Handbook of Experimental Pharmacology

Volume 267

The *Handbook of Experimental Pharmacology* is one of the most authoritative and influential book series in pharmacology. It provides critical and comprehensive discussions of the most significant areas of pharmacological research, written by leading international authorities. Each volume in the series represents the most informative and contemporary account of its subject available, making it an unrivalled reference source.

HEP is indexed in PubMed and Scopus.

More information about this series at http://www.springer.com/series/164

Nikita Gamper • KeWei Wang
Editors

Pharmacology of Potassium Channels

Springer

Editors
Nikita Gamper
Faculty of Biological Sciences
University of Leeds
Leeds, UK

KeWei Wang
School of Pharmacy
Qingdao University
Qingdao, China

ISSN 0171-2004 ISSN 1865-0325 (electronic)
Handbook of Experimental Pharmacology
ISBN 978-3-030-84051-8 ISBN 978-3-030-84052-5 (eBook)
https://doi.org/10.1007/978-3-030-84052-5

This Springer imprint is published by the registered company Springer Nature Switzerland AG.
The registered company address is: Gewerbestrasse 11, 6330 Cham, Switzerland

Preface

We are happy to see this volume of the *Handbook of Experimental Pharmacology*, entitled "Pharmacology of Potassium Channels" to finally roll off the press, despite all the setbacks and havoc of the COVID-19 global pandemic. We would like to express our sincere gratitude to all the contributing authors whose commitment, hard work, and enthusiasm for the potassium channel research have made this book possible. With the ongoing worldwide efforts towards global vaccination, there is finally a hope for the return to life, which gives room to communication, interaction, and collaboration necessary to the scientific progress and development of new therapeutics – the areas of human endeavor this book series is striving to promote.

The purpose of this volume is to provide academic and industrial scientists, students, and interested public with a comprehensive introduction to the field of potassium (K^+) channels with specific focus on pharmacology and therapeutic applications. K^+ channels are the most ubiquitously expressed type of ion channel in mammals and they are present in virtually all living organisms. Their functions range from maintenance of resting membrane potential and regulation of excitability in neurons and muscle cells, to secretion, cell volume regulation, metabolic control, and cell cycle regulation. Genetic mutations in K^+ channel genes or compromised function of K^+-channel-forming protein subunits causes a spectrum of severe human diseases, including cardiac arrhythmias, epilepsies, ataxias, diabetes, deafness, and many others. K^+ channels are in a sharp focus of current drug discovery and cardiovascular safety liability, thus understanding of K^+ channel pharmacology is a critical step within these processes. This knowledge is also important for the development of basic understanding of human physiology and cell biology, as well as for the identification and validation of future drug targets.

This volume aims to cover most members of K^+ channel superfamily, including voltage-gated K^+ channels, Ca^{2+}-activated K^+ channels, inwardly rectifying K^+ channels, and tandem pore domain K^+ channels. Information on these major subfamilies of K^+ channels is organized according to the general themes used in pharmacology classes, academic and industrial research communities: classification, characterization of properties, atomic structures, channelopathies, and pharmacological modulation by small molecules, toxins, and antibodies.

All chapters in this book are contributed by experts in their respective fields and contain both the historic perspectives and current state of the art with a specific focus on emerging pharmacology and therapeutic targeting. Some chapters also include the most recently solved K^+ channel structures and discuss how this emerging wealth of structural information can be used in drug design and development. A chapter reviewing major classes of K^+ channel auxiliary subunits and the ways these affect channel properties is also included in the volume.

Although we aimed to provide both broad and in-depth coverage of the field, due to the space limitations and time constrains not all the exciting recent developments in K^+ channel pharmacology were included in this edition. We hope that this leaves us with an opportunity to update and revise the field in the future editions. Comments about the current volume, as well as suggestions for the future volumes are always welcome and should be directed to the publisher (general information about the Handbook of Experimental Pharmacology and contact details can be found at https://www.springer.com/series/164).

Finally, we wish to acknowledge all the great help and efforts of the editorial staff at the Handbook of Experimental Pharmacology, with specific cordial thanks to Ms. Susanne Dathe, Ms. Alamelu Damodharan, and Mr. Arunkumar Kathiravan.

Leeds, UK
Qingdao, China

Nikita Gamper
KeWei Wang

Contents

Comparison of K^+ Channel Families

Jaume Taura, Daniel M. Kircher, Isabel Gameiro-Ros, and Paul A. Slesinger

Contents

Abstract

K^+ channels enable potassium to flow across the membrane with great selectivity. There are four K^+ channel families: voltage-gated K (K_v), calcium-activated (K_{Ca}), inwardly rectifying K (K_{ir}), and two-pore domain potassium (K_{2P}) channels. All four K^+ channels are formed by subunits assembling into a classic tetrameric (4x1P = 4P for the K_v, K_{Ca}, and K_{ir} channels) or tetramer-like (2x2P = 4P for the K_{2P} channels) architecture. These subunits can either be the same (homomers) or different (heteromers), conferring great diversity to these channels. They share a highly conserved selectivity filter within the pore but show different gating mechanisms adapted for their function. K^+ channels play essential roles in controlling neuronal excitability by shaping action potentials, influencing the resting membrane potential, and responding to diverse

Jaume Taura, Daniel M. Kircher, and Isabel Gameiro-Ros contributed equally to this work.

J. Taura · D. M. Kircher · I. Gameiro-Ros · P. A. Slesinger (✉)
Nash Family Department of Neuroscience, Icahn School of Medicine at Mount Sinai, New York, NY, USA
e-mail: paul.slesinger@mssm.edu

N. Gamper, K. Wang (eds.), *Pharmacology of Potassium Channels*,
Handbook of Experimental Pharmacology 267, https://doi.org/10.1007/164_2021_460

physicochemical stimuli, such as a voltage change (K_v), intracellular calcium oscillations (K_{Ca}), cellular mediators (K_{ir}), or temperature (K_{2P}).

Keywords

Calcium-activated · Conductivity · Gating · Inwardly rectifying K · Ion channel · Potassium channel · Selectivity · Two-pore domain potassium · Voltage-gated K

Abbreviations

EKG	Electrocardiogram
GIRK	G protein-gated inwardly rectifying potassium channel
K_{ATP}	ATP-sensitive inwardly rectifying potassium channel
TALK	Two-pore ALkaline-activated K^+ channel
TASK	Two-pore acid-sensitive K^+ channel
THIK	Two-pore halothane-inhibited K^+ channel
TRAAK	TWIK-related arachidonic acid-stimulated K^+ channel
TREK	TWIK-related K^+ channel
TRESK	TWIK-related spinal-cord K^+ channel
TWIK	Two-pore weak inward-rectifying K^+ channel

1 Overview

A key property of all K^+ channels is their ability to selectively allow permeation of K^+ across the membrane at near diffusion limited rates. That is, they discriminate between K^+ and other monovalent cations and anions, with high fidelity, providing a conduit for K^+ to flow in and out of cells. Built on the framework of K^+ selectivity, K^+ channels have evolved different gating mechanisms (i.e., opening and closing) and functions in a variety of cell types. In this chapter, we compare some of the essential features of K^+ channels across the different families and subfamilies.

The voltage-gated K^+ channels (K_v) form the largest gene family in the K^+ channel group, first described by Hodgkin and Huxley (1945) and cloned 36 years ago (Noda et al. 1984). In mammals, K_v channels are encoded by 40 genes, with each gene encoding a corresponding α subunit. Traditionally, K_v channels play a role in cell excitability, where channel opening helps to repolarize excitable cells via efflux of K^+, such as during the action potential (Hille 1986). The K_v channel family is divided into 12 subfamilies (A-González and Castrillo 2011; Abbott et al. 2001) (Fig. 1), based on analyses of the hydrophobic domain containing the six transmembrane segments (S1-S6).

The first evidence of K^+ currents activated by calcium was described by Gardos over 60 years ago, who observed the activation of K^+ selective conductance by

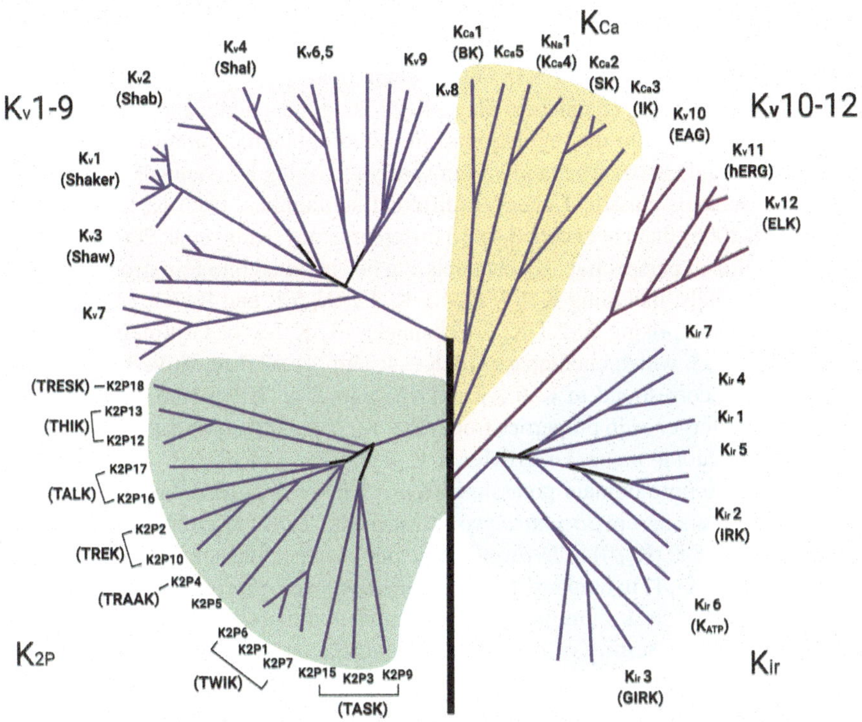

Fig. 1 Potassium channel tree. Dendrogram of the different families of potassium channels

intracellular free calcium in red blood cells (Gardos 1958). The family of calcium-activated (K_{Ca}) channels encompasses a group of K^+ channels with different physiological and pharmacological properties. The calcium sensitivity characteristic of K_{Ca} channels allows them to couple membrane potential changes during the action potential with elevations in intracellular Ca^{2+} concentration ($[Ca^{2+}]_i$), contributing to activation of the afterhyperpolarization (AHP) and regulation of the action potential (Berkefeld et al. 2010). Based on their single-channel conductance, K_{Ca} channels can be classified as small conductance (SK) K_{Ca} channels, which are $K_{Ca}2.1$-$K_{Ca}2.3$ (4–14 pS) (Kohler et al. 1996), intermediate conductance (IK) K_{Ca} channel, also named as $K_{Ca}3.1$ (32–39 pS) (Ishii et al. 1997), and large conductance (*big* conductance, BK) K_{Ca} channel, also known as $K_{Ca}1.1$ or Slo1 (200–300 pS) (Butler et al. 1993; Kshatri et al. 2018). With a fair degree of sequence homology, the K_{Ca} channel family includes the sodium-activated K^+ channels $K_{Ca}4.1$ and $K_{Ca}4.2$ (also called Slack/Slo2.1 and Slick/Slo2.2, respectively), as well as the pH-dependent $K_{Ca}5.1$ channels (also named as Slo3) (Fig. 1) (Wei et al. 2005).

Inwardly rectifying K (K_{ir}) channels were first described over 70 years ago in skeletal muscle fibers by Bernard Katz (1949), who observed an "anomalous" rectifier inward current in the presence of different extracellular K^+ concentrations. K_{ir} gating appeared to shift with the Nernst potential for K^+. Years later, several studies explained that the property of inward rectification arises from an asymmetric blockade of the open channel pore by intracellular Mg^{2+} (Matsuda et al. 1987) and polyamines (Lopatin et al. 1994; Oliver et al. 2000). This property of inward rectification enables K_{ir} channels to play a key role in the maintenance of the resting membrane potential and the regulation of the action potential duration in excitable cells (Hibino et al. 2010). The family of inwardly rectifying K^+ channels comprises a variety of channels classified in seven different subfamilies, from $K_{ir}1.x$ to $K_{ir}7.1$ (Kubo et al. 2005) that are encoded by 15 different genes (Kubo et al. 2005) (Fig. 1). From a functional perspective, K_{ir} channels can be classified into four groups: (1) K^+ transport channels, including $K_{ir}1.1$, $K_{ir}4.1$-$K_{ir}4.2$, $K_{ir}5.1$, and $K_{ir}7.1$; (2) Classical K_{ir} channels, comprising $K_{ir}2.1$-$K_{ir}2.4$ channels; (3) $K_{ir}3.x$ or G-protein-gated K_{ir} channels (GIRK), which encompass GIRK1–4; and (4) ATP-sensitive K^+ channels (K_{ATP}), which correspond to $K_{ir}6.1$-$K_{ir}6.2$ (Hibino et al. 2010; Kubo et al. 2005). Due to their divergence in properties from other K_{ir} channels, some are often referred to by their functional name, i.e. GIRK for $K_{ir}3$ and K_{ATP} for $K_{ir}6.2$.

The first two-pore domain potassium (K_{2P}) channel ever described was discovered only 25 years ago in *Saccharomyces cerevisiae*, TOK1 (Ketchum et al. 1995). A year later, dORK ($K_{2P}0$) in *Drosophila melanogaster* (Goldstein et al. 1996)) and finally TWIK1 ($K_{2P}1$) in humans (Lesage et al. 1996) were discovered. K_{2P} channels contribute to a K^+ leak current in excitable and non-excitable cells (Czirjak and Enyedi 2002). This resting or background conductance is critical in motoneurons (Berg et al. 2004; Talley et al. 2000), dorsal root ganglion neurons (Kang and Kim 2006; Pereira et al. 2014), or cerebellar granule neurons (Plant et al. 2002). Whereas "leakage current" typically refers to a non-selective current following membrane damage, K_{2P} channels support a K^+-selective leak that is fairly voltage-independent. At rest, open K_{2P} channels enable K^+ efflux due to K^+ concentration gradient, making the intracellular more negative, limiting further K^+ efflux and suppressing depolarization. Under physiological conditions, neurons display a resting membrane potential (V_m) of −60 to −90 mV, while the equilibrium potential for K^+ (E_K) is approx. −90 mV, with a K^+ concentration of 5 mM outside and 140 mM inside at 37°C. Nevertheless, K^+ leakage contributes more to the V_m. K_{2P} channels affect the frequency, duration, and amplitude of action potentials. K_{2P} are tightly regulated by splicing, post-translational modifications (phosphorylation, sumoylation, glycosylation) and numerous chemical (phospholipid composition, GPCR activation, second messengers) and physical agents (extracellular and intracellular pH, mechanical stretch, temperature) (Gada and Plant 2019; Goldstein et al. 2001; Niemeyer et al. 2016). Currently, the K_{2P} family is composed of 15 different subunits ($K_{2P}1$–15) and encoded by genes numbered KCNK1–18 (no expression has been found for KCNK8, KCNK11, or KCNK14). They have been historically grouped according to structural and functional relatedness in six subfamilies: TREK1-TREK2-TRAAK ($K_{2P}2$-$K_{2P}10$-$K_{2P}4$), TALK1-TALK2 ($K_{2P}16$-$K_{2P}17$), TWIK1-TWIK2 ($K_{2P}1$-$K_{2P}6$), TASK1-TASK2-TASK3-TASK5 ($K_{2P}3$-$K_{2P}5$-$K_{2P}9$-$K_{2P}15$), THIK1-THIK2 ($K_{2P}12$-$K_{2P}13$), and TRESK ($K_{2P}18$). Like K_{ir} channels, K_{2P} channels are also commonly referred to via their functional name.

2 Subunits/Assembly/Topology

Potassium channels share many similarities when it comes to their topology, assembly, and subunit composition. However, there are some key differences, which we will explore here. For voltage-gated potassium channels each K_v channel is a tetramer composed of similar or identical pore-forming α subunits, and in some cases also contains auxiliary β subunits which can alter channel localization and/or function (A-González and Castrillo 2011; Abbott et al. 2006). The α subunits are arranged around a central axis that forms a pore (Coetzee et al. 1999). Each α subunit is a polypeptide with 6 transmembrane domains (S1-S6) and five loops connecting the segments (Fig. 2). The N- and C-terminal regions are cytoplasmic. The pore-forming region of the channel is produced by the S5-S6 linker (P-loop) and contains the K^+ selectivity filter (Heginbotham et al. 1994). The voltage-sensing domain (VSD) is formed by segments S1-S4 that control pore opening via the S4-S5 intracellular loop that is connected to the pore domain (Bezanilla 2000; Cui 2016; Gandhi and Isacoff 2002; Schmidt and Mackinnon 2008). Within each subfamily both homomeric and heteromeric channels may form with a range of biophysical properties (Abbott et al. 2006; Albrecht et al. 1995), leading to a large diversity of channels.

K_{Ca} channels basic topology is similar to that of K_v channels; in fact, both families belong to the 6/7TM group of K^+ channels (Gutman et al. 2005; Wei et al. 2005). Cryo-EM structures of the full length $K_{Ca}1.1$ channel have provided extensive information about its structure and gating (Hite et al. 2017; Tao et al. 2017) (PDB: 5TJ6 and 5TJI, respectively). Importantly, small and large conductance subfamilies of K_{Ca} channels have two main differences in their structure: $K_{Ca}1.1$ channels have an additional transmembrane domain, S0, and their S4 transmembrane domain is voltage-sensitive (Fig. 2) (Kshatri et al. 2018). Due to the S0, the N-terminus is extracellular, and the large C-terminal domain that comprises around two-thirds of the protein is intracellular (Meera et al. 1997). Like the K_v, the S1-S4 transmembrane segments of the $K_{Ca}1.1$ channel form the VSD (Diaz et al. 1998; Ma et al. 2006) and the S5-S6 segments contain the P-loop with the K^+ selectivity filter (Meera et al. 1997). Another major difference between $K_{Ca}1.1$ channels and $K_{Ca}2.x$/ $K_{Ca}3.1$ is the cytoplasmic C-terminal domain, which in $K_{Ca}1.1$ channels contains two regulating conductance of K^+ (RCK) domains, RCK1 and RCK2 (Yuan et al. 2010). X-ray crystal structures of the isolated C-terminus of $K_{Ca}1.1$ have provided valuable structural information about RCK1 and RCK2 (PDB: 3NAF) (Yuan et al. 2011). Both RCK domains possess a high affinity Ca^{2+}-binding site: a string of negatively-charged aspartate residues located at RCK2, labeled as the Ca^{2+}-bowl (Schreiber and Salkoff 1997), and a site containing the residues D362 and D376 in RCK1 (Yuan et al. 2010). Four of these pore-forming subunits of $K_{Ca}1.1$ (α subunit) assemble to form functional homotetramers (Shen et al. 1994).

$K_{Ca}2.1$-$K_{Ca}2.3$ and $K_{Ca}3.1$ channels share with K_v channels the six TM domain (S1-S6) topology (Kohler et al. 1996), with the S5 and S6 TM pore-forming

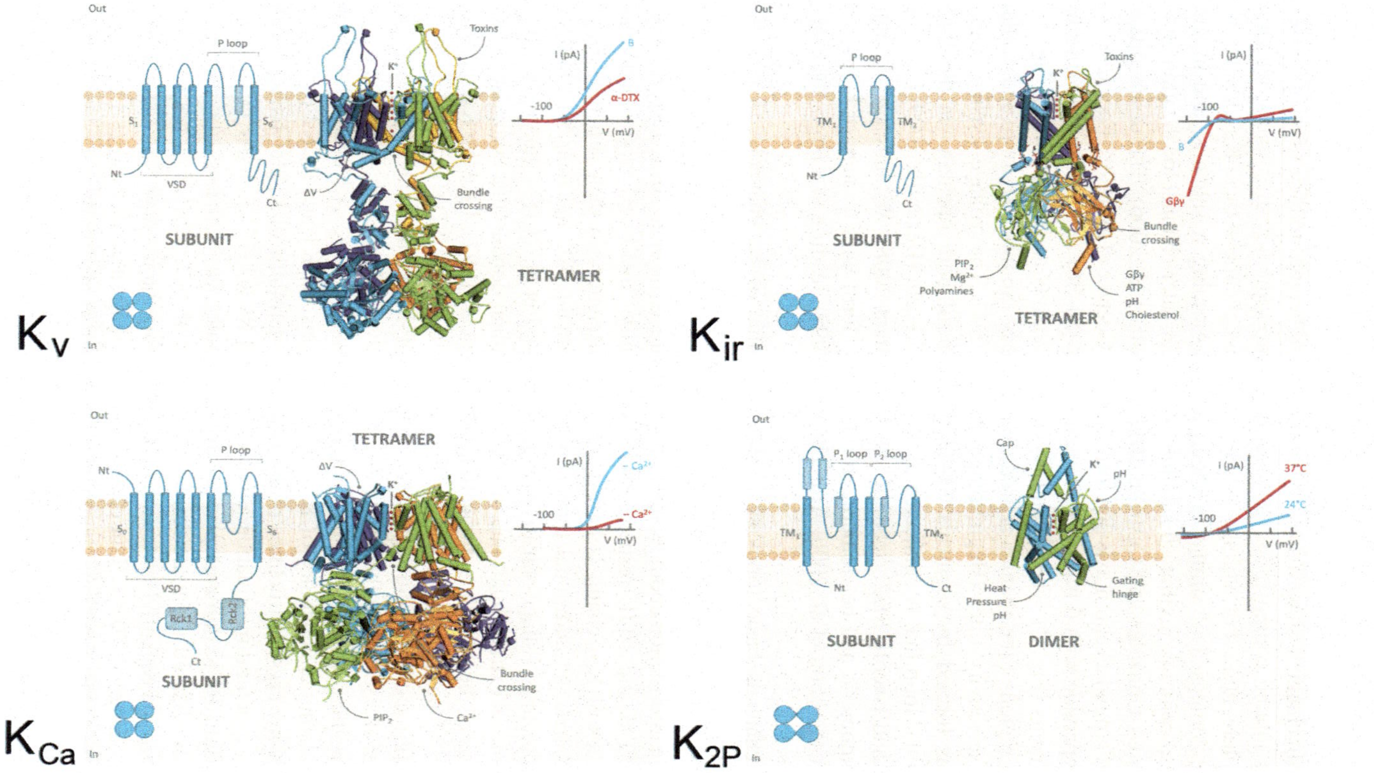

Fig. 2 Potassium channel structure and function. The membrane topology, atomic structure, and typical current-voltage (I-V) plots are shown for the K_V (basal vs α-DTX), K_{Ca} (± Ca^{2+}), K_{ir} (± Gβγ), and K_{2P} (37 ° C vs 24 ° C) channels. Adapted from Yu and Catterall (2004)

domains, and the N- and C-terminal domains facing the cytosol (Kshatri et al. 2018). Unlike K_v and $K_{Ca}1.1$ channels, the S4 domain lacks voltage sensitivity, and therefore gating is membrane potential independent (Hirschberg et al. 1999). In the cytosolic C-terminal domain of small conductance K_{Ca} channels, a calmodulin (CaM) binding domain (CaMBD) (Adelman 2016; Fanger et al. 1999) is located that indirectly confers Ca^{2+} sensitivity (Xia et al. 1998). In general, $K_{Ca}2.1$-$K_{Ca}2.3$ and $K_{Ca}3.1$ assemble in homotetramers to form functional channels (Kohler et al. 1996; Sforna et al. 2018), but $K_{Ca}2.1$-$K_{Ca}2.3$ can also arrange in heterotetramers (Strassmaier et al. 2005).

K_{ir} channels share a relatively simple topology, as compared to K_v and K_{Ca} channels (Fig. 2). They contain two transmembrane domains, TM1 and TM2, separated by a linking pore-forming P-loop sequence that includes the K^+ selectivity filter (Heginbotham et al. 1994). The cytoplasmic N- and C-terminal domains form a characteristic cytoplasmic extended pore structure (Fig. 2) (Nishida et al. 2007). Four subunits associate to form functional homotetramers or heterotetramers. While $K_{ir}1.1$ and $K_{ir}7.1$ can only form homotetramers (Kumar and Pattnaik 2014; Leng et al. 2006), the majority of K_{ir} channels assemble with subunits within the same subfamily (Hibino et al. 2010). $K_{ir}4.x$ forms homotetramers (Pessia et al. 2001), while the formation of functional $K_{ir}5.1$ homotetramers has not been described yet (Hibino et al. 2010). However, $K_{ir}5.1$ can associate with $K_{ir}4.1$-$K_{ir}4.2$ to form functional channels ($K_{ir}4.1$-$K_{ir}5.1$ or $K_{ir}4.2$-$K_{ir}5.1$) (Pessia et al. 2001).

Like K_{ir} channels, K_{2P} channels also contain a simplified topology, compared to K_v and K_{Ca} channels (Fig. 2). K_{2P} channels contain the two pore-forming P loops (P1, P2), where the K^+ selectivity filter can be found, and four transmembrane helices (TM1-TM4). The first pore-forming P-loop sequence (P1) is located in between TM1 and TM2, while the second one (P2) is found between TM3 and TM4. The 2P/4TM topology (2P/8TM for TOK1) is unique among other K^+ channels (1P/6TM for K_v or 1P/2TM for K_{ir}). However, despite differences in the topology, the overall K_{2P} channel structure does not differ much from K_v, K_{Ca}, and K_{ir} channels, due to its pseudo tetrameric architecture. Each protomer (2P) will assemble to form a dimer (2x2P = 4P) to recreate a classic tetrameric K^+ channels configuration (4x1P = 4P) (Gada and Plant 2019; Goldstein et al. 2001; Niemeyer et al. 2016). The P1 and P2 loops share high homology with the K_v channel P-loop (Fig. 3). K_{2P} channels also possess an extracellular cap domain, constituted by the external loop located in between TM1 and P1. The two subunits assemble their helical caps to generate two lateral tunnels where the ions move from the exterior to the pore (extracellular ion pathway). This assembly is stabilized in most K_{2P} channels by a disulfide bond (Lesage et al. 1996; Niemeyer et al. 2003). The cap impedes direct ion transport between the pore and the extracellular medium.

K^+ ions. Despite some differences, K_{2P} channels exhibit similar sequences to the T-I/V-G-Y/F-G consensus sequence (Schewe et al. 2016) (Fig. 3). Interestingly, the specific sequence composition can confer the channel a change in ion selectivity under certain conditions. For example, the presence of a second threonine (Thr118) within the selectivity filter of TWIK1 enables the channel to support Na^+ leak currents (Ma et al. 2011).

4 Gating Mechanisms

One of the areas where there is a large divergence in K^+ channels is through their different gating mechanisms. The K_v class of channels are voltage-dependent and have evolved to use the power of the electric field that exists across excitable membranes to move charged groups of ions crosswise across the membrane (Ishida et al. 2015; Schmidt and Mackinnon 2008). Voltage-dependent gating of K_v channels involves several molecular processes including: (1) detection of changes in voltage across the membrane by the voltage-sensing domain (VSD). VSD activation results in conformational rearrangement leading to (2) propagation of VSD movements to the ion conduction pore via a helical linker (Khalili-Araghi et al. 2006). Rearrangement of the pore, via VSD-pore coupling, results in (3) pore opening and ion conduction into the cell (Cui 2016). The VSD can adopt a stable conformation in the absence of the rest of the channel (PD) (Chakrapani et al. 2008; Krepkiy et al. 2009). One of the first VSDs defined at the atomic level and considered a native, non-altered structure was part of the mammalian $K_v1.2$ channel (PDB: 2A79) (Long 2005). This structure showed that the VSD interacts with the pore domain of an adjacent subunit where the voltage sensor is latched around the pore of an adjacent subunit and voltage sensing in one subunit would affect the pore region of another subunit (Long et al. 2007).

A key characteristic of the VSD is the presence of basic amino acids, including positively charged arginine and lysine amino acids in the S4 segment, in a repeating motif of one positive charge separated by two hydrophobic residues. The number of positive charges is variable, with some Shaker related channels having as many as seven (Zhou et al. 2001a). Most models of voltage-dependent gating suggest that as transmembrane voltage changes polarity during depolarization, with the cytoplasmic side becoming positive, the energy exerted on the S4 charges is altered and moves the S4 segment. The S4 segment appears to translate and rotate counterclockwise (Ahern and Horn 2005; Larsson et al. 1996). The C-terminal portion of the S4 segment is accessible to the intracellular solution at rest and as depolarization occurs, the charged residues become less accessible to the intracellular solution and instead become more accessible to the extracellular solution, resulting in the S4 segment moving to an external position (Ahern and Horn 2004, 2005; Baker et al. 1998; Broomand and Elinder 2008; Gandhi and Isacoff 2002; Larsson et al. 1996; Lin et al. 2011). The result is channel activation, leading to opening of the water-filled channel and the flow of K^+ down the electrochemical gradient out of the cell. As the membrane potential repolarizes, the VSD returns to the resting state, which in turn closes the channel and terminates K^+ permeability.

K_v channels can exhibit various types of inactivation, each involving distinct mechanisms. In some channels, inactivation occurs soon after the channel is activated. This fast inactivation, or N-type inactivation, is mainly due to an intracellular block of the channel by the intracellular N-terminus, often referred to as the inactivation particle (Aldrich 2001). A relatively slower form of inactivation termed C-type occurs after tens or hundreds of milliseconds have elapsed following channel activation (Pau et al. 2017). It appears that the pore structure of this class of channels and the permeating ions play a pivotal role in this process; however, it remains the subject of investigation (Hoshi and Armstrong 2013). Modulation of the inactivation process conveys the ability to control the cellular availability of K_v channel currents. In some cases, inactivation is sensitive to the cellular redox environment (Sahoo et al. 2014). A structural component within the N-terminus has been identified that serves as a sensor for the cytoplasmic redox potential (e.g., exposure to oxidizing agents) and leads to inactivation of the channel (Finol-Urdaneta et al. 2006).

Two different gating mechanisms can be observed in K_{Ca} channels: voltage and calcium-dependent gating for $K_{Ca}1.1$ channels, and calcium-dependent for small and intermediate conductance K_{Ca} channels (Fakler and Adelman 2008). Large conductance $K_{Ca}1.1$ channels exhibit voltage sensitivity similar to K_v channels; membrane depolarization and intracellular Ca^{2+} combine allosterically to activate the channel and open the inner pore (Horrigan and Aldrich 2002). Four RCK1-RCK2 intracellular domains of the $K_{Ca}1.1$ tetrameric assembly comprise the gating ring (Hite et al. 2017; Yuan et al. 2011). Upon Ca^{2+} binding to the Ca^{2+}-sensitive sites in the gating ring, an aspartate string in RCK1 and the Ca^{2+} bowl in RCK2, the RCK1-RCK2 tandems rearrange to open the gating ring (Yuan et al. 2011). When Ca^{2+} binding to the intracellular domain of the channel combines with the activation of the VSD by membrane depolarization, the chemical and the electrical energies released additively fuel the conformational change of the PD from the closed to the open state (Hite et al. 2017; Horrigan and Aldrich 2002), in a process that requires the interaction of the gating ring with the VSD (Hite et al. 2017). As the membrane voltage depolarizes, the intracellular Ca^{2+} concentration ($[Ca^{2+}]_i$) required to activate $K_{Ca}1.1$ decreases (Cui et al. 1997), ranging from 0.5 to 50 mM (Xia et al. 2002). A regulatory Mg^{2+}-binding site, located in RCK1, has also been described for $K_{Ca}1.1$ channels (Shi and Cui 2001), through which Mg^{2+} contributes to channel activation (Xia et al. 2002; Yang et al. 2008).

In contrast to $K_{Ca}1.1$ channels, gating is voltage-independent in $K_{Ca}2.x$ and $K_{Ca}3.1$ channels (Hirschberg et al. 1999). Ca^{2+} activates $K_{Ca}2.x$ and $K_{Ca}3.1$ channels through binding to the highly Ca^{2+}-sensitive protein calmodulin (CaM), which is constitutively associated to α subunits of the channel (Fanger et al. 1999; Sforna et al. 2018; Xia et al. 1998). The binding of Ca^{2+} to $K_{Ca}2.x$ and $K_{Ca}3.1$ channels through CaM accounts for their elevated Ca^{2+} sensitivity compared to submicromolar sensitivity of $K_{Ca}1.1$ channels (Adelman et al. 2012; Xia et al. 1998). $K_{Ca}2.x$ and $K_{Ca}3.1$ channels interact with CaM through a highly conserved CaM binding domain (CaMBD) (Adelman 2016; Fanger et al. 1999) in each α subunit of the channel, and it has been confirmed by cryo-EM (Lee and MacKinnon 2018). Ca^{2+} binding to the EF-hand domains in the N-lobe of CaM promotes the rearrangement of two CaM-CaMBD dimers into a "dimer of dimers," that leads to the conformational change of the helices forming the pore required for channel opening (Lee and MacKinnon 2018; Schumacher et al. 2001).

For K_{ir} channels, the inward rectification is their most distinctive feature. In fact, different levels of inward rectification can be described in K_{ir} channels, ranging from strong inward rectifiers, such as $K_{ir}2.1$-$K_{ir}2.4$, to medium, e.g. GIRK1-GIRK4, and to weak, such as $K_{ir}1.1$ and $K_{ir}6.1$-$K_{ir}6.2$ channels (Hibino et al. 2010; Walsh 2020). Although K_{ir} channels are not intrinsically voltage-dependent, since they lack the voltage-sensing S4 domain (Hibino et al. 2010), the inward rectification shows an apparent voltage sensitivity. Inward rectification is mediated by intracellular Mg^{2+} (Lu and MacKinnon 1994; Matsuda et al. 1987) and naturally occurring polyamines (e.g., putrescine^{2+}, spermidine^{3+}, and spermine^{4+}) (Lopatin et al. 1994; Nichols and Lee 2018). At membrane potentials positive to the equilibrium potential of K^+ ($E_K \approx -95$ mV), Mg^{2+} and polyamines occlude the inner vestibule, only allowing a small outward current. In contrast, at potentials negative to E_K, Mg^{2+} and polyamines flow out of the channel into the cell, allowing a large inward K^+ current (Lopatin et al. 1995). The affinity of Mg^{2+} and the polyamines for binding sites in the pore-forming TM2 helix (Stanfield et al. 1994; Wible et al. 1994) and the cytoplasmic domain of K_{ir} channels (Kubo and Murata 2001; Taglialatela et al. 1995) dictates the strength of the inward rectification (Baronas and Kurata 2014; Clarke et al. 2010). Besides Mg^{2+} and polyamines, K_{ir} channels K^+ conductance is also influenced by extracellular K^+ concentrations ($[K^+]_o$) (Lopatin and Nichols 1996), being this conductance higher at increasing $[K^+]_o$ (Hibino et al. 2010). $K_{ir}7.1$ is an exception and exhibits only a slight dependence on $[K^+]_o$ due to the presence of a methionine at position 125 in the pore domain, instead of the conserved arginine found in the majority of K_{ir} channels (Doring et al. 1998). The intrinsic gating of K_{ir} channels is controlled by two gating structures: the bundle-crossing region in the TM2 of the transmembrane domain (Sadja et al. 2001; Yi et al. 2001) and the G loop in the cytoplasmic domain (Pegan et al. 2005). The first K_{ir} channel structures to be resolved, involving bacterial K_{ir} channels, such as KscA (Doyle et al. 1998) (PDB: 1BL8) and K_{ir}Bac1.1 (Kuo et al. 2003) (PDB: 1P7B), pointed at TM1 and TM2 as key players in the gating of K_{ir} channels. The gating of some K_{ir} channels depends on other regulators, apart from Mg^{2+}, polyamines, and $[K^+]_o$, such as pH, Na^+, ATP, and/or G proteins (Hibino et al. 2010). For instance, changes in the intracellular pH alter the gating of $K_{ir}1.1$ (Schulte and Fakler 2000), $K_{ir}4.1$-$K_{ir}4.2$ (Pessia et al. 2001), and $K_{ir}5.1$ (Tucker et al. 2000), which are closed upon intracellular acidification, while $K_{ir}7.1$ shows maximal response at pH 7.0 (Yuan et al. 2003). In fact, homomeric $K_{ir}4.1$ and $K_{ir}4.1/K_{ir}5.1$ channels exhibit different pH sensitivities (Casamassima et al. 2003). In the case of $K_{ir}2.1$-$K_{ir}2.4$ channels, intracellular alkalization activates $K_{ir}2.4$ (Hughes et al. 2000), while either extracellular or intracellular alkalization enhances $K_{ir}2.3$ activity (Zhu et al. 1999). $K_{ir}6.1$-$K_{ir}6.2$ channels, also called K_{ATP}, are regulated by intracellular ATP, which leads to the inactivation of the channel (Terzic et al. 1995), while intracellular nucleoside diphosphates, such as ADP, activate the channel through the interaction with SUR, the auxiliary subunits of $K_{ir}6.1$-$K_{ir}6.2$ channels (Hibino et al. 2010; Matsuoka et al. 2000). G-protein gated K_{ir} channels (GIRK) are opened by an interaction of the Gβγ subunit with the βL-βM sheets in the cytoplasmic C-domain of GIRK (PDB: 4KFM) (Finley et al. 2004; He et al. 1999; Ivanina et al. 2003; Whorton and MacKinnon 2013), producing a conformational change that opens the channel pore in a PIP_2-dependent process (Huang et al. 1998; Whorton and MacKinnon 2013). Moreover, the gating of GIRK channels containing GIRK2 or GIRK4 subunits is also influenced by intracellular Na^+ (Ho and Murrell-Lagnado 1999), which promotes the binding of PIP_2 to the channel and activation (Rosenhouse-Dantsker et al. 2008). The structural binding site for Na^+ has been identified in a GIRK2 X-ray structure (PDB: 3SYA) (Whorton and MacKinnon 2011). In this way, the intracellular Na^+ increase after cell depolarization enhances the activity of GIRK channels, bringing the cell back to the resting state.

Like K_v, K_{Ca}, and K_{ir} channels, K_{2P} possesses a gating hinge (Brohawn et al. 2012; Miller and Long 2012; Niemeyer et al. 2016). TM1 and TM3 are located on the outer pore, while the inner helices, TM2 and TM4, play a crucial role in channel activation. The TM4 helix motion, up and down (closer and farther TM2 helix), is a pivotal determinant of the open-close configuration. The interfacial C helix is adjacent to TM4 and movement is transferred to the TM2-TM4 hinge to support pore widening and ion conduction (Brohawn et al. 2012; Miller and Long 2012; Niemeyer et al. 2016). K_{2P} channels are sensitive not only to cytosolic factors, but also to membrane components (and/or alterations). K_{2P} channels also possess intramembrane openings that confer connections between lipid membrane and ion pore. These openings, termed fenestrations, have been named for analogous side portals present in prokaryotic voltage-dependent Na^+ channels (Payandeh et al. 2011). They are located in between the TM2 of one protomer and the TM4 of the other one. The two transmembrane cavities can accommodate acyl chains and influence the channel conductivity (Brohawn et al. 2014).

Although many factors can modulate K_{2P} gating, the extracellular pH (pH_o) is probably the best characterized. Many K_{2P} channels have a histidine located at the entrance of the selectivity filter (TM1-P1) that is protonated upon pH_o decrease. In TASK1 (His98), TASK3 (His98), TWIK1 (His122), and TREK1 (His126), histidine protonation prevents the ion passage. Thus, channel closure is similar to C-type inactivation in K_v channels (Chatelain et al. 2012; Cohen et al. 2008; Kim et al. 2000; Lopes et al. 2001; Rajan et al. 2000). Uniquely, extracellular acidification induces channel activation in TREK2 (His151) and involves a region of the P2-TM4 extracellular loop (Sandoz et al. 2009). Interestingly, TWIK1 switches ion selectivity upon a decrease in pH_o (Ma et al. 2012). Histidine is not the only basic residue in K_{2P} channels that can operate as H^+ sensor. TASK2 lacks the histidine sensor but has an arginine (Arg224) at the second pore domain that confers selectivity filter pH_o-sensing as well. TASK2 is inhibited by acidic pH_o, and surprisingly activated when pH_o increases. In this case, protonation/deprotonation of the side chain of the residue alters the electrostatic stability on the selectivity filter (Niemeyer et al. 2007; Zuniga et al. 2011). Additionally, some K_{2P} channels respond to intracellular pH (pH_i)

alterations. Thereby, K_{2P} channels such as TREK1 or TASK2 switch from low to high activity upon intracellular acidification. The mechanism does not involve any residue in the selectivity filter, but an acidic amino acid in the interfacial C helix (i.e., Glu306 in TREK1), working as an activation gate (Bagriantsev et al. 2011).

In addition to pH, the intracellular C-terminus also supports different gating mechanisms. TREK1 possesses a group of positively charged residues in the C helix that confers the channel the capacity to respond to phospholipids (Chemin et al. 2007). TREK1 also contains a phosphorylation site (Ser348) in the C-terminus that alters channel gating properties, switching the channel from a voltage-independent into a voltage-dependent phenotype (Bockenhauer et al. 2001). The interfacial C helix is also modulated by GPCR activation. TASK2 is closed by direct G protein βγ subunits binding at the Lysine 245 (Anazco et al. 2013; Niemeyer et al. 2016). TASK1 and TASK3 are also closed by G protein α subunit induced diacylglycerol (DAG) generation that is believed to bind the C-terminal domain (Wilke et al. 2014). Physical stimuli such as pressure and temperature also influence K_{2P} gating. Little is known about the specific mechanism but surely involves the intracellular C-terminus (Bagriantsev et al. 2011). For instance, partial deletion of the interfacial C helix in TREK1 lowers heat-induced activation (Maingret et al. 2000). Besides the C-terminus domain that operates as cytosolic gate, K_{2P} channels show an inner gate that modulates the pore conductivity by membrane composition. The current hypothesis sustains that intramembrane fenestration determine TM4 position over TM2, working as a gating hinge and affecting the selectivity filter. Thus, two possible conformations exist: 1) when the lipid acyl chains penetrate into the fenestration up to the cavity located below the selectivity filter, the TM4 helix is in the down conformation and the ion transit is hindered, and 2) in contrast, when the lipid fenestration is empty, TM4 moves up towards TM2 (up conformation), closing the fenestration and releasing the selectivity filter, which can accommodate an extra ion and facilitate ion conduction (Brohawn et al. 2014).

Recent structural determinations support some of these gating mechanisms. TREK2 structure resolved with the inhibitor fluoxetine exhibits a down conformation (PDB: 4XDJ) (Dong et al. 2015). On the other hand, TRAAK crystallization in the presence of the activator trichloroethanol shows the up conformation (PDB: 3UM7) (Brohawn et al. 2014). Moreover, artificially trapping TRAAK into the up state, by a disulfide bridge between TM4 and TM2, induces the channel to a reversible low activity profile. The least understood gating mechanism is the effect of membrane voltage. Although K_{2P} channels lack a specific voltage-sensing domain (i.e., S4 in K_v) and the first leak K^+ channels were initially described as a voltage-independent outward rectifier K^+ channels (Goldstein et al. 2001), some K_{2P} (except for the unphosphorylated TWIK1) can unequivocally alter their activity in response to membrane potential changes. It has been recently proposed that a one-way "check valve" mechanism, in which the selectivity filter acts as a voltage-gate, takes place. Depolarization induces filter opening and outward K^+ flow, whereas at membrane potentials below E_K the non-return valve promotes filter inactivation. The second threonine (i.e., Thr157 in TREK1 or Thr103 in TRAAK) in the selectivity filter of P1 plays a major role in this mechanism. Thus, mutagenesis experiments in TREK1 and

TRAAK turn them into a leak mode (Schewe et al. 2016). Interestingly, TREK1 voltage-gated mode is abolished upon pH_i, pressure and PIP_2 activation (Chemin et al. 2005).

5 Role of Lipids/PIP_2

It has been widely recognized that the lipid bilayer can modulate the function of K^+ channels (Forte et al. 1981; Van Dalen and De Kruijff 2004). One such role is for inactivation of K_v channels, where interaction with the membrane causes prolonged channel closing (Schmidt et al. 2009; Schmidt and Mackinnon 2008). Using $K_v1.2$ as an example, the VSDs are embedded in the membrane, with S4 being mostly shielded away from lipids (Long 2005). The top gating charges found in S4 have been modeled to interact with lipid headgroups, making stable electrostatic interactions with their negatively-charged phosphates (Cuello 2004; Lee and Mackinnon 2004; Long et al. 2007). The mechanical properties of the membrane are dictated by lipid composition, and interaction with the headgroups can facilitate sensor movement and subsequently pore opening.

A version of the K_v channel, lacking a sensor region (PDB: 1K4C), exhibits four immobilized lipids filling and surrounding a crevice between subunits on the extracellular surface of the channel, suggesting affinity for lipids at this region (Santos et al. 2012). Inclusion of lipids with headgroups that coat the extracellular membrane-solution interface with hydroxyl groups (e.g., glycerol and phosphoinositol) drastically increases the probability of finding the channel open (Syeda et al. 2014), suggesting that the pore-forming region of the K_v channels may be transformed into an open conductor of K^+ through interaction with lipid modulators that target either the bundle gate, via direct interaction, or the filter gate, by destabilization of water structure. So, not only are lipids critical for proper protein folding (Valiyaveetil et al. 2002), they also allow for modulation of channel properties. An example of this would be the K_v7 channel, where channel opening requires the membrane lipid PIP_2, which serves as a cofactor that mediates coupling of VSD with the pore gate (Zaydman and Cui 2014).

For K_{Ca} channels, several membrane and cholesterol-related lipids have been shown to modulate the activity of some of these channels. For instance, the membrane lipid phosphatidylinositol-4,5-bisphosphate (PIP_2) influences the activity of small and large conductance K_{Ca} channels. $K_{Ca}1.1$ channels can be either activated or inhibited by PIP_2, depending on the β auxiliary subunits to which they are associated (Tian et al. 2015). Particularly, PIP_2 has an inhibitory effect on $K_{Ca}1.1$ in the absence of auxiliary subunits and when they are in complex with $\gamma 1$ subunits, while PIP_2 activates the channel when it is complexed with $\beta 1$ or $\beta 4$ subunits (Tian et al. 2015). Small conductance K_{Ca} channels are also modulated by PIP_2 (Zhang et al. 2014). This phospholipid acts as a cofactor for $K_{Ca}2.x$ channels activation by CaM upon Ca^{2+} binding, whilst PIP_2 removal leads to channel inhibition (Zhang et al. 2014). Moreover, the regulation of $K_{Ca}2.x$ by PIP_2 is dependent on CaM phosphorylation by casein Kinase 2 (CK2), which phosphorylates the amino acid

T80 in CaM weakening the affinity of PIP_2 for the CaM-$K_{Ca}2.x$ complex (Zhang et al. 2014).

Cholesterol is another membrane lipid that modulates $K_{Ca}1.1$ activity (Dopico and Bukiya 2017). The cytosolic C-terminal domain of $K_{Ca}1.1$ channels presents a cholesterol recognition amino acid consensus motif (CRAC4) that confers cholesterol sensitivity to the channel (Singh et al. 2012). Cholesterol has shown to inhibit $K_{Ca}1.1$ channels in heterologous expression systems (Wu et al. 2013), although the in vivo effects of cholesterol enrichment or depletion on $K_{Ca}1.1$ channels activity are in some cases contradictory, depending on the tissue where the channel is expressed (Dopico and Bukiya 2017). $K_{Ca}1.1$ channel activity is modulated as well by certain steroid hormones such as 17β-estradiol or dehydroepiandrosterone (DHEA). When co-expressed with β1 auxiliary subunits, $K_{Ca}1.1$ channels are activated by 17-β-estradiol, which exerts no effect on $K_{Ca}1.1$ α subunits alone (Valverde et al. 1999), nor when they are associated with β2 subunits (King et al. 2006). However, the adrenal androgen DHEA is able to activate β2-associated $K_{Ca}1.1$ channels, an effect also exerted by corticosterone (King et al. 2006). The bile acid lithocholate and the non-steroid leukotriene LTB4 are also potentiators of $K_{Ca}1.1$ channels activity in a β1-dependent manner (Bukiya et al. 2007, 2014). Lastly, the omega-3 lipid docosahexaenoic acid activates β1 and β4-associated $K_{Ca}1.1$ channels, exhibiting no effect when the pore-forming α subunits are in complex with β2 or γ1 subunits (Hoshi et al. 2013). The auxiliary subunit-dependent activation of $K_{Ca}1.1$ channels exerted by some of these lipids could be contributing to vascular smooth muscle relaxation and consequently to vasodilation (Latorre et al. 2017).

The activation of all K_{ir} channels is also dependent on PIP_2 (Hibino et al. 2010; Rohacs et al. 2003). The structural mechanism of PIP_2 binding has been elucidated in two X-ray structures for $K_{ir}2.1$ and GIRK2 (PDB: 3SPI and 3SYA, respectively) (Hansen et al. 2011; Whorton and MacKinnon 2011). The presence of this membrane phospholipid in the inner surface of the plasma membrane is essential for K_{ir} activation (Huang et al. 1998; Li et al. 1999), as well as for the activation mediated by the different endogenous gating regulators of K_{ir} channels (Du et al. 2004), like $K_{ir}6.x$ activation by ATP (Baukrowitz et al. 1998) and GIRK activation by Na^+ and Gβγ (Huang et al. 1998; Rosenhouse-Dantsker et al. 2008). Cholesterol is another membrane lipid that modulates the activity of some K_{ir} channels, such as $K_{ir}2.x$, that become inactive at increasing concentrations of cholesterol (Romanenko et al. 2004), and GIRK channels, which in contrast are activated by cholesterol in a PIP_2-dependent and G-protein-independent manner (Glaaser and Slesinger 2017). Other lipids are also involved in K_{ir} activity modulation. For instance, arachidonic acid has shown to increase the current flow through $K_{ir}2.3$ containing channels (Liu et al. 2001), and the intracellular increase of long-chain CoA esters has an opposite effect on $K_{ir}2.1$ and $K_{ir}6.x$ channels, inhibiting the former and activating the latter (Shumilina et al. 2006).

K_{2P} channels are also influenced by surrounding lipids, most likely through the two lateral portals or fenestrations. Many K_{2P} are activated by PIP_2 (i.e., TASK1-TASK3, TREK1, and TRAAK) leading to a leak K^+ conductance mode. Stimulation of G_q/G_{11} coupled receptors such as muscarinic M1 induces PIP_2 hydrolysis and a subsequent inhibition of TASK1-TASK3, TREK1, and TREK2 (Lopes et al. 2005). However, the relationship between phospholipids and K_{2P} channels can be quite complex in some cases. For example, PIP_2 exerts a dual regulation in TREK1. In transiently transfected cells, intracellular PIP_2 stimulates TREK1 currents in half of the patches and inhibits currents in the other half. Interestingly, pressure, intracellular acidification, and arachidonic acid induced activation are all blocked by the presence of PIP_2. The removal of the C-terminal domain abolished PIP_2-inhibitory capacity, suggesting the implication of this region on the PIP_2-induced gating regulation (Chemin et al. 2007).

6 Trafficking and Accessory Subunits

K_v channels exhibit subfamily-specific patterns of localization within cells (Vacher et al. 2008). For example, in neurons K_v1 channels are expressed at the axon initial segment (AIS). The AIS plays an important role in generating axonal action potentials. K_v1 channels regulate action potential initiation and propagation (Kole and Stuart 2012). Within this channel, specific amino acid sequence motifs act as trafficking determinants (TDs) and direct the initiation, continuation of expression, and localization of these channels to the AIS. TDs are located within the C-terminal domain and act on different interacting proteins (Magidovich et al. 2006). The C-terminal domains are highly conserved in mammalian channels and include a specific motif within the extracellular loop between TM segments S1 and S2 (McKeown et al. 2008) and an acidic motif in the C-terminus of K_v1 α subunits (Manganas et al. 2001). Some K_v subfamily members contain TDs with lower or higher affinity for interacting proteins. The degree of affinity that different subfamily members have for interacting protein leads to trafficking characteristics that are sensitive to co-assembly (Manganas and Trimmer 2000), where localization depends on the TD-interacting protein coupling. TDs also play a role in trafficking from the endoplasmic reticulum (ER). The composition and stoichiometric assembly of K_v1 heterotetrameric channels produces interaction with different proteins and controls ER export of the channel to different loci (Vacher et al. 2007). For example, $K_v1.1$ contains an ER retention signal in its extracellular pore domain that inhibits export from the ER (Manganas and Trimmer 2000; Manganas et al. 2001; Zhu et al. 2001). The retention signal overlaps with the binding site for the neurotoxin α dendrotoxin (DTX), suggesting that $K_v1.1$ retention is due to a DTX-like prototoxin. Phosphorylation also regulates trafficking of $K_v1.2$, with phosphorylation of specific C-terminal tyrosine residues triggering endocytosis of the channels (Nesti et al. 2004). In addition, phosphorylation at a different C-terminal tyrosine residue regulates $K_v1.2$ clustering (Gu and Gu 2011; Smith et al. 2012) and serine phosphorylation sites regulate biogenic trafficking (Yang et al. 2007).

Initial biochemical studies on native K_v1 channels indicated the presence of stoichiometric amounts of copurifying protein components that were initially proposed to be β subunits (Parcej and Dolly 1989). In fact, the majority of K_v1 channels in mammalian brain are associated with $K_v\beta$ subunits (Coleman et al. 1999; Rhodes et al. 1995, 1996, 1997). There are three genes that encode $K_v\beta$ subunits ($K_v\beta1$-$K_v\beta3$), with various alternative splicing leading to a larger number of functionally distinct isoforms (Pongs et al. 1999). Certain $K_v\beta$ subunits contain a domain in the N-terminus region that confers the rapid "N-type" inactivation to K_v channels (Rettig et al. 1994). The N-terminus region acts like the inactivation particle found in some K_v channels that works to occlude the pore of the activated K_v1 channels.

For K_{Ca} channels, the stable association of the pore-forming α subunits with auxiliary subunits confers versatility in their different physiological roles (Berkefeld et al. 2010). The activity of $K_{Ca}1.1$ channel is regulated by several β and γ subunits, which are expressed in different tissues and modify the biophysical and pharmacological properties of the channel (Latorre et al. 2017; Li and Yan 2016). Four different β subunits, β1-β4, have been identified, all of them composed of two transmembrane domains linked by an extracellular loop and with intracellular C- and N-termini (Latorre et al. 2017). The stoichiometry of the association between α and β subunits is generally considered to be 1:1 (Latorre et al. 2017), as revealed by cryo-EM studies (PDB: 6 V22) (Tao and MacKinnon 2019). Interestingly, functional channels can operate with less than four β subunits, exhibiting a proportional modification of the channel properties as the number of β subunits increases (Wang et al. 2002). β1 subunits are mainly expressed in the vascular smooth muscle (Knaus et al. 1994a; Latorre et al. 2017) and enhance the apparent voltage and Ca^{2+}-sensitivity of the channel (McManus et al. 1995) and slow down activation and deactivation kinetics (Dworetzky et al. 1996). In addition, β subunits provide $K_{Ca}1.1$ channel with distinct pharmacological properties, depending on the β subunit to which they are associated (Latorre et al. 2017; Li and Yan 2016). Similarly to β1, β2 subunits increase the apparent Ca^{2+}-sensitivity in the $K_{Ca}1.1$ channel (Brenner et al. 2000a) and decrease the gating kinetics rate (Brenner et al. 2000a). Moreover, β2 subunits are responsible for $K_{Ca}1.1$ channel inactivation (Wallner et al. 1999; Xia et al. 2003), in a process where the N-terminal domain of β2 behaves like a peptide ball that occludes the $K_{Ca}1.1$ channel (Bentrop et al. 2001; Wallner et al. 1999), resembling the ball-and-chain inactivation of K_v channels. In the case of β3 subunits, four different isoforms have been identified (β3a-d) (Uebele et al. 2000), which do not affect $K_{Ca}1.1$ channel Ca^{2+}-sensitivity (Latorre et al. 2017). Alternately, β3a, β3b, and β3c isoforms exert a partial inactivation of the channel (Uebele et al. 2000), while β3b is also responsible for conferring an outward rectification via the extracellular loop (Uebele et al. 2000; Zeng et al. 2003). On the other hand, β4 subunits are mainly expressed in the brain (Weiger et al. 2000) and slow down the activation and deactivation kinetics of $K_{Ca}1.1$ channels (Behrens et al. 2000; Weiger et al. 2000). β4 subunits display a dual effect on Ca^{2+}-sensitivity of $K_{Ca}1.1$ channels: it is reduced in the presence of β4 at low $[Ca^{2+}]_i$, while β4 enhances channel Ca^{2+}-sensitivity at high $[Ca^{2+}]_i$ conditions (Brenner et al. 2000a; Wang et al. 2006).

Four γ subunits have been described (γ1-γ4) (Latorre et al. 2017; Li and Yan 2016), each with a single transmembrane domain and large extracellular leucine-rich repeat N-terminal domain (Yan and Aldrich 2012). Although the stoichiometry of γ association with $K_{Ca}1.1$ subunits has not been yet fully elucidated (Latorre et al. 2017), experiments investigating different ratios indicate a single γ subunit can modulate the activity of the α-homotetramer (Gonzalez-Perez et al. 2014). γ1 is the first subunit identified (Yan and Aldrich 2010) and produces a negative shift in the voltage dependence of the activation of the channel (∿140 mV shift) (Yan and Aldrich 2010), leading to the accelerated activation and slower deactivation kinetics, with no effect on Ca^{2+}-sensitivity (Yan and Aldrich 2010). The γ2, γ3, and γ4 subunits also shift the voltage dependence of activation to more negative potentials, although less intensively than γ1 (∿101 mV shift for γ2, ∿51 mV for γ3, and ∿19 mV for γ4) (Yan and Aldrich 2012). Lastly, it has been recently shown that both β2 and γ1 subunits can simultaneously assemble with $K_{Ca}1.1$ homotetramers, endowing the channel with unique gating properties, being active at resting potentials (Gonzalez-Perez et al. 2015).

Apart from the co-assembly with β and/or γ subunits, $K_{Ca}1.1$ channels co-localize in the cell membrane with Ca_V channels forming multiprotein complexes (Berkefeld et al. 2006, 2010; Grunnet and Kaufmann 2004). Given the Ca^{2+}-sensitivity of these channels, in the micromolar range (Berkefeld et al. 2006), they localize close to Ca^{2+} sources in the membrane to achieve these Ca^{2+} concentrations (Augustine et al. 2003; Berkefeld et al. 2006).

For $K_{Ca}2.x$ and $K_{Ca}3.1$ channels, CaM is considered the β subunit of these channels (Berkefeld et al. 2010). CaM is constitutively bound to the pore-forming subunits of the channel with a 1:1 stoichiometry (Fanger et al. 1999) and has been visualized with cryo-EM (PDB: 6CNM) (Lee and MacKinnon 2018). CaM is responsible for K_{Ca} elevated Ca^{2+}-sensitivity (Xia et al. 1998) and has also been implicated in channel assembly and membrane trafficking (Joiner et al. 2001). Apart from CaM, an additional pair of proteins assembles with the $K_{Ca}2.x$-CaM complexes in the membrane to modulate channel activity: CK2 and protein phosphatase 2A (PP2A) (Adelman et al. 2012; Berkefeld et al. 2010). CK2 and PP2A are constitutively bound to $K_{Ca}2.x$ channels, co-assembling with both the CaMBD and the $K_{Ca}2.x$ intracellular N-terminal domain, thus forming together with CaM a multiprotein complex (Allen et al. 2007; Bildl et al. 2004). Instead of phosphorylating the $K_{Ca}2.x$ subunits, CK2 phosphorylates the amino acid T80 of CaM when the channel is closed, decreasing Ca^{2+}-sensitivity of the channel (Allen et al. 2007). On the other hand, PP2A dephosphorylates CaM in the open state of the channel, allowing it to recover its Ca^{2+}-sensitivity (Allen et al. 2007). In terms of kinase modulation, $K_{Ca}3.1$ activity is influenced by 5'AMP-activated protein kinase (AMPK), which interacts through its γ1 subunit with a leucine zipper domain located in the C-terminus of the channel (Klein et al. 2009).

K_{ir} channels typically associate as homo or heterotetramers without accessory subunits. However, $K_{ir}6.x$ channels function as octamers, composed of four pore-forming $K_{ir}6.x$ subunits, and four auxiliary subunits of the sulfonylurea receptor (SUR1, SUR2A, or SUR2A) (Clement et al. 1997; Shyng and Nichols 1997). The combinations of $K_{ir}6.x$ and SUR auxiliary subunits found in different tissues account for the distinct functional and pharmacological properties of the native channels (Aguilar-Bryan et al. 1998). The intracellular trafficking of some K_{ir} channels is also subjected to protein regulation. Particularly, GIRK channel subunits present trafficking motifs that result in different trafficking patterns of the homo or heterotetramers

(Ma et al. 2002). For instance, GIRK2a (a splicing variant of GIRK2) and GRIK4 present an ER export motif in the N-terminal region and a surface-promoting motif in the C-terminal domain that guide the endosomal exportation of GIRK channels containing these subunits to the cell surface (Ma et al. 2002). In contrast, the lysosomal targeting signal present in GIRK3 downregulates the membrane expression of GIRK3-containing channels (Ma et al. 2002). Sorting nexin 27 (SNX27) also regulates the trafficking of GIRK channels, through the interaction of its PDZ domain with the C-terminal domain of GIRK2c and GIRK3, promoting channel trafficking (Lunn et al. 2007; Munoz and Slesinger 2014). Other PDZ-containing proteins play a role in regulating the localization of some K_{ir} receptors in polarized cells, as is the case of $K_{ir}1.1$ (Yoo et al. 2004), $K_{ir}2.3$ (Olsen et al. 2002), and $K_{ir}4.1$ (Tanemoto et al. 2005).

Few associated proteins have been identified for K_{2P}. The first one is AKAP150, which interacts with TREK channels, both TREK1 and 2. Interestingly, AKAP150 is an A-kinase-anchoring protein, key in the native TREK1 environment that can transform small outward TREK1 currents into big leak K^+ conductance no longer responsive to pressure, intracellular acidification and arachidonic acid induced (Sandoz et al. 2006). New advances in proteomics will surely bring up new K_2P accessory proteins that might be key for its regulation at the activity and trafficking levels. Regulation of K^+ channels by auxiliary subunits is described in more detail in chapter "Control of Biophysical and Pharmacological Properties of K^+ Channels by Auxiliary Subunits".

7 Pharmacology: Blockers and Modulators

The structural and functional diversity among K^+ channels accounts for the wide variety of toxins and small molecules that modulate the activity of these channels. Various toxins exploit different K_v channel characteristics to exert their actions on the channel. One common class of toxins work by occluding the narrow pore of the channel from its extracellular side preventing ion flow and are referred to as "pore blockers." Many of these toxins are composed of a positive lysine and a hydrophobic tyrosine/phenylalanine in a dyad motif (Eriksson and Roux 2002; Gao and Garcia 2003; Miller 1995). With this arrangement the lysine residue occludes the K_v channel selectivity filter and prevents K^+ ions from entering the channel, at the same time the hydrophobic portion of the dyad aids docking and toxin binding to the channel (Dauplais et al. 1997; Gilquin et al. 2002; Savarin et al. 1998; Srinivasan et al. 2002). Examples of this class of toxins include κM-RIIIK and ConK-S1 from cone snails (Al-Sabi et al. 2004; Jouirou et al. 2004) and ShK from a sea anemone (Finol-Urdaneta et al. 2020). Another distinct mechanism is exhibited by "gating modifiers," which bind to the extracellular exposed linker between the TM segments S3 and S4 within the VSD. These toxins inhibit channel function, increasing the energy required to open the channel by shifting the voltage dependence to more depolarized potentials raising the activation threshold. An example of this class is the HaTx toxin from spiders (Tudor et al. 1996). δ-dendrotoxin (DTX), isolated from the green mamba snake venom, is another well-known blocker of K_v channels (Harvey and Anderson 1985; Harvey and Robertson 2004). Various toxins, in particular conotoxins produced by marine cone snails, have been employed as molecular tools for the study of K_v channels in mammalian targets (Teichert et al. 2015). You can read more on K^+ channel toxins in chapter "Peptide Toxins Targeting K_v Channels".

In addition to toxins, work is currently being conducted to identify small molecule modulators of K_v channels which could prove useful for treating various brain disorders. For example, K_v channel activators could be used to dampen hyperexcitability for treating epilepsy or attention deficit disorder (Wulff et al. 2009). K_v channel inhibitors, on the other hand, could be used to increase excitability in disorders involving reduced neuronal activity, such as multiple sclerosis (MS). For a detailed review of the potential therapeutic utility of K_v modulators, see (Wulff et al. 2009). One example is 4-aminopyridine (4-AP) which is a non-selective K_v channel inhibitor (Wu et al. 2009) which has undergone phase III clinical trials for the treatment of MS (Goodman et al. 2009; Korenke et al. 2008). Another example is dofetilide, a class III antiarrhythmic and inhibitor of K_v11.1, that is efficacious in reverting and preventing atrial fibrillation of the heart (Kamath and Mittal 2008).

For K_{Ca} channels, there are different modulators for large, intermediate, and small conductance channels (Kshatri et al. 2018). For example, K_{Ca}1.1 channels are blocked by tetraethylammonium (TEA) (Blatz and Magleby 1984; Villarroel et al. 1988), like many K_v channels (Bretschneider et al. 1999), whereas K_{Ca}2.x and K_{Ca}3.1 are not affected by this quaternary amine. K_{Ca} subtypes also exhibit different sensitivities to toxins (Kshatri et al. 2018). K_{Ca}1.1 channels are classically inhibited by the scorpion venom peptide iberiotoxin (IbTX) (Galvez et al. 1990) and charybdotoxin (ChTX) (Miller et al. 1985). The selectivity of ChTX and IbTX for K_{Ca}1.1 channels depends on the type of β subunit associated with the α pore-forming subunit, demonstrating how auxiliary subunits can modify the pharmacological properties of the channel (Latorre et al. 2017). For example, association of K_{Ca}1.1 channels in complex with β2/3 or β4 subunits decreases the affinity for ChTX (Meera et al. 2000; Xia et al. 1999), while channels associated with β1 are highly sensitive to ChTX (Hanner et al. 1997). In the case of IbTX, β4-associated K_{Ca}1.1 channels are resistant to the blockade by this toxin (Meera et al. 2000). Slotoxin (Garcia-Valdes et al. 2001) and martentoxin (Shi et al. 2008), two scorpion venom toxins closely related to ChTX and IbTX, are potent K_{Ca}1.1 blockers. Their affinity for the channel is also dependent on the β subunit composition, with slotoxin weakly blocking K_{Ca}1.1 channels assembled with β4 subunits (Garcia-Valdes et al. 2001), while martentoxin exhibits an opposite behavior and selectively blocks $\alpha + \beta 4$ K_{Ca}1.1 (Shi et al. 2008). Other natural toxins that inhibit K_{Ca}1.1 channels are the scorpion venom toxin BmP_{09} (Yao et al. 2005), and the fungal alkaloids paxilline, panitrem, and lolitrem B, which have shown to block the channel at low nanomolar concentrations (Imlach et al. 2009; Knaus et al. 1994b). K_{Ca}2.x channels are characterized by their sensitivity to inhibition with the bee venom toxin apamin (to which K_{Ca}1.1 are insensitive) (Grunnet et al. 2001; Weatherall et al. 2010). K_{Ca}2.x channels are also inhibited by the scorpion venom toxin tamapin (Pedarzani et al. 2002). Lastly, like K_{Ca}1.1 channels, K_{Ca}3.1 are blocked by ChTX (Sforna et al. 2018; Wei et al. 2005) and by another scorpion venom peptide toxin, maurotoxin

(Castle et al. 2003), the latter having a high affinity and selectivity for this subfamily of K_{Ca} channels (Castle et al. 2003).

For small molecule modulators of K_{Ca} channels, several activators and inhibitors have been described for the different subfamilies. For example, $K_{Ca}1.1$ channels are activated by the synthetic compounds NS1608 (Strobaek et al. 1996) and BMS-204352 (Gribkoff et al. 2001), which show promise in in vivo models for the treating fragile X syndrome (Hebert et al. 2014). For $K_{Ca}2.x$ channels, the inhibitors UCL1684 (Strobaek et al. 2000) and NS8593 (Strobaek et al. 2006) have been described. Regarding $K_{Ca}2.x$ activators, CyPPA (Hougaard et al. 2007), NS13001 (selective for $K_{Ca}2.2/K_{Ca}2.3$) (Kasumu et al. 2012), and NS309 (Strobaek et al. 2004) are of notice, the latter acting by increasing the Ca^{2+}-sensitivity of the channel. Moreover, the $K_{Ca}2.x$ activator EBIO (Devor et al. 1996) has shown in vivo efficacy as an anticonvulsant (Anderson et al. 2006). Chlorzoxazone is a $K_{Ca}2.2$ activator (Cao et al. 2001) and muscle relaxant that has been approved for the treatment of severe spasticity (Losin and McKean 1966). For $K_{Ca}3.1$ channels, the antifungal drug clotrimazole is a classical small molecule blocker (Wulff et al. 2000), and has been used as scaffold for the development of $K_{Ca}3.1$ inhibitors such as TRAM-34, which exhibits high selectivity for $K_{Ca}3.1$ (Wulff et al. 2000). On the other hand, some $K_{Ca}2.x$ activators also act on $K_{Ca}3.1$ channels to enhance their activity, such as for the benzimidazolones EBIO (Devor et al. 1996), DCEBIO (Singh et al. 2001), NS309 (Strobaek et al. 2004), and SKA-31 (Sankaranarayanan et al. 2009). In short, the strategies that pursue the activation of $K_{Ca}1.1$ and $K_{Ca}2.x$ channels, or the inhibition of $K_{Ca}3.1$ channels, are the most important when it comes to the treatment of diseases involving K_{Ca} channels (Kshatri et al. 2018).

While the biophysical features of K_{ir} channels have been thoroughly studied, their pharmacological modulation remains largely unexplored. Initially, inorganic cations like Ba^{2+} and Cs^{+} were found to block the majority of K_{ir} channels (Hagiwara et al. 1976, 1978), in a voltage- and $[K^{+}]_{o}$-dependent manner (Quayle et al. 1993). Nevertheless, $K_{ir}7.1$ shows a much lower sensitivity to the blockade by these cations (Krapivinsky et al. 1998). Interestingly, the K_{v} channel blockers tetraethylammonium (TEA) and 4-aminopyridine (4AP) have little effect on K_{ir} channels (Hagiwara et al. 1976; Oonuma et al. 2002). Several naturally occurring toxins have been described as blockers of some K_{ir} channels (Doupnik 2017). The bee venom peptide toxin tertiapin is a $K_{ir}1.1$ and GIRK channel blocker (Jin and Lu 1998; Kanjhan et al. 2005), as well as its synthetic oxidation-resistant derivative tertiapin-Q (Jin and Lu 1999). Tertiapins are not effective blockers of $K_{ir}2.1$ channels (Jin and Lu 1998). The scorpion venom peptide toxin Lq2, an isoform of charybdotoxin (ChTX), is also a $K_{ir}1.1$ blocker, and again has no inhibitory effect on $K_{ir}2.1$ (Lu and MacKinnon 1997). δ-dendrotoxin (DTX), isolated from the green mamba snake venom, is a well-known blocker of K_{v} channels (Harvey and Anderson 1985; Harvey and Robertson 2004) that is also a potent $K_{ir}1.1$ inhibitor (Imredy et al. 1998).

In addition to toxins, small chemical modulators have been isolated that activate some K_{ir} channels. GIRK channels are activated by small molecules such as the ureas ML297 and GiGA1, which have shown promising anticonvulsant (Kaufmann et al. 2013; Zhao et al. 2020) and anxiolytic effects (Wydeven et al. 2014). Ethanol, as well as other short-chain alcohols, also activates GIRK channels (Kobayashi et al. 1999), implicating these channels in alcohol motivational and addictive effects (Rifkin et al. 2017). Activators and inhibitors of $K_{ir}6.x$, through the interaction with their SUR auxiliary subunits, have therapeutic applications in human (Hibino et al. 2010). For example, sulfonylureas, such as tolbutamide (Ashfield et al. 1999), glibenclamide (Schmid-Antomarchi et al. 1987), or glimepiride, block the $K_{ir}6.2$/SUR1 channel, stabilizing a conformation of SUR1 that prevents the pore opening (Doyle and Egan 2003). The recent structural determination of the $K_{ir}6.2$/SUR1 complex by cryo-EM (Lee et al. 2017; Martin et al. 2017) has helped in the identification of the sulfonylurea glibenclamide binding site (PDB: 5TWV) (Martin et al. 2017). These drugs have an important clinical use in the treatment of diabetes mellitus II, since the blockade of $K_{ir}6.2$/SUR1 expressed in pancreatic β cells promotes insulin secretion (Ashcroft 2005). Potassium channel openers (KCO), such as nicorandil (Horinaka 2011) and pinacidil (Muiesan et al. 1985), activate $K_{ir}6.x$ channels upon SUR binding and are used in the treatment of myocardial infarction, ischemia-reperfusion injury, and hypertension (Grover and Garlid 2000; Mannhold 2004).

For K_{2P} channels, many different halogenated anesthetics, such as isoflurane or sevoflurane, stimulate the channels (i.e., TREK1, TASK1-TASK3, TRAAK, and TRESK). These volatile anesthetics increase K_{2P} channel open probability and K^+ conductance, resulting in membrane hyperpolarization (Patel et al. 1999; Plant 2012). The mechanism of action, however, is not fully understood but some evidence suggests it involves the C-terminal domain. In addition, halogenated anesthetics could disrupt the inhibitory influence of the G_q/G_{11} (Chen et al. 2006).

Selective serotonin reuptake inhibitors (SSRI) fluoxetine and norfluoxetine inhibit TREK1-TREK2 throughout the lateral portals, as visualized in the TREK2 structure (Dong et al. 2015). Interestingly, some clinical studies have described analgesic activity as a side effect of these antidepressants, which can be explained by their influence on K_{2P} channels (Kennard et al. 2005), which is puzzling to explain since the inhibition of K_{2P} channels is expected to increase pain. For example, fenamates are nonsteroidal anti-inflammatory drugs that selectively activate lipid-sensitive mechano-gated K_{2P} channels, which is, together with the inhibition of pro-excitatory ion channels, the mechanism of analgesic action (Takahira et al. 2005). Fenamates exert their influence by interacting with the N-terminus of K_{2P} channels (Veale et al. 2014). Finally, new compounds, like arylsulfonamide, ML335 and the thiophene-carboxamide ML402, have been reported as activators of TREK1,2/TRAAK (Lolicato et al. 2017).

8 Physiology and Function

Potassium channels support a wide array of functions within the body and the brain. Due to the extensive diversity of the K^+ channel family, a discussion of the roles all these channels play in physiology and function is beyond the scope of this chapter. Instead, we will describe examples for each of the four general K^+ channel families.

K_v channels most notably play a role in the excitability of neurons and help shape action potentials. K_v1.1 is one of the most abundant K_v1 subunits expressed in mammalian brain (Trimmer and Rhodes 2004) and often exists as part of a heteromeric channel complex (Rhodes et al. 1997). K_v1.1 associates with K_v1.2 at axon initial segments (Dodson et al. 2002; Inda et al. 2006; Van Wart et al. 2007), where they control synaptic efficacy via modulation of the action potential (Kole et al. 2007). The role of these channels in controlling neuronal excitability was revealed using venom toxins such as dendrotoxin, which can elicit seizures in rodents (Bagetta et al. 1992). In addition, mice lacking K_v1.1 are predisposed to seizures and exhibit spontaneous seizures and changes in CNS structure (Smart et al. 1998). In humans, several loss-of-function mutations have been identified that have been linked to episodic ataxia, myokymia disorders, and partial seizures (Zuberi et al. 1999). In addition to direct loss of K_v1.1 channel function, mutations in a protein that co-expresses with K_v1.1, the leucine-rich glioma-inactivated protein 1 (LGI1), have been associated with temporal lobe epilepsy (Schulte et al. 2006). In these examples, errant changes in action potential firing frequency can lead to various neurological and psychological disorders such as epilepsy. Targeting K_v1.1 subunit containing channels with some form of intervention could rescue the increased likelihood of seizures and epileptiform activity observed in humans with loss-of-function mutations.

K_v11.1 channels, often referred to as human ether-a-go-go (hERG) channels (Kaplan and Trout 3rd 1969), are particularly important in heart tissue. There are two kinetically distinct components of the delayed rectifier potassium current observed in cardiac myocytes, referred to as the rapid delayed rectifier (I_{Kr}) and the slow delayed rectifier (I_{Ks}) (Noble and Tsien 1969a, b). These two components are sufficient to account for cardiac repolarization (Noble and Tsien 1969b). I_{Kr} is mediated by K_v11.1 and displays a telltale "hook" characteristic of these channels when being recorded during deactivation (Shibasaki 1987). In cardiac cells, the slow activation and deactivation kinetics of K_v11.1 coupled with rapid voltage-dependent inactivation and recovery from inactivation make the current that passes through the channels ideal for determining the duration of plateau phase of atrial and ventricular myocyte action potentials (Sanguinetti et al. 1995; Smith et al. 1996) (Fig. 4). The maintenance of this plateau is critical for ensuring sufficient time for Ca^{2+} release from the sarcoplasm to enable cardiac contraction, and K_v11.1 current contributes to pacemaking activity of the sinoatrial and atrioventricular node cells (Clark et al. 2004; Furukawa et al. 1999; Mitcheson and Hancox 1999). K_v11.1 is the molecular target for most drugs that cause drug-induced arrhythmias (Sanguinetti et al. 1995), many of which require channel opening prior to gaining access to receptor site within the inner cavity of the channel pore (Carmeliet 1992; Kiehn et al. 1996; Yang et al.

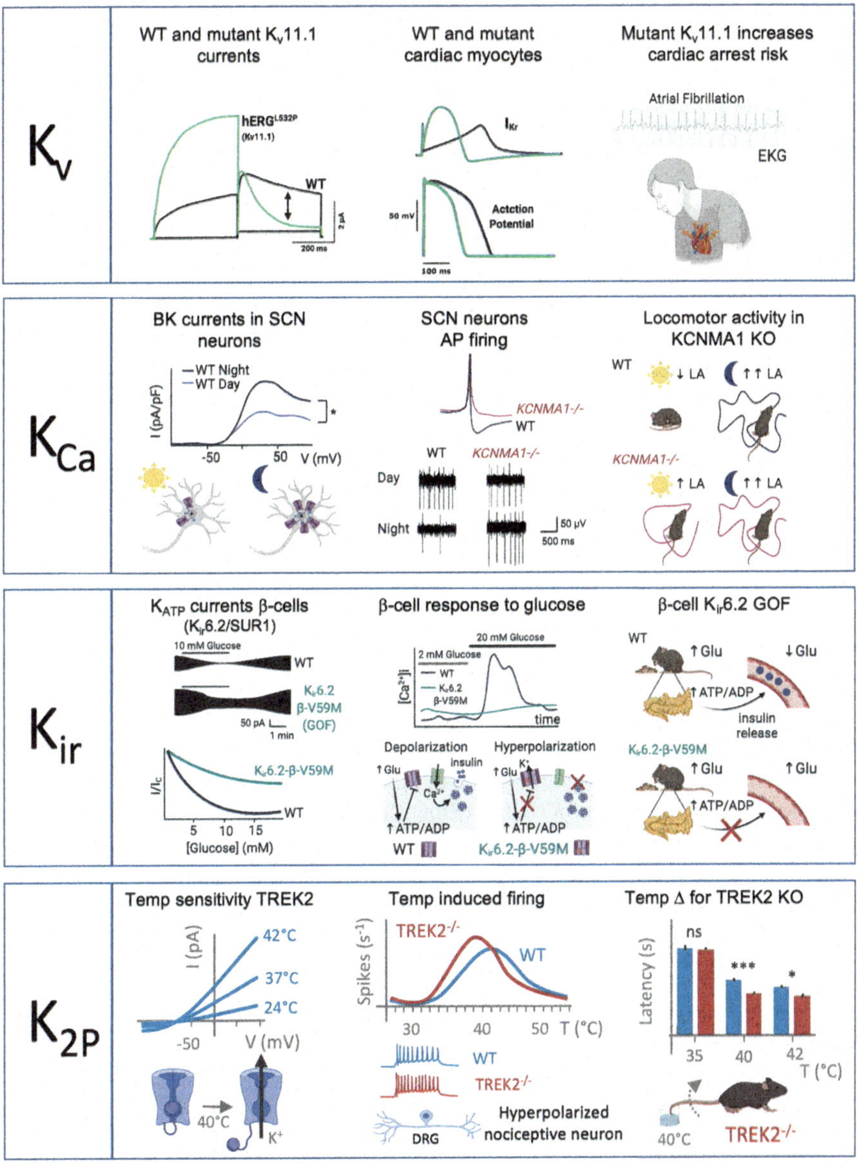

Fig. 4 Examples of physiological role of K_V, K_{Ca}, K_{ir}, and K_{2P} channels. K_V Mutant $K_v11.1$ (hERG) in cardiac myocytes (green) exhibits a rapid repolarization as compared to wild type (black) $K_v11.1$ channels. Expression of Mutant $K_v11.1$ results in a change in I_{Kr} and a less sustained action potential in cardiac tissue, an indication of short QT syndrome (SQTS). SQTS produced by the expression of mutant $K_v11.1$ channels can lead to atrial fibrillation and sudden cardiac arrest. K_{Ca} $K_{Ca}1.1$ (BK) channels in the suprachiasmatic nuclei (SCN) are essential for the maintenance of the circadian rhythm. The expression of BK channels is enhanced at night, decreasing SNC neuronal activity to keep an activity pattern (high SCN neuronal activity during daytime, and low at nighttime) responsible for circadian rhythmicity. In *KCNMA1*$^{-/-}$ mice this neuronal activity pattern is lost, with SNC neurons similarly active during day and night. This leads to an increased locomotor activity of *KCNMA1*$^{-/-}$ mice during the day, due to the altered SCN pacemaker function

1995). All clinical compounds developed need to be screened for off-target activity on $K_v11.1$ channel, which could potentially result in arrhythmias. Mutations in the *KCNH2* gene that encodes $K_v11.1$ underlies chromosome 7-associated long QT syndrome (LQTS type 2), accounting for roughly 40% of cases of genetically confirmed LQTS (Fig. 4).

In all genotypes that produce long QT intervals there are some common traits: they occur at an early onset and LQTS carries an increased risk for sudden cardiac arrest (SCA) (Priori et al. 2003). Individuals with LQTS exhibit disruptions in T-wave morphology that are characteristic for different subtypes of LQTS (type 1, 2, and 3) (Moss et al. 1995; Zhang et al. 2000). Disruptions in T-waves may not be apparent at rest, especially in individuals with type 2 LQTS, who develop a bifid or notched T-wave appearance during exercise (Takenaka et al. 2003). It is still unknown why reduced $K_v11.1$ function presents with this bifid or notched T-wave pattern; however, some initial evidence suggests this phenotype may be due to an increase in transmural dispersion of repolarization in cardiac myocytes when $K_v11.1$ current is reduced (Shimizu and Antzelevitch 2000). With LQTS different behaviors are most associated with negative cardiac events or SCA. For type 1, exercise is the primary risk factor (62% of individuals), arousal is the primary trigger (43%) for type 2, and rest or sleep is the most common trigger (49%) for type 3 (Schwartz et al. 2001). While risk factors have been well defined and tools like EKG (Fig. 4) can be used to diagnose individuals, there is still limited understanding of the influence common polymorphisms of the *KCNH2* gene can play in the disorder.

In contrast to LQTS, short QT syndrome (SQTS) is a disorder with a shortened duration of the QT interval on an electrocardiogram. This disorder is usually accompanied by atrial fibrillation (Patel et al. 2010). Like LQTS, SQTS appears to arise from mutations in the *KCNH2* (Zhang et al. 2011) which result in reduced inactivation and a greater current flow during the plateau of the cardiac potential (Fig. 4), leading to the ventricular and atrial action potential having a shorter duration and a shortening of the QT interval and predispose the individual to sudden cardiac death (Brugada et al. 2004). More on cardiac K^+ channels can be found in

---

Fig. 4 (continued) (Meredith et al. 2006; Montgomery et al. 2013). *K_{ir}* $K_{ir}6.2$ is expressed in pancreatic β cells together with its auxiliary subunit SUR1, where it plays a key role in glucose homeostasis: upon food intake and subsequent glucose and ATP increase, the latter inhibits $K_{ir}6.2$/SUR1 channels, promoting cell depolarization, Ca^{2+} intracellular increase, and insulin secretion. The human gain of function mutation β-V59M in $K_{ir}6.2$ leads to a type of neonatal diabetes. Mice bearing this mutation reproduce the human phenotype, exhibiting decreased K_{ATP} currents and glucose response (measured as intracellular Ca^{2+} increase) in β cells (Girard et al. 2009) and impaired insulin secretion upon food intake (Brereton and Ashcroft 2013). *K_{2P}* TREK2 increases K^+ outflow in response to heat, within the 24–42°C range. Nociceptive expresses TREK2, which regulates non-aversive warmth perception. In a wild-type individual, heat-sensitive c-fibers increase their firing activity gradually as they approach 42°C. The lack of TREK2 in mice significantly increases the number of action potentials by 30% in the 30–40°C range compared to wild type. $TREK2^{-/-}$ mice exhibit hyperalgesia, since tail flick latencies upon 40 and 42°C bath immersion are reduced chapters "Cardiac K^+ Channels and Channelopathies" and "Cardiac hERG K^+ Channel as Safety and Pharmacological Target".

K_{Ca} channels are widely expressed in both neuronal and non-neuronal tissues, where they play a diversity of physiological roles that are based on their ability to couple membrane potential and the intracellular Ca^{2+} concentration (Berkefeld et al. 2010). Increases in $[Ca^{2+}]_i$ lead to an outward K^+ flux through K_{Ca} channels that contributes to cell hyperpolarization. This helps to maintain Ca^{2+} homeostasis, limiting Ca^{2+} influx either through voltage-gated Ca^{2+} channel inactivation or increasing the activity of Na^+/Ca^{2+} exchangers (Fakler and Adelman 2008). $K_{Ca}1.1$ channels (BK channels) are mainly expressed with β1 subunits in the vascular smooth muscle (Latorre et al. 2017), where these play a key role in the regulation of the vascular tone (Brenner et al. 2000b; Latorre et al. 2017). Ca^{2+} release from the sarcoplasmic reticulum forms Ca^{2+} sparks that activate BK channels, inducing vasodilation (Latorre et al. 2017; Pluger et al. 2000). The dysfunction of $K_{Ca}1.1$ channels or β1 mutations is involved in altered vasoregulation, such as hypertension (Dogan et al. 2019; Latorre et al. 2017). $K_{Ca}1.1$ expression has also been described in the intercalated cells of the kidney, in complex with either β1 or β4 subunits, where they participate in K^+ secretion (Holtzclaw et al. 2011). Importantly, $K_{Ca}1.1$ channels are abundant and broadly found in the CNS (Sausbier et al. 2006), where they mainly associate with β4 subunits (Weiger et al. 2000). In neurons, $K_{Ca}1.1$ channels contribute to different processes involved in neuronal excitability, such as AP repolarization (Storm 1987), mediation of the fast phase of the AHP (Gu et al. 2007; Lancaster and Nicoll 1987; Storm 1987) and shaping the Ca^{2+} dendritic spikes (Golding et al. 1999), as well as to the modulation of neurotransmitter release (Griguoli et al. 2016; Yazejian et al. 2000).

Of particular interest is the role of $K_{Ca}1.1$ channels expressed in the suprachiasmatic nuclei (SCN) in the regulation of the circadian rhythm (Fig. 4) (Meredith et al. 2006; Montgomery et al. 2013; Whitt et al. 2016). $K_{Ca}1.1$ expression and outward currents are increased during nighttime (Montgomery et al. 2013), decreasing SCN neuronal activity at night. This decrease in activity is essential to maintain the high amplitude of the neural activity pattern in the SCN that restricts locomotor activity to the appropriate phase (night) (Montgomery et al. 2013). In *KCNMA1* (gene encoding $K_{Ca}1.1$) knockout mice, the SCN neural activity amplitude is lost, altering the SCN pacemaker function, and making mice more active during daytime (Meredith et al. 2006) (Fig. 4).

$K_{Ca}3.1$ expression has been observed in diverse set of cells, including epithelial, vascular endothelial, vascular smooth muscle cells (Wulff and Castle 2010), hematopoietic cells, such as erythrocytes, lymphocytes, monocytes, and macrophages (Logsdon et al. 1997), and in CNS-resident immune cells, namely microglia (D'Alessandro et al. 2018; Ferreira et al. 2014). $K_{Ca}3.1$ plays an essential role in regulating cellular volume (Sforna et al. 2018), mediating the Ca^{2+}-dependent K^+ efflux that is part of the regulatory volume decrease (RVD) that occurs upon cell swelling (Sforna et al. 2018; Vandorpe et al. 1998). This regulation of cellular volume links $K_{Ca}3.1$ channels with cell migration, since volume increases in one edge for protrusion and decreases for retraction (D'Alessandro et al. 2018).

Interestingly, this role of $K_{Ca}3.1$ channels also accounts for their involvement in glioblastoma multiforme, where $K_{Ca}3.1$ is necessary for cell infiltration, and which expression correlates with worse prognosis (D'Alessandro et al. 2018; Turner et al. 2014).

K_{ir} channels are widely expressed throughout the organism, playing a variety of roles in different cells and tissues. Their characteristic inward rectification accounts for their contribution not only to the maintenance of the resting membrane potential in excitable cells (Hibino et al. 2010), but also to the preservation of ionic gradients in renal tissues (Welling 2016). The ATP sensitivity of $K_{ir}6.x$ channels (Terzic et al. 1995) accounts for their physiological role, coupling the cellular metabolism with the excitability of the membrane (Tinker et al. 2018). Several $K_{ir}6.x$ and SUR subunits combinations are expressed in different tissues (Hibino et al. 2010). Cardiac myocytes and skeletal muscle express $K_{ir}6.2$/SUR2A, where they play a protective role against ischemia-reperfusion (Suzuki et al. 2002) and as linkers to glucose metabolism (Weik and Neumcke 1989), respectively. In vascular smooth muscle, the predominant isoform is $K_{ir}6.1$/SUR2B (Aziz et al. 2014), which participates in the regulation of the vascular tone (Aziz et al. 2014). $K_{ir}6.2$/SUR1 channels have been described in hypothalamic neurons (Ashford et al. 1990), where they play a role in coupling glucose metabolism to glucagon secretion (Miki et al. 2001), and also in pancreatic β cells (Fig. 4). In these cells, $K_{ir}6.2$/SUR1 are key players in glucose homeostasis, linking glucose metabolism to insulin secretion (Ashcroft et al. 1984). An increase in glucose levels elevates intracellular ATP, which binds to SUR1 closing $K_{ir}6.2$ channel pore (Fig. 4) and promoting β cell depolarization, with the subsequent increase of intracellular Ca^{2+} that leads to insulin secretion (Hibino et al. 2010). Importantly, mutations in $K_{ir}6.2$ and SUR1 lead to a range of insulin secretion disorders (Fig. 4) (Remedi and Koster 2010). Gain of function mutations are responsible for different types of neonatal diabetes (Gloyn et al. 2004) (Tinker et al. 2018), while loss of function mutations in both $K_{ir}6.2$ and SUR1 cause congenital hyperinsulinism and hypoglycemia (Nestorowicz et al. 1997; Tinker et al. 2018). Sulfonylureas block $K_{ir}6.2$ channels through their interaction with the SUR subunits and are commonly used for the treatment of diabetes (Ashcroft 2005).

In the case of $K_{ir}4.x$ and $K_{ir}5.1$ channels, $K_{ir}4.1$ homotetramers and $K_{ir}4.1/K_{ir}5.1$ heterotetramers are abundantly expressed in astrocytes (Hibino et al. 2004a) and in retinal Müller glial cells (Ishii et al. 2003), where they play an essential role in the spatial buffering extracellular K^+, helping maintain the osmotic balance (Hibino et al. 2004a; Ishii et al. 2003). $K_{ir}4.1/K_{ir}5.1$ and $K_{ir}4.2/K_{ir}5.1$ channels have also been found in the kidney, particularly in the basolateral surface of renal epithelial cells (Lourdel et al. 2002; Tanemoto et al. 2000), where they contribute to the maintenance of the driving force required for Na^+ reabsorption by recycling K^+ across the basolateral membrane (Huang et al. 2007; Palygin et al. 2017). Moreover, $K_{ir}4.1/5.1$ is also expressed in the cochlea of the inner ear (Hibino et al. 1997, 2004b), contributing to the generation of the endocochlear potential of the inner ear endolymph (Hibino and Kurachi 2006). Mutations in $K_{ir}4.1$ lead to the SeSAME syndrome (Scholl et al. 2009), with a symptomatology characterized by seizures, sensorineural deafness, ataxia, mental retardation, and electrolyte imbalance (SeSAME), that correlates with $K_{ir}4.1$ expression in the organism (Scholl et al. 2009).

K_{2P} channels are expressed in motoneurons (Berg et al. 2004; Talley et al. 2000), dorsal root ganglion (DRG) neurons (Kang and Kim 2006; Pereira et al. 2014), cortical, hippocampal, hypothalamic neurons (Fink et al. 1996; Medhurst et al. 2001), cerebellar granule neurons (Plant et al. 2002) and cortical astrocytes (Hwang et al. 2014). In particular, K_{2P} channels in the DRG control the generation of an action potential through thermal-gating of TREK2 (Fig. 4). TREK2 increases K^+ outflow in response to heat, within the 24–42°C range (Kang et al. 2005). Nociceptive neurons in the DRGs that innervate most of the body surface express TREK2, which regulates non-aversive warmth perception. In a wild-type individual, heat-sensitive c-fibers increase their firing activity gradually as they approach 42°C, due to the activation of mainly thermosensitive transient receptor potential (TRP) channels (Caterina et al. 1997). However, in mice lacking TREK2, the number of action potentials significantly increases by 30% in the 30–40°C range compared to wild type. Additionally, $TREK2^{-/-}$ mice exhibit hyperalgesia, since tail flick latencies upon 40 and 42°C bath immersion are reduced (Pereira et al. 2014) (Fig. 4). Overall, the channel, neuron, and animal behavior indicate the following: TREK2 by being active with heat contributes to a hyperpolarizing environment in the nociceptive neurons which dampen nociceptive signals upon non-aversive warmth.

K_2P malfunctioning has been extensively associated to different pain manifestations such as neuropathic pain or migraine. TRESK also contributes to background current in DRG neurons (Plant 2012; Tulleuda et al. 2011). TRESK is downregulated in spared nerve injury (SNI) model of chronic pain in rats. Interestingly, the hyperalgesia and gliocytes activation are reduced after inducing recombinant TRESK gene overexpression (Zhou et al. 2017). TWIK1 and TASK3 are also reduced in SNI. However, their levels are restored after weeks in the case of TWIK1 and months for TASK3 (Pollema-Mays et al. 2013). Multiple TRESK mutations in humans have been associated to migraine with aura (Lafreniere et al. 2010; Rainero et al. 2014). Proximal point mutations in regions close to the pore (i.e., A34V and C110R) lead to smaller TRESK currents (Andres-Enguix et al. 2012). TRESK deletions display dominant-negative phenotype in heterologous systems when is co-expressed with wild-type TRESK (Lafreniere and Rouleau 2011). Together, these studies suggest TRESK as a target for new analgesics.

A noteworthy contribution of the K_2P channels in the brain is their implication on glutamate release from astrocytes. Classically, neurons have been exclusively attributed for fast glutamatergic synaptic transmission. However, recent studies have shown astrocytes induce slow and fast glutamate release, involving mechanisms of neuron-like exocytosis or transporter/channel mediated. Astrocytes display a leaky membrane with a low resistance, attributed to primarily the outwardly rectifier TREK1 (Fink et al. 1996). Activation of TREK1, either directly or upon CB1 activation Gβγ to the N-terminal, induces astrocytic glutamate release (Woo et al. 2012). TREK1 downregulation eliminates glutamate release fast mode but does not affect the slow mode. Interestingly, the "non-functional" TWIK1 is also expressed in astrocytes and forms a functional heteromer with TREK1 (Hwang et al. 2014).

Acknowledgements Figures created with BioRender.com

References

Abbott GW, Butler MH, Bendahhou S, Dalakas MC, Ptacek LJ, Goldstein SAN (2001) MiRP2 forms potassium channels in skeletal muscle with Kv3.4 and is associated with periodic paralysis. Cell 104:217–231

Abbott GW, Butler MH, Goldstein SAN (2006) Phosphorylation and protonation of neighboring MiRP2 sites: function and pathophysiology of MiRP2-Kv3.4 potassium channels in periodic paralysis. FASEB J 20:293–301

Adelman JP (2016) SK channels and calmodulin. Channels (Austin) 10:1–6

Adelman JP, Maylie J, Sah P (2012) Small-conductance Ca2+−activated K+ channels: form and function. Annu Rev Physiol 74:245–269

A-González N, Castrillo A (2011) Liver X receptors as regulators of macrophage inflammatory and metabolic pathways. Biochim Biophys Acta (BBA) Mol Basis Dis 1812:982–994

Aguilar-Bryan L, Clement JPt, Gonzalez G, Kunjilwar K, Babenko A, Bryan J (1998) Toward understanding the assembly and structure of KATP channels. Physiol Rev 78:227–245

Ahern CA, Horn R (2004) Specificity of charge-carrying residues in the voltage sensor of potassium channels. J Gen Physiol 123:205–216

Ahern CA, Horn R (2005) Focused electric field across the voltage sensor of potassium channels. Neuron 48:25–29

Albrecht B, Weber K, Pongs O (1995) Characterization of a voltage-activated K-channel gene cluster on human chromosome 12p13. Receptors Channels 3:213–220

Aldrich RW (2001) Fifty years of inactivation. Nature 411:643–644

Allen D, Fakler B, Maylie J, Adelman JP (2007) Organization and regulation of small conductance Ca2+−activated K+ channel multiprotein complexes. J Neurosci 27:2369–2376

Al-Sabi A, Lennartz D, Ferber M, Gulyas J, Rivier JEF, Olivera BM, Carlomagno T, Terlau H (2004) κM-conotoxin RIIIK, structural and functional novelty in a K+channel antagonist†. Biochemistry 43:8625–8635

Anazco C, Pena-Munzenmayer G, Araya C, Cid LP, Sepulveda FV, Niemeyer MI (2013) G protein modulation of K2P potassium channel TASK-2: a role of basic residues in the C terminus domain. Pflügers Arch 465:1715–1726

Anderson NJ, Slough S, Watson WP (2006) In vivo characterisation of the small-conductance KCa (SK) channel activator 1-ethyl-2-benzimidazolinone (1-EBIO) as a potential anticonvulsant. Eur J Pharmacol 546:48–53

Andres-Enguix I, Shang L, Stansfeld PJ, Morahan JM, Sansom MS, Lafreniere RG, Roy B, Griffiths LR, Rouleau GA, Ebers GC et al (2012) Functional analysis of missense variants in the TRESK (KCNK18) K channel. Sci Rep 2:237

Åqvist J, Luzhkov V (2000) Ion permeation mechanism of the potassium channel. Nature 404:881–884

Ashcroft FM (2005) ATP-sensitive potassium channelopathies: focus on insulin secretion. J Clin Invest 115:2047–2058

Ashcroft FM, Harrison DE, Ashcroft SJ (1984) Glucose induces closure of single potassium channels in isolated rat pancreatic beta-cells. Nature 312:446–448

Ashfield R, Gribble FM, Ashcroft SJ, Ashcroft FM (1999) Identification of the high-affinity tolbutamide site on the SUR1 subunit of the K(ATP) channel. Diabetes 48:1341–1347

Ashford ML, Boden PR, Treherne JM (1990) Glucose-induced excitation of hypothalamic neurones is mediated by ATP-sensitive K+ channels. Pflugers Arch 415:479–483

Augustine GJ, Santamaria F, Tanaka K (2003) Local calcium signaling in neurons. Neuron 40:331–346

Aziz Q, Thomas AM, Gomes J, Ang R, Sones WR, Li Y, Ng KE, Gee L, Tinker A (2014) The ATP-sensitive potassium channel subunit, Kir6.1, in vascular smooth muscle plays a major role in blood pressure control. Hypertension 64:523–529

Bagetta G, Nistico G, Dolly JO (1992) Production of seizures and brain damage in rats by alpha-dendrotoxin, a selective K+ channel blocker. Neurosci Lett 139:34–40

Bagriantsev SN, Peyronnet R, Clark KA, Honore E, Minor DL Jr (2011) Multiple modalities converge on a common gate to control K2P channel function. EMBO J 30:3594–3606
Baker OS, Larsson HP, Mannuzzu LM, Isacoff EY (1998) Three transmembrane conformations and sequence-dependent displacement of the S4 domain in shaker K+ channel gating. Neuron 20:1283–1294
Baronas VA, Kurata HT (2014) Inward rectifiers and their regulation by endogenous polyamines. Front Physiol 5:325
Baukrowitz T, Schulte U, Oliver D, Herlitze S, Krauter T, Tucker SJ, Ruppersberg JP, Fakler B (1998) PIP2 and PIP as determinants for ATP inhibition of KATP channels. Science 282:1141–1144
Behrens R, Nolting A, Reimann F, Schwarz M, Waldschutz R, Pongs O (2000) hKCNMB3 and hKCNMB4, cloning and characterization of two members of the large-conductance calcium-activated potassium channel beta subunit family. FEBS Lett 474:99–106
Bentrop D, Beyermann M, Wissmann R, Fakler B (2001) NMR structure of the "ball-and-chain" domain of KCNMB2, the beta 2-subunit of large conductance Ca2+– and voltage-activated potassium channels. J Biol Chem 276:42116–42121
Berg AP, Talley EM, Manger JP, Bayliss DA (2004) Motoneurons express heteromeric TWIK-related acid-sensitive K+ (TASK) channels containing TASK-1 (KCNK3) and TASK-3 (KCNK9) subunits. J Neurosci 24:6693–6702
Berkefeld H, Sailer CA, Bildl W, Rohde V, Thumfart JO, Eble S, Klugbauer N, Reisinger E, Bischofberger J, Oliver D et al (2006) BKCa-Cav channel complexes mediate rapid and localized Ca2+–activated K+ signaling. Science 314:615–620
Berkefeld H, Fakler B, Schulte U (2010) Ca2+–activated K+ channels: from protein complexes to function. Physiol Rev 90:1437–1459
Bernèche S, Roux B (2001) Energetics of ion conduction through the K+ channel. Nature 414:73–77
Bezanilla F (2000) The voltage sensor in voltage-dependent ion channels. Physiol Rev 80:555–592
Bichet D, Haass FA, Jan LY (2003) Merging functional studies with structures of inward-rectifier K (+) channels. Nat Rev Neurosci 4:957–967
Bildl W, Strassmaier T, Thurm H, Andersen J, Eble S, Oliver D, Knipper M, Mann M, Schulte U, Adelman JP et al (2004) Protein kinase CK2 is coassembled with small conductance ca(2+)-activated K+ channels and regulates channel gating. Neuron 43:847–858
Blatz AL, Magleby KL (1984) Ion conductance and selectivity of single calcium-activated potassium channels in cultured rat muscle. J Gen Physiol 84:1–23
Bockenhauer D, Zilberberg N, Goldstein SA (2001) KCNK2: reversible conversion of a hippocampal potassium leak into a voltage-dependent channel. Nat Neurosci 4:486–491
Brenner R, Jegla TJ, Wickenden A, Liu Y, Aldrich RW (2000a) Cloning and functional characterization of novel large conductance calcium-activated potassium channel beta subunits, hKCNMB3 and hKCNMB4. J Biol Chem 275:6453–6461
Brenner R, Perez GJ, Bonev AD, Eckman DM, Kosek JC, Wiler SW, Patterson AJ, Nelson MT, Aldrich RW (2000b) Vasoregulation by the beta1 subunit of the calcium-activated potassium channel. Nature 407:870–876
Brereton MF, Ashcroft FM (2013) Mouse models of beta-cell KATP channel dysfunction. Drug Discov Today Dis Models 10:e101–e109
Bretschneider F, Wrisch A, Lehmann-Horn F, Grissmer S (1999) External tetraethylammonium as a molecular caliper for sensing the shape of the outer vestibule of potassium channels. Biophys J 76:2351–2360
Brohawn SG, del Marmol J, MacKinnon R (2012) Crystal structure of the human K2P TRAAK, a lipid- and mechano-sensitive K+ ion channel. Science 335:436–441
Brohawn SG, Campbell EB, MacKinnon R (2014) Physical mechanism for gating and mechanosensitivity of the human TRAAK K+ channel. Nature 516:126–130
Broomand A, Elinder F (2008) Large-scale movement within the voltage-sensor paddle of a potassium channel-support for a helical-screw motion. Neuron 59:770–777

Brugada R, Hong K, Dumaine R, Cordeiro J, Gaita F, Borggrefe M, Menendez TM, Brugada J, Pollevick GD, Wolpert C et al (2004) Sudden death associated with short-QT syndrome linked to mutations in HERG. Circulation 109:30–35

Bukiya AN, Liu J, Toro L, Dopico AM (2007) Beta1 (KCNMB1) subunits mediate lithocholate activation of large-conductance Ca2+−activated K+ channels and dilation in small, resistance-size arteries. Mol Pharmacol 72:359–369

Bukiya AN, McMillan J, Liu J, Shivakumar B, Parrill AL, Dopico AM (2014) Activation of calcium- and voltage-gated potassium channels of large conductance by leukotriene B4. J Biol Chem 289:35314–35325

Butler A, Tsunoda S, McCobb DP, Wei A, Salkoff L (1993) mSlo, a complex mouse gene encoding "maxi" calcium-activated potassium channels. Science 261:221–224

Cao Y, Dreixler JC, Roizen JD, Roberts MT, Houamed KM (2001) Modulation of recombinant small-conductance ca(2+)-activated K(+) channels by the muscle relaxant chlorzoxazone and structurally related compounds. J Pharmacol Exp Ther 296:683–689

Carmeliet E (1992) Voltage- and time-dependent block of the delayed K+ current in cardiac myocytes by dofetilide. J Pharmacol Exp Ther 262:809–817

Casamassima M, D'Adamo MC, Pessia M, Tucker SJ (2003) Identification of a heteromeric interaction that influences the rectification, gating, and pH sensitivity of Kir4.1/Kir5.1 potassium channels. J Biol Chem 278:43533–43540

Castle NA, London DO, Creech C, Fajloun Z, Stocker JW, Sabatier JM (2003) Maurotoxin: a potent inhibitor of intermediate conductance Ca2+−activated potassium channels. Mol Pharmacol 63:409–418

Caterina MJ, Schumacher MA, Tominaga M, Rosen TA, Levine JD, Julius D (1997) The capsaicin receptor: a heat-activated ion channel in the pain pathway. Nature 389:816–824

Chakrapani S, Cuello LG, Cortes DM, Perozo E (2008) Structural dynamics of an isolated voltage-sensor domain in a lipid bilayer. Structure 16:398–409

Chatelain FC, Bichet D, Douguet D, Feliciangeli S, Bendahhou S, Reichold M, Warth R, Barhanin J, Lesage F (2012) TWIK1, a unique background channel with variable ion selectivity. Proc Natl Acad Sci U S A 109:5499–5504

Chemin J, Patel AJ, Duprat F, Lauritzen I, Lazdunski M, Honore E (2005) A phospholipid sensor controls mechanogating of the K+ channel TREK-1. EMBO J 24:44–53

Chemin J, Patel AJ, Duprat F, Sachs F, Lazdunski M, Honore E (2007) Up- and down-regulation of the mechano-gated K(2P) channel TREK-1 by PIP (2) and other membrane phospholipids. Pflügers Arch 455:97–103

Chen X, Talley EM, Patel N, Gomis A, McIntire WE, Dong B, Viana F, Garrison JC, Bayliss DA (2006) Inhibition of a background potassium channel by Gq protein alpha-subunits. Proc Natl Acad Sci U S A 103:3422–3427

Clark RB, Mangoni ME, Lueger A, Couette B, Nargeot J, Giles WR (2004) A rapidly activating delayed rectifier K+ current regulates pacemaker activity in adult mouse sinoatrial node cells. Am J Physiol Heart Circ Physiol 286:H1757–H1766

Clarke OB, Caputo AT, Hill AP, Vandenberg JI, Smith BJ, Gulbis JM (2010) Domain reorientation and rotation of an intracellular assembly regulate conduction in Kir potassium channels. Cell 141:1018–1029

Clement JPt, Kunjilwar K, Gonzalez G, Schwanstecher M, Panten U, Aguilar-Bryan L, Bryan J (1997) Association and stoichiometry of K(ATP) channel subunits. Neuron 18:827–838

Coetzee WA, Amarillo Y, Chiu J, Chow A, Lau D, McCormack T, Morena H, Nadal MS, Ozaita A, Pountney D et al (1999) Molecular diversity of K+ channels. Ann N Y Acad Sci 868:233–255

Cohen A, Ben-Abu Y, Hen S, Zilberberg N (2008) A novel mechanism for human K2P2.1 channel gating. Facilitation of C-type gating by protonation of extracellular histidine residues. J Biol Chem 283:19448–19455

Coleman SK, Newcombe J, Pryke J, Dolly JO (1999) Subunit composition of Kv1 channels in human CNS. J Neurochem 73:849–858

Cuello LG (2004) Molecular architecture of the KvAP voltage-dependent K+ channel in a lipid bilayer. Science 306:491–495
Cui J (2016) Voltage-dependent gating: novel insights from KCNQ1 channels. Biophys J 110:14–25
Cui J, Cox DH, Aldrich RW (1997) Intrinsic voltage dependence and Ca2+ regulation of mslo large conductance ca-activated K+ channels. J Gen Physiol 109:647–673
Czirjak G, Enyedi P (2002) Formation of functional heterodimers between the TASK-1 and TASK-3 two-pore domain potassium channel subunits. J Biol Chem 277:5426–5432
D'Alessandro G, Limatola C, Catalano M (2018) Functional roles of the Ca2+−activated K+ channel, KCa3.1, in brain tumors. Curr Neuropharmacol 16:636–643
Dauplais M, Lecoq A, Song J, Cotton J, Jamin N, Gilquin B, Roumestand C, Vita C, De Medeiros CLC, Rowan EG et al (1997) On the convergent evolution of animal toxins. J Biol Chem 272:4302–4309
Devor DC, Singh AK, Frizzell RA, Bridges RJ (1996) Modulation of cl- secretion by benzimidazolones. I. Direct activation of a ca(2+)-dependent K+ channel. Am J Phys 271: L775–L784
Diaz L, Meera P, Amigo J, Stefani E, Alvarez O, Toro L, Latorre R (1998) Role of the S4 segment in a voltage-dependent calcium-sensitive potassium (hSlo) channel. J Biol Chem 273:32430–32436
Dodson PD, Barker MC, Forsythe ID (2002) Two heteromeric Kv1 potassium channels differentially regulate action potential firing. J Neurosci 22:6953–6961
Dogan MF, Yildiz O, Arslan SO, Ulusoy KG (2019) Potassium channels in vascular smooth muscle: a pathophysiological and pharmacological perspective. Fundam Clin Pharmacol 33:504–523
Dong YY, Pike AC, Mackenzie A, McClenaghan C, Aryal P, Dong L, Quigley A, Grieben M, Goubin S, Mukhopadhyay S et al (2015) K2P channel gating mechanisms revealed by structures of TREK-2 and a complex with Prozac. Science 347:1256–1259
Dopico AM, Bukiya AN (2017) Regulation of ca(2+)-sensitive K(+) channels by cholesterol and bile acids via distinct channel subunits and sites. Curr Top Membr 80:53–93
Doring F, Derst C, Wischmeyer E, Karschin C, Schneggenburger R, Daut J, Karschin A (1998) The epithelial inward rectifier channel Kir7.1 displays unusual K+ permeation properties. J Neurosci 18:8625–8636
Doupnik CA (2017) Venom-derived peptides inhibiting Kir channels: past, present, and future. Neuropharmacology 127:161–172
Doyle ME, Egan JM (2003) Pharmacological agents that directly modulate insulin secretion. Pharmacol Rev 55:105–131
Doyle DA, Morais Cabral J, Pfuetzner RA, Kuo A, Gulbis JM, Cohen SL, Chait BT, MacKinnon R (1998) The structure of the potassium channel: molecular basis of K+ conduction and selectivity. Science 280:69–77
Du X, Zhang H, Lopes C, Mirshahi T, Rohacs T, Logothetis DE (2004) Characteristic interactions with phosphatidylinositol 4,5-bisphosphate determine regulation of kir channels by diverse modulators. J Biol Chem 279:37271–37281
Dworetzky SI, Boissard CG, Lum-Ragan JT, McKay MC, Post-Munson DJ, Trojnacki JT, Chang CP, Gribkoff VK (1996) Phenotypic alteration of a human BK (hSlo) channel by hSlobeta subunit coexpression: changes in blocker sensitivity, activation/relaxation and inactivation kinetics, and protein kinase A modulation. J Neurosci 16:4543–4550
Eriksson MAL, Roux B (2002) Modeling the structure of agitoxin in complex with the shaker K+ channel: a computational approach based on experimental distance restraints extracted from thermodynamic mutant cycles. Biophys J 83:2595–2609
Fakler B, Adelman JP (2008) Control of K(ca) channels by calcium nano/microdomains. Neuron 59:873–881

Fanger CM, Ghanshani S, Logsdon NJ, Rauer H, Kalman K, Zhou J, Beckingham K, Chandy KG, Cahalan MD, Aiyar J (1999) Calmodulin mediates calcium-dependent activation of the intermediate conductance KCa channel. IKCa1 J Biol Chem 274:5746–5754

Ferreira R, Lively S, Schlichter LC (2014) IL-4 type 1 receptor signaling up-regulates KCNN4 expression, and increases the KCa3.1 current and its contribution to migration of alternative-activated microglia. Front Cell Neurosci 8:183

Fink M, Duprat F, Lesage F, Reyes R, Romey G, Heurteaux C, Lazdunski M (1996) Cloning, functional expression and brain localization of a novel unconventional outward rectifier K+ channel. EMBO J 15:6854–6862

Finley M, Arrabit C, Fowler C, Suen KF, Slesinger PA (2004) betaL-betaM loop in the C-terminal domain of G protein-activated inwardly rectifying K(+) channels is important for G(betagamma) subunit activation. J Physiol 555:643–657

Finol-Urdaneta RK, StrüVer N, Terlau H (2006) Molecular and functional differences between heart mKv1.7 channel isoforms. J Gen Physiol 128:133–145

Finol-Urdaneta RK, Belovanovic A, Micic-Vicovac M, Kinsella GK, McArthur JR, Al-Sabi A (2020) Marine toxins targeting Kv1 channels: pharmacological tools and therapeutic scaffolds. Mar Drugs 18:173

Forte M, Satow Y, Nelson D, Kung C (1981) Mutational alteration of membrane phospholipid composition and voltage-sensitive ion channel function in paramecium. Proc Natl Acad Sci 78:7195–7199

Furukawa Y, Miyashita Y, Nakajima K, Hirose M, Kurogouchi F, Chiba S (1999) Effects of verapamil, zatebradine, and E-4031 on the pacemaker location and rate in response to sympathetic stimulation in dog hearts. J Pharmacol Exp Ther 289:1334–1342

Gada K, Plant LD (2019) Two-pore domain potassium channels: emerging targets for novel analgesic drugs: IUPHAR review 26. Br J Pharmacol 176:256–266

Galvez A, Gimenez-Gallego G, Reuben JP, Roy-Contancin L, Feigenbaum P, Kaczorowski GJ, Garcia ML (1990) Purification and characterization of a unique, potent, peptidyl probe for the high conductance calcium-activated potassium channel from venom of the scorpion Buthus tamulus. J Biol Chem 265:11083–11090

Gandhi CS, Isacoff EY (2002) Molecular models of voltage sensing. J Gen Physiol 120:455–463

Gao Y-D, Garcia ML (2003) Interaction of agitoxin2, charybdotoxin, and iberiotoxin with potassium channels: selectivity between voltage-gated and maxi-K channels. Proteins Struct Funct Genet 52:146–154

Garcia-Valdes J, Zamudio FZ, Toro L, Possani LD (2001) Slotoxin, alphaKTx1.11, a new scorpion peptide blocker of MaxiK channels that differentiates between alpha and alpha+beta (beta1 or beta4) complexes. FEBS Lett 505:369–373

Gardos G (1958) The function of calcium in the potassium permeability of human erythrocytes. Biochim Biophys Acta 30:653–654

Gilquin B, Racapé J, Wrisch A, Visan V, Lecoq A, Grissmer S, Ménez A, Gasparini S (2002) Structure of the BgK-Kv1.1 complex based on distance restraints identified by double mutant cycles. J Biol Chem 277:37406–37413

Girard CA, Wunderlich FT, Shimomura K, Collins S, Kaizik S, Proks P, Abdulkader F, Clark A, Ball V, Zubcevic L et al (2009) Expression of an activating mutation in the gene encoding the KATP channel subunit Kir6.2 in mouse pancreatic beta cells recapitulates neonatal diabetes. J Clin Invest 119:80–90

Glaaser IW, Slesinger PA (2017) Dual activation of neuronal G protein-gated inwardly rectifying potassium (GIRK) channels by cholesterol and alcohol. Sci Rep 7:4592

Gloyn AL, Pearson ER, Antcliff JF, Proks P, Bruining GJ, Slingerland AS, Howard N, Srinivasan S, Silva JM, Molnes J et al (2004) Activating mutations in the gene encoding the ATP-sensitive potassium-channel subunit Kir6.2 and permanent neonatal diabetes. N Engl J Med 350:1838–1849

Golding NL, Jung HY, Mickus T, Spruston N (1999) Dendritic calcium spike initiation and repolarization are controlled by distinct potassium channel subtypes in CA1 pyramidal neurons. J Neurosci 19:8789–8798
Goldstein SA, Price LA, Rosenthal DN, Pausch MH (1996) ORK1, a potassium-selective leak channel with two pore domains cloned from Drosophila melanogaster by expression in Saccharomyces cerevisiae. Proc Natl Acad Sci U S A 93:13256–13261
Goldstein SA, Bockenhauer D, O'Kelly I, Zilberberg N (2001) Potassium leak channels and the KCNK family of two-P-domain subunits. Nat Rev Neurosci 2:175–184
Gonzalez-Perez V, Xia XM, Lingle CJ (2014) Functional regulation of BK potassium channels by gamma1 auxiliary subunits. Proc Natl Acad Sci U S A 111:4868–4873
Gonzalez-Perez V, Xia XM, Lingle CJ (2015) Two classes of regulatory subunits coassemble in the same BK channel and independently regulate gating. Nat Commun 6:8341
Goodman AD, Brown TR, Krupp LB, Schapiro RT, Schwid SR, Cohen R, Marinucci LN, Blight AR (2009) Sustained-release oral fampridine in multiple sclerosis: a randomised, double-blind, controlled trial. Lancet 373:732–738
Gribkoff VK, Starrett JE Jr, Dworetzky SI, Hewawasam P, Boissard CG, Cook DA, Frantz SW, Heman K, Hibbard JR, Huston K et al (2001) Targeting acute ischemic stroke with a calcium-sensitive opener of maxi-K potassium channels. Nat Med 7:471–477
Griguoli M, Sgritta M, Cherubini E (2016) Presynaptic BK channels control transmitter release: physiological relevance and potential therapeutic implications. J Physiol 594:3489–3500
Grover GJ, Garlid KD (2000) ATP-sensitive potassium channels: a review of their cardioprotective pharmacology. J Mol Cell Cardiol 32:677–695
Grunnet M, Kaufmann WA (2004) Coassembly of big conductance Ca2+−activated K+ channels and L-type voltage-gated Ca2+ channels in rat brain. J Biol Chem 279:36445–36453
Grunnet M, Jensen BS, Olesen SP, Klaerke DA (2001) Apamin interacts with all subtypes of cloned small-conductance Ca2+−activated K+ channels. Pflugers Arch 441:544–550
Gu C, Gu Y (2011) Clustering and activity tuning of Kv1 channels in myelinated hippocampal axons. J Biol Chem 286:25835–25847
Gu N, Vervaeke K, Storm JF (2007) BK potassium channels facilitate high-frequency firing and cause early spike frequency adaptation in rat CA1 hippocampal pyramidal cells. J Physiol 580:859–882
Gutman GA, Chandy KG, Grissmer S, Lazdunski M, McKinnon D, Pardo LA, Robertson GA, Rudy B, Sanguinetti MC, Stuhmer W et al (2005) International Union of Pharmacology. LIII. Nomenclature and molecular relationships of voltage-gated potassium channels. Pharmacol Rev 57:473–508
Hagiwara S, Miyazaki S, Rosenthal NP (1976) Potassium current and the effect of cesium on this current during anomalous rectification of the egg cell membrane of a starfish. J Gen Physiol 67:621–638
Hagiwara S, Miyazaki S, Moody W, Patlak J (1978) Blocking effects of barium and hydrogen ions on the potassium current during anomalous rectification in the starfish egg. J Physiol 279:167–185
Hanner M, Schmalhofer WA, Munujos P, Knaus HG, Kaczorowski GJ, Garcia ML (1997) The beta subunit of the high-conductance calcium-activated potassium channel contributes to the high-affinity receptor for charybdotoxin. Proc Natl Acad Sci U S A 94:2853–2858
Hansen SB, Tao X, MacKinnon R (2011) Structural basis of PIP2 activation of the classical inward rectifier K+ channel Kir2.2. Nature 477:495–498
Harvey AL, Anderson AJ (1985) Dendrotoxins: snake toxins that block potassium channels and facilitate neurotransmitter release. Pharmacol Ther 31:33–55
Harvey AL, Robertson B (2004) Dendrotoxins: structure-activity relationships and effects on potassium ion channels. Curr Med Chem 11:3065–3072
He C, Zhang H, Mirshahi T, Logothetis DE (1999) Identification of a potassium channel site that interacts with G protein bg subunits to mediate agonist-induced signaling. J Biol Chem 274:12517–12524

Hebert B, Pietropaolo S, Meme S, Laudier B, Laugeray A, Doisne N, Quartier A, Lefeuvre S, Got L, Cahard D et al (2014) Rescue of fragile X syndrome phenotypes in Fmr1 KO mice by a BKCa channel opener molecule. Orphanet J Rare Dis 9:124
Heginbotham L, Lu Z, Abramson T, MacKinnon R (1994) Mutations in the K+ channel signature sequence. Biophys J 66:1061–1067
Hibino H, Kurachi Y (2006) Molecular and physiological bases of the K+ circulation in the mammalian inner ear. Physiology (Bethesda) 21:336–345
Hibino H, Horio Y, Inanobe A, Doi K, Ito M, Yamada M, Gotow T, Uchiyama Y, Kawamura M, Kubo T et al (1997) An ATP-dependent inwardly rectifying potassium channel, KAB-2 (Kir4. 1), in cochlear stria vascularis of inner ear: its specific subcellular localization and correlation with the formation of endocochlear potential. J Neurosci 17:4711–4721
Hibino H, Fujita A, Iwai K, Yamada M, Kurachi Y (2004a) Differential assembly of inwardly rectifying K+ channel subunits, Kir4.1 and Kir5.1, in brain astrocytes. J Biol Chem 279:44065–44073
Hibino H, Higashi-Shingai K, Fujita A, Iwai K, Ishii M, Kurachi Y (2004b) Expression of an inwardly rectifying K+ channel, Kir5.1, in specific types of fibrocytes in the cochlear lateral wall suggests its functional importance in the establishment of endocochlear potential. Eur J Neurosci 19:76–84
Hibino H, Inanobe A, Furutani K, Murakami S, Findlay I, Kurachi Y (2010) Inwardly rectifying potassium channels: their structure, function, and physiological roles. Physiol Rev 90:291–366
Hille B (1986) Ionic channels: molecular pores of excitable membranes. Harvey Lect 82:47–69
Hirschberg B, Maylie J, Adelman JP, Marrion NV (1999) Gating properties of single SK channels in hippocampal CA1 pyramidal neurons. Biophys J 77:1905–1913
Hite RK, Tao X, MacKinnon R (2017) Structural basis for gating the high-conductance ca(2+)-activated K(+) channel. Nature 541:52–57
Ho IH, Murrell-Lagnado RD (1999) Molecular determinants for sodium-dependent activation of G protein-gated K+ channels. J Biol Chem 274:8639–8648
Hodgkin AL, Huxley AF (1945) Resting and action potentials in single nerve fibres. J Physiol 104:176–195
Holtzclaw JD, Grimm PR, Sansom SC (2011) Role of BK channels in hypertension and potassium secretion. Curr Opin Nephrol Hypertens 20:512–517
Horinaka S (2011) Use of nicorandil in cardiovascular disease and its optimization. Drugs 71:1105–1119
Horrigan FT, Aldrich RW (2002) Coupling between voltage sensor activation, Ca2+ binding and channel opening in large conductance (BK) potassium channels. J Gen Physiol 120:267–305
Horvath GA, Zhao Y, Tarailo-Graovac M, Boelman C, Gill H, Shyr C, Lee J, Blydt-Hansen I, Drogemoller BI, Moreland J et al (2018) Gain-of-function KCNJ6 mutation in a severe hyperkinetic movement disorder phenotype. Neuroscience 384:152–164
Hoshi T, Armstrong CM (2013) C-type inactivation of voltage-gated K+ channels: pore constriction or dilation? J Gen Physiol 141:151–160
Hoshi T, Wissuwa B, Tian Y, Tajima N, Xu R, Bauer M, Heinemann SH, Hou S (2013) Omega-3 fatty acids lower blood pressure by directly activating large-conductance ca(2)(+)-dependent K (+) channels. Proc Natl Acad Sci U S A 110:4816–4821
Hougaard C, Eriksen BL, Jorgensen S, Johansen TH, Dyhring T, Madsen LS, Strobaek D, Christophersen P (2007) Selective positive modulation of the SK3 and SK2 subtypes of small conductance Ca2+−activated K+ channels. Br J Pharmacol 151:655–665
Huang CL, Feng S, Hilgemann DW (1998) Direct activation of inward rectifier potassium channels by PIP2 and its stabilization by Gbetagamma. Nature 391:803–806
Huang C, Sindic A, Hill CE, Hujer KM, Chan KW, Sassen M, Wu Z, Kurachi Y, Nielsen S, Romero MF et al (2007) Interaction of the Ca2+−sensing receptor with the inwardly rectifying potassium channels Kir4.1 and Kir4.2 results in inhibition of channel function. Am J Physiol Renal Physiol 292:F1073–F1081

Hughes BA, Kumar G, Yuan Y, Swaminathan A, Yan D, Sharma A, Plumley L, Yang-Feng TL, Swaroop A (2000) Cloning and functional expression of human retinal kir2.4, a pH-sensitive inwardly rectifying K(+) channel. Am J Physiol Cell Physiol 279:C771–C784

Hwang EM, Kim E, Yarishkin O, Woo DH, Han KS, Park N, Bae Y, Woo J, Kim D, Park M et al (2014) A disulphide-linked heterodimer of TWIK-1 and TREK-1 mediates passive conductance in astrocytes. Nat Commun 5:3227

Imlach WL, Finch SC, Dunlop J, Dalziel JE (2009) Structural determinants of lolitrems for inhibition of BK large conductance Ca2+−activated K+ channels. Eur J Pharmacol 605:36–45

Imredy JP, Chen C, MacKinnon R (1998) A snake toxin inhibitor of inward rectifier potassium channel ROMK1. Biochemistry 37:14867–14874

Inda MC, Defelipe J, Munoz A (2006) Voltage-gated ion channels in the axon initial segment of human cortical pyramidal cells and their relationship with chandelier. Cell 103:2920–2925

Ishida IG, Rangel-Yescas GE, Carrasco-Zanini J, Islas LD (2015) Voltage-dependent gating and gating charge measurements in the Kv1.2 potassium channel. J Gen Physiol 145:345–358

Ishii TM, Silvia C, Hirschberg B, Bond CT, Adelman JP, Maylie J (1997) A human intermediate conductance calcium-activated potassium channel. Proc Natl Acad Sci U S A 94:11651–11656

Ishii M, Fujita A, Iwai K, Kusaka S, Higashi K, Inanobe A, Hibino H, Kurachi Y (2003) Differential expression and distribution of Kir5.1 and Kir4.1 inwardly rectifying K+ channels in retina. Am J Physiol Cell Physiol 285:C260–C267

Ivanina T, Rishal I, Varon D, Mullner C, Frohnwieser-Steinecke B, Schreibmayer W, Dessauer CW, Dascal N (2003) Mapping the Gβγ-binding sites in GIRK1 and GIRK2 subunits of the G protein-activated K^+ channel. J Biol Chem 278:29174–29183

Jin W, Lu Z (1998) A novel high-affinity inhibitor for inward-rectifier K+ channels. Biochemistry 37:13291–13299

Jin W, Lu Z (1999) Synthesis of a stable form of tertiapin: a high-affinity inhibitor for inward-rectifier K+ channels. Biochemistry 38:14286–14293

Joiner WJ, Khanna R, Schlichter LC, Kaczmarek LK (2001) Calmodulin regulates assembly and trafficking of SK4/IK1 Ca2+−activated K+ channels. J Biol Chem 276:37980–37985

Jouirou B, Mouhat S, Andreotti N, De Waard M, Sabatier J-M (2004) Toxin determinants required for interaction with voltage-gated K+ channels. Toxicon 43:909–914

Kamath GS, Mittal S (2008) The role of antiarrhythmic drug therapy for the prevention of sudden cardiac death. Prog Cardiovasc Dis 50:439–448

Kang D, Kim D (2006) TREK-2 (K2P10.1) and TRESK (K2P18.1) are major background K+ channels in dorsal root ganglion neurons. Am J Physiol Cell Physiol 291:C138–C146

Kang D, Choe C, Kim D (2005) Thermosensitivity of the two-pore domain K+ channels TREK-2 and TRAAK. J Physiol 564:103–116

Kanjhan R, Coulson EJ, Adams DJ, Bellingham MC (2005) Tertiapin-Q blocks recombinant and native large conductance K+ channels in a use-dependent manner. J Pharmacol Exp Ther 314:1353–1361

Kaplan WD, Trout WE 3rd (1969) The behavior of four neurological mutants of Drosophila. Genetics 61:399–409

Kasumu AW, Hougaard C, Rode F, Jacobsen TA, Sabatier JM, Eriksen BL, Strobaek D, Liang X, Egorova P, Vorontsova D et al (2012) Selective positive modulator of calcium-activated potassium channels exerts beneficial effects in a mouse model of spinocerebellar ataxia type 2. Chem Biol 19:1340–1353

Katz B (1949) Les Constantes Electriques De La Membrane Du Muscle. Arch Sci Physiol 3:285–300

Kaufmann K, Romaine I, Days E, Pascual C, Malik A, Yang L, Zou B, Du Y, Sliwoski G, Morrison RD et al (2013) ML297 (VU0456810), the first potent and selective activator of the GIRK potassium channel, displays antiepileptic properties in mice. ACS Chem Neurosci 4:1278–1286

Kennard LE, Chumbley JR, Ranatunga KM, Armstrong SJ, Veale EL, Mathie A (2005) Inhibition of the human two-pore domain potassium channel, TREK-1, by fluoxetine and its metabolite norfluoxetine. Br J Pharmacol 144:821–829

Ketchum KA, Joiner WJ, Sellers AJ, Kaczmarek LK, Goldstein SA (1995) A new family of outwardly rectifying potassium channel proteins with two pore domains in tandem. Nature 376:690–695

Khalili-Araghi F, Tajkhorshid E, Schulten K (2006) Dynamics of K+ ion conduction through Kv1.2. Biophys J 91:L72–L74

Kiehn J, Wible B, Lacerda AE, Brown AM (1996) Mapping the block of a cloned human inward rectifier potassium channel by dofetilide. Mol Pharmacol 50:380–387

Kim Y, Bang H, Kim D (2000) TASK-3, a new member of the tandem pore K(+) channel family. J Biol Chem 275:9340–9347

King JT, Lovell PV, Rishniw M, Kotlikoff MI, Zeeman ML, McCobb DP (2006) Beta2 and beta4 subunits of BK channels confer differential sensitivity to acute modulation by steroid hormones. J Neurophysiol 95:2878–2888

Klein H, Garneau L, Trinh NT, Prive A, Dionne F, Goupil E, Thuringer D, Parent L, Brochiero E, Sauve R (2009) Inhibition of the KCa3.1 channels by AMP-activated protein kinase in human airway epithelial cells. Am J Physiol Cell Physiol 296:C285–C295

Knaus HG, Folander K, Garcia-Calvo M, Garcia ML, Kaczorowski GJ, Smith M, Swanson R (1994a) Primary sequence and immunological characterization of beta-subunit of high conductance ca(2+)-activated K+ channel from smooth muscle. J Biol Chem 269:17274–17278

Knaus HG, McManus OB, Lee SH, Schmalhofer WA, Garcia-Calvo M, Helms LM, Sanchez M, Giangiacomo K, Reuben JP, Smith AB 3rd et al (1994b) Tremorgenic indole alkaloids potently inhibit smooth muscle high-conductance calcium-activated potassium channels. Biochemistry 33:5819–5828

Kobayashi T, Ikeda K, Kojima H, Niki H, Yano R, Yoshioka T, Kumanishi T (1999) Ethanol opens G-protein-activated inwardly rectifying K+ channels. Nat Neurosci 2:1091–1097

Kohler M, Hirschberg B, Bond CT, Kinzie JM, Marrion NV, Maylie J, Adelman JP (1996) Small-conductance, calcium-activated potassium channels from mammalian brain. Science 273:1709–1714

Kole MHP, Stuart GJ (2012) Signal processing in the axon initial segment. Neuron 73:235–247

Kole MH, Letzkus JJ, Stuart GJ (2007) Axon initial segment Kv1 channels control axonal action potential waveform and synaptic efficacy. Neuron 55:633–647

Korenke AR, Rivey MP, Allington DR (2008) Sustained-release fampridine for symptomatic treatment of multiple sclerosis. Ann Pharmacother 42:1458–1465

Krapivinsky G, Medina I, Eng L, Krapivinsky L, Yang Y, Clapham DE (1998) A novel inward rectifier K+ channel with unique pore properties. Neuron 20:995–1005

Krepkiy D, Mihailescu M, Freites JA, Schow EV, Worcester DL, Gawrisch K, Tobias DJ, White SH, Swartz KJ (2009) Structure and hydration of membranes embedded with voltage-sensing domains. Nature 462:473–479

Kshatri AS, Gonzalez-Hernandez A, Giraldez T (2018) Physiological roles and therapeutic potential of ca(2+) activated potassium channels in the nervous system. Front Mol Neurosci 11:258

Kubo Y, Murata Y (2001) Control of rectification and permeation by two distinct sites after the second transmembrane region in Kir2.1 K+ channel. J Physiol 531:645–660

Kubo Y, Adelman JP, Clapham DE, Jan LY, Karschin A, Kurachi Y, Lazdunski M, Nichols CG, Seino S, Vandenberg CA (2005) International Union of Pharmacology. LIV. Nomenclature and molecular relationships of inwardly rectifying potassium channels. Pharmacol Rev 57:509–526

Kumar M, Pattnaik BR (2014) Focus on Kir7.1: physiology and channelopathy. Channels (Austin) 8:488–495

Kuo A, Gulbis JM, Antcliff JF, Rahman T, Lowe ED, Zimmer J, Cuthbertson J, Ashcroft FM, Ezaki T, Doyle DA (2003) Crystal structure of the potassium channel KirBac1.1 in the closed state. Science 300:1922–1926

Lafreniere RG, Rouleau GA (2011) Migraine: role of the TRESK two-pore potassium channel. Int J Biochem Cell Biol 43:1533–1536

Lafreniere RG, Cader MZ, Poulin JF, Andres-Enguix I, Simoneau M, Gupta N, Boisvert K, Lafreniere F, McLaughlan S, Dube MP et al (2010) A dominant-negative mutation in the TRESK potassium channel is linked to familial migraine with aura. Nat Med 16:1157–1160

Lancaster B, Nicoll RA (1987) Properties of two calcium-activated hyperpolarizations in rat hippocampal neurones. J Physiol 389:187–203

Larsson HP, Baker OS, Dhillon DS, Isacoff EY (1996) Transmembrane movement of the shaker K+ channel S4. Neuron 16:387–397

Latorre R, Castillo K, Carrasquel-Ursulaez W, Sepulveda RV, Gonzalez-Nilo F, Gonzalez C, Alvarez O (2017) Molecular determinants of BK channel functional diversity and functioning. Physiol Rev 97:39–87

Lee S-Y, Mackinnon R (2004) A membrane-access mechanism of ion channel inhibition by voltage sensor toxins from spider venom. Nature 430:232–235

Lee CH, MacKinnon R (2018) Activation mechanism of a human SK-calmodulin channel complex elucidated by cryo-EM structures. Science 360:508–513

Lee KPK, Chen J, MacKinnon R (2017) Molecular structure of human KATP in complex with ATP and ADP. eLife 6

Leng Q, MacGregor GG, Dong K, Giebisch G, Hebert SC (2006) Subunit-subunit interactions are critical for proton sensitivity of ROMK: evidence in support of an intermolecular gating mechanism. Proc Natl Acad Sci U S A 103:1982–1987

Lesage F, Reyes R, Fink M, Duprat F, Guillemare E, Lazdunski M (1996) Dimerization of TWIK-1 K+ channel subunits via a disulfide bridge. EMBO J 15:6400–6407

Li Q, Yan J (2016) Modulation of BK channel function by auxiliary beta and gamma subunits. Int Rev Neurobiol 128:51–90

Li Q, Zhang M, Duan Z, Stamatoyannopoulos G (1999) Structural analysis and mapping of DNase I hypersensitivity of HS5 of the beta-globin locus control region. Genomics 61:183–193

Lin MC, Hsieh JY, Mock AF, Papazian DM (2011) R1 in the shaker S4 occupies the gating charge transfer center in the resting state. J Gen Physiol 138:155–163

Liu Y, Liu D, Heath L, Meyers DM, Krafte DS, Wagoner PK, Silvia CP, Yu W, Curran ME (2001) Direct activation of an inwardly rectifying potassium channel by arachidonic acid. Mol Pharmacol 59:1061–1068

Liu S, Focke PJ, Matulef K, Bian X, Moënne-Loccoz P, Valiyaveetil FI, Lockless SW (2015) Ion-binding properties of a K + channel selectivity filter in different conformations. Proc Natl Acad Sci U S A 112:15096–15100

Logsdon NJ, Kang J, Togo JA, Christian EP, Aiyar J (1997) A novel gene, hKCa4, encodes the calcium-activated potassium channel in human T lymphocytes. J Biol Chem 272:32723–32726

Lolicato M, Arrigoni C, Mori T, Sekioka Y, Bryant C, Clark KA, Minor DL Jr (2017) K2P2.1 (TREK-1)-activator complexes reveal a cryptic selectivity filter binding site. Nature 547:364–368

Long SB (2005) Crystal structure of a mammalian voltage-dependent shaker family K+ channel. Science 309:897–903

Long SB, Tao X, Campbell EB, Mackinnon R (2007) Atomic structure of a voltage-dependent K+ channel in a lipid membrane-like environment. Nature 450:376–382

Lopatin AN, Nichols CG (1996) [K+] dependence of open-channel conductance in cloned inward rectifier potassium channels (IRK1, Kir2.1). Biophys J 71:682–694

Lopatin AN, Makhina EN, Nichols CG (1994) Potassium channel block by cytoplasmic polyamines as the mechanism of intrinsic rectification. Nature 372:366–369

Lopatin AN, Makhina EN, Nichols CG (1995) The mechanism of inward rectification of potassium channels: "long-pore plugging" by cytoplasmic polyamines. J Gen Physiol 106:923–955

Lopes CM, Zilberberg N, Goldstein SA (2001) Block of Kcnk3 by protons. Evidence that 2-P-domain potassium channel subunits function as homodimers. J Biol Chem 276:24449–24452

Lopes CM, Rohacs T, Czirjak G, Balla T, Enyedi P, Logothetis DE (2005) PIP2 hydrolysis underlies agonist-induced inhibition and regulates voltage gating of two-pore domain K+ channels. J Physiol 564:117–129

Losin S, McKean CM (1966) Chlorzoxazone (paraflex) in the treatment of severe spasticity. Dev Med Child Neurol 8:768–769

Lourdel S, Paulais M, Cluzeaud F, Bens M, Tanemoto M, Kurachi Y, Vandewalle A, Teulon J (2002) An inward rectifier K(+) channel at the basolateral membrane of the mouse distal convoluted tubule: similarities with Kir4-Kir5.1 heteromeric channels. J Physiol 538:391–404

Lu Z, MacKinnon R (1994) Electrostatic tuning of Mg^{2+} affinity in an inward-rectifier K^+ channel. Nature 371:243–246

Lu Z, MacKinnon R (1997) Purification, characterization, and synthesis of an inward-rectifier K+ channel inhibitor from scorpion venom. Biochemistry 36:6936–6940

Lunn ML, Nassirpour R, Arrabit C, Tan J, McLeod I, Arias CM, Sawchenko PE, Yates JR 3rd, Slesinger PA (2007) A unique sorting nexin regulates trafficking of potassium channels via a PDZ domain interaction. Nat Neurosci 10:1249–1259

Ma D, Zerangue N, Raab-Graham K, Fried SR, Jan YN, Jan LY (2002) Diverse trafficking patterns due to multiple traffic motifs in G protein-activated inwardly rectifying potassium channels from brain and heart. Neuron 33:715–729

Ma Z, Lou XJ, Horrigan FT (2006) Role of charged residues in the S1-S4 voltage sensor of BK channels. J Gen Physiol 127:309–328

Ma L, Zhang X, Chen H (2011) TWIK-1 two-pore domain potassium channels change ion selectivity and conduct inward leak sodium currents in hypokalemia. Sci Signal 4:ra37

Ma L, Zhang X, Zhou M, Chen H (2012) Acid-sensitive TWIK and TASK two-pore domain potassium channels change ion selectivity and become permeable to sodium in extracellular acidification. J Biol Chem 287:37145–37153

Magidovich E, Fleishman SJ, Yifrach O (2006) Intrinsically disordered C-terminal segments of voltage-activated potassium channels: a possible fishing rod-like mechanism for channel binding to scaffold proteins. Bioinformatics 22:1546–1550

Maingret F, Lauritzen I, Patel AJ, Heurteaux C, Reyes R, Lesage F, Lazdunski M, Honore E (2000) TREK-1 is a heat-activated background K(+) channel. EMBO J 19:2483–2491

Makary SM, Claydon TW, Dibb KM, Boyett MR (2006) Base of pore loop is important for rectification, activation, permeation, and block of Kir3.1/Kir3.4. Biophys J 90:4018–4034

Manganas LN, Trimmer JS (2000) Subunit composition determines Kv1 potassium channel surface expression. J Biol Chem 275:29685–29693

Manganas LN, Wang Q, Scannevin RH, Antonucci DE, Rhodes KJ, Trimmer JS (2001) Identification of a trafficking determinant localized to the Kv1 potassium channel pore. Proc Natl Acad Sci 98:14055–14059

Mannhold R (2004) KATP channel openers: structure-activity relationships and therapeutic potential. Med Res Rev 24:213–266

Martin GM, Yoshioka C, Rex EA, Fay JF, Xie Q, Whorton MR, Chen JZ, Shyng SL (2017) Cryo-EM structure of the ATP-sensitive potassium channel illuminates mechanisms of assembly and gating. eLife 6

Matsuda H, Saigusa A, Irisawa H (1987) Ohmic conductance through the inwardly rectifying K channel and blocking by internal Mg2+. Nature 325:156–159

Matsuoka T, Matsushita K, Katayama Y, Fujita A, Inageda K, Tanemoto M, Inanobe A, Yamashita S, Matsuzawa Y, Kurachi Y (2000) C-terminal tails of sulfonylurea receptors control ADP-induced activation and diazoxide modulation of ATP-sensitive K(+) channels. Circ Res 87:873–880

McKeown L, Burnham MP, Hodson C, Jones OT (2008) Identification of an evolutionarily conserved extracellular threonine residue critical for surface expression and its potential coupling of adjacent voltage-sensing and gating domains in voltage-gated potassium channels. J Biol Chem 283:30421–30432

McManus OB, Helms LM, Pallanck L, Ganetzky B, Swanson R, Leonard RJ (1995) Functional role of the beta subunit of high conductance calcium-activated potassium channels. Neuron 14:645–650

Medhurst AD, Rennie G, Chapman CG, Meadows H, Duckworth MD, Kelsell RE, Gloger II, Pangalos MN (2001) Distribution analysis of human two pore domain potassium channels in tissues of the central nervous system and periphery. Brain Res Mol Brain Res 86:101–114

Meera P, Wallner M, Song M, Toro L (1997) Large conductance voltage- and calcium-dependent K + channel, a distinct member of voltage-dependent ion channels with seven N-terminal transmembrane segments (S0-S6), an extracellular N terminus, and an intracellular (S9-S10) C terminus. Proc Natl Acad Sci U S A 94:14066–14071

Meera P, Wallner M, Toro L (2000) A neuronal beta subunit (KCNMB4) makes the large conductance, voltage- and Ca2+−activated K+ channel resistant to charybdotoxin and iberiotoxin. Proc Natl Acad Sci U S A 97:5562–5567

Meredith AL, Wiler SW, Miller BH, Takahashi JS, Fodor AA, Ruby NF, Aldrich RW (2006) BK calcium-activated potassium channels regulate circadian behavioral rhythms and pacemaker output. Nat Neurosci 9:1041–1049

Miki T, Liss B, Minami K, Shiuchi T, Saraya A, Kashima Y, Horiuchi M, Ashcroft F, Minokoshi Y, Roeper J et al (2001) ATP-sensitive K+ channels in the hypothalamus are essential for the maintenance of glucose homeostasis. Nat Neurosci 4:507–512

Miller C (1995) The charybdotoxin family of K+ channel-blocking peptides. Neuron 15:5–10

Miller AN, Long SB (2012) Crystal structure of the human two-pore domain potassium channel K2P1. Science 335:432–436

Miller C, Moczydlowski E, Latorre R, Phillips M (1985) Charybdotoxin, a protein inhibitor of single Ca2+−activated K+ channels from mammalian skeletal muscle. Nature 313:316–318

Mitcheson JS, Hancox JC (1999) An investigation of the role played by the E-4031-sensitive (rapid delayed rectifier) potassium current in isolated rabbit atrioventricular nodal and ventricular myocytes. Pflugers Arch 438:843–850

Montgomery JR, Whitt JP, Wright BN, Lai MH, Meredith AL (2013) Mis-expression of the BK K (+) channel disrupts suprachiasmatic nucleus circuit rhythmicity and alters clock-controlled behavior. Am J Physiol Cell Physiol 304:C299–C311

Moss AJ, Zareba W, Benhorin J, Locati EH, Hall WJ, Robinson JL, Schwartz PJ, Towbin JA, Vincent GM, Lehmann MH (1995) ECG T-wave patterns in genetically distinct forms of the hereditary long QT syndrome. Circulation 92:2929–2934

Muiesan G, Fariello R, Muiesan ML, Christensen OE (1985) Effect of pinacidil on blood pressure, plasma catecholamines and plasma renin activity in essential hypertension. Eur J Clin Pharmacol 28:495–499

Munoz MB, Slesinger PA (2014) Sorting nexin 27 regulation of G protein-gated inwardly rectifying K+ channels attenuates *in vivo* cocaine response. Neuron 82:659–669

Nesti E, Everill B, Morielli AD (2004) Endocytosis as a mechanism for tyrosine kinase-dependent suppression of a voltage-gated potassium channel. Mol Biol Cell 15:4073–4088

Nestorowicz A, Inagaki N, Gonoi T, Schoor KP, Wilson BA, Glaser B, Landau H, Stanley CA, Thornton PS, Seino S et al (1997) A nonsense mutation in the inward rectifier potassium channel gene, Kir6.2, is associated with familial hyperinsulinism. Diabetes 46:1743–1748

Nichols CG, Lee SJ (2018) Polyamines and potassium channels: a 25-year romance. J Biol Chem 293:18779–18788

Niemeyer MI, Cid LP, Valenzuela X, Paeile V, Sepulveda FV (2003) Extracellular conserved cysteine forms an intersubunit disulphide bridge in the KCNK5 (TASK-2) K+ channel without having an essential effect upon activity. Mol Membr Biol 20:185–191

Niemeyer MI, Gonzalez-Nilo FD, Zuniga L, Gonzalez W, Cid LP, Sepulveda FV (2007) Neutralization of a single arginine residue gates open a two-pore domain, alkali-activated K+ channel. Proc Natl Acad Sci U S A 104:666–671

Niemeyer MI, Cid LP, Gonzalez W, Sepulveda FV (2016) Gating, regulation, and structure in K2P K+ channels: in varietate concordia? Mol Pharmacol 90:309–317

Nishida M, Cadene M, Chait BT, MacKinnon R (2007) Crystal structure of a Kir3.1-prokaryotic Kir channel chimera. EMBO J 26:4005–4015

Noble D, Tsien RW (1969a) Outward membrane currents activated in the plateau range of potentials in cardiac Purkinje fibres. J Physiol 200:205–231

Noble D, Tsien RW (1969b) Reconstruction of the repolarization process in cardiac Purkinje fibres based on voltage clamp measurements of membrane current. J Physiol 200:233–254

Noda M, Shimizu S, Tanabe T, Takai T, Kayano T, Ikeda T, Takahashi H, Nakayama H, Kanaoka Y, Minamino N et al (1984) Primary structure of electrophorus electricus sodium channel deduced from cDNA sequence. Nature 312:121–127

Noskov SY, Bernèche S, Roux B (2004) Control of ion selectivity in potassium channels by electrostatic and dynamic properties of carbonyl ligands. Nature 431:830–834

Oliver D, Baukrowitz T, Fakler B (2000) Polyamines as gating molecules of inward-rectifier K+ channels. Eur J Biochem 267:5824–5829

Olsen O, Liu H, Wade JB, Merot J, Welling PA (2002) Basolateral membrane expression of the Kir 2.3 channel is coordinated by PDZ interaction with Lin-7/CASK complex. Am J Physiol Cell Physiol 282:C183–C195

Oonuma H, Iwasawa K, Iida H, Nagata T, Imuta H, Morita Y, Yamamoto K, Nagai R, Omata M, Nakajima T (2002) Inward rectifier K(+) current in human bronchial smooth muscle cells: inhibition with antisense oligonucleotides targeted to Kir2.1 mRNA. Am J Respir Cell Mol Biol 26:371–379

Palygin O, Pochynyuk O, Staruschenko A (2017) Role and mechanisms of regulation of the basolateral Kir 4.1/Kir 5.1K(+) channels in the distal tubules. Acta Physiol 219:260–273

Parcej DN, Dolly JO (1989) Elegance persists in the purification of K+ channels. Biochem J 264:623–624

Patel AJ, Honore E, Lesage F, Fink M, Romey G, Lazdunski M (1999) Inhalational anesthetics activate two-pore-domain background K+ channels. Nat Neurosci 2:422–426

Patel C, Yan GX, Antzelevitch C (2010) Short QT syndrome: from bench to bedside. Circ Arrhythm Electrophysiol 3:401–408

Patil N, Cox DR, Bhat D, Faham M, Myers RM, Peterson AS (1995) A potassium channel mutation in weaver mice implicates membrane excitability in granule cell differentiation. Nat Genet 11:126–129

Pau V, Zhou Y, Ramu Y, Xu Y, Lu Z (2017) Crystal structure of an inactivated mutant mammalian voltage-gated K+ channel. Nat Struct Mol Biol 24:857–865

Payandeh J, Scheuer T, Zheng N, Catterall WA (2011) The crystal structure of a voltage-gated sodium channel. Nature 475:353–358

Pedarzani P, D'Hoedt D, Doorty KB, Wadsworth JD, Joseph JS, Jeyaseelan K, Kini RM, Gadre SV, Sapatnekar SM, Stocker M et al (2002) Tamapin, a venom peptide from the Indian red scorpion (Mesobuthus tamulus) that targets small conductance Ca2+−activated K+ channels and afterhyperpolarization currents in central neurons. J Biol Chem 277:46101–46109

Pegan S, Arrabit C, Zhou W, Kwiatkowski W, Collins A, Slesinger PA, Choe S (2005) Cytoplasmic domain structures of Kir2.1 and Kir3.1 show sites for modulating gating and rectification. Nat Neurosci 8:279–287

Pereira V, Busserolles J, Christin M, Devilliers M, Poupon L, Legha W, Alloui A, Aissouni Y, Bourinet E, Lesage F et al (2014) Role of the TREK2 potassium channel in cold and warm thermosensation and in pain perception. Pain 155:2534–2544

Pessia M, Imbrici P, D'Adamo MC, Salvatore L, Tucker SJ (2001) Differential pH sensitivity of Kir4.1 and Kir4.2 potassium channels and their modulation by heteropolymerisation with Kir5.1. J Physiol 532:359–367

Plant LD (2012) A role for K2P channels in the operation of somatosensory nociceptors. Front Mol Neurosci 5:21

Plant LD, Kemp PJ, Peers C, Henderson Z, Pearson HA (2002) Hypoxic depolarization of cerebellar granule neurons by specific inhibition of TASK-1. Stroke 33:2324–2328

Pluger S, Faulhaber J, Furstenau M, Lohn M, Waldschutz R, Gollasch M, Haller H, Luft FC, Ehmke H, Pongs O (2000) Mice with disrupted BK channel beta1 subunit gene feature abnormal ca(2+) spark/STOC coupling and elevated blood pressure. Circ Res 87:E53–E60

Pollema-Mays SL, Centeno MV, Ashford CJ, Apkarian AV, Martina M (2013) Expression of background potassium channels in rat DRG is cell-specific and down-regulated in a neuropathic pain model. Mol Cell Neurosci 57:1–9

Pongs O, Leicher T, Berger M, Roeper J, Bahring R, Wray D, Giese KP, Silva AJ, Storm JF (1999) Functional and molecular aspects of voltage-gated K+ channel beta subunits. Ann N Y Acad Sci 868:344–355

Priori SG, Schwartz PJ, Napolitano C, Bloise R, Ronchetti E, Grillo M, Vicentini A, Spazzolini C, Nastoli J, Bottelli G et al (2003) Risk stratification in the long-QT syndrome. N Engl J Med 348:1866–1874

Quayle JM, McCarron JG, Brayden JE, Nelson MT (1993) Inward rectifier K+ currents in smooth muscle cells from rat resistance-sized cerebral arteries. Am J Phys 265:C1363–C1370

Rainero I, Rubino E, Gallone S, Zavarise P, Carli D, Boschi S, Fenoglio P, Savi L, Gentile S, Benna P et al (2014) KCNK18 (TRESK) genetic variants in Italian patients with migraine. Headache 54:1515–1522

Rajan S, Wischmeyer E, Xin Liu G, Preisig-Muller R, Daut J, Karschin A, Derst C (2000) TASK-3, a novel tandem pore domain acid-sensitive K+ channel. An extracellular histiding as pH sensor. J Biol Chem 275:16650–16657

Remedi MS, Koster JC (2010) K(ATP) channelopathies in the pancreas. Pflugers Arch 460:307–320

Rettig J, Heinemann SH, Wunder F, Lorra C, Parcej DN, Oliver Dolly J, Pongs O (1994) Inactivation properties of voltage-gated K+ channels altered by presence of β-subunit. Nature 369:289–294

Rhodes K, Keilbaugh S, Barrezueta N, Lopez K, Trimmer J (1995) Association and colocalization of K+ channel alpha- and beta-subunit polypeptides in rat brain. J Neurosci 15:5360–5371

Rhodes KJ, Monaghan MM, Barrezueta NX, Nawoschik S, Bekele-Arcuri Z, Matos MF, Nakahira K, Schechter LE, Trimmer JS (1996) Voltage-gated K+channel β subunits: expression and distribution of Kvβ1 and Kvβ2 in adult rat brain. J Neurosci 16:4846–4860

Rhodes KJ, Strassle BW, Monaghan MM, Bekele-Arcuri Z, Matos MF, Trimmer JS (1997) Association and colocalization of the Kvβ1 and Kvβ2 β-subunits with Kv1 α-subunits in mammalian brain K+channel complexes. J Neurosci 17:8246–8258

Rifkin RA, Moss SJ, Slesinger PA (2017) G protein-gated potassium channels: a link to drug addiction. Trends Pharmacol Sci 38:378–392

Rohacs T, Lopes CM, Jin T, Ramdya PP, Molnar Z, Logothetis DE (2003) Specificity of activation by phosphoinositides determines lipid regulation of Kir channels. Proc Natl Acad Sci U S A 100:745–750

Romanenko VG, Fang Y, Byfield F, Travis AJ, Vandenberg CA, Rothblat GH, Levitan I (2004) Cholesterol sensitivity and lipid raft targeting of Kir2.1 channels. Biophys J 87:3850–3861

Rosenhouse-Dantsker A, Sui JL, Zhao Q, Rusinova R, Rodriguez-Menchaca AA, Zhang Z, Logothetis DE (2008) A sodium-mediated structural switch that controls the sensitivity of Kir channels to PtdIns(4,5)P(2). Nat Chem Biol 4:624–631

Roux B (2005) Ion conduction and selectivity in K+ channels. Ann Rev Biophys Biomol Struct 34:153–171

Sadja R, Smadja K, Alagem N, Reuveny E (2001) Coupling Gbetagamma-dependent activation to channel opening via pore elements in inwardly rectifying potassium channels. Neuron 29:669–680

Sahoo N, Hoshi T, Heinemann SH (2014) Oxidative modulation of voltage-gated potassium channels. Antioxidants Redox Signal 21:933–952

Sandoz G, Thummler S, Duprat F, Feliciangeli S, Vinh J, Escoubas P, Guy N, Lazdunski M, Lesage F (2006) AKAP150, a switch to convert mechano-, pH- and arachidonic acid-sensitive TREK K (+) channels into open leak channels. EMBO J 25:5864–5872

Sandoz G, Douguet D, Chatelain F, Lazdunski M, Lesage F (2009) Extracellular acidification exerts opposite actions on TREK1 and TREK2 potassium channels via a single conserved histidine residue. Proc Natl Acad Sci U S A 106:14628–14633

Sanguinetti MC, Jiang C, Curran ME, Keating MT (1995) A mechanistic link between an inherited and an acquired cardiac arrhythmia: HERG encodes the IKr potassium channel. Cell 81:299–307

Sankaranarayanan A, Raman G, Busch C, Schultz T, Zimin PI, Hoyer J, Kohler R, Wulff H (2009) Naphtho[1,2-d]thiazol-2-ylamine (SKA-31), a new activator of KCa2 and KCa3.1 potassium channels, potentiates the endothelium-derived hyperpolarizing factor response and lowers blood pressure. Mol Pharmacol 75:281–295

Santos JS, Asmar-Rovira GA, Han GW, Liu W, Syeda R, Cherezov V, Baker KA, Stevens RC, Montal M (2012) Crystal structure of a voltage-gated K+channel pore module in a closed state in lipid membranes. J Biol Chem 287:43063–43070

Sausbier U, Sausbier M, Sailer CA, Arntz C, Knaus HG, Neuhuber W, Ruth P (2006) Ca2+ −activated K+ channels of the BK-type in the mouse brain. Histochem Cell Biol 125:725–741

Savarin P, Guenneugues M, Gilquin B, Lamthanh H, Gasparini S, Zinn-Justin S, Ménez A (1998) Three-dimensional structure of κ-conotoxin PVIIA, a novel potassium channel-blocking toxin from cone snails†,‡. Biochemistry 37:5407–5416

Schewe M, Nematian-Ardestani E, Sun H, Musinszki M, Cordeiro S, Bucci G, de Groot BL, Tucker SJ, Rapedius M, Baukrowitz T (2016) A non-canonical voltage-sensing mechanism controls gating in K2P K(+) channels. Cell 164:937–949

Schmid-Antomarchi H, de Weille J, Fosset M, Lazdunski M (1987) The antidiabetic sulfonylurea glibenclamide is a potent blocker of the ATP-modulated K+ channel in insulin secreting cells. Biochem Biophys Res Commun 146:21–25

Schmidt D, Mackinnon R (2008) Voltage-dependent K+ channel gating and voltage sensor toxin sensitivity depend on the mechanical state of the lipid membrane. Proc Natl Acad Sci 105:19276–19281

Schmidt D, Cross SR, Mackinnon R (2009) A gating model for the archeal voltage-dependent K+ channel KvAP in DPhPC and POPE:POPG decane lipid bilayers. J Mol Biol 390:902–912

Scholl UI, Choi M, Liu T, Ramaekers VT, Hausler MG, Grimmer J, Tobe SW, Farhi A, Nelson-Williams C, Lifton RP (2009) Seizures, sensorineural deafness, ataxia, mental retardation, and electrolyte imbalance (SeSAME syndrome) caused by mutations in KCNJ10. Proc Natl Acad Sci U S A 106:5842–5847

Schreiber M, Salkoff L (1997) A novel calcium-sensing domain in the BK channel. Biophys J 73:1355–1363

Schulte U, Fakler B (2000) Gating of inward-rectifier K+ channels by intracellular pH. Eur J Biochem 267:5837–5841

Schulte U, Thumfart JO, Klocker N, Sailer CA, Bildl W, Biniossek M, Dehn D, Deller T, Eble S, Abbass K et al (2006) The epilepsy-linked Lgi1 protein assembles into presynaptic Kv1 channels and inhibits inactivation by Kvbeta1. Neuron 49:697–706

Schumacher MA, Rivard AF, Bachinger HP, Adelman JP (2001) Structure of the gating domain of a Ca2+−activated K+ channel complexed with Ca2+/calmodulin. Nature 410:1120–1124

Schwartz PJ, Priori SG, Spazzolini C, Moss AJ, Vincent GM, Napolitano C, Denjoy I, Guicheney P, Breithardt G, Keating MT et al (2001) Genotype-phenotype correlation in the long-QT syndrome: gene-specific triggers for life-threatening arrhythmias. Circulation 103:89–95

Sforna L, Megaro A, Pessia M, Franciolini F, Catacuzzeno L (2018) Structure, gating and basic functions of the Ca2+−activated K channel of intermediate conductance. Curr Neuropharmacol 16:608–617

Shen KZ, Lagrutta A, Davies NW, Standen NB, Adelman JP, North RA (1994) Tetraethylammonium block of slowpoke calcium-activated potassium channels expressed in Xenopus oocytes: evidence for tetrameric channel formation. Pflugers Arch 426:440–445

Shi J, Cui J (2001) Intracellular mg(2+) enhances the function of BK-type ca(2+)-activated K(+) channels. J Gen Physiol 118:589–606

Shi J, He HQ, Zhao R, Duan YH, Chen J, Chen Y, Yang J, Zhang JW, Shu XQ, Zheng P et al (2008) Inhibition of martentoxin on neuronal BK channel subtype (alpha+beta4): implications for a novel interaction model. Biophys J 94:3706–3713

Shibasaki T (1987) Conductance and kinetics of delayed rectifier potassium channels in nodal cells of the rabbit heart. J Physiol 387:227–250
Shimizu W, Antzelevitch C (2000) Differential effects of beta-adrenergic agonists and antagonists in LQT1, LQT2 and LQT3 models of the long QT syndrome. J Am Coll Cardiol 35:778–786
Shin N, Soh H, Chang S, Kim DH, Park CS (2005) Sodium permeability of a cloned small-conductance calcium-activated potassium channel. Biophys J 89:3111–3119
Shrivastava IH, Peter Tieleman D, Biggin PC, Sansom MSP (2002) K+ versus Na+ ions in a K channel selectivity filter: a simulation study. Biophys J 83:633–645
Shumilina E, Klocker N, Korniychuk G, Rapedius M, Lang F, Baukrowitz T (2006) Cytoplasmic accumulation of long-chain coenzyme A esters activates KATP and inhibits Kir2.1 channels. J Physiol 575:433–442
Shyng S, Nichols CG (1997) Octameric stoichiometry of the KATP channel complex. J Gen Physiol 110:655–664
Singh S, Syme CA, Singh AK, Devor DC, Bridges RJ (2001) Benzimidazolone activators of chloride secretion: potential therapeutics for cystic fibrosis and chronic obstructive pulmonary disease. J Pharmacol Exp Ther 296:600–611
Singh AK, McMillan J, Bukiya AN, Burton B, Parrill AL, Dopico AM (2012) Multiple cholesterol recognition/interaction amino acid consensus (CRAC) motifs in cytosolic C tail of Slo1 subunit determine cholesterol sensitivity of Ca2+– and voltage-gated K+ (BK) channels. J Biol Chem 287:20509–20521
Slesinger PA, Patil N, Liao YJ, Jan YN, Jan LY, Cox DR (1996) Functional effects of the mouse weaver mutation on G protein-gated inwardly rectifying K+ channels. Neuron 16:321–331
Smart SL, Lopantsev V, Zhang CL, Robbins CA, Wang H, Chiu SY, Schwartzkroin PA, Messing A, Tempel BL (1998) Deletion of the K(V)1.1 potassium channel causes epilepsy in mice. Neuron 20:809–819
Smith PL, Baukrowitz T, Yellen G (1996) The inward rectification mechanism of the HERG cardiac potassium channel. Nature 379:833–836
Smith SEP, Xu L, Kasten MR, Anderson MP (2012) Mutant LGI1 inhibits seizure-induced trafficking of Kv4.2 potassium channels. J Neurochem 120:611–621
Srinivasan KN, Sivaraja V, Huys I, Sasaki T, Cheng B, Kumar TKS, Sato K, Tytgat J, Yu C, San BCC et al (2002) κ-Hefutoxin1, a novel toxin from the scorpionheterometrus fulvipes with unique structure and function. J Biol Chem 277:30040–30047
Stanfield PR, Davies NW, Shelton PA, Sutcliffe MJ, Khan IA, Brammar WJ, Conley EC (1994) A single aspartate residue is involved in both intrinsic gating and blockage by Mg2+ of the inward rectifier, IRK1. J Physiol 478(Pt 1):1–6
Storm JF (1987) Action potential repolarization and a fast after-hyperpolarization in rat hippocampal pyramidal cells. J Physiol 385:733–759
Strassmaier T, Bond CT, Sailer CA, Knaus HG, Maylie J, Adelman JP (2005) A novel isoform of SK2 assembles with other SK subunits in mouse brain. J Biol Chem 280:21231–21236
Strobaek D, Christophersen P, Holm NR, Moldt P, Ahring PK, Johansen TE, Olesen SP (1996) Modulation of the ca(2+)-dependent K+ channel, hslo, by the substituted diphenylurea NS 1608, paxilline and internal Ca2+. Neuropharmacology 35:903–914
Strobaek D, Jorgensen TD, Christophersen P, Ahring PK, Olesen SP (2000) Pharmacological characterization of small-conductance ca(2+)-activated K(+) channels stably expressed in HEK 293 cells. Br J Pharmacol 129:991–999
Strobaek D, Teuber L, Jorgensen TD, Ahring PK, Kjaer K, Hansen RS, Olesen SP, Christophersen P, Skaaning-Jensen B (2004) Activation of human IK and SK Ca2+ –activated K+ channels by NS309 (6,7-dichloro-1H-indole-2,3-dione 3-oxime). Biochim Biophys Acta 1665:1–5
Strobaek D, Hougaard C, Johansen TH, Sorensen US, Nielsen EO, Nielsen KS, Taylor RD, Pedarzani P, Christophersen P (2006) Inhibitory gating modulation of small conductance Ca2 +–activated K+ channels by the synthetic compound (R)-N-(benzimidazol-2-yl)-1,2,3,4-

tetrahydro-1-naphtylamine (NS8593) reduces afterhyperpolarizing current in hippocampal CA1 neurons. Mol Pharmacol 70:1771–1782

Suzuki M, Sasaki N, Miki T, Sakamoto N, Ohmoto-Sekine Y, Tamagawa M, Seino S, Marban E, Nakaya H (2002) Role of sarcolemmal K(ATP) channels in cardioprotection against ischemia/reperfusion injury in mice. J Clin Invest 109:509–516

Syeda R, Santos JS, Montal M (2014) Lipid bilayer modules as determinants of K+channel gating. J Biol Chem 289:4233–4243

Tabcharani JA, Misler S (1989) Ca2+-activated K+ channel in rat pancreatic islet B cells: permeation, gating and blockade by cations. Biochim Biophys Acta 982:62–72

Taglialatela M, Ficker E, Wible BA, Brown AM (1995) C-terminus determinants for Mg2+ and polyamine block of the inward rectifier K+ channel IRK1. EMBO J 14:5532–5541

Takahira M, Sakurai M, Sakurada N, Sugiyama K (2005) Fenamates and diltiazem modulate lipid-sensitive mechano-gated 2P domain K(+) channels. Pflügers Arch 451:474–478

Takenaka K, Ai T, Shimizu W, Kobori A, Ninomiya T, Otani H, Kubota T, Takaki H, Kamakura S, Horie M (2003) Exercise stress test amplifies genotype-phenotype correlation in the LQT1 and LQT2 forms of the long-QT syndrome. Circulation 107:838–844

Talley EM, Lei Q, Sirois JE, Bayliss DA (2000) TASK-1, a two-pore domain K+ channel, is modulated by multiple neurotransmitters in motoneurons. Neuron 25:399–410

Tanemoto M, Kittaka N, Inanobe A, Kurachi Y (2000) In vivo formation of a proton-sensitive K+ channel by heteromeric subunit assembly of Kir5.1 with Kir4.1. J Physiol 525(Pt 3):587–592

Tanemoto M, Abe T, Ito S (2005) PDZ-binding and di-hydrophobic motifs regulate distribution of Kir4.1 channels in renal cells. J Am Soc Nephrol 16:2608–2614

Tao X, MacKinnon R (2019) Molecular structures of the human Slo1 K(+) channel in complex with beta4. eLife 8

Tao X, Hite RK, MacKinnon R (2017) Cryo-EM structure of the open high-conductance ca(2+)-activated K(+) channel. Nature 541:46–51

Teichert RW, Schmidt EW, Olivera BM (2015) Constellation pharmacology: a new paradigm for drug discovery. Ann Rev Pharmacol Toxicol 55:573–589

Terzic A, Jahangir A, Kurachi Y (1995) Cardiac ATP-sensitive K+ channels: regulation by intracellular nucleotides and K+ channel-opening drugs. Am J Phys 269:C525–C545

Tian Y, Ullrich F, Xu R, Heinemann SH, Hou S, Hoshi T (2015) Two distinct effects of PIP2 underlie auxiliary subunit-dependent modulation of Slo1 BK channels. J Gen Physiol 145:331–343

Tinker A, Aziz Q, Li Y, Specterman M (2018) ATP-sensitive potassium channels and their physiological and pathophysiological roles. Compr Physiol 8:1463–1511

Tong Y, Wei J, Zhang S, Strong JA, Dlouhy SR, Hodes ME, Ghetti B, Yu L (1996) The weaver mutation changes the ion selectivity of the affected inwardly rectifying potassium channel GIRK2. FEBS Lett 390:63–68

Trimmer JS, Rhodes KJ (2004) Localization of voltage-gated ion channels in mammalian brain. Annu Rev Physiol 66:477–519

Tucker SJ, Imbrici P, Salvatore L, D'Adamo MC, Pessia M (2000) pH dependence of the inwardly rectifying potassium channel, Kir5.1, and localization in renal tubular epithelia. J Biol Chem 275:16404–16407

Tudor JE, Pallaghy PK, Pennington MW, Norton RS (1996) Solution structure of ShK toxin, a novel potassium channel inhibitor from a sea anemone. Nat Struct Biol 3:317–320

Tulleuda A, Cokic B, Callejo G, Saiani B, Serra J, Gasull X (2011) TRESK channel contribution to nociceptive sensory neurons excitability: modulation by nerve injury. Mol Pain 7:30

Turner KL, Honasoge A, Robert SM, McFerrin MM, Sontheimer H (2014) A proinvasive role for the ca(2+) -activated K(+) channel KCa3.1 in malignant glioma. Glia 62:971–981

Uebele VN, Lagrutta A, Wade T, Figueroa DJ, Liu Y, McKenna E, Austin CP, Bennett PB, Swanson R (2000) Cloning and functional expression of two families of beta-subunits of the large conductance calcium-activated K+ channel. J Biol Chem 275:23211–23218

Vacher H, Mohapatra DP, Misonou H, Trimmer JS (2007) Regulation of Kvl channel trafficking by the mamba snake neurotoxin dendrotoxin K. FASEB J 21:906–914

Vacher H, Mohapatra DP, Trimmer JS (2008) Localization and targeting of voltage-dependent ion channels in mammalian central neurons. Physiol Rev 88:1407–1447

Valiyaveetil FI, Zhou Y, Mackinnon R (2002) Lipids in the structure, folding, and function of the KcsA K+channel. Biochemistry 41:10771–10777

Valverde MA, Rojas P, Amigo J, Cosmelli D, Orio P, Bahamonde MI, Mann GE, Vergara C, Latorre R (1999) Acute activation of maxi-K channels (hSlo) by estradiol binding to the beta subunit. Science 285:1929–1931

Van Dalen A, De Kruijff B (2004) The role of lipids in membrane insertion and translocation of bacterial proteins. Biochim Biophys Acta (BBA) Mol Cell Res 1694:97–109

Van Wart A, Trimmer JS, Matthews G (2007) Polarized distribution of ion channels within microdomains of the axon initial segment. J Comp Neurol 500:339–352

Vandorpe DH, Shmukler BE, Jiang L, Lim B, Maylie J, Adelman JP, de Franceschi L, Cappellini MD, Brugnara C, Alper SL (1998) cDNA cloning and functional characterization of the mouse Ca2+−gated K+ channel, mIK1. Roles in regulatory volume decrease and erythroid differentiation. J Biol Chem 273:21542–21553

Varma S, Rogers DM, Pratt LR, Rempe SB (2011) Design principles for K+ selectivity in membrane transport. J Gen Physiol 137:479–488

Veale EL, Al-Moubarak E, Bajaria N, Omoto K, Cao L, Tucker SJ, Stevens EB, Mathie A (2014) Influence of the N terminus on the biophysical properties and pharmacology of TREK1 potassium channels. Mol Pharmacol 85:671–681

Villarroel A, Alvarez O, Oberhauser A, Latorre R (1988) Probing a Ca2+−activated K+ channel with quaternary ammonium ions. Pflugers Arch 413:118–126

Wallner M, Meera P, Toro L (1999) Molecular basis of fast inactivation in voltage and Ca2 +−activated K+ channels: a transmembrane beta-subunit homolog. Proc Natl Acad Sci U S A 96:4137–4142

Walsh KB (2020) Screening technologies for inward rectifier potassium channels: discovery of new blockers and activators. SLAS Discov 25:420–433

Wang YW, Ding JP, Xia XM, Lingle CJ (2002) Consequences of the stoichiometry of Slo1 alpha and auxiliary beta subunits on functional properties of large-conductance Ca2+−activated K+ channels. J Neurosci 22:1550–1561

Wang B, Rothberg BS, Brenner R (2006) Mechanism of beta4 subunit modulation of BK channels. J Gen Physiol 127:449–465

Weatherall KL, Goodchild SJ, Jane DE, Marrion NV (2010) Small conductance calcium-activated potassium channels: from structure to function. Prog Neurobiol 91:242–255

Wei AD, Gutman GA, Aldrich R, Chandy KG, Grissmer S, Wulff H (2005) International Union of Pharmacology. LII. Nomenclature and molecular relationships of calcium-activated potassium channels. Pharmacol Rev 57:463–472

Weiger TM, Holmqvist MH, Levitan IB, Clark FT, Sprague S, Huang WJ, Ge P, Wang C, Lawson D, Jurman ME et al (2000) A novel nervous system beta subunit that downregulates human large conductance calcium-dependent potassium channels. J Neurosci 20:3563–3570

Weik R, Neumcke B (1989) ATP-sensitive potassium channels in adult mouse skeletal muscle: characterization of the ATP-binding site. J Membr Biol 110:217–226

Welling PA (2016) Roles and regulation of renal K channels. Annu Rev Physiol 78:415–435

Whitt JP, Montgomery JR, Meredith AL (2016) BK channel inactivation gates daytime excitability in the circadian clock. Nat Commun 7:10837

Whorton MR, MacKinnon R (2011) Crystal structure of the mammalian GIRK2 K^+ channel and gating regulation by G proteins, PIP_2, and sodium. Cell 147:199–208

Whorton MR, MacKinnon R (2013) X-ray structure of the mammalian GIRK2-betagamma G-protein complex. Nature 498:190–197

Wible BA, Taglialatela M, Ficker E, Brown AM (1994) Gating of inwardly rectifying K^+ channels localized to a single negatively charged residue. Nature 371:246–249

Wilke BU, Lindner M, Greifenberg L, Albus A, Kronimus Y, Bunemann M, Leitner MG, Oliver D (2014) Diacylglycerol mediates regulation of TASK potassium channels by Gq-coupled receptors. Nat Commun 5:5540

Wischmeyer E, Doring F, Karschin A (2000) Stable cation coordination at a single outer pore residue defines permeation properties in Kir channels. FEBS Lett 466:115–120

Woo DH, Han KS, Shim JW, Yoon BE, Kim E, Bae JY, Oh SJ, Hwang EM, Marmorstein AD, Bae YC et al (2012) TREK-1 and Best1 channels mediate fast and slow glutamate release in astrocytes upon GPCR activation. Cell 151:25–40

Wu ZZ, Li DP, Chen SR, Pan HL (2009) Aminopyridines potentiate synaptic and neuromuscular transmission by targeting the voltage-activated calcium channel beta subunit. J Biol Chem 284:36453–36461

Wu W, Wang Y, Deng XL, Sun HY, Li GR (2013) Cholesterol down-regulates BK channels stably expressed in HEK 293 cells. PLoS One 8:e79952

Wulff H, Castle NA (2010) Therapeutic potential of KCa3.1 blockers: recent advances and promising trends. Expert Rev Clin Pharmacol 3:385–396

Wulff H, Miller MJ, Hansel W, Grissmer S, Cahalan MD, Chandy KG (2000) Design of a potent and selective inhibitor of the intermediate-conductance Ca2+−activated K+ channel, IKCa1: a potential immunosuppressant. Proc Natl Acad Sci U S A 97:8151–8156

Wulff H, Castle NA, Pardo LA (2009) Voltage-gated potassium channels as therapeutic targets. Nat Rev Drug Discov 8:982–1001

Wydeven N, Marron Fernandez de Velasco E, Du Y, Benneyworth MA, Hearing MC, Fischer RA, Thomas MJ, Weaver CD, Wickman K (2014) Mechanisms underlying the activation of G-protein-gated inwardly rectifying K+ (GIRK) channels by the novel anxiolytic drug, ML297. Proc Natl Acad Sci U S A 111:10755–10760

Xia XM, Fakler B, Rivard A, Wayman G, Johnson-Pais T, Keen JE, Ishii T, Hirschberg B, Bond CT, Lutsenko S et al (1998) Mechanism of calcium gating in small-conductance calcium-activated potassium channels. Nature 395:503–507

Xia XM, Ding JP, Lingle CJ (1999) Molecular basis for the inactivation of Ca2+− and voltage-dependent BK channels in adrenal chromaffin cells and rat insulinoma tumor cells. J Neurosci 19:5255–5264

Xia XM, Zeng X, Lingle CJ (2002) Multiple regulatory sites in large-conductance calcium-activated potassium channels. Nature 418:880–884

Xia XM, Ding JP, Lingle CJ (2003) Inactivation of BK channels by the NH2 terminus of the beta2 auxiliary subunit: an essential role of a terminal peptide segment of three hydrophobic residues. J Gen Physiol 121:125–148

Yan J, Aldrich RW (2010) LRRC26 auxiliary protein allows BK channel activation at resting voltage without calcium. Nature 466:513–516

Yan J, Aldrich RW (2012) BK potassium channel modulation by leucine-rich repeat-containing proteins. Proc Natl Acad Sci U S A 109:7917–7922

Yang T, Snyders DJ, Roden DM (1995) Ibutilide, a methanesulfonanilide antiarrhythmic, is a potent blocker of the rapidly activating delayed rectifier K+ current (IKr) in AT-1 cells. Concentration-, time-, voltage-, and use-dependent effects. Circulation 91:1799–1806

Yang JW, Vacher H, Park KS, Clark E, Trimmer JS (2007) Trafficking-dependent phosphorylation of Kv1.2 regulates voltage-gated potassium channel cell surface expression. Proc Natl Acad Sci 104:20055–20060

Yang H, Shi J, Zhang G, Yang J, Delaloye K, Cui J (2008) Activation of Slo1 BK channels by Mg2 + coordinated between the voltage sensor and RCK1 domains. Nat Struct Mol Biol 15:1152–1159

Yao J, Chen X, Li H, Zhou Y, Yao L, Wu G, Chen X, Zhang N, Zhou Z, Xu T et al (2005) BmP09, a "long chain" scorpion peptide blocker of BK channels. J Biol Chem 280:14819–14828

Yazejian B, Sun XP, Grinnell AD (2000) Tracking presynaptic Ca2+ dynamics during neurotransmitter release with Ca2+−activated K+ channels. Nat Neurosci 3:566–571

Yi BA, Lin YF, Jan YN, Jan LY (2001) Yeast screen for constitutively active mutant G protein-activated potassium channels. Neuron 29:657–667
Yoo D, Flagg TP, Olsen O, Raghuram V, Foskett JK, Welling PA (2004) Assembly and trafficking of a multiprotein ROMK (Kir 1.1) channel complex by PDZ interactions. J Biol Chem 279:6863–6873
Yu FH, Catterall WA (2004) The VGL-chanome: a protein superfamily specialized for electrical signaling and ionic homeostasis. Sci Signal 2004:re15
Yuan Y, Shimura M, Hughes BA (2003) Regulation of inwardly rectifying K+ channels in retinal pigment epithelial cells by intracellular pH. J Physiol 549:429–438
Yuan P, Leonetti MD, Pico AR, Hsiung Y, MacKinnon R (2010) Structure of the human BK channel Ca2+−activation apparatus at 3.0 A resolution. Science 329:182–186
Yuan P, Leonetti MD, Hsiung Y, MacKinnon R (2011) Open structure of the Ca2+ gating ring in the high-conductance Ca2+−activated K+ channel. Nature 481:94–97
Zaydman MA, Cui J (2014) PIP2 regulation of KCNQ channels: biophysical and molecular mechanisms for lipid modulation of voltage-dependent gating. Front Physiol 5:195
Zeng XH, Xia XM, Lingle CJ (2003) Redox-sensitive extracellular gates formed by auxiliary beta subunits of calcium-activated potassium channels. Nat Struct Biol 10:448–454
Zhang L, Timothy KW, Vincent GM, Lehmann MH, Fox J, Giuli LC, Shen J, Splawski I, Priori SG, Compton SJ et al (2000) Spectrum of ST-T-wave patterns and repolarization parameters in congenital long-QT syndrome: ECG findings identify genotypes. Circulation 102:2849–2855
Zhang YH, Colenso CK, Sessions RB, Dempsey CE, Hancox JC (2011) The hERG K+ channel S4 domain L532P mutation: characterization at 37°C. Biochim Biophys Acta (BBA) Biomembr 1808:2477–2487
Zhang M, Meng XY, Cui M, Pascal JM, Logothetis DE, Zhang JF (2014) Selective phosphorylation modulates the PIP2 sensitivity of the CaM-SK channel complex. Nat Chem Biol 10:753–759
Zhao Y, Ung PM, Zahoranszky-Kohalmi G, Zakharov AV, Martinez NJ, Simeonov A, Glaaser IW, Rai G, Schlessinger A, Marugan JJ et al (2020) Identification of a G-protein-independent activator of GIRK channels. Cell Rep 31:107770
Zhou M, Morais-Cabral JH, Mann S, Mackinnon R (2001a) Potassium channel receptor site for the inactivation gate and quaternary amine inhibitors. Nature 411:657–661
Zhou Y, Morais-Cabral JH, Kaufman A, Mackinnon R (2001b) Chemistry of ion coordination and hydration revealed by a K+ channel–fab complex at 2.0 Å resolution. Nature 414:43–48
Zhou J, Chen H, Yang C, Zhong J, He W, Xiong Q (2017) Reversal of TRESK downregulation alleviates neuropathic pain by inhibiting activation of gliocytes in the spinal cord. Neurochem Res 42:1288–1298
Zhu G, Chanchevalap S, Cui N, Jiang C (1999) Effects of intra- and extracellular acidifications on single channel Kir2.3 currents. J Physiol 516(Pt 3):699–710
Zhu J, Watanabe I, Gomez B, Thornhill WB (2001) Determinants involved in Kv1 potassium channel folding in the endoplasmic reticulum, glycosylation in the golgi, and cell surface expression. J Biol Chem 276:39419–39427
Zuberi SM, Eunson LH, Spauschus A, De Silva R, Tolmie J, Wood NW, McWilliam RC, Stephenson JB, Kullmann DM, Hanna MG (1999) A novel mutation in the human voltage-gated potassium channel gene (Kv1.1) associates with episodic ataxia type 1 and sometimes with partial epilepsy. Brain 122(Pt 5):817–825
Zuniga L, Marquez V, Gonzalez-Nilo FD, Chipot C, Cid LP, Sepulveda FV, Niemeyer MI (2011) Gating of a pH-sensitive K(2P) potassium channel by an electrostatic effect of basic sensor residues on the selectivity filter. PLoS One 6:e16141

High-Resolution Structures of K$^+$ Channels

Qiu-Xing Jiang

Contents

Abstract

Potassium channels are present in every living cell and essential to setting up a stable, non-zero transmembrane electrostatic potential which manifests the off-equilibrium livelihood of the cell. They are involved in other cellular activities

Q.-X. Jiang (✉)
Laboratory of Molecular Physiology and Biophysics and the Cryo-EM Center, Hauptmann-Woodward Medical Research Institute, Buffalo, NY, USA

Department of Medicinal Chemistry, University of Florida, Gainesville, FL, USA

Departments of Materials Design and Invention and Physiology and Biophysics, University of Buffalo (SUNY), Buffalo, NY, USA
e-mail: qxjiang@hwi.buffalo.edu

N. Gamper, K. Wang (eds.), *Pharmacology of Potassium Channels*,
Handbook of Experimental Pharmacology 267, https://doi.org/10.1007/164_2021_454

and regulation, such as the controlled release of hormones, the activation of T-cells for immune response, the firing of action potential in muscle cells and neurons, etc. Pharmacological reagents targeting potassium channels are important for treating various human diseases linked to dysfunction of the channels. High-resolution structures of these channels are very useful tools for delineating the detailed chemical basis underlying channel functions and for structure-based design and optimization of their pharmacological and pharmaceutical agents. Structural studies of potassium channels have revolutionized biophysical understandings of key concepts in the field – ion selectivity, conduction, channel gating, and modulation, making them multi-modality targets of pharmacological regulation. In this chapter, I will select a few high-resolution structures to illustrate key structural insights, proposed allostery behind channel functions, disagreements still open to debate, and channel–lipid interactions and co-evolution. The known structural consensus allows the inference of conserved molecular mechanisms shared among subfamilies of K^+ channels and makes it possible to develop channel-specific pharmaceutical agents.

Keywords

Activation, deactivation, and inactivation · Co-evolution of channels and lipids · Energetics and allostery · Ligand-gated K^+ channels · Lipid-dependent gating · Pharmacological regulators and small molecule compounds · Structure-based drug design · Voltage-gated K^+ channels (Kv)

1 Importance of High-Resolution Structures to Pharmacology of K^+ Channels

Potassium channels are ubiquitously present in all three kingdoms of life because all living cells utilize negative transmembrane electrostatic potential for material transport and signal transduction (Iwasa and Marshall 2015). Setting up of negative resting transmembrane potential relies on potassium channels that keep it close to the Nernst potential of K^+, which is set by a gradient of high intracellular $[K^+]$ and low extracellular $[K^+]$ across plasma membranes maintained by ionic transporters. Continued consumption of chemical energy drives such an off-equilibrium state at a cost of net increase in total entropy. The need of K^+ channels in plasma membranes is thus intrinsic to all cells. Because cells vary in their lipid environments and their functions, K^+ channels must diversify and adapt to such changes in cellular environments (Jiang 2018). With the cellular environments changing in space and time, a cell may alter the expression level of different types of K^+ channels in order to satisfy its needs. Such a diversification of K^+ channels in adaptation to cellular environments makes them suitable targets for developing specific pharmaceutical reagents that may treat human diseases caused by malfunctions of K^+ channels (Ashcroft 2000).

A specific type of cells may express their own subsets of K^+ channels in their membrane systems in order to satisfy their functional needs. For example, cerebellar

Purkinje neurons are morphologically specialized in order to accept input from thousands of neuronal cells, integrate signals from them, and send out uniquely coded signals (Martina et al. 2007; Masoli et al. 2015). The expression of K^+ channels in a Purkinje neuron is spatially and temporally regulated such that the cell may fire a specific pattern of action potentials in order to coordinate or balance the activities in its incumbent neural circuit between central and peripheral nervous systems. Each Purkinje neuron expresses dozens of different types of ion channels, including multiple K^+ channels (Masoli et al. 2015). Pharmacological manipulation of these channels may allow cell-specific regulation or modulation of physiological activities at the systems level.

Structures of K^+ channels are important not only to the basic understandings of their functional determinants, but also to the development of channel-specific pharmacological reagents including small molecule compounds. Due to the conserved pore domains and diversified regulatory domains among K^+ channels, it might be preferential to develop chemical reagents targeting the regulatory domains to achieve high channel-specificity. Understanding the structural basis defining the functions of the pore and regulatory domains becomes an essential component for pharmacology of K^+ channels. Even though structural studies often only reveal static snapshots of the channels ensnared in energetically favored states and it is still debated whether some of the fundamental allostery behind the structural rearrangements for channel functions is truly defined by structural snapshots or is biased by structural interpretations, specific sites of action for channel pharmacology are usually valid and channel structures in complex with the reagents can be obtained for structure-guided drug design or optimization. In the next sections, I will highlight some of the key chemical underpinnings for different types of K^+ channels and reveal conserved and diversified domains that may become suitable sites of action for pharmacological regulation of K^+ channels. A word of caution is that there is always a gap between high-resolution structures and functional states of the channels in their native environments even though structural biologists may always have the best expectations on a genuine mapping between the two. Cell biologists and molecular geneticists sometimes may argue whether the biophysical mechanisms revealed by structural studies of K^+ channels might be out of context, if not purely artificial, or may state that structural studies of either soluble or membrane proteins in crystallization conditions or by cryo-EM in many cases, if not leading to biophysical artifacts, do not open new areas in biology. It suffices to say that most, if not all, molecular structures reflect good approximations to physical states that are frequently visited by functional channels in their native environments, which should be sufficient for structure-based design or development of pharmaceutical agents and justifies cautious tests of these agents for practical applications.

2 Phylogenetic Subfamilies of K^+ Channels and Their Typical Structural Models

A typical tetrameric K^+ channel has one pore domain in each subunit, four of which, homotypically or heterotypically, together form the ion-conducting pore (Fig. 1a.I), and whose selectivity segments are responsible for achieving high ion selectivity. Regulatory domains of a K^+ channel may be directly linked to either terminus of the pore domain or in an accessory subunit (Fig. 1a.II and IV). A small number of K^+ channels contain two pore domains in each subunit (Fig. 1a.III) such that two subunits suffice to form a functional channel. Based on the number of pore domains per subunit and the regulatory factors that modulate them, the K channels may be classified into different phylogenetic subgroups as in Table 1.

Typical topology of the transmembrane domains of K^+ channels is depicted in Fig. 1a, which defines four main subfamilies. The pore domain of a K^+ channel is responsible for selective permeation of K^+ ions at a diffusion-limited speed, which means that the chemical structure for selecting K^+ ions against other cations does not introduce any energetic trap or barrier to slow down the crossing of ions through the pore (Hille 2001). The regulatory domains of K^+ channels include the

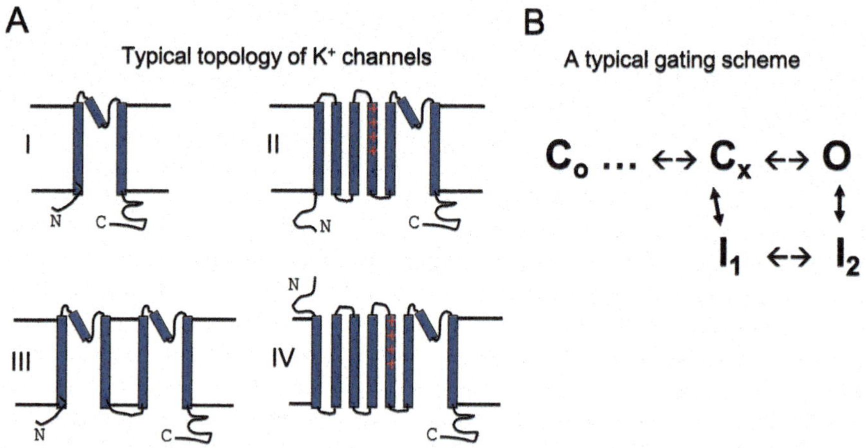

Fig. 1 Transmembrane topology and a simple gating scheme for K^+ channels. (**a**) Each subunit may have only the pore domain which is made of two transmembrane helices and a pore loop made of a short pore helix and a selectivity loop afterwards (I). This is sometimes referred to as 1P topology, whose duplication leads to the 2P topology in III. A typical Kv channel has a voltage-sensing domain (VSD) attached to N-terminal side of a 1P domain (II). The Ca^{2+}-activated K channels may have an extra transmembrane helix (TM0 or S0) before its voltage-sensor domain (IV). Other structural or functional domains may be attached to the N- and/or C-terminus or the loops between transmembrane segments in some channels. (**b**) A channel can be switched among closed states before reaching the open state (O), and both closed states ($\mathbf{C_o} \ldots \mathbf{C_x}$) and the open state (**O**) can switch into the inactivated states ($\mathbf{I_1}$ and $\mathbf{I_2}$). Some channels may have multiple open states and multiple routes to switch from closed states to the open ones. The structural studies presumably catch channels in some of these gating states, which may leave uncertainty in the interpretation of structural data

Table 1 Subfamilies of K channels

# Pore domains	Regulatory factors	Typical channels	Typical structures (PDB codes)
Two (2P)	Temperature, mechanical stretch, lipids, anesthetic agents, etc.	K2P subfamily include TWIK; TREK; TASK; TRAAK, TALK; THIK; TRESK, etc.	3UKM, 4RUE, 4RUEF, 4I9W, 6V36, 6CQ6, 3UM7
One (1P)	Voltage-sensor domain (VSD)	Kv1–Kv12	Kv1.2chim (2R9R)
	** Accessory subunits for Kv channels	Kvβ, minK, MiRP, KCNE, KCNIP	2I2R, 2NZ0, 2EB3
	Calcium activated (K_{Ca})	BK, SK, IK, MthK	6V3G, 6V38, 6A6E, 6UWN,6U5R, 6U6D, 6AEJ, 6AEF
	ATP-sensitive	K_{ATP}	6TWV, 6PZ9, 5WUA, 6JB1, 63CO,
	GPCR activation	GIRK	3SYA, 3SYP, 3SYO, 3SYQ,4KFM
	PIP_2	ROMK (K_{ir}1.x); K_{ir}2-6, Kv7	2QKS, 2WLN, 6M84, 3SPH, 6O9T
	Nucleotide	HERG (Kv11.1), Eag1,	5K7L,5VA1
	Nonphospholipids	Kv2, Kv4.3, KvAP, Kir	6UWM, 1ORQ
	Others (temperature, pH, alcohol, stretch, etc.)	KcsA, BK, ROMK, NaK, Kv, etc.	1BL8, 1K4C, 1R3J, 2IH1, 2HVJ, 2HVK, 3TEF, 6H8P, 3FB5, 3FB6, 3GB7, 2NFU

** Accessory subunits have no pore domains

voltage-sensing domains (VSDs) in voltage-gated potassium (Kv) channels, Ca^{2+}-sensitive domains, ATP/ADP-sensitive domains, Gβγ, PIP_2-binding sites, etc. Some ion channels are known to be sensitive to lipids or lipid-soluble components, such as nonphospholipids, alcohol, heme, etc. (Jiang 2018, 2019). There are many high-resolution structures of K^+ channels available and Table 1 lists the PDB codes for only a fraction of them. Instead of being inundated by detailed structural features, I will discuss key structural components for the four typical subfamilies, which may serve as the main themes for K-channel pharmacology. More detailed introductions on using these structural components for developing pharmaceutical reagents will be covered in later chapters.

3 A Simplified Gating Model for Different K+ Channels

A channel may essentially be in conductive or nonconductive states, quick switching between them leading to intermediate states. A sodium pump is generally used by eukaryotic cells to exchange sides of three sodium and two potassium ions per ATP. For a typical K^+ channel with a single channel conductance of say 20 pS, a 1-millisecond opening at 0 mV in a typical cell under physiological conditions would allow 10,000 K^+ to move out of the cell, which would require the consumption of 5,000 ATP molecules to be moved back. It is therefore naturally important for cells to let the K^+ channels stay in the conductive state only transiently at transmembrane potentials deviating from the K^+ Nernst potential. A process used to switch a channel between conductive and nonconductive states is called *gating*. The open (conductive) states of many channels are not stable and can go through a process called inactivation to become nonconductive. It is thus possible to use a generic, simplified model with multiple closed (nonconductive) states, one open state and two nonconductive, inactivated states to describe schematically the gating of various K^+ channels and correlate their structures with these functional states (Fig. 1b) (Jiang 2018).

Pharmacological reagents may recognize channels in these states or in gating states resembling any of them in order to maintain the channel pore either conductive (channel openers or agonists) or nonconductive (channel blockers, inhibitors, or antagonists). Structures of Kv channels in complex with these modulators will offer accurate chemical insights to refine their binding poses and redesign them in order to enhance their efficacies.

4 Structural Basis for Ion Selectivity of K^+ Channels

The pore structure revealed the exquisite chemical coordination of K^+ ions in the pore domain. The first crystal structure of the bacterial KcsA channel made a deep mark in history (Doyle et al. 1998; Morais-Cabral et al. 2001; Zhou et al. 2001) (Fig. 2a) and revolutionized our understandings of the pore structure for the K^+ channel family (Fig. 2b). The selectivity loops (TTVGYG) from four subunits jointly form the structure of the selectivity filter (SF), which contains four main K^+-binding sites (called K1, K2, K3, and K4 sites; Fig. 2b). Each K^+ in every one of the four binding sites is stabilized by eight carbonyl oxygens of the peptide bonds in the pore loops, whose partial negative charges neutralize at least in part the positive charge of the K^+ ion and stabilize it in the binding lattice. Structures in other potassium channels corroborated the SF structure and the K^+ binding sites (Table **1** and next). Studies of the nonselective NaK and its mutants suggested all four binding sites are required to achieve high selectivity (Alam and Jiang 2009a, b; Linn et al. 2010; Derebe et al. 2011a). Molecular dynamics analysis also suggested that destabilizing the K2 site would impair the selectivity significantly (Aqvist and Luzhkov 2000; Berneche and Roux 2000). Together, these data showed that the geometric arrangement of atoms in the SF and the chemical interactions are chiefly responsible for K^+ selectivity against smaller ions like Na^+ and Li^+ or larger ions like Cs^+ or Rb^+. On the other hand, Tl^+, being very close to K^+ in radius (1.4 Å versus 1.33 Å), can move across the channel equally well as K^+ does (Zhou and MacKinnon 2003). These suggest that the geometric factor be a decisive one in defining selectivity, even though other factors, such as the electric field or the stability of the SF structure, may be important as well.

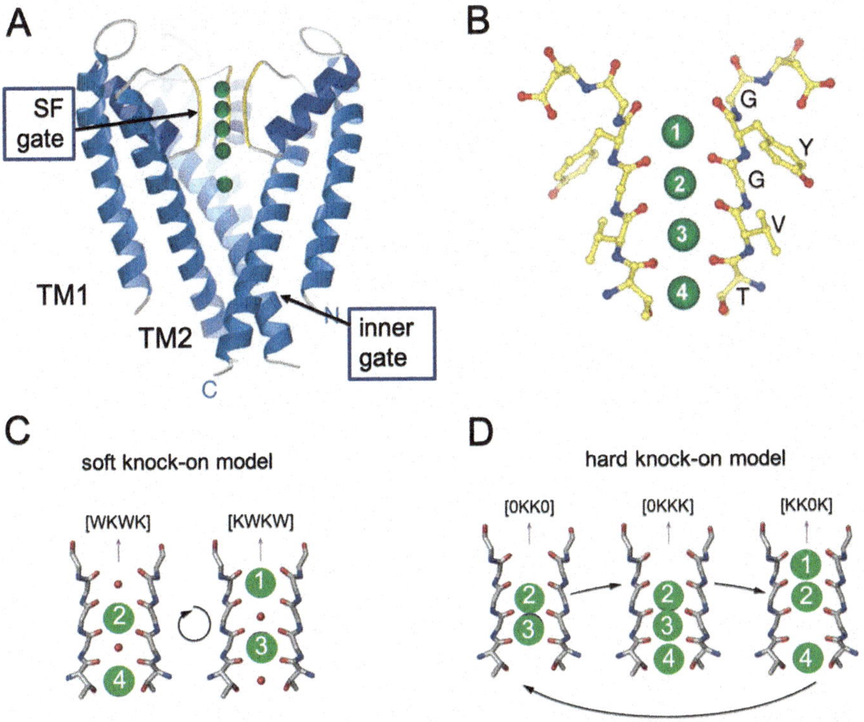

Fig. 2 Structural basis for ion selectivity and fast permeation in a K^+ channel. (**a**) Structural model of the KcsA channel (PDB ID 1K4C), showing two transmembrane helices and the pore helix of each subunit as blue ribbons and the selectivity loops in yellow. One subunit in the front is taken away for presentation. K^+ ions in the selectivity filter (SF) and the cavity are presented as green balls. (**b**) Electronegative carbonyl oxygens (red) of the selectivity loops coordinate and stabilize the K^+ ions in four binding sites (K1-K4; here not using the usual S1-S4 labels). (**c**) Throughput cycle of K^+ ions switching between two different configurations in SF ([K2, K4] and [K1, K3]) for outward flux. Always two K^+ and two water molecules, denoted as green and small red balls, respectively, are in the SF. This is named "soft knock-on" because of the separation of two neighboring K^+ ions by a water molecule. (**d**) Throughput cycle by a "hard knock-on" model through three distinct configurations in SF for outward flux. The three depicted states all have one pair of K^+ ions in two neighboring binding sites without an intervening water. The knock-on of a K^+ from the bottom of the SF (cavity site) into the two-ion state (left) would knock off the water molecule in the K4 site such that no water flow would accompany the K^+ flux. Figure panels adapted from references (Morais-Cabral et al. 2001; Zhou et al. 2001; Kopfer et al. 2014; Kopec et al. 2018) with permissions in accord with the PMC open access policy

The refined occupancies of the four K^+ binding sites, long-run molecular dynamics simulations, and other data suggest two distinct models for K^+ permeation across the SF in a single profile fashion (Zhou and MacKinnon 2003; Kopfer et al. 2014; Coates 2020). The main difference is on whether two K^+ ions are able to occupy two neighboring binding sites without an intervening water molecule (Fig. 2c vs. d). First, the chemical nature of the pore structure at atomic resolutions and the consideration of strong Coulombic repulsion between K^+ led to a "throughput cycle" of K^+ and water through the SF (Fig. 2c), which is also called the "soft knock-on," "water-mediated knock-on," or "indirect knock-on" (Kopec et al. 2018). Further, refinement of B-factor and ion occupancy from the anomalous data of Tl^+ in the KcsA crystals gave rise to a calibrated occupancy of ~0.5 for all four binding sites in the SF, accounted for by the equal probability of K^+ in [WKWK] and [KWKW] configurations (Aqvist and Luzhkov 2000; Zhou and MacKinnon 2003). This model predicted two water molecules in the SF and water permeation together with K^+, both of which were supported by experimentally measured streaming potential (Alcayaga et al. 1989). IR spectroscopy and solid-state NMR analysis supported the hydration of SF and mutational studies can catch the intermediate state of the throughput cycle (Blasic et al. 2015; Kratochvil et al. 2016). Such ample amount of evidence has cemented the "soft knock-on model" pretty well, which has been generally accepted and is a textbook presentation (Iwasa and Marshall 2015).

However, in 2014 the de Groot group and their collaborators performed very long molecular dynamics calculations to catch thousands of permeation events across the SF of the KcsA and NaK2K channels and re-did the occupancy analysis of the Tl^+ anomalous signal reported by Zhou et al. in 2003 (Kopfer et al. 2014; Kopec et al. 2018; Shi et al. 2018). They found that the absolute occupancy of K^+ in KcsA (PDB ID 1r3j) and MthK (PDB ID 3ldc) is nearly unity for all four sites. The intermediate permeation steps suggested that there are always a pair of K^+ ions in two neighboring sites (Fig. 2d), and the knocking of a dehydrated K^+ into the SF removes the H_2O (at the fourth site) and leads to a 0KKK configuration of three K^+ ions in a row. The resulted throughput cycle is called a "direct knock-on" mechanism for ion permeation. It is also called "hard knock-on." Brownian dynamics simulation and the energetic profiling due to Coulombic interactions in the SF could account for the high rate of ion permeation, much faster than the water-mediated "soft knock-on" (Kopec et al. 2018). Further, reanalysis of the IR spectroscopy seemed to suggest that the "direct knock-on mechanism" can explain the data equally well without water permeation coupled with K^+ crossing. Solid state NMR studies of the NaK2K channel in membrane suggested that the SF was not hydrated, contradicting the neutron diffraction/solid-state NMR data that supported the soft knock-on (Blasic et al. 2015; Kopec et al. 2018; Oster et al. 2019).

In contrast, a careful check of the different populations of ion configurations in the SF – 0KKK (47%), KK0K (28%), KKKK (14%), 0KK0 (5.0%), K0KK (4.0%), and KKK0 (3.0%) from the simulated events reported by Kopec and colleagues [see Fig. S3 in (Kopec et al. 2018)] suggested the relative occupancies of the four sites (K1-K4) to be *0.49: 0.96: 0.73: 0.93*, which contradict with the absolute occupancies of near unity (*0.92: 0.80: 1.0: 1.0*) derived from the SHELXL/SHELXD analysis of the anomalous signal from Tl^+-soaked crystals (Kopfer et al. 2014). The contradictions between the two models for ion permeation, the opposite interpretations of the IR spectroscopic data, the anomalous X-ray scattering signals and the NMR data, and the contradicting ion occupancies between the simulated permeation events and the analysis of anomalous X-ray scattering data will need to be fully examined and resolved before the "direct knock-on" mechanism will become a dominant model in the field because of more overwhelming evidence for the soft knock-on (Fig. 2).

The pore of the K channels is believed to contain two different gating mechanisms: one at the SF (called the SF gate) and the other at the intracellular bottom of the inner helices (called the inner gate; Fig. 2a). Gating by definition means a change in the pore that favors or disfavors energetically ion passing through it. The gate could be a physical (geometric) block or a chemical block due to environmental changes, such as increased hydrophobicity (dewetting or hydrophobic gate) or electro-positivity (electrostatic gate). In KcsA the inner gate is linked to its pH-sensor in the intracellular side. In MthK, the inner gate is coupled to the allostery of the RCK domains inside. In Kv channels, the inner gate is linked or coupled to the structural rearrangements in the voltage-sensor domains (VSDs). In BK channels, both the VSDs and the RCK domains may exert control to the energetic changes of hydrated K^+ ions near the inner gate. Pharmacological agents may act on or around the SF or the inner gate in order to keep the pore open (opener) or keep it closed (blocker). We will discuss the SF gate in the section on channel inactivation.

5 Structural Diversity Among Voltage-Gated Potassium (Kv) Channels and Differences in the Physical Movements of the Voltage-Sensor Domains (VSDs)

Voltage-gated ion channels (VGICs) are present in all living cells (Hille 2001). Their dysfunction causes severe diseases (Poolos 2005; Bjerregaard et al. 2006; Cox et al. 2006; Estevez 2006; Howard et al. 2007; Kordasiewicz and Gomez 2007; Rajakulendran et al. 2007; Pietrobon 2010; Tremblay and Hamet 2010; Xie et al. 2010; Baig et al. 2011; Poolos and Johnston 2012; Shribman et al. 2013). Except voltage-gated H^+ (Hv) channels which have their ion-conducting pores within the VSDs, all known VGICs contain four subunits/domains, each of which is made of 6 transmembrane α-helices (TMs; S1–S6 in Fig. 1a.II), and consist of a central pore and four flanking VSDs (Fig. 3a). The first four TMs (S1–S4) form a helical bundle and act as the voltage-sensing domain (VSD). The S5 and S6 of a Kv channel correspond to the TM1 and TM2 of the KcsA channel, respectively (Fig. 2a). Their gating means the switch of the VSDs or the pore between "down and up" or "closed and open" states, respectively. A 4th TM (S4) in a VSD often contains four evenly-spaced positively-charged residues (R1-R4) believed to move inside a *gating pore* (Fig. 3b) that is divided into outer (or upper) and inner (or lower) crevices by a "hydrophobic gasket" or the gating charge transfer center (Tao et al. 2010; Li et al. 2014a). Structures of multiple VSDs assigned to the "up" or "intermediate closed" states reveal significant structural variations among them and in the coupling of VSDs to PDs (Payandeh et al. 2011; Wu et al. 2015; Guo et al. 2016; Kintzer and Stroud 2016; Whicher and MacKinnon 2016; Hite and MacKinnon 2017; Lee and MacKinnon 2017; Wang and MacKinnon 2017b). Recently reported Kv structures by cryo-EM further testified the structural diversity (Whicher and MacKinnon 2016;

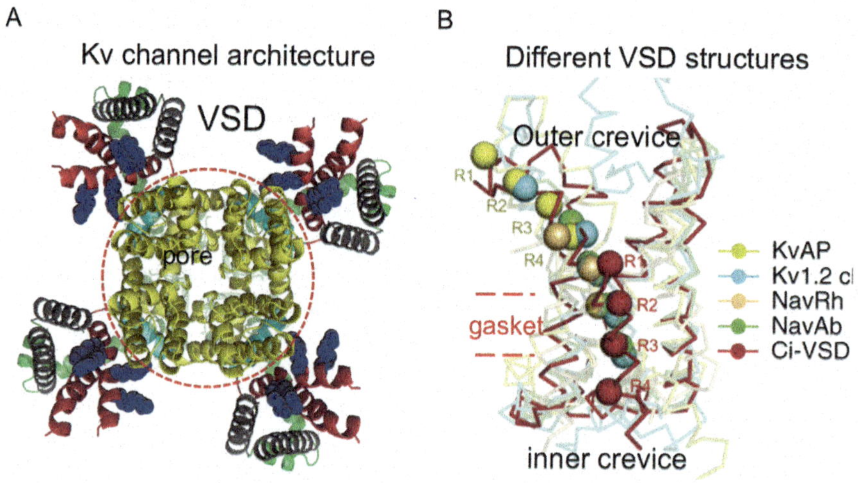

Fig. 3 Architecture of a Kv channel and structural variations among VSD structures. (**a**) A KvAP structure modeled after the Kv1.2/2.1 chimera structure (PDB ID 2R9R) shows a central pore domain (red circle) and four flanking VSDs. The S1/S2 are in gray and the S3/S4 in red. Four Arg residues in the S4 are showed in blue balls. (**b**) Five different VSD structures aligned with S4 Arg residues across the gating pores. Outer (upper) and inner (lower) crevices separated by a hydrophobic gasket (red dashed lines). In a resting (down) state, all four Arg residues (R1–R4) are expected to be in the inner crevice and drive the pore domain into the closed state. Panels modified from references (Long et al. 2007; Li et al. 2014a) with permissions following the PMC open access policy

Lee and MacKinnon 2017; Sun and MacKinnon 2017; Wang and MacKinnon 2017a), which appears to have evolved for their diversified needs to function in varying native lipid environments.

Past studies of VGICs in different membrane systems have revealed very important mechanistic insights. Key residues for voltage sensing and chemical basis for ion selectivity are established, revealing a range of gating charges per channel (Schoppa et al. 1992; Heginbotham and MacKinnon 1993; Heginbotham et al. 1994; Aggarwal and MacKinnon 1996; Seoh et al. 1996; Schoppa and Sigworth 1998; Islas and Sigworth 1999, Islas and Sigworth 2001; Starace and Bezanilla 2001; Zhou and MacKinnon 2003, 2004; Payandeh et al. 2011; Tang et al. 2014). Despite structural variations among VSDs (Fig. 3b), congruent structural features in PDs ensure fast ion conduction and high ion selectivity. But coupling between VSDs and PDs may vary markedly (Jiang et al. 2003a; Lee et al. 2005; Long et al. 2007; Clayton et al. 2009; Tao et al. 2010; Payandeh et al. 2011, 2012; Zhang et al. 2012; Catterall 2014; Takeshita et al. 2014; Tang et al. 2014; Whicher and MacKinnon 2016; Hite et al. 2017; Lee and MacKinnon 2017; Sun and MacKinnon 2017; Tao et al. 2017; Wang and MacKinnon 2017a, b). The VSD in a *Ciona* voltage-sensitive phosphatase (*Ci*-VSP) is not coupled to a PD, but instead to an enzyme. Its two states showed a one-register movement (~5 Å translation and ~60° rotation) of its S4 and minor changes in the other three α-helices, largely in accord with an intermediate "down" states reported in structures of NavAb and TPCs (Payandeh et al. 2011; Li et al. 2014b; Guo et al. 2016; Kintzer and Stroud 2016) and agreeing with a short vertical dislocation of the KvAP VS paddle in the gating pore (not in lipids), a 6–10 Å S4 shift in the focused-field model, a 3_{10}-helix sliding in the Rosetta modeling of NavAb, or a toxin-induced motion of Nav1.7 VSD (Jiang et al. 2003a, b; Ahern and Horn 2005; Yarov-Yarovoy et al. 2006; Vargas et al. 2012; Xu et al. 2019). Despite these appealing models, a structure of a canonical VGIC in the fully resting state with four VSDs in a "down" state remains unknown, disallowing direct comparisons of VSDs of the same channel in the "up" and "down" states, which still await further structural investigations. Similarly, structures of Kv's expected to be in steady-state inactivation failed to exhibit such a conformation, probably because experimental conditions did not enrich proteins in the right state (Shin et al. 2004; Kurata and Fedida 2006; Long et al. 2007; Cuello et al. 2010a; Cuello et al. 2010b; Li et al. 2017a, 2018; Matthies et al. 2018). These two aspects highlight not only some uncertainty of structural studies in state assignment, but also the importance of obtaining more structural information of the same VGICs in distinct gating states in near-native environments.

The main open question on voltage-dependent gating is how the VSDs change their structures in response to change in transmembrane electrostatic potential. There are clear and sometimes strong structural variations among VSDs (Fig. 3b), which might be a source for the observed differences in VSD rearrangements by different groups. Adding more to structural variations, the spatial arrangement between VSDs and PDs may be *domain-swapped or non-swapped* (Tao and MacKinnon 2019) and the coupling between the two domains through the S4-S5 linkers could vary substantially. Almost all the known VSD structures of various Kv channels differ from each other, albeit they all are 4-helix bundles. All structural changes resolved so far by X-ray and cryo-EM have only showed small local rearrangement of one or two Arg residues. Although the structural variations may be difficult to reconcile in order to reveal coherent structural basis for the VSD movement in voltage-gating, such differences among VSD structures suggest a strikingly rich source and a solid structural basis for VSD-targeting pharmacological agents that can modulate the voltage-gating of a specific subtype of Kv channels.

6 Structural Basis Underlying the N-Type Inactivation

N-type inactivation refers to the fast inactivation of the channel (Fig. 1b) empowered by a short peptide that inserts into the cavity of an open pore from the intracellular side. It is also called the ball-and-chain inactivation (Zagotta et al. 1990; Keynes 1992; Gomez-Lagunas and Armstrong 1994). It was first demonstrated in the *Shaker* K^+ channel whose N-terminal sequence serves the ball (Zagotta et al. 1990; MacKinnon et al. 1993). The inactivation peptide can also be presented by accessory subunits (De Biasi et al. 1997; Yellen 1998). For the Kv-type channels the ball-and-chain action on the channel pore has not been physically observed in structural studies. Recently, in a MthK channel structure by single particle cryo-EM analysis

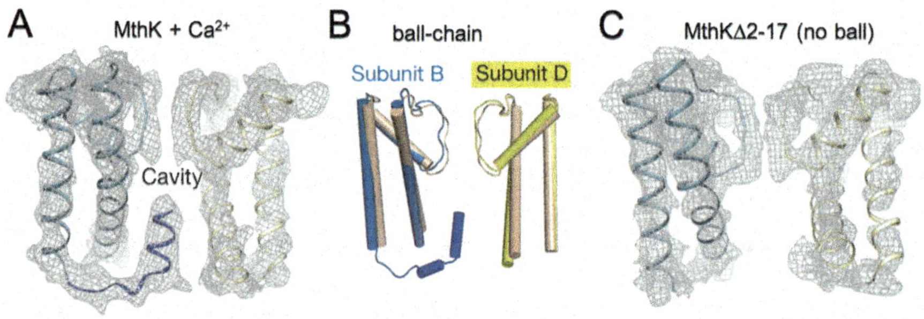

Fig. 4 Structural evidence for the N-type ball-and-chain inactivation. (**a**) A cryo-EM structure of the transmembrane part of the Ca^{2+}-activated MthK channel in the open state with an N-terminal ball peptide inside the cavity. Only two subunits are shown with the structural models as blue and yellow ribbons in the map (gray). (**b**) Schematic cartoon shows the ball-and-chain relative to the two subunits of the transmembrane pore region. The model from cryo-EM structure is in blue (subunit B) and yellow (subunit D), and the overlaying model in beige is from the crystal structure (PDB ID 3LDC). (**c**) The cryo-EM structure co-responding to two subunits of the transmembrane domain of the deletion mutant lacking the N-terminal ball peptide. Figure panels were reproduced from Fan et al. (2020) with permissions following PMC open access policy

(Fan et al. 2020), it was possible to catch the structure with the N-terminal peptide inserted in the open pore (Fig. 4a, b), which was missing in the structure of a mutant MthKΔ2-17 channel lacking the peptide (Fig. 4c).

7 Structure Determinants for Channels in a Steady-State Inactivation

Besides the N-type inactivation, many channels undergo a steady-state inactivation that was named the C-type inactivation because initially it was found to be contributed by the C-terminal region, around the S6 helix (Hoshi et al. 1991; Boland et al. 1994; Baukrowitz and Yellen 1995). The C-type inactivation reflects an allosteric rearrangement of the pore structure in the open state that leads to the closure of the ion-conducting pathway. It is believed by many in the field that there are at least two different pathways leading to the C-type inactivation – one goes from the open state and the other from the closed state(s) (Klemic et al. 2001; Kurata et al. 2005; Kurata and Fedida 2005). It is unclear whether these two pathways (Fig. 1b) ultimately lead to the same structure of the inactivated channel. Although multiple Kv channels at zero mV were expected to transit into an inactivated state, their structures did not reveal distinct features of such a state, e.g., see Long et al. (2007); Wang and MacKinnon (2017b). Instead, the structures were assigned to specific states based on a generic gating mechanism for *Shaker*-like channels. For example, Kv1.2 is expected to be inactivated, but its structures in detergents and phospholipid nanodiscs are the same, both reflecting probably the open state (Long et al. 2007; Matthies et al. 2018). The lipid-dependent gating of Kv channels suggests that these

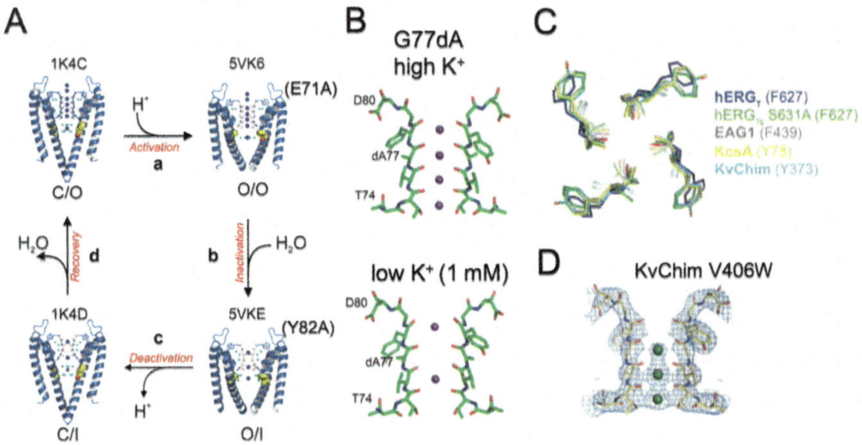

Fig. 5 Possible structural basis for C-type inactivation. (**a**) Gating cycle of KcsA. The PDB codes (1K4C, 1K4D, 5VK6, and 5VKE) represent structures of the wild-type channel in high (1K4C) and low (1K4D) K^+ solutions and its two mutants – (E71A) and Y82A. The latter two were assigned to two different functional states. The C/O, O/O, O/I, and C/I pairs denote the state of the inner gate (C or O in the first position) and that of the SF gate (O or I in the second position), respectively. The collapse of the SF, affecting the K2 and K3 ion-binding sites in KcsA, correlates with SF inactivation. (**b**) G77dA KcsA pore does not collapse in high or low K^+, but it still inactivates as the wild-type channel, suggesting that the SF collapse is not essential for inactivation. (**c**) SFs of the hERG, hERG S631A, EAG1 and Kv1.2/2.1 chimera (KvChim) compared at a level corresponding to the KcsA Y78. hERG C-type inactivation might result from a subtle movement of F627 in the SF, which causes a slight reorientation of the carbonyl oxygens in the SF and may affect the K1 site. (**d**) KvChim mutant V406W has low occupancy of K1 and inactivates quicker than the wild-type, suggesting that small changes that affect the energetics of K^+ ions in the site are sufficient to affect inactivation. Panels were adapted from references (Cuello et al. 2010a; Matulef et al. 2016; Pau et al. 2017; Wang and MacKinnon 2017a) with permissions following PMC public access policy

channels may require nonphospholipid-rich environments for them to be at proper energetic levels and in right states [(Jiang 2018, 2019); see Sect. 9].

In the transmembrane part, there are at two different ways to cause pore closure by deactivation or inactivation – one at the SF and the other at or close to the inner gate region (Fig. 2a). We still don't have structures reflecting the putative VSD-related inactivation, which likely is linked to structural changes at or near the inner gate region. Instead, the structural basis for the pH-triggered activation and the SF-based inactivation has been studied extensively in the pore domain of KcsA (Cuello et al. 2010a; Ostmeyer et al. 2013; Li et al. 2018). Crystallographic studies of KcsA suggested that the channel inactivates through the collapse of the SF (Cordero-Morales et al. 2007; Cuello et al. 2010a) (Fig. 5). The two gates in combination give rise to four different states, which essentially suggests that the opening of the inner gate during activation is coupled to the collapsing (constriction) of the SF and induces an inactivated state at the filter region through the flipping of the peptide backbone and the destabilization of some of the ion-binding sites (Zhou et al. 2001;

Derebe et al. 2011b). For example, in Fig. 5a, the second and third ion-binding sites were less stable in the O/I state as suggested by the KcsA Y82A structure.

The idea of the collapse or constriction of the SF region for C-type inactivation faced some challenges. (1) In a semisynthetic channel – G77dA KcsA, which contains a D-alanine at G77 position, the collapse of the SF was never experimentally observed (Fig. 5b), even in 1.0 mM K^+, but the channel inactivates normally as the wild-type (Valiyaveetil et al. 2004, 2006; Devaraneni et al. 2013; Matulef et al. 2016). (2) The E71A KcsA showed two X-ray structures with normal and G77 flipped configures in the SF, but the E71A channel in membrane is almost always open (Cordero-Morales et al. 2006, 2007). Given that crystallography tends to keep the channels in the most stable states, it is puzzling that E71A does not visit the collapsed state in membranes. (3) So far, all other reported structures of K^+ channels in high [K^+] without introducing mutations to the SF regions have not showed the collapsed configuration, although multiple of them were expected to inactivate (Figs. 3a and 4a, c).

On the other hand, molecular dynamics simulation has suggested that the collapsed state might be less stable under the cryogenic temperature, but could be visited by the channel in membrane at physiological temperatures (Li et al. 2017a, 2018). Further, the HERG K^+ channel has strong closed-state inactivation (Fig. 1b), but its cryo-EM structure revealed no constriction of SF (Wang and MacKinnon 2017a). Subtle changes in the conserved F627, equivalent to Y82 of KcsA, were proposed to destabilize the K1 ion-binding site and accelerate inactivation (Fig. 5c), which was observed in a mutant Kv1.2 chimera (Fig. 5d) (Pau et al. 2017). Furthermore, these subtle changes might still need more testing because destabilization of the K1 site was expected to impair ion selectivity (Fig. 2c, d) (Derebe et al. 2011a; Matulef et al. 2016). The various mutants destabilizing the SF itself are reminiscent of the W434F mutant in the *Shaker* channel, which introduced a faster C-type-like inactivation (Yang et al. 1997). Similar to ion permeation across the SF and the motion of the VSDs for voltage-gating, the disagreements on the C-type inactivation in KcsA-like channels and VGICs await further experimental evidence to settle.

8 Ligand-Gated Potassium Channels and the Structural Basis for Their Gating

Ligand-gated K^+ channels play critical regulatory roles among different cell types. A lot of studies have been done, and multiple structures have been published. Due to space limitation, I will only briefly describe key structural features and point the readers to other cited articles that provided more detailed insights or analysis of structural features. More details will also be discussed by others in this volume. The K2P channels may form a separate subfamily, but are grouped here with the ligand-gated channels because of their roles in integrating different regulatory signals.

8.1 Ca^{2+}-Activated Potassium (K_{Ca}) Channels

The importance of K_{Ca} resides in its connection to Ca^{2+} signaling inside different cells, especially in cardiovascular system (McCobb et al. 1995; Cox et al. 1997; Saito et al. 1997; Salkoff et al. 2006; Pyott et al. 2007; Singh et al. 2012; Zhou et al. 2017; Lingle et al. 2019). Some of these channels have an extra transmembrane segment that sends its N-terminus to the extracellular side (Fig. 1a.IV). They have functional voltage-sensor domains that harbor intrinsic gating charges much smaller than the *Shaker*-like Kv channels (Contreras et al. 2012). The structural studies of this family started with the crystal structures of the MthK (Jiang et al. 2002), a bacterial Ca^{2+}-regulated channel that has no VSDs, but contains relatively-conserved intracellular C-terminal regulatory domains, called RCK domains, that bear similarity to the C-terminal domains of eukaryotic BK channels and are known to bind Ca^{2+}. Initially the open pore structure of the MthK was proposed to close at the inner gate. But in eukaryotic BK channels, mounting evidence suggested that the inner gate remains largely open during different gating modes (Zhou et al. 2017). The recent cryo-EM structures of the acBK and Slo2.2 channels in both Ca^{2+}-free and Ca^{2+}-bound states showed that the inner gates remained open, despite some minor changes at the interfaces between the RCK domains and the transmembrane domain (Hite et al. 2015, 2017; Hite and MacKinnon 2017; Tao et al. 2017) (Fig. 6a).

Certainly, lack of physical closure is not equivalent to lack of gating. Environmental changes at the inner gate and/or inside the cavity might create a chemical barrier (Jia et al. 2018) (Fig. 6b). The conformational changes in the SF might also contribute to the gating process. Indeed, in recent cryo-EM structures of MthK in nanodiscs, the channel was closed at the inner gate due to hydrophobic residues (L95 and I99; Fig. 6c) (Fan et al. 2020). A similar hydrophobic gating (dewetting) was proposed for the Ca^{2+}-free BK channel, whose cavity region is likely to become more hydrophobic and prohibitive for K^+ permeation (Fig. 6d). These data are not sufficient to rule out a possible coupling of the slight inner helix motion to the closure of the SF gate, which would make the pore nonconductive as proposed for KcsA (Fig. 5a).

The K_{Ca} channels are good druggable targets for pharmacological modulations of the physiological activities in their residing cells. Certain toxins have been identified and extensively studied on these channels (Kaczorowski and Garcia 2016). It is expected that the pore, the VSDs, and the RCK domains may all be druggable. High-resolution structures by cryo-EM will become useful for structure-based drug design in the near future.

8.2 Structures of K_{ATP} Channels

ATP is the energy currency in all cells. ATP-gated K^+ channels (K_{ATP}) connect cellular metabolism and energy level to the change in transmembrane potential, cellular excitability and signaling. It is thus well expected that K_{ATP} plays important

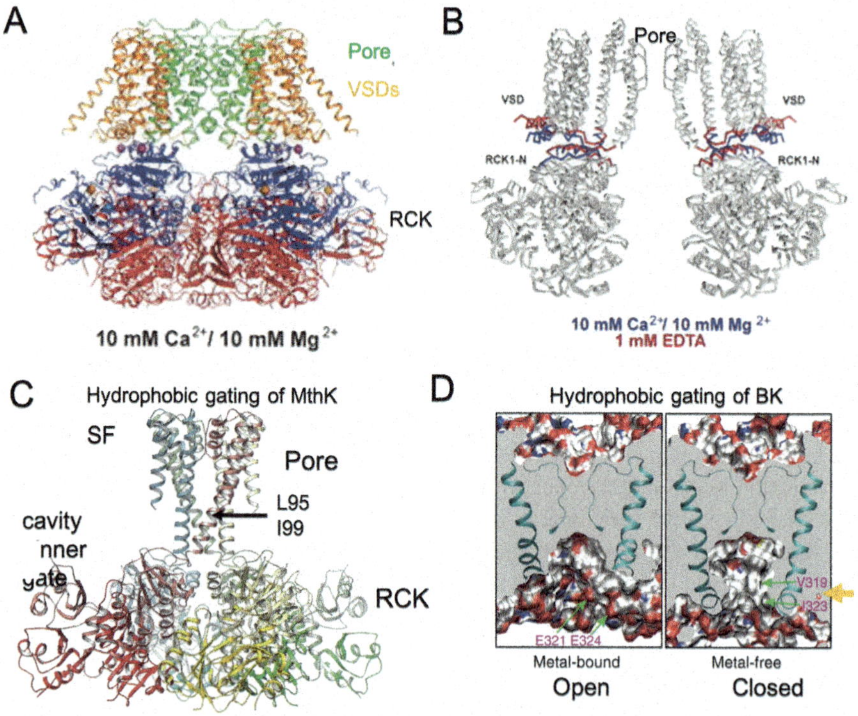

Fig. 6 Structures of Ca^{2+}-activated K channels and hydrophobic gating at the inner gate. (**a**) Structure of acBK channel in high Ca^{2+}/Mg^{2+}. (**b**) Minor structural rearrangements of acBK from the metal-bound (blue) to metal-free (red) state, showing that the inner gate is wide-open, and the RCK domains experience some subtle changes that lead to small rearrangements at the interface between the RCK domains (gating ring) and the transmembrane part, which are not enough to move the inner gate. (**c**) Structure of the Ca^{2+}-free MthK in nanodisc, showing a closed state at the inner gate due to hydrophobic residues and the physical motion of the inner gate. (**d**) The proposed hydrophobic gating of BK due to the switching of side chains of apolar residues (yellow arrow) lining the cavity to make it less favorable for hydrated K^+ ions to pass. Panels were adapted from references (Hite et al. 2017; Tao et al. 2017; Jia et al. 2018; Fan et al. 2020) with permissions following the PMC open access policy

roles in different cells, especially in neuronal cells, which are exquisitely sensitive to ATP-deficiency and O_2-level. By cryo-EM three groups have worked on the structures of Kir/SUR complexes and obtained 3D reconstructions at intermediate to near-atomic resolutions (Lee et al. 2017; Li et al. 2017b; Martin et al. 2017; Ding et al. 2019; Martin et al. 2019). Figure 7a shows a 3.3 Å map with the channel pore surrounded by four SUR subunits. The structure was stabilized by ATPγS and repaglinide (RPG). RPG is an inhibitor. In membrane, the SUR1 interacts with the Kir6.2 via a hydrophobic interface (TMD0-L0) (Fig. 7b, c). The whole model for the complex is presented in Fig. 7d. Under the experimental conditions the channel structures showed a closed pore. An ATP-binding pocket on the surface of the Kir part is located at the interface made of an N-terminal loop of one Kir subunit, the

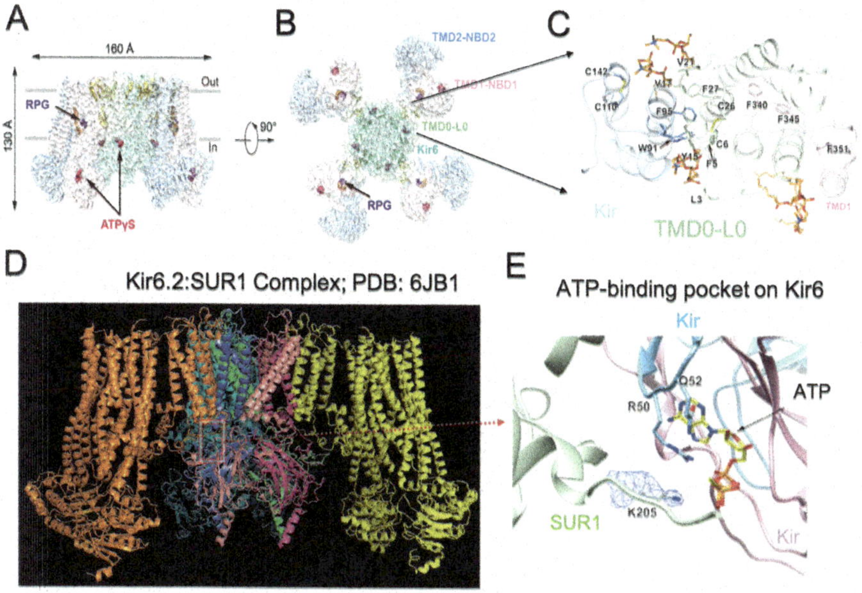

Fig. 7 Structure of the K_{ATP} channel. (**a**) Cryo-EM map of the Kir/SUR1 complex at 3.3. Å showing the arrangement of density and binding sites for ATPγS and RPG relative to the membrane. (**b**). Subunit arrangement viewed from the intracellular side. RPG binds inside the SUR1. ATPγS binds to the NBD1 of the SUR1 as well as at the channel/SUR1 interface. (**c**). The TMD0-L0 region of the SUR interfaces with the Kir subunit through hydrophobic interactions. (**d**) The molecular model of the whole octameric complex with one SUR1 subunit close to the reader removed for clarity. The white box locates the ATP-binding site at the Kir/SUR interface and is expanded in (**e**) to show the key residues important for the ATP-binding sites. Panels adapted from original publications (Li et al. 2017b; Ding et al. 2019; Martin et al. 2019) with permissions in accord with the PMC public access policy

C-terminal (cytoplasmic) domain (CTD) of a neighboring Kir and a loop of the SUR1 subunit (Fig. 7e). Key positively-charged residues contribute to ATP-binding, and this ATP-binding site is right below the PIP_2-binding site. The channel activation needs PIP_2 to open the inner gate (Baukrowitz et al. 1998). The ATP-binding makes the pore nonconductive, probably through a gate at the top of the CTD (similar to what is shown in Fig. 8). The Mg-ADP/Mg-ATP binding to the canonical binding site in the NBD1 of the SUR1 leads to partial reactivation of the channel. The RPG-binding to the middle tunnel of the TMD1 domain of the SUR1 subunit removes the Mg-ADP-induced partial reactivation and keeps the pore nonconductive, which then depolarizes the transmembrane potential in pancreatic β-cells and augments the release of insulin-secretory granules to the blood stream (Gribble et al. 1998; Proks et al. 1999). Due to such physiological roles, K_{ATP} is connected genetically to diabetes and has been a major drug target.

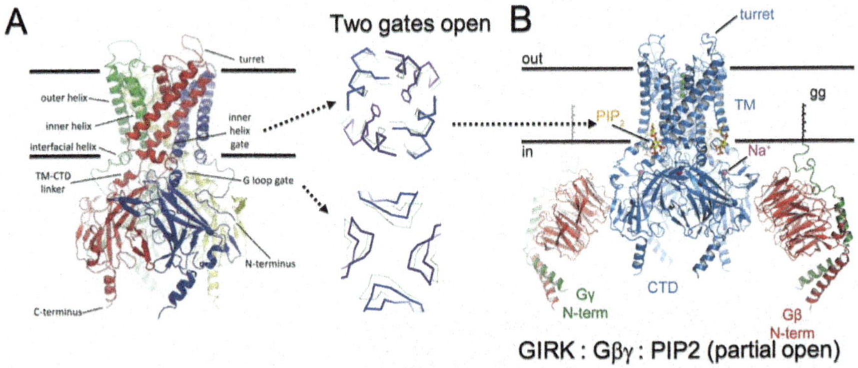

Fig. 8 Structural basis for GIRK gating. (**a**) The structure of GIRK2 in a closed state (left; Kir3.2; PDB ID 3SYO). The transmembrane domain forms the pore structure, resembling KcsA, which would have both an SF gate and an inner helix gate. The beta-rich CTD domains form a tetramer containing an ion-conducting tunnel in the middle that is pinched closed at the top of CTD with the G-loop gate right under the presumed inner helix gate. Right: The two lower gates presumably may open by splaying laterally the inner helices (top; PDB ID 3SYA) or the G-loops (bottom; PDB ID 3SYQ). (**b**) Opening of the two lower gates by PIP_2 and Gβγ, respectively (PDB ID 4KFM). The PIP_2-binding drags the interfacial and inner helices laterally to pen the inner helix gate. The beta-wheel of the Gβγ attached to the CTD laterally through both hydrophobic and electrostatic interactions, causing the opening of the G-loop gate in a stochastic fashion and thus a higher P_o than in the absence of Gβγ. Panels adapted from original publications (Whorton and MacKinnon 2011, 2013) with permissions under the PMC public access

8.3 GPCR-Coupled K^+ Channels

Similar to K_{ATP}, other Kir channels form their own subgroups and may be regulated by various factors. As an example, the GIRK channels are discussed here, partly due to their possible connections to a large number of G-protein-coupled receptors (GPCRs) (Logothetis et al. 1987). These channels integrate extracellular signals with the change in transmembrane potential and cellular excitability. The inward rectification of these channels is due to the intracellular cations, such as polyamines and Mg^{2+}, that can enter into the pore from the intracellular side when carried by an outward flow. It is known that multiple Kir channels need PIP_2 binding from the inner leaflet in order to open the inner gate (Fig. 8a) (Whorton and MacKinnon 2011). The GIRK has a gate right underneath the PIP_2-binding interface, called the G-loop gate, which could be open transiently in the presence of Gβγ (Fig. 8b) (Whorton and MacKinnon 2013). The binding of four Gβγ subunits causes a counterclockwise rotation of the cytoplasmic domain (CTD) relative the transmembrane domain (viewed from inside), which pulls the G-loops, leading to a higher probability for the gate to open ($P_o \sim 0.15$) (Mullner et al. 2003). There is a Na^+-binding pocket at the interface next to the G-loop gate, which could explain the increased opening of GIRK when local Na^+ concentration increases due to cell excitation and Na^+ influx (not showed in Fig. 8) (Ho and Murrell-Lagnado 1999). The role of SF gate in the GIRK channels is not clear, although it might also contribute partially to the low single channel open probability in the presence of both PIP_2 and G-protein activation by the coupling of the inner gate opening to a possible closure of the SF gate (Fig. 5a).

8.4 Two-Pore K^+ Channels (K2P)

The K2P channels (KCNK subfamily with 15 known genes) contain two concatenated pore domains with four helices and two pore loops such that two copies of them together can form a functional channel (Figs. 1a.III and 9) (Ketchum et al. 1995; Brohawn et al. 2012, 2014a; Miller and Long 2012; Lolicato et al. 2014, 2017; Bayliss et al. 2019). They are outward-rectifying probably due to high [K^+] inside and low [K^+] outside the cell. So far, we have crystal structures of K2P1, K2P2, and K2P4. There is a domain-swapping between the two subunits resolved in the Fab-bound TRAAK (K2P4) structure (Brohawn et al. 2013, 2014a). These channels have a cap region that is helix-rich and stabilized by disulfide bonds. The apposition of the cap to the outer orifice of the SF (Fig. 9a) makes it necessary for the outflowing K^+ ions to move laterally (two red arrows) and also renders it impossible for conventional pore blockers to approach the SF from right above. TEA, pore-binding toxins, etc. thus fails to block K2P channels. The crystal structures reveal

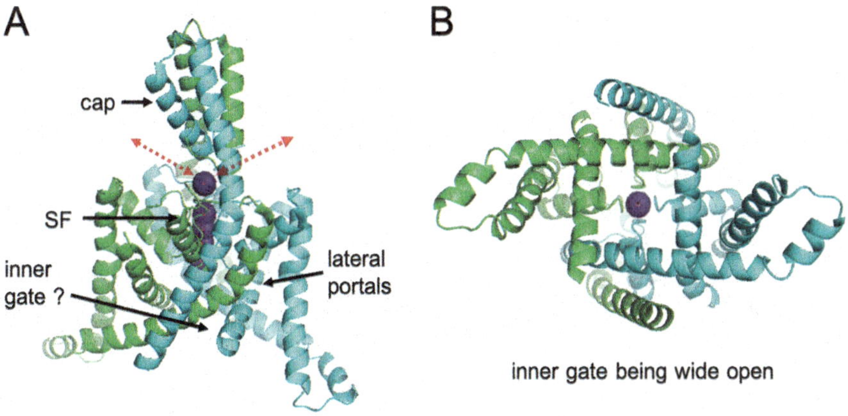

Fig. 9 Structural features of the K2P channels. (**a**) The crystal structure of a K2P4 channel shows the helical bundle in the cap region right above the SF (the top K0 ion-binding site), leading to two lateral flux pathways marked by red arrows (PDB IDs: 4I9W, 4RUE, 4RUF). The SF is the same as in Fig. 2a. The inner helices at the inner gate region are far away from each other such that the inner gate might not be used to close the pore. There is domain-swapping between the two subunits in the cap region. At the dimer interfaces there are two lateral portals that allow hydrophobic molecules to access the cavity and block the pore. Upward movement of the inner helix 2 is likely to block the lateral portals and relieves the channel from lipid-soluble blockers. (**b**) A view of the inner gate region and the cavity from the intracellular side along the central axis. K^+ ions are shown as purple balls. Panels adapted from original publications (Brohawn et al. 2014a) with permissions under the PMC public access policy

two lateral portals at the two dimer interfaces, which allow acyl chains or hydrophobic lipid-chain-analogs to reach the cavity region from the inner leaflet of the membrane and block ion conduction. The inner helix gate is wide-open (Fig. 9b), making it less likely to gate the pore via physical motion. The arachidonic acid might activate the channel by competing away the hydrophobic chains in the lateral portals, but not blocking the channel pore. Alternatively, it might introduce changes in mechanical stretch in the bilayer to open the channel (Brohawn et al. 2014a, b). The SF gate might be coupled to the inner helix movement such that the up-down motion of the inner helices (IH2) could lead to the C-type-like inactivation. The K2P channels are thought to be regulated by lipids, neurotransmitter-activated second messenger pathways, anesthetics, phosphorylation, SUMOylation, lipophilic drugs, mechanical stretch in bilayer membranes as well as thermal fluctuations. How these changes are integrated by different K2P channels remains unclear or controversial. Recently, it was reported that the TRAAK channels are preferentially distributed to a majority (~80%) of mammalian nodes of Ranvier in myelinated axons in both central and peripheral nervous systems and serve as the long-known "leak" K^+ conductance that is important for action potential firing at the nodes (Brohawn et al. 2019). Whether the channels at the nodes of Ranvier are able to sense and integrate various different regulatory factors remains a question for future investigation.

9 Lipid-Dependent Gating of Kv Channels, and Lipid-Soluble Pharmacological Agents

Kv channels function in cell membranes that contain both phospholipids (Group I) and nonphospholipids (Group II). All major human lipid metabolic defects alter nonphospholipid distribution in plasma membranes (Kolter and Sandhoff 2006) and may cause severe neurological defects and early death (Bellettato and Scarpa 2010; Bolsover et al. 2014; Cheng 2014). Some cases of neurodegenerative diseases are linked to lipidomic changes (Mesa-Herrera et al. 2019). Chemical depletion of cholesterol (CHOL) inhibits or stimulates VGICs (Hajdu et al. 2003; Bowles et al. 2004; Pouvreau et al. 2004; Xia et al. 2004; Heaps et al. 2005; Abi-Char et al. 2007; Balijepalli et al. 2007; Pottosin et al. 2007; Guo et al. 2008; Chun et al. 2010; Finol-Urdaneta et al. 2010; Huang et al. 2011; Purcell et al. 2011; Coyan et al. 2014; Rudakova et al. 2015; Balajthy et al. 2016), probably due to complicated (intra)-cellular effects of the chemical depletion and the heterogeneous lipid domains in cell membranes. But chemical loading of CHOL *invariably exerts inhibitory effects* to VGICs (Jiang 2019). My group has consistently observed that nonphospholipids in homogeneous membranes favor KvAP in a resting state with its VSDs in the "down" conformation. This state apparently is equivalent to the physiological "resting" state driven by hyper-polarization of transmembrane electrostatic potential (Zheng et al. 2012). We named such a non-phospholipid-induced conformational change of the VSDs as a steady-state "lipid-dependent gating." The lipid-dependent gating can account for all results from my lab as well as the partial immobilization of gating charges in different Kv channels after sphingomyelinase treatment of cell membranes (Xu et al. 2008), the lack of activity from Kv channels clustered in CHOL-rich domains (O'Connell et al. 2010), the gating charge immobilization of Kv4.3 in midbrain dopamine neurons treated by endocannabinoids (besides change in inactivation) (Gantz and Bean 2017), CHOL-induced change of Nav1.9 activity in DRG neurons and pain sensation (Amsalem et al. 2018), and the allosteric changes of a bacterial Nav channel (MVP) in Group II lipids (Randich et al. 2014). We hence proposed that lipid-dependent gating may mediate more general inhibitory effects of nonphospholipids on VGICs. Lipid-dependent gating may be less strong for certain eukaryotic voltage-gated Na^+ (Nav) or Ca^{2+} (Cav) channels (Hajdu et al. 2003; Bowles et al. 2004; Pouvreau et al. 2004; Xia et al. 2004; Heaps et al. 2005; Abi-Char et al. 2007; Balijepalli et al. 2007; Pottosin et al. 2007; Guo et al. 2008; Chun et al. 2010; Finol-Urdaneta et al. 2010; Huang et al. 2011; Purcell et al. 2011; Coyan et al. 2014; Rudakova et al. 2015; Balajthy et al. 2016), probably because their four VSDs differ in function and could be shielded (at least partially) from annular lipids by auxiliary transmembrane subunits (Catterall 1988; Calhoun and Isom 2014; Hofmann et al. 2015; Wu et al. 2015).

These observations support the idea that in membranes of different lipid compositions, the voltage-gated channels may change their VSD conformations and switch into different gating states (Fig. 1b). Further, different Kv channels are known to be distributed into specific locations in a cell, say a neuron with many processes, and function in a local environment with specific lipid composition or lipid domains. In this sense, we believe that the K^+ channels have evolved together with the lipid organization/distribution in cell membranes where these channels are delivered to function. The distribution of Kv2.1, Nav1.6, and TRAAK in the paranodes and nodes of Ranvier is a recent example (Brohawn et al. 2019). The lipid compositions of the axonal membranes in the nodes, the paranodes, and the juxtaparanodes are pretty different such that the K^+ channels are foreseeably regulated by the local lipid molecules.

Following the same line of thinking, we expect that unconventional lipid molecules, which are not well-studied or characterized in different membrane domains, may have specific gating effects on K^+ channels. PIP_2 and its analogs are well demonstrated to be important for multiple Kir and Kv7 channels. Gangliosides, glycosphingolipids, plasmalogen lipids, lysolipids, long-chain fatty acids, cannabinoids, and analogs, etc. are less studied examples.

10 Unresolved Questions on the Structural Bases of K^+ Channels and Future Directions

Structural analysis of the DNA double helix led the way to modern Molecular Cell Biology and has raised many new biological and biomedical questions, whose solutions drove scientists to achieve better mechanistic understandings and offer more accurate physical descriptions of biological processes. It is probably safe to say that the same has happened to the K^+ channels in the past two decades or so. The pioneering studies of KcsA have provided a succinct, precise description of the conserved structural features for high K^+ selectivity against other smaller or larger cations and in the meantime achieving high permeation rate. The studies of Kv, ligand-gated K^+ channels, and K2P channels have unlocked many secrets on the gating and regulation of these channels. There are still multiple open questions on the differences between the soft knock-on and hard knock-on models, the discrepancies on whether SF constriction and destabilization of ion-binding sites are an essential and universal mechanism behind C-type inactivation, and the proposed roles of environmental changes at or near the inner gate region besides the physical constrictions (closure) being the conventional (activation-) gating of the channels. The voltage-dependent gating has encountered equal or more disagreements because different biophysical methods have revealed different models for the VSD movement in membranes, and different calibrations have yielded varying dimensional changes from a few to dozens of angstroms. Almost all VSD structures from different proteins have taken slight or much more variations (Fig. 3b) and the obtained structures assigned to different states have revealed local rearrangements of one or two charged residues, rather than what was expected from the canonical displacement of all four charged S4 residues, from one side of the gating pore to the other. Lack of the same Kv channel in three or more distinct conformational states as depicted in Fig. 1b has been a major limiting factor (Tao et al. 2010). It might be essential to implement new strategies to satisfy this need or utilize reconstituted Kv channels in vesicles to enable voltage-jump and recapitulate the gating process for cryo-EM studies (Jiang et al. 2001; Jensen et al. 2016). The gating of ligand-gated K^+ channels and the K2P channels still have a lot of unknowns. Reconstituting the gating processes on cryo-EM grids other than in the milieu for growing 3D crystals might be a better strategy for addressing these open questions (Lee et al. 2013; Llaguno et al. 2014).

With the development of new digital electron counting devices, better energy filters and phase plates, cryo-EM studies of all four main groups of K^+ channels (Table 1) will be advanced to approach truly atomic resolutions (~1.2 Å or better). Using chemically functionalized ultrathin carbon (ChemiC) films (Llaguno et al. 2014), we expect to be able to control and synchronize K^+ channels along the reaction coordinates for their gating process such that individual channel molecules will be frozen in motion and their structures would be easier to be matched with particular gating states.

From a pharmaceutical standpoint, gaining deeper structural insights and performing structure-based drug design all need preferentially structures at atomic resolutions and in well-defined physiological states. The structural studies of the K^+ channels together with their molecular modulators will expectedly unleash the secrets needed for developing better treatment for many human diseases caused by dysfunctions of K^+ channels. Many investigators have been working in these areas as exemplified in the other chapters of this volume.

Acknowledgements The research programs in the Jiang laboratory over the past years have been supported by NIH (R21GM131231, R01GM111367, R01GM093271 & R01GM088745), AHA

(12IRG9400019), CF Foundation (JIANG15G0), Welch Foundation (I-1684), CPRIT (RP120474) and intramural funds at UT Southwestern, the University of Florida and the Hauptman-Woodward Medical Research Institute. I am indebted to many colleagues in the ion channel field for their valuable suggestions and advice to my research programs. Due to space limitation, I had to omit a lot of valuable work by many colleagues. I apologize for such omission due to personal selection.

Declaration The author claims no conflict of interest.

References

Abi-Char J, Maguy A, Coulombe A et al (2007) Membrane cholesterol modulates Kv1.5 potassium channel distribution and function in rat cardiomyocytes. J Physiol 582:1205–1217. https://doi.org/10.1113/jphysiol.2007.134809

Aggarwal SK, MacKinnon R (1996) Contribution of the S4 segment to gating charge in the Shaker K+ channel. Neuron 16:1169–1177

Ahern CA, Horn R (2005) Focused electric field across the voltage sensor of potassium channels. Neuron 48:25–29

Alam A, Jiang Y (2009a) High-resolution structure of the open NaK channel. Nat Struct Mol Biol 16:30–34. https://doi.org/10.1038/nsmb.1531

Alam A, Jiang Y (2009b) Structural analysis of ion selectivity in the NaK channel. Nat Struct Mol Biol 16:35–41. https://doi.org/10.1038/nsmb.1537

Alcayaga C, Cecchi X, Alvarez O, Latorre R (1989) Streaming potential measurements in Ca2+−activated K+ channels from skeletal and smooth muscle. Coupling of ion and water fluxes. Biophys J 55:367–371. https://doi.org/10.1016/S0006-3495(89)82814-0

Amsalem M, Poilbout C, Ferracci G, Delmas P, Padilla F (2018) Membrane cholesterol depletion as a trigger of Nav1.9 channel-mediated inflammatory pain. EMBO J 37(8):e97349. https://doi.org/10.15252/embj.201797349

Aqvist J, Luzhkov V (2000) Ion permeation mechanism of the potassium channel. Nature 404:881–884. https://doi.org/10.1038/35009114

Ashcroft FM (2000) Ion channels and disease 95:95

Baig SM, Koschak A, Lieb A et al (2011) Loss of Ca(v)1.3 (CACNA1D) function in a human channelopathy with bradycardia and congenital deafness. Nat Neurosci 14:77–84. doi: nn.2694 [pii] 111038/nn.2694

Balajthy A, Somodi S, Petho Z et al (2016) 7DHC-induced changes of Kv1.3 operation contributes to modified T cell function in Smith-Lemli-Opitz syndrome. Pflugers Arch. https://doi.org/10.1007/s00424-016-1851-4

Balijepalli RC, Delisle BP, Balijepalli SY, Foell JD, Slind JK, Kamp TJ, January CT (2007) Kv11.1 (ERG1) K+ channels localize in cholesterol and sphingolipid enriched membranes and are modulated by membrane cholesterol. Channels (Austin) 1:263–272. doi: 4946 [pii]

Baukrowitz T, Yellen G (1995) Modulation of K+ current by frequency and external [K+]: a tale of two inactivation mechanisms. Neuron 15:951–960

Baukrowitz T, Schulte U, Oliver D et al (1998) PIP2 and PIP as determinants for ATP inhibition of KATP channels. Science 282:1141–1144. https://doi.org/10.1126/science.282.5391.1141

Bayliss DA, Czirjak G, Enyedi P et al (2019) Two P domain potassium channels (version 2019.4) in the IUPHAR/BPS guide to pharmacology database. IUPHAR/BPS guide to pharmacology CITE 4. https://doi.org/10.2218/gtopdb/F79/2019.4

Bellettato CM, Scarpa M (2010) Pathophysiology of neuropathic lysosomal storage disorders. J Inherit Metab Dis 33:347–362. https://doi.org/10.1007/s10545-010-9075-9

Berneche S, Roux B (2000) Molecular dynamics of the KcsA K(+) channel in a bilayer membrane. Biophys J 78:2900–2917. https://doi.org/10.1016/S0006-3495(00)76831-7

Bjerregaard P, Jahangir A, Gussak I (2006) Targeted therapy for short QT syndrome. Expert Opin Ther Targets 10:393–400. https://doi.org/10.1517/14728222.10.3.393

Blasic JR, Worcester DL, Gawrisch K, Gurnev P, Mihailescu M (2015) Pore hydration states of KcsA potassium channels in membranes. J Biol Chem 290:26765–26775. https://doi.org/10.1074/jbc.M115.661819

Boland LM, Jurman ME, Yellen G (1994) Cysteines in the Shaker K+ channel are not essential for channel activity or zinc modulation. Biophys J 66:694–699

Bolsover FE, Murphy E, Cipolotti L, Werring DJ, Lachmann RH (2014) Cognitive dysfunction and depression in Fabry disease: a systematic review. J Inherit Metab Dis 37:177–187. https://doi.org/10.1007/s10545-013-9643-x

Bowles DK, Heaps CL, Turk JR, Maddali KK, Price EM (2004) Hypercholesterolemia inhibits L-type calcium current in coronary macro-, not microcirculation. J Appl Physiol 96:2240–2248. https://doi.org/10.1152/japplphysiol.01229.2003

Brohawn SG, del Marmol J, MacKinnon R (2012) Crystal structure of the human K2P TRAAK, a lipid- and mechano-sensitive K+ ion channel. Science 335:436–441. https://doi.org/10.1126/science.1213808

Brohawn SG, Campbell EB, MacKinnon R (2013) Domain-swapped chain connectivity and gated membrane access in a Fab-mediated crystal of the human TRAAK K+ channel. Proc Natl Acad Sci U S A 110:2129–2134. https://doi.org/10.1073/pnas.1218950110

Brohawn SG, Campbell EB, MacKinnon R (2014a) Physical mechanism for gating and mechanosensitivity of the human TRAAK K+ channel. Nature 516:126–130. https://doi.org/10.1038/nature14013

Brohawn SG, Su Z, MacKinnon R (2014b) Mechanosensitivity is mediated directly by the lipid membrane in TRAAK and TREK1 K+ channels. Proc Natl Acad Sci U S A 111:3614–3619. https://doi.org/10.1073/pnas.1320768111

Brohawn SG, Wang W, Handler A, Campbell EB, Schwarz JR, MacKinnon R (2019) The mechanosensitive ion channel TRAAK is localized to the mammalian node of Ranvier. eLife:8. https://doi.org/10.7554/eLife.50403

Calhoun JD, Isom LL (2014) The role of non-pore-forming beta subunits in physiology and pathophysiology of voltage-gated sodium channels. Handb Exp Pharmacol 221:51–89. https://doi.org/10.1007/978-3-642-41588-3_4

Catterall WA (1988) Molecular properties of voltage-sensitive sodium and calcium channels. Braz J Med Biol Res 21:1129–1144

Catterall WA (2014) Structure and function of voltage-gated sodium channels at atomic resolution. Exp Physiol 99:35–51. https://doi.org/10.1113/expphysiol.2013.071969

Cheng SH (2014) Gene therapy for the neurological manifestations in lysosomal storage disorders. J Lipid Res 55:1827–1838. https://doi.org/10.1194/jlr.R047175

Chun YS, Shin S, Kim Y et al (2010) Cholesterol modulates ion channels via down-regulation of phosphatidylinositol 4,5-bisphosphate. J Neurochem 112:1286–1294. https://doi.org/10.1111/j.1471-4159.2009.06545.x

Clayton GM, Aller SG, Wang J, Unger V, Morais-Cabral JH (2009) Combining electron crystallography and X-ray crystallography to study the MlotiK1 cyclic nucleotide-regulated potassium channel. J Struct Biol 167:220–226. doi: S1047-8477(09)00154-3 [pii]. https://doi.org/10.1016/j.jsb.2009.06.012

Coates L (2020) Ion permeation in potassium ion channels. Acta Crystallogr D Struct Biol 76:326–331. https://doi.org/10.1107/S2059798320003599

Contreras GF, Neely A, Alvarez O, Gonzalez C, Latorre R (2012) Modulation of BK channel voltage gating by different auxiliary beta subunits. Proc Natl Acad Sci U S A 109:18991–18996. https://doi.org/10.1073/pnas.1216953109

Cordero-Morales JF, Cuello LG, Zhao Y, Jogini V, Cortes DM, Roux B, Perozo E (2006) Molecular determinants of gating at the potassium-channel selectivity filter. Nat Struct Mol Biol 13:311–318. doi: nsmb1069 [pii] 381038/nsmb1069

Cordero-Morales JF, Jogini V, Lewis A, Vasquez V, Cortes DM, Roux B, Perozo E (2007) Molecular driving forces determining potassium channel slow inactivation. Nat Struct Mol Biol 14:1062–1069. doi: nsmb1309 [pii]. https://doi.org/10.1038/nsmb1309

Cox DH, Cui J, Aldrich RW (1997) Allosteric gating of a large conductance Ca-activated K+ channel. J Gen Physiol 110:257–281

Cox JJ, Reimann F, Nicholas AK et al (2006) An SCN9A channelopathy causes congenital inability to experience pain. Nature 444:894–898. doi: nature05413 [pii]. https://doi.org/10.1038/nature05413

Coyan FC, Abderemane-Ali F, Amarouch MY et al (2014) A long QT mutation substitutes cholesterol for phosphatidylinositol-4,5-bisphosphate in KCNQ1 channel regulation. PLoS One 9:e93255. https://doi.org/10.1371/journal.pone.0093255

Cuello LG, Jogini V, Cortes DM et al (2010a) Structural basis for the coupling between activation and inactivation gates in K(+) channels. Nature 466:272–275. doi: nature09136 [pii]. https://doi.org/10.1038/nature09136

Cuello LG, Jogini V, Cortes DM, Perozo E (2010b) Structural mechanism of C-type inactivation in K(+) channels. Nature 466:203–208. doi: nature09153 [pii]. https://doi.org/10.1038/nature09153

De Biasi M, Wang Z, Accili E, Wible B, Fedida D (1997) Open channel block of human heart hKv1.5 by the beta-subunit hKv beta 1.2. Am J Phys 272:H2932–H2941. https://doi.org/10.1152/ajpheart.1997.272.6.H2932

Derebe MG, Sauer DB, Zeng W, Alam A, Shi N, Jiang Y (2011a) Tuning the ion selectivity of tetrameric cation channels by changing the number of ion binding sites. Proc Natl Acad Sci U S A 108:598–602. https://doi.org/10.1073/pnas.1013636108

Derebe MG, Zeng W, Li Y, Alam A, Jiang Y (2011b) Structural studies of ion permeation and Ca2+ blockage of a bacterial channel mimicking the cyclic nucleotide-gated channel pore. Proc Natl Acad Sci U S A 108:592–597. https://doi.org/10.1073/pnas.1013643108

Devaraneni PK, Komarov AG, Costantino CA, Devereaux JJ, Matulef K, Valiyaveetil FI (2013) Semisynthetic K+ channels show that the constricted conformation of the selectivity filter is not the C-type inactivated state. Proc Natl Acad Sci U S A 110:15698–15703. https://doi.org/10.1073/pnas.1308699110

Ding D, Wang M, Wu JX, Kang Y, Chen L (2019) The structural basis for the binding of Repaglinide to the pancreatic KATP Channel. Cell Rep 27:1848–1857 e1844. https://doi.org/10.1016/j.celrep.2019.04.050

Doyle DA, Morais CJ, Pfuetzner RA et al (1998) The structure of the potassium channel: molecular basis of K+ conduction and selectivity. Science 280(5360):69–77

Estevez M (2006) Invertebrate modeling of a migraine channelopathy. Headache 46(Suppl 1):S25–S31

Fan C, Sukomon N, Flood E, Rheinberger J, Allen TW, Nimigean CM (2020) Ball-and-chain inactivation in a calcium-gated potassium channel. Nature 580:288–293. https://doi.org/10.1038/s41586-020-2116-0

Finol-Urdaneta RK, McArthur JR, Juranka PF, French RJ, Morris CE (2010) Modulation of KvAP unitary conductance and gating by 1-alkanols and other surface active agents. Biophys J 98:762–772. doi: S0006-3495(09)01733-0 [pii]. https://doi.org/10.1016/j.bpj.2009.10.053

Gantz SC, Bean BP (2017) Cell-autonomous excitation of midbrain dopamine neurons by endocannabinoid-dependent lipid signaling. Neuron 93:1375–1387 e1372. https://doi.org/10.1016/j.neuron.2017.02.025

Gomez-Lagunas F, Armstrong CM (1994) The relation between ion permeation and recovery from inactivation of ShakerB K+ channels. Biophys J 67:1806–1815. https://doi.org/10.1016/S0006-3495(94)80662-9

Gribble FM, Tucker SJ, Haug T, Ashcroft FM (1998) MgATP activates the beta cell KATP channel by interaction with its SUR1 subunit. Proc Natl Acad Sci U S A 95:7185–7190. https://doi.org/10.1073/pnas.95.12.7185

Guo J, Chi S, Xu H, Jin G, Qi Z (2008) Effects of cholesterol levels on the excitability of rat hippocampal neurons. Mol Membr Biol 25:216–223. doi: 791444084 [pii]. https://doi.org/10.1080/09687680701805541

Guo J, Zeng W, Chen Q et al (2016) Structure of the voltage-gated two-pore channel TPC1 from Arabidopsis thaliana. Nature 531:196–201. https://doi.org/10.1038/nature16446

Hajdu P, Varga Z, Pieri C, Panyi G, Gaspar R Jr (2003) Cholesterol modifies the gating of Kv1.3 in human T lymphocytes. Pflugers Arch 445:674–682

Heaps CL, Tharp DL, Bowles DK (2005) Hypercholesterolemia abolishes voltage-dependent K+ channel contribution to adenosine-mediated relaxation in porcine coronary arterioles. Am J Physiol Heart Circ Physiol 288:H568–H576. https://doi.org/10.1152/ajpheart.00157.2004

Heginbotham L, MacKinnon R (1993) Conduction properties of the cloned Shaker K+ channel. Biophys J 65(5):2089–2096

Heginbotham L, Lu Z, Abramson T, MacKinnon R (1994) Mutations in the K+ channel signature sequence. Biophys J 66(4):1061–1067 1131

Hille B (2001) Ion channels of excitable membranes. Sinauer Associates, Inc., Sunderland

Hite RK, MacKinnon R (2017) Structural titration of Slo2.2, a Na+-dependent K+ channel. Cell 168:390–399 e311. https://doi.org/10.1016/j.cell.2016.12.030

Hite RK, Yuan P, Li Z, Hsuing Y, Walz T, MacKinnon R (2015) Cryo-electron microscopy structure of the Slo2.2 Na(+)-activated K(+) channel. Nature 527:198–203. https://doi.org/10.1038/nature14958

Hite RK, Tao X, MacKinnon R (2017) Structural basis for gating the high-conductance Ca2+−activated K+ channel. Nature 541:52–57. https://doi.org/10.1038/nature20775

Ho IH, Murrell-Lagnado RD (1999) Molecular mechanism for sodium-dependent activation of G protein-gated K+ channels. J Physiol 520(Pt 3):645–651. https://doi.org/10.1111/j.1469-7793.1999.00645.x

Hofmann F, Belkacemi A, Flockerzi V (2015) Emerging alternative functions for the auxiliary subunits of the voltage-gated calcium channels. Curr Mol Pharmacol 8:162–168

Hoshi T, Zagotta WN, Aldrich RW (1991) Two types of inactivation in Shaker K+ channels: effects of alterations in the carboxy-terminal region. Neuron 7:547–556. https://doi.org/10.1016/0896-6273(91)90367-9

Howard RJ, Clark KA, Holton JM, Minor DL Jr (2007) Structural insight into KCNQ (Kv7) channel assembly and channelopathy. Neuron 53:663–675. doi: S0896-6273(07)00109-2 [pii] 711016/j.neuron.2007.02.010

Huang CW, Wu YJ, Wu SN (2011) Modification of activation kinetics of delayed rectifier K+ currents and neuronal excitability by methyl-beta-cyclodextrin. Neuroscience 176:431–441. https://doi.org/10.1016/j.neuroscience.2010.10.060

Islas LD, Sigworth FJ (1999) Voltage sensitivity and gating charge in Shaker and Shab family potassium channels. J Gen Physiol 114:723–742

Islas LD, Sigworth FJ (2001) Electrostatics and the gating pore of Shaker potassium channels. J Gen Physiol 117:69–89

Iwasa J, Marshall W (2015) Karp's cell and molecular biology. Wiley, Hoboken

Jensen KH, Brandt SS, Shigematsu H, Sigworth FJ (2016) Statistical modeling and removal of lipid membrane projections for cryo-EM structure determination of reconstituted membrane proteins. J Struct Biol 194:49–60. https://doi.org/10.1016/j.jsb.2016.01.012

Jia Z, Yazdani M, Zhang G, Cui J, Chen J (2018) Hydrophobic gating in BK channels. Nat Commun 9:3408. https://doi.org/10.1038/s41467-018-05970-3

Jiang QX (2018) Lipid-dependent gating of ion channels. Protein-lipid interactions: perspectives, techniques and challenges, vol 1, p 196

Jiang QX (2019) Cholesterol-dependent gating effects on ion channels. Adv Exp Med Biol 1115:167–190. https://doi.org/10.1007/978-3-030-04278-3_8

Jiang QX, Chester DW, Sigworth FJ (2001) Spherical reconstruction: a method for structure determination of membrane proteins from cryo-EM images. J Struct Biol 133(2–3):119–131

Jiang Y, Lee A, Chen J, Cadene M, Chait BT, MacKinnon R (2002) The open pore conformation of potassium channels. Nature 417:523–526
Jiang Y, Lee A, Chen J, Ruta V, Cadene M, Chait BT, MacKinnon R (2003a) X-ray structure of a voltage-dependent K+ channel. Nature 423:33–41
Jiang Y, Ruta V, Chen J, Lee A, MacKinnon R (2003b) The principle of gating charge movement in a voltage-dependent K+ channel. Nature 423:42–48
Kaczorowski GJ, Garcia ML (2016) Developing molecular pharmacology of BK channels for therapeutic benefit. Int Rev Neurobiol 128:439–475. https://doi.org/10.1016/bs.irn.2016.02.013
Ketchum KA, Joiner WJ, Sellers AJ, Kaczmarek LK, Goldstein SA (1995) A new family of outwardly rectifying potassium channel proteins with two pore domains in tandem. Nature 376:690–695. https://doi.org/10.1038/376690a0
Keynes RD (1992) A new look at the mechanism of activation and inactivation of voltage-gated ion channels. Proc Biol Sci 249:107–112
Kintzer AF, Stroud RM (2016) Structure, inhibition and regulation of two-pore channel TPC1 from Arabidopsis thaliana. Nature 531:258–262. https://doi.org/10.1038/nature17194
Klemic KG, Kirsch GE, Jones SW (2001) U-type inactivation of Kv3.1 and Shaker potassium channels. Biophys J 81:814–826. https://doi.org/10.1016/S0006-3495(01)75743-8
Kolter T, Sandhoff K (2006) Sphingolipid metabolism diseases. Biochim Biophys Acta 1758:2057–2079. https://doi.org/10.1016/j.bbamem.2006.05.027
Kopec W, Kopfer DA, Vickery ON, Bondarenko AS, Jansen TLC, de Groot BL, Zachariae U (2018) Direct knock-on of desolvated ions governs strict ion selectivity in K(+) channels. Nat Chem 10:813–820. https://doi.org/10.1038/s41557-018-0105-9
Kopfer DA, Song C, Gruene T, Sheldrick GM, Zachariae U, de Groot BL (2014) Ion permeation in K(+) channels occurs by direct coulomb knock-on. Science 346:352–355. https://doi.org/10.1126/science.1254840
Kordasiewicz HB, Gomez CM (2007) Molecular pathogenesis of spinocerebellar ataxia type 6. Neurotherapeutics 4:285–294. doi: S1933-7213(07)00004-9 [pii]. https://doi.org/10.1016/j.nurt.2007.01.003
Kratochvil HT, Carr JK, Matulef K et al (2016) Instantaneous ion configurations in the K+ ion channel selectivity filter revealed by 2D IR spectroscopy. Science 353:1040–1044. https://doi.org/10.1126/science.aag1447
Kurata HT, Fedida D (2005) A structural interpretation of voltage-gated potassium channel inactivation. Prog Biophys Mol Biol:1562
Kurata HT, Fedida D (2006) A structural interpretation of voltage-gated potassium channel inactivation. Prog Biophys Mol Biol 92:185–208
Kurata HT, Doerksen KW, Eldstrom JR, Rezazadeh S, Fedida D (2005) Separation of P/C- and U-type inactivation pathways in Kv1.5 potassium channels. J Physiol 568:31–46. https://doi.org/10.1113/jphysiol.2005.087148
Lee CH, MacKinnon R (2017) Structures of the human HCN1 hyperpolarization-Activated Channel. Cell 168:111–120 e111. https://doi.org/10.1016/j.cell.2016.12.023
Lee SY, Lee A, Chen J, MacKinnon R (2005) Structure of the KvAP voltage-dependent K+ channel and its dependence on the lipid membrane. Proc Natl Acad Sci U S A 102:15441–15446
Lee S, Zheng H, Shi L, Jiang QX (2013) Reconstitution of a Kv channel into lipid membranes for structural and functional studies. J Vis Exp:e50436. https://doi.org/10.3791/50436
Lee KPK, Chen J, MacKinnon R (2017) Molecular structure of human KATP in complex with ATP and ADP. eLife 6. https://doi.org/10.7554/eLife.32481
Li Q, Wanderling S, Paduch M et al (2014a) Structural mechanism of voltage-dependent gating in an isolated voltage-sensing domain. Nat Struct Mol Biol 21:244–252. https://doi.org/10.1038/nsmb.2768
Li Q, Wanderling S, Sompornpisut P, Perozo E (2014b) Structural basis of lipid-driven conformational transitions in the KvAP voltage-sensing domain. Nat Struct Mol Biol 21:160–166. https://doi.org/10.1038/nsmb.2747

Li J, Ostmeyer J, Boulanger E, Rui H, Perozo E, Roux B (2017a) Chemical substitutions in the selectivity filter of potassium channels do not rule out constricted-like conformations for C-type inactivation. Proc Natl Acad Sci U S A 114:11145–11150. https://doi.org/10.1073/pnas.1706983114

Li N, Wu JX, Ding D, Cheng J, Gao N, Chen L (2017b) Structure of a pancreatic ATP-sensitive potassium channel. Cell 168:101–110 e110. https://doi.org/10.1016/j.cell.2016.12.028

Li J, Ostmeyer J, Cuello LG, Perozo E, Roux B (2018) Rapid constriction of the selectivity filter underlies C-type inactivation in the KcsA potassium channel. J Gen Physiol 150:1408–1420. https://doi.org/10.1085/jgp.201812082

Lingle CJ, Martinez-Espinosa PL, Yang-Hood A et al (2019) LRRC52 regulates BK channel function and localization in mouse cochlear inner hair cells. Proc Natl Acad Sci U S A 116:18397–18403. https://doi.org/10.1073/pnas.1907065116

Linn KM, Derebe MG, Jiang Y, Valiyaveetil FI (2010) Semisynthesis of NaK, a Na(+) and K(+) conducting ion channel. Biochemistry 49:4450–4456. https://doi.org/10.1021/bi100413z

Llaguno MC, Xu H, Shi L, Huang N, Zhang H, Liu Q, Jiang QX (2014) Chemically functionalized carbon films for single molecule imaging. J Struct Biol 185:405–417

Logothetis DE, Kurachi Y, Galper J, Neer EJ, Clapham DE (1987) The beta gamma subunits of GTP-binding proteins activate the muscarinic K+ channel in heart. Nature 325:321–326. https://doi.org/10.1038/325321a0

Lolicato M, Riegelhaupt PM, Arrigoni C, Clark KA, Minor DL Jr (2014) Transmembrane helix straightening and buckling underlies activation of mechanosensitive and thermosensitive K (2P) channels. Neuron 84:1198–1212. https://doi.org/10.1016/j.neuron.2014.11.017

Lolicato M, Arrigoni C, Mori T, Sekioka Y, Bryant C, Clark KA, Minor DL Jr (2017) K2P2.1 (TREK-1)-activator complexes reveal a cryptic selectivity filter binding site. Nature 547:364–368. https://doi.org/10.1038/nature22988

Long SB, Tao X, Campbell EB, MacKinnon R (2007) Atomic structure of a voltage-dependent K+ channel in a lipid membrane-like environment. Nature 450:376–382

MacKinnon R, Aldrich RW, Lee AW (1993) Functional stoichiometry of Shaker potassium channel inactivation. Science 262(5134):757–759 1836

Martin GM, Kandasamy B, DiMaio F, Yoshioka C, Shyng SL (2017) Anti-diabetic drug binding site in a mammalian KATP channel revealed by Cryo-EM. eLife:6. https://doi.org/10.7554/eLife.31054

Martin GM, Sung MW, Yang Z et al (2019) Mechanism of pharmacochaperoning in a mammalian KATP channel revealed by cryo-EM. eLife:8. https://doi.org/10.7554/eLife.46417

Martina M, Metz AE, Bean BP (2007) Voltage-dependent potassium currents during fast spikes of rat cerebellar Purkinje neurons: inhibition by BDS-I toxin. J Neurophysiol 97:563–571. doi: 00269.2006 [pii] 1171152/jn.00269.2006

Masoli S, Solinas S, D'Angelo E (2015) Action potential processing in a detailed Purkinje cell model reveals a critical role for axonal compartmentalization. Front Cell Neurosci 9:47. https://doi.org/10.3389/fncel.2015.00047

Matthies D, Bae C, Toombes GE, Fox T, Bartesaghi A, Subramaniam S, Swartz KJ (2018) Single-particle cryo-EM structure of a voltage-activated potassium channel in lipid nanodiscs. eLife:7. https://doi.org/10.7554/eLife.37558

Matulef K, Annen AW, Nix JC, Valiyaveetil FI (2016) Individual ion binding sites in the K(+) channel play distinct roles in C-type inactivation and in recovery from inactivation. Structure 24:750–761. https://doi.org/10.1016/j.str.2016.02.021

McCobb DP, Fowler NL, Featherstone T, Lingle CJ, Saito M, Krause JE, Salkoff L (1995) A human calcium-activated potassium channel gene expressed in vascular smooth muscle. Am J Phys 269:H767–H777. https://doi.org/10.1152/ajpheart.1995.269.3.H767

Mesa-Herrera F, Taoro-Gonzalez L, Valdes-Baizabal C, Diaz M, Marin R (2019) Lipid and lipid raft alteration in aging and neurodegenerative diseases: a window for the development of new biomarkers. Int J Mol Sci 20. https://doi.org/10.3390/ijms20153810

Miller AN, Long SB (2012) Crystal structure of the human two-pore domain potassium channel K2P1. Science 335:432–436. https://doi.org/10.1126/science.1213274
Morais-Cabral JH, Zhou Y, MacKinnon R (2001) Energetic optimization of ion conduction rate by the K+ selectivity filter. Nature 414:37–42. https://doi.org/10.1038/35102000
Mullner C, Yakubovich D, Dessauer CW, Platzer D, Schreibmayer W (2003) Single channel analysis of the regulation of GIRK1/GIRK4 channels by protein phosphorylation. Biophys J 84:1399–1409. https://doi.org/10.1016/S0006-3495(03)74954-6
O'Connell KM, Loftus R, Tamkun MM (2010) Localization-dependent activity of the Kv2.1 delayed-rectifier K+ channel. Proc Natl Acad Sci U S A 107:12351–12356. https://doi.org/10.1073/pnas.1003028107
Oster C, Hendriks K, Kopec W et al (2019) The conduction pathway of potassium channels is water free under physiological conditions. Sci Adv 5:eaaw6756. https://doi.org/10.1126/sciadv.aaw6756
Ostmeyer J, Chakrapani S, Pan AC, Perozo E, Roux B (2013) Recovery from slow inactivation in K + channels is controlled by water molecules. Nature 501:121–124. https://doi.org/10.1038/nature12395
Pau V, Zhou Y, Ramu Y, Xu Y, Lu Z (2017) Crystal structure of an inactivated mutant mammalian voltage-gated K(+) channel. Nat Struct Mol Biol 24:857–865. https://doi.org/10.1038/nsmb.3457
Payandeh J, Scheuer T, Zheng N, Catterall WA (2011) The crystal structure of a voltage-gated sodium channel. Nature 475:353–358. nature10238 [pii]. https://doi.org/10.1038/nature10238
Payandeh J, Gamal El-Din TM, Scheuer T, Zheng N, Catterall WA (2012) Crystal structure of a voltage-gated sodium channel in two potentially inactivated states. Nature 486:135–139. https://doi.org/10.1038/nature11077
Pietrobon D (2010) CaV2.1 channelopathies. Pflugers Arch 460:375–393. https://doi.org/10.1007/s00424-010-0802-8
Poolos NP (2005) The h-channel: a potential channelopathy in epilepsy? Epilepsy Behav 7:51–56
Poolos NP, Johnston D (2012) Dendritic ion channelopathy in acquired epilepsy. Epilepsia 53 (Suppl 9):32–40. https://doi.org/10.1111/epi.12033
Pottosin II, Valencia-Cruz G, Bonales-Alatorre E, Shabala SN, Dobrovinskaya OR (2007) Methyl-beta-cyclodextrin reversibly alters the gating of lipid rafts-associated Kv1.3 channels in Jurkat T lymphocytes. Pflugers Arch 454:235–244
Pouvreau S, Berthier C, Blaineau S, Amsellem J, Coronado R, Strube C (2004) Membrane cholesterol modulates dihydropyridine receptor function in mice fetal skeletal muscle cells. J Physiol 555:365–381
Proks P, Gribble FM, Adhikari R, Tucker SJ, Ashcroft FM (1999) Involvement of the N-terminus of Kir6.2 in the inhibition of the KATP channel by ATP. J Physiol 514(Pt 1):19–25. https://doi.org/10.1111/j.1469-7793.1999.019af.x
Purcell EK, Liu L, Thomas PV, Duncan RK (2011) Cholesterol influences voltage-gated calcium channels and BK-type potassium channels in auditory hair cells. PLoS One 6:e26289. https://doi.org/10.1371/journal.pone.0026289. PONE-D-11-13726 [pii]
Pyott SJ, Meredith AL, Fodor AA, Vazquez AE, Yamoah EN, Aldrich RW (2007) Cochlear function in mice lacking the BK channel alpha, beta1, or beta4 subunits. J Biol Chem 282:3312–3324. https://doi.org/10.1074/jbc.M608726200
Rajakulendran S, Schorge S, Kullmann DM, Hanna MG (2007) Episodic ataxia type 1: a neuronal potassium channelopathy. Neurotherapeutics 4:258–266. doi: S1933-7213(07)00011-6 [pii]. https://doi.org/10.1016/j.nurt.2007.01.010
Randich AM, Cuello LG, Wanderling SS, Perozo E (2014) Biochemical and structural analysis of the hyperpolarization-activated K(+) channel MVP. Biochemistry 53:1627–1636. https://doi.org/10.1021/bi4014243
Rudakova E, Wagner M, Frank M, Volk T (2015) Localization of Kv4.2 and KChIP2 in lipid rafts and modulation of outward K+ currents by membrane cholesterol content in rat left ventricular myocytes. Pflugers Arch 467:299–309. https://doi.org/10.1007/s00424-014-1521-3

Saito M, Nelson C, Salkoff L, Lingle CJ (1997) A cysteine-rich domain defined by a novel exon in a slo variant in rat adrenal chromaffin cells and PC12 cells. J Biol Chem 272:11710–11717. https://doi.org/10.1074/jbc.272.18.11710

Salkoff L, Butler A, Ferreira G, Santi C, Wei A (2006) High-conductance potassium channels of the SLO family. Nat Rev Neurosci 7:921–931

Schoppa NE, Sigworth FJ (1998) Activation of Shaker potassium channels. III. An activation gating model for wild-type and V2 mutant channels. J Gen Physiol 111(2):313–342 2661

Schoppa NE, McCormack K, Tanouye MA, Sigworth FJ (1992) The size of gating charge in wild-type and mutant Shaker potassium channels. Science 255:1712–1715

Seoh SA, Sigg D, Papazian DM, Bezanilla F (1996) Voltage-sensing residues in the S2 and S4 segments of the Shaker K+ channel. Neuron 16:1159–1167

Shi C, He Y, Hendriks K et al (2018) A single NaK channel conformation is not enough for non-selective ion conduction. Nat Commun 9:717. https://doi.org/10.1038/s41467-018-03179-y

Shin KS, Maertens C, Proenza C, Rothberg BS, Yellen G (2004) Inactivation in HCN channels results from reclosure of the activation gate: desensitization to voltage. Neuron 41:737–744

Shribman S, Patani R, Deeb J, Chaudhuri A (2013) Voltage-gated potassium channelopathy: an expanding spectrum of clinical phenotypes. BMJ Case Rep:2013. https://doi.org/10.1136/bcr-2012-007742

Singh H, Stefani E, Toro L (2012) Intracellular BK(Ca) (iBK(Ca)) channels. J Physiol 590:5937–5947. https://doi.org/10.1113/jphysiol.2011.215533

Starace DM, Bezanilla F (2001) Histidine scanning mutagenesis of basic residues of the S4 segment of the shaker k+ channel. J Gen Physiol 117:469–490

Sun J, MacKinnon R (2017) Cryo-EM structure of a KCNQ1/CaM complex reveals insights into congenital Long QT syndrome. Cell 169:1042–1050 e1049. https://doi.org/10.1016/j.cell.2017.05.019

Takeshita K, Sakata S, Yamashita E et al (2014) X-ray crystal structure of voltage-gated proton channel. Nat Struct Mol Biol 21:352–357. https://doi.org/10.1038/nsmb.2783

Tang L, Gamal El-Din TM, Payandeh J et al (2014) Structural basis for Ca2+ selectivity of a voltage-gated calcium channel. Nature 505:56–61. https://doi.org/10.1038/nature12775

Tao X, MacKinnon R (2019) Cryo-EM structure of the KvAP channel reveals a non-domain-swapped voltage sensor topology. eLife:8. https://doi.org/10.7554/eLife.52164

Tao X, Lee A, Limapichat W, Dougherty DA, MacKinnon R (2010) A gating charge transfer center in voltage sensors. Science 328:67–73

Tao X, Hite RK, MacKinnon R (2017) Cryo-EM structure of the open high-conductance Ca2+–activated K+ channel. Nature 541:46–51. https://doi.org/10.1038/nature20608

Tremblay J, Hamet P (2010) Genetics of pain, opioids, and opioid responsiveness. Metabolism 59 (Suppl 1):S5–S8. S0026-0495(10)00235-0 [pii]. https://doi.org/10.1016/j.metabol.2010.07.015

Valiyaveetil FI, Sekedat M, MacKinnon R, Muir TW (2004) Glycine as a D-amino acid surrogate in the K(+)-selectivity filter. Proc Natl Acad Sci U S A 101(49):17045–17049 13151

Valiyaveetil FI, Leonetti M, Muir TW, Mackinnon R (2006) Ion selectivity in a semisynthetic K+ channel locked in the conductive conformation. Science 314:1004–1007. https://doi.org/10.1126/science.1133415

Vargas E, Yarov-Yarovoy V, Khalili-Araghi F et al (2012) An emerging consensus on voltage-dependent gating from computational modeling and molecular dynamics simulations. J Gen Physiol 140:587–594. https://doi.org/10.1085/jgp.201210873

Wang W, MacKinnon R (2017a) Cryo-EM structure of the open human ether-a-go-go-related K(+) channel hERG. Cell 169:422–430 e410. https://doi.org/10.1016/j.cell.2017.03.048

Wang W, MacKinnon R (2017b) Cryo-EM structure of the open human ether-a-go-go-related K+ channel hERG. Cell 169:422–430 e410. https://doi.org/10.1016/j.cell.2017.03.048

Whicher JR, MacKinnon R (2016) Structure of the voltage-gated K(+) channel Eag1 reveals an alternative voltage sensing mechanism. Science 353:664–669. https://doi.org/10.1126/science.aaf8070

Whorton MR, MacKinnon R (2011) Crystal structure of the mammalian GIRK2 K+ channel and gating regulation by G proteins, PIP2, and sodium. Cell 147:199–208. https://doi.org/10.1016/j.cell.2011.07.046

Whorton MR, MacKinnon R (2013) X-ray structure of the mammalian GIRK2-betagamma G-protein complex. Nature 498:190–197. https://doi.org/10.1038/nature12241

Wu J, Yan Z, Li Z, Yan C, Lu S, Dong M, Yan N (2015) Structure of the voltage-gated calcium channel Cav1.1 complex. Science 350:aad2395. https://doi.org/10.1126/science.aad2395

Xia F, Gao X, Kwan E et al (2004) Disruption of pancreatic beta-cell lipid rafts modifies Kv2.1 channel gating and insulin exocytosis. J Biol Chem 279:24685–24691

Xie G, Harrison J, Clapcote SJ, Huang Y, Zhang JY, Wang LY, Roder JC (2010) A new Kv1.2 channelopathy underlying cerebellar ataxia. J Biol Chem 285:32160–32173. doi: M110.153676 [pii]. https://doi.org/10.1074/jbc.M110.153676

Xu Y, Ramu Y, Lu Z (2008) Removal of phospho-head groups of membrane lipids immobilizes voltage sensors of K+ channels. Nature 451:826–829. doi: nature06618 [pii]. https://doi.org/10.1038/nature06618

Xu H, Li T, Rohou A et al (2019) Structural basis of Nav1.7 inhibition by a gating-modifier spider toxin. Cell 176:702–715 e714. https://doi.org/10.1016/j.cell.2018.12.018

Yang Y, Yan Y, Sigworth FJ (1997) How does the W434F mutation block current in Shaker potassium channels? J Gen Physiol 109:779–789. https://doi.org/10.1085/jgp.109.6.779

Yarov-Yarovoy V, Baker D, Catterall WA (2006) Voltage sensor conformations in the open and closed states in ROSETTA structural models of K(+) channels. Proc Natl Acad Sci U S A 103:7292–7297. doi: 0602350103 [pii]. https://doi.org/10.1073/pnas.0602350103

Yellen G (1998) The moving parts of voltage-gated ion channels. Q Rev Biophys 31:239–295

Zagotta WN, Hoshi T, Aldrich RW (1990) Restoration of inactivation in mutants of Shaker potassium channels by a peptide derived from ShB. Science 250:568–571

Zhang X, Ren W, DeCaen P et al (2012) Crystal structure of an orthologue of the NaChBac voltage-gated sodium channel. Nature 486:130–134. https://doi.org/10.1038/nature11054

Zheng H, Liu W, Anderson LY, Jiang QX (2012) Lipid-dependent gating of a voltage-gated potassium channel. Nat Commun 2:250. doi: ncomms1254 [pii]. https://doi.org/10.1038/ncomms1254

Zhou Y, MacKinnon R (2003) The occupancy of ions in the K+ selectivity filter: charge balance and coupling of ion binding to a protein conformational change underlie high conduction rates. J Mol Biol 333(5):965–975 3587

Zhou Y, MacKinnon R (2004) Ion binding affinity in the cavity of the KcsA potassium channel. Biochemistry 43(17):4978–4982 3588

Zhou Y, Morais-Cabral JH, Kaufman A, MacKinnon R (2001) Chemistry of ion coordination and hydration revealed by a K+ channel-Fab complex at 2.0 A resolution. Nature 414:43–48. https://doi.org/10.1038/35102009

Zhou Y, Yang H, Cui J, Lingle CJ (2017) Threading the biophysics of mammalian Slo1 channels onto structures of an invertebrate Slo1 channel. J Gen Physiol 149:985–1007. https://doi.org/10.1085/jgp.201711845

Pharmacological Approaches to Studying Potassium Channels

Alistair Mathie, Emma L. Veale, Alessia Golluscio, Robyn G. Holden, and Yvonne Walsh

Contents

A. Mathie (✉)
Medway School of Pharmacy, University of Kent, Kent, UK

Medway School of Pharmacy, University of Greenwich, London, UK

School of Engineering, Arts, Science and Technology, University of Suffolk, Ipswich, UK
e-mail: a.a.mathie@kent.ac.uk

E. L. Veale · A. Golluscio · R. G. Holden · Y. Walsh
Medway School of Pharmacy, University of Kent, Kent, UK

Medway School of Pharmacy, University of Greenwich, London, UK

N. Gamper, K. Wang (eds.), *Pharmacology of Potassium Channels*,
Handbook of Experimental Pharmacology 267, https://doi.org/10.1007/164_2021_502

Abstract

In this review, we consider the pharmacology of potassium channels from the perspective of these channels as therapeutic targets. Firstly, we describe the three main families of potassium channels in humans and disease states where they are implicated. Secondly, we describe the existing therapeutic agents which act on potassium channels and outline why these channels represent an under-exploited therapeutic target with potential for future drug development. Thirdly, we consider the evidence desired in order to embark on a drug discovery programme targeting a particular potassium channel. We have chosen two "case studies": activators of the two-pore domain potassium (K_{2P}) channel TREK-2 ($K_{2P}10.1$), for the treatment of pain and inhibitors of the voltage-gated potassium channel $K_V1.3$, for use in autoimmune diseases such as multiple sclerosis. We describe the evidence base to suggest why these are viable therapeutic targets. Finally, we detail the main technical approaches available to characterise the pharmacology of potassium channels and identify novel regulatory compounds. We draw particular attention to the Comprehensive in vitro Proarrhythmia Assay initiative (CiPA, https://cipaproject.org) project for cardiac safety, as an example of what might be both desirable and possible in the future, for ion channel regulator discovery projects.

Keywords

CiPA · hERG · Kv1.3 · Patch-clamp electrophysiology · Potassium channel · TREK-2

1 Potassium Channel Families

Potassium selective ion channels are pore-forming proteins that allow the flow of potassium ions across membranes, primarily, but not exclusively, the plasma membrane. Potassium channels regulate cell excitability, control cell resting membrane potential and determine the shape of the action potential waveform in cells that use action potentials such as neurons and cardiac cells. However, potassium channels are present in virtually all cells within the body, influencing a wide range of diverse processes.

Potassium channels are the largest class of mammalian ion channel proteins. The human genome contains over 75 genes that encode for the primary (alpha)-subunits of potassium channel proteins. These genes are divided into either three or four (Taura et al. 2021) distinct families in mammals (see Fig. 1) that encode pore-forming subunits based on their structural and functional properties.

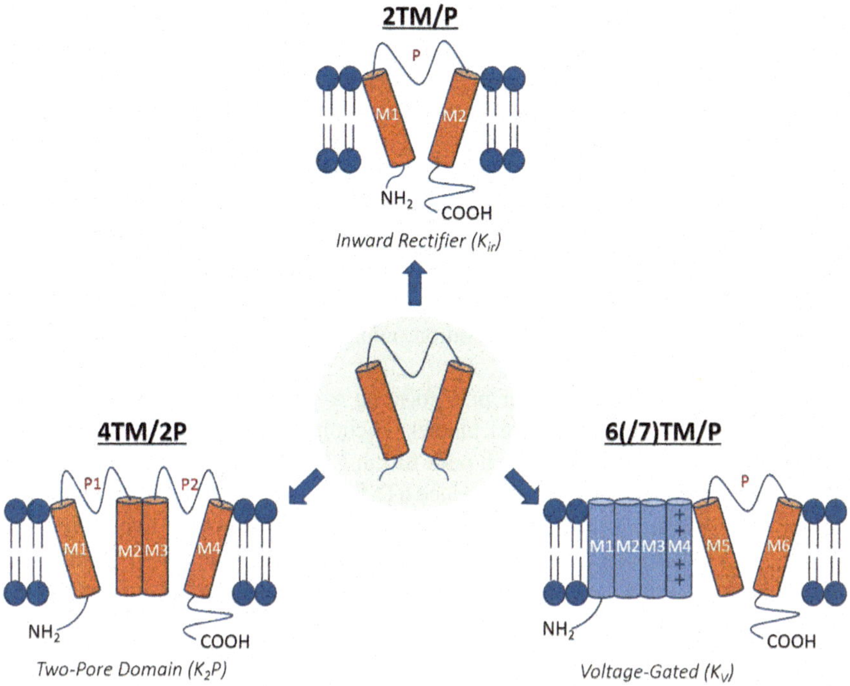

Fig. 1 Potassium channel families

The number of different, functioning ion channel proteins is, potentially, much greater than this because of the formation of heteromeric channel subunit combinations. Diversity is further enhanced by subunit variation due to alternative splicing, alternative translation initiation (ATI) and by co-assembly with accessory proteins. A standardised nomenclature for potassium channels has been proposed by NC-IUPHAR (Adelman et al. 2019; Aldrich et al. 2019; Attali et al. 2019; Bayliss et al. 2019). Nomenclature of K channels, however, remains a divisive topic. Whilst formal classification exists (K_V1.x, $K_{ir}1$.x, $K_{2P}1$.x, etc.,), there is an established and entrenched literature which utilises the more familiar potassium channel names such as hERG, BK_{Ca}, K_{ATP}, TWIK, KCNQ1, etc., that are often much more recognisable to researchers both within and outside the field. In this review, we have attempted to accommodate both positions by using either nomenclature where appropriate.

An introduction to potassium channels is given in Taura et al. (2021) and a comprehensive description of each channel subtype and its pharmacology is given by NC-IUPHAR's Guide to Pharmacology: (http://www.guidetopharmacology.org/GRAC/FamilyDisplayForward?familyId=696).

An up-to-date snapshot of current K channel pharmacology, in particular a tabulated summary of the important properties of each subfamily of K channels, is provided by the related Concise Guide to Pharmacology (Alexander et al. 2019).

1.1 6TM Potassium Channels

The six transmembrane domain (6TM), family of K channels is the largest of the K channel families and is made up of the voltage-gated K_V subfamilies, the KCNQ subfamily (which includes KCNQ1 channels, $K_V7.1$), the EAG subfamily (which includes hERG channels, $K_V11.1$), the Ca^{2+} activated K_{Ca} subfamilies and the Na^+ activated K_{Na} subfamilies. Sometimes, the latter two subfamilies are classified in a separate family (Taura et al. 2021). There are 40 human voltage-gated potassium channel genes belonging to 12 "K_V" subfamilies. In nerve cells and cardiac cells, K_V channels regulate the waveform and firing pattern of action potentials. They may also regulate the cell volume, proliferation, and migration of a wide range of cell types.

Each K_V channel contains four pore-forming α-subunits and may also contain auxiliary β-subunits that affect the channel function and/or localisation (Gonzalez et al. 2012; Vacher et al. 2008). Each pore-forming subunit of K_V channels contains six transmembrane segments (S1-S6, hence 6TM), with the first four transmembrane segments (S1-S4) constituting the voltage sensor and the last two transmembrane segments flanking a pore loop (S5-P-S6) as the pore (P) domain. The P domain for all K channels contains the sequence T/SxxTxGxG which is termed the K selectivity sequence. Within each of the K_V subfamilies, homomeric and heteromeric channels may form with a range of functional properties (González et al. 2012; Johnston et al. 2010).

K_V channel modulators may inhibit channel activity either by occluding the channel permeation pathway, as in the case of pore-blocking toxins and inner pore blockers, or through interacting with the voltage sensor to stabilise the closed state of the channel, as in the case of gating modifier toxins. Some small molecules act by binding to the gating machinery as gating modifiers, or by interacting with the interface between the α- and β-subunits to alter channel activity (Wulff et al. 2009). Mutations of K_V channel genes may cause neurological diseases such as episodic ataxia and epilepsies, heart diseases and deafness (Lehmann-Horn and Jurkat-Rott 1999; Kullmann and Hanna 2002; Abriel and Zaklyazminskaya 2013; Villa and Combi 2016).

1.2 2TM Potassium Channels

The 2TM domain family of K channels is also known as the inward-rectifier K channel family (K_{ir}). Seven structurally distinct subfamilies of the K_{ir} family have been identified in mammals. This family includes the strong inward-rectifier K channels ($K_{ir}2.x$) that are constitutively active, the G-protein-activated inward-rectifier K channels ($K_{ir}3.x$) and the ATP-sensitive K channels ($K_{ir}6.x$, which combines with sulphonylurea receptors to form the functional channel complex). The pore-forming α subunits form tetramers, and heteromeric channels may be formed within subfamilies (e.g. $K_{ir}3.2$ with $K_{ir}3.3$). K_{IR} channels play central roles in control of cellular excitability and K^+ ion homeostasis. K_{IR} channels possess only two membrane-spanning helices and have evolved distinct voltage-independent mechanisms for opening and closing, including gating by G proteins, pH and ATP.

Mutations in K_{ir} channels cause a range of diseases including Bartter syndrome (Simon et al. 1996), Andersen-Tawil syndrome leading to ventricular arrhythmias (Plaster et al. 2001), epilepsy (Buono et al. 2004; Ferraro et al. 2004), vasospastic angina (Miki et al. 2002) and neonatal diabetes (Gloyn et al. 2004). Loss of $K_{ir}4.1$ expression abolishes endo-cochlear membrane potential and causes deafness in Pendred syndrome (Wangemann et al. 2010).

1.3 4TM Potassium Channels

The 4TM family of K channels (K_{2P}) underlies background currents, which are expressed throughout the body. They are open across the physiological voltage-range and are regulated by many neurotransmitters and physiological mediators. The pore-forming α-subunit contains two pore loop (P) domains and so two subunits assemble as dimers to form one pore lined by four P domains in the functional channel. Some of the K_{2P} subunits can form heterodimers across subfamilies (e.g. $K_{2P}3.1$ with $K_{2P}9.1$).

K_{2P} channel subtypes have emerging roles in a multitude of physiological responses and pathological conditions (Mathie et al. 2021a), including action potential propagation in myelinated axons (Brohawn et al. 2019; Kanda et al. 2019), microglial surveillance (Madry et al. 2018), inflammation (Bittner et al. 2009), cancer (Pei et al. 2003; Mu et al. 2003; Sun et al. 2016), cardiac arrhythmias (Decher et al. 2017) and pain (Alloui et al. 2006; Devilliers et al. 2013; Vivier et al. 2017).

2 Classical Pharmacology of Potassium Channels

At least some of us will have taken, or will take in the future, drugs that produce their effects through an action on potassium channels. These include certain oral hypoglycaemic agents, such as gliclazide which block ATP-sensitive K channels ($K_{ir}6.x$) in pancreatic beta cells; openers of the same ATP-sensitive K channels such as nicorandil which hyperpolarise vascular smooth muscle cells leading to vasodilation and are used in the treatment of angina; and blockers of $K_V11.1$ (hERG) channels such as amiodarone which delay the repolarisation phase of cardiac action potentials and are used in the treatment of atrial and ventricular fibrillation. All of the above compounds are listed in the WHO model list of essential medicines: (https://www.who.int/groups/expert-committee-on-selection-and-use-of-essential-medicines/essential-medicines-lists).

Each of these drugs has been in clinical use for decades but, perhaps surprisingly, there are no recent additions to this list of essential drugs which activate or block potassium channels as their primary mechanism. However, a small number of new drugs which target potassium channels, such as fampridine, a formulation of

4-aminopyridine (4-AP, see below) used in multiple sclerosis, have been introduced. These older drugs are not particularly potent, nor are they particularly selective for their primary target. As a result, the doses used can be high and can be associated with undesirable side effects.

It is difficult to obtain precise numbers regarding the relative importance of different proteins as current drug targets, not least because it is often unclear what the primary target of an existing therapeutic agent actually is. Nevertheless, relatively recent estimates suggest that existing drugs target just a few hundred proteins (667 human plus 189 pathogen proteins, or protein families, Overington et al. 2006; Santos et al. 2017 but see Imming et al. 2006). Nineteen per cent of these (or around 160 proteins) are ion channels. To consider this from the perspective of existing drugs, there are around 1,600 distinct approved drugs on the market today of which 18% (or around 300 drugs) target ion channels, making them the second largest gene family targeted by existing drugs, behind G protein coupled receptors (GPCRs).

Early identified pharmacological agents that act on K channels are the "classical" K channel blockers, tetraethylammonium ions (TEA) and 4-AP, which block a range of K_V channels to different degrees but are largely ineffective against K_{ir} or K_{2P} channels. There are also a number of naturally occurring toxins which bind potently and selectively to particular potassium channels, such as charybdotoxin, iberiotoxin and apamin (certain K_{Ca} potassium channels), ShK toxin (from the sea anemone, *Stichodactyla helianthus*; $K_V1.3$ channels) and guangxitoxin ($K_V2.1$ and $K_V2.2$ channels) (Matsumura et al. 2021). These toxins have provided considerable insight, both into how channels function and into potential sites of action for the development of novel therapeutics. Furthermore, naturally occurring compounds such as hydroxy-α-sanshool (from Szechuan peppers), which blocks certain two-pore domain potassium channels, including TRESK channels (Bautista et al. 2008; see also Mathie 2010) have revealed the potential importance of these channels as therapeutic targets in the treatment of pain.

The pharmaceutical industry has not yet exploited potassium channels fully as a drug target despite the introduction of novel, faster screening techniques for compounds acting on potassium channels, although there are several compounds currently in different stages of clinical trials (https://www.ionchannellibrary.com/drugs-in-clinical-trials/). The amino acid sequence homology among potassium channel subfamily members can be quite high (>70%), and this can often make it challenging to both identify and develop subtype-selective compounds. Nevertheless, there is ample opportunity to discover and develop novel "first-in class" molecules targeting potassium channels, as considered later in this review.

3 Identifying K Channel Pharmacological Targets

Embarking on a drug discovery programme focussing on a novel ion channel target is an expensive and time-consuming operation. It is not something that is entered into lightly or without the potential of a profitable (drug-to-market) outcome. A pharmaceutical company would expect to see a portfolio of evidence from a range of different scientific and technical approaches in order to be persuaded that a particular potassium channel is a viable therapeutic target.

At a very fundamental level, knowledge of the exact subunit composition of potassium channels in different cell types and tissues is incomplete and this can be complicated further when channel subunit expression levels are altered in disease. As such the precise subunit composition of potassium channels and what auxiliary subunits may (or may not) be associated with them is often unknown for particular disease states.

Even if one is confident of the potassium channel subtype underlying a particular disease phenotype, there is a paucity of existing pharmacological tools to aid the development of useful assays. Furthermore, despite the resolution of some potassium channel structures and the increased understanding of their function that has resulted from this, there is a lack of resolved structures of potassium channels in complex with channel activators or inhibitors (but see Dong et al. 2015; Lolicato et al. 2017) which restricts identification of pharmacophores for rational drug design (McGivern and Ding 2020). When selective pharmacological tools or structural insights are not available, additional information can be obtained from human genetic data, which show correlations between loss or gain of function mutations in ion channel subunits and disease phenotypes, when selecting which targets to pursue. However, genetic association of a mutation with a disease does not prove causation of the disease (McGivern and Ding 2020).

In the sections below, we outline the key information that might be needed for identifying a disease-associated ion channel target. In most cases, the availability of this information is fragmented or incomplete and one has to make a judgement call as to whether the channel is indeed a viable target. To illustrate this, we pick two examples of different K channels and outline the existing evidence (and evidence gaps) that suggest they may be viable therapeutic targets.

3.1 Physiological and Pathophysiological Role of the Channel of Interest

A detailed knowledge of the physiological and pathophysiological roles of the potassium channel of interest is critical but this is often lacking detail. Fundamental questions such as "do we want to enhance or inhibit the current?" rely on an understanding of the physiological roles of particular ion channels and how this might change in pathophysiology. For example, there is good evidence that $K_V2.1$ channels are important in regulating the neuronal firing in motoneurons where they can either maintain or suppress repetitive firing depending on the existing activity of the neurons (Romer et al. 2019; see also Liu and Bean 2014). Loss of function of these channels can lead to decreased firing and neuro-developmental epileptic disorders (de Kovel et al. 2017). Some mutations of $K_V2.1$ channels cause a loss of ion selectivity for K^+ ions over Na^+ ions which will have complex effects on firing, depending on the degree of synaptic activity. Furthermore, gain of function mutations can, paradoxically, still lead to epilepsy phenotypes (Niday and

Tzingounis 2018). So, depending on both the underlying pathophysiology and the level of tonic firing activity (Romer et al. 2019) opposing therapeutic approaches might be desirable.

Perhaps an alternative way to consider this is to seek to restore optimum/normal levels of activity of the channel target. For example, there is a strong genetic linkage between loss of function mutations of TASK-1 K_{2P} channels and the development of pulmonary hypertension (Ma et al. 2013; Olschewski et al. 2017; Cunningham et al. 2019). More recent studies suggest that reduced TASK-1 channel activity is a more general condition seen not just in genetic pulmonary hypertension but also in the idiopathic condition (Antigny et al. 2016). As such, restoration of "normal" channel function may be a more widely applicable therapeutic intervention in this and other situations.

3.2 Knowledge of Distribution Channel mRNA and Protein Expression in the Appropriate Places in the Body

A fundamental component of the evidence base for suggesting a particular ion channel target is the knowledge that the channels are actually expressed (and functional) in the cells or organs of the body that are relevant. This is exemplified in the case studies below for TREK-2 and $K_V1.3$ channels as therapeutic targets, where there is good evidence that these channels are both expressed and functional in the cells that matter and, indeed, that expression and/or function is then altered in disease states.

3.3 Are There Species Differences? This May Be Important for Extrapolation from Preclinical Physiology and/or Pharmacology Studies

Since much preclinical research occurs on rodent models, it is extremely important to consider species differences between channel protein structures and whether this alters responsiveness to drugs. One example of this is where the respiratory stimulant, doxapram, was found to be a more selective inhibitor of TASK-1 than TASK-3 K2P channels in rodents (Cotten et al. 2006), but this channel specificity is lost for human TASK-1 and TASK-3 channels where doxapram is an equipotent inhibitor of both channel types (Cunningham et al. 2020).

3.4 Evidence from Diseases States of Channel Up- or Down-Regulation

Channelopathies provide important information about the roles of ion channels and have contributed to identification of novel ion channel targets in disease. Mutations in different ion channel genes can give rise to disease states such as episodic ataxias, epilepsy, diabetes, cardiac arrhythmias, Birk Barel mental retardation syndrome and cystic fibrosis (see Ashcroft 2006; Catterall et al. 2008, Veale et al. 2014b). Many of these channelopathies result from mutations in the coding regions, leading to a gain or loss of channel function or from mutations in the promoter regions leading to over- or under-expression of ion channels. Similarly, genetic ablation studies with knockout animals have confirmed the potential importance of ion channels in disease states but have also provided novel and surprising insights into new roles for specific ion channels in physiological processes, which might be targeted in the future (see Mathie 2010).

3.5 Case Study 1: TREK-2 Channel Activators for Pain

Over 1.5 billion people worldwide suffer from chronic pain. Existing first-line drugs for treatment of chronic pain, including cyclooxygenase inhibitors (non-steroidal anti-inflammatory agents, NSAIDs) and opioid receptor modulators, do not alleviate pain completely and, in certain situations including neuropathic pain, do not work well except at high doses (Mathie and Veale 2015). Furthermore, tolerance and addiction to opioids, particularly if used chronically, is a major current health issue.

Chronic pain conditions are often characterised by persistent over-excitability of peripheral nociceptors brought about by changes in ion channel organisation and/or activity. Increasing evidence points to an important contribution from a number of different potassium channels (see Du and Gamper 2013; Tsantoulas and McMahon 2014; Waxman and Zamponi 2014), including K_{2P} channels (Alloui et al. 2006; Woolf and Ma 2007; Noel et al. 2009; Mathie et al. 2010; Gada and Plant 2018), in pain processing. Among K_{2P} channels, the strongest body of evidence from expression and functional studies highlights the importance of TREK-1, TREK-2 (and also TRESK) channels (see Mathie and Veale 2015).

TREK channels are expressed in both human (mRNA) and rodent (mRNA/protein) small/medium sized dorsal root ganglion (DRG) neurons which relay painful stimuli to the CNS. In these neurons TREKs are often co-localised with excitatory TRPV1 channels (Maingret et al. 2000; Medhurst et al. 2001; Talley et al. 2001; Alloui et al. 2006; Dedman et al. 2008; Loucif et al. 2018). Recent immunohistochemistry studies, using selective commercially available antibodies, also show the expression of all three TREK channels (TREK-1, TREK-2 and TRAAK) in small sized rat DRG neurons (Viatchenko-Karpinski et al. 2018, Acosta et al. 2014). Given the high expression levels of these channels in nociceptive neurons, activation of these channels would be predicted to reduce nociceptor firing, thereby reducing pain.

Despite its broad CNS expression, studies of TREK-2 KO mice did not observe any major behavioural abnormalities (Mirkovic et al. 2012). In humans, initial TaqMan RT-PCR assays showed that there are very low levels of expression of TREK-2 (or TREK-1) channel mRNA in the heart (Medhurst et al. 2001), suggesting low risk of cardiac side effects in humans.

It has been suggested that TREK-2 contributes 69% (Kang and Kim 2006) and 59% (Acosta et al. 2014) of the resting potassium current in neonatal rat small DRGs and rat small $IB4^+$ DRG neurons, respectively. In contrast, TREK-1 and TRESK channels contributed just 12% and 16%, respectively (Kang and Kim 2006). TREK-2 channel activity therefore has a significant impact on the resting membrane potential – and so excitability – of these cells.

TREK-2 channels have been implicated in regulating somatosensory nociceptive neuron excitability. After CFA-induced inflammation, spontaneous pain, measured as spontaneous foot lifting, was increased by siRNA-induced TREK-2 knockdown in vivo (Acosta et al. 2014); furthermore, TREK-2 KO mice were more sensitive to von Frey filaments (Pereira et al. 2014).

There is a wealth of structural information from a variety of different potassium channel subunits, which has formed the basis of our understanding of how all ion channels function. Following the original publications of crystal structures of TWIK-1 and TRAAK (Brohawn et al. 2012; Miller and Long 2012), more recent work has identified a number of different structural forms of TRAAK and TREK-2, which purport to correspond to different open and closed conformations of the TREK family of K_{2P} channels (Brohawn et al. 2014; Lolicato et al. 2014; Dong et al. 2015).

It has been shown that the TREK family of channels are subject to a process known as alternative translation initiation (ATI). For TREK-2 there are three potential translational (ATG or methionine) sites in its N-terminus, thus allowing for the possible generation of three different TREK-2 proteins (Staudacher et al. 2011; Simkin et al. 2008). Furthermore, it has been shown that different isoforms of TREK channels generated by ATI can cause an alteration in drug sensitivity (Eckert et al. 2011; Kisselbach et al. 2014; Veale et al. 2014a).

As well as ATI, the TREK family has also been shown to be subject to alternative exon splicing. For TREK-2, the three isoforms identified do not differ hugely, in terms of their functional properties; however, they are differentially expressed in various tissues. TREK-2a is strongly expressed in the brain, pancreas and kidney, whilst TREK-2b is expressed in the proximal tubule of the kidney and in the pancreas and TREK-2c is expressed mainly in the brain (Gu et al. 2002). It is not known which subtype predominates in DRG neurons.

At least in expression systems, there is accumulating evidence to suggest that all three TREK channels (TREK-1, TREK-2 and TRAAK) can form functional heterodimeric channels with each other (Blin et al. 2016; Lengyel et al. 2016; Levitz et al. 2016) resulting in unique functional properties. Functional recordings of currents with properties suggestive of the formation of TREK-1/TREK-2 heteromeric channels were also demonstrated in native DRG neurons indicating that hetero-dimerisation of TREK channels may occur in native cells to provide greater diversity of leak potassium conductances (Lengyel et al. 2016). A recent study provides convincing evidence for TREK-2 and TRESK heterodimer channels in native trigeminal ganglion neurons (Lengyel et al. 2020).

There are, however, several gaps in the evidence in favour of TREK-2 channels as a therapeutic target for pain. Firstly, there is no direct genetic linkage at this stage between TREK-2 channels and pain. Therefore, it is not possible to select patients on the basis of genotype. It is also not practically possible to select patients based upon TREK-2 expression levels in neurons. Nevertheless, this is not a unique problem as, in general, the pain therapeutic field suffers from a lack of predictive/efficacy and patient stratification biomarkers.

Secondly, there has not been satisfactory pharmacological validation of TREK-2 activators in vivo due to the non-selectivity of tool compounds. Therefore, it is not clear whether TREK-2 activation alone will have sufficient analgesic effects or co-activation of TREK-1/TRAAK will also be required. Thirdly, the side effects of TREK-2 channel activation are not known and therefore, it is not clear whether selectivity for TREK-2 over TREK-1 will avoid drug related side effects.

3.6 Case Study 2: $K_V1.3$ Channel Blockers in Autoimmune Disorders

The voltage-gated potassium channel, $K_V1.3$ regulates the membrane potential of lymphocytes (DeCoursey et al. 1984; Wulff and Zhorov 2008; Chiang et al. 2017) which is critical for the activation of these immune cells (Veytia-Bucheli et al. 2018).

Several studies have confirmed that $K_V1.3$ is highly expressed in macrophages, microglia and T cells, suggesting that $K_V1.3$ plays a crucial role in immune and inflammatory responses to human diseases such as multiple sclerosis (MS), rheumatoid arthritis, type 1 diabetes mellitus and asthma (Toldi et al. 2010; Huang et al. 2017; Tanner et al. 2017; Zhou et al. 2018). In these conditions, the expression levels of $K_V1.3$ channels are significantly elevated (Rangaraju et al. 2009).

The pathogenesis of autoimmune diseases involves activation and proliferation of effector memory T cells (TEM cells) and persistence of autoantigens (Devarajan and Chen 2013). During the activation of TEM cells, the expression of the $K_V1.3$ channel was up-regulated significantly (Cahalan and Chandy 2009). Accumulated data for $K_V1.3$ showed higher expression levels in myelin-reactive T cells from the peripheral blood (PB) of MS patients compared to healthy controls (Wulff et al. 2003). Also, in an animal model of experimental autoimmune encephalomyelitis (EAE), it has been confirmed that expression of $K_V1.3$ is significantly elevated (Rus et al. 2005).

Taken together, these data suggest that $K_V1.3$ channels are an attractive therapeutic target for immunomodulation (Beeton et al. 2006; Wulff and Zhorov 2008). In patients with multiple sclerosis (MS), type 1 diabetes mellitus, rheumatoid arthritis, psoriasis, or chronic asthma, disease-associated T cells are $K_V1.3$-dependent TEM cells. Selective $K_V1.3$ inhibitors suppress the proliferation and cytokine production of these cells (Cahalan and Chandy 2009). There is also a growing body of evidence suggesting that $K_V1.3$ channel blockers have beneficial therapeutic effect on rheumatoid arthritis, autoimmune encephalitis and other autoimmune diseases (see Chang et al. 2018, Wang et al. 2020).

$K_V1.3$ channel blockers have been found to alleviate disease symptoms in animal autoimmune diseases, chronic inflammatory diseases and metabolic disease models without obvious side effects (Perez-Verdaguer et al. 2016). Importantly, positive results have been shown in preclinical trials (see below, also Prentis et al. 2018). Therefore, $K_V1.3$ channel blockers have the potential to be developed as effective drugs for the treatment of MS and EAE (Rangaraju et al. 2009).

Toxin blockers of $K_V1.3$ and other potassium channels have been found in the venom of numerous animals, including the venom of sea anemone (Wulff et al. 2019). In 1995, an effective blocker was extracted from the Caribbean sea anemone (*Stichodactyla helianthus*) by Castaneda et al. (1995) subsequently named ShK toxin (Wulff and Zhorov 2008). ShK and subsequent derivatives of it (see below and Lanigan et al. 2001) reduce the inflammatory response of autoimmune diseases by maintaining the integrity of the blood brain barrier (BBB) (Huang et al. 2017), reducing activation of TEM cells (Beeton et al. 2006) and eliminating respiratory bursts in activated microglia and subsequent secondary damage of neurons by microglia (Fordyce et al. 2005).

The structures of several ShK and ShK-like toxins have been determined, all of which have been reported to modulate ion channels (Chhabra et al. 2014). Like many K channel blocking toxins, they are polypeptides of between 34 and 75 amino acids, characterised by six cysteine motifs (forming 3 disulphide bonds) and a functional lysine residue which enters and blocks the pore of the channels at the selectivity filter. The novel analog ShK-186 (dalazatide) has a 100-fold improvement of selectivity for $K_V1.3$ over $K_V1.1$ and 1,000-fold over $K_V1.4$ as well as $K_V1.6$ (Pennington et al. 2009) compared to the natural ShK toxin. ShK-186 and its analogs had strong therapeutic actions in animal models of human autoimmune diseases including MS and rheumatoid arthritis (Beeton et al. 2001). Unexpectedly, ShK-186 was found to have a long half-life if given by sub-cutaneous injection leading to a sustained concentration, at pM levels, in plasma. This, in turn, gave a prolonged therapeutic efficacy (Tarcha et al. 2012). ShK-186 completed Phase 1a and 1b trials in healthy volunteers in 2016 and was believed to enter phase 2 trials in 2018/2019 (Wang et al. 2020).

Despite these encouraging data, there are still challenges for ShK to be used to the treatment of neuroinflammatory diseases. The first problem underlying the application of these peptides is that they cannot be taken orally, mainly because they have difficulty penetrating the intestinal mucosa. Furthermore, due to the polypeptide's molecular size, hydrophobicity and polarity, the cell membrane penetration of ShK is poor. The second obstacle is that ShK cannot cross the BBB in most neuroinflammatory diseases where this barrier is not compromised by the condition (Li et al. 2015). Thus, there is scope for the development of novel, pharmacologically active agents, either polypeptides or small molecules, to selectively target and block $K_V1.3$ channels in humans.

4 Techniques to Study the Pharmacology of K Channels

Ion channel drug discovery traditionally involves the use of model assay systems, often utilising a recombinant channel expressed in cell lines (see Mathie et al. 2021b). These assays support the identification of active compounds and the characterisation and optimisation of these lead series. Promising lead candidate molecules can then be tested in models of disease before entering clinical trials (see McGivern and Ding 2020; Walsh 2020). The preferred ion channel assay is often a cell line that transiently or stably expresses the appropriate combination of channel α- and auxiliary subunits (if known) and that yields relevant functional and pharmacological responses on (semi) high throughput assay platforms. There is an underlying assumption that the channel of interest expressed in an assay cell line should perform consistently over time, when tested using different methodological approaches and when used in different laboratories (see CiPA study below).

The gold standard approach for studying ion channels remains the manual patch-clamp method. This is, however, an extremely low-throughput technique where single cells are studied one at a time but, with appropriate expertise, this technique allows the collection of both high-quality data and potential mechanistic insight, unsurpassed by other methods. Used in isolation, this technique is unlikely to generate sufficient novel drug candidates to satisfy the demands of either the pharmaceutical industry or therapeutic need (McGivern and Ding 2020).

Accordingly, a range of different assay technologies have been developed and refined to address, primarily, the issue of high throughput. An excellent review of the range of available technologies, from the perspective of K_{ir} channels has recently been published by Walsh (2020, see also McGivern and Ding 2020) and has helped shape our thinking in this section of this review. Walsh (2020) analyses the pros and cons of a number of assay technologies that have been used for screening K_{ir} channels, including membrane potential-based fluorescent dye assays, using either fast-response fluorescence resonance energy transfer (FRET) probes or slow-response probes, thallium-sensitive fluorescent assays, radiometric and nonradiometric rubidium (Rb^+) flux assays, automated patch-clamp assays, a cell-free ion flux assay that uses K_{ir} channel-containing liposomes, and a K^+-dependent yeast growth assay.

In the sections below, we describe some of the key technological approaches that can be utilised to advance discovery of new therapeutic agents targeting K channels. Although these technologies have been used successfully for screening compound libraries against different K channels, each technology has its own limitations in parameters such as sensitivity, specificity, cost and physiological relevance (Walsh 2020). Therefore, there is a need for continued development of ion channel screening technologies.

4.1 Improved Structural Information for Ion Channels

High resolution elucidation of the structure of ion channels using X-ray crystallography was led by the outstanding work on bacterial, and later mammalian, potassium channels by MacKinnon and colleagues (e.g. Doyle et al. 1998; Jiang et al. 2002, 2003). These structures have led to rapid advances in our understanding of ion channel function and the identification of regions of channels that are important for determining channel function and drug binding. Knowing regions of the channel important for drug binding, or pharmacophores, it is then possible to use these pharmacophores as templates for the design of novel therapeutic agents (Mathie et al. 2021a).

The fact that ion channels are present in lipid membranes coupled with the dynamic nature of ion channel proteins has made resolution of structures difficult. Consequently still relatively few ion channels have been resolved using crystallography-based techniques. However, recent advancements in the field of Cryo-EM have led to significant improvements in the observed resolution and a substantial increase in the number of published structures. The recent development of direct electron detector cameras alongside improvements in computational processing has allowed greater resolution of structures, down to the atomic level (see Nakane et al. 2020). For example, many tens of Cryo-EM structures for the transient receptor potential (TRP) channel family have been published in the last few years (Renaud et al. 2018). The recently published human β3 $GABA_A$ receptor structure (Nakane et al. 2020) is (at 1.7 Å), the highest resolution membrane-spanning protein structure obtained to date from cryo-EM studies, which is significantly better than the accepted resolution of at least 2.5 Å required to model drug–protein interactions (Renaud et al. 2018). Improved structural images from cryo-EM techniques will undoubtedly lead to significant advances in the development of selective ion channel blockers and activators and, hopefully, an increase in useful drugs targeting ion channels. Readers are also directed to (Jiang 2021) for discussion of recent progress in structural biology of K^+ channels.

4.2 Flux Assays

Ion channel high throughput screening programs have mainly used either radiometric Rb^+ assays (e.g. Loucif et al. 2018) or fluorescence-based assays of ion flux in stable cell lines (using, for example, Tl^+, Ca^{2+} or Na^+ dyes) to measure permeation of the channel.

4.2.1 Thallium (Tl^+) Flux Assay

Perhaps the most common of these flux-assay approaches, at present, is the thallium (Tl^+) flux assay which takes advantage of the high permeability of K channels to T^{l+} ions in order to measure Tl^+ flux as a surrogate of ion channel activity. The Tl^+-sensitive, fluorescent-based assay for multi-well plate screening of K channels was introduced by Weaver et al. (2004). In this high throughput assay, cells are loaded

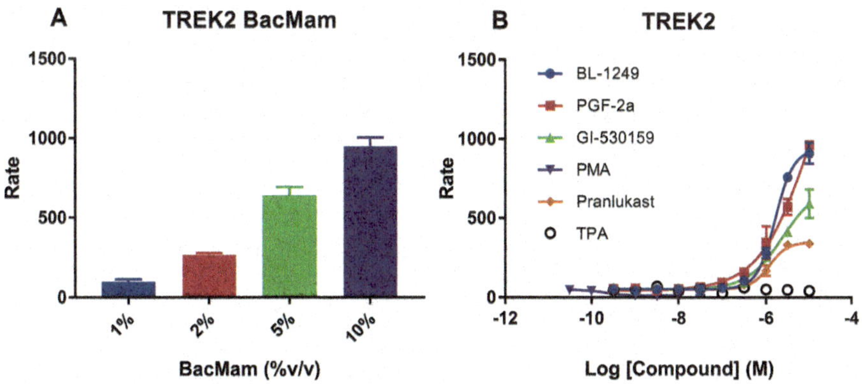

Fig. 2 Thallium flux assays of TREK-2 channel activators using "BacMam" baculovirus transfection. (**a**) Titration of BacMam. As BacMam levels increase (%v/v present in media), there is an increase in the rate of fluorescence change between 13 and 19 s after the addition of thallium, error bars denote S.E.M. This is proportional to increased TREK-2 activity in the assay. (**b**) Pharmacology of TREK-2 activators. Exemplar data showing activity of BL-1249, 11-deoxy prostaglandin F2a (PGF-2a), GI-530159, PMA, Pranlukast and TPA (adapted from Wright PD, McCoull D, Walsh Y, Large JM, Hadrys BW, Gauritcikaite E et al. Pranlukast Is a novel small molecule activator of the two-pore domain potassium channel TREK2. Biochem Biophys Res Commun. 2019; 520: 35–40 with permission)

with a Tl^+-sensitive, membrane-permeant reporter dye. K channel modulators can be rapidly screened by monitoring changes in the Tl^+-induced fluorescent signal.

In our own research, in collaboration with colleagues at LifeArc (Wright et al. 2017, 2019), cells transiently expressing K channels of interest were generated by using U-2 OS cells and a BacMam baculovirus transfection method. Generation of cell lines stably expressing ion channels can be challenging for a number of reasons including inherent toxicity. Moreover, the ability to identify channel activators can be compromised by systems in which the target is over-expressed at high levels due, for example, to saturation of the dye. Conversely, ion channel inhibitors can be hard to detect when expression, and therefore signal, is low. To minimise these issues BacMam can be used to deliver ion channels into mammalian cells. BacMam confers a number of advantages, including safety and reduced time compared to generating stable cell lines but, primarily, it allows the precise titratable expression of the gene of interest (Wright et al. 2013, 2017, 2019, see Fig. 2). In our work with K_{2P} channels this enabled us to generate cell systems in which we were able to intricately and robustly select a level of K channel expression, functionally optimised for the identification of channel activators. Another advantage of the BacMam system is that it allows screening against parental cells in the absence of the heterologous expressed ion channel. The use of a high throughput, fluorescence-based assay also allows the possibility of a "target class" approach to drug discovery (Barnash et al. 2017; McCoull et al. 2021). This approach allows the simultaneous prosecution of multiple targets from a target family. Knowledge accumulated from one program can be leveraged across multiple potential therapeutic targets (McCoull et al. 2021).

One major limitation to the Tl^+ assay is that some cells contain endogenous Tl^+ transport pathways that can interfere with the K channel efflux under study and cause a higher than acceptable rate of "false-positive" hits. Furthermore, ion flux fluorescent assays often require expensive instrumentation such as FLIPR or FDSS workstations that combine sophisticated liquid handling microfluidics with fluorescent imaging. In addition, biosafety and disposal of a toxic heavy metal such as Tl^+ is an issue.

A further technical drawback of plate-based, fluorescence assays is that the activating or inhibiting agent, once added, cannot be removed easily, so it is difficult to study the reversibility of the compound's effects on channel activity. Furthermore, potencies of ion channel modulators from fluorescence assays do not always correlate with manual or automated patch-clamp electrophysiological measurements.

4.2.2 Liposome Flux Assay

The liposome flux assay (LFA) described by Su et al. (2016) is a cell-free assay which provides a putative alternative high throughput screening system for K channel drug discovery. In this procedure, K channels are purified and reconstituted into lipid vesicles. A concentration gradient for K^+ efflux is established by adding the vesicles into a NaCl solution. K^+ efflux is initiated by the addition of a proton ionophore, such as carbonyl cyanide m-chlorophenylhydrazone (CCCP), which allows the influx of protons to balance the efflux of K^+ (Walsh 2020). The efflux of K^+ is quantified indirectly by monitoring the proton-dependent quenching of a fluorescent dye. In the original description of LFA, a library of 100,000 compounds was screened to identify both inhibitors and activators of GIRK2 ($K_{ir}3.2$) channels, hERG channels, TRAAK K_{2P} channels and K_{Ca} channels (Su et al. 2016). LFA potentially provides an efficient and low-cost method for K channel screening. However, the technique is not yet widely used and drug potencies obtained using LFA will need to be compared with values obtained using more established ion flux and automated patch-clamp systems before LFA becomes a widely used K channel high throughput method (Walsh 2020).

4.3 Electrophysiological Approaches

Until quite recently, the study and development of new drugs that act on ion channels has been restricted by the lack of high throughput screens that measure current directly passing through the ion channels using electrophysiological approaches. Manual patch-clamp recording techniques (Hamill et al. 1981) allow exquisite resolution of electrical activity but are restricted by the number of recordings that can be made in a given time and the level of technical expertise required to achieve these recordings. They do, however, remain critical for exploring the complex biophysical properties of ion channels and, in detail, the mechanisms of how lead compounds might work. For example, information arising from structural studies can be harnessed to explore the site and mechanism of drug action through site-directed mutagenesis of the ion channel under investigation.

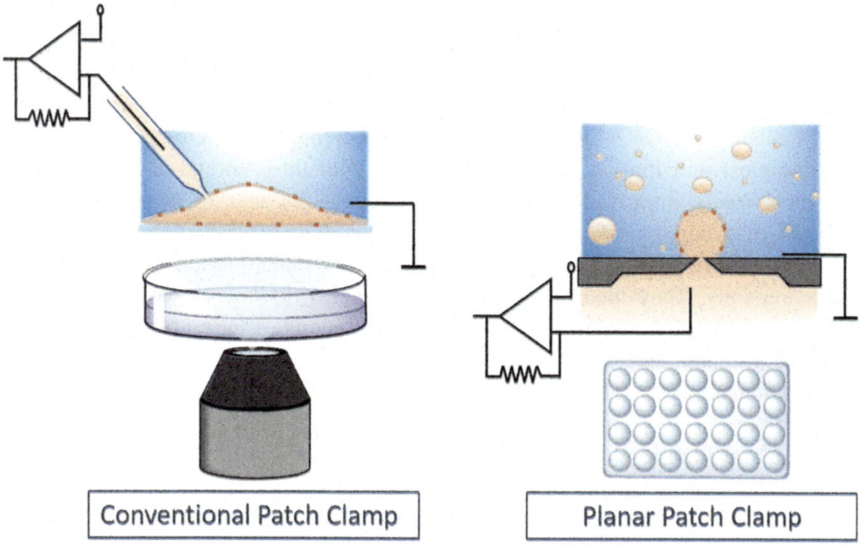

Fig. 3 Schematic representation of conventional patch clamp recording versus high throughput automated patch recording. Adapted and redrawn from an original schematic by Nanion Technologies: https://www.nanion.de/en/technology/technology-explained.html

Now, however, in large part driven by the need for pharmaceutical companies to screen compounds against hERG channels (see below), higher throughput screening technologies have been introduced and continue to be refined and optimised. Automated patch-clamp increases throughput and ease of use compared to the conventional patch-clamp technique making it accessible to a wider audience. The automation of the patch-clamp method was made possible with the development of chip-based planar patch-clamp technology (Fig. 3). Multi-well planar arrays give higher throughput screens by allowing multiple recordings in parallel. Most automated patch-clamp (APC) systems utilise a planar array which contains micron-sized apertures in silicon-based chips. Cells are added in suspension to a multi-well recording plate and, in a similar fashion to forming a seal using a microelectrode in manual parch-clamp recordings, negative pressure is applied to attract cells to the apertures. Further application of negative pressure causes the patch of membrane immediately beneath the aperture to rupture, thus establishing the whole-cell configuration. Alternatively, electrical access to the cell can be obtained using a pore-forming antibiotic such as nystatin or amphotericin B.

A range of different systems now exist, and an excellent analyses of the advantages and disadvantages of the evolving versions of each system have been provided by Dunlop et al. (2008) and, more recently, by Bell and Dallas (2018). Each system allows significant up-scaling of activity when compared with conventional patch-clamp electrophysiological methods or two-electrode voltage clamp from oocytes, increasing output from single digit to hundreds or even thousands of drug data points per day (Bell and Dallas 2018).

In the last few years, and in particular following the development of high-quality miniaturised amplifiers, automated, 384-well plate-based electrophysiology platforms have emerged, ranging from the perforated-patch-based IonWorks Barracuda (Molecular Devices, Sunnyvale, CA) to the latest generation, giga-seal-based instruments such as Qube (Sophion Bioscience), SyncroPatch (Nanion Technologies) and Ionflux (Fluxion). Coupled with this, storage and analysis of data have vastly improved, enabling large amounts of data to be recorded and analysed in a comparatively short space of time (Obergrussberger et al. 2020). These recent whole-cell patch-clamp platforms offer much higher throughput than earlier versions. They incorporate microfluidic networks and temperature control and are capable of recording currents from populations of individual cells, which increases the likelihood of obtaining useful data from each well of an assay plate. This latter feature also allows K^+ current averaging to compensate for cell-to-cell variations in current amplitudes and kinetics. They have been used in screening campaigns for libraries approaching 200,000 compounds.

Automated electrophysiology now provides high-quality data-rich information for driving structure–activity relationship (SAR) and an ability to explore mechanism of action early in screening. Initially developed and used for recombinant ion channels, either stably or transiently transfected into cell lines such as CHO cells and HEK293 cells, the use of these techniques is now being expanded into mammalian cells such as neurons or smooth muscle cells, either freshly isolated or maintained in primary culture (Milligan et al. 2009). In the near future, there is interest in applying high throughput, automated patch-clamp to native cells and human induced pluripotent stem cells (see below). The development of dynamic clamp on an APC instrument, where "currents" of choice may be introduced electronically (Goversen et al. 2018), may prove to be a major breakthrough in the use of this technology to study more physiologically relevant channel properties and regulation (Obergrussberger et al. 2020).

There are still areas for consideration. For example, for the most part, stably transfected cell lines work best in these systems and there can be issues with expression levels and cost associated with this. Furthermore, the sheer volume of data generated requires consideration when deciding how best to analyse and display this. While many APC systems now provide high-quality voltage clamp data that approach traditional whole-cell patch-clamp currents, some instruments have sacrificed data quality in exchange for higher screening throughput (Obergrussberger et al. 2018). In addition, as in the case with fluorescent plate readers (above), the high initial start-up price and consumable costs involved in using APC systems often limit their application to large academic and pharmaceutical research facilities. As such, there remains considerable value in the continued development of lower-throughput, higher fidelity automated systems.

Impressively, the major commercial suppliers of these devices (Sophion, Nanion, Fluxion) have developed advanced programmes to reach out to academic and pharmaceutical researchers in order to develop and expand the range of devices and resources available to make them appropriate to as many needs as possible. The

CiPA initiative described below exemplifies the potential value of such open and constructive collaboration.

5 Quantification and Standardisation of Drug Action on K Channels

For pharmacologists, there is a desire to quantify the effects of agents which regulate receptor or ion channel activity in order to fully define their selectivity and their usefulness as tool compounds or therapeutic agents (Mathie et al. 2021a). It is also critical that measurements can be obtained that are as consistent as possible between different platforms and different research groups. In this regard, there are considerations around quantification and characterisation of pharmacologically active compounds on potassium channels which might enhance our understanding of how these channels function and are regulated and enable us to obtain more consistent and reliable measurements across studies.

Practically, for all of the techniques described above, there are issues around resting K channel activity. The activity and regulation of K channels can vary depending on the recording method used (flux assay versus patch-clamp electrophysiology; mammalian cell versus oocyte expression system). It may also vary between different channel isoforms (generated through splice variants or alternative translation initiation) or between wild type and mutated channels or channels from different species (see above, also Mathie et al. 2021a). It is important, therefore, that these variables are documented carefully between studies and varied as little as possible from one study to another.

When it comes to quantification of effect, again as much standardisation as possible is desirable. For example, for K channel activating compounds, should efficacy (measured as the percentage increase or absolute maximum increase in current evoked) or potency (measured as a calculated EC_{50} value) or both, be used as a quantitative measure of drug activity? If different measurements are used by different groups, comparisons are difficult. One possible solution, regardless of how the measurement is obtained or quantified, is to compare the effects of novel compounds with a known, well-characterised "standard". For example, in our studies of TREK-2 channel activators (see above), we routinely use BL-1249 as a standard activator of these channels and express the effects of other compounds both in absolute terms and in comparison with the effects of a defined concentration of BL-1249. However, this may not be so appropriate for compounds with a different site of action where the degree of tonic channel activity may influence the activity of a compound acting at one site more than a compound acting at a different site on the channel.

The need for consistent and robust quantification of the effects of drugs on ion channels has, perhaps paradoxically, been addressed best by the need to minimise the risks caused by drugs modifying the activity of ion channels in the heart.

5.1 hERG Channels and the CiPA Initiative

Inhibition of hERG ($K_V11.1$) potassium channels by drugs can lead to a concentration-dependent prolongation of the QT interval, a condition described as long QT syndrome and associated with an increased risk of cardiac proarrhythmia (Sanguinetti and Tristani-Firouzi 2006). A number of drugs have been withdrawn from use or had their indications severely limited in many countries because of their propensity to inhibit hERG channels. These include astemizole (an antihistamine), terfenadine (also an antihistamine) and cisapride (which stimulates gastrointestinal motility) (Mathie 2010). hERG channels are also inhibited by a wide range of commonly used drugs such as chlorpromazine, imipramine and amitriptyline (Redfern et al. 2003). As such, pharmaceutical companies have been compelled to introduce preclinical testing of all new chemical entities for hERG-channel blocking activity during preclinical trials (Redfern et al. 2003, but see Kramer et al. 2013). This includes the incorporation of higher throughput screens for ion channels.

More recently, based on a more comprehensive understanding of cardiac electrophysiology and cellular mechanisms underlying the drug-induced abnormal heartbeat, torsade de pointes (TdP), the Comprehensive in vitro Proarrhythmia Assay (CiPA) initiative was established with the aim of integrating multi-ion channel pharmacology measured in vitro into in silico models to assess TdP risk (Sager et al. 2014; Brinkwirth et al. 2020; Kramer et al. 2020; Ridder et al. 2020). If successful, this programme may facilitate the use of nonclinical data as part of an integrated risk assessment strategy to inform clinical decision making, improve accuracy and reduce costs in predicting cardiac liability of new drug candidates. The programme has three primary strands. The first is to characterise and quantify, in vitro, the effects of drugs on multiple cardiac ion channels, the second to accurately reconstruct the cardiac action potential in silico. The third strand is to confirm predictions made using the in silico model using human stem cell derived cardiac myocytes (Su et al. 2021).

As a result, this initiative has brought together a consortium of commercial and academic laboratories, including several CROs, to extend and integrate the use of electrophysiology-based cardiac ion channel screening on six primary cardiac ion channels, including four potassium channels: hERG ($K_V11.1$), KCNQ1 ($K_V7.1$) + KCNE1, $K_{ir}2.1$ and $K_V4.3$ (see Fig. 4).

An important component of this initiative was an attempt to quantify cross-site and cross-platform variability in the study of hERG ($K_V11.1$) channels (Kramer et al. 2020). It is clear that despite best attempts to minimise variability, differences in experimental protocols, instruments and procedures introduce variability and affect IC_{50} values characterising the effects of drugs. The study utilised 12 centrally-supplied, blinded drugs and tested them against hERG and two other non-K channels involved in the cardiac action potential at multiple (17) sites using five APC platforms but using, as far as possible, a standard operating protocol. Whilst many of the results were similar across sites and platforms, there were some notable differences. For example, there was a 10.4-fold variance in IC_{50} for dofetilide block of hERG channels (12–103 nM) across 14 sites, four platforms. This is a

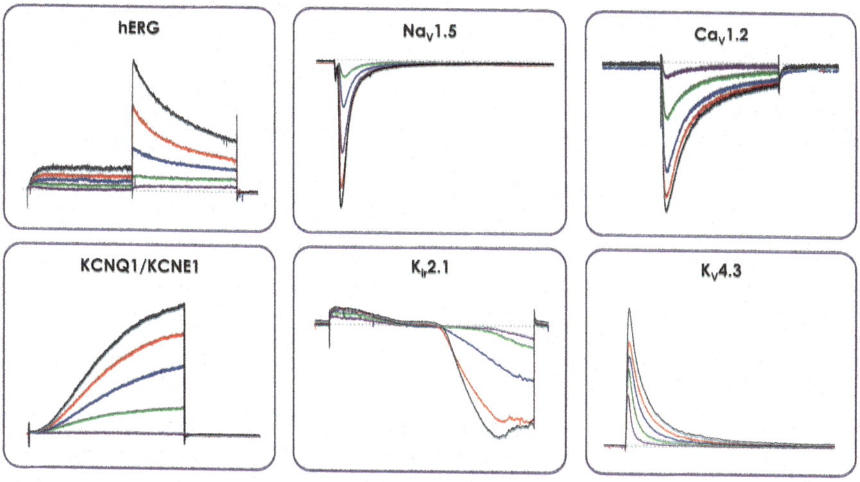

Fig. 4 Ion channel currents investigated in the CiPA protocol. Four different potassium channel currents are studied as part of this protocol: hERG (K_V11.1), KCNQ1 (K_V7.1) co-expressed with KCNE1, K_{ir}2.1 and K_V4.3. Other channels that form part of the study are the voltage-gated sodium channel, Na_V1.5 and the voltage-gated calcium channel, Ca_V1.2. Figure kindly provided by Dr. John Ridley, Dr. Andrew Southan, Dr. Robert Kirby, Dr. Marc Rogers and their colleagues at Metrion Biosciences

potential issue because the whole tenet of the CiPA programme is to develop in silico modelling based on reliable and reproducible IC_{50} values from APCs.

There are a number of potential explanations for variability in the data. The kinetics of block are often important when considering hERG blockers. This depends on the protocols chosen which may have shown subtle variability across platforms, despite the fact that certain quality control parameters (Rm, leak current, baseline current) were applied across all platforms and sites and a standard fitting process used for all concentration response relationships. Other suggested sources of inconsistency include equilibration times, variability in drug forms and concentrations, expressed versus background currents, intracellular buffer and temperature. Temperature, in particular, may be a wider problem for the CiPA programme as experiments are primarily done at room temperature. For example, Lo and Kuo (2019) have described the temperature dependence of amiodarone block of hERG channels and found it to be much more voltage sensitive at 37°C compared to 22°C. This may be an important consideration when extrapolating the CiPA in vitro data for drugs on ion channels collected at room temperature compared to the situation in patients at body temperature.

Despite these sources of variability, the CiPA project represents by far the best attempt to date to obtain reliable and consistent quantitative data of the effects of drugs on ion channels. The project consistently measures and collates these differences and makes strenuous efforts to eliminate or minimise them or at least take them into consideration when evaluating drug action. As such, in our view, the CiPA project, despite having the primary aim of minimising the adverse effects of drugs, sets an outstanding example for future research models aimed at developing novel therapeutic drugs acting on ion channels.

Acknowledgements AM and ELV's research in this area has been supported by grants from the BBSRC and the Royal Society. AM and ELV are currently supported by a LifeArc Centre for Therapeutics Discovery Award. YW is supported by the Leverhulme Trust. Special thanks to Dr. John Ridley, Dr. Andrew Southan, Dr. Robert Kirby, Dr. Marc Rogers and their colleagues at Metrion Biosciences for providing Fig. 4 and for helpful, constructive comments on the manuscript.

References

Abriel H, Zaklyazminskaya EV (2013) Cardiac channelopathies: genetic and molecular mechanisms. Gene 517:1–11

Acosta C, Djouhri L, Watkins R, Berry C, Bromage K, Lawson SN (2014) TREK2 expressed selectively in IB4-binding C-fiber nociceptors hyperpolarizes their membrane potentials and limits spontaneous pain. J Neurosci 34:1494–1509

Adelman JP, Clapham DE, Hibino H, Inanobe A, Jan LY, Karschin A et al (2019) Inwardly rectifying potassium channels (version 2019.4) in the IUPHAR/BPS guide to pharmacology database. IUPHAR/BPS Guide Pharmacol CITE 2019(4). https://doi.org/10.2218/gtopdb/F74/2019.4

Aldrich R, Chandy KG, Grissmer S, Gutman GA, Kaczmarek LK, Wei AD, Wulff H (2019) Calcium- and sodium-activated potassium channels (version 2019.4) in the IUPHAR/BPS guide to pharmacology database. IUPHAR/BPS Guide Pharmacol CITE 2019(4). https://doi.org/10.2218/gtopdb/F69/2019.4

Alexander SPH, Mathie A, Peters JA, Veale EL, Striessnig J, Kelly E et al (2019) The concise guide to pharmacology 2019/20: ion channels. Br J Pharmacol 176(S1):S142–S228

Alloui A, Zimmermann K, Mamet J, Duprat F, Noel J, Chemin YN (2006) TREK-1, a K+ channel involved in polymodal pain perception. EMBO J 25:2368–2376

Antigny F, Hautefort A, Meloche J, Belacel-Ouari M, Manoury B, Rucker-Martin C et al (2016) Potassium channel subfamily K member 3 (KCNK3) contributes to the development of pulmonary arterial hypertension. Circulation 133:1371–1385

Ashcroft FM (2006) From molecule to malady. Nature 440:440–447

Attali B, Chandy KG, Giese MH, Grissmer S, Gutman GA, Jan LY et al (2019) Voltage-gated potassium channels (version 2019.4) in the IUPHAR/BPS guide to pharmacology database. IUPHAR/BPS Guide Pharmacol CITE 2019(4). https://doi.org/10.2218/gtopdb/F81/2019.4

Barnash KD, James LI, Frye SV (2017) Target class drug discovery. Nat Chem Biol 13:1053–1056

Bautista DM, Sigal YM, Milstein AD, Garrison JL, Zorn JA, Tsuruda PR et al (2008) Pungent agents from Szechuan peppers excite sensory neurons by inhibiting two-pore potassium channels. Nat Neurosci 11:772–779

Bayliss DA, Czirják G, Enyedi P, Goldstein SA, Lesage F, Minor DL Jr et al (2019) Two P domain potassium channels (version 2019.4) in the IUPHAR/BPS guide to pharmacology database. IUPHAR/BPS Guide Pharmacol CITE 2019(4). https://doi.org/10.2218/gtopdb/F79/2019.4

Beeton C, Wulff H, Barbaria J, Clot-Faybesse O, Pennington M, Bernard D et al (2001) Selective blockade of T lymphocyte K(+) channels ameliorates experimental autoimmune encephalomyelitis, a model for multiple sclerosis. Proc Natl Acad Sci U S A 98:13942–13947

Beeton C, Wulff H, Standifer NE, Azam P, Mullen KM, Pennington MW et al (2006) Kv1.3 channels are a therapeutic target for T cell-mediated autoimmune diseases. Proc Natl Acad Sci U S A 103:17414–17419

Bell DC, Dallas ML (2018) Using automated patch clamp electrophysiology platforms in pain-related ion channel research: insights from industry and academia. Br J Pharmacol 175:2312–2321

Bittner S, Meuth SG, Göbel K, Melzer N, Herrmann AM, Ole J et al (2009) TASK1 modulates inflammation and neurodegeneration in autoimmune inflammation of the central nervous system. Brain 132:2501–2516

Blin S, Soussia BI, Kim EJ, Brau F, Kang D, Lesage F et al (2016) Mixing and matching TREK/TRAAK subunits generate heterodimeric K2P channels with unique properties. Proc Natl Acad Sci U S A 113:4200–4205

Brinkwirth N, Takasuna K, Doi M, Becker N, Obergrussberger A, Friis S et al (2020) Reliable identification of cardiac liability in drug discovery using automated patch clamp: benchmarking best practices and calibration standards for improved proarrhythmic assessment. J Pharmacol Toxicol Methods 105:106884. https://doi.org/10.1016/j.vascn.2020.106884

Brohawn SG, del Mármol J, MacKinnon R (2012) Crystal structure of the human K2P TRAAK, a lipid- and mechano-sensitive K+ ion channel. Science 335:436–441

Brohawn SG, Campbell EB, MacKinnon R (2014) Physical mechanism for gating and mechanosensitivity of the human TRAAK K+ channel. Nature 516:126–130

Brohawn SG, Wang W, Handler A, Campbell EB, Schwarz JR, MacKinnon R (2019) The mechanosensitive ion channel TRAAK is localized to the mammalian node of Ranvier. Elife 8:e50403

Buono RJ, Lohoff FW, Sander T, Sperling MR, O'Connor MJ, Dlugos DJ et al (2004) Association between variation in the human KCNJ10 potassium ion channel gene and seizure susceptibility. Epilepsy 58:175–183

Cahalan MD, Chandy KG (2009) The functional network of ion channels in T lymphocytes. Immunol Rev 231:59–87

Castaneda O, Sotolongo V, Amor AM, Stocklin R, Anderson AJ, Harvey AL et al (1995) Characterization of a potassium channel toxin from the Caribbean Sea anemone *Stichodactyla helianthus*. Toxicon 33:603–613

Catterall WA, Sulayman D-H, Meisler MH, Pietrobon D (2008) Inherited neuronal ion channelopathies: new windows on complex neurological diseases. J Neurosci 28:11768–11777

Chang SC, Bajaj S, Chandy G (2018) ShK toxin: history, structure and therapeutic applications for autoimmune diseases. WikiJournal Sci 1:3. https://doi.org/10.15347/wjs/003

Chhabra S, Chang SC, Nguyen HM, Huq R, Tanner MR, Londono LM et al (2014) Kv1.3 channel-blocking immunomodulatory peptides from parasitic worms: implications for autoimmune diseases. FASEB J 28:3952–3964

Chiang EY, Li T, Jeet S, Peng I, Zhang J, Lee WP et al (2017) Potassium channels Kv1.3 and KCa3.1 cooperatively and compensatorily regulate antigen specific memory T cell functions. Nat Commun 8:14644. https://doi.org/10.1038/ncomms14644

Cotten JF, Keshavaprasad B, Laster MJ, Eger EI 2nd, Yost CS (2006) The ventilatory stimulant doxapram inhibits TASK tandem pore (K2P) potassium channel function but does not affect minimum alveolar anesthetic concentration. Anesth Analg 102:779–785

Cunningham KP, Holden RG, Escribano-Subias PM, Cogolludo A, Veale EL, Mathie A (2019) Characterization and regulation of wild-type and mutant TASK-1 two pore domain potassium channels indicated in pulmonary arterial hypertension. J Physiol 597:1087–1101

Cunningham KP, MacIntyre DE, Mathie A, Veale EL (2020) Effects of the ventilatory stimulant, doxapram on human TASK-3 (KCNK9, K2P9.1) channels and TASK-1 (KCNK3, K2P3.1) channels. Acta Physiol 228(2):e13361

de Kovel CGF, Syrbe S, Brilstra EH, Verbeek N, Kerr B, Dubbs H (2017) Neurodevelopmental disorders caused by de novo variants in KCNB1 genotypes and phenotypes. JAMA Neurol 74:1228–1236

Decher N, Ortiz-Bonnin B, Friedrich C, Schewe M, Kiper AK, Rinné S et al (2017) Sodium permeable and "hypersensitive" TREK-1 channels cause ventricular tachycardia. EMBO Mol Med 9:403–414

DeCoursey TE, Chandy KG, Gupta S, Cahalan MD (1984) Voltage-gated K+ channels in human T lymphocytes: a role in mitogenesis? Nature 307:465–468. https://doi.org/10.1038/307465a0

Dedman A, Sharif-Naeini R, Folgering JH, Duprat F, Patel A, Honoré E (2008) The mechano-gated K(2P) channel TREK-1. Eur Biophys J 38:293–303

Devarajan P, Chen Z (2013) Autoimmune effector memory T cells: the bad and the good. Immunol Res 57:12–22

Devilliers M, Busserolles J, Lolignier S, Deval E, Pereira V, Alloui A et al (2013) Activation of TREK-1 by morphine results in analgesia without adverse side effects. Nat Commun 4:2941

Dong YY, Pike AC, Mackenzie A, McClenaghan C, Aryal P, Dong L et al (2015) K2P channel gating mechanisms revealed by structures of TREK-2 and a complex with Prozac. Science 347:1256–1259

Doyle DA, Cabral JM, Pfuetzner RA, Kuo A, Gulbis JM, Cohen SL et al (1998) The structure of the potassium channel: molecular basis of K+ conduction and selectivity. Science 280:69–77

Du X, Gamper N (2013) Potassium channels in peripheral pain pathways: expression, function and therapeutic potential. Curr Neuropharmacol 11:621–640

Dunlop J, Bowlby M, Peri R, Vasilyev D, Arias R (2008) High-throughput electrophysiology: an emerging paradigm for ion-channel screening and physiology. Nat Rev Drug Discov 7:358–368

Eckert M, Egenberger B, Doring F, Wischmeyer E (2011) TREK-1 isoforms generated by alternative translation initiation display different susceptibility to the antidepressant fluoxetine. Neuropharmacology 61:918–923

Ferraro TN, Golden GT, Smith GG, Martin JF, Lohoff FW, Gieringer TA et al (2004) Fine mapping of a seizure susceptibility locus on mouse chromosome 1: nomination of Kcnj10 as a causative gene. Mamm Genome 15:239–251

Fordyce CB, Jagasia R, Zhu X, Schlichter LC (2005) Microglia Kv1.3 channels contribute to their ability to kill neurons. J Neurosci 25:7139–7149

Gada K, Plant LD (2018) Two-pore domain potassium channels: emerging targets for novel analgesic drugs: IUPHAR review 26. Br J Pharmacol 176:256–266

Gloyn AL, Pearson ER, Antcliff JF, Proks P, Bruining GJ, Slingerland AS (2004) Activating mutations in the gene encoding the ATP-sensitive potassium-channel subunit $K_{ir}6.2$ and permanent neonatal diabetes. N Engl J Med 350:1838–1849

González C, Baez-Nieto D, Valencia I, Oyarzún I, Rojas P, Naranjo D et al (2012) K(+) channels: function-structural overview. Compr Physiol 2:2087–2149

Goversen B, Becker N, Stoelzle-Feix S, Obergrussberger A, Vos MA, van Veen TAB et al (2018) A hybrid model for safety pharmacology on an automated patch clamp platform: using dynamic clamp to join iPSC-derived cardiomyocytes and simulations of Ik1 ion channels in real-time. Front Physiol 8:1094

Gu W, Schlichthorl G, Hirsch JR, Engels H, Karschin C, Karschin A et al (2002) Expression pattern and functional characteristics of two novel splice variants of the two-pore-domain potassium channel TREK-2. J Physiol 539:657–668

Hamill OP, Marty A, Neher E, Sakmann B, Sigworth FJ (1981) Improved patch-clamp techniques for high resolution current recording from cells and cell-free membrane patches. Pflugers Arch 391:85–100

Huang J, Han S, Sun Q, Zhao Y, Liu J, Yuan X et al (2017) Kv1.3 channel blocker (ImKTx88) maintains blood-brain barrier in experimental autoimmune encephalomyelitis. Cell Biosci 7:3

Imming P, Sinning C, Meyer A (2006) Drugs, their targets and the nature and number of drug targets. Nat Rev Drug Discov 5:821–834

Jiang Y, Lee A, Chen J, Cadene M, Chait BT, MacKinnon R (2002) Crystal structure and mechanism of a calcium-gated potassium channel. Nature 417:515–522

Jiang Y, Lee A, Chen J, Ruta V, Cadene M, Chait BT, MacKinnon R (2003) X-ray structure of a voltage-dependent K+ channel. Nature 423:33–41

Jiang QX (2021) High-resolution structures of K+ channels. Handb Exp Pharmacol. In press. https://doi.org/10.1007/164_2021_454

Johnston J, Forsythe ID, Kopp-Scheinpflug C (2010) Going native: voltage-gated potassium channels controlling neuronal excitability. J Physiol 588:3187–3200

Kanda H, Ling J, Tonomura S, Noguchi K, Matalon S, Gu JG (2019) TREK-1 and TRAAK are principal K+ channels at the nodes of Ranvier for rapid action potential conduction on mammalian myelinated afferent nerves. Neuron 104:960–971

Kang D, Kim D (2006) TREK-2 (K2P10.1) and TRESK (K2P18.1) are major background K+ channels in dorsal root ganglion neurons. Am J Physiol Cell Physiol 291:C138–C146

Kisselbach J, Seyler C, Schweizer PA, Gerstberger R, Becker R, Katus HA et al (2014) Modulation of K2P 2.1 and K2P 10.1 K(+) channel sensitivity to carvedilol by alternative mRNA translation initiation. Br J Pharmacol 171:5182–5194

Kramer J, Obejero-Paz CA, Myatt G, Kuryshev YA, Bruening-Wright A, Verducci JS, Brown AM (2013) MICE models: superior to the HERG model in predicting torsade de pointes. Sci Rep 3:2100

Kramer J, Himmel HM, Lindqvist A, Stoelzle-Feix S, Chaudhary KW, Li D, Bohme GA et al (2020) Cross-site and cross-platform variability of automated patch clamp assessments of drug effects on human cardiac currents in recombinant cells. Sci Rep 10:5627. Erratum in: Sci Rep. 2020 Jul 14;10(1):11884

Kullmann DM, Hanna MG (2002) Neurological disorders caused by inherited ion-channel mutations. Lancet Neurol 1:157–166

Lanigan MD, Pennington MW, Lefievre Y, Rauer H, Norton RS (2001) Designed peptide analogues of the potassium channel blocker ShK toxin. Biochemistry 40:15528–15537

Lehmann-Horn F, Jurkat-Rott K (1999) Voltage-gated ion channels and hereditary disease. Physiol Rev 79:1317–1372

Lengyel M, Czirják G, Enyedi P (2016) Formation of functional heterodimers by TREK-1 and TREK-2 two-pore domain potassium channel subunits. J Biol Chem 291:13649–13661

Lengyel M, Czirják G, Jacobson DA, Enyedi P (2020) TRESK and TREK-2 two-pore-domain potassium channel subunits form functional heterodimers in primary somatosensory neurons. J Biol Chem 295:12408–12425

Levitz J, Royal P, Comoglio Y, Wdziekonski B, Schaub S, Clemens DM et al (2016) Heterodimerization within the TREK channel subfamily produces a diverse family of highly regulated potassium channels. Proc Natl Acad Sci U S A 113:4194–4199

Li S, Hou H, Mori T, Sawmiller D, Smith A, Tian J et al (2015) Swedish mutant APP-based BACE1 binding site peptide reduces APP β-cleavage and cerebral Aβ levels in Alzheimer's mice. Sci Rep 5:11322

Liu PW, Bean BP (2014) Kv2 channel regulation of action potential repolarization and firing patterns in superior cervical ganglion neurons and hippocampal CA1 pyramidal neurons. J Neurosci 34:4991–5002

Lo YC, Kuo CC (2019) Temperature dependence of the biophysical mechanisms underlying the inhibition and enhancement effect of amiodarone on hERG channels. Mol Pharmacol 96:330–344

Lolicato M, Riegelhaupt PM, Arrigoni C, Clark KA, Minor DL Jr (2014) Transmembrane helix straightening and buckling underlies activation of mechanosensitive and thermosensitive K (2P) channels. Neuron 84:1198–1212

Lolicato M, Arrigoni C, Mori T, Sekioka Y, Bryant C, Clark KA et al (2017) K2P2.1 (TREK-1)-activator complexes reveal a cryptic selectivity filter binding site. Nature 547:364–368

Loucif AJC, Saintot PP, Liu J, Antonio BM, Zellmer SG, Yoger K et al (2018) GI-530159, a novel, selective, mechanosensitive two-pore-domain potassium (K2P) channel opener, reduces rat dorsal root ganglion neuron excitability. Br J Pharmacol 175:2272–2283

Ma L, Roman-Campos D, Austin ED, Eyries M, Sampson KS, Soubrier F et al (2013) A novel channelopathy in pulmonary arterial hypertension. N Engl J Med 369:351–361

Madry C, Kyrargyri V, Arancibia-Cárcamo IL, Jolivet R, Kohsaka S, Bryan RM et al (2018) Microglial ramification, surveillance, and interleukin-1β release are regulated by the two-pore domain K(+) channel THIK-1. Neuron 97:299–312

Maingret F, Lauritzen I, Patel AJ, Heurteaux C, Reyes R, Lesage F et al (2000) TREK-1 is a heat-activated background K(+) channel. EMBO J 19:2483–2491

Mathie A (2010) Ion channels as novel therapeutic targets in the treatment of pain. J Pharm Pharmacol 62:1089–1095

Mathie A, Veale EL (2015) Two-pore domain potassium channels: potential therapeutic targets for the treatment of pain. Pflugers Arch 467:931–943

Mathie A, Al-Moubarak E, Veale EL (2010) Gating of two pore domain potassium channels. J Physiol 588:3149–3156

Mathie A, Veale EL, Cunningham KP, Holden RG, Wright PD (2021a) Two-pore domain potassium channels as drug targets: anesthesia and beyond. Annu Rev Pharmacol Toxicol 61:401–420

Mathie A, Veale EL, Holden RG (2021b) Heterologous expression of ion channels in mammalian cell lines. In: Dallas M, Bell D (eds) Patch clamp electrophysiology: methods and protocols, Methods in molecular biology, vol 2188, pp 51–65

Matsumura K, Yokogawa M, Osawa M (2021) Peptide toxins targeting KV channels. Handb Exp Pharmacol. In press. https://doi.org/10.1007/164_2021_500

McCoull D, Ococks E, Large JM, Tickle DC, Mathie A, Jerman J et al (2021) A "target-class" screen to identify activators of two-pore domain potassium (K2P) channels. SLAS Discov 26:428–438

McGivern JG, Ding M (2020) Ion channels and relevant drug screening approaches. SLAS Discov 25:413–419

Medhurst AD, Rennie G, Chapman CG, Meadows H, Duckworth MD, Kelsell RE et al (2001) Distribution analysis of human two pore domain potassium channels in tissues of the central nervous system and periphery. Brain Res Mol Brain Res 86:101–114

Miki T, Suzuki M, Shibasaki T, Uemura H, Sato T, Yamaguchi K (2002) Mouse model of Prinzmetal angina by disruption of the inward rectifier $K_{ir}6.1$. Nat Med 8:466–472

Miller AN, Long SB (2012) Crystal structure of the human two-pore domain potassium channel K2P1. Science 335:432–436

Milligan CJ, Li J, Sukumar P, Majeed Y, Dallas ML, English A et al (2009) Robotic multiwell planar patch-clamp for native and primary mammalian cells. Nat Protoc 4:244–255

Mirkovic K, Palmersheim J, Lesage F, Wickman K (2012) Behavioral characterization of mice lacking Trek channels. Front Behav Neurosci 6:60

Mu D, Chen L, Zhang X, See LH, Koch CM, Yen C et al (2003) Genomic amplification and oncogenic properties of the KCNK9 potassium channel gene. Cancer Cell 3:297–302

Nakane T, Kotecha A, Sente A, McMullan G, Masiulis S, Brown PMGE et al (2020) Single-particle cryo-EM at atomic resolution. Nature 587:152–156

Niday Z, Tzingounis AV (2018) Potassium channel gain of function in epilepsy: an unresolved paradox. Neuroscientist 24:368–380

Noel J, Zimmermann K, Busserolles J, Deval E, Alloui A, Diochot S (2009) The mechano-activated K+ channels TRAAK and TREK-1 control both warm and cold perception. EMBO J 28:1308–1318

Obergrussberger A, Goetze TA, Brinkwirth N, Becker N, Friis S et al (2018) An update on the advancing high-throughput screening techniques for patch clamp-based ion channel screens: implications for drug discovery. Expert Opin Drug Discovery 3:269–277

Obergrussberger A, Friis S, Brüggemann A, Fertig N (2020) Automated patch clamp in drug discovery: major breakthroughs and innovation in the last decade. Expert Opin Drug Discovery:1–5

Olschewski A, Veale EL, Nagy BM, Nagaraj C, Kwapiszewska G, Antigny F et al (2017) TASK-1 (KCNK3) channels in the lung: from cell biology to clinical implications. Eur Respir J 50:1700754

Overington JP, Al-Lazikani B, Hopkins AL (2006) How many drug targets are there? Nat Rev Drug Discov 5:993–996

Pei L, Wiser O, Slavin A, Mu D, Powers S, Jan LY et al (2003) Oncogenic potential of TASK3 (Kcnk9) depends on K+ channel function. Proc Natl Acad Sci U S A 100:7803–7807

Pennington MW, Beeton C, Galea CA, Smith BJ, Chi V, Monaghan KP et al (2009) Engineering a stable and selective peptide blocker of the Kv1.3 channel in T lymphocytes. Mol Pharmacol 75:762–773

Pereira V, Busserolles J, Christin M, Devilliers M, Poupon L, Legha W (2014) A. et al. Role of the TREK2 potassium channel in cold and warm thermosensation and in pain perception. Pain 155:2534–2544

Perez-Verdaguer M, Capera J, Serrano-Novillo C, Estadella I, Sastre D, Felipe A (2016) The voltage-gated potassium channel Kv1.3 is a promising multitherapeutic target against human pathologies. Expert Opin Ther Targets 20:577–591

Plaster NM, Tawil R, Tristani-Firouzi M, Canun S, Bendahhou S, Tsunoda A et al (2001) Mutations in $K_{ir}2.1$ cause the developmental and episodic electrical phenotypes of Andersen's syndrome. Cell 105:511–519

Prentis PJ, Pavasovic A, Norton RS (2018) Sea anemones: quiet achievers in the field of peptide toxins. Toxins 10:36

Rangaraju S, Chi V, Pennington MW, Chandy KG (2009) Kv1.3 potassium channels as a therapeutic target in multiple sclerosis. Expert Opin Ther Targets 13:909–924

Redfern WS, Carlsson L, Davis AS, Lynch WG, MacKenzie I, Palethorpe S et al (2003) Relationships between preclinical cardiac electrophysiology, clinical QT interval prolongation and torsade de pointes for a broad range of drugs: evidence for a provisional safety margin in drug development. Cardiovasc Res 58:32–45

Renaud JP, Chari A, Ciferri C, Liu W-T, Rémigy H-W, Stark H et al (2018) Cryo-EM in drug discovery: achievements, limitations and prospects. Nat Rev Drug Discov 17:471–492

Ridder BJ, Leishman DJ, Bridgland-Taylor M, Samieegohar M, Han X, Wu WW et al (2020) A systematic strategy for estimating hERG block potency and its implications in a new cardiac safety paradigm. Toxicol Appl Pharmacol 394:114961. Erratum in: Toxicol Appl Pharmacol. 2020; 395:114983

Romer SH, Deardorff AS, Fyffe REW (2019) A molecular rheostat: Kv2.1 current maintain or suppress repetitive firing in motoneurons. J Physiol 597:3769–3786

Rus H, Pardo CA, Hu L, Darrah E, Cudrici C, Niculescu T et al (2005) The voltage-gated potassium channel Kv1.3 is highly expressed on inflammatory infiltrates in multiple sclerosis brain. Proc Natl Acad Sci U S A 102:11094–11099

Sager PT, Gintant G, Turner JR, Pettit S, Stockbridge N (2014) Rechanneling the cardiac proarrhythmia safety paradigm: a meeting report from the cardiac safety research consortium. Am Heart J 167:292–300

Sanguinetti MC, Tristani-Firouzi M (2006) hERG potassium channels and cardiac arrhythmia. Nature 440:463–469

Santos R, Ursu O, Gaulton A, Bento AP, Donadi RS, Bologa CG (2017) A comprehensive map of molecular drug targets. Nat Rev Drug Discov 16:19–34

Simkin D, Cavanaugh EJ, Kim D (2008) Control of the single channel conductance of K2P10.1 (TREK-2) by the amino-terminus: role of alternative translation initiation. J Physiol 586:5651–5663

Simon DB, Karet FE, Rodriguez-Soriano J, Hamdan JH, DiPietro A, Trachtman H et al (1996) Genetic heterogeneity of Bartter's syndrome revealed by mutations in the K^+ channel. ROMK Nat Genet 14:152–156

Staudacher K, Baldea I, Kisselbach J, Staudacher I, Rahm A-K, Schweizer PA et al (2011) Alternative splicing determines mRNA translation initiation and function of human K(2P) 10.1 K+ channels. J Physiol 589:3709–3720

Su Z, Brown EC, Wang W, Mackinnon R (2016) Novel cell-free high-throughput screening method for pharmacological tools targeting K+ channels. Proc Natl Acad Sci U S A 113:5748–5753

Su S, Sun J, Wang Y, Xu Y (2021) Cardiac hERG K+ channel as safety and pharmacological target. Handbk Exp Pharmacol. https://doi.org/10.1007/164_2021_455

Sun H, Luo L, Lal B, Ma X, Chen L, Hann CL et al (2016) A monoclonal antibody against KCNK9 K(+) channel extracellular domain inhibits tumour growth and metastasis. Nat Commun 7:10339

Talley EM, Solorzano G, Lei Q, Kim D, Bayliss DA (2001) Cns distribution of members of the two-pore-domain (KCNK) potassium channel family. J Neurosci 21:7491–7505

Tanner MR, Tajhya RB, Huq R, Gehrmann EJ, Rodarte KE, Atik MA et al (2017) Prolonged immunomodulation in inflammatory arthritis using the selective Kv1.3 channel blocker HsTX1 [R14A] and its PEGylated analog. Clin Immunol 180:45–57

Tarcha EJ, Chi V, Munoz-Elias EJ, Bailey D, Londono LM, Upadhyay SK et al (2012) Durable pharmacological responses from the peptide ShK-186, a specific Kv1.3 channel inhibitor that suppresses T cell mediators of autoimmune disease. J Pharmacol Exp Ther 342:642–653

Taura J, Kircher DM, Gameiro-Ros I, Slesinger PA (2021) Comparison of K+ channel families. Handb Exp Pharmacol (in this volume)

Toldi G, Vasarhelyi B, Kaposi A, Meszaros G, Panczel P, Hosszufalusi N et al (2010) Lymphocyte activation in type 1 diabetes mellitus: the increased significance of Kv1.3 potassium channels. Immunol Lett 133:35–41

Tsantoulas C, McMahon SB (2014) Opening paths to novel analgesics: the role of potassium channels in chronic pain. Trends Neurosci 37:146–158

Vacher H, Mohapatra DP, Trimmer JS (2008) Localization and targeting of voltage-dependent ion channels in mammalian central neurons. Physiol Rev 88:1407–1447

Veale EL, Al-Moubarak E, Bajaria N, Omoto K, Cao L, Tucker SJ, Stevens EB, Mathie A (2014a) Influence of the N terminus on the biophysical properties and pharmacology of TREK1 potassium channels. Mol Pharmacol 85:671–681

Veale EL, Hassan M, Walsh Y, Al Moubarak E, Mathie A (2014b) Recovery of current through mutated TASK3 potassium channels underlying Birk Barel syndrome. Mol Pharmacol 85:397–407

Veytia-Bucheli JI, Jimenez-Vargas JM, Melchy-Perez EI, Sandoval-Hernandez MA, Possani LD, Rosenstein Y (2018) Kv1.3 channel blockade with the Vm24 scorpion toxin attenuates the CD4 (+) effector memory T cell response to TCR stimulation. Cell Commun Signal 16:45

Viatchenko-Karpinski V, Ling J, Gu JG (2018) Characterization of temperature-sensitive leak K+ currents and expression of TRAAK, TREK-1, and TREK2 channels in dorsal root ganglion neurons of rats. Mol Brain 11:40

Villa C, Combi R (2016) Potassium channels and human epileptic phenotypes: an updated overview. Front Cell Neurosci 10:81

Vivier D, Soussia IB, Rodrigues N, Lolignier S, Devilliers M, Chatelain FC et al (2017) Development of the first two-pore domain potassium channel TWIK-related K+ channel 1-selective agonist possessing in vivo antinociceptive activity. J Med Chem 260:1076–1088

Walsh KB (2020) Screening technologies for inward rectifier potassium channels: discovery of new blockers and activators. SLAS Discov 25:420–433

Wang X, Li G, Guo J, Zhang Z, Zhang S, Zhu Y et al (2020) Kv1.3 channel as a key therapeutic target for neuroinflammatory diseases: state of the art and beyond. Front Neurosci 13:1393

Wangemann P, Itza EM, Albrecht B, Wu T, Jabba SV, Maganti RJ et al (2010) Loss of KCNJ10 protein expression abolishes endocochlear potential and causes deafness in Pendred syndrome mouse model. BMC Med 2:30

Waxman SG, Zamponi GW (2014) Regulating excitability of peripheral afferents: emerging ion channel targets. Nat Neurosci 17:153–163

Weaver CD, Harden D, Dworetzky SI, Robertson B, Knox RJ (2004) A thallium-sensitive, fluorescence-based assay for detecting and characterizing potassium channel modulators in mammalian cells. J Biomol Screen 9:671–677

Woolf CJ, Ma Q (2007) Nociceptors–noxious stimulus detectors. Neuron 55:353–364

Wright PD, Weir G, Cartland J, Tickle D, Kettleborough C, Cader ZM et al (2013) Cloxyquin (5-chloroquinolin-8-ol) is an activator of the two-pore domain potassium channel TRESK. Biochem Biophys Res Commun 441:463–468

Wright PD, Veale EL, McCoull D, Tickle DC, Large JM, Ococks E et al (2017) Terbinafine is a novel and selective activator of the two-pore domain potassium channel TASK3. Biochem Biophys Res Commun 493:444–450

Wright PD, McCoull D, Walsh Y, Large JM, Hadrys BW, Gauritcikaite E et al (2019) Pranlukast is a novel small molecule activator of the two-pore domain potassium channel TREK2. Biochem Biophys Res Commun 520:35–40

Wulff H, Zhorov BS (2008) K+ channel modulators for the treatment of neurological disorders and autoimmune diseases. Chem Rev 108:1744–1773

Wulff H, Calabresi PA, Allie R, Yun S, Pennington M, Beeton C et al (2003) The voltage-gated Kv1.3 K(+) channel in effector memory T cells as new target for MS. J Clin Invest 111:1703–1713

Wulff H, Castle NA, Pardo LA (2009) Voltage-gated potassium channels as therapeutic targets. Nat Rev Drug Discov 8:982–1001

Wulff H, Christophersen P, Colussi P, Chandy KG, Yarov-Yarovoy V (2019) Antibodies and venom peptides: new modalities for ion channels. Nat Rev Drug Discov 18:339–357

Zhou QL, Wang TY, Li M, Shang YX (2018) Alleviating airway inflammation by inhibiting ERK-NF-κB signaling pathway by blocking Kv1.3 channels. Int Immunopharmacol 63:110–118

Cardiac K^+ Channels and Channelopathies

Julian A. Schreiber and Guiscard Seebohm

Contents

Abstract

The physiological heart function is controlled by a well-orchestrated interplay of different ion channels conducting Na^+, Ca^{2+} and K^+. Cardiac K^+ channels are key players of cardiac repolarization counteracting depolarizating Na^+ and Ca^{2+} currents. In contrast to Na^+ and Ca^{2+}, K^+ is conducted by many different channels that differ in activation/deactivation kinetics as well as in their contribution to different phases of the action potential. Together with modulatory subunits these K^+ channel α-subunits provide a wide range of repolarizing currents with specific

J. A. Schreiber
Institute of Pharmaceutical and Medicinal Chemistry, University of Münster, Münster, Germany

Department of Cardiovascular Medicine, Institute for Genetics of Heart Diseases (IfGH), University Hospital Münster, Münster, Germany

G. Seebohm (✉)
Department of Cardiovascular Medicine, Institute for Genetics of Heart Diseases (IfGH), University Hospital Münster, Münster, Germany
e-mail: Guiscard.Seebohm@ukmuenster.de

N. Gamper, K. Wang (eds.), *Pharmacology of Potassium Channels*,
Handbook of Experimental Pharmacology 267, https://doi.org/10.1007/164_2021_513

characteristics. Moreover, due to expression differences, K^+ channels strongly influence the time course of the action potentials in different heart regions. On the other hand, the variety of different K^+ channels increase the number of possible disease-causing mutations. Up to now, a plethora of gain- as well as loss-of-function mutations in K^+ channel forming or modulating proteins are known that cause severe congenital cardiac diseases like the long-QT-syndrome, the short-QT-syndrome, the Brugada syndrome and/or different types of atrial tachyarrhythmias. In this chapter we provide a comprehensive overview of different K^+ channels in cardiac physiology and pathophysiology.

Keywords

Disease · Heart · Inherited · Mutation · Rhythm

1 Function of Cardiac Potassium Channels

The physiological cardiac function is controlled by a tightly balanced electrochemical cycle resulting from the interplay of Na^+, Ca^{2+}, and K^+ ion channels. The mainly voltage-controlled operation of divergent channels produces a periodic change of membrane potential characteristic of the cardiac action potentials (AP). Depending on the cell type and the localization in the heart the AP of a cell has a specific appearance. The sequential activation, inactivation, and deactivation of ion channels lead to a recurring course of the membrane potential. The shape of this course and the duration of the different phases are determined by the ion channel expression pattern characteristic for the respective cardiac cell subtypes and differ among sinoatrial, atrial or the ventricular myocardium (Rudy 2008). While only a few specific subtypes of Na^+ and Ca^{2+} channels are involved, a broad variety of different potassium channels contribute to different phases of the AP. Potassium channels are built by pore forming α-subunits and they can be modulated by auxiliary subunits (β-subunits) as well as by intracellular interacting proteins (Pongs et al. 1999). The potassium channel derived currents can be subdivided into two groups: outward rectifying currents like the slow, rapid and ultra-rapid currents I_{Ks}, I_{Kr} or I_{Kur}, that are controlled by the opening and closing of voltage-gated potassium channels at distinct membrane potentials, and inward rectifying currents like the I_{K1} (see below) (Deal et al. 1996). In addition to outward and inward rectifying channels, background K^+ currents conducted by K2P channels like TASK1 were identified as important factors in atria as well (Kim et al. 1999).

In general, the ventricular AP can be subdivided in phases from 0 to 4, that are characterized by specific inward and outward currents (I) resulting from ion channel opening (Fig. 1) (Shih 1994). Phase 0 is dominated by a fast rise of the membrane potential to more positive voltages mostly caused by the inward directed I_{Na} through voltage-gated $Na_v1.5$ channels (Maroni et al. 2019). In Phase 1 the rise of membrane potential is stopped and partially reversed by the opposed directed I_{Tof} and I_{Tos}, that result from the fast and slow transient outward K^+ currents through voltage-gated $K_v4.2$, $K_v4.3$/KChIP (I_{Tof}) (An et al. 2000) and $K_v1.4$ (I_{Tos}) channels (Johnson et al.

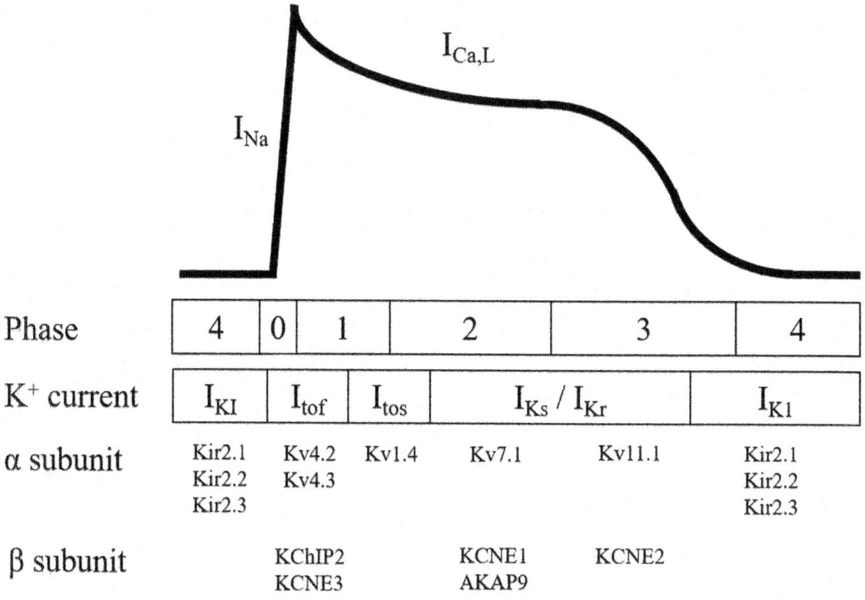

Fig. 1 Schematic depiction of a ventricular AP subdivided into Phases 0–4. The dominating potassium current as well as the contributing pore forming (α-subunits) and modulating proteins (β-subunit) is named for each phase of the AP

2018). The depolarized membrane potential leads to the opening of voltage-gated L-Type Ca^{2+} channels (α_1-subunit $Ca_v1.2$). In addition to the α-subunit, the L-type Ca^{2+} channels contain 4 different subunits, the α_2, δ, β and γ subunits (Bodi et al. 2005). The α_2, δ and β subunits modulate channel trafficking and biophysical properties of the α_1 subunit. Together with opening of the classical components of I_K $K_v7.1$/KCNE1 (I_{Ks}) (Barhanin et al. 1996; Sanguinetti et al. 1996) and $K_v11.1$/KCNE2 (I_{Kr}) channels (Abbott et al. 1999), a plateau phase is generated by balanced activity of the inward directed $I_{Ca,L}$ and the outward directed delayed rectifier K^+ currents I_{Ks} and I_{Kr} (Phase 2) (Sanguinetti and Jurkiewicz 1990). The return to negative potential is driven by the subsequent inactivation of Ca^{2+} channels, while delayed rectifier K^+ channels $K_v7.1$/KCNE1 and $K_v11.1$/KCNE2 stay open or even activate to a larger extent (Phase 3). Further opening of inward rectifying K^+ channels producing the I_{K1} ($K_{ir}2.1$, $K_{ir}2.2$, $K_{ir}2.3$) additionally repolarizes the membrane, until the resting potential is reached again (Phase 4) (Liu et al. 2001; Grunnet 2010).

Compared to the classical AP of ventricular cells, the AP of atrial cells in the sinoatrial node (SAN) shows a spontaneous depolarization in the Phase 4 that is caused by the auto-depolarization of the so-called pacemaker cells (Fig. 2) (Fenske et al. 2020). Moreover, the AP does not have a Phase 1 or 2 and the Phase 0 is mostly caused by the activation of L-Type Ca^{2+} channels and the subsequent release of Ca^{2+} from the sarcoplasmic reticulum instead of opening $Na_v1.5$ channels (Lyashkov

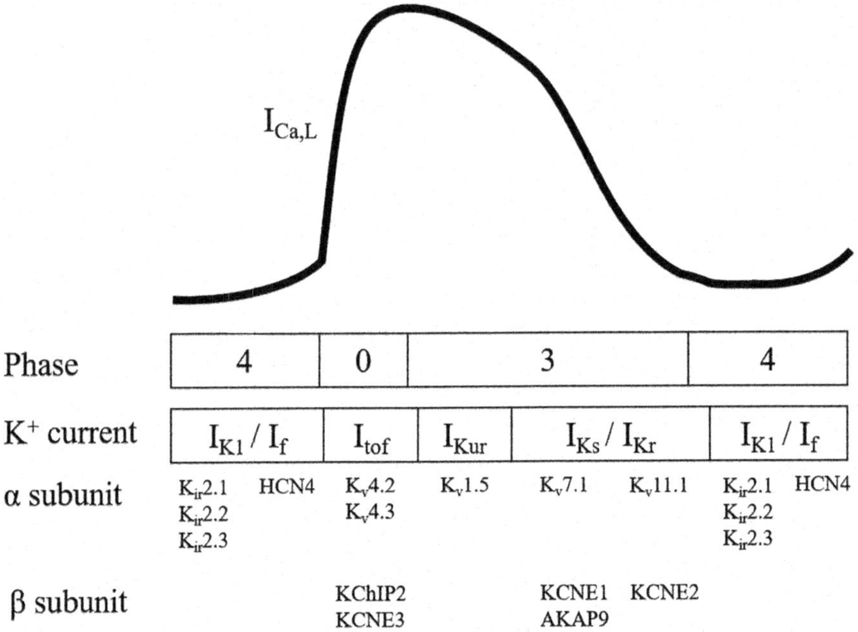

Fig. 2 Schematic depiction of a sinoatrial AP subdivided into Phases 0–4. The dominating potassium current as well as the contributing pore forming (α-subunits) and modulating proteins (β-subunit) is presented for each phase of the AP

et al. 2018). In addition to the I_{Ks} and I_{Kr} forming K^+ channels, the Phase 3 of atrial cells is dominated by the ultra-rapid outward current I_{Kur} produced by $K_v1.5$ channels (Wettwer et al. 2004). The most obvious change in the action of pacemaker cells from the atrial myocard is the reduced resting potential and the auto-depolarization in phase 4 caused by reduced expression of I_{K1} and high expression of hyperpolarization-activated cyclic nucleotide-gated channels (HCN), that generate the "funny" current I_f. Although four members of HCN channels are known (HCN1-4), HCN4 is mostly responsible for the I_f in human SAN pacemaker cells (Li et al. 2015). This nonselective cation channel conducts Na^+ and K^+ at hyperpolarized membrane potentials leading to the slow spontaneous depolarization, that is needed for the activation of $Ca_v1.2$-channel complex (Sartiani et al. 2017). HCN4 may associate with KCNE2 to form the cardiac channel (Decher et al. 2003).

To summarize, cardiac potassium channels contribute to nearly all phases of the action potential and become dominant in the later AP phases. Therefore, inherited dysfunctions of cardiac potassium channels can cause severe and life-threatening diseases. Depending on the affected channel type as well as the resulting increase or decrease of the affected current, characteristic phenotypes can be observed. In this chapter the most frequently observed inherited as well as acquired (drug induced) potassium channelopathies are discussed.

2 Characteristics and History of Congenital Long QT Syndromes

The long QT syndrome (LQTS) is named by the prolongation of the QT interval in a 12-lead electrocardiogram (ECG) (Fig. 3). The ECG of a healthy person shows characteristic signals that are named P wave, QRS complex, T wave and U wave. These deflections from the isoelectric line are caused by the sequential depolarization starting from the sinoatrial node in the atrium (P wave), the repolarization of atria and depolarization of ventricles (QRS complex) and the subsequent repolarization of the ventricle (T wave). After the T wave another wave of low amplitude (U wave) can be seen in a few ECGs at lower heart rate (Draghici and Taylor 2016). Together these components from the P to the U wave represent a single heartbeat that is controlled by the sequential excitation of neighbouring cells from the sinoatrial node through the atrium to the atrioventricular node that transfers the excitation through the bundles of HIS to the Purkinje fibres through the whole ventricle (Stephenson et al. 2017). The direction of normal cardiac excitation is fixed by a refractory phase, whereby cardiomyocytes that were just excited cannot be re-stimulated. However, if repolarization time is longer than this refractory phase, re-entry or circular excitation and arrhythmias are possible (Kimrey et al. 2020).

Prolongation of the ventricular depolarization/repolarization process (QT interval) can lead to re-entry or circular excitation in the ventricle that subsequently reduces the cardiac efficiency and causes severe ventricular arrhythmia *Torsades de Pointes* (*TdP*) which can eventually result in sudden cardiac death (Wilders and Verkerk 2018). *TdP*s are often observed after physical or emotional cardiovascular stress that leads to an increased heart rate. *TdPs are* characterized by undulating electrical activity around the isoelectric baseline. Under physiological conditions the heart rate is increased by a shortening of the depolarization/repolarization process that facilitates the possibility of circular excitation especially in patients with altered ion channel functions that result in delayed repolarization. A prolonged QT interval exists if the heart rate corrected Q wave to T wave duration is ≥450 ms (male) or ≥460 ms (female) (Rautaharju et al. 2009).

The long QT syndrome (LQTS) can be caused by the use of different drugs or electrolyte disturbance (acquired forms; aLQTS) or as a result of genetic variants

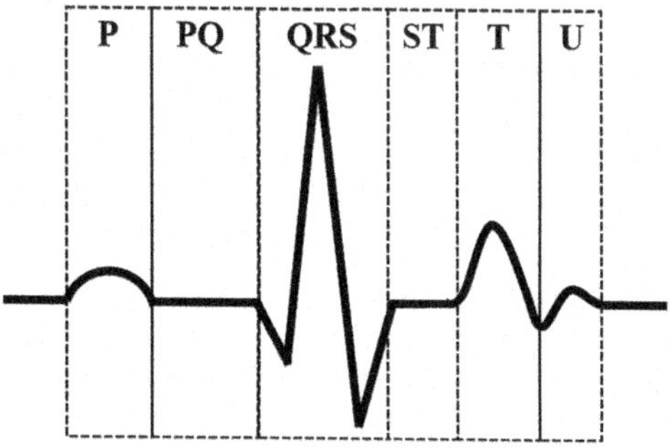

Fig. 3 Schematic depiction of an ECG subdivided into P wave, PQ interval, QRS complex, ST interval, T wave and U wave

Table 1 LQTS diagnostic Schwartz score (Priori et al. 2013)

Electrocardiographic findings		
A	Corrected QT (Bazett formula; male)	
	≥480 ms	3
	460–479 ms	2
	450–459 ms	1
B	Corrected QT fourth minute of recovery from exercise stress test ≥480 ms	1
C	*Torsades-de-pointes*	2
D	T-wave alternans	1
E	Notched T wave in 3 leads	1
F	Low heart rate for age	0.5
Clinical history		
A	Syncope with stress	2
	Syncope without stress	1
B	Congenital deafness	0.5
Family history		
A	Family members with definite LQTS	1
B	Unexplained sudden cardiac death younger than age 30 among immediate family members	0.5

(congenital forms; cLQTS) (Sugrue et al. 2017). The congenital forms of the LQT syndrome can manifest as a life-threatening disease that can cause severe pathophysiological symptoms up to sudden cardiac death in early lifetime. The prevalence of cLQTS is approximately 1:2000 (Schwartz et al. 2009). However, the symptoms as well as the severity of the disease are remarkably diverse. Therefore, clinical diagnostic is based on the ECG, the corrected QT time as well as the clinical Schwartz score (Table 1; Priori et al. 2013). LQTS is diagnosed in presence of a Schwartz risk score ≥3.5 or/and genetic analysis identifying a pathogenic mutation, that is associated with LQTS.

The first congenital LQTS was discovered in 1957 by Anton Jervell and Fred Lange-Nielsen, which described a combination of congenital deafness, fainting attacks (syncope) and prolonged QT intervals in four siblings, whereas three of them died at young age of sudden cardiac death (Jervell and Lange-Nielsen 1957). With a small delay, Cesarino Romano and Owen Conor Ward report similar symptoms among patients without deafness (Romano et al. 1963; Ward 1964). These findings represent the starting point for the characterization of the autosomal recessive Jervell-Lange-Nielsen syndrome (JLNS) as well as the autosomal dominant Romano-Ward syndrome. The Jervell-Lange-Nielson syndrome (JLNS) is created by mutations in the *KCNQ1* (JLNS1) or *KCNE1* (JLNS2) genes. These mutations lead to a loss of function of the Kv7.1/KCNE1 channel, which conducts the repolarizing outward I_{Ks} current. In contrast to the JLNS, the Romano-Ward syndrome (RWS) is an autosomal dominant disease that can be caused by mutations in the genes encoding for KCNQ1, KCNE1, hERG or $Na_v1.5$ (Faridi et al. 2018).

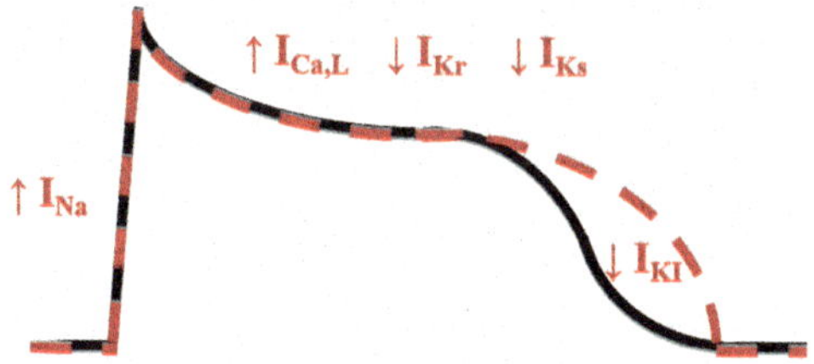

Fig. 4 Schematic depiction of a normal ventricular AP (black) vs. a typical ventricular AP from patients with LQTS. The prolonged Phase 2 can be caused by reduced K^+ channel activity or increased Na^+- or Ca^{2+} channel activity

Beside the classical JLNS and RWS other LQTS are known. The nomenclature is based on the mutation's location in genes coding for ion channel pore forming or modulating proteins. These mutations lead to a reduction of the repolarizing I_{Ks}, I_{Kr}, or I_{K1} or an increment of I_{Na} or $I_{Ca,L}$ prolonging the depolarizing Phase 2 of the action potential (Fig. 4) (Giudicessi et al. 2018). Approximately 80% of the mutations underlying clinical LQTS cases are located in the genes *KCNQ1* (LQT1), *KCNH2* (LQT2) and *SCN5A* (LQT3) (Altmann et al. 2015). These genes are coding for pore forming α subunits of the channels $K_v7.1$, $K_v11.1$ and $Na_v1.5$, respectively. LQT1 and LQT2 are caused by loss of function mutations resulting in reduced I_{Ks} or I_{Kr} currents, which lead to a prolonged Phase 2 of the AP. On the contrary, LQT3 is caused by gain of function mutations resulting in elevated I_{Na} currents, that lead to stronger depolarization in Phase 0 and subsequently prolonged repolarization time (Wang et al. 1995). After the identification of the three major LQTS even more mutations in other ion channel pore-forming or modulatory proteins were identified (LQT4–LQT16). However, LQT4–LQT16 are called minor LQT syndromes since their prevalence is extremely low (Table 2) (Schwartz et al. 2012).

In LQTS both, early afterdepolarizations (EADs) and delayed afterdepolarizations (DADs) occur. DADs occur after full repolarization. They emerge spontaneously and are characterized by a relatively small amplitude. DADs bear the potential to initiate ectopic firing. DADs are facilitated in a setting of abnormal Ca^{2+} handling, such as anomalous spontaneous or excessive Ca^{2+} release from the sarcoplasmic reticulum (SR) via Ryanodine receptors 2 (RyR2) during AP phase 4. The excessive intracellular Ca^{2+} activates the Na^+/Ca^{2+} exchanger NCX which in turn increases Na^+ influx and thus creates a characteristic depolarizing current producing the DAD (Nattel et al. 2008; Dobrev et al. 2011; Wakili et al. 2011). EADs are produced by extended cardiac APs observed in LQTS; these allow L-type Ca^{2+} channels to reactivate and produce a depolarization during the late action potential phase 2/early action potential phase 3. These EADs can trigger ectopic activity as well (Zeng and Rudy 1995; Nattel et al. 2008).

Table 2 Nomenclature of LQT and JLN syndromes (Giudicessi et al. 2018)

	Gene	Protein	Functional effect	Freq.
LQT1	*KCNQ1*	K_v7.1	Reduced I_{Ks}	30–35%
LQT2	*KCNH2*	K_v11.1	Reduced I_{Kr}	25–30%
LQT3	*SCN5A*	Na_v1.5	Increased I_{Na}	5–10%
LQT4	*ANK2*	Ankyrin B	Aberrant ion channel/transporter localization	<1%
LQT5	*KCNE1*	KCNE1	Reduced I_{Ks}	<1%
LQT6	*KCNE2*	KCNE2	Reduced I_{Kr}	<1%
LQT7	*KCNJ2*	Kir2.1	Reduced I_{K1}	<1%
LQT8	*CACNA1C*	Ca_v1.2	Increased $I_{Ca,L}$	<1%
LQT9	*CAV3*	Caveolin 3	Increased I_{Na}	<1%
LQT10	*SCN4B*	Na_v1.5/β4-subunit	Increased I_{Na}	<1%
LQT11	*AKAP9*	Yotiao	Reduced I_{Ks}	<1%
LQT12	*SNTA1*	Syntrophin-α1	Increased I_{Na}	<1%
LQT13	*KCNJ5*	Kir3.4	Reduced $I_{K,Ach}$	<1%
LQT14	*CALM1*	Calmodulin 1	Increased $I_{Ca,L}$	<1%
LQT15	*CALM2*	Calmodulin 2	Increased $I_{Ca,L}$	<1%
LQT16	*CALM3*	Calmodulin 3	Increased $I_{Ca,L}$	<1%
JLNS1	*KCNQ1*	K_v7.1	Reduced I_{Ks}	<1%
JLNS2	*KCNE1*	KCNE1	Reduced I_{Ks}	<1%

3 Congenital Long QT Syndromes Associated with I_{Ks}: LQT1, 5, 11

Loss of function mutations influencing the I_{Ks} current are the most common basis of cLQTS (Kannampuzha et al. 2018). To generate a physiological I_{Ks} current, the association of the homotetrameric K_v7.1 channel with one or more modulating β-subunits KCNE1 as well as the association with the A-kinase anchoring protein 9 (AKAP9) is required. The slow activated potassium outward current (I_{Ks}) is named by this typical behaviour of the K_v7.1/KCNE1 channel complex: The depolarization in phase 1 of the cardiac action potential leads to opening of the channel complex with a slowly rising current amplitude that reached its maximum at the end of phase 3. A reduction or complete loss of this outward potassium current prolongs the plateau in phase 2 and stretches repolarization phase 3 of the AP dramatically (Kojima et al. 2020). A reduction of I_{Ks} can be caused by two separate mechanisms: First, mutated α- or β-subunits can assemble as I_{Ks} producing channels with reduced conductance or altered kinetics (Duggal et al. 1998; Seebohm et al. 2001; Henrion et al. 2009). Second, the assembly of the K_v7.1/KCNE1 complex can be disturbed leading to insufficient expression of functional I_{Ks} channels at the plasma membrane surface (Moss et al. 2007). Mutants with altered trafficking were reported for $K_v7.1$ as well as KCNE1 (Harmer and Tinker 2007; Seebohm et al. 2008).

Among the clinically confirmed congenital LQT syndrome cases approximately 35% are associated with mutations in the pore forming α-subunit of the $K_v7.1$ channel (LQT1). The α-subunit encompasses 676 amino acids and contains an N-terminal domain, six transmembrane domains (S1–S6), that can be subdivided into the outer voltage sensor domain (VSD, S1–S4) and the inner pore forming domain (PD, S5–S6) and a C-terminal domain (Wang et al. 2012). Reader can find additional information about KCNQ channels in chapter "Kv7 Channels and Excitability Disorders" (Jones et al. 2021). More than 250 different mutations of the *KCNQ1* gene causing LQT1 are described in literature (Hedley et al. 2009). These mutations are distributed over the complete protein. However, LQT1 causing mutations are more frequently located in the transmembrane region or in the C-terminal region rather than in the N-Terminal-region (Kapa et al. 2009). The location inside the protein sequence can influence the severity of the LQT symptoms. It was found that mutations within the transmembrane domains more often cause pronounced cQT prolongation than mutations in other domains leading to more cardiac events at younger age (Shimizu et al. 2004). Especially mutations in the pore region or mutations near the cytoplasmatic loops are known to cause severe loss of functions and subsequently long cQT intervals (Splawski et al. 2000; Barsheshet et al. 2012).

It is not surprising that beside mutations in the pore forming α-subunit (LQT1), reduction of I_{Ks} can also be caused by mutations within the modulating β-subunit KCNE1 (LQT5). Mutations in KCNE1 causing LQT5 are exceedingly rare and belong to the minor LQTS (LQT4–LQT16). According to recent studies, the prevalence of LQT5 is approximately 1–2% of the diagnosed LQTS patients (Roberts et al. 2020). However, the results also suggest that KCNE1 loss-of-function variants are only weakly penetrant and that LQT5 patients usually possess additional risk factors for prolonged cQT intervals.

Often LQTS1 is exacerbated under stress conditions as the I_{Ks} channel is facilitated under acute (β-adrenergic) or chronic (cortisol mediated) physiological stress to counteract excessive calcium influx via $I_{Ca,L}$ (Kurokawa et al. 2004; Seebohm et al. 2007; Dvir et al. 2014). AKAP9 binds to the C-terminus of KCNQ1 and mediates channel modulation by several stress-related effector proteins like the protein kinase A (PKA) or protein phosphatase 1 (PP1) or even PIP_2 (Dvir et al. 2014). Thus, it is not surprising that mutations within the AKAP9 (Yotiao) cause another form of LQTS (LQTS11) (Kurokawa et al. 2004).

Marx and colleagues showed that stimulation of G_s-coupled β_1 receptors activates PKA and PP1 that subsequently phosphorylate the Kv7.1 subunit mediated by the Yotiao protein (Marx et al. 2002). The phosphorylation of the channel leads to an increased I_{Ks} shortening the plateau phase of the AP and, thereby, allowing accelerated heart rate. The stimulation of I_{Ks} is due to kinetic effects and increased RAB4 dependent exocytosis leading to increased plasma membrane abundance of I_{Ks} (Piccini et al. 2017). However, mutations like Yotiao S1570L disturb the phosphorylation of KCNQ1 by the PKA/PP1 system resulting in prolonged cQT intervals especially under β-adrenergic stimulations (Chen et al. 2007). The β-adrenergic adaption of the I_{Ks} channel can be disturbed by loss of function mutations in the KCNQ1 or KCNE1 subunit as well. Beside acute physical or emotional stress, chronic stress mediated by cortisol leads to RAB11 mediated I_{Ks} channel exocytosis that is disrupted in specific mutations contributing to loss of I_{Ks} function (Seebohm et al. 2007, 2012).

Recent studies suggest that directed and selective modulation of the I_{Ks} or I_{Kr} is a promising strategy for pharmacological rescue especially in JLNS. While patients with normal LQTS are often responsive to β-blocker therapy, the treatment of JLNS patients with β-blockers is often not sufficient. These patients could benefit from a selective activation of I_{Kr} compensating the underdeveloped I_{Ks} component in the AP (Zhang et al. 2014a).

4 I_{Kr} Associated Congenital Long QT Syndromes: LQT2 and LQT6

Like I_{Ks}, I_{Kr} is generated by the heteromeric assembly of an α-subunit (K_v11.1) with its associated β-subunit KCNE2 (MiRP1) (Abbott et al. 1999). The K_v11.1 is a voltage-gated potassium channel comprised from four α-subunits that are encoded by the gene *KCNH2* also known as human Ether-à-go-go related gene (*hERG*) (Ng et al. 2020). The overall structure of the channel is similar to other K_v channels containing a transmembrane region with a voltage sensor domain (S1–S4; VSD) and a pore domain (S5S6) as well as intracellular N- and C-terminal domains (Wang and MacKinnon 2017). At the N-terminus, the channel possesses a Per-Arnt-Sim domain (PASD), which impacts channel gating by slowing the rate of deactivation (Cabral et al. 1998). Loss-of-function mutations in *KCNH2* are distributed over the complete structure and can cause LQT2, that is the second common form of LQTS (Smith et al. 2016; Adler et al. 2020). Similar to mutations in the *KCNQ1* gene, mutations inside regions coding for the pore loop or the transmembrane region of K_v11.1 are known to facilitate cardiac events more than mutations in the PASD or C-terminal domain (Kim et al. 2010). In contrast to loss-of-function, gain-of-function mutations in the *KCNH2* gene are the most common reason for inherited short QT syndrome (SQT1) (Brugada et al. 2004).

Nearly 500 *KCNH2* mutations are linked to LQT2 and the mechanisms of QT prolongation are quite diverse (Anderson et al. 2014). The most prominent reason for I_{Kr} reduction in LQT2 is a reduced number of K_v11.1 channels at the membrane surface (Anderson et al. 2014). This can be caused by either missense mutations that lead to misfolding and subsequent reduced intracellular transport and trafficking or nonsense mutations that caused abnormal transcription/translation of the channel (Anderson et al. 2006; Gong et al. 2007). Around 90% of the LQT2-associated mutations show reduced trafficking to the cell surface (Smith et al. 2016). However, some mutations lead to altered gating behaviour or reduced ion permeability as well. For example, mutations in the PASD lead to accelerated channel deactivation and cause reduction of the I_{Kr} (Chen et al. 1999). More about Kv11.1/hERG channels can be found in chapter "Cardiac hERG K^+ Channel as Safety and Pharmacological Target" (Su et al. 2021).

Expression of $K_v11.1$ without KCNE2 in heterologous systems generate currents that are not completely comparable with the physiologic I_{Kr} (Abbott et al. 1999). Thus, the modulating β-subunit may be required for proper current modulation. Association of the $K_v11.1$ channel with KCNE2 may lead to current amplitude reduction by slowed activation as well as accelerated deactivation (Abbott et al. 1999). Like the closely related KCNE1, KCNE2 is a single helical transmembrane subunit that also alters the function of HCN channels as well as voltage-gated Ca^{2+} channels (Abbott 2015). Mutations located in the *KCNE2* gene can cause LQT6 that belongs to the minor LQTS, since the prevalence is rather low (<1%) (Priori et al. 2013). Very recent studies suggest that LQT6 is not only a rare variant of LQTS but also a variant with low risk for cardiac events (Roberts et al. 2017). Moreover the authors suggest that LQT6 patients develop proarrhythmic susceptibility only with additional environmental or congenital risk factors (Roberts et al. 2017).

5 Mutations in Kir2.1 Cause the LQT7/Andersen-Tawil Syndrome

The Anderson-Tawil syndrome (LQT7) belongs to the minor group of LQTS (LQT4–LQT16) and can be subdivided into two types (ATS1, ATS2) (Donaldson et al. 2004). It is named by Ellen Anderson (Andersen et al. 1971), who firstly described the typical symptoms, as well as Rabi Tawil (Tawil et al. 1994), who contributes significantly to the understanding of the disease. Contrary to most other LQTS, ATS is a multiorgan disease and it is characterized by a triad of periodic muscle weakness (up to paralysis), ventricular arrhythmias as well as dysmorphic features (Yoon et al. 2006). The prevalence of this autosomal dominant disease is rare (approximately 1:1,000,000) (Rajakulendran et al. 2010). ATS1 is caused by loss-of-function mutations in the inward rectifier potassium channel $K_{ir}2.1$, a main component of the I_{K1} current (Donaldson et al. 2004). Only a few patients are known with mutations in the *KCNJ5* gene coding for the $K_{ir}3.4$ channel (Kokunai et al. 2014). However, nearly 40% of the ATS patients do not have mutations in the associated genes suggesting that additional causative factors for ATS exist (Donaldson et al. 2004).

The $K_{ir}2$ channels can be formed as homo- or heterotetramers and they are expressed in the heart and skeletal muscles (Preisig-Muller et al. 2002). A single subunit of $K_{ir}2.1$ (*KCNJ2*) contains two transmembrane helices (M1, M2) that are connected by a pore forming loop, as well as large intracellular N- and C-terminal domains (Pegan et al. 2006). The inward rectification of K_{ir} channels is based on cytosolic interactions with polyamines as well as Mg^{2+} ions leading to a preferable influx of potassium ions especially at hyperpolarized voltages (Pegan et al. 2005). The influx of potassium can be blocked by Ba^{2+} or Cs^+ ions (Bradley et al. 1999). Several ATS1-causing mutations of *KCNJ2* lead to reduced K^+ influx at the late phase 3 and 4 of the cardiac AP and prolonged repolarization (Seemann et al. 2007). The ECG often shows a prominent U wave, while QT interval in the most cases is not prolonged (Tristani-Firouzi et al. 2002).

ATS2 (not *KCNJ2* associated) is much rarer than ATS1 and it is caused, for example, by a mutation in the *KCNJ5* gene, that encodes for the $K_{ir}3.4$ channel, a G protein coupled inward rectifying potassium channel (GIRK) (Kokunai et al. 2014). It is activated by the βγ subunit of the G_i coupled M_2 receptor and contributes to the $I_{K,Ach}$ that shortens the action potential in the atrium and slows down the heart rate (Moreno-Galindo et al. 2016). It was suggested that mutated $K_{ir}3.4$ form heterodimers with $K_{ir}2.1$ and subsequently inhibit the normal function of the $K_{ir}2.1$ subunit (Kokunai et al. 2014). Clearly, further experimental data are needed to clarify the mechanism of action. More on physiology and pharmacology of K_{ir} channels can be found in chapter "Kir Channel Molecular Physiology, Pharmacology and Therapeutic Implications" (Lin et al. 2021).

6 Short QT Syndrome (SQTS)

Contrary to the LQTS, patients with short QT syndrome (SQTS) show cQT intervals of $\leq$340 ms in the electrocardiogram (ECG). SQTS is also diagnosed when cQT intervals of $\leq$360 ms appear together with family history of diagnosed SQTS or sudden cardiac death at the age < 40 (Priori et al. 2015). While the diversity of symptoms in LQTS patients is high, most SQTS patients described so far suffer from severe ventricular arrhythmia and early cardiac death. Thus, SQTS is characterized as a highly malignant disease with high lethality, especially in infants (Gaita et al. 2003). However, with only a few hundred cases described worldwide congenital SQTS is extremely rare compared to congenital LQTS (cLQTS) (Campuzano et al. 2018). More recent studies suggest that not all patients with SQTS have severe symptoms and that previously described riskiness of SQTS is influenced by the low number of diagnosed cases (Akdis et al. 2018).

As compared to LQTS, opposite mechanisms are responsible for the congenital SQTS: shortening the Phase 2 of the AP by increasing I_{Kr}, I_{Ks} or I_{K1} or decreasing the $I_{Ca,L}$ (Brugada et al. 2004; Bellocq et al. 2004; Priori et al. 2005; Antzelevitch et al. 2007). The SQTS was first described in a case report of a family in 2000, where three members suffered from several episodes of paroxysmal atrial fibrillation requiring electrical cardioversion (Gussak et al. 2000). From then to now mutations in at least seven different genes were described as cause for cSQTS, that is transmitted in an autosomal dominant fashion (Campuzano et al. 2018). Most frequently observed are gain-of-function mutations in the gene *KCNH2* (SQT1) that can increase the current amplitude of $K_v11.1$ channel and subsequently accelerate repolarization in Phase 2 of the AP (Sun et al. 2011). Also, gain-of-function mutations in *KCNQ1* (I_{Ks}, SQT2) and *KCNJ2* (I_{K1}, SQT3) are known, however, these are much less common (Moreno et al. 2015; Fernández-Falgueras et al. 2017). The mechanisms for increased I_{Ks} and I_{K1} are diverse and range from accelerated activation kinetics (I_{Ks}) (Bellocq et al. 2004) to shifted voltage dependency (I_{K1}) (Hattori et al. 2012) or enhanced membrane expression of the channel (I_{K1}) (Ambrosini et al. 2014).

Even more rare are SQTS patients, who possess a loss-of-function mutation in genes coding for $I_{Ca,L}$ generating proteins (Antzelevitch et al. 2007). Thus, SQT4 is caused by mutations in the gene *CACNA1C* that can reduce the trafficking of the L-type α1C-subunit ($Ca_v1.2$) to the membrane surface (Antzelevitch et al. 2007). On the contrary, mutations in the gene *CACNB2b* (SQT5), encoding for the β2-subunit of the Ca^{2+} channel, were found to decrease the $I_{Ca,L}$ without changing trafficking (Antzelevitch et al. 2007). ECG from SQT4 and SQT5 patients also show elevation of the ST interval, which is a characteristic phenotype for Burgada syndrome (Antzelevitch et al. 2007). In 2011 another loss-of-function mutation was reported for the gene *CACNA2D1* (SQT6) encoding for the α2δ1-subunit of the voltage gated Ca^{2+} channel that also led to reduced $I_{Ca,L}$ without affecting membrane expression (Templin et al. 2011).

Very recently, another gene was associated with SQTS. Contrary to all the other genes discussed above, *SLC4A3* encodes for a membrane transport protein (anion exchange protein 3) instead of an ion channel (Thorsen et al. 2017). However, the exact molecular mechanism of QT shortening in this case remains elusive (Thorsen et al. 2017).

7 Brugada Syndrome

The Brugada syndrome (BrS) is a rare autosomal dominant cardiac disease that was firstly described in 1992 by Pedro and Joseph Brugada (Brugada and Brugada 1992). Patients with BrS show elevated ST segments in the ECG and can suffer from ventricular arrhythmias up to sudden cardiac death (Brugada and Brugada 1992). The worldwide prevalence for BgS is estimated 1:2000 (Vutthikraivit et al. 2018). While BrS patients are usually asymptomatic, trigger factors like fever can unmask the syndrome (Roomi et al. 2020). The diagnosis of BrS is difficult since characteristic ECG alterations are not always present (Wilde et al. 2002). Beside the ECG, class I antiarrhythmics like flecainide or ajmaline can be used for diagnosis (Wilde et al. 2002). The administration of these drugs blocking the Na^+ channel can provoke or quench the typical BrS symptoms in the ECG depending on the ionic current underlying the pathology (Wilde et al. 2002).

The syndrome can be caused by decreased I_{Na} or $I_{Ca,L}$ or increased I_{to} (Garcia-Elias and Benito 2018). Most patients with an associated mutation have a reduced I_{Na} caused by mutations in the gene *SCNA5* (Watanabe and Minamino 2016). Up to now more than 300 loss-of-function mutations in the *SCNA5* gene were identified that cause BrS by influencing the trafficking or the gating behaviour of the $Na_v1.5$ channel (Kapplinger et al. 2010). Especially mutations leading to truncated proteins lead to a pronounced reduction of I_{Na} and subsequently to severe BrS symptoms (Kapplinger et al. 2010).

Beside the reduced I_{Na}, also reduction of the $I_{Ca,L}$ is described in a few cases to cause BrS. Loss-of-function mutations in the genes *CACNA1C*, *CACNB2b* and *CACNA2D1* can reduce the $I_{Ca,L}$ dramatically (Antzelevitch et al. 2007; Templin et al. 2011). Surprisingly, the same genes are also associated with the development of short QT syndrome (SQTS) leading to mixed BrS and SQTS symptomatic in these patients (Campuzano et al. 2018).

In contrast to the above-mentioned loss-of-function mutations, increment of the I_{to} by several gain-of-function mutations in genes coding for potassium channel forming and modulating proteins are known (You et al. 2015). The I_{tof} is generated by the $K_v4.2$ (*KCND2*) and $K_v4.3$ (*KCND3*) together with the modulating subunits KChiP2 and KCNE3 (Panama et al. 2016). The α-subunits of $K_v4.2$ and $K_v4.3$ show the typical structure of voltage gated potassium channels with six transmembrane helices and form homo- or heterotetramers with other K_v4 channels (Jerng et al. 2004). Compared to the I_{Ks} and I_{Kr} generating potassium channels, $K_v4.2$ and $K_v4.3$ are rapidly activating and also rapidly deactivating channels resulting in the typical notch of the cardiac AP (Jerng et al. 2004). Experimental data expressing $K_v4.3$ alone show that a modulation by KChiP2 is needed to form the physiological I_{tof} (Deschênes and Tomaselli 2002). KChiP2 belongs to the potassium channel interacting proteins and associates at the cytosolic site with pore forming subunit accelerating the channel activation and increasing the current density (Deschênes and Tomaselli 2002). In contrast, KCNE3 interacts with $K_v4.2$ and $K_v4.3$ and alters the inactivation of these channels (Abbott 2016). BrS causing mutations were found in these two interacting proteins and other modulatory proteins (KCNAB2, KCNE5), as well as in the pore forming α-subunits. These mutations typically result in enhanced I_{tof} by increasing surface expression of the channel complex or/and by increased conductance (Abbott 2016; Campuzano et al. 2020).

8 Ion Channels in Atrial Tachyarrhythmias

Alterations in cardiac electrophysiology are crucial for initiation, progression and maintenance of common atrial arrhythmias called atrial fibrillations (AF). This ion channel remodelling has been of particular interest in AF research. Cellular mechanisms believed to be responsible for ectopic activity – the first step of AF generation – include enhanced atrial automaticity, EAD or DAD (Nattel et al. 2008). Sustained ectopic high-rate activity can trigger atrial re-entry, by acting on vulnerable electrical substrate, thereby leading to AF that may become manifest AF (Heijman et al. 2014). Re-entry is promoted and maintained by a number of factors: atrial refractory period shortening and/or slowed conduction, resulting from atrial electrical and structural remodelling (Wakili et al. 2011).

AF patients display shorter atrial refractoriness, due to abbreviated atrial action potential durations (APD; Fig. 5) (Daoud et al. 1996; Franz et al. 1997). Alterations in ion channel function underlie the so-called electrical remodelling and can occur within hours of AF onset, while changes in gene expression occur much slower (Allessie et al. 2002). The shortening of APD is believed to be caused by impairment of L-type Ca^{2+} channels/$I_{Ca,L}$, whose rate of recovery from inactivation is reduced in AF patients, which contributes to reduced Ca^{2+} influx at high frequencies (Van Wagoner et al. 1999; Bosch et al. 1999; Skasa et al. 2001). Altered functions of repolarizing K^+ currents I_{to}, I_{K1}, I_{Kur} and $I_{K,ACh}$ have also been reported in AF (Schotten et al. 2011). Thus, as a compensatory mechanism, I_{K1} (via $K_{ir}2.1$) was shown to increase in AF patients, which is likely to contribute to the observed APD

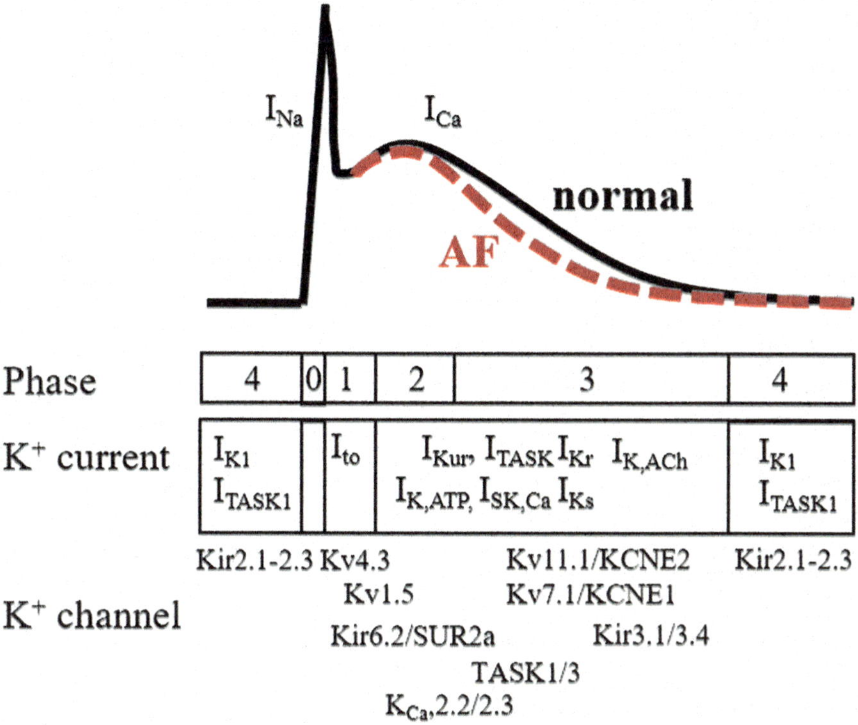

Fig. 5 Schematic depiction of an atrial AP subdivided into Phases 0–4. The dominating potassium current as well as the contributing K channel forming proteins is named for each phase of the AP. Shortening of atrial APs as it occurs in AF patients is indicated as red dotted line

(ZHANG et al. 2005; Gaborit et al. 2005). Additionally, $I_{K,ACh}$ was found to be facilitated by hyperphosphorylation leading to constitutively active channels (Dobrev et al. 2005; Voigt et al. 2007) and I_{Kur} (Kv1.5) was potentially modulated in AF (Van Wagoner et al. 1997; Workman et al. 2001). Due to suggested predominantly atrial distribution of these currents, blockers of $I_{K,ACh}$ or I_{Kur} were considered as promising AF drug candidates (Schotten et al. 2011). However promising preclinical results were not confirmed the subsequent clinical trials.

Another K^+ channel upregulated in AF is TASK-1, a K2P channel which shows certain atrial enrichment. This upregulation may contribute to pathological shortening of the atrial APD as well. Pharmacologic blockade of this $I_{TASK\text{-}1}$ currents could revert APD shortening in cardiomyocytes of AF patients (Schmidt et al. 2015). Thus, $I_{TASK\text{-}1}$ inhibition may represent another approach for AF therapy (Schmidt et al. 2014; Wiedmann et al. 2016). Pharmacological inhibition of atrial $I_{TASK\text{-}1}$ showed acute cardioversion in a porcine model of paroxysmal AF, an effect in agreement with in silico predictions. These experiments indicate the potential therapeutic use of $I_{TASK\text{-}1}$ inhibition in AF treatment (Wiedmann et al. 2020). It will be interesting to see if this concept holds true for future AF therapeutics.

9 Calmodulinopathy Leads a Path to a Potential SK Channel-Based AF Therapy

Over a decade ago, small conductance Ca^{2+}-activated K^+ (SK) channels were reported to be functionally expressed in both human and mouse hearts (Xu et al. 2003; Skibsbye et al. 2014). Using a two-hybrid approach, Xia and colleagues showed that the C-terminal domain of SK channels interacts with calmodulin (CaM) which operates as the intracellular Ca^{2+} sensor (Xia et al. 1998). A flexible linker connects the N-and C-terminal lobes providing CaM with high conformational plasticity allowing it to interact with the SK channels. In each terminal lobe a pair of EF hands are found, enabling a total of four Ca^{2+} ions to interact with one CaM (Marshall et al. 2015). Four CaM molecules are tethered in one functional SK channel. Upon Ca^{2+} binding two CaMBD dimer complexes are formed, which induce a rotary movement that transmits directly to the S6 segment to open the channel (Schumacher et al. 2001; Lee and MacKinnon 2018). The thus Ca^{2+}-activated SK channels are activated during the repolarization phase 3 of the AP in cardiomyocytes when the intracellular Ca^{2+} concentrations are increased (Gu et al. 2018). The expression patterns of SK channels are species specific. In the human, levels of the three SK channel subtypes vary in the four chambers of the heart, whereas the SK channel expression and function is much higher in atria, compared to ventricles. The first CaM mutation associated with heart disease was reported by Nyegaard et al. in 2012 (Nyegaard et al. 2012). Subsequently, about 18 CaM mutations have been identified that associate with catecholaminergic polymorphic ventricular tachycardia (CPVT), long QT syndrome (LQTS) and idiopathic ventricular fibrillation (IVF) (Jensen et al. 2018). The clinical manifestations of CaM gene mutations cause life-threatening arrhythmias and are termed calmodulinopathies (Kotta et al. 2018). Taking the crucial role of CaM in function of several ion channels and the lethality of misfunction into account, it is not surprising that most of these are de novo mutations discovered in infants or very young children and only three are found to be inherited.

SK channels as one of the major CaM interaction partners are apparently atrial selectively expressed. Therefore, SK channels represent as an interesting target for treatment of AF and are the latest addition to the atrial-selective targets is the SK channel. A common SK channel gene variant was associated with AF and, consistent with an SK channel key function in AF, SK2-knockout mice represented prolonged atrial APD, EAD and increased occurrence of AF (Ozgen et al. 2007). Numerous animal studies have shown similar results with relevance for AF (Ozgen et al. 2007; Diness et al. 2010, 2017, 2020; Skibsbye et al. 2011, 2018; Chua et al. 2011; Hsieh et al. 2013; Chang et al. 2013; Zhang et al. 2014b; Kirchhoff et al. 2015; Haugaard et al. 2015; Hundahl et al. 2017; Lubberding et al. 2019). Therefore, cardiac SK channels are promising atrial-selective druggable targets as atrial anti-arrhythmic drugs in otherwise healthy patients. However, in pathophysiological setting of heart failure ventricular SK channels might be of importance (Bonilla et al. 2014).

Atrial fibrosis is a second line of AF that occurs after electrical remodelling implicated in the maintenance of AF by promoting re-entry and conduction slowing.

Furthermore, the degree of atrial structural remodelling and fibrosis is highly correlated with the burden and progression of AF. To be successful, an ion channel targeting anti-AF strategy should therefore be applied as fast as possible after occurrence of AF. Initial upregulation of SK channel currents in patients with persistent AF has been reported, but is followed by SK channel downregulation in patients with permanent AF (Yu et al. 2012; Ling et al. 2013; Skibsbye et al. 2014). Possibly, SK channels are initially upregulated to allow for sufficient repolarization reserve, but are then downregulated in the persistent disease setting because of substantial atrial remodelling (Skibsbye et al. 2014). Thus, precise timing of a potential SK-targeting therapy is currently hard to forecast. Currently, an SK channel inhibitor is undergoing clinical trial, which might disclose the potential of SK block in treatment of AF (Gal et al. 2020).

References

Abbott GW (2015) The KCNE2 K+ channel regulatory subunit: ubiquitous influence, complex pathobiology. Gene 569:162–172. https://doi.org/10.1016/j.gene.2015.06.061

Abbott GW (2016) Regulation of human cardiac potassium channels by full-length KCNE3 and KCNE4. Sci Rep 6:38412. https://doi.org/10.1038/srep38412

Abbott GW, Sesti F, Splawski I et al (1999) MiRP1 forms I Kr potassium channels with HERG and is associated with cardiac arrhythmia. Cell 97:175–187. https://doi.org/10.1016/S0092-8674(00)80728-X

Adler A, Novelli V, Amin AS et al (2020) An international, multicentered, evidence-based reappraisal of genes reported to cause congenital long QT syndrome. Circulation 141:418–428. https://doi.org/10.1161/CIRCULATIONAHA.119.043132

Akdis D, Saguner AM, Medeiros-Domingo A et al (2018) Multiple clinical profiles of families with the short QT syndrome. EP Eur 20:f113–f121. https://doi.org/10.1093/europace/eux186

Allessie M, Ausma J, Schotten U (2002) Electrical, contractile and structural remodeling during atrial fibrillation. Cardiovasc Res 54:230–246. https://doi.org/10.1016/S0008-6363(02)00258-4

Altmann HM, Tester DJ, Will ML et al (2015) Homozygous/compound heterozygous Triadin mutations associated with autosomal-recessive long-QT syndrome and pediatric sudden cardiac arrest. Circulation 131:2051–2060. https://doi.org/10.1161/CIRCULATIONAHA.115.015397

Ambrosini E, Sicca F, Brignone MS et al (2014) Genetically induced dysfunctions of Kir2.1 channels: implications for short QT3 syndrome and autism–epilepsy phenotype. Hum Mol Genet 23:4875–4886. https://doi.org/10.1093/hmg/ddu201

An WF, Bowlby MR, Betty M et al (2000) Modulation of A-type potassium channels by a family of calcium sensors. Nature 403:553–556. https://doi.org/10.1038/35000592

Andersen ED, Krasilnikoff PA, Overvad H (1971) Intermittent muscular weakness, extrasystoles, and multiple developmental anomalies. Acta Paediatr 60:559–564. https://doi.org/10.1111/j.1651-2227.1971.tb06990.x

Anderson CL, Delisle BP, Anson BD et al (2006) Most LQT2 mutations reduce Kv11.1 (hERG) current by a class 2 (trafficking-deficient) mechanism. Circulation 113:365–373. https://doi.org/10.1161/CIRCULATIONAHA.105.570200

Anderson CL, Kuzmicki CE, Childs RR et al (2014) Large-scale mutational analysis of Kv11.1 reveals molecular insights into type 2 long QT syndrome. Nat Commun 5:1–13. https://doi.org/10.1038/ncomms6535

Antzelevitch C, Pollevick GD, Cordeiro JM et al (2007) Loss-of-function mutations in the cardiac calcium channel underlie a new clinical entity characterized by ST-segment elevation, short QT

intervals, and sudden cardiac death. Circulation 115:442–449. https://doi.org/10.1161/CIRCULATIONAHA.106.668392
Barhanin J, Lesage F, Guillemare E et al (1996) KvLQT1 and IsK (minK) proteins associate to form the IKS cardiac potassium current. Nature 384:78–80. https://doi.org/10.1038/384078a0
Barsheshet A, Goldenberg I, O-Uchi J et al (2012) Mutations in cytoplasmic loops of the KCNQ1 channel and the risk of life-threatening events. Circulation 125:1988–1996. https://doi.org/10.1161/CIRCULATIONAHA.111.048041
Bellocq C, van Ginneken ACG, Bezzina CR et al (2004) Mutation in the KCNQ1 gene leading to the short QT-interval syndrome. Circulation 109:2394–2397. https://doi.org/10.1161/01.CIR.0000130409.72142.FE
Bodi I, Mikala G, Koch SE et al (2005) The L-type calcium channel in the heart: the beat goes on. J Clin Invest 115:3306–3317. https://doi.org/10.1172/JCI27167
Bonilla IM, Long VP, Vargas-Pinto P et al (2014) Calcium-activated potassium current modulates ventricular repolarization in chronic heart failure. PLoS One 9:e108824. https://doi.org/10.1371/journal.pone.0108824
Bosch R, Zeng X, Grammer JB et al (1999) Ionic mechanisms of electrical remodeling in human atrial fibrillation. Cardiovasc Res 44:121–131. https://doi.org/10.1016/S0008-6363(99)00178-9
Bradley KK, Jaggar JH, Bonev AD et al (1999) K ir 2.1 encodes the inward rectifier potassium channel in rat arterial smooth muscle cells. J Physiol 515:639–651. https://doi.org/10.1111/j.1469-7793.1999.639ab.x
Brugada P, Brugada J (1992) Right bundle branch block, persistent ST segment elevation and sudden cardiac death: a distinct clinical and electrocardiographic syndrome. J Am Coll Cardiol 20:1391–1396. https://doi.org/10.1016/0735-1097(92)90253-J
Brugada R, Hong K, Dumaine R et al (2004) Sudden death associated with short-QT syndrome linked to mutations in HERG. Circulation 109:30–35. https://doi.org/10.1161/01.CIR.0000109482.92774.3A
Cabral JHM, Lee A, Cohen SL et al (1998) Crystal structure and functional analysis of the HERG potassium channel N terminus. Cell 95:649–655. https://doi.org/10.1016/S0092-8674(00)81635-9
Campuzano O, Sarquella-Brugada G, Cesar S et al (2018) Recent advances in short QT syndrome. Front Cardiovasc Med 5:1–7. https://doi.org/10.3389/fcvm.2018.00149
Campuzano O, Sarquella-Brugada G, Cesar S et al (2020) Update on genetic basis of Brugada syndrome: monogenic, polygenic or oligogenic? Int J Mol Sci 21:1–10. https://doi.org/10.3390/ijms21197155
Chang P-C, Hsieh Y-C, Hsueh C-H et al (2013) Apamin induces early afterdepolarizations and torsades de pointes ventricular arrhythmia from failing rabbit ventricles exhibiting secondary rises in intracellular calcium. Hear Rhythm 10:1516–1524. https://doi.org/10.1016/j.hrthm.2013.07.003
Chen J, Zou A, Splawski I et al (1999) Long QT syndrome-associated mutations in the per-arnt-sim (PAS) domain of HERG potassium channels accelerate channel deactivation. J Biol Chem 274:10113–10118. https://doi.org/10.1074/jbc.274.15.10113
Chen L, Marquardt ML, Tester DJ et al (2007) Mutation of an A-kinase-anchoring protein causes long-QT syndrome. Proc Natl Acad Sci 104:20990–20995. https://doi.org/10.1073/pnas.0710527105
Chua S-K, Chang P-C, Maruyama M et al (2011) Small-conductance calcium-activated potassium channel and recurrent ventricular fibrillation in failing rabbit ventricles. Circ Res 108:971–979. https://doi.org/10.1161/CIRCRESAHA.110.238386
Daoud EG, Bogun F, Goyal R et al (1996) Effect of atrial fibrillation on atrial refractoriness in humans. Circulation 94:1600–1606. https://doi.org/10.1161/01.CIR.94.7.1600
Deal KK, England SK, Tamkun MM (1996) Molecular physiology of cardiac potassium channels. Physiol Rev 76:49–67. https://doi.org/10.1152/physrev.1996.76.1.49

Decher N, Bundis F, Vajna R, Steinmeyer K (2003) KCNE2 modulates current amplitudes and activation kinetics of HCN4: influence of KCNE family members on HCN4 currents. Pflügers Arch Eur J Physiol 446:633–640. https://doi.org/10.1007/s00424-003-1127-7

Deschênes I, Tomaselli GF (2002) Modulation of Kv4.3 current by accessory subunits. FEBS Lett 528:183–188. https://doi.org/10.1016/S0014-5793(02)03296-9

Diness JG, Sørensen US, Nissen JD et al (2010) Inhibition of small-conductance ca 2+ -activated K + channels terminates and protects against atrial fibrillation. Circ Arrhythmia Electrophysiol 3:380–390. https://doi.org/10.1161/CIRCEP.110.957407

Diness JG, Skibsbye L, Simó-Vicens R et al (2017) Termination of Vernakalant-resistant atrial fibrillation by inhibition of small-conductance ca 2+ -activated K + channels in pigs. Circ Arrhythmia Electrophysiol 10:1–13. https://doi.org/10.1161/CIRCEP.117.005125

Diness JG, Kirchhoff JE, Speerschneider T et al (2020) The KCa2 channel inhibitor AP30663 selectively increases atrial refractoriness, converts Vernakalant-resistant atrial fibrillation and prevents its reinduction in conscious pigs. Front Pharmacol 11:1–9. https://doi.org/10.3389/fphar.2020.00159

Dobrev D, Friedrich A, Voigt N et al (2005) The G protein–gated potassium current I K,ACh is constitutively active in patients with chronic atrial fibrillation. Circulation 112:3697–3706. https://doi.org/10.1161/CIRCULATIONAHA.105.575332

Dobrev D, Voigt N, Wehrens XHT (2011) The ryanodine receptor channel as a molecular motif in atrial fibrillation: pathophysiological and therapeutic implications. Cardiovasc Res 89:734–743. https://doi.org/10.1093/cvr/cvq324

Donaldson M, Yoon G, Fu Y, Ptacek L (2004) Andersen-Tawil syndrome: a model of clinical variability, pleiotropy, and genetic heterogeneity. Ann Med 36:92–97. https://doi.org/10.1080/17431380410032490

Draghici AE, Taylor JA (2016) The physiological basis and measurement of heart rate variability in humans. J Physiol Anthropol 35:22. https://doi.org/10.1186/s40101-016-0113-7

Duggal P, Vesely MR, Wattanasirichaigoon D et al (1998) Mutation of the gene for I sK associated with both Jervell and Lange-Nielsen and Romano-Ward forms of long-QT syndrome. Circulation 97:142–146. https://doi.org/10.1161/01.CIR.97.2.142

Dvir M, Strulovich R, Sachyani D et al (2014) Long QT mutations at the interface between KCNQ1 helix C and KCNE1 disrupt IKS regulation by PKA and PIP2. J Cell Sci 127:3943–3955. https://doi.org/10.1242/jcs.147033

Faridi R, Tona R, Brofferio A et al (2018) Mutational and phenotypic spectra of KCNE1 deficiency in Jervell and Lange-Nielsen syndrome and Romano-Ward syndrome. Hum Mutat 40: humu.23689. https://doi.org/10.1002/humu.23689

Fenske S, Hennis K, Rötzer RD et al (2020) cAMP-dependent regulation of HCN4 controls the tonic entrainment process in sinoatrial node pacemaker cells. Nat Commun 11. https://doi.org/10.1038/s41467-020-19304-9

Fernández-Falgueras A, Sarquella-Brugada G, Brugada J et al (2017) Cardiac channelopathies and sudden death: recent clinical and genetic advances. Biology (Basel) 6:7. https://doi.org/10.3390/biology6010007

Franz MR, Karasik PL, Li C et al (1997) Electrical remodeling of the human atrium: similar effects in patients with chronic atrial fibrillation and atrial flutter 11To discuss this article on-line, visit the ACC home page at www.acc.org/members and click on the JACC forum. J Am Coll Cardiol 30:1785–1792. https://doi.org/10.1016/S0735-1097(97)00385-9

Gaborit N, Steenman M, Lamirault G et al (2005) Human atrial ion channel and transporter subunit gene-expression remodeling associated with Valvular heart disease and atrial fibrillation. Circulation 112:471–481. https://doi.org/10.1161/CIRCULATIONAHA.104.506857

Gaita F, Giustetto C, Bianchi F et al (2003) Short QT syndrome. Circulation 108:965–970. https://doi.org/10.1161/01.CIR.0000085071.28695.C4

Gal P, Klaassen ES, Bergmann KR et al (2020) First clinical study with AP30663 – a K Ca 2 channel inhibitor in development for conversion of atrial fibrillation. Clin Transl Sci 13 (6):1336–1344. https://doi.org/10.1111/cts.12835

Garcia-Elias A, Benito B (2018) Ion channel disorders and sudden cardiac death. Int J Mol Sci 19:692. https://doi.org/10.3390/ijms19030692

Giudicessi JR, Wilde AAM, Ackerman MJ (2018) The genetic architecture of long QT syndrome: a critical reappraisal. Trends Cardiovasc Med 28:453–464. https://doi.org/10.1016/j.tcm.2018.03.003

Gong Q, Zhang L, Vincent GM et al (2007) Nonsense mutations in hERG cause a decrease in mutant mRNA transcripts by nonsense-mediated mRNA decay in human long-QT syndrome. Circulation 116:17–24. https://doi.org/10.1161/CIRCULATIONAHA.107.708818

Grunnet M (2010) Repolarization of the cardiac action potential. Does an increase in repolarization capacity constitute a new anti-arrhythmic principle? Acta Physiol 198:1–48. https://doi.org/10.1111/j.1748-1716.2009.02072.x

Gu M, Zhu Y, Yin X, Zhang D-M (2018) Small-conductance Ca2+−activated K+ channels: insights into their roles in cardiovascular disease. Exp Mol Med 50:23. https://doi.org/10.1038/s12276-018-0043-z

Gussak I, Brugada P, Brugada J et al (2000) Idiopathic short QT interval: a new clinical syndrome? Cardiology 94:99–102. https://doi.org/10.1159/000047299

Harmer SC, Tinker A (2007) The role of abnormal trafficking of KCNE1 in long QT syndrome 5. Biochem Soc Trans 35:1074–1076. https://doi.org/10.1042/BST0351074

Hattori T, Makiyama T, Akao M et al (2012) A novel gain-of-function KCNJ2 mutation associated with short-QT syndrome impairs inward rectification of Kir2.1 currents. Cardiovasc Res 93:666–673. https://doi.org/10.1093/cvr/cvr329

Haugaard MM, Hesselkilde EZ, Pehrson S et al (2015) Pharmacologic inhibition of small-conductance calcium-activated potassium (SK) channels by NS8593 reveals atrial antiarrhythmic potential in horses. Hear Rhythm 12:825–835. https://doi.org/10.1016/j.hrthm.2014.12.028

Hedley PL, Jørgensen P, Schlamowitz S et al (2009) The genetic basis of long QT and short QT syndromes: a mutation update. Hum Mutat 30:1486–1511. https://doi.org/10.1002/humu.21106

Heijman J, Voigt N, Nattel S, Dobrev D (2014) Cellular and molecular electrophysiology of atrial fibrillation initiation, maintenance, and progression. Circ Res 114:1483–1499. https://doi.org/10.1161/CIRCRESAHA.114.302226

Henrion U, Strutz-Seebohm N, Duszenko M et al (2009) Long QT syndrome-associated mutations in the voltage sensor of IKs channels. Cell Physiol Biochem 24:11–16. https://doi.org/10.1159/000227828

Hsieh Y-C, Chang P-C, Hsueh C-H et al (2013) Apamin-sensitive potassium current modulates action potential duration restitution and arrhythmogenesis of failing rabbit ventricles. Circ Arrhythmia Electrophysiol 6:410–418. https://doi.org/10.1161/CIRCEP.111.000152

Hundahl LA, Sattler SM, Skibsbye L et al (2017) Pharmacological blockade of small conductance Ca2+−activated K+ channels by ICA reduces arrhythmic load in rats with acute myocardial infarction. Pflügers Arch Eur J Physiol 469:739–750. https://doi.org/10.1007/s00424-017-1962-6

Jensen HH, Brohus M, Nyegaard M, Overgaard MT (2018) Human calmodulin mutations. Front Mol Neurosci 11. https://doi.org/10.3389/fnmol.2018.00396

Jerng HH, Pfaffinger PJ, Covarrubias M (2004) Molecular physiology and modulation of somatodendritic A-type potassium channels. Mol Cell Neurosci 27:343–369. https://doi.org/10.1016/j.mcn.2004.06.011

Jervell A, Lange-Nielsen F (1957) Congenital deaf-mutism, functional heart disease with prolongation of the Q-T interval, and sudden death. Am Heart J 54:59–68. https://doi.org/10.1016/0002-8703(57)90079-0

Johnson EK, Springer SJ, Wang W et al (2018) Differential expression and remodeling of transient outward potassium currents in human left ventricles. Circ Arrhythmia Electrophysiol 11:1–16. https://doi.org/10.1161/CIRCEP.117.005914

Jones F, Gamper N, Gao H (2021) Kv7 channels and excitability disorders. Handb Exp Pharmacol. https://doi.org/10.1007/164_2021_457. Epub ahead of print. PMID: 33860384

Kannampuzha JA, Sengodan P, Avula S et al (2018) Non-sustained microvolt level T-wave alternans in congenital long QT syndrome types 1 and 2. J Electrocardiol 51:303–308. https://doi.org/10.1016/j.jelectrocard.2017.11.008
Kapa S, Tester DJ, Salisbury BA et al (2009) Genetic testing for long-QT syndrome. Circulation 120:1752–1760. https://doi.org/10.1161/CIRCULATIONAHA.109.863076
Kapplinger JD, Tester DJ, Alders M et al (2010) An international compendium of mutations in the SCN5A-encoded cardiac sodium channel in patients referred for Brugada syndrome genetic testing. Hear Rhythm 7:33–46. https://doi.org/10.1016/j.hrthm.2009.09.069
Kim Y, Bang H, Kim D (1999) TBAK-1 and TASK-1, two-pore K + channel subunits: kinetic properties and expression in rat heart. Am J Physiol Circ Physiol 277:H1669–H1678. https://doi.org/10.1152/ajpheart.1999.277.5.H1669
Kim JA, Lopes CM, Moss AJ et al (2010) Trigger-specific risk factors and response to therapy in long QT syndrome type 2. Hear Rhythm 7:1797–1805. https://doi.org/10.1016/j.hrthm.2010.09.011
Kimrey J, Vo T, Bertram R (2020) Canard analysis reveals why a large Ca2+ window current promotes early afterdepolarizations in cardiac myocytes. PLoS Comput Biol 16:e1008341. https://doi.org/10.1371/journal.pcbi.1008341
Kirchhoff JE, Goldin Diness J, Sheykhzade M et al (2015) Synergistic antiarrhythmic effect of combining inhibition of Ca2+−activated K+ (SK) channels and voltage-gated Na+ channels in an isolated heart model of atrial fibrillation. Hear Rhythm 12:409–418. https://doi.org/10.1016/j.hrthm.2014.12.010
Kojima A, Fukushima Y, Itoh H et al (2020) A computational analysis of the effect of sevoflurane in a human ventricular cell model of long QT syndrome: importance of repolarization reserve in the QT-prolonging effect of sevoflurane. Eur J Pharmacol 883:173378. https://doi.org/10.1016/j.ejphar.2020.173378
Kokunai Y, Nakata T, Furuta M et al (2014) A Kir3.4 mutation causes Andersen-Tawil syndrome by an inhibitory effect on Kir2.1. Neurology 82:1058–1064. https://doi.org/10.1212/WNL.0000000000000239
Kotta M-C, Sala L, Ghidoni A et al (2018) Calmodulinopathy: a novel, life-threatening clinical entity affecting the young. Front Cardiovasc Med 5:1–10. https://doi.org/10.3389/fcvm.2018.00175
Kurokawa J, Motoike HK, Rao J, Kass RS (2004) Regulatory actions of the A-kinase anchoring protein Yotiao on a heart potassium channel downstream of PKA phosphorylation. Proc Natl Acad Sci 101:16374–16378. https://doi.org/10.1073/pnas.0405583101
Lee C-H, MacKinnon R (2018) Activation mechanism of a human SK-calmodulin channel complex elucidated by cryo-EM structures. Science 360:508–513. https://doi.org/10.1126/science.aas9466
Li N, Csepe TA, Hansen BJ et al (2015) Molecular mapping of sinoatrial node HCN channel expression in the human heart. Circ Arrhythmia Electrophysiol 8:1219–1227. https://doi.org/10.1161/CIRCEP.115.003070
Ling T-Y, Wang X-L, Chai Q et al (2013) Regulation of the SK3 channel by microRNA-499 – potential role in atrial fibrillation. Hear Rhythm 10:1001–1009. https://doi.org/10.1016/j.hrthm.2013.03.005
Liu GX, Derst C, Schlichthörl G et al (2001) Comparison of cloned Kir2 channels with native inward rectifier K + channels from Guinea-pig cardiomyocytes. J Physiol 532:115–126. https://doi.org/10.1111/j.1469-7793.2001.0115g.x
Lin Y, Grinter SZ, Lu Z, Xu X, Wang HZ, Liang H, Hou P, Gao J, Clausen C, Shi J, Zhao W, Ma Z, Liu Y, White KM, Zhao L, Kang PW, Zhang G, Cohen IS, Zou X, Cui J (2021) Modulating the voltage sensor of a cardiac potassium channel shows antiarrhythmic effects. Proc Natl Acad Sci U S A 118(20):e2024215118. https://doi.org/10.1073/pnas.2024215118. PMID: 33990467; PMCID: PMC8157969

Lubberding AF, Sattler SM, Grunnet M et al (2019) Arrhythmia development during inhibition of small-conductance calcium-activated potassium channels in acute myocardial infarction in a porcine model. EP Eur 21:1584–1593. https://doi.org/10.1093/europace/euz223

Lyashkov AE, Beahr J, Lakatta EG et al (2018) Positive feedback mechanisms among local ca releases, NCX, and ICaL ignite pacemaker action potentials. Biophys J 114:1176–1189. https://doi.org/10.1016/j.bpj.2017.12.043

Maroni M, Körner J, Schüttler J et al (2019) β1 and β3 subunits amplify mechanosensitivity of the cardiac voltage-gated sodium channel Nav1.5. Pflugers Arch Eur J Physiol 471:1481–1492. https://doi.org/10.1007/s00424-019-02324-w

Marshall CB, Nishikawa T, Osawa M et al (2015) Calmodulin and STIM proteins: two major calcium sensors in the cytoplasm and endoplasmic reticulum. Biochem Biophys Res Commun 460:5–21. https://doi.org/10.1016/j.bbrc.2015.01.106

Marx SO, Kurokawa J, Reiken S et al (2002) Requirement of a macromolecular signaling complex for beta adrenergic receptor modulation of the KCNQ1-KCNE1 potassium channel. Science 295:496–499. https://doi.org/10.1126/science.1066843

Moreno C, Oliveras A, de la Cruz A et al (2015) A new KCNQ1 mutation at the S5 segment that impairs its association with KCNE1 is responsible for short QT syndrome. Cardiovasc Res 107:613–623. https://doi.org/10.1093/cvr/cvv196

Moreno-Galindo EG, Alamilla J, Sanchez-Chapula JA et al (2016) The agonist-specific voltage dependence of M2 muscarinic receptors modulates the deactivation of the acetylcholine-gated K + current (I KACh). Pflügers Arch Eur J Physiol 468:1207–1214. https://doi.org/10.1007/s00424-016-1812-y

Moss AJ, Shimizu W, Wilde AAM et al (2007) Clinical aspects of type-1 long-QT syndrome by location, coding type, and biophysical function of mutations involving the KCNQ1 gene. Circulation 115:2481–2489. https://doi.org/10.1161/CIRCULATIONAHA.106.665406

Nattel S, Burstein B, Dobrev D (2008) Atrial remodeling and atrial fibrillation. Circ Arrhythmia Electrophysiol 1:62–73. https://doi.org/10.1161/CIRCEP.107.754564

Ng C-A, Perry MD, Liang W et al (2020) High-throughput phenotyping of heteromeric human ether-à-go-go-related gene potassium channel variants can discriminate pathogenic from rare benign variants. Hear Rhythm 17:492–500. https://doi.org/10.1016/j.hrthm.2019.09.020

Nyegaard M, Overgaard MT, Søndergaard MT et al (2012) Mutations in calmodulin cause ventricular tachycardia and sudden cardiac death. Am J Hum Genet 91:703–712. https://doi.org/10.1016/j.ajhg.2012.08.015

Ozgen N, Dun W, Sosunov E et al (2007) Early electrical remodeling in rabbit pulmonary vein results from trafficking of intracellular SK2 channels to membrane sites. Cardiovasc Res 75:758–769. https://doi.org/10.1016/j.cardiores.2007.05.008

Panama BK, Korogyi AS, Aschar-Sobbi R et al (2016) Reductions in the cardiac transient outward K + current I to caused by chronic β-adrenergic receptor stimulation are partly rescued by inhibition of nuclear factor κB. J Biol Chem 291:4156–4165. https://doi.org/10.1074/jbc.M115.694984

Pegan S, Arrabit C, Zhou W et al (2005) Cytoplasmic domain structures of Kir2.1 and Kir3.1 show sites for modulating gating and rectification. Nat Neurosci 8:279–287. https://doi.org/10.1038/nn1411

Pegan S, Arrabit C, Slesinger PA, Choe S (2006) Andersen's syndrome mutation effects on the structure and assembly of the cytoplasmic domains of Kir2.1. Biochemistry 45:8599–8606. https://doi.org/10.1021/bi060653d

Piccini I, Fehrmann E, Frank S et al (2017) Adrenergic stress protection of human iPS cell-derived cardiomyocytes by fast Kv7.1 recycling. Front Physiol 8:1–13. https://doi.org/10.3389/fphys.2017.00705

Pongs O, Leicher T, Berger M et al (1999) Functional and molecular aspects of voltage-gated K+ channel beta subunits. Ann N Y Acad Sci 868:344–355. https://doi.org/10.1111/j.1749-6632.1999.tb11296.x

Preisig-Muller R, Schlichthorl G, Goerge T et al (2002) Heteromerization of Kir2.x potassium channels contributes to the phenotype of Andersen's syndrome. Proc Natl Acad Sci 99:7774–7779. https://doi.org/10.1073/pnas.102609499

Priori SG, Pandit SV, Rivolta I et al (2005) A novel form of short QT syndrome (SQT3) is caused by a mutation in the KCNJ2 gene. Circ Res 96:800–807. https://doi.org/10.1161/01.RES.0000162101.76263.8c

Priori SG, Wilde AA, Horie M et al (2013) HRS/EHRA/APHRS expert consensus statement on the diagnosis and management of patients with inherited primary arrhythmia syndromes. Hear Rhythm 10:1932–1963. https://doi.org/10.1016/j.hrthm.2013.05.014

Priori SG, Blomström-Lundqvist C, Mazzanti A et al (2015) 2015 ESC guidelines for the management of patients with ventricular arrhythmias and the prevention of sudden cardiac death. Eur Heart J 36:2793–2867. https://doi.org/10.1093/eurheartj/ehv316

Rajakulendran S, Tan SV, Hanna MG (2010) Muscle weakness, palpitations and a small chin: the Andersen-Tawil syndrome. Pract Neurol 10:227–231. https://doi.org/10.1136/jnnp.2010.217794

Rautaharju PM, Surawicz B, Gettes LS (2009) AHA/ACCF/HRS recommendations for the standardization and interpretation of the electrocardiogram. Part IV: the ST segment, T and U waves, and the QT interval a scientific statement from the American Heart Association Electrocardiography and Arrhythmias Co. J Am Coll Cardiol 53:982–991. https://doi.org/10.1016/j.jacc.2008.12.014

Roberts JD, Krahn AD, Ackerman MJ et al (2017) Loss-of-function KCNE2 variants. Circ Arrhythmia Electrophysiol 10:1–11. https://doi.org/10.1161/CIRCEP.117.005282

Roberts JD, Asaki SY, Mazzanti A et al (2020) An international multicenter evaluation of type 5 long QT syndrome. Circulation 141:429–439. https://doi.org/10.1161/CIRCULATIONAHA.119.043114

Romano C, Gemme G, Pongiglione R (1963) Rare cardiac arrythmias of the pediatric age. II. Syncopal attacks due to paroxysmal ventricular fibrillation. Clin Pediatr 45:656–683

Roomi SS, Ullah W, Abbas H et al (2020) Brugada syndrome unmasked by fever: a comprehensive review of literature. J Commun Hosp Intern Med Perspect 10:224–228. https://doi.org/10.1080/20009666.2020.1767278

Rudy Y (2008) Molecular basis of cardiac action potential repolarization. Ann N Y Acad Sci 1123:113–118. https://doi.org/10.1196/annals.1420.013

Sanguinetti MC, Jurkiewicz NK (1990) Two components of cardiac delayed rectifier K+ current. Differential sensitivity to block by class III antiarrhythmic agents. J Gen Physiol 96:195–215. https://doi.org/10.1085/jgp.96.1.195

Sanguinetti MC, Curran ME, Zou A et al (1996) Coassembly of KVLQT1 and minK (IsK) proteins to form cardiac IKS potassium channel. Nature 384:80–83. https://doi.org/10.1038/384080a0

Sartiani L, Mannaioni G, Masi A et al (2017) The hyperpolarization-activated cyclic nucleotide-gated channels: from biophysics to pharmacology of a unique family of ion channels. Pharmacol Rev 69:354–395. https://doi.org/10.1124/pr.117.014035

Schmidt C, Wiedmann F, Langer C et al (2014) Cloning, functional characterization, and remodeling of K2P3.1 (TASK-1) potassium channels in a porcine model of atrial fibrillation and heart failure. Hear Rhythm 11:1798–1805. https://doi.org/10.1016/j.hrthm.2014.06.020

Schmidt C, Wiedmann F, Voigt N et al (2015) Upregulation of K 2P 3.1 K + current causes action potential shortening in patients with chronic atrial fibrillation. Circulation 132:82–92. https://doi.org/10.1161/CIRCULATIONAHA.114.012657

Schotten U, Verheule S, Kirchhof P, Goette A (2011) Pathophysiological mechanisms of atrial fibrillation: a translational appraisal. Physiol Rev 91:265–325. https://doi.org/10.1152/physrev.00031.2009

Schumacher MA, Rivard AF, Bächinger HP, Adelman JP (2001) Structure of the gating domain of a Ca^{2+}-activated K^{+} channel complexed with Ca ^{2+}/calmodulin. Nature 410:1120–1124

Schwartz PJ, Stramba-Badiale M, Crotti L et al (2009) Prevalence of the congenital long-QT syndrome. Circulation 120:1761–1767. https://doi.org/10.1161/CIRCULATIONAHA.109.863209

Schwartz PJ, Crotti L, Insolia R (2012) Long-QT syndrome: from genetics to management. Circ Arrhythm Electrophysiol 5:868–877. https://doi.org/10.1161/CIRCEP.111.962019

Seebohm G, Scherer CR, Busch AE, Lerche C (2001) Identification of specific pore residues mediating KCNQ1 inactivation. J Biol Chem 276:13600–13605. https://doi.org/10.1074/jbc.M008373200

Seebohm G, Strutz-Seebohm N, Birkin R et al (2007) Regulation of endocytic recycling of KCNQ1/KCNE1 potassium channels. Circ Res 100:686–692. https://doi.org/10.1161/01.RES.0000260250.83824.8f

Seebohm G, Strutz-Seebohm N, Ureche ON et al (2008) Long QT syndrome–associated mutations in KCNQ1 and KCNE1 subunits disrupt normal endosomal recycling of I Ks channels. Circ Res 103:1451–1457. https://doi.org/10.1161/CIRCRESAHA.108.177360

Seebohm G, Strutz-Seebohm N, Ursu ON et al (2012) Altered stress stimulation of inward rectifier potassium channels in Andersen-Tawil syndrome. FASEB J 26:513–522. https://doi.org/10.1096/fj.11-189126

Seemann G, Sachse FB, Weiss DL et al (2007) Modeling of I K1 mutations in human left ventricular myocytes and tissue. Am J Physiol Circ Physiol 292:H549–H559. https://doi.org/10.1152/ajpheart.00701.2006

Shih HT (1994) Anatomy of the action potential in the heart. Tex Hear Inst J 21:30–41

Shimizu W, Horie M, Ohno S et al (2004) Mutation site-specific differences in arrhythmic risk and sensitivity to sympathetic stimulation in the LQT1 form of congenital long QT syndrome. J Am Coll Cardiol 44:117–125. https://doi.org/10.1016/j.jacc.2004.03.043

Skasa M, Jüngling E, Picht E et al (2001) L-type calcium currents in atrial myocytes from patients with persistent and non-persistent atrial fibrillation. Basic Res Cardiol 96:151–159. https://doi.org/10.1007/s003950170065

Skibsbye L, Diness JG, Sørensen US et al (2011) The duration of pacing-induced atrial fibrillation is reduced in vivo by inhibition of small conductance Ca2+−activated k+ channels. J Cardiovasc Pharmacol 57:672–681. https://doi.org/10.1097/FJC.0b013e318217943d

Skibsbye L, Poulet C, Diness JG et al (2014) Small-conductance calcium-activated potassium (SK) channels contribute to action potential repolarization in human atria. Cardiovasc Res 103:156–167. https://doi.org/10.1093/cvr/cvu121

Skibsbye L, Bengaard AK, Uldum-Nielsen AM et al (2018) Inhibition of small conductance calcium-activated potassium (SK) channels prevents arrhythmias in rat atria during β-adrenergic and muscarinic receptor activation. Front Physiol 9:1–13. https://doi.org/10.3389/fphys.2018.00510

Smith JL, Anderson CL, Burgess DE et al (2016) Molecular pathogenesis of long QT syndrome type 2. J Arrhythmia 32:373–380. https://doi.org/10.1016/j.joa.2015.11.009

Splawski I, Shen J, Timothy KW et al (2000) Spectrum of mutations in long-QT syndrome genes. Circulation 102:1178–1185. https://doi.org/10.1161/01.CIR.102.10.1178

Stephenson RS, Atkinson A, Kottas P et al (2017) High resolution 3-dimensional imaging of the human cardiac conduction system from microanatomy to mathematical modeling. Sci Rep 7:7188. https://doi.org/10.1038/s41598-017-07694-8

Su S, Sun J, Wang Y, Xu Y (2021) Cardiac hERG K+ channel as safety and pharmacological target. Handb Exp Pharmacol. https://doi.org/10.1007/164_2021_455. Epub ahead of print. PMID: 33829343

Sugrue A, Noseworthy PA, Kremen V et al (2017) Automated T-wave analysis can differentiate acquired QT prolongation from congenital long QT syndrome. Ann Noninvasive Electrocardiol 22:e12455. https://doi.org/10.1111/anec.12455

Sun Y, Quan X-Q, Fromme S et al (2011) A novel mutation in the KCNH2 gene associated with short QT syndrome. J Mol Cell Cardiol 50:433–441. https://doi.org/10.1016/j.yjmcc.2010.11.017

Tawil R, Ptacek LJ, Pavlakis SG et al (1994) Andersen's syndrome: potassium-sensitive periodic paralysis, ventricular ectopy, and dysmorphic features. Ann Neurol 35:326–330. https://doi.org/10.1002/ana.410350313
Templin C, Ghadri J-R, Rougier J-S et al (2011) Identification of a novel loss-of-function calcium channel gene mutation in short QT syndrome (SQTS6). Eur Heart J 32:1077–1088. https://doi.org/10.1093/eurheartj/ehr076
Thorsen K, Dam VS, Kjaer-Sorensen K et al (2017) Loss-of-activity-mutation in the cardiac chloride-bicarbonate exchanger AE3 causes short QT syndrome. Nat Commun 8:1696. https://doi.org/10.1038/s41467-017-01630-0
Tristani-Firouzi M, Jensen JL, Donaldson MR et al (2002) Functional and clinical characterization of KCNJ2 mutations associated with LQT7 (Andersen syndrome). J Clin Invest 110:381–388. https://doi.org/10.1172/JCI15183
Van Wagoner DR, Pond AL, McCarthy PM et al (1997) Outward K + current densities and Kv1.5 expression are reduced in chronic human atrial fibrillation. Circ Res 80:772–781. https://doi.org/10.1161/01.RES.80.6.772
Van Wagoner DR, Pond AL, Lamorgese M et al (1999) Atrial L-type Ca 2+ currents and human atrial fibrillation. Circ Res 85:428–436. https://doi.org/10.1161/01.RES.85.5.428
Voigt N, Friedrich A, Bock M et al (2007) Differential phosphorylation-dependent regulation of constitutively active and muscarinic receptor-activated IK,ACh channels in patients with chronic atrial fibrillation. Cardiovasc Res 74:426–437. https://doi.org/10.1016/j.cardiores.2007.02.009
Vutthikraivit W, Rattanawong P, Putthapiban P et al (2018) Worldwide prevalence of Brugada syndrome: a systematic review and meta-analysis. Acta Cardiol Sin 34:267–277. https://doi.org/10.6515/ACS.201805_34(3).20180302B
Wakili R, Voigt N, Kääb S et al (2011) Recent advances in the molecular pathophysiology of atrial fibrillation. J Clin Invest 121:2955–2968. https://doi.org/10.1172/JCI46315
Wang W, MacKinnon R (2017) Cryo-EM structure of the open human ether-à-go-go -related K + channel hERG. Cell 169:422–430.e10. https://doi.org/10.1016/j.cell.2017.03.048
Wang Q, Shen J, Splawski I et al (1995) SCN5A mutations associated with an inherited cardiac arrhythmia, long QT syndrome. Cell 80:805–811. https://doi.org/10.1016/0092-8674(95)90359-3
Wang Y, Zhang M, Xu Y et al (2012) Probing the structural basis for differential KCNQ1 modulation by KCNE1 and KCNE2. J Gen Physiol 140:653–669. https://doi.org/10.1085/jgp.201210847
Ward OC (1964) A new familial cardiac syndrome in children. J Irish Med Assoc 54:103–106
Watanabe H, Minamino T (2016) Genetics of Brugada syndrome. J Hum Genet 61:57–60. https://doi.org/10.1038/jhg.2015.97
Wettwer E, Hála O, Christ T et al (2004) Role of I Kur in controlling action potential shape and contractility in the human atrium. Circulation 110:2299–2306. https://doi.org/10.1161/01.CIR.0000145155.60288.71
Wiedmann F, Schmidt C, Lugenbiel P et al (2016) Therapeutic targeting of two-pore-domain potassium (K2P) channels in the cardiovascular system. Clin Sci 130:643–650. https://doi.org/10.1042/CS20150533
Wiedmann F, Beyersdorf C, Zhou X et al (2020) Pharmacologic TWIK-related acid-sensitive K+ channel (TASK-1) potassium channel inhibitor A293 facilitates acute cardioversion of paroxysmal atrial fibrillation in a porcine large animal model. J Am Heart Assoc 9:e015751. https://doi.org/10.1161/JAHA.119.015751
Wilde AAM, Antzelevitch C, Borggrefe M et al (2002) Proposed diagnostic criteria for the Brugada syndrome. Circulation 106:2514–2519. https://doi.org/10.1161/01.CIR.0000034169.45752.4A
Wilders R, Verkerk AO (2018) Long QT syndrome and sinus bradycardia–a mini review. Front Cardiovasc Med 5:1–7. https://doi.org/10.3389/fcvm.2018.00106

Workman AJ, Kane KA, Rankin AC (2001) The contribution of ionic currents to changes in refractoriness of human atrial myocytes associated with chronic atrial fibrillation. Cardiovasc Res 52:226–235. https://doi.org/10.1016/S0008-6363(01)00380-7
Xia X-M, Fakler B, Rivard A et al (1998) Mechanism of calcium gating in small-conductance calcium-activated potassium channels. Nature 395:503–507. https://doi.org/10.1038/26758
Xu Y, Tuteja D, Zhang Z et al (2003) Molecular identification and functional roles of a Ca 2+ -activated K + channel in human and mouse hearts. J Biol Chem 278:49085–49094. https://doi.org/10.1074/jbc.M307508200
Yoon G, Oberoi S, Tristani-Firouzi M et al (2006) Andersen-Tawil syndrome: prospective cohort analysis and expansion of the phenotype. Am J Med Genet Part A 140A:312–321. https://doi.org/10.1002/ajmg.a.31092
YOU T, MAO W, CAI B et al (2015) Two novel Brugada syndrome-associated mutations increase KV4.3 membrane expression and function. Int J Mol Med 36:309–315. https://doi.org/10.3892/ijmm.2015.2223
Yu T, Deng C, Wu R et al (2012) Decreased expression of small-conductance Ca2+−activated K+ channels SK1 and SK2 in human chronic atrial fibrillation. Life Sci 90:219–227. https://doi.org/10.1016/j.lfs.2011.11.008
Zeng J, Rudy Y (1995) Early afterdepolarizations in cardiac myocytes: mechanism and rate dependence. Biophys J 68:949–964. https://doi.org/10.1016/S0006-3495(95)80271-7
Zhang H, Garratt C, Zhu J, Holden A (2005) Role of up-regulation of in action potential shortening associated with atrial fibrillation in humans. Cardiovasc Res 66:493–502. https://doi.org/10.1016/j.cardiores.2005.01.020
Zhang M, D'Aniello C, Verkerk AO et al (2014a) Recessive cardiac phenotypes in induced pluripotent stem cell models of Jervell and Lange-Nielsen syndrome: disease mechanisms and pharmacological rescue. Proc Natl Acad Sci 111:E5383–E5392. https://doi.org/10.1073/pnas.1419553111
Zhang X-D, Timofeyev V, Li N et al (2014b) Critical roles of a small conductance Ca2+−activated K+ channel (SK3) in the repolarization process of atrial myocytes. Cardiovasc Res 101:317–325. https://doi.org/10.1093/cvr/cvt262

Cardiac hERG K^+ Channel as Safety and Pharmacological Target

Shi Su, Jinglei Sun, Yi Wang, and Yanfang Xu

Contents

Abstract

The human *ether-á-go-go related* gene (*hERG*, *KCNH2*) encodes the pore-forming subunit of the potassium channel responsible for a fast component of the cardiac delayed rectifier potassium current (I_{Kr}). Outward I_{Kr} is an important determinant of cardiac action potential (AP) repolarization and effectively

S. Su · J. Sun · Y. Wang · Y. Xu (✉)
Department of Pharmacology, Hebei Medical University, The Key Laboratory of New Drug Pharmacology and Toxicology, Hebei, China
e-mail: yanfangxu@hebmu.edu.cn

N. Gamper, K. Wang (eds.), *Pharmacology of Potassium Channels*,
Handbook of Experimental Pharmacology 267, https://doi.org/10.1007/164_2021_455

controls the duration of the QT interval in humans. Dysfunction of hERG channel can cause severe ventricular arrhythmias and thus modulators of the channel, including hERG inhibitors and activators, continue to attract intense pharmacological interest. Certain inhibitors of hERG channel prolong the action potential duration (APD) and effective refractory period (ERP) to suppress premature ventricular contraction and are used as class III antiarrhythmic agents. However, a reduction of the hERG/I_{Kr} current has been recognized as a predominant mechanism responsible for the drug-induced delayed repolarization known as acquired long QT syndromes (LQTS), which is linked to an increased risk for "torsades de pointes" (TdP) ventricular arrhythmias and sudden cardiac death. Many drugs of different classes and structures have been identified to carry TdP risk. Hence, assessing hERG/I_{Kr} blockade of new drug candidates is mandatory in the drug development process according to the regulatory agencies. In contrast, several hERG channel activators have been shown to enhance I_{Kr} and shorten the APD and thus might have potential antiarrhythmic effects against pathological LQTS. However, these activators may also be proarrhythmic due to excessive shortening of APD and the ERP.

Keywords

Activator · Arrhythmia · hERG · Inhibitor · Long QT syndromes

1 Introduction

Cardiac arrhythmias are one of the major causes of cardiovascular disease-related deaths worldwide. Ion channels are pore-forming proteins that provide pathways for the transmembrane movement of ions and thus control the cardiac action potential (AP) generation and propagation, resulting in the release of Ca^{2+} from intracellular stores and triggering cardiac muscle contraction. Abnormalities in cardiac ion channel function may lead to arrhythmias and sudden cardiac death (Keating and Sanguinetti 2001). The human *ether-á-go-go related* gene (hERG, *KCNH2*) encodes the pore-forming subunit (Kv11.1) of the channel that in cardiac myocytes conducts the rapidly activating delayed rectifier potassium current (I_{Kr}). Outward I_{Kr} is a critical current in the phase 3 AP repolarization in the human ventricle and effectively controls the QT interval of the electrocardiogram (Sanguinetti et al. 1995). Inhibition of I_{Kr} results in the prolongation of repolarization, which has been described as an antiarrhythmic mechanism of Class III antiarrhythmic agents (Singh and Vaughan Williams 1970). However, these drugs have also been found to be associated with an increased risk of arrhythmias. In addition to antiarrhythmic agents, a wide variety of different classes of non-antiarrhythmic pharmaceuticals have the potential to inhibit hERG/I_{Kr} current and, thus, can pose a threat of the drug-induced form of acquired long QT syndromes (LQTS) associated with an increased risk of an unusual life-threatening form of arrhythmia known as torsades de pointes (TdP) (Sanguinetti and Tristani-Firouzi 2006; Vandenberg et al. 2012). Consequently, assessing potential I_{Kr}/hERG inhibition of drug candidates has become a major requirement in new drug development process (Hancox et al. 2008; Sanguinetti and Mitcheson 2005). Considerable effort has been made to understand the molecular basis underlying the susceptibility of hERG channel to pharmacological inhibition. A recent cryoelectron microscopy (cryo-EM) structure of hERG (Wang and MacKinnon 2017) has provided opportunities to better understand hERG channel gating and pharmacology (Butler et al. 2019). This review briefly describes hERG channel as a pharmacological and safety target for antiarrhythmic/proarrhythmic actions of drugs.

2 Structure of hERG Channel

Like other Kv channels, hERG channel is formed by co-assembly of four α subunits. Each α subunit has six transmembrane spanning α-helical segments (S1–S6) along with the intracellularly located N- and C-terminus. The voltage sensor domain (VSD) that senses transmembrane potential is formed by S1–S4 helices (Piper et al. 2003; Subbiah et al. 2004). S4 helix contains positively charged amino acids mainly separated by hydrophobic residues. S5–S6 segments along with the intervening pore loop contribute to the pore domains. S5 is connected to S6 by an extracellular helix, followed by the pore helix (PH) and the K$^+$ selective filter (SF) (Jiang et al. 2005). The SF of the hERG channel adopts a unique signature sequence of Ser-Val-Gly-Phe-Gly (Doyle et al. 1998). It has been supposed that below the SF the pore widens to form a water-filled central cavity that is lined by residues from the S6 helices (Perry et al. 2010). However, recently solved cryo-EM structure of the hERG channel in the open state reveals that four deep cylindrical hydrophobic pockets below the SF extend out from the central pore cavity (Fig. 1a) (Wang and MacKinnon 2017). These pockets exclusively exist in hERG channel since the S6 inner helix of hERG is displaced to create a separation between the PH and S6 helix (Wang and MacKinnon 2017).

hERG channel has a unique kinetic behavior that is characterized by slow deactivation but very fast, voltage-dependent inactivation (Vandenberg et al. 2004). This unusual combination of kinetics gives rise to an apparent inward rectification that is crucial for maintaining a prolonged plateau phase of the cardiac AP. The channel opens following membrane depolarization as a result of its VSD's response to the voltage; however, the channel almost immediately inactivates, limiting K$^+$ passage until the start of the repolarization phase of the AP (due to the rapid recovery from inactivation). In addition, hERG deactivates very slowly so that the outward K$^+$ current is passed even as the membrane potential returns toward the resting potential (Fig. 1b, c). Therefore, the unique kinetics makes the hERG current ideally suited for determining the duration of the plateau phase of the AP (Smith et al. 1996; Sanguinetti and Tristani-Firouzi 2006). Maintenance of plateau is crucial for ensuring sufficient time for calcium release from the sarcoplasmic reticulum to enable cardiac contraction. The gating kinetics of hERG also enables the channel to generate rapid transient currents late in AP repolarization/early diastole, to protect against arrhythmogenic premature depolarizations.

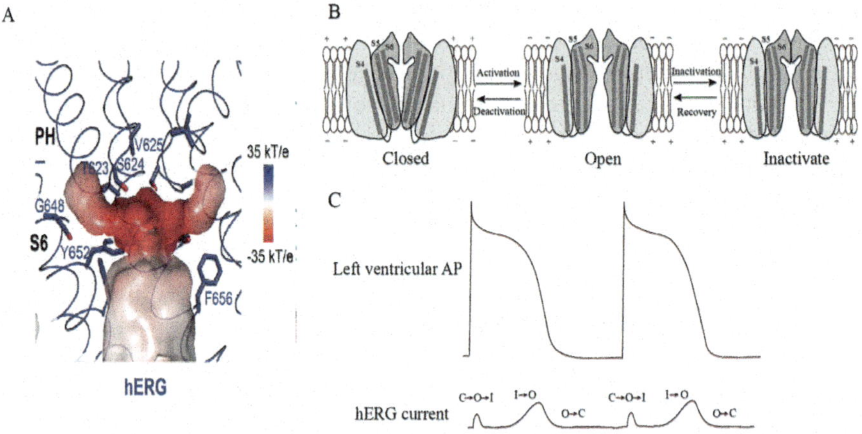

Fig. 1 The structure and gating of hERG channel. (**a**) The structure of pore cavity; adopted from (Wang and MacKinnon 2017) with permission. The central cavity has an atypically small central volume surrounded by four deep hydrophobic pockets. Internal molecular surface around the central cavity is represented as translucent surface colored by electrostatic potential according to the scale shown. Residues related to drug binding are shown as sticks on the otherwise ribbon representation of the channel. (**b**) hERG channel exists in closed, open, or inactivated states; transitions between these states are voltage dependent. (**c**) hERG current response (bottom) to the AP voltage waveform (top). hERG channel opens following membrane depolarization and then rapidly inactivates. During repolarization of the AP waveform, the current increases due to the recovery from inactivation and then slowly decreases again as the electrochemical gradient for K^+ efflux decreases

3 Mechanisms of Arrhythmias

Cardiac arrhythmias are commonly believed to arise primarily from abnormal automaticity, reentrant excitation, or the combination of both. Abnormal automaticity may occur as a result of enhanced automaticity or triggered activity (Wit 1990). The triggered activity and reentrant excitation are highly associated with hERG dysfunction-induced tachycardial ventricular arrhythmias. It is generally accepted that tachyarrhythmic events are obligated depending on two phenomena: a triggering event for initiation and a reentry substrate for sustainability (Schmitt et al. 2014). Triggered activity results from the premature activation of cardiac tissues by afterdepolarizations, which are oscillations in membrane potential that follow the primary depolarization phase (0) of an AP. If afterdepolarizations develop before full repolarization, corresponding to phase 2 or phase 3 of the cardiac AP, they are classified as early afterdepolarizations (EADs) and those originating from phase 4 of AP are classified as delayed afterdepolarizations (DADs) (Fig. 2a). EADs are usually but not exclusively associated with prolonged action potential durations (APD). It is generally considered that EADs occur primarily due to the reactivation of the voltage-gated $Ca_V1.2$ channels (L-type Ca^{2+} channels) (January and Riddle 1989). If the change in membrane potential brought about by the EAD is large enough to

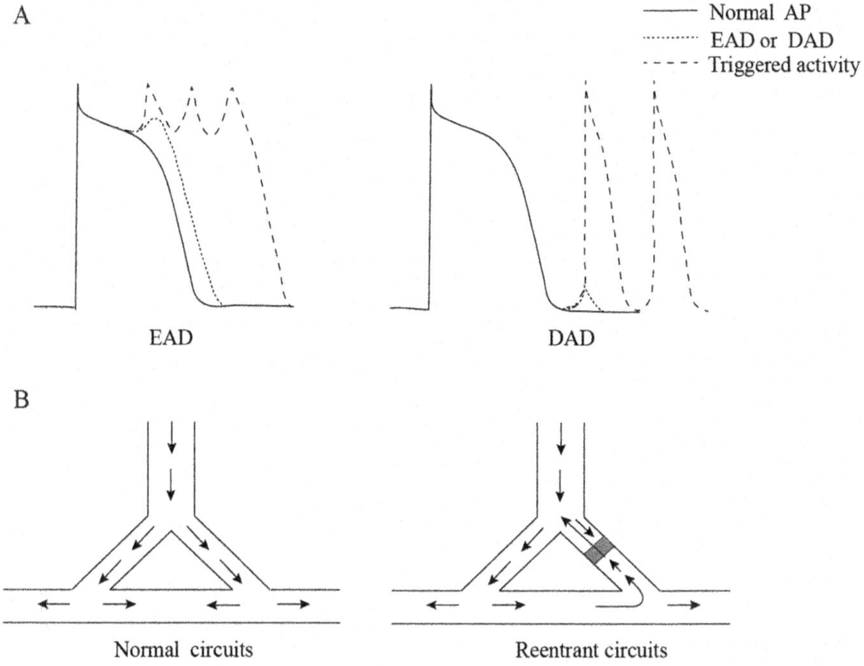

Fig. 2 The mechanisms of arrhythmias. (**a**) The afterdepolarizations developing before full repolarization, corresponding to phase 2 or phase 3 of the cardiac AP are classified as EADs (left,······), and those originating from phase 4 of AP are classified as DADs (right,······). When afterdepolarizations reach the threshold potential, a new AP is generated, leading to the triggered activity (-----). (**b**) Propagation of normal AP (left) and conditions for a reentrant excitation (right). Under normal conditions, the electrical signals travel down each branch of Purkinje fiber with equal velocity, and the signals will not progress if the two branches are connected. However, if one branch exhibits a unidirectional block, the electrical signal will travel down only one branch and may back-propagate until the point of blocking. If a retrogradely progressing impulse encounters excitable tissue, a reentry is set up

reach the threshold potential for initiation of APs, it will cause triggered activity (Fig. 2a). EADs and their resulting triggered activity are thought to underlie the arrhythmogenesis observed in LQTS (Maruyama et al. 2011). DADs usually occur under conditions of intracellular calcium overload and involve spontaneous release of calcium from the sarcoplasmic reticulum.

In order for sustained arrhythmias to occur, the triggering events must subsequently initiate a self-sustained episode of APD propagation, which is known as reentry-based arrhythmia (where reentry denotes an ongoing loop of unintended electrical signaling). A normally-propagating AP usually encounters neighboring tissue with equal conducting velocity and completely extinguish (Fig. 2b left). If an impulse is blocked in a specific area of the tissue but not elsewhere and the retrograde conduction is still possible, a unidirectional blocking is said to have occurred. If a retrogradely conducting impulse encounters excitable tissue, a reentry is being set up (Fig. 2b right). Such electrophysiological blocks may result from an anatomical or functional obstacle under pathological conditions such as myocardial infarction or inflammation or altered electrophysiologic properties due to electrolyte imbalance or ischemia. Another important factor forming arrhythmic substrates is electrophysiological heterogeneity of the myocardium. The APD diverges in different parts of the myocardium, and there is a significant heterogeneity among cardiac cells along several axes including the transmural, left-right, and apicobasal axes (Boukens et al. 2009). The dispersion is increased in the conditions with inherited ion channelopathies and after unintended inhibition of I_{Kr} by cardiac and non-cardiac drugs (Antzelevitch 2007, 2008). This amplification of spatial dispersion of repolarization can form substrates for reentry loops and thus contribute to life-threatening arrhythmias (Antzelevitch 2007; Keating and Sanguinetti 2001).

4 hERG Inhibitors

4.1 hERG Inhibitors as Antiarrhythmic Agents

Class III antiarrhythmic agents include nonselective K^+ channel blockers ambasilide, amiodarone, and dronedarone and selective I_{Kr} blockers dofetilide, ibutilide, and sotalol (Lei et al. 2018). The supposed mechanism of antiarrhythmic effects of these compounds is the inhibition of reentry-based arrhythmias through prolongation of the effective refractory period (ERP). However, inhibition of I_{Kr} by these compounds has also been found to be associated with an increased risk of arrhythmias and sudden cardiac death (Vandenberg et al. 2001). The proarrhythmic effect of class III compounds results from excessive prolongation of APD, especially an extended and slowly decaying phase 3-repolarization (triangulation), which could promote reactivation of L-type Ca^{2+} channels and, thus, lead to EADs. According to the aforementioned arrhythmogenic mechanisms, increased dispersion of repolarizations form reentry substrates can, in turn, result in TdP, which may ultimately degenerate to ventricular fibrillation.

4.2 hERG Inhibition by Structurally Diverse Drugs

In 1922, syncope and sudden death were firstly reported in patients treated with the quinidine (Levy 1922). These phenomena were further revealed in 1964, when Selzer and Wray (1964) observed TdP on electrocardiograms from patients with quinidine-related syncope, which was resulted from prolongation of cardiac repolarization due to hERG channel blockage. Since then, more and more drugs with miscellaneous structures are discovered to block hERG channel and, thus, carry the TdP risk. Antiarrhythmic, antihistamine, antimicrobial, antipsychotic, and antidepressant drugs are important classes associated with proarrhythmic risk (Rampe and Brown 2013). Hitherto, several drugs have been withdrawn from the market or given strict limitation for use because of TdP risk, including terfenadine, lidoflazine, astemizole, sertindole, levomethadyl, droperidol, cisapride, and grepafloxacin (Table 1). A database is available for drugs with the risk of TdP, which is categorized into three classes: drugs with known risk of TdP, possible risk of TdP, and conditional risk of TdP. Drugs with known risk of TdP related to hERG channel inhibition are listed in Table 1. Updated information about drug-associated TdP risk can be found at www.crediblemeds.org.

4.3 Molecular Basis Underlying hERG Channel Inhibition

The question of why the hERG channel is so susceptible to "nonspecific" block by such a wide variety of medications has attracted intense interest. Much effort has been made to explore the structural basis underlying this unusual susceptibility to inhibition, with approaches ranging from electrophysiology to, protein structure solution and *in silico* modeling. It is generally considered that there are at least two important structural features of hERG channel that are responsible for the above property. Firstly, many drugs bind to hERG channel by being trapped in its inner cavity, which appears to be much larger than in any other voltage-gated K^+ channel. Thus, the large inner cavity of hERG channel can accommodate and trap large molecules that other K^+ channels cannot trap (Mitcheson et al. 2000). Recently, the cryo-EM structure of hERG has been solved (Wang and MacKinnon 2017), it provides a valuable insight into the channel structure with regard to the drug binding. It has been demonstrated that there are four unique elongated, relatively hydrophobic pockets that extend from the central cavity (Wang and MacKinnon 2017) (Fig. 1a). Drugs are proposed to occupy the center of the cavity and insert a functional group into the hydrophic pockets. The central cavity of the channel in the region just below the SF is slightly narrower than that seen in *Shaker-like* voltage-gated K^+ channel structures. As a consequence, there is a greater negative electrostatic potential in this region of the cavity (Vandenberg et al. 2017), which attracts cations (e.g., metal ions or positively charged drugs) to form a more stable structure. Secondly, it is believed that a number of aromatic residues in a specific hERG channel region can form binding sites for inhibitory drugs. The electrons of the aromatic ring may form π-cation or π-π interactions with the drug molecule *via* charged nitrogen or aromatic ring, respectively (Fernandez et al. 2004; Stansfeld et al. 2007). Mutagenesis screening has demonstrated that residues on the S6 helix (Y652, F656, G648) and residues at the base of the SF (T623, S624, and V625) are critical to binding for a range of hERG blockers (Kamiya et al. 2006; Lees-Miller et al. 2000; Mitcheson et al. 2000; Perry et al. 2004). Among these, the two aromatic residues on the S6 helices (Y652, F656) are highly conserved in hERG channel orthologs, but not in other voltage-dependent K^+ channels (Shealy et al. 2003). Substantial evidence has shown that channel blockage by almost all hERG blocking drugs tested is dramatically attenuated by mutations of one or both of these two key residues (Y652 and F656) that form much of the lining of the K^+ conductance pathway. In addition, *in silico* hERG blocking studies have also demonstrated that Y652 and F656 in the hERG S6 domain play critical roles in drug binding (Hyang-Ae et al. 2018). These

Table 1 Drugs with a known risk of TdP due to hERG inhibition

Drugs	Drug class	References
Amiodarone	Antiarrhythmic	(Kamiya et al. 2001; Kiehn et al. 1999)
Arsenic trioxide[a]	Anticancer	(Ficker et al. 2004)
Astemizole[b]	Antihistamine	(Suessbrich et al. 1996; Zhou et al. 1999)
Azithromycin	Antibiotic	(Yang et al. 2017; Zhi et al. 2015)
Bepridil[b]	Antianginal	(Chouabe et al. 1998, 2000)
Chloroquine	Antimalarial	(Sánchez-Chapula et al. 2002; Traebert et al. 2004)
Chlorpromazine	Antipsychotic/ antiemetic	(Lee et al. 2004; Thomas et al. 2003b)
Ciprofloxacin	Antibiotic	(Bischoff et al. 2000; Kang et al. 2001)
Cisapride[b]	GI stimulant	(Mohammad et al. 1997; Rampe et al. 1997)
Citalopram[a]	Antidepressant, SSRI	(Chae et al. 2014; Witchel et al. 2002)
Clarithromycin	Antibiotic	(Stanat et al. 2003; Volberg et al. 2002)
Cocaine	Local anesthetic	(Guo et al. 2006; Zhang et al. 2001)
Disopyramide	Antiarrhythmic	(Paul et al. 2001; Yang et al. 2001)
Dofetilide	Antiarrhythmic	(Kiehn et al. 1995; Yang et al. 2001)
Domperidone	Antiemetic	(Claassen and Zünkler 2005; Drolet et al. 2000)
Donepezil[a]	Cholinesterase inhibitor	(Chae et al. 2015)
Dronedarone	Antiarrhythmic	(Ridley et al. 2004; Thomas et al. 2003a)
Droperidol	Antipsychotic/ antiemetic	(Drolet et al. 1999; Luo et al. 2008)
Erythromycin	Antibiotic	(Duncan et al. 2006; Stanat et al. 2003)
Escitalopram[a]	Antidepressant, SSRI	(Chae et al. 2014)
Flecainide	Antiarrhythmic	(Paul et al. 2002)
Fluconazole[a]	Antifungal	(Han et al. 2011)
Gatifloxacin[b]	Antibiotic	(Kang et al. 2001)
Grepafloxacin[b]	Antibiotic	(Bischoff et al. 2000; Kang et al. 2001)
Halofantrine	Antimalarial	(Tie et al. 2000; Traebert et al. 2004)
Haloperidol	Antipsychotic	(Shuba et al. 2001; Suessbrich et al. 1997)
Ibogaine	Psychedelic	(Koenig et al. 2013; Thurner et al. 2014)
Ibutilide	Antiarrhythmic	(Kodirov et al. 2019; Yang et al. 2001)
Levofloxacin	Antibiotic	(Kang et al. 2001)
Levomethadyl acetate[b]	Opioid agonist	(Katchman et al. 2002)
Mesoridazine[b]	Antipsychotic	(Su et al. 2004)
Methadone	Opioid agonist	(Katchman et al. 2002)
Moxifloxacin	Antibiotic	(Bischoff et al. 2000; Kang et al. 2001)
Nifekalant	Antiarrhythmic	(Kushida et al. 2002)
Ondansetron	Antiemetic	(Kuryshev et al. 2000)
Papaverine HCl (Intracoronary)	Vasodilator, coronary	(Kim et al. 2007, 2008)

(continued)

Table 1 (continued)

Drugs	Drug class	References
Pentamidine[a]	Antifungal	(Kuryshev et al. 2005; Tanaka et al. 2014)
Pimozide	Antipsychotic	(Kang et al. 2000)
Probucol[a,b]	Antilipemic	(Guo et al. 2007, 2011)
Procainamide	Antiarrhythmic	(Yang et al. 2001)
Propofol	Anesthetic, general	(Han et al. 2016)
Quinidine	Antiarrhythmic	(Sǎnchez-Chapula et al. 2003; Yang et al. 2001)
Roxithromycin[a]	Antibiotic	(Han et al. 2013; Volberg et al. 2002)
Sertindole[b]	Antipsychotic	(Rampe et al. 1998)
Sevoflurane	Anesthetic, general	(Yamada et al. 2006)
Sotalol	Antiarrhythmic	(Numaguchi et al. 2000; Sanguinetti and Jurkiewicz 1990)
Sparfloxacin[b]	Antibiotic	(Bischoff et al. 2000; Kang et al. 2001)
Sulpiride	Antipsychotic, atypical	(Lee et al. 2009)
Terfenadine[b]	Antihistamine	(Suessbrich et al. 1996; Tanaka et al. 2014)
Terodiline	Muscle relaxant	(Martin et al. 2006)
Thioridazine	Antipsychotic	(Kim and Kim 2005; Milnes et al. 2006)
Vandetanib	Anticancer	(Lee et al. 2018)

[a]Drug with effect of trafficking inhibition
[b]Drug withdrawn from market by FDA. Data acquired from www.crediblemeds.org in May, 2020

two aromatic residues in each subunit were originally proposed to face into the inner cavity so as to provide a total of eight binding sites for drugs (Mitcheson et al. 2000). However, recent cryo-EM structure of hERG channel in open state revealed that Y652 projects towards K^+ permeation pathway, while F656 side chains projects away from the permeation pathway towards the outer PH (Fig. 1a). This structure is not consistent with the original hypothesis that drugs directly bind to F656 within the permeation pathway. The molecular basis for this discrepancy is not yet fully understood. One possibility is that inactivation in hERG is associated with repositioning of Y652 and (especially) F656 side chains into a configuration that promotes interaction with blockers in the pore since drugs prefer to bind to the hERG channel in its inactivated state (Chen et al. 2002). This might involve a small clockwise rotation of the inner S6 helix containing these side chains (Chen et al. 2002; Helliwell et al. 2018). Comprehensive reviews with the detailed information about molecular basis of hERG drug binding can be found in the references (Butler et al. 2019; Dickson et al. 2020; Helliwell et al. 2018; Vandenberg et al. 2017; Wacker et al. 2017; Wang and MacKinnon 2017).

In addition to the direct inhibition of channel activity, forward trafficking impairment can reduce hERG current through a reduction in the number of hERG channels on cell membrane. Experiments indicate that arsenic trioxide (Ficker et al. 2004), pentamidine (Kuryshev et al. 2005), and probucol (Guo et al. 2007) disrupt hERG trafficking at concentrations known to cause QT prolongation and arrhythmia without direct channel block. Some other drugs such as fluoxetine and ketoconazole both can acutely block hERG channel and reduce hERG plasma membrane protein abundance following long-term exposure by inhibiting trafficking (Rajamani et al. 2006; Takemasa et al. 2008). It is important to consider impaired trafficking as an alternative mechanism for drug-induced QT prolongation, as conventional compound screening methods for hERG block liability may not detect reductions in channel abundance.

4.4 Methodology of hERG Assays

Since hERG channel plays an important role in cardiac repolarization and is susceptible to inhibition by a wide variety of compounds, evaluation of the potential hERG blocking effect of new compounds for identifying potential risk of proarrhythmic side effects is a necessary step in a drug discovery process. The International Conference on Harmonisation of Technical Requirements for Registration of Pharmaceuticals for Human Use (ICH) adopted a guideline S7B putting forward requirements in assessing hERG blocking of new drugs for the cardiac safety in 2005.

A variety of technologies have been applied to evaluate effects of hERG channel blocking based on multiple test systems including heterologous hERG expression in Xenopus oocyte and mammalian cells such as HEK293 cells and CHO cells, and native cardiomyocytes with I_{Kr} current. Because the cardiomyocytes of adult mice and rats heart lack the I_{Kr} current component, native cardiomyocytes for testing are commonly derived from the hearts of larger animals such as guinea pigs, rabbits, and dogs. Evaluation technologies include direct electrophysiological measurement (i.e., patch clamp), and indirect non-electrophysiological measurements such as competitive radioligand binding assays, ion flux assays, fluorescence-based assays, and *in silico* modeling.

Patch clamp technique remains a gold standard to directly assess hERG blocking liability of compounds (Hancox et al. 2008). It provides accurate and physiologically relevant data of ion channel function at the single cell or single channel level. However, traditional manual patch clamp has been limited in drug screening due to low throughput and a requirement for highly skilled operators. Recently developed automated patch clamp approach, which offers high-throughput electrophysiological data acquisition, has transformed the situation (Guo and Guthrie 2005; Jones et al. 2009). At present, both manual patch clamp and automated patch clamp are widely used in evaluation of hERG safety (Danker and Möller 2014; Lindqvist 2019). Non-electrophysiological measurements are also widely used, these assess the potency of drugs for hERG blocking by measuring the hERG channel related indicators. The competitive radioligand binding assays determine displacement of specific radiolabeled hERG ligands such as [^{3}H]dofetilide (Diaz et al. 2004; Finlayson et al. 2001a, b), [^{3}H]astemizole (Chiu et al. 2004), [^{35}S]-MK-499 (Raab et al. 2006), and [^{125}I]-BeKm1 (Angelo et al. 2003) to reflect the binding affinity of test drugs. The ion flux-based assays (often in combination with fluorescence-based approaches) measure the amount of ions such as Rubidium (Rb^+) (Terstappen 1999) and Thallium (Tl^+) (Titus et al. 2009; Weaver et al. 2004) permeating through the hERG channel and thus indirectly reflect the alterations of hERG function under the action of drugs. In recent years, *in silico* models of hERG channel were developed for predicting the action of hERG modulators. *In silico* models are based on structural properties of the hERG channel and incorporate the information of channel gating and ligand binding kinetics. The aim of such modeling is to characterize the interactions of compounds with the hERG channel by computer simulations (Lee et al. 2016; Pearlstein et al. 2016; Zhang and Hancox 2004). However, the electrophysiological measurements remain necessary to confirm data obtained by such modeling.

The potency of compounds for producing hERG inhibition, usually indicated by the compound's IC_{50} (concentration of half-maximal inhibition), can be normalized to the clinically relevant concentrations of the given compound, such as $C_{max,free}$ (free plasma concentration) to calculate the safety margin, as proposed by the S7B guideline. According to relevant studies, the closer the hERG IC_{50} value is to the $C_{max,free}$ the higher is the risk of QT interval prolongation (Redfern et al. 2003; van Noord et al. 2011). A 30-fold margin between $C_{max,free}$ and hERG IC_{50} has been considered as a cardiac safety value in many cases (Redfern et al. 2003; van Noord et al. 2011). However, it is also recognized that an increase in the margin should be considered, especially for drug candidates aimed for non-debilitating diseases (Redfern et al. 2003).

However, due to the lack of standardization for measuring hERG modulator potency, there are often significant differences in measured IC_{50} values reported by different laboratories for the same compounds. For instance, the difference in IC_{50} of cisapride reported by different laboratories exceeds 60-fold (Potet et al. 2001; Rezazadeh et al. 2004). The essential factors that contribute to such variability generally include differences in test systems and recording conditions such as temperature and voltage protocols.

Using different test systems, such as native cardiac myocytes and cell lines heterologously expressing hERG can lead to significant discrepancy of IC_{50} values. As much as 50-fold difference of E-4031 IC_{50} has been observed between native cardiac myocytes (Sanguinetti and Jurkiewicz 1990) and transfected cells (Zhou et al. 1998). This discrepancy may result from the differences in the composition of hERG channel. In the native cardiomyocytes, in addition to the dominant hERG1a isoform, the hERG1b isoform is also expressed (although at much lower level) and can contribute to the composition of heteromeric channel (McNally et al. 2017). Indeed, a study has shown that a homomeric hERG1a channel expressed in HEK293 cells is blocked by E-4031 more rapidly than with a heteromeric channel containing both hERG1a and hERG1b (Sale et al. 2008). A similar trend has been found for dofetilide (Abi-Gerges et al. 2011).

Additional complications arise from the state-dependent binding of some compounds to the hERG. Substantial evidence indicates that different hERG blockers have a high-affinity binding to the activated or inactivated channel (Stork et al. 2007; Walker et al. 1999). The channel state can be modulated by temperature and voltage protocols including voltage pattern, duration, and pulse frequency (Lee et al. 2019; Stork et al. 2007). Thus, it is not difficult to understand why there are significant differences in measured IC_{50} values under distinct temperature and voltage protocols. In addition, temperature and voltage protocols have an influence on drug binding kinetics and trapping (Kirsch et al. 2004; Stork et al. 2007). These factors also lead to discrepancies in the reported potency parameters.

4.5 A New CiPA Paradigm to Evaluate Drug-Induced TdP

Although no approved drugs have been withdrawn from the market because of the TdP risk since the ICH S7B Guideline was implemented (Sager et al. 2014), the hERG safety remains a necessary phase in drug discovery. Yet, limitations of only assessing hERG blockage have been recognized. The cardiac AP is coordinated by multiple ion currents and requires relative balance between inward and outward currents. It is therefore insufficient to focus on a single component in predicting the risk of delayed repolarization and TdP. For example, verapamil has been shown to inhibit hERG current with high potency (Zhang et al. 1999), but it does not lead to QT interval prolongation and does not increase the TdP risk because of the concomitant inhibition on inward I_{CaL} (Winters et al. 1985). A recent study based on 30 drugs of different risk categories (high, intermediate, and low) has shown that blocking inward currents such as sodium and calcium current may reduce proarrhythmic effect of hERG current inhibition (Crumb et al. 2016). Thus, assessing hERG blockage alone carries a risk for false-positive predictions and leads to potentially valuable new compounds being discarded early in drug discovery. A study has indicated that as many as 60% of new molecular entities developed as potential therapeutic agents are abandoned early due to hERG inhibition (Ponti 2008). Therefore, a new parardigm, a Comprehensive Invitro Proarrhythmia Assay (CiPA) has been proposed in the field of cardiac safety; CiPA presents a more comprehensive approach to predicting proarrhythmic risk (Sager et al. 2014).

There are three preclinical components in CiPA paradigm: (1) drug effects on multiple human cardiac currents; (2) *in silico* reconstruction of human ventricular electrophysiology, and (3) *in vitro* effects on human stem-cell derived ventricular myocytes. Specific study groups have been established to refine the approaches and benchmarks within each of these components.

In CiPA paradigm, hERG blocking is no longer the unique indicator; instead, a more comprehensive *in vitro* set of ion current assays is used to explore the effects of drugs on multiple potassium, sodium, and calcium currents. A recent study has shown that, under the premise of evaluation of hERG, incorporating $Na_V1.5$ or $Ca_V1.2$ in particularly into the evaluation system has significantly improved the TdP predictability (Kramer et al. 2013). The ion channel working group of CiPA has developed a series of protocols to test the effects of compounds on the main cardiac ion channels including hERG, L-type calcium, and fast and late inward sodium currents, hoping to provide standardized protocols to be used in different patch clamp facilities of the academic and industrial research institutions (Fermini et al.

2016; Huang et al. 2017; Windley et al. 2017). In the next step, *in silico* reconstruction of ventricular APs assesses the effects of compounds more intimately on the basis of electrophysiological data. Finally, cadiomyocytes such as human induced pluripotent stem cell-derived cardiac myocytes (hiPSC-CMs) would be used to provide an assessment of the integrated electrophysiological response to a drug (Sager et al. 2014; Wallis et al. 2018). The updated information about the progress of CiPA groups is available at www.cipaproject.org. Hopefully, the CiPA paradigm can provide more precise and comprehensive information for assessment of hERG inhibition to predict the risk of drug-induced arrhythmia.

5 hERG Activators

In contrast to numerous hERG channel blockers, some compounds have been discovered to increase hERG channel currents during the course of screening for hERG channel-blocking activity early in preclinical safety evaluation (Grunnet et al. 2008). Thus, Kang and colleagues reported the first synthetic activator of hERG channel, RPR260243 (Kang et al. 2005). Since then several other hERG activators have been identified, including PD118057 (Zhou et al. 2005), NS1643 (Casis et al. 2006; Hansen et al. 2006a), NS3623 (Hansen et al. 2006b), Mallotoxin (Zeng et al. 2006), PD307243 (Gordon et al. 2008; Xu et al. 2008), A935142 (Su et al. 2009), ICA-105574 (Gerlach et al. 2010), KB130015 (Gessner et al. 2010), etc. These compounds shorten cardiac APD and have been proposed as a new therapeutic approach for the treatment of acquired or congenital LQTS (reviewed in (Sanguinetti 2014; Szabó et al. 2011; Vandenberg et al. 2012; Zhou et al. 2011)).

5.1 Mechanisms of Action of hERG Channel Activators

Different to hERG blockers that simply block K^+ conduction and have little influence on channel gating, hERG activators primarily exert their effects by modulating channel gating. Four distinct mechanisms have been described: (1) slowing the rate of channel deactivation; (2) attenuation of C-type inactivation; (3) negative shift of voltage dependence of activation; (4) increase in channel open probability (Sanguinetti 2014) (Fig. 3). Accordingly, depending on the predominant mechanism of action, hERG activators can be categorized in four types (although most hERG activators have multiple mechanisms of action). Here, we will give a brief review on the gating modulation by several known activators. More detailed information on these mechanisms can be found in several previous reviews (Perry et al. 2010; Sanguinetti 2014; Szabó et al. 2011; Zhou et al. 2011). The chemical structures of major hERG activators are shown in Fig. 4.

5.1.1 Slowing the Deactivation

RPR260243 is the first compound designed as a type 1 hERG channel activator (Kang et al. 2005). This small molecule enhances current by attenuating inactivation

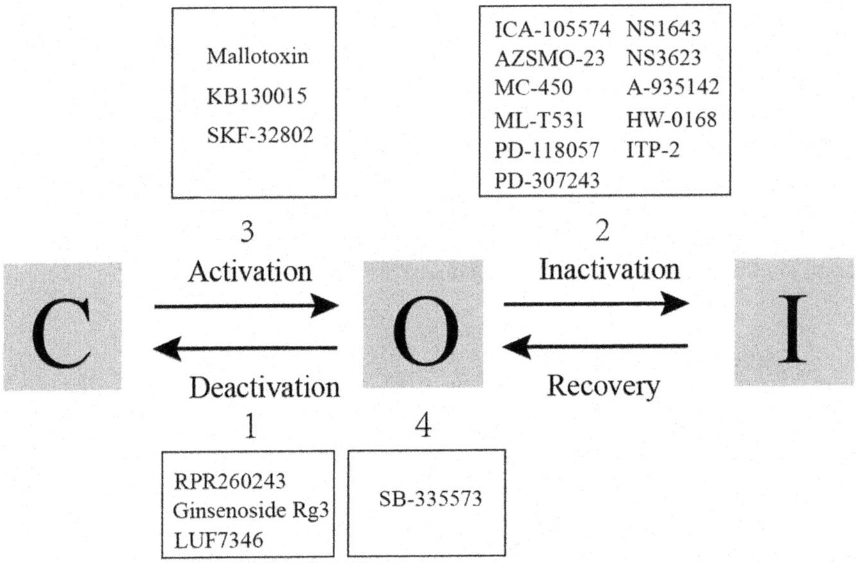

Fig. 3 The action of hERG channel activators. hERG activators primarily exert their effects by modulating channel gating. There are four distinct mechanisms including slowing of channel deactivation (1), attenuation of C-type inactivation (2), negative shift of voltage dependence of activation (3), and increase in channel open probability (4). Known hERG activators are assigned to types 1-4, according to the predominant mechanism of action

and severely slowing the rate of channel deactivation (Kang et al. 2005; Perry et al. 2007). Another compound, Ginsenoside Rg3, an alkaloid isolated from the root of *Panax ginseng* plants, increases current magnitude primarily by slowing the rate of hERG deactivation (Choi et al. 2011). More recently, compound LUF7346 has been identified as a type 1 hERG channel activator, which increases hERG current by slowing deactivation and positively shifting voltage dependence of inactivation (Sala et al. 2016).

Scanning mutagenesis has identified the putative binding site for RPR260243, which is located near the cytoplasmic ends of the S5 and S6 helices of the hERG subunit, a region of the channel that is important for activation and deactivation. Hence, it is proposed that binding of RPR260243 to a single subunit may directly constrain movement of the S6 domains to slow the rate of channel closure (Perry et al. 2007).

5.1.2 Attenuation of C-Type Inactivation

As mentioned, one of the most important gating features of hERG channel is its fast C-type inactivation. Attenuation of C-type inactivation is produced by some of the hERG channel activators, an effect resulting in an enhancement of hERG current. Up to now, more than ten compounds such as PD118057 (Zhou et al. 2005), PD307243 (Gordon et al. 2008), NS1643 (Casis et al. 2006), NS3623 (Hansen et al. 2006b), A-935142 (Su et al. 2009), ICA-105574 (Gerlach et al. 2010), ML-T531 (Zhang

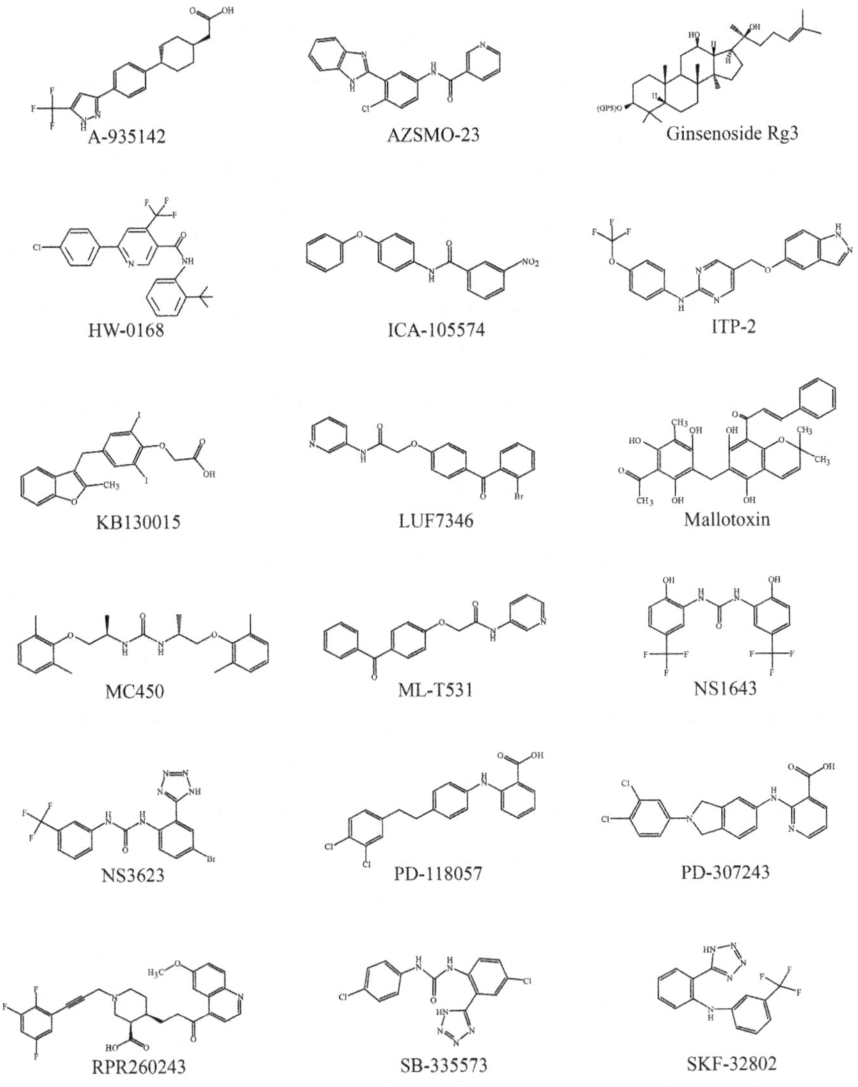

Fig. 4 Chemical structures of major hERG channel activators

et al. 2012), AZSMO-23 (Mannikko et al. 2015), ITP-2 (Sale et al. 2017), MC-450 (Gualdani et al. 2017), and HW-0168 (Dong et al. 2019) have been identified to enhance hERG current primarily through attenuating the channel inactivation and are thus classified as type 2 activators (Perry et al. 2010). However, most of those activators may have multiple mechanisms of action. The mechanistic and structural basis underlying the fast inactivation of hERG channel is not fully understood. It is believed to be caused by a subtle voltage-dependent conformational changes in the SF of the outer pore domain (for reviews, see Ref. Vandenberg et al. 2012).

Experimental evidence has shown that the binding sites of many type 2 activators are located closer to the SF (Garg et al. 2011; Gerlach et al. 2010; Perry et al. 2009). Scanning mutagenesis combined with molecular modeling studies have revealed that PD118057 interacts with residues located in the PH of one hERG subunit and the N-terminal half of the S6 helix in an adjacent subunit to attenuate inactivation (Perry et al. 2009). Similarly, the residues interacting with ICA-105574, another potent type 2 activator (Gerlach et al. 2010), are located in the PH and the base of the SF and S6 segments (Garg et al. 2011). A recent study has proposed a common mechanism to prevent C-type inactivation by a group of negatively charged activators such as PD-118057 (Schewe et al. 2019). This type of activators may directly stabilize the SF in its active state through binding to similar sites below the SF (Schewe et al. 2019). In line with this hypothesis, a molecular dynamics simulation has demonstrated that ICA-105574 increases the stability of the SF to attenuate channel inactivation (Zangerl-Plessl et al. 2020). However, whether other type 2 activators with distinct chemical structures share the same molecular mechanism remains uncertain.

5.1.3 Negative Shift of Voltage Dependence of Activation

Previous experimental findings indicate that both Mallotoxin and KB130015 increase hERG current amplitude primarily by causing a hyperpolarizing shift in the voltage dependence of channel activation (Zeng et al. 2006; Gessner et al. 2010). Mallotoxin also accelerates the rate of activation and slows the rate of deactivation (Zeng et al. 2006). KB130015 is a derivative of the hERG blocker, amiodarone, and presumably binds to the hERG pore from the cytosolic side and functionally competes with amiodarone (Gessner et al. 2010). SKF-32802, a structural analog of NS3623, induces a leftward shift in the voltage dependence of activation. The above compounds are identified as the type 3 activators (Donovan et al. 2018).

5.1.4 Increase in Channel Open Probability

Similar to SKF-32802, SB-335573 is also a structural analog of NS3623. However, it enhances hERG current through increasing open probability without affecting the voltage dependence of activation and, thus, identified as a type 4 activator (Donovan et al. 2018). In addition, PD-118057 has been reported to increase single hERG channel open probability (Perry et al. 2009).

5.2 Potential Antiarrhythmic Effect of hERG Channel Activators

Several hERG activators have been tested for their antiarrhythmic effectiveness in inherited or drug-induced acquired LQTS. Thirteen subtypes of inherited LQTS have been identified, with the most prevalent forms being LQTS1, 2, and 3 (Schwartz et al. 2012). The underlying channelopathies are loss-of-function mutations in I_{Ks} (type 1) and in I_{Kr} (type 2) and increased sustained I_{Na} current (type 3). Theoretically, the LQTS phenotype could be rescued by the compensatory effect of hERG channel activators if I_{Kr} current is not completely lost. Experimental evidence obtained in cardiac myocytes, especially in hiPSC-CMs derived from LQTS patients and in transgenic animals, supports this notion. A study has demonstrated that NS1643 significantly shortens APD and QT interval in a rabbit model of inherited LQTS1 (Bentzen et al. 2011). Type 2 activator ML-T531 normalizes the prolonged APD by selectively enhancing I_{Kr} in hiPSC-CMs derived from LQTS1 patient (Zhang et al. 2012). NS1643 and ICA-105574 effectively restore hERG current from heterozygous LQTS2 mutant channels in heterologous expression systems (Huo et al. 2017; Perry et al. 2020). Several activators, including NS1643, ICA-105574, and LUF-7346, have been shown to reverse the prolonged repolarization in hiPSC-CMs derived from LQTS2 patients carrying different mutations (Duncan et al. 2017; Perry et al. 2020; Sala et al. 2016). In addition, both NS3623 and Mallotoxin show the antiarrhythmic potential in a cellular model of LQTS3 (Diness et al. 2009).

Many hERG activators with different gating modulation mechanisms have been demonstrated to counteract the inhibition by hERG blockers either in heterologous expression systems or in native cardiac myocytes (review, Ref. (Szabó et al. 2011)). However, only few of those activators have been tested *in vivo* or in intact hearts for their effectiveness of suppressing drug-induced arrhythmias. An experiment has demonstrated that *in vivo* administration of NS3623 results in shortening of the QT interval as well as reversal of a pharmacologically induced QT prolongation in both anesthetized and conscious guinea pigs (Hansen et al. 2008). NS1643 completely suppresses arrhythmic activity caused by I_{Kr} inhibitor dofetilide in the *in vivo* rabbit models of TdP (Diness et al. 2008). ICA-105574 effectively prevents ventricular arrhythmias caused by I_{Kr} or I_{Ks} inhibitors in intact guinea-pig hearts (Meng et al. 2013). Recent experiments demonstrates that LUF7244 and RPR260243 counteract dofetilide-induced arrhythmias in a chronic atrioventricular block model in dogs (Qile et al. 2019) and in whole organ zebrafish hearts (Shi et al. 2020), respectively. These findings support the notion that hERG activators may provide an effective antiarrhythmic approach in drug-induced, disease-induced, or gene mutation-linked LQTS.

5.3 Proarrhythmic Risk of hERG Channel Activators

The fact that congenital short QT syndromes (SQT) (Crotti et al. 2010) may lead to susceptibility to arrhythmias raises concerns that QT-shortening drugs could also lead to arrhythmias. Several reports have revealed the potential proarrhythmic risk of some hERG activators including mallotoxin, NS1643, ICA-105574, and PD-118057 in experimental and *in silico* models (Bentzen et al. 2011; Lu et al. 2008; Peitersen et al. 2008; Schewe et al. 2019) and, thus, those hERG activators have been used to create drug-induced SQT models. The arrhythmogenesis of these activators may result from a decrease of ERP and an increase of the transmural dispersion of repolarization (TDR). Amplification of the spatial dispersion of repolarization in the form of TDR is the basis for the development of life-threatening ventricular arrhythmias (Antzelevitch 2007). In addition, ICA-105574 causes temporal redistribution of the peak I_{Kr} to much earlier in the plateau phase of the AP and, thus, results in early repolarization (Perry et al. 2020; Qiu et al. 2019), which, in turn, my result in the development of phase 2 reentry and ventricular tachycardia/ventricular fibrillation.

6 Conclusion

The shape of the cardiac AP depends on a fine balance between various depolarizing and repolarizing ionic currents. The unique gating kinetic properties of hERG channel make it ideal for determining the morphology and duration of the cardiac AP repolarization. Consequently, alterations of hERG channel function by inhibitors or activators may result in either prolongation or shortening of APD, which can counteract abnormal electroactivity under specific pathological condition. However, unintended disturbance or overcorrection of hERG channel function may result in arrhythmogenesis. Thus, hERG channel becomes an important pharmacological and safety target for antiarrhythmic/proarrhythmic actions of drugs.

References

Abi-Gerges N, Holkham H, Jones EM, Pollard CE, Valentin JP, Robertson GA (2011) hERG subunit composition determines differential drug sensitivity. Br J Pharmacol 164:419–432

Angelo K, Korolkova YV, Grunnet M, Grishin EV, Pluzhnikov KA, Klaerke DA, Knaus HG, Moller M, Olesen SP (2003) A radiolabeled peptide ligand of the hERG channel, [125I]-BeKm-1. Pflugers Arch 447:55–63

Antzelevitch C (2007) Role of spatial dispersion of repolarization in inherited and acquired sudden cardiac death syndromes. Am J Physiol Heart Circ Physiol 293:H2024–H2038

Antzelevitch C (2008) Drug-induced spatial dispersion of repolarization. Cardiol J 15:100–121

Bentzen BH, Bahrke S, Wu K, Larsen AP, Odening KE, Franke G, Storm vańs Gravesande K, Biermann J, Peng X, Koren G, Zehender M, Bode C, Grunnet M, Brunner M (2011) Pharmacological activation of Kv11.1 in transgenic long QT-1 rabbits. J Cardiovasc Pharmacol 57:223–230

Bischoff U, Schmidt C, Netzer R, Pongs O (2000) Effects of fluoroquinolones on HERG currents. Eur J Pharmacol 406:341–343

Boukens BJ, Christoffels VM, Coronel R, Moorman AF (2009) Developmental basis for electrophysiological heterogeneity in the ventricular and outflow tract myocardium as a substrate for life-threatening ventricular arrhythmias. Circ Res 104:19–31

Butler A, Helliwell MV, Zhang Y, Hancox JC, Dempsey CE (2019) An update on the structure of hERG. Front Pharmacol 10:1572

Casis O, Olesen SP, Sanguinetti MC (2006) Mechanism of action of a novel human ether-a-go-go-related gene channel activator. Mol Pharmacol 69:658–665

Chae YJ, Jeon JH, Lee HJ, Kim IB, Choi JS, Sung KW, Hahn SJ (2014) Escitalopram block of hERG potassium channels. Naunyn Schmiedebergs Arch Pharmacol 387:23–32

Chae YJ, Lee HJ, Jeon JH, Kim IB, Choi JS, Sung KW, Hahn SJ (2015) Effects of donepezil on hERG potassium channels. Brain Res 1597:77–85

Chen J, Seebohm G, Sanguinetti MC (2002) Position of aromatic residues in the S6 domain, not inactivation, dictates cisapride sensitivity of HERG and eag potassium channels. Proc Natl Acad Sci U S A 99:12461–12466

Chiu PJ, Marcoe KF, Bounds SE, Lin CH, Feng JJ, Lin A, Cheng FC, Crumb WJ, Mitchell R (2004) Validation of a [3H]astemizole binding assay in HEK293 cells expressing HERG K+ channels. J Pharmacol Sci 95:311–319

Choi SH, Shin TJ, Hwang SH, Lee BH, Kang J, Kim HJ, Jo SH, Choe H, Nah SY (2011) Ginsenoside Rg(3) decelerates hERG K(+) channel deactivation through Ser631 residue interaction. Eur J Pharmacol 663:59–67

Chouabe C, Drici MD, Romey G, Barhanin J, Lazdunski M (1998) HERG and KvLQT1/IsK, the cardiac K+ channels involved in long QT syndromes, are targets for calcium channel blockers. Mol Pharmacol 54:695–703

Chouabe C, Drici MD, Romey G, Barhanin J (2000) Effects of calcium channel blockers on cloned cardiac K+ channels IKr and IKs. Therapie 55:195–202

Claassen S, Zünkler BJ (2005) Comparison of the effects of metoclopramide and domperidone on HERG channels. Pharmacology 74:31–36

Crotti L, Taravelli E, Girardengo G, Schwartz PJ (2010) Congenital short QT syndrome. Indian Pacing Electrophysiol J 10:86–95

Crumb WJ Jr, Vicente J, Johannesen L, Strauss DG (2016) An evaluation of 30 clinical drugs against the comprehensive in vitro proarrhythmia assay (CiPA) proposed ion channel panel. J Pharmacol Toxicol Methods 81:251–262

Danker T, Möller C (2014) Early identification of hERG liability in drug discovery programs by automated patch clamp. Front Pharmacol 5:203

Diaz GJ, Daniell K, Leitza ST, Martin RL, Su Z, McDermott JS, Cox BF, Gintant GA (2004) The [3H]dofetilide binding assay is a predictive screening tool for hERG blockade and proarrhythmia: Comparison of intact cell and membrane preparations and effects of altering [K+]o. J Pharmacol Toxicol Methods 50:187–199

Dickson CJ, Velez-Vega C, Duca JS (2020) Revealing molecular determinants of hERG blocker and activator binding. J Chem Inf Model 60:192–203

Diness TG, Yeh YH, Qi XY, Chartier D, Tsuji Y, Hansen RS, Olesen SP, Grunnet M, Nattel S (2008) Antiarrhythmic properties of a rapid delayed-rectifier current activator in rabbit models of acquired long QT syndrome. Cardiovasc Res 79:61–69

Diness JG, Hansen RS, Nissen JD, Jespersen T, Grunnet M (2009) Antiarrhythmic effect of IKr activation in a cellular model of LQT3. Heart Rhythm 6:100–106

Dong X, Liu Y, Niu H, Wang G, Dong L, Zou A, Wang K (2019) Electrophysiological characterization of a small molecule activator on human ether-a-go-go-related gene (hERG) potassium channel. J Pharmacol Sci 140:284–290

Donovan BT, Bandyopadhyay D, Duraiswami C, Nixon CJ, Townsend CY, Martens SF (2018) Discovery and electrophysiological characterization of SKF-32802: a novel hERG agonist found through a large-scale structural similarity search. Eur J Pharmacol 818:306–327

Doyle DA, Morais Cabral J, Pfuetzner RA, Kuo A, Gulbis JM, Cohen SL, Chait BT, MacKinnon R (1998) The structure of the potassium channel: molecular basis of K+ conduction and selectivity. Science 280:69–77

Drolet B, Zhang S, Deschênes D, Rail J, Nadeau S, Zhou Z, January CT, Turgeon J (1999) Droperidol lengthens cardiac repolarization due to block of the rapid component of the delayed rectifier potassium current. J Cardiovasc Electrophysiol 10:1597–1604

Drolet B, Rousseau G, Daleau P, Cardinal R, Turgeon J (2000) Domperidone should not be considered a no-risk alternative to cisapride in the treatment of gastrointestinal motility disorders. Circulation 102:1883–1885

Duncan RS, Ridley JM, Dempsey CE, Leishman DJ, Leaney JL, Hancox JC, Witchel HJ (2006) Erythromycin block of the HERG K+ channel: accessibility to F656 and Y652. Biochem Biophys Res Commun 341:500–506

Duncan G, Firth K, George V, Hoang MD, Staniforth A, Smith G, Denning C (2017) Drug-mediated shortening of action potentials in LQTS2 human induced pluripotent stem cell-derived cardiomyocytes. Stem Cells Dev 26:1695–1705

Fermini B, Hancox JC, Abi-Gerges N, Bridgland-Taylor M, Chaudhary KW, Colatsky T, Correll K, Crumb W, Damiano B, Erdemli G, Gintant G, Imredy J, Koerner J, Kramer J, Levesque P, Li Z, Lindqvist A, Obejero-Paz CA, Rampe D, Sawada K, Strauss DG, Vandenberg JI (2016) A new perspective in the field of cardiac safety testing through the comprehensive in vitro proarrhythmia assay paradigm. J Biomol Screen 21:1–11

Fernandez D, Ghanta A, Kauffman GW, Sanguinetti MC (2004) Physicochemical features of the hERG channel drug binding site. J Biol Chem 279:10120–10127

Ficker E, Kuryshev YA, Dennis AT, Obejero-Paz C, Wang L, Hawryluk P, Wible BA, Brown AM (2004) Mechanisms of arsenic-induced prolongation of cardiac repolarization. Mol Pharmacol 66:33–44

Finlayson K, Pennington AJ, Kelly JS (2001a) [3H]dofetilide binding in SHSY5Y and HEK293 cells expressing a HERG-like K+ channel? Eur J Pharmacol 412:203–212

Finlayson K, Turnbull L, January CT, Sharkey J, Kelly JS (2001b) [3H]dofetilide binding to HERG transfected membranes: a potential high throughput preclinical screen. Eur J Pharmacol 430:147–148

Garg V, Stary-Weinzinger A, Sachse F, Sanguinetti MC (2011) Molecular determinants for activation of human ether-à-go-go-related gene 1 potassium channels by 3-nitro-n-(4-phenoxyphenyl) benzamide. Mol Pharmacol 80:630–637

Gerlach AC, Stoehr SJ, Castle NA (2010) Pharmacological removal of human ether-à-go-go-related gene potassium channel inactivation by 3-nitro-N-(4-phenoxyphenyl) benzamide (ICA-105574). Mol Pharmacol 77:58–68

Gessner G, Macianskiene R, Starkus JG, Schönherr R, Heinemann SH (2010) The amiodarone derivative KB130015 activates hERG1 potassium channels via a novel mechanism. Eur J Pharmacol 632:52–59

Gordon E, Lozinskaya IM, Lin Z, Semus SF, Blaney FE, Willette RN, Xu X (2008) 2-[2-(3,4-dichloro-phenyl)-2,3-dihydro-1H-isoindol-5-ylamino]-nicotinic acid (PD-307243) causes instantaneous current through human ether-a-go-go-related gene potassium channels. Mol Pharmacol 73:639–651

Grunnet M, Hansen RS, Olesen SP (2008) hERG1 channel activators: a new anti-arrhythmic principle. Prog Biophys Mol Biol 98:347–362

Gualdani R, Cavalluzzi MM, Tadini-Buoninsegni F, Lentini G (2017) Discovery of a new mexiletine-derived agonist of the hERG K(+) channel. Biophys Chem 229:62–67

Guo L, Guthrie H (2005) Automated electrophysiology in the preclinical evaluation of drugs for potential QT prolongation. J Pharmacol Toxicol Methods 52:123–135

Guo J, Gang H, Zhang S (2006) Molecular determinants of cocaine block of human ether-á-go-go-related gene potassium channels. J Pharmacol Exp Ther 317:865–874

Guo J, Massaeli H, Li W, Xu J, Luo T, Shaw J, Kirshenbaum LA, Zhang S (2007) Identification of IKr and its trafficking disruption induced by probucol in cultured neonatal rat cardiomyocytes. J Pharmacol Exp Ther 321:911–920

Guo J, Li X, Shallow H, Xu J, Yang T, Massaeli H, Li W, Sun T, Pierce GN, Zhang S (2011) Involvement of caveolin in probucol-induced reduction in hERG plasma-membrane expression. Mol Pharmacol 79:806–813

Han S, Zhang Y, Chen Q, Duan Y, Zheng T, Hu X, Zhang Z, Zhang L (2011) Fluconazole inhibits hERG K(+) channel by direct block and disruption of protein trafficking. Eur J Pharmacol 650:138–144

Han SN, Yang SH, Zhang Y, Duan YY, Sun XY, Chen Q, Fan TL, Ye ZK, Huang CZ, Hu XJ, Zhang Z, Zhang LR (2013) Blockage of hERG current and the disruption of trafficking as induced by roxithromycin. Can J Physiol Pharmacol 91:1112–1118

Han SN, Jing Y, Yang LL, Zhang Z, Zhang LR (2016) Propofol inhibits hERG K(+) channels and enhances the inhibition effects on its mutations in HEK293 cells. Eur J Pharmacol 791:168–178

Hancox JC, McPate MJ, El Harchi A, Zhang YH (2008) The hERG potassium channel and hERG screening for drug-induced torsades de pointes. Pharmacol Ther 119:118–132

Hansen RS, Diness TG, Christ T, Demnitz J, Ravens U, Olesen SP, Grunnet M (2006a) Activation of human ether-a-go-go-related gene potassium channels by the diphenylurea 1,3-bis-(2-hydroxy-5-trifluoromethyl-phenyl)-urea (NS1643). Mol Pharmacol 69:266–277

Hansen RS, Diness TG, Christ T, Wettwer E, Ravens U, Olesen SP, Grunnet M (2006b) Biophysical characterization of the new human ether-a-go-go-related gene channel opener NS3623 [N-(4-bromo-2-(1H-tetrazol-5-yl)-phenyl)-N'-(3'-trifluoromethylphenyl)urea]. Mol Pharmacol 70:1319–1329

Hansen RS, Olesen SP, Rønn LC, Grunnet M (2008) In vivo effects of the IKr agonist NS3623 on cardiac electrophysiology of the guinea pig. J Cardiovasc Pharmacol 52:35–41

Helliwell MV, Zhang Y, El Harchi A, Du C, Hancox JC, Dempsey CE (2018) Structural implications of hERG K(+) channel block by a high-affinity minimally structured blocker. J Biol Chem 293:7040–7057

Huang H, Pugsley MK, Fermini B, Curtis MJ, Koerner J, Accardi M, Authier S (2017) Cardiac voltage-gated ion channels in safety pharmacology: review of the landscape leading to the CiPA initiative. J Pharmacol Toxicol Methods 87:11–23

Huo J, Guo X, Lu Q, Qiang H, Liu P, Bai L, Huang CL, Zhang Y, Ma A (2017) NS1643 enhances ionic currents in a G604S-WT hERG co-expression system associated with long QT syndrome 2. Clin Exp Pharmacol Physiol 44:1125–1133

Hyang-Ae L, Sung-Ae H, Byungjin B, Jong-Hak C, Ki-Suk K, Shang-Zhong XJPO (2018) Electrophysiological mechanisms of vandetanib-induced cardiotoxicity: Comparison of action potentials in rabbit Purkinje fibers and pluripotent stem cell-derived cardiomyocytes. PLoS One 13:e0195577

January CT, Riddle JM (1989) Early afterdepolarizations: mechanism of induction and block. A role for L-type Ca2+ current. Circ Res 64:977–990

Jiang M, Zhang M, Maslennikov IV, Liu J, Wu DM, Korolkova YV, Arseniev AS, Grishin EV, Tseng GN (2005) Dynamic conformational changes of extracellular S5-P linkers in the hERG channel. J Physiol 569:75–89

Jones KA, Garbati N, Zhang H, Large CH (2009) Automated patch clamping using the QPatch. Methods Mol Biol 565:209–223

Kamiya K, Nishiyama A, Yasui K, Hojo M, Sanguinetti MC, Kodama I (2001) Short- and long-term effects of amiodarone on the two components of cardiac delayed rectifier K(+) current. Circulation 103:1317–1324

Kamiya K et al (2006) Molecular determinants of hERG channel block. Mol Pharmacol 69:1709–1716

Kang J, Wang L, Cai F, Rampe D (2000) High affinity blockade of the HERG cardiac K(+) channel by the neuroleptic pimozide. Eur J Pharmacol 392:137–140

Kang J, Wang L, Chen XL, Triggle DJ, Rampe D (2001) Interactions of a series of fluoroquinolone antibacterial drugs with the human cardiac K+ channel HERG. Mol Pharmacol 59:122–126

Kang J, Chen XL, Wang H, Ji J, Cheng H, Incardona J, Reynolds W, Viviani F, Tabart M, Rampe D (2005) Discovery of a small molecule activator of the human ether-a-go-go-related gene (HERG) cardiac K+ channel. Mol Pharmacol 67:827–836

Katchman AN, McGroary KA, Kilborn MJ, Kornick CA, Manfredi PL, Woosley RL, Ebert SN (2002) Influence of opioid agonists on cardiac human ether-a-go-go-related gene K(+) currents. J Pharmacol Exp Ther 303:688–694

Keating MT, Sanguinetti MC (2001) Molecular and cellular mechanisms of cardiac arrhythmias. Cell 104:569–580

Kiehn J, Wible B, Ficker E, Taglialatela M, Brown AM (1995) Cloned human inward rectifier K+ channel as a target for class III methanesulfonanilides. Circ Res 77:1151–1155

Kiehn J, Thomas D, Karle CA, Schöls W, Kübler W (1999) Inhibitory effects of the class III antiarrhythmic drug amiodarone on cloned HERG potassium channels. Naunyn Schmiedebergs Arch Pharmacol 359:212–219

Kim KS, Kim EJ (2005) The phenothiazine drugs inhibit hERG potassium channels. Drug Chem Toxicol 28:303–313

Kim CS, Lee N, Son SJ, Lee KS, Kim HS, Kwak YG, Chae SW, Lee SD, Jeon BH, Park JB (2007) Inhibitory effects of coronary vasodilator papaverine on heterologously-expressed HERG currents in Xenopus oocytes. Acta Pharmacol Sin 28:503–510

Kim YJ, Hong HK, Lee HS, Moh SH, Park JC, Jo SH, Choe H (2008) Papaverine, a vasodilator, blocks the pore of the HERG channel at submicromolar concentration. J Cardiovasc Pharmacol 52:485–493

Kirsch GE, Trepakova ES, Brimecombe JC, Sidach SS, Erickson HD, Kochan MC, Shyjka LM, Lacerda AE, Brown AM (2004) Variability in the measurement of hERG potassium channel inhibition: effects of temperature and stimulus pattern. J Pharmacol Toxicol Methods 50:93–101

Kodirov SA, Zhuravlev VL, Brachmann J (2019) Prevailing effects of ibutilide on fast delayed rectifier K(+) channel. J Membr Biol 252:609–616

Koenig X, Kovar M, Rubi L, Mike AK, Lukacs P, Gawali VS, Todt H, Hilber K, Sandtner W (2013) Anti-addiction drug ibogaine inhibits voltage-gated ionic currents: a study to assess the drug's cardiac ion channel profile. Toxicol Appl Pharmacol 273:259–268

Kramer J, Obejero-Paz CA, Myatt G, Kuryshev YA, Bruening-Wright A, Verducci JS, Brown AM (2013) MICE models: superior to the HERG model in predicting torsade de pointes. Sci Rep 3:2100

Kuryshev YA, Brown AM, Wang L, Benedict CR, Rampe D (2000) Interactions of the 5-hydroxytryptamine 3 antagonist class of antiemetic drugs with human cardiac ion channels. J Pharmacol Exp Ther 295:614–620

Kuryshev YA, Ficker E, Wang L, Hawryluk P, Dennis AT, Wible BA, Brown AM, Kang J, Chen XL, Sawamura K, Reynolds W, Rampe D (2005) Pentamidine-induced long QT syndrome and block of hERG trafficking. J Pharmacol Exp Ther 312:316–323

Kushida S, Ogura T, Komuro I, Nakaya H (2002) Inhibitory effect of the class III antiarrhythmic drug nifekalant on HERG channels: mode of action. Eur J Pharmacol 457:19–27

Lee SY, Choi SY, Youm JB, Ho WK, Earm YE, Lee CO, Jo SH (2004) Block of HERG human K (+) channel and IKr of guinea pig cardiomyocytes by chlorpromazine. J Cardiovasc Pharmacol 43:706–714

Lee HA, Kim KS, Park SJ, Kim EJ (2009) Cellular mechanism of the QT prolongation induced by sulpiride. Int J Toxicol 28:207–212

Lee W, Mann SA, Windley MJ, Imtiaz MS, Vandenberg JI, Hill AP (2016) In silico assessment of kinetics and state dependent binding properties of drugs causing acquired LQTS. Prog Biophys Mol Biol 120:89–99

Lee HA, Hyun SA, Byun B, Chae JH, Kim KS (2018) Electrophysiological mechanisms of vandetanib-induced cardiotoxicity: comparison of action potentials in rabbit Purkinje fibers and pluripotent stem cell-derived cardiomyocytes. PLoS One 13:e0195577

Lee W, Windley MJ, Perry MD, Vandenberg JI, Hill AP (2019) Protocol-dependent differences in IC50 values measured in human ether-a-go-go-related gene assays occur in a predictable way and can be used to quantify state preference of drug binding. Mol Pharmacol 95:537–550

Lees-Miller JP, Duan YJ, Teng GQ, Duff HJ (2000) Molecular determinant of high-affinity dofetilide binding toHERG1 expressed in xenopus oocytes: involvement of S6 sites. Mol Pharmacol 57:367–374

Lei M, Wu L, Terrar DA, Huang CL-H (2018) Modernized classification of cardiac antiarrhythmic drugs. Circulation 138:1879–1896

Levy RL (1922) Clinical studies of quinidin: IV. The clinical toxicology of quinidin. JAMA 79:1108–1113

Lindqvist A (2019) Estimating hERG drug binding using temperature-controlled high-throughput automated patch-clamp. J Pharmacol Toxicol Methods 99:106595

Lu HR, Vlaminckx E, Hermans AN, Rohrbacher J, Van Ammel K, Towart R, Pugsley M, Gallacher DJ (2008) Predicting drug-induced changes in QT interval and arrhythmias: QT-shortening drugs point to gaps in the ICHS7B guidelines. Br J Pharmacol 154:1427–1438

Luo T, Luo A, Liu M, Liu X (2008) Inhibition of the HERG channel by droperidol depends on channel gating and involves the S6 residue F656. Anesth Analg 106:1161–1170. table of contents

Mannikko R, Bridgland-Taylor MH, Pye H, Swallow S, Abi-Gerges N, Morton MJ, Pollard CE (2015) Pharmacological and electrophysiological characterization of AZSMO-23, an activator of the hERG K(+) channel. Br J Pharmacol 172:3112–3125

Martin RL, Su Z, Limberis JT, Palmatier JD, Cowart MD, Cox BF, Gintant GA (2006) In vitro preclinical cardiac assessment of tolterodine and terodiline: multiple factors predict the clinical experience. J Cardiovasc Pharmacol 48:199–206

Maruyama M, Lin SF, Xie Y, Chua SK, Joung B, Han S, Shinohara T, Shen MJ, Qu Z, Weiss JN, Chen PS (2011) Genesis of phase 3 early afterdepolarizations and triggered activity in acquired long-QT syndrome. Circ Arrhythm Electrophysiol 4:103–111

McNally BA, Pendon ZD, Trudeau MC (2017) hERG1a and hERG1b potassium channel subunits directly interact and preferentially form heteromeric channels. J Biol Chem 292:21548–21557

Meng J, Shi C, Li L, Du Y, Xu Y (2013) Compound ICA-105574 prevents arrhythmias induced by cardiac delayed repolarization. Eur J Pharmacol 718:87–97

Milnes JT, Witchel HJ, Leaney JL, Leishman DJ, Hancox JC (2006) hERG K+ channel blockade by the antipsychotic drug thioridazine: an obligatory role for the S6 helix residue F656. Biochem Biophys Res Commun 351:273–280

Mitcheson JS, Chen J, Lin M, Culberson C, Sanguinetti MC (2000) A structural basis for drug-induced long QT syndrome. Proc Natl Acad Sci 97:12329–12333

Mohammad S, Zhou Z, Gong Q, January CT (1997) Blockage of the HERG human cardiac K+ channel by the gastrointestinal prokinetic agent cisapride. Am J Physiol 273:H2534–H2538

Numaguchi H, Mullins FM, Johnson JP Jr, Johns DC, Po SS, Yang IC, Tomaselli GF, Balser JR (2000) Probing the interaction between inactivation gating and Dd-sotalol block of HERG. Circ Res 87:1012–1018

Paul AA, Witchel HJ, Hancox JC (2001) Inhibition of HERG potassium channel current by the class 1a antiarrhythmic agent disopyramide. Biochem Biophys Res Commun 280:1243–1250

Paul AA, Witchel HJ, Hancox JC (2002) Inhibition of the current of heterologously expressed HERG potassium channels by flecainide and comparison with quinidine, propafenone and lignocaine. Br J Pharmacol 136:717–729

Pearlstein RA, MacCannell KA, Erdemli G, Yeola S, Helmlinger G, Hu QY, Farid R, Egan W, Whitebread S, Springer C, Beck J, Wang HR, Maciejewski M, Urban L, Duca JS (2016) Implications of dynamic occupancy, binding kinetics, and channel gating kinetics for hERG blocker safety assessment and mitigation. Curr Top Med Chem 16:1792–1818

Peitersen T, Grunnet M, Benson AP, Holden AV, Holstein-Rathlou NH, Olesen SP (2008) Computational analysis of the effects of the hERG channel opener NS1643 in a human ventricular cell model. Heart Rhythm 5:734–741

Perry M, de Groot MJ, Helliwell R, Leishman D, Tristani-Firouzi M, Sanguinetti MC, Mitcheson J (2004) Structural determinants of HERG channel block by clofilium and ibutilide. Mol Pharmacol 66:240–249

Perry M, Sachse FB, Sanguinetti MC (2007) Structural basis of action for a human ether-a-go-go-related gene 1 potassium channel activator. Proc Natl Acad Sci U S A 104:13827–13832

Perry M, Sachse FB, Abbruzzese J, Sanguinetti MC (2009) PD-118057 contacts the pore helix of hERG1 channels to attenuate inactivation and enhance K+ conductance. Proc Natl Acad Sci U S A 106:20075–20080

Perry M, Sanguinetti M, Mitcheson J (2010) Revealing the structural basis of action of hERG potassium channel activators and blockers. J Physiol 588:3157–3167

Perry MD, Ng CA, Mangala MM, Ng TYM, Hines AD, Liang W, Xu MJO, Hill AP, Vandenberg JI (2020) Pharmacological activation of IKr in models of long QT Type 2 risks overcorrection of repolarization. Cardiovasc Res 116:1434–1445

Piper DR, Varghese A, Sanguinetti MC, Tristani-Firouzi M (2003) Gating currents associated with intramembrane charge displacement in HERG potassium channels. Proc Natl Acad Sci U S A 100(18):10534–10539

Ponti FD (2008) Pharmacological and regulatory aspects of QT prolongation. Wiley-VCH Verlag GmbH & Co, KGaA

Potet F, Bouyssou T, Escande D, Baro I (2001) Gastrointestinal prokinetic drugs have different affinity for the human cardiac human ether-a-gogo K(+) channel. J Pharmacol Exp Ther 299:1007–1012

Qile M, Beekman HDM, Sprenkeler DJ, Houtman MJC, van Ham WB, Stary-Weinzinger A, Beyl S, Hering S, van den Berg DJ, de Lange ECM, Heitman LH, IJzerman AP, Vos MA, van der MAG H (2019) LUF7244, an allosteric modulator/activator of K(v) 11.1 channels, counteracts dofetilide-induced torsades de pointes arrhythmia in the chronic atrioventricular block dog model. Br J Pharmacol 176:3871–3885

Qiu B, Wang Y, Li C, Guo H, Xu Y (2019) Utility of the JT peak interval and the JT area in determining the proarrhythmic potential of QT-shortening agents. J Cardiovasc Pharmacol Ther 24:160–171

Raab CE, Butcher JW, Connolly TM, Karczewski J, Yu NX, Staskiewicz SJ, Liverton N, Dean DC, Melillo DG (2006) Synthesis of the first sulfur-35-labeled hERG radioligand. Bioorg Med Chem Lett 16:1692–1695

Rajamani S, Eckhardt LL, Valdivia CR, Klemens CA, Gillman BM, Anderson CL, Holzem KM, Delisle BP, Anson BD, Makielski JC, January CT (2006) Drug-induced long QT syndrome: hERG K+ channel block and disruption of protein trafficking by fluoxetine and norfluoxetine. Br J Pharmacol 149:481–489

Rampe D, Brown AM (2013) A history of the role of the hERG channel in cardiac risk assessment. J Pharmacol Toxicol Methods 68:13–22

Rampe D, Roy ML, Dennis A, Brown AM (1997) A mechanism for the proarrhythmic effects of cisapride (Propulsid): high affinity blockade of the human cardiac potassium channel HERG. FEBS Lett 417:28–32

Rampe D, Murawsky MK, Grau J, Lewis EW (1998) The antipsychotic agent sertindole is a high affinity antagonist of the human cardiac potassium channel HERG. J Pharmacol Exp Ther 286:788–793

Redfern WS, Carlsson L, Davis AS, Lynch WG, MacKenzie I, Palethorpe S, Siegl PK, Strang I, Sullivan AT, Wallis R, Camm AJ, Hammond TG (2003) Relationships between preclinical cardiac electrophysiology, clinical QT interval prolongation and torsade de pointes for a broad range of drugs: evidence for a provisional safety margin in drug development. Cardiovasc Res 58:32–45

Rezazadeh S, Hesketh JC, Fedida D (2004) Rb+ flux through hERG channels affects the potency of channel blocking drugs: correlation with data obtained using a high-throughput Rb+ efflux assay. J Biomol Screen 9:588–597

Ridley JM, Milnes JT, Witchel HJ, Hancox JC (2004) High affinity HERG K(+) channel blockade by the antiarrhythmic agent dronedarone: resistance to mutations of the S6 residues Y652 and F656. Biochem Biophys Res Commun 325:883–891

Sager PT, Gintant G, Turner JR, Pettit S, Stockbridge N (2014) Rechanneling the cardiac proarrhythmia safety paradigm: a meeting report from the Cardiac Safety Research Consortium. Am Heart J 167:292–300

Sala L, Yu Z, Ward-van Oostwaard D, van Veldhoven JP, Moretti A, Laugwitz KL, Mummery CL, IJzerman AP, Bellin M (2016) A new hERG allosteric modulator rescues genetic and drug-induced long-QT syndrome phenotypes in cardiomyocytes from isogenic pairs of patient induced pluripotent stem cells. EMBO Mol Med 8:1065–1081

Sale H, Wang J, O'Hara TJ, Tester DJ, Phartiyal P, He JQ, Rudy Y, Ackerman MJ, Robertson GA (2008) Physiological properties of hERG 1a/1b heteromeric currents and a hERG 1b-specific mutation associated with Long-QT syndrome. Circ Res 103:e81–e95

Sale H, Roy S, Warrier J, Thangathirupathy S, Vadari Y, Gopal SK, Krishnamurthy P, Ramarao M (2017) Modulation of K(v) 11.1 (hERG) channels by 5-(((1H-indazol-5-yl)oxy)methyl)-N-(4-(trifluoromethoxy)phenyl)pyrimidin-2-amine (ITP-2), a novel small molecule activator. Br J Pharmacol 174:2484–2500
Sánchez-Chapula JA, Navarro-Polanco RA, Culberson C, Chen J, Sanguinetti MC (2002) Molecular determinants of voltage-dependent human ether-a-go-go related gene (HERG) K+ channel block. J Biol Chem 277:23587–23595
Sănchez-Chapula JA, Ferrer T, Navarro-Polanco RA, Sanguinetti MC (2003) Voltage-dependent profile of human ether-a-go-go-related gene channel block is influenced by a single residue in the S6 transmembrane domain. Mol Pharmacol 63:1051–1058
Sanguinetti MC (2014) HERG1 channel agonists and cardiac arrhythmia. Curr Opin Pharmacol 15:22–27
Sanguinetti MC, Jurkiewicz NK (1990) Two components of cardiac delayed rectifier K+ current. Differential sensitivity to block by class III antiarrhythmic agents. J Gen Physiol 96:195–215
Sanguinetti MC, Mitcheson JS (2005) Predicting drug-hERG channel interactions that cause acquired long QT syndrome. Trends Pharmacol Sci 26:119–124
Sanguinetti MC, Tristani-Firouzi M (2006) hERG potassium channels and cardiac arrhythmia. Nature 440:463–469
Sanguinetti MC, Jiang C, Curran ME, Keating MT (1995) A mechanistic link between an inherited and an acquired cardiac arrhythmia: HERG encodes the IKr potassium channel. Cell 81:299–307
Schewe M, Sun H, Mert Ü, Mackenzie A, Pike ACW, Schulz F, Constantin C, Vowinkel KS, Conrad LJ, Kiper AK, Gonzalez W, Musinszki M, Tegtmeier M, Pryde DC, Belabed H, Nazare M, de Groot BL, Decher N, Fakler B, Carpenter EP, Tucker SJ, Baukrowitz T (2019) A pharmacological master key mechanism that unlocks the selectivity filter gate in K(+) channels. Science 363:875–880
Schmitt N, Grunnet M, Olesen SP (2014) Cardiac potassium channel subtypes: new roles in repolarization and arrhythmia. Physiol Rev 94:609–653
Schwartz PJ, Crotti L, Insolia R (2012) Long-QT syndrome: from genetics to management. Circ Arrhythm Electrophysiol 5:868–877
Selzer A, Wray HWJC (1964) Quinidine syncope. Paroxysmal ventricular fibrillation occurring during treatment of chronic atrial arrhythmias. Circulation 30:17–26
Shealy RT, Murphy AD, Ramarathnam R, Jakobsson E, Subramaniam S (2003) Sequence-function analysis of the K+-selective family of ion channels using a comprehensive alignment and the KcsA channel structure. Biophys J 84:2929–2942
Shi YP, Pang Z, Venkateshappa R, Gunawan M, Kemp J, Truong E, Chang C, Lin E, Shafaattalab S, Faizi S, Rayani K, Tibbits GF, Claydon VE, Claydon TW (2020) The hERG channel activator, RPR260243, enhances protective I(Kr) current early in the refractory period reducing arrhythmogenicity in zebrafish hearts. Am J Physiol Heart Circ Physiol 319:H251–h261
Shuba YM, Degtiar VE, Osipenko VN, Naidenov VG, Woosley RL (2001) Testosterone-mediated modulation of HERG blockade by proarrhythmic agents. Biochem Pharmacol 62:41–49
Singh BN, Vaughan Williams EM (1970) A third class of anti-arrhythmic action. Effects on atrial and ventricular intracellular potentials, and other pharmacological actions on cardiac muscle, of MJ 1999 and AH 3474. Br J Pharmacol 39:675–687
Smith PL, Baukrowitz T, Yellen G (1996) The inward rectification mechanism of the HERG cardiac potassium channel. Nature 379:833–836
Stanat SJ, Carlton CG, Crumb WJ Jr, Agrawal KC, Clarkson CW (2003) Characterization of the inhibitory effects of erythromycin and clarithromycin on the HERG potassium channel. Mol Cell Biochem 254:1–7
Stansfeld PJ, Gedeck P, Gosling M, Cox B, Mitcheson JS, Sutcliffe MJ (2007) Drug block of the hERG potassium channel: insight from modeling. Proteins 68:568–580

Stork D, Timin EN, Berjukow S, Huber C, Hohaus A, Auer M, Hering S (2007) State dependent dissociation of HERG channel inhibitors. Br J Pharmacol 151:1368–1376

Su Z, Martin R, Cox BF, Gintant G (2004) Mesoridazine: an open-channel blocker of human ether-a-go-go-related gene K+ channel. J Mol Cell Cardiol 36:151–160

Su Z, Limberis J, Souers A, Kym P, Mikhail A, Houseman K, Diaz G, Liu X, Martin RL, Cox BF, Gintant GA (2009) Electrophysiologic characterization of a novel hERG channel activator. Biochem Pharmacol 77:1383–1390

Subbiah RN, Clarke CE, Smith DJ, Zhao J, Campbell TJ, Vandenberg JI (2004) Molecular basis of slow activation of the human ether-a-go-go related gene potassium channel. J Physiol 558:417–431

Suessbrich H, Waldegger S, Lang F, Busch AE (1996) Blockade of HERG channels expressed in Xenopus oocytes by the histamine receptor antagonists terfenadine and astemizole. FEBS Lett 385:77–80

Suessbrich H, Schönherr R, Heinemann SH, Attali B, Lang F, Busch AE (1997) The inhibitory effect of the antipsychotic drug haloperidol on HERG potassium channels expressed in Xenopus oocytes. Br J Pharmacol 120:968–974

Szabó G, Farkas V, Grunnet M, Mohácsi A, Nánási PP (2011) Enhanced repolarization capacity: new potential antiarrhythmic strategy based on HERG channel activation. Curr Med Chem 18:3607–3621

Takemasa H, Nagatomo T, Abe H, Kawakami K, Igarashi T, Tsurugi T, Kabashima N, Tamura M, Okazaki M, Delisle BP, January CT, Otsuji Y (2008) Coexistence of hERG current block and disruption of protein trafficking in ketoconazole-induced long QT syndrome. Br J Pharmacol 153:439–447

Tanaka H, Takahashi Y, Hamaguchi S, Iida-Tanaka N, Oka T, Nishio M, Ohtsuki A, Namekata I (2014) Effect of terfenadine and pentamidine on the HERG channel and its intracellular trafficking: combined analysis with automated voltage clamp and confocal microscopy. Biol Pharm Bull 37:1826–1830

Terstappen GC (1999) Functional analysis of native and recombinant ion channels using a high-capacity nonradioactive rubidium efflux assay. Anal Biochem 272:149–155

Thomas D, Kathofer S, Zhang W, Wu K, Wimmer AB, Zitron E, Kreye VA, Katus HA, Schoels W, Karle CA, Kiehn J (2003a) Acute effects of dronedarone on both components of the cardiac delayed rectifier K+ current, HERG and KvLQT1/minK potassium channels. Br J Pharmacol 140:996–1002

Thomas D, Wu K, Kathöfer S, Katus HA, Schoels W, Kiehn J, Karle CA (2003b) The antipsychotic drug chlorpromazine inhibits HERG potassium channels. Br J Pharmacol 139:567–574

Thurner P, Stary-Weinzinger A, Gafar H, Gawali VS, Kudlacek O, Zezula J, Hilber K, Boehm S, Sandtner W, Koenig X (2014) Mechanism of hERG channel block by the psychoactive indole alkaloid ibogaine. J Pharmacol Exp Ther 348:346–358

Tie H, Walker BD, Singleton CB, Valenzuela SM, Bursill JA, Wyse KR, Breit SN, Campbell TJ (2000) Inhibition of HERG potassium channels by the antimalarial agent halofantrine. Br J Pharmacol 130:1967–1975

Titus SA, Beacham D, Shahane SA, Southall N, Xia M, Huang R, Hooten E, Zhao Y, Shou L, Austin CP, Zheng W (2009) A new homogeneous high-throughput screening assay for profiling compound activity on the human ether-a-go-go-related gene channel. Anal Biochem 394:30–38

Traebert M, Dumotier B, Meister L, Hoffmann P, Dominguez-Estevez M, Suter W (2004) Inhibition of hERG K+ currents by antimalarial drugs in stably transfected HEK293 cells. Eur J Pharmacol 484:41–48

van Noord C, Sturkenboom MC, Straus SM, Witteman JC, Stricker BH (2011) Non-cardiovascular drugs that inhibit hERG-encoded potassium channels and risk of sudden cardiac death. Heart 97:215–220

Vandenberg JI, Walker BD, Campbell TJ (2001) HERG K+ channels: friend and foe. Trends Pharmacol Sci 22:240–246

Vandenberg JI, Torres AM, Campbell TJ, Kuchel PW (2004) The HERG K+ channel: progress in understanding the molecular basis of its unusual gating kinetics. Eur Biophys J 33:89–97

Vandenberg JI, Perry MD, Perrin MJ, Mann SA, Ke Y, Hill AP (2012) hERG K(+) channels: structure, function, and clinical significance. Physiol Rev 92:1393–1478

Vandenberg JI, Perozo E, Allen TW (2017) Towards a structural view of drug binding to hERG K (+) channels. Trends Pharmacol Sci 38:899–907

Volberg WA, Koci BJ, Su W, Lin J, Zhou J (2002) Blockade of human cardiac potassium channel human ether-a-go-go-related gene (HERG) by macrolide antibiotics. J Pharmacol Exp Ther 302:320–327

Wacker S, Noskov SY, Perissinotti LL (2017) Computational models for understanding of structure, function and pharmacology of the cardiac potassium channel Kv11.1 (hERG). Curr Top Med Chem 17:2681–2702

Walker BD, Singleton CB, Bursill JA, Wyse KR, Valenzuela SM, Qiu MR, Breit SN, Campbell TJ (1999) Inhibition of the human ether-a-go-go-related gene (HERG) potassium channel by cisapride: affinity for open and inactivated states. Br J Pharmacol 128:444–450

Wallis R, Benson C, Darpo B, Gintant G, Kanda Y, Prasad K, Strauss DG, Valentin JP (2018) CiPA challenges and opportunities from a non-clinical, clinical and regulatory perspectives. An overview of the safety pharmacology scientific discussion. J Pharmacol Toxicol Methods 93:15–25

Wang W, MacKinnon R (2017) Cryo-EM structure of the open human ether-à-go-go-related K(+) channel hERG. Cell 169:422–430.e10

Weaver CD, Harden D, Dworetzky SI, Robertson B, Knox RJ (2004) A thallium-sensitive, fluorescence-based assay for detecting and characterizing potassium channel modulators in mammalian cells. J Biomol Screen 9:671–677

Windley MJ, Abi-Gerges N, Fermini B, Hancox JC, Vandenberg JI, Hill AP (2017) Measuring kinetics and potency of hERG block for CiPA. J Pharmacol Toxicol Methods 87:99–107

Winters SL, Schweitzer P, Kupersmith J, Gomes JA (1985) Verapamil-induced polymorphous ventricular tachycardia. J Am Coll Cardiol 6:257–259

Wit AL (1990) Cellular electrophysiologic mechanisms of cardiac arrhythmias. Cardiol Clin 8:393–409

Witchel HJ, Pabbathi VK, Hofmann G, Paul AA, Hancox JC (2002) Inhibitory actions of the selective serotonin re-uptake inhibitor citalopram on HERG and ventricular L-type calcium currents. FEBS Lett 512:59–66

Xu X, Recanatini M, Roberti M, Tseng GN (2008) Probing the binding sites and mechanisms of action of two human ether-a-go-go-related gene channel activators, 1,3-bis-(2-hydroxy-5-trifluoromethyl-phenyl)-urea (NS1643) and 2-[2-(3,4-dichloro-phenyl)-2,3-dihydro-1H-isoindol-5-ylamino]-nicotinic acid (PD307243). Mol Pharmacol 73:1709–1721

Yamada M, Hatakeyama N, Malykhina AP, Yamazaki M, Momose Y, Akbarali HI (2006) The effects of sevoflurane and propofol on QT interval and heterologously expressed human ether-a-go-go related gene currents in Xenopus oocytes. Anesth Analg 102:98–103

Yang T, Snyders D, Roden DM (2001) Drug block of I(kr): model systems and relevance to human arrhythmias. J Cardiovasc Pharmacol 38:737–744

Yang Z, Prinsen JK, Bersell KR, Shen W, Yermalitskaya L, Sidorova T, Luis PB, Hall L, Zhang W, Du L, Milne G, Tucker P, George AL Jr, Campbell CM, Pickett RA, Shaffer CM, Chopra N, Yang T, Knollmann BC, Roden DM, Murray KT (2017) Azithromycin causes a novel proarrhythmic syndrome. Circ Arrhythm Electrophysiol 10(4):e003560

Zangerl-Plessl EM, Berger M, Drescher M, Chen Y, Wu W, Maulide N, Sanguinetti M, Stary-Weinzinger A (2020) Toward a structural view of hERG activation by the small-molecule activator ICA-105574. J Chem Inf Model 60:360–371

Zeng H, Lozinskaya IM, Lin Z, Willette RN, Brooks DP, Xu X (2006) Mallotoxin is a novel human ether-a-go-go-related gene (hERG) potassium channel activator. J Pharmacol Exp Ther 319:957–962

Zhang H, Hancox JC (2004) In silico study of action potential and QT interval shortening due to loss of inactivation of the cardiac rapid delayed rectifier potassium current. Biochem Biophys Res Commun 322:693–699

Zhang S, Zhou Z, Gong Q, Makielski JC, January CT (1999) Mechanism of block and identification of the verapamil binding domain to HERG potassium channels. Circ Res 84:989–998

Zhang S, Rajamani S, Chen Y, Gong Q, Rong Y, Zhou Z, Ruoho A, January CT (2001) Cocaine blocks HERG, but not KvLQT1+minK, potassium channels. Mol Pharmacol 59:1069–1076

Zhang H, Zou B, Yu H, Moretti A, Wang X, Yan W, Babcock JJ, Bellin M, McManus OB, Tomaselli G, Nan F, Laugwitz KL, Li M (2012) Modulation of hERG potassium channel gating normalizes action potential duration prolonged by dysfunctional KCNQ1 potassium channel. Proc Natl Acad Sci U S A 109:11866–11871

Zhi D, Feng PF, Sun JL, Guo F, Zhang R, Zhao X, Li BX (2015) The enhancement of cardiac toxicity by concomitant administration of Berberine and macrolides. Eur J Pharm Sci 76:149–155

Zhou Z, Gong Q, Ye B, Fan Z, Makielski JC, Robertson GA, January CT (1998) Properties of HERG channels stably expressed in HEK 293 cells studied at physiological temperature. Biophys J 74:230–241

Zhou Z, Vorperian VR, Gong Q, Zhang S, January CT (1999) Block of HERG potassium channels by the antihistamine astemizole and its metabolites desmethylastemizole and norastemizole. J Cardiovasc Electrophysiol 10:836–843

Zhou J, Augelli-Szafran CE, Bradley JA, Chen X, Koci BJ, Volberg WA, Sun Z, Cordes JS (2005) Novel potent human ether-a-go-go-related gene (hERG) potassium channel enhancers and their in vitro antiarrhythmic activity. Mol Pharmacol 68:876–884

Zhou PZ, Babcock J, Liu LQ, Li M, Gao ZB (2011) Activation of human ether-a-go-go related gene (hERG) potassium channels by small molecules. Acta Pharmacol Sin 32:781–788

Pharmacology of A-Type K^+ Channels

Jamie Johnston

Contents

Abstract

Transient outward potassium currents were first described nearly 60 years ago, since then major strides have been made in understanding their molecular basis and physiological roles. From the large family of voltage-gated potassium channels members of 3 subfamilies can produce such fast-inactivating A-type potassium currents. Each subfamily gives rise to currents with distinct biophysical properties and pharmacological profiles and a simple workflow is provided to aid the identification of channels mediating A-type currents in native cells. Their unique properties and regulation enable A-type K^+ channels to perform varied roles in excitable cells including repolarisation of the cardiac action potential,

J. Johnston (✉)
Faculty of Biological Sciences, University of Leeds, Leeds, UK
e-mail: J.Johnston@leeds.ac.uk

N. Gamper, K. Wang (eds.), *Pharmacology of Potassium Channels*,
Handbook of Experimental Pharmacology 267, https://doi.org/10.1007/164_2021_456

controlling spike and synaptic timing, regulating dendritic integration and long-term potentiation as well as being a locus of neural plasticity.

Keywords

A-type K^+ channel · Kv1.4 · Kv3.4 · Kv4 · Potassium channel · Voltage-clamp

1 Introduction

A-type refers to neuronal K^+ currents generating a transient outward current in response to depolarising voltage-clamp steps. This stereotypical phenotype was first described in neurons from sea slug *Onchidium* (Hagiwara et al. 1961). These early experiments described the now stereotypical hallmarks of A-type currents. They identified a transient outward voltage-gated K^+ current activated by depolarising steps from a hyperpolarised potential (Fig. 1c, d). In current clamp when hyperpolarising steps were used to remove A-current inactivation, they observed an increase in the latency to the first spike in response to depolarising current steps (Fig. 1a, b). Such transient outward currents were later designated as I_A, a term which has persisted (Connor and Stevens 1971). Work over the subsequent decades has described A-type channels in the heart (the corresponding current fraction is called I_{to} in the heart; for more details see chapter "Cardiac K^+ Channels and Channelopathies") where they contribute to the early repolarisation phase (Dixon et al. 1996; Yeola and Snyders 1997) and in many different types of neurons where they contribute to such varied functions as controlling spiking, neurotransmitter release, dendritic integration and long-term potentiation (Molineux et al. 2005; Imai et al. 2019; Watanabe et al. 2002; Frick et al. 2004; Chen et al. 2006; Shevchenko et al. 2004; Kim et al. 2005; Kuo et al. 2017).

The transient nature of A-type currents is due to rapid voltage-dependent inactivation, with inactivation decay time constants typically in the 10s of milliseconds. The subunit composition and presence of accessory subunits or other modulatory factors determine the exact biophysics and pharmacological profile of these currents. Alpha subunits from three subfamilies of voltage-gated K^+ channel generate rapid inactivating K^+ currents when expressed as homomers: Kv1.4, Kv3.3, Kv3.4 and all three Kv4 members, Kv4.1, Kv4.2 & Kv4.3 (Coetzee et al. 1999). The presence of the ß subunit Kvß1 can also confer A-type kinetics to other, normally delayed rectifier subunits from the Kv1 family (Rettig et al. 1994). These different subfamilies have distinct biophysical properties, enabling the subfamily mediating a native A-type current to be narrowed down with simple voltage-clamp protocols and confirmed with their pharmacological profiles (Fig. 1).

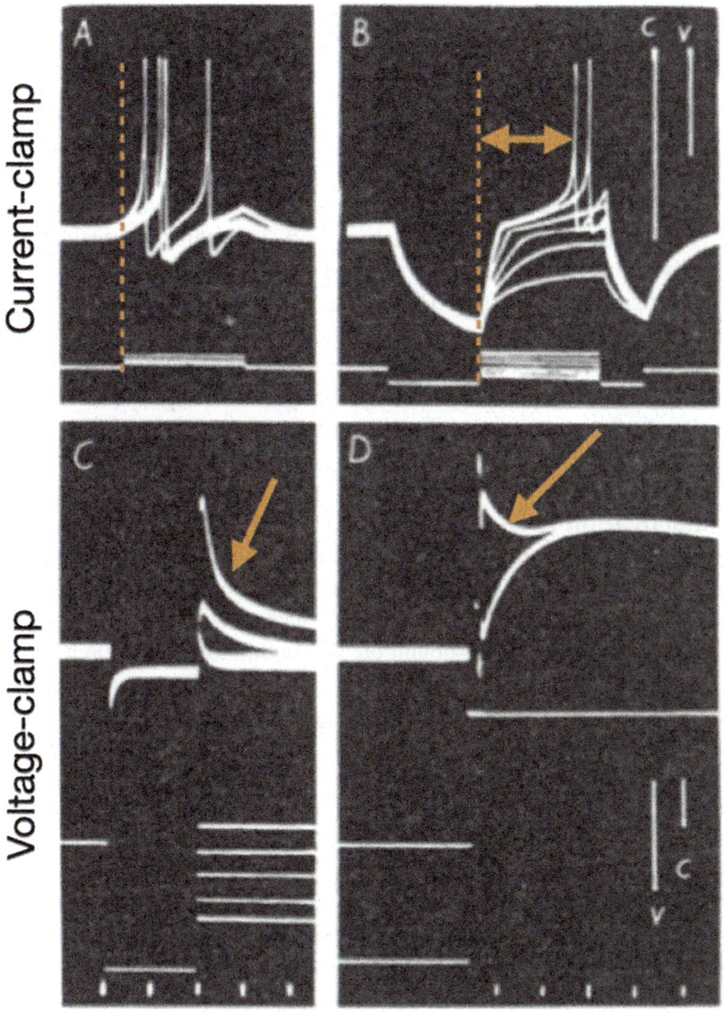

Fig. 1 Initial observations of A-type currents in *Onchidium.* (**a**) Current clamp recording of action potentials evoked from rest with short spike latencies. (**b**) A preceding hyperpolarising step removes steady-state inactivation of the A-type current and results in a lag to the first spike as the A-current inactivates (yellow arrow). (**c**) Voltage-clamp recording showing a fast-inactivating outward current activated at voltage around resting membrane potentials after hyperpolarising steps, yellow arrow. (**d**) The A-current is present at more positive potentials only with a hyperpolarising pre-step, yellow arrow. The delayed rectifier can be isolated by inactivating the A-current. Adapted from Hagiwara et al. (1961) reproduced with permission

2 Identifying Native A-Type Channels

2.1 Biophysics

2.1.1 Activation Range

Canonical A-type K^+ currents have sub-threshold activation ranges, i.e. they begin to activate at voltages more negative than that required to generate an action potential (Fig. 1). A hallmark of the Kv3 subfamily is their supra-threshold activation range, they typically do not begin to activate until voltages greater than −30 mV with half activation voltages around +10 mV (Rudy and McBain 2001). This property restricts their physiological activation window to that of the action potential, in which Kv3 channels act to rapidly repolarise the cell to the resting state (Rudy and McBain 2001; Johnston et al. 2010). A-type K^+ currents mediated by channels containing Kv3.3 and/or Kv3.4 can therefore be distinguished from other A-type subunits by their positively shifted activation curve (Fig. 2b).

2.1.2 Inactivation Properties

The molecular basis of rapid K^+ channel inactivation has been extensively studied with Kv1.4 used as the prototype. Two forms of inactivation have been demonstrated; fast N-type or "ball and chain" inactivation involves the N-terminal inactivation peptide of the alpha subunit binding within the central cavity to occlude the inner pore (Hoshi et al. 1990; Zhou et al. 2001). The rate of this type of inactivation is directly related to the number of N-terminal peptides, e.g. Kv1.4 homomers, possessing 4 N-terminal peptides, show faster inactivation than Kv1.4/Kv1.1 heteromers, which have 3 of less depending on the stoichiometry (Ruppersberg et al. 1990; MacKinnon et al. 1993). The presence of Kvß1 subunits confers a similar N-terminal inactivation peptide (one per subunit), which further accelerates the inactivation rate of Kv1.4 or can endow Kv1.1 with fast A-type inactivation (Rettig et al. 1994). Kv3.3 and Kv3.4 are thought to inactivate through a similar mechanism showing a similar dose dependency of N-terminal inactivating peptide on inactivation rate (Rudy et al. 1999; Weiser et al. 1994). In addition to the described N-type inactivation, Kv1.4 channels also inactivate by a slower C-type inactivation attributed to partial collapse of the selectivity filter/outer pore. N and C-type inactivation in Kv1.4 are allosterically coupled (Hoshi et al. 1990, 1991; Bett and Rasmusson 2004) and the rate of recovery from inactivation is controlled by the slower C-type inactivation (Rasmusson et al. 1995). As a result Kv1.4 channels recover from inactivation with a time constant of seconds (Bett and Rasmusson 2004; Roeper et al. 1997).

In contrast, the molecular details concerning inactivation of Kv4 channels have not been determined with such precision. Kv4 subunits use neither the N nor C-type inactivation processes identified for Kv1.4, instead inactivation seems to involve the inner vestibule of the pore and readily occurs through the closed state (Jerng et al. 1999; Bähring et al. 2001; Beck and Covarrubias 2001; Shahidullah and Covarrubias 2003). Kv4 channels possess an N-terminal inactivating domain but rather than contributing to inactivation this region is the binding site of the ß-subunits of Kv4

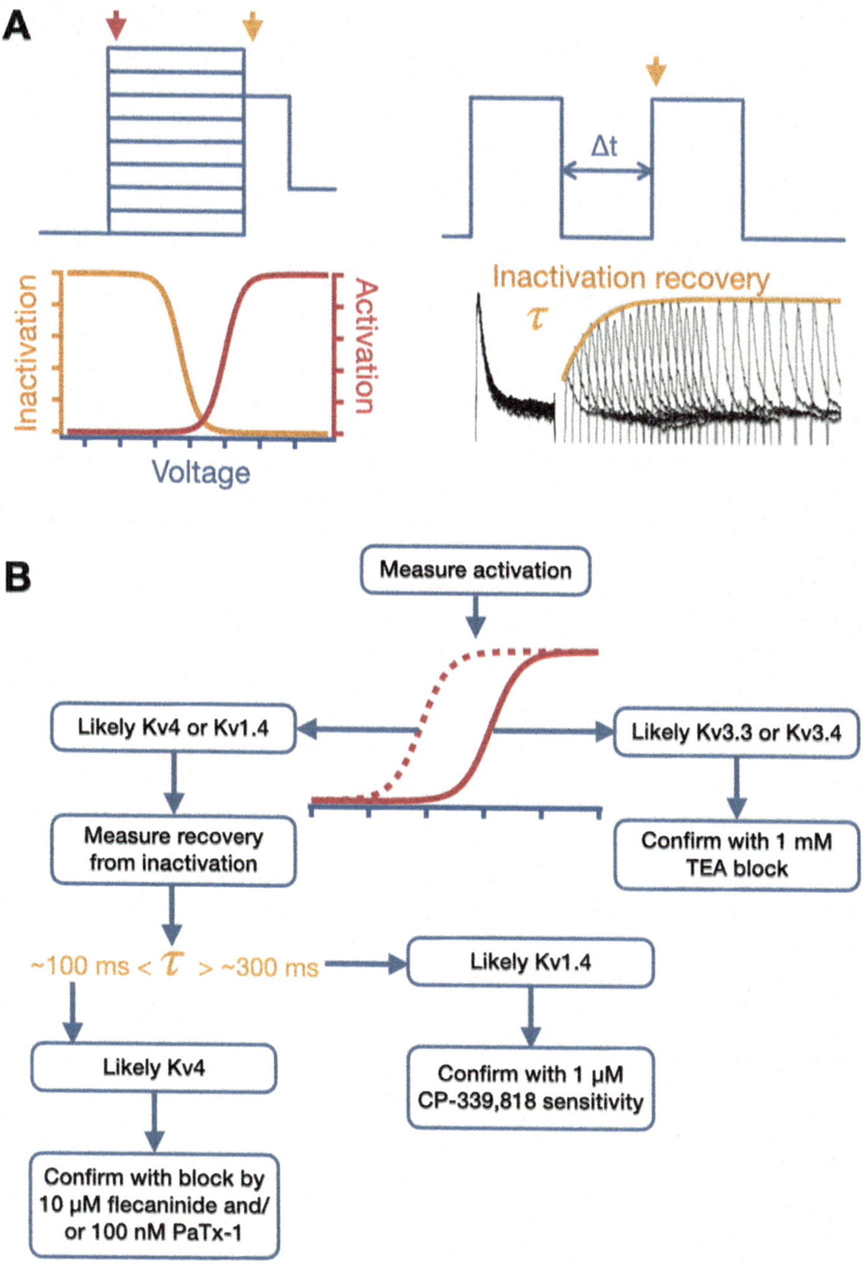

Fig. 2 Biophysical and pharmacological identification of channels mediating native A-type K^+ currents. (**a**) *Left:* Voltage protocol to determine the activation and inactivation curves for A-type channels. The activation curve can be constructed by measuring the peak current soon after the test step (red arrow) and correcting it for the non-linear driving force (see text). The steady-state inactivation curve can be obtained by measuring the current at the yellow arrow as a function of the preceding test step. Note: the duration of the initial hyperpolarising pre-pulse and the test step should be adjusted depending on the kinetics of that current being measured, e.g. the duration

channels known as K^+ channel interacting proteins (KChIPs). These accessory proteins appear to be obligatory subunits for native channels (Rhodes et al. 2004). KChIPs and DPPX, another group of accessory proteins, dramatically increase the surface expression of Kv4 subunits and alter the kinetics of and recovery from inactivation of Kv4 subunits (Beck et al. 2002; Jerng et al. 2004).

The kinetics of inactivation cannot reliably distinguish between Kv1.4 and Kv4 subunits as their decay time constants can overlap, largely depending on the composition of accessory subunits. However, the recovery from inactivation of Kv4 channels is usually much more rapid, with time constants around 10s of milliseconds (Kim et al. 2005; Beck et al. 2002; Johnston et al. 2008; Amadi et al. 2007), compared to Kv1.4 with recovery rates of 100 s of milliseconds to seconds (Ruppersberg et al. 1990; Rasmusson et al. 1995; Roeper et al. 1997).

Two simple voltage protocols can therefore help to determine the subfamily responsible for native A-type currents. First, the activation curve should be attained with a simple I/V protocol with test steps from ~−100 mV, to ensure removal of inactivation (Fig. 2a). The peak current soon after the test step (red arrow in Fig. 2a) should be corrected for the non-linear driving force of K+ before being normalised and fit with a Boltzmann function (Clay 2000). With the protocol shown in Fig. 2a the steady-state inactivation curve can also be constructed by plotting the current amplitudes measured at the yellow arrow vs the voltage of the preceding test step. The activation curve will distinguish between Kv3 mediated A-type currents with positive shifted activation ranges and Kv4 or Kv1 mediated A-type currents which have more negative sub-threshold activation ranges (Fig. 2b). If necessary, the A-type current can be isolated from other endogenous currents with a subtraction protocol as shown in Johnston et al. (2008). If the A-current has a sub-threshold activation range, the rate of recovery from inactivation can be determined with a protocol similar to that shown in Fig. 2b right; rapid recovery from inactivation, a tau <~100 ms, is indicative of Kv4 subunits whereas currents displaying time constants of several hundreds of milliseconds or longer are likely to be mediated by Kv1.4 subunits.

2.2 Pharmacology

Pharmacology of ion channels can serve multiple purposes as illustrated by the holy grail of pharmacology, a selective antagonist, a molecule that will antagonise one

Fig. 2 (continued) pre-pulse should ensure complete recovery from inactivation. *Right:* The rate of recovery from inactivation can be determined with a pair of test steps of sufficient duration to allow complete inactivation of the A-current and separated by a variable delay (Δt). The time constant of recovery can then be determined by fitting an exponential to the current amplitude of the second step as a function of Δt. Note the current amplitude of the first step should be constant. Adapted from Johnston et al. (2008). (**b**) A biophysical and pharmacological procedure to aid identification of the channel mediating a native A-current (see text for further description) kind of channel and no other. Such a molecule can be used to unambiguously determine the presence of a channel, its contribution to total current and the role of this channel in normal physiology. There are some striking examples of this with biologically derived toxins, e.g. Dendrotoxin-K is selective for any channel containing Kv1.1 subunits and blocks such channels in the 10 nM range (Wang et al. 1999; Robertson et al. 1996; Dodson et al. 2002). Unfortunately, such molecules are not always available particularly for the large diversity of Kv channels. However, less selective antagonists can still serve a useful role; when combined with other antagonists or biophysical characteristics they can aid identification of the subfamily mediating an A-type current. With these considerations in mind the following discussion largely focuses on molecules that fit these criteria.

2.2.1 Kv3 A-Type Currents

All Kv3 channels are blocked by 1 mM tetraethylammonium (TEA) which is readily washed on and off tissue (Johnston et al. 2010). TEA has an IC50 for Kv3.3 and Kv3.4 in the 100–200 μM range and acts by occluding the pore (De Miera et al. 1992; Rettig et al. 1992). TEA sensitivity in combination with a positive activation range provides a strong indication that a Kv3 subunit mediates an A-type current (Fig. 2b). Block of an A-type current by 1 mM TEA rules out Kv4 subunits which are insensitive to even 100 mM (Johnston et al. 2008, 2010), though the presence of DPPX can increase their sensitivity to TEA somewhat (Colinas et al. 2008). Kv1.4 homomers are also insensitive to TEA (Ruppersberg et al. 1990); however, heteromeric Kv1.1/Kv1.4 channels are endowed with a mix of properties from their homomeric counterparts; they display A-type kinetics with slower inactivation rates and have moderate TEA sensitivity with an IC50 of ~10 mM (Ruppersberg et al. 1990). Although 1 mM TEA can be used to unambiguously identify a high-voltage activated A-type currents, it is of limited use in exploring their functional role as many other non-A-type channels are sensitive to TEA, e.g. some Kv7 subunits and BK channels (Johnston et al. 2010). The sea anemone toxins BD-I and BDS-II were thought to be selective antagonists of Kv3.4 channels which would provide a powerful tool for elucidating the functional role of Kv3.4 A-type currents in physiology (Diochot et al. 1998). However, later detailed investigation of the interaction of BDS toxins with Kv3 subunits revealed them to act as gating modifiers and with little selectivity between Kv3 subunits (Yeung et al. 2005). BDS-I is also highly potent for Nav1.7 channels, slowing inactivation rates with a consequent broadening of action potentials and increased resurgent currents during repolarisation (Liu et al. 2012). Consequently, BDS toxins have limited potential for revealing the physiological role of Kv3 A-type currents. This also highlights a pitfall of using poorly characterised pharmacological tools; it took 14 years after their first characterisation as selective Kv3.4 blockers to realise a more complete pharmacological profile of BDS toxins. To date there are no sufficiently selective antagonists of Kv3.3 or Kv3.4 channels suitable for studying their physiological roles.

2.2.2 Kv1.4 A-Type Currents

Kv1.4 channels are relatively insensitive to TEA (Ruppersberg et al. 1990) and have a lower sensitivity (IC50 ~ 12.5 mM) to the broad spectrum K^+ channel blocker 4-AP (Stühmer et al. 1989). In fact, there is a paucity of pharmacological tools available for Kv1.4 channels, they are somewhat unique in their resistance to biological toxins which are very potent against many of the other Kv1.x members. At 1 μM the small molecule CP-339,818 shows comparative selectivity for Kv1.4 and Kv1.3 over other Kv channels, including Kv3 and Kv4 subfamilies (Nguyen et al. 1996). CP-339,818 preferentially binds to the C-type inactivated state and therefore shows a use dependent block. At concentrations greater than 1 μM off-target effects become a problem with many other channels being inhibited, e.g. Kv2.1, HCN1, Kv3.2 (Lee et al. 2008; Nguyen et al. 1996; Sforna et al. 2015). Another small molecule inhibitor, UK-78,282, has a very similar pharmacological profile to CP-339,818 (Hanson et al. 1999). To date there are no well-established selective antagonists of Kv1.4 channels suitable for studying its physiological roles, though if a contribution from Kv1.3 can be ruled out, CP-339,818 may serve this purpose.

2.2.3 Kv4 A-Type Currents

The anti-arrhythmic drug, flecainide, at 10 μM inhibits Kv4.2 & Kv4.3 leaving Kv1.4 unaffected and has been used to determine the contribution of these channels to the cardiac I_{to} current (Dixon et al. 1996; Yeola and Snyders 1997). Although flecainide can be used to distinguish between A-type currents mediated by Kv1.4 and Kv4 subunits, it also inhibits a number of other channels including the cardiac Nav1.7 channel (Ramos and O'leary 2004), ether-a-go-go related gene (ERG) K^+ channels (Paul et al. 2002) and ryanodine receptors (Mehra et al. 2014). The spider toxins, phrixotoxin 1 & 2, are potent and selective inhibitors of Kv4.2 and Kv4.3 with IC50s between 5–70 nM (Diochot et al. 1999), these toxins are gating modifiers stabilising the closed state (Chagot et al. 2004) which shifts the activation curve to more positive potentials (Diochot et al. 1999).

Kv4.2 and Kv4.3 show strong expression in many areas of the brain and heart (Serodio et al. 1996) and consequently have received the most attention. The first member of this family to be cloned, Kv4.1, shows much weaker expression in both heart and brain (Serodio et al. 1996) and as a result the pharmacological profiles of many molecules have mostly omitted this subunit. However, Kv4.1 is expressed in the suprachiasmatic nucleus (Hermanstyne et al. 2017), striatum (Song et al. 1998), basolateral amygdala (Dabrowska and Rainnie 2010) and nociceptive neurons in the dorsal root ganglion (Phuket and Covarrubias 2009). The spider toxin Heteroscodra maculata 1 (HmTx-1) is reported to inhibit all 3 Kv4 members but also inhibits Kv2.1 (Escoubas et al. 2002). Phrixotoxin-1, although only initially tested on Kv4.2 and Kv4.3, also inhibits Kv4.1 but with 3 times lower affinity; this block seems to be voltage-independent (Yunoki et al. 2014). At present phrioxtoxin-1 provides a useful tool for identifying the presence of Kv4 currents and exploring their physiological roles, without being able to distinguish between different Kv4 subunits.

3 Physiological Roles of A Currents

3.1 Neural Excitability and Spike Properties

A-type channels are rapidly activated upon depolarisation and therefore contribute to action potential repolarisation (Kim et al. 2005; Carrasquillo et al. 2012; Rudy and McBain 2001), including in the heart (Dixon et al. 1996; Yeola and Snyders 1997). Kv4 subunits are predominantly located in the somato-dendritic compartments of neurons (Trimmer 2015), which regulates back-propagation of action potentials into the dendritic tree (Fig. 3b). The sub-threshold activation range of Kv4 and Kv1.4 channels enables them to regulate intrinsic firing/bursting of neurons in many areas including the: suprachiasmatic nucleus (Hermanstyne et al. 2017), hypothalamus (Mendonça et al. 2018; Imai et al. 2019), dorsal root ganglion (Zemel et al. 2018), hippocampus (Bourdeau et al. 2007), substantia nigra (Liss et al. 2001) and cortex (Carrasquillo et al. 2012). A key factor determining the ability of A-type channels to exert influence is the availability of channels in their non-inactivated state, for example fast spiking will result in more inactivated A-type channels and consequent spike broadening (Kim et al. 2005). The steady-state inactivation curve (Fig. 2a yellow) describes the proportion of channels available at resting membrane potentials. This property is influenced by the presence of accessory subunits (Jerng et al. 1999) and can be dynamically regulated by phosphorylation (Rosenkranz et al. 2009) and Ca^{2+} (Anderson et al. 2010a). Cav3 T-type channels which are also activated at sub-threshold potentials form complexes with Kv4 channels and provide a source of Ca^{2+} that shifts the inactivation curve of Kv4 channels to more positive potentials (Anderson et al. 2010a, b), increasing their availability. The presence of Cav3 channels can therefore alter the contribution that A-type channels make to neural excitability (Heath et al. 2014) (Fig. 3).

Due to the lack of specific inhibitors of Kv1.4, Kv3.3 and Kv3.4 there are few concrete demonstrations of their physiological roles. Throughout the brain Kv1.4 is often found in axons and presynaptic terminals (Veh et al. 1995; Cooper et al. 1998) and may therefore play a role in regulating transmitter release. Kv1.4's slow recovery from inactivation means that with repetitive firing the repolarisation rate of presynaptic action potentials will reduce as Kv1.4 accumulates in its inactivated state. Slower repolarisation of the action potential causes larger presynaptic Ca^{2+} influx and an increase in release probability (Yang and Wang 2006), suggesting that Kv1.4 may play a role in short-term synaptic plasticity. Consistent with this idea, knockdown of Kv1.4 in the hippocampus reduced paired pulse facilitation (Meiri et al. 1998), a form of presynaptic short-term plasticity. Ca^{2+} can also regulate the rate of recovery from inactivation of Kv1.4 in a CamKinase II dependent manner (Roeper et al. 1997), providing scope for regulation of such short-term plasticity. Kv3.3 and Kv3.4 are also located in axons and synaptic terminals (Laube et al. 1996; Ishikawa et al. 2003; Trimmer 2015) and could have similar roles.

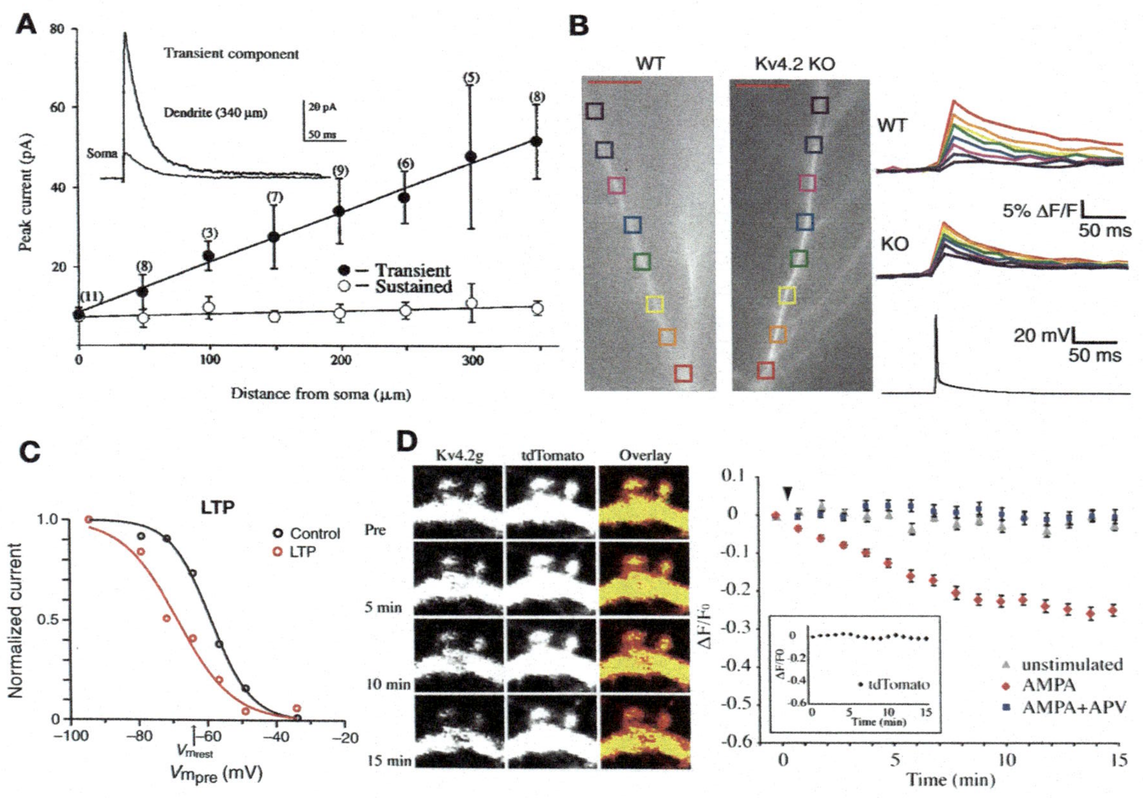

Fig. 3 Dendritic Kv4 A-type channels regulate and are a locus of plasticity in hippocampal CA1 neurons. (**a**) Dendritic recordings of Kv4 A-type K^+ currents demonstrate a gradient of current that increases with distance from soma adapted from Hoffman et al. (1997) with permission. (**b**) Imaging of fura-2 loaded CA1

neurons reveal attenuation of back-propagating action potential evoked Ca^{2+} transients. This attenuation is dramatically reduced when Kv4.2 is absent, adapted from Chen et al. (2006, Copyright 2006, Society for Neuroscience) with permission. (**c**) The steady-state inactivation curve of dendritically located Kv4 channels is shifted to more hyperpolarised potentials after induction of LTP with theta burst stimuli. Note that this decreases the available Kv4 current at resting membrane potentials by ~50%, adapted from Frick et al. (2004) with permission. (**d**) Fluorescently tagged A-type channels (Kv4.2 g) are trafficked from the membrane following activation of synaptic receptors, adapted from Kim et al. (2007) with permission

3.2 Dendritic Integration and LTP

In hippocampal pyramidal neurons the density of Kv4 channels increases from the soma towards the distal dendrites (Fig. 3a), this gradient depends on expression of the accessory subunit DPP6 (Sun et al. 2011). The gradient of Kv4 channels controls the ability of action potentials to back-propagate and invade the dendritic tree (Hoffman et al. 1997; Johnston et al. 2000) (Fig. 3b), a phenomenon required for induction of synaptic plasticity in these cells (Magee and Johnston 1997; Buchanan and Mellor 2007). Consistent with this idea, action potentials more readily back-propagate into the dendritic tree when Kv4.2 channels are knocked down (Fig. 3b) and consequently the threshold for LTP induction is lowered (Chen et al. 2006). Kv4 channels are also a locus of plasticity in CA1 dendrites being targeted by mitogen-activated protein kinase (Rosenkranz et al. 2009). Induction of LTP results in a shift in the steady-state inactivation of Kv4 channels to more negative-potentials (Fig. 3c) (Frick et al. 2004) and internalisation of channels (Fig. 3d) (Kim et al. 2007), effects which are localised to the site of synaptic input. Both of these changes reduce the available A-type current and consequently increase dendritic excitability and therefore the likelihood of further plasticity.

A similar attenuation of back-propagating action potentials by A-type currents is observed in the lateral dendrites of mitral cells of the olfactory bulb (Margrie et al. 2001; Christie and Westbrook 2003), this influences the extent of lateral dendrodendritic interactions with inhibitory granule cells. A-type K^+ channels also influence the inhibitory side of the reciprocal synapse between granule and mitral cell dendrites. Granule cells receive excitation from mitral cell dendrites via AMPA and NMDA receptors and provide GABAergic feedback inhibition. The A-type current in granule cells attenuates excitation for the fast AMPA component; spiking and feedback inhibition then occur with a delay, following the time course of the NMDA component (Schoppa and Westbrook 1999).

The varied biophysics, sub-cellular distributions and dynamic modulation of A-type K^+ channels enable them to perform a multitude of roles in regulating cellular excitability. Large strides have been made in the 60 years since their discovery but much still remains to be elucidated.

References

Amadi CC, Brust RD, Skerritt MR, Campbell DL (2007) Regulation of Kv4.3 closed state inactivation and recovery by extracellular potassium and intracellular KChIP2b. Channels (Austin) 1:305–314

Anderson D, Mehaffey WH, Iftinca M, Rehak R, Engbers JD, Hameed S, Zamponi GW, Turner RW (2010a) Regulation of neuronal activity by Cav3-Kv4 channel signaling complexes. Nat Neurosci 13:333–337

Anderson D, Rehak R, Hameed S, Mehaffey WH, Zamponi GW, Turner RW (2010b) Regulation of the KV4.2 complex by CaV3.1 calcium channels. Channels (Austin) 4:163–167

Bähring R, Boland LM, Varghese A, Gebauer M, Pongs O (2001) Kinetic analysis of open-and closed-state inactivation transitions in human Kv4. 2 A-type potassium channels. J Physiol 535:65

Beck EJ, Covarrubias M (2001) Kv4 channels exhibit modulation of closed-state inactivation in inside-out patches. Biophys J 81:867–883

Beck EJ, Bowlby M, An WF, Rhodes KJ, Covarrubias M (2002) Remodelling inactivation gating of Kv4 channels by KChIP1, a small-molecular-weight calcium-binding protein. J Physiol 538:691–706

Bett GCL, Rasmusson RL (2004) Inactivation and recovery in Kv1.4 K+ channels: lipophilic interactions at the intracellular mouth of the pore. J Physiol Lond 556:109–120

Bourdeau ML, Morin F, Laurent CE, Azzi M, Lacaille JC (2007) Kv4.3-mediated A-type K+ currents underlie rhythmic activity in hippocampal interneurons. J Neurosci 27:1942–1953

Buchanan KA, Mellor JR (2007) The development of synaptic plasticity induction rules and the requirement for postsynaptic spikes in rat hippocampal CA1 pyramidal neurones. J Physiol 585:429–445

Carrasquillo Y, Burkhalter A, Nerbonne JM (2012) A-type K+ channels encoded by Kv4.2, Kv4.3 and Kv1.4 differentially regulate intrinsic excitability of cortical pyramidal neurons. J Physiol 590:3877–3890

Chagot B, Escoubas P, Villegas E, Bernard C, Ferrat G, Corzo G, Lazdunski M, Darbon H (2004) Solution structure of Phrixotoxin 1, a specific peptide inhibitor of Kv4 potassium channels from the venom of the theraphosid spider Phrixotrichus auratus. Protein Sci 13:1197–1208

Chen X, Yuan L-L, Zhao C, Birnbaum SG, Frick A, Jung WE, Schwarz TL, Sweatt JD, Johnston D (2006) Deletion of Kv4.2 gene eliminates dendritic A-type K+ current and enhances induction of long-term potentiation in hippocampal CA1 pyramidal neurons. J Neurosci 26:12143–12151

Christie JM, Westbrook GL (2003) Regulation of backpropagating action potentials in mitral cell lateral dendrites by A-type potassium currents. J Neurophysiol 89:2466–2472

Clay JR (2000) Determining K+ channel activation curves from K+ channel currents. Eur Biophys J 29:555–557

Coetzee WA, Amarillo Y, Cu JOANNA, Chow ALAN, Lau DAVID, McCormack TOM, Morena HERMAN, Nadal MARCELAS, Ozaita ANDER, Pountney DAVID, Saganich MICHAEL, de Miera ELEAZARVEGA-SAENZ, Rudy BERNARDO (1999) Molecular diversity of K+ channels. Ann N Y Acad Sci 868:233–255

Colinas O, Pérez-Carretero FD, López-López JR, Pérez-García MT (2008) A role for DPPX modulating external TEA sensitivity of Kv4 channels. J Gen Physiol 131:455–471

Connor JA, Stevens CF (1971) Voltage clamp studies of a transient outward membrane current in gastropod neural somata. J Physiol 213:21

Cooper EC, Milroy A, Jan YN et al (1998) Presynaptic localization of Kv1. 4-containing A-type potassium channels near excitatory synapses in the hippocampus. J Neurosci 18:965–974

Dabrowska J, Rainnie DG (2010) Expression and distribution of Kv4 potassium channel subunits and potassium channel interacting proteins in subpopulations of interneurons in the basolateral amygdala. Neuroscience 171:721–733

De Miera EV-S, Moreno H, Fruhling D, Kentros C, Rudy B (1992) Cloning of ShIII (Shaw-like) cDNAs encoding a novel high-voltage-activating, TEA-sensitive, type-A K+ channel. Proc R Soc Lond Ser B Biol Sci 248:9–18

Diochot S, Schweitz H, Béress L, Lazdunski M (1998) Sea anemone peptides with a specific blocking activity against the fast inactivating potassium channel Kv3.4. J Biol Chem 273:6744–6749

Diochot S, Drici M, Moinier D, Fink M, Lazdunski M (1999) Effects of phrixotoxins on the Kv4 family of potassium channels and implications for the role of Ito1 in cardiac electrogenesis. Br J Pharmacol 126:251–263

Dixon JE, Shi W, Wang HS, McDonald C, Yu H, Wymore RS, Cohen IS, McKinnon D (1996) Role of the Kv4.3 K+ channel in ventricular muscle. A molecular correlate for the transient outward current. Circ Res 79:659–668

Dodson PD, Barker MC, Forsythe ID (2002) Two heteromeric Kv1 potassium channels differentially regulate action potential firing. J Neurosci 22:6953–6961

Escoubas P, Diochot S, Celerier M-L, Nakajima T, Lazdunski M (2002) Novel tarantula toxins for subtypes of voltage-dependent potassium channels in the Kv2 and Kv4 subfamilies. Mol Pharmacol 62:48–57

Frick A, Magee J, Johnston D (2004) LTP is accompanied by an enhanced local excitability of pyramidal neuron dendrites. Nat Neurosci 7:126–135

Hagiwara S, Kusano K, Saito N (1961) Membrane changes of Onchidium nerve cell in potassium-rich media. J Physiol 155:470

Hanson DC, Nguyen A, Mather RJ, Rauer H, Koch K, Burgess LE, Rizzi JP, Donovan CB, Bruns MJ, Canniff PC (1999) UK-78,282, a novel piperidine compound that potently blocks the Kv1. 3 voltage-gated potassium channel and inhibits human T cell activation. Br J Pharmacol 126:1707–1716

Heath NC, Rizwan AP, Engbers JD, Anderson D, Zamponi GW, Turner RW (2014) The expression pattern of a Cav3-Kv4 complex differentially regulates spike output in cerebellar granule cells. J Neurosci 34:8800–8812

Hermanstyne TO, Granados-Fuentes D, Mellor RL, Herzog ED, Nerbonne JM (2017) Acute knockdown of Kv4.1 regulates repetitive firing rates and clock gene expression in the suprachiasmatic nucleus and daily rhythms in locomotor behavior. eNeuro 4

Hoffman DA, Magee JC, Colbert CM, Johnston D (1997) K+ channel regulation of signal propagation in dendrites of hippocampal pyramidal neurons. Nature 387:869–875

Hoshi T, Zagotta WN, Aldrich RW (1990) Biophysical and molecular mechanisms of shaker potassium channel inactivation. Science 250:533–538

Hoshi T, Zagotta WN, Aldrich RW (1991) Two types of inactivation in shaker K+ channels: effects of alterations in the carboxy-terminal region. Neuron 7:547–556

Imai R, Yokota S, Horita S, Ueta Y, Maejima Y, Shimomura K (2019) Excitability of oxytocin neurons in paraventricular nucleus is regulated by voltage-gated potassium channels Kv4.2 and Kv4.3. Biosci Biotechnol Biochem 83:202–211

Ishikawa T, Nakamura Y, Saitoh N, Li W-B, Iwasaki S, Takahashi T (2003) Distinct roles of Kv1 and Kv3 potassium channels at the Calyx of held presynaptic terminal. J Neurosci 23:10445–10453

Jerng HH, Shahidullah M, Covarrubias M (1999) Inactivation gating of Kv4 potassium channels: molecular interactions involving the inner vestibule of the pore. J Gen Physiol 113:641–660

Jerng HH, Pfaffinger PJ, Covarrubias M (2004) Molecular physiology and modulation of somatodendritic A-type potassium channels. Mol Cell Neurosci 27:343–369

Johnston D, Hoffman DA, Magee JC, Poolos NP, Watanabe S, Colbert CM, Migliore M (2000) Dendritic potassium channels in hippocampal pyramidal neurons. J Physiol 525(Pt 1):75–81

Johnston J, Griffin SJ, Baker C, Forsythe ID (2008) Kv4 (A-type) potassium currents in the mouse medial nucleus of the trapezoid body. Eur J Neurosci 27:1391–1399

Johnston J, Forsythe ID, Kopp-Scheinpflug C (2010) Going native: voltage-gated potassium channels controlling neuronal excitability. J Physiol 588:3187–3200

Kim J, Wei D-S, Hoffman DA (2005) Kv4 potassium channel subunits control action potential repolarization and frequency-dependent broadening in rat hippocampal CA1 pyramidal neurones. J Physiol 569:41–57

Kim J, Jung SC, Clemens AM, Petralia RS, Hoffman DA (2007) Regulation of dendritic excitability by activity-dependent trafficking of the A-type K+ channel subunit Kv4.2 in hippocampal neurons. Neuron 54:933–947

Kuo Y-L, Cheng J-K, Hou W-H, Chang Y-C, Du P-H, Jian J-J, Rau R-H, Yang J-H, Lien C-C, Tsaur M-L (2017) K+ channel modulatory subunits KChIP and DPP participate in Kv4-mediated mechanical pain control. J Neurosci 37:4391–4404

Laube G, Röper J, Pitt JC, Sewing S, Kistner U, Garner CC, Pongs O, Veh RW (1996) Ultrastructural localization of shaker-related potassium channel subunits and synapse-associated protein 90 to septate-like junctions in rat cerebellar Pinceaux. Brain Res Mol Brain Res 42:51–61

Lee YT, Vasilyev DV, Shan QJ, Dunlop J, Mayer S, Bowlby MR (2008) Novel pharmacological activity of loperamide and CP-339,818 on human HCN channels characterized with an automated electrophysiology assay. Eur J Pharmacol 581:97–104

Liss B, Franz O, Sewing S, Bruns R, Neuhoff H, Roeper J (2001) Tuning pacemaker frequency of individual dopaminergic neurons by Kv4.3L and KChip3.1 transcription. EMBO J 20:5715–5724

Liu P, Jo S, Bean BP (2012) Modulation of neuronal sodium channels by the sea anemone peptide BDS-I. J Neurophysiol 107:3155–3167

MacKinnon R, Aldrich RW, Lee AW (1993) Functional stoichiometry of shaker potassium channel inactivation. Science 262:757–759

Magee JC, Johnston D (1997) A synaptically controlled, associative signal for Hebbian plasticity in hippocampal neurons. Science 275:209–213

Margrie TW, Sakmann B, Urban NN (2001) Action potential propagation in mitral cell lateral dendrites is decremental and controls recurrent and lateral inhibition in the mammalian olfactory bulb. Proc Natl Acad Sci U S A 98:319–324

Mehra D, Imtiaz MS, van Helden DF, Knollmann BC, Laver DR (2014) Multiple modes of ryanodine receptor 2 inhibition by flecainide. Mol Pharmacol 86:696–706

Meiri N, Sun MK, Segal Z, Alkon DL (1998) Memory and long-term potentiation (LTP) dissociated: normal spatial memory despite CA1 LTP elimination with Kv1.4 antisense. Proc Natl Acad Sci U S A 95:15037–15042

Mendonça PRF, Kyle V, Yeo SH, Colledge WH, Robinson HPC (2018) Kv4.2 channel activity controls intrinsic firing dynamics of arcuate kisspeptin neurons. J Physiol 596:885–899

Molineux ML, Fernandez FR, Mehaffey WH, Turner RW (2005) A-type and T-type currents interact to produce a novel spike latency-voltage relationship in cerebellar stellate cells. J Neurosci 25:10863–10873

Nguyen A, Kath JC, Hanson DC, Biggers MS, Canniff PC, Donovan CB, Mather RJ, Bruns MJ, Rauer H, Aiyar J, Lepple-Wienhues A, Gutman GA, Grissmer S, Cahalan MD, Chandy KG (1996) Novel nonpeptide agents potently block the C-type inactivated conformation of Kv1.3 and suppress T cell activation. Mol Pharmacol 50:1672–1679

Paul AA, Witchel HJ, Hancox JC (2002) Inhibition of the current of heterologously expressed HERG potassium channels by flecainide and comparison with quinidine, propafenone and lignocaine. Br J Pharmacol 136:717–729

Phuket TR, Covarrubias M (2009) Kv4 channels underlie the subthreshold-operating A-type K-current in nociceptive dorsal root ganglion neurons. Front Mol Neurosci 2:3

Ramos E, O'leary ME (2004) State-dependent trapping of flecainide in the cardiac sodium channel. J Physiol 560:37–49

Rasmusson RL, Morales MJ, Castellino RC, Zhang Y, Campbell DL, Strauss HC (1995) C-type inactivation controls recovery in a fast inactivating cardiac K+ channel (Kv1. 4) expressed in Xenopus oocytes. J Physiol 489:709–721

Rettig J, Wunder F, Stocker M, Lichtinghagen R, Mastiaux F, Beckh S, Kues W, Pedarzani P, Schröter KH, Ruppersberg JP (1992) Characterization of a Shaw-related potassium channel family in rat brain. EMBO J 11:2473–2486

Rettig J, Heinemann SH, Wunder F, Lorra C, Parcej DN, Dolly JO, Pongs O (1994) Inactivation properties of voltage-gated K+ channels altered by presence of β-subunit. Nature 369:289–294

Rhodes KJ, Carroll KI, Sung MA, Doliveira LC, Monaghan MM, Burke SL, Strassle BW, Buchwalder L, Menegola M, Cao J, An WF, Trimmer JS (2004) KChIPs and Kv4 alpha subunits as integral components of A-type potassium channels in mammalian brain. J Neurosci 24:7903–7915

Robertson B, Owen D, Stow J, Butler C, Newland C (1996) Novel effects of dendrotoxin homologues on subtypes of mammalian Kv1 potassium channels expressed in Xenopus oocytes. FEBS Lett 383:26–30

Roeper J, Lorra C, Pongs O (1997) Frequency-dependent inactivation of mammalian A-type K+ channel KV1.4 regulated by Ca2+/calmodulin-dependent protein kinase. J Neurosci 17:3379–3391

Rosenkranz JA, Frick A, Johnston D (2009) Kinase-dependent modification of dendritic excitability after long-term potentiation. J Physiol 587:115–125

Rudy B, McBain CJ (2001) Kv3 channels: voltage-gated K+ channels designed for high-frequency repetitive firing. Trends Neurosci 24:517–526

Rudy B, Chow A, Lau D, Amarillo Y, Ozaita A, Saganich M, Moreno H, Nadal MS, Hernandez-Pineda RICARDO, Hernandez-Cruz ARTURO (1999) Contributions of Kv3 channels to neuronal excitability. Ann N Y Acad Sci 868:304–343

Ruppersberg JP, Schröter KH, Sakmann B, Stocker M, Sewing S, Pongs O (1990) Heteromultimeric channels formed by rat brain potassium-channel proteins. Nature 345:535–537

Schoppa NE, Westbrook GL (1999) Regulation of synaptic timing in the olfactory bulb by an A-type potassium current. Nat Neurosci 2:1106–1113

Serodio P, Vega-Saenz de Miera E, Rudy B (1996) Cloning of a novel component of A-type K+ channels operating at subthreshold potentials with unique expression in heart and brain. J Neurophysiol 75:2174–2179

Sforna L, D'Adamo MC, Servettini I, Guglielmi L, Pessia M, Franciolini F, Catacuzzeno L (2015) Expression and function of a CP339,818-sensitive K+ current in a subpopulation of putative nociceptive neurons from adult mouse trigeminal ganglia. J Neurophysiol 113:2653–2665

Shahidullah M, Covarrubias M (2003) The link between ion permeation and inactivation gating of Kv4 potassium channels. Biophys J 84:928–941

Shevchenko T, Teruyama R, Armstrong WE (2004) High-threshold, Kv3-like potassium currents in magnocellular neurosecretory neurons and their role in spike repolarization. J Neurophysiol 92:3043–3055

Song W-J, Tkatch T, Baranauskas G, Ichinohe N, Kitai ST, Surmeier DJ (1998) Somatodendritic depolarization-activated potassium currents in rat neostriatal cholinergic interneurons are predominantly of the A type and attributable to coexpression of Kv4. 2 and Kv4. 1 subunits. J Neurosci 18:3124–3137

Stühmer W, Ruppersberg JP, Schröter KH, Sakmann B, Stocker M, Giese KP, Perschke A, Baumann A, Pongs O (1989) Molecular basis of functional diversity of voltage-gated potassium channels in mammalian brain. EMBO J 8:3235–3244

Sun W, Maffie JK, Lin L, Petralia RS, Rudy B, Hoffman DA (2011) DPP6 establishes the A-type K (+) current gradient critical for the regulation of dendritic excitability in CA1 hippocampal neurons. Neuron 71:1102–1115

Trimmer JS (2015) Subcellular localization of K+ channels in mammalian brain neurons: remarkable precision in the midst of extraordinary complexity. Neuron 85:238–256

Veh RW, Lichtinghagen R, Sewing S, Wunder F, Grumbach IM, Pongs O (1995) Immunohistochemical localization of five members of the Kv1 channel subunits: contrasting subcellular locations and neuron-specific co-localizations in rat brain. Eur J Neurosci 7:2189–2205

Wang FC, Parcej DN, Dolly JO (1999) Subunit compositions of Kv1.1-containing K+channel subtypes fractionated from rat brain using dendrotoxins. Eur J Biochem 263:230–237

Watanabe S, Hoffman DA, Migliore M, Johnston D (2002) Dendritic K+ channels contribute to spike-timing dependent long-term potentiation in hippocampal pyramidal neurons. Proc Natl Acad Sci U S A 99:8366–8371

Weiser M, de Miera EV-S, Kentros C, Moreno H, Franzen L, Hillman D, Baker H, Rudy B (1994) Differential expression of Shaw-related K+ channels in the rat central nervous system. J Neurosci 14:949–972

Yang Y-M, Wang L-Y (2006) Amplitude and kinetics of action potential-evoked Ca2+ current and its efficacy in triggering transmitter release at the developing Calyx of held Synapse. J Neurosci 26:5698–5708. https://doi.org/10.1523/JNEUROSCI.4889-05.2006

Yeola SW, Snyders DJ (1997) Electrophysiological and pharmacological correspondence between Kv4.2 current and rat cardiac transient outward current. Cardiovasc Res 33:540–547
Yeung SY, Thompson D, Wang Z, Fedida D, Robertson B (2005) Modulation of Kv3 subfamily potassium currents by the sea anemone toxin BDS: significance for CNS and biophysical studies. J Neurosci 25:8735–8745
Yunoki T, Takimoto K, Kita K, Funahashi Y, Takahashi R, Matsuyoshi H, Naito S, Yoshimura N (2014) Differential contribution of Kv4-containing channels to A-type, voltage-gated potassium currents in somatic and visceral dorsal root ganglion neurons. J Neurophysiol 112:2492–2504
Zemel BM, Ritter DM, Covarrubias M, Muqeem T (2018) A-type K_V channels in dorsal root ganglion neurons: diversity, function, and dysfunction. Front Mol Neurosci 11:253
Zhou M, Morais-Cabral JH, Mann S, MacKinnon R (2001) Potassium channel receptor site for the inactivation gate and quaternary amine inhibitors. Nature 411:657–661

Kv7 Channels and Excitability Disorders

Frederick Jones, Nikita Gamper, and Haixia Gao

Contents

F. Jones
Faculty of Biological Sciences, University of Leeds, Leeds, UK

N. Gamper (✉)
Faculty of Biological Sciences, University of Leeds, Leeds, UK

Department of Pharmacology, Hebei Medical University, Shijiazhuang, China
e-mail: n.gamper@leeds.ac.uk

H. Gao (✉)
Department of Pharmacology, Hebei Medical University, Shijiazhuang, China
e-mail: gaohx686@hebmu.edu.cn

N. Gamper, K. Wang (eds.), *Pharmacology of Potassium Channels*,
Handbook of Experimental Pharmacology 267, https://doi.org/10.1007/164_2021_457

Abstract

Kv7.1-Kv7.5 (KCNQ1–5) K^+ channels are voltage-gated K^+ channels with major roles in neurons, muscle cells and epithelia where they underlie physiologically important K^+ currents, such as neuronal M current and cardiac I_{Ks}. Specific biophysical properties of Kv7 channels make them particularly well placed to control the activity of excitable cells. Indeed, these channels often work as 'excitability breaks' and are targeted by various hormones and modulators to regulate cellular activity outputs. Genetic deficiencies in all five KCNQ genes result in human excitability disorders, including epilepsy, arrhythmias, deafness and some others. Not surprisingly, this channel family attracts considerable attention as potential drug targets. Here we will review biophysical properties and tissue expression profile of Kv7 channels, discuss recent advances in the understanding of their structure as well as their role in various neurological, cardiovascular and other diseases and pathologies. We will also consider a scope for therapeutic targeting of Kv7 channels for treatment of the above health conditions.

Keywords

Channelopathy · Epilepsy · KCNQ · Kv7 channel · M current · Pain

1 Introduction

The family of Kv7 (KCNQ) K^+ channels encompasses voltage-gated K^+ channels produced by homo- or heterotetrameric assembly of Kv7 subunits encoded by *KCNQ* genes. As a general rule, these channels conduct K^+ currents with a very negative activation threshold (~−60 mV), slow activation and deactivation kinetics and no inactivation (Delmas and Brown 2005; Du et al. 2018; Gamper and Shapiro 2015). These features make these channels an exquisite instrument for controlling cellular excitability. Some of the currents produced by these channels were functionally described long before the genes were identified. Thus, a slow, non-inactivating K^+ current fraction inhibited by *m*uscarinic acetylcholine receptors (mAChRs) was discovered by David Brown and colleagues in sympathetic neurons and named '*M* current' (Brown and Adams 1980; Constanti and Brown 1981). Due to its biophysical properties, M current (I_M) serves as an effective brake on neuronal excitability and its inhibition has proven to be one of the major mechanisms of the excitatory effect of acetylcholine (Brown et al. 1997). A K^+ current with even slower kinetics was also identified in the heart and termed 'I_{Ks}'; it mediates initial repolarization of the cardiac action potential (which is much slower than a neuronal one). I_{ks} is potentiated by protein kinase A (PKA), a property which is partly responsible for the speeding of the heart rate upon adrenergic stimulation (Walsh and Kass 1988). Additionally, a small-conductance K^+ current sensitive to chromanol 293B was identified in the basolateral membrane of some epithelial cells, where it provides the driving force for electrogenic Cl^- secretion (Lohrmann et al. 1995; Warth et al. 1996). Found in different types of cells (neuronal, cardiac and epithelial) these currents were not initially thought to represent related phenomena, but the advent of molecular cloning and genetics unveiled the true genetic identity of the underlying ion channels as belonging to the same protein family: the Kv7 channel family.

In 1996–1997, an inherited form of human cardiac arrhythmia, the 'long-QT syndrome' (LQTS) was linked to mutations in a novel K^+ channel gene, named KvLQT1 (Shalaby et al. 1997; Wang et al. 1996; Yang et al. 1997). The gene was identified by positional cloning approach and localized to chromosome 11p15.5 (Wang et al. 1996). The KvLQT1 gene predicted to code for a K^+ channel subunit orthologous to the *Shaker* K^+ channel with predicted 6 membrane-spanning domains (S1-S6), a voltage sensor residing in S4, and a long carboxyl terminus. Subsequent studies revealed that native I_{Ks} to be composed of KvLQT1, as the pore-forming subunit, in complex with an auxiliary β-subunit, called KCNE1 (Kaczmarek and Blumenthal 1997; Romey et al. 1997; Sanguinetti et al. 1996). Orthologs of human KvLQT1 (KQT1–2) were then identified in *C. elegans* (Wei et al. 1996). The next step was the discovery of two other human KvLQT1-related genes, mapped to chromosomes 20q1.3 and 8q24 (Biervert et al. 1998; Singh et al. 1998). Mutations in these novel genes were linked to a form of inherited epilepsy in humans, benign familial neonatal convulsions (BFNC) (Charlier et al. 1998; Yang et al. 1998). When exogenously overexpressed in the cell lines these gene products neatly recapitulated all major properties of neuronal I_M, including kinetics, pharmacology and sensitivity to mAChRs (Selyanko et al. 2000; Shapiro et al. 2000; Wang et al. 1998). The new gene family was named *KCNQ*, with KvLQT1 receiving a new name – *KCNQ1*, and the two epilepsy-associated genes named *KCNQ2* and *KCNQ3*. The protein subunits encoded by *KCNQ* genes were grouped into the Kv7 family of voltage-gated K^+ channels, although 'KCNQ' is also commonly used to describe channel subunits in the literature. In the meantime, the basolateral K^+ channel of epithelial cells was pinpointed to arise from co-assembly of KCNQ1 α-subunit and another member of KCNE β-subunit family, KCNE3 (Bleich and Warth 2000; Schroeder et al. 2000b). And soon, two other genes encoding final Kv7 subunits were identified: *KCNQ4* encoding Kv7.4 channel, predominantly expressed in the inner ear (Wangemann 2002) and auditory nuclei in the brain (Coucke et al. 1999; Kharkovets et al. 2000; Kubisch et al. 1999; Sogaard et al. 2001) and *KCNQ5*, coding for another neuronal M channel subunit, Kv7.5 (Lerche et al. 2000; Passmore et al. 2003; Schroeder et al.

2000a). These two subunits were later found expressed in various smooth muscle cells, where these control contractility (Brueggemann et al. 2007, 2012; Mackie et al. 2008; Ohya et al. 2003). Thus, the *KCNQ* gene family includes five genes in humans, *KCNQ1–5,* giving rise to five types of α-subunits, Kv7.1-Kv7.5 (each with a number of splice variants). In the next two chapters we will discuss properties, roles and therapeutic potential of the Kv7 channels. Here we will focus on the general properties of these channels, their tissue expression, regulation and role in excitability disorders. Chapter "Pharmacological Activation of Neuronal Voltage-Gated Kv7/KCNQ/M-Channels for Potential Therapy of Epilepsy and Pain" will specifically explore the Kv7 channel pharmacology and prospects of targeting these channels for therapy of epilepsy and pain.

2 Biophysical Properties of Kv7 Channels as a Guide to Their Role in Cellular Excitability

Some key biophysical properties of individual Kv7 channels are summarized in Table 1. Most Kv7 subunits assemble to conduct slow, M-like voltage-gated K^+ currents with a very negative activation threshold (~−60 mV but can be as negative as −80 mV in some cases (Huang and Trussell 2011)) and no inactivation (Fig. 1a). Two exceptions to this rule are the Kv7.1 homomers, which do express modest inactivation (Tristani-Firouzi and Sanguinetti 1998) and Kv7.1/KCNE3 channel complex, which is constitutively-open and, hence, voltage-independent (Schroeder et al. 2000b). The single-channel conductance (γ) varies between the channel isoforms but is generally small (Table 1). Kv7.1 homotetramers have a very low γ (1–4 pS), but assembly with KCNE1 increases it considerably (Pusch 1998; Sesti and Goldstein 1998; Werry et al. 2013) and removes inactivation (Tristani-Firouzi and Sanguinetti 1998); both effects resulting in considerably higher macroscopic current amplitudes as compared to homomeric Kv7.1 channels (Jespersen et al. 2005). Other Kv7 channels also have relatively low γ, with Kv7.3 having the largest and Kv7.4 and Kv7.5 the smallest (Table 1; (Hernandez et al. 2008a; Li et al. 2004a; Selyanko and Brown 1999; Selyanko et al. 2001)). Most Kv7 channels saturate at around 0 mV, yet, the open probability at saturating voltages ($P_{o,\,max}$) of different channels is remarkably different and reflects their differential affinity for membrane phosphoinositide, phosphatidylinositol 4,5-bisphosphate (PIP_2) (Li et al. 2004a; Selyanko et al. 2001; Tatulian and Brown 2003). Thus, Kv7.3 overexpressed in Chinese hamster ovary (CHO) cells has $P_{o,\,max}$ near 1, while Kv7.2, Kv7.4 and Kv7.5 have their $P_{o,\,max}$ values at 0.2 or lower. Consistently, single-channel studies revealed that Kv7.3 has ~30 timed higher apparent PIP_2 affinity, as compared to Kv7.2 or Kv7.4 (Li et al. 2005). Heteromeric Kv7.2/Kv7.3 channels have intermediate $P_{o,\,max}$ (~0.3) and intermediate PIP_2 affinity (Hernandez et al. 2008a; Li et al. 2004a, 2005). PIP_2 serves as a co-factor necessary for Kv7 channel activity, thus, at tonic PIP_2 levels in cells, Kv7.3 homomers are saturated with PIP_2, permitting high $P_{o,\,max}$, while channels consisting of other subunits are not saturated and, hence, have lower $P_{o,\,max}$.

Table 1 Properties of Kv7 channels

Channel/ subunit	$V_{1/2}$ (mV)	τ_{act} (0 mV, ms)	τ_{deact} (−60 mV, ms)	γ (pS)	Inacti-vation	β-subunits	Tissue localization (major)
Kv7.1	~15	150–200	200–450	<2	Yes	KCNE1–5, CaM	Heart, inner ear, epithelia, smooth muscle
Kv7.1/ KCNE1	2–8	>500	>200	3–6, varied	No	CaM	Heart, epithelia
Kv7.2	−15 to −30	~200	150–280	~6	No	CaM	Brain, autonomic and sensory ganglia
Kv7.3	−35 to −50	100–150	100–200	~9	No	CaM	Brain, autonomic and sensory ganglia, gut
Kv7.2/ Kv7.3	−20 to −27	130–170	75–100	9.0	No	CaM	Brain, autonomic and sensory ganglia
Kv7.4	−20 to −30	125–200	~100	~2	No	KCNE1–4 CaM	Cochlea, auditory nuclei, mechanosensitive neurons, smooth muscle
Kv7.5	−25 to −44	~150	~125	~2	No	KCNE1 KCNE3 CaM	Brain, autonomic and sensory ganglia, smooth muscle
Kv7.4/ Kv7.5	−38	?	~125	?	No	CaM(?)	Smooth muscle

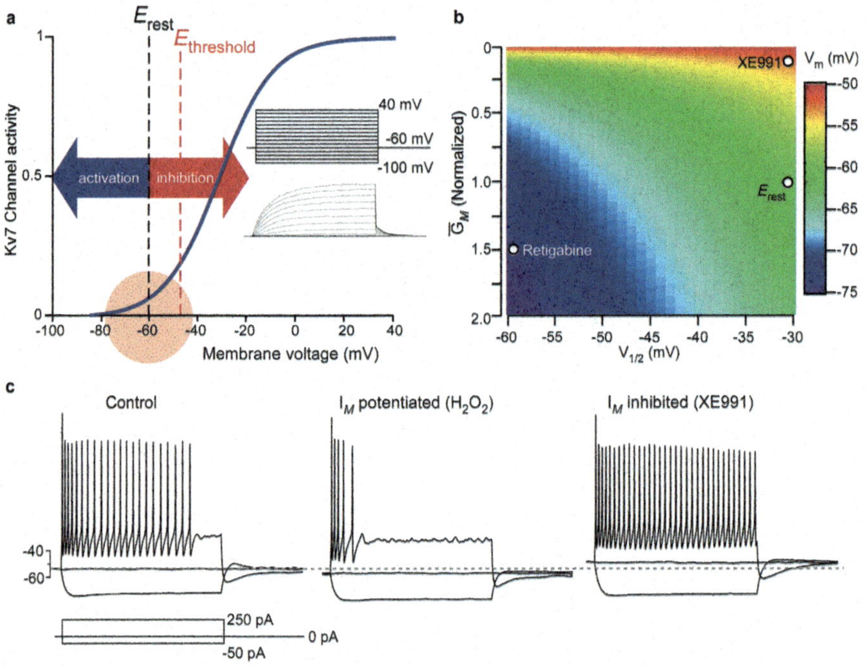

Fig. 1 Kv7 channels and neuronal excitability. (**a**) Voltage-dependence of a Kv7.2/Kv7.3 heteromeric 'M channel' with relation to the resting membrane potential and firing threshold of a generalized somatosensory neuron. Inset shows current traces recorded in a CHO cell, overexpressing Kv7.2 and Kv7.3 (voltage protocol is depicted above the traces). (**b**) Simulation of the effect of modulation of the Kv7.2/Kv7.3 heteromeric channel activity with an 'opener', retigabine, and blocker, XE991, on steady-state membrane potential (Vm). Varying M channel voltage dependence and maximum conductance (G_M) according to the effect of the compounds on channel activity shifts Vm (leak currents were kept constant). Panels a and b are reproduced from Du et al. (2018) with permission. (**c**) effect of M current potentiation (with H_2O_2) or inhibition (with XE991) on firing of cultured sympathetic neuron recorded in a current clamp mode; reproduced from Gamper et al. (2006) with permission

Kv7s display slow activation and deactivation kinetics with time constants (τ) for both processes (at 0 mV or −60 mV, respectively) in the range of 100–200 ms (Table 1). The gating process underlying these kinetics is rather complex and not yet fully elucidated; at least two open states and multiple closed states with multiple transitions between states have been reported (Owen et al. 1990; Selyanko and Brown 1999; Selyanko et al. 2001; Stansfeld et al. 1993; Telezhkin et al. 2012). The above biophysical properties, particularly, the negative threshold activation voltage and lack of inactivation endow Kv7 channels with the ability to control resting membrane potential and, hence, the excitability threshold of cells (Fig. 1b). Slow activation kinetics prevent Kv7 channels from being able to repolarize individual action potentials in neurons, yet, a gradual channel activation increases the threshold for each subsequent action potential during the burst and, thus, underlies the so-called burst accommodation. Thus, Kv7 channels impose strong and dynamic control over the excitability and contractility parameters of neurons and muscle cells (Fig. 1c). The fact that Kv7 channels have a very rich repertoire of physiological modulators (see below) makes these a powerful and versatile tuner of an excitable cell output.

3 Structural Insights into Kv7 Channel Assembly and Function

Extensive previous work, including recent cryo-EM structures, provided a wealth of understanding of structural features of a Kv7 channel. The overall architecture of Kv7 channel complex is consistent with that of the other Kv and TRP channel families: individual subunits contain six transmembrane domains (S1-S6) with a K^+-selective pore in the S5-S6 region and a voltage sensing domain containing S1-S4. All Kv7s have an extended carboxy-terminus (comprising roughly half of the protein) responsible for multimerization and interaction with multiple regulatory molecules (see below). There are four helical regions within the C-terminus of a Kv7 subunit, helixes A-D (Fig. 2a). Helixes A and B bind calmodulin (CaM; (Gamper and Shapiro 2003; Sun and MacKinnon 2017, 2020; Tobelaim et al. 2017; Wen and Levitan 2002; Yus-Najera et al. 2002)) and are involved in channel regulation, whereas helices C and D form tandem coiled coils which are necessary for subunit assembly and tetramerization ((Howard et al. 2007; Maljevic et al. 2003; Schwake et al. 2006; Schwake et al. 2003; Tobelaim et al. 2017), reviewed in Choveau and Shapiro (2012)). Kv7 subunits form heteromers but not at random. Thus, Kv7.3 co-assembles with either Kv7.2 (Shapiro et al. 2000; Wang et al. 1998), Kv7.4 (Bal et al. 2008) or Kv7.5 (Wickenden et al. 2001); Kv7.4 and Kv7.5 also form heteromeric channels with each other (Brueggemann et al. 2014c; Chadha et al. 2014) but Kv7.2 does not multimerize with Kv7.4 (Sogaard et al. 2001) or Kv7.5 (Lerche et al. 2000). KCNQ1 is not known to co-assemble with any other Kv7 subunits (Maljevic et al. 2003; Schenzer et al. 2005; Schwake et al. 2003).

All Kv7 channels require a co-factor, PIP_2, for their activity and also incorporate CaM as an obligatory subunit (reviewed in (Abbott 2020; Gamper and Shapiro 2015; Nunez et al. 2020)). The latter is necessary for channel assembly and trafficking (Etxeberria et al. 2008; Ghosh et al. 2006; Liu and Devaux 2014; Shamgar et al. 2006) and also for its modulation by Ca^{2+} (Gamper et al. 2003, 2005; Kosenko and Hoshi 2013; Shamgar et al. 2006; Sihn et al. 2016; Tobelaim et al. 2017). Recent work indicates that both PIP_2 and CaM might target the same or overlapping parts of the Kv7 protein exerting a complex control over the channel function. Although the details of this intricate molecular interplay are only beginning to emerge, recent structural work in this area does offer some insights. Several X-ray crystallographic structures of proximal C-terminal domains of Kv7.1 (Sachyani et al. 2014), Kv7.4 (Archer et al. 2019; Chang et al. 2018; Xu et al. 2013) and Kv7.5 (Chang et al. 2018) in complex with CaM, as well as an NMR structure of a proximal C-terminus of Kv7.2 in complex with CaM (Bernardo-Seisdedos et al. 2018) were obtained recently. In addition, two high-resolution cryo-EM structures of the Kv7.1 (Sun and MacKinnon 2017, 2020) and Kv7.2 (Li et al. 2021) in complex with CaM were

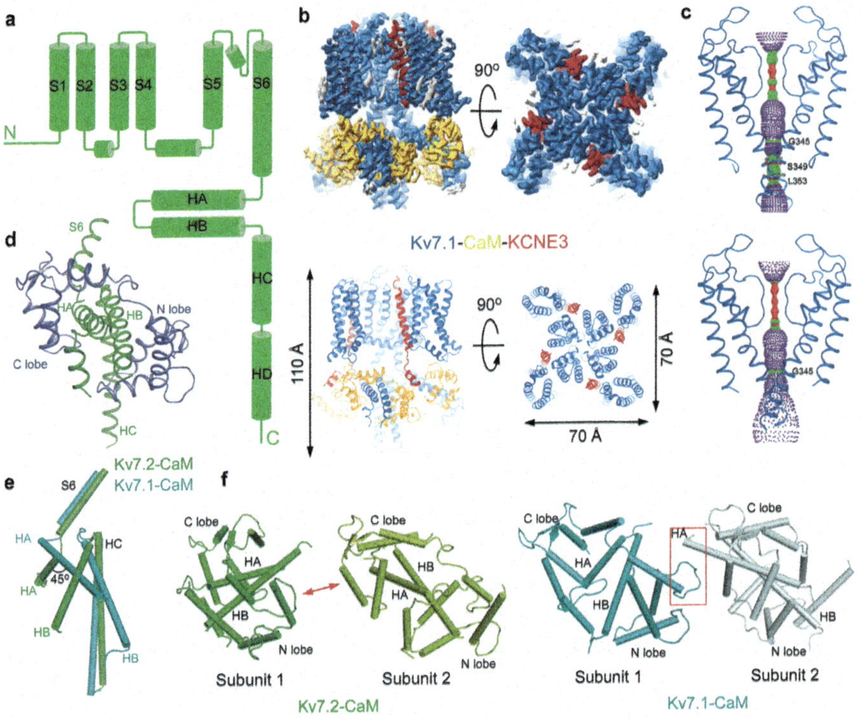

Fig. 2 Structural insights into the Kv7 channel complex. (**a**) Secondary structure of a Kv7 channel. (**b**) Top: side and top view of the human Kv7.1 (blue), CaM (yellow) and KCNE3 (red) complex as a cryo-EM density map. Bottom: structure model of the same complex. (**c**) View of ion-conducting pathways for PIP_2-free (top) and PIP_2-bound (bottom) states of the Kv7.1 channel (front and back subunits excluded for clarity). Panels b-c are reproduced from Sun and MacKinnon (2020) with permission. (**d**) Structural comparisons of CaM in Kv7.2 (green) and Kv7.1 (cyan) channel complexes. (**e**) Alignment of interacting domains of Kv7.2-CaM (green) and Kv7.1-CaM (cyan) structures. (**f**) Difference in the position of CaM in Kv7.2 (left) and Kv7.1 (right) structures: the two adjacent HA-HB-CaM domains are separated from each other in Kv7.2 while they form a continuous structure in Kv7.1. Panels d-f are reproduced from Li et al. (2021) with permission

also released (Fig. 2b-f). Despite some differences, all structures agree that CaM binds simultaneously to A and B helixes of the Kv7 C-terminus which, when bound to CaM, adopts anti-parallel coiled-coil (Fig. 2d). The other contact region identified in more complete Cryo-EM structures is formed between the S2-S3 loop of the channel and the third (out of four in total) EF hand of CaM (Sun and MacKinnon 2017, 2020). Concurrent interaction of CaM with both, the voltage sensor region and the C-terminus of the Kv7 channels is likely to explain its role in both the assembly/trafficking and Ca^{2+}-dependent modulation of the channel. Clarity with how exactly CaM is involved in both these processes is yet to be achieved.

For the case of Ca^{2+} modulation, the matter is complicated by the fact that Kv7.1 channels are activated by Ca^{2+} (Sachyani et al. 2014; Shamgar et al. 2006; Tobelaim et al. 2017), while the rest of the family are rather inhibited (Chang et al. 2018; Gamper et al. 2005; Gamper and Shapiro 2003; Kosenko and Hoshi 2013; Selyanko and Brown 1996; Sihn et al. 2016). Several hypotheses were suggested in which CaM rearranges itself around the A-B helical domain upon binding Ca^{2+}, either by releasing the helix A (Chang et al. 2018; Tobelaim et al. 2017) or by clamping both helixes together (Archer et al. 2019), these rearrangements, in turn, would bring about modulation of gating. Whatever the exact trajectory of these rearrangements is, these are likely to affect channel relationships with its activity-permitting co-factor, PIP_2. Indeed, the phosphoinositide interacts with several channel regions which are also involved in the interaction with CaM, including the A-B region (Choveau et al. 2018; Hernandez et al. 2008a; Tobelaim et al. 2017; Zaydman and Cui 2014; Zaydman et al. 2013) and S2-S3 linker (but additionally also the S4-S5 loop) (Choveau et al. 2018; Sun and MacKinnon 2020; Zaydman and Cui 2014; Zaydman et al. 2013). PIP_2 is necessary for Kv7 channel activity and the depletion of its plasma membrane pool by phospholipase C (PLC) underlies inhibition of I_M by mAChRs and other G protein coupled receptors linked to the PLC signalling cascade (see below) (Ford et al. 2003, 2004; Linley et al. 2008; Liu et al. 2010; Loussouarn et al. 2003; Suh and Hille 2002; Zhang et al. 2003). The mechanism of Kv7 channel modulation by PIP_2 is under intense scrutiny and several reviews are available covering this topic (Gamper and Shapiro 2015; Villalba-Galea 2020; Wang et al. 2020b; Zaydman and Cui 2014). Thus far the most direct insight into structural consequences of PIP_2 binding to a Kv7 channel was obtained by the cryo-EM study from MacKinnon group, who reported PIP_2-bound and PIP_2-free structures of Kv7.1 (in complex with CaM and KCNE3) (Sun and MacKinnon 2020). These authors discovered that PIP_2 occupies a site within the inner membrane leaflet, formed largely by the S2-S3 and the S4-S5 linkers. Binding of PIP_2 causes a large conformational changes in the channel protein, which results in widening of the pore (Fig. 2c). Two helixes separated by a loop: the S6 and C-terminal A helix join into a single continuous helix upon PIP_2 binding, this causes CaM to rotate almost 180° and release its interaction with the voltage sensor.

Of note, KCNE3 causes the Kv7.1 channel to lose its voltage dependence, in addition the N and most of the C-termini of the Kv7.1 were truncated in the protein used for the structure determination. These circumstances make Kv7.1/KCNE3/CaM/PIP_2 structure (Sun and MacKinnon 2020) considerably different from either the native I_{Ks} or the M channel. The truncation of C-terminal domains perhaps explains the fact that interaction of PIP_2 within the A-B region, reported by several labs (Choveau et al. 2018; Hernandez et al. 2008a; Tobelaim et al. 2017; Zaydman and Cui 2014; Zaydman et al. 2013) was not detected. However, the latter interaction may hold the key to understanding differential action of Ca^{2+}/CaM on Kv7.1 (potentiation) vs. other Kv7s (inhibition). Indeed, it was suggested that CaM substitutes for PIP_2 at helix B in Kv7.1 (Tobelaim et al. 2017) while it competes PIP_2 off its C-terminal binding sites in other Kv7s (Archer et al. 2019; Kosenko and Hoshi 2013). Interestingly, recent Kv7.2/CaM structure (Li et al. 2021) revealed several differences in the way CaM integrates into the channel structure (as compared to Kv7.1 channel complex): in Kv7.2, CaM surrounding the Helixes A and B is shifted by about 10–15 Å relative to that in Kv7.1 (Fig. 2d-f). In addition, in the Kv7.2/CaM complex, the four CaM molecules are separated from each other, whereas in Kv7.1/CaM complex they form a continuous cytosolic ring (Fig. 2f; (Sun and MacKinnon 2017, 2020)) further highlighting the difference between the 'cardiac' and 'neuronal' Kv7 channels.

4 Auxiliary Subunits of Kv7 Channels

4.1 KCNEs

A family of small single transmembrane domain proteins formerly known as minK and 'minK-related peptides' (MiRP1–4) now is referred to as KCNE family of auxiliary subunits (KCNE1–5) (McCrossan and Abbott 2004). KCNE1 is perhaps the most studied of the five isoforms due to the fact that it is a part of the cardiac I_{Ks} channel complex. KCNE1 and Kv7.1 are often found co-expressed in the heart, epithelia and other relevant tissues and several KCNE mutations are associated with LQTS and hearing loss (see below). KCNE1 has several profound and well-characterized effects on Kv7.1 gating: ~1,000-fold slowing of activation, positive shift of the voltage dependence of activation (20–50 mV), threefold to sevenfold increase of single-channel conductance, stabilizing the open state and removal of inactivation (Barhanin et al. 1996; Hou et al. 2017; Meisel et al. 2018; Pusch 1998; Sanguinetti et al. 1996; Sesti and Goldstein 1998; Tristani-Firouzi and Sanguinetti 1998; Werry et al. 2013; Yang and Sigworth 1998) (reviewed in (Wang et al. 2020b)). In addition, KCNE1 reduces the Kv7.1 sensitivity to XE991 (Wang et al. 2000) while increasing sensitivity to the chromanol 293B and azimilide (Busch et al. 1997).

The stoichiometry of KCNQ1/KCNE1 assembly has been a matter of some debate as both fixed and variable stoichiometry models have been suggested (reviewed in (Wang et al. 2020b; Wrobel et al. 2012)). Co-expression of WT KCNE1 with a lethal KCNE1 mutant at different ratios reveals a stoichiometry of two subunits per functional channel (Wang and Goldstein 1995). Experiments with forced stoichiometry (linked Kv7.1-KCNE1 constructs) and unnatural amino acid photo-crosslinking (Murray et al. 2016; Nakajo et al. 2010; Westhoff et al. 2017) established that two proteins could indeed assemble at variable stoichiometry (1:4 to 4:4) with channel activation becoming progressively slower as more KCNE1 subunits are added to the complex. Interestingly, the recent Kv7.1/KCNE3 structure reports 4:4 stoichiometry (Fig. 2b; (Sun and MacKinnon 2020)) and, given the structural similarities between KCNE subunits, this perhaps should be interpreted as the stoichiometry at saturating levels of a given KCNE subunit. Solution NMR, mutagenesis and simulations proposed several models of KCNQ1-KCNE1 interactions with most of them assuming that KCNE1 is piercing through the channel complex aligning along the Kv7.1 channel pore (Kang et al. 2008; Strutz-Seebohm et al. 2011; Tian et al. 2007), this idea is further supported by the recent Kv7.1/KCNE3 structure (Fig. 2b; (Sun and MacKinnon 2020)).

The mechanisms of gating modifications imposed on Kv7.1 by KCNE1 have been intensively investigated (reviewed in Wang et al. (2020b); Wrobel et al. (2012).

KCNE1 was suggested to stabilize resting (closed) states of Kv7.1 (Nakajo and Kubo 2007), retard pore opening (Rocheleau and Kobertz 2008), decelerate the movement of the S4 segment of the voltage sensing domain of KCNQ1 (Nakajo and Kubo 2007; Ruscic et al. 2013) or disturb coupling between the voltage sensor movement and the pore opening (Barro-Soria et al. 2014; Nakajo and Kubo 2014; Osteen et al. 2010, 2012; Ruscic et al. 2013; Westhoff et al. 2019; Zaydman et al. 2014).

Kv7.1 co-assembles with KCNE2 or KCNE3 resulting in constitutively-open, voltage-independent channels (Bendahhou et al. 2005; Schroeder et al. 2000b; Tinel et al. 2000a); KCNE2 decreases the macroscopic current by over threefold (Bendahhou et al. 2005; Tinel et al. 2000a), whilst KCNE3 produces over tenfold increase (Bendahhou et al. 2005; Schroeder et al. 2000b). The properties of constitutively-open Kv7.1/KCNE3 and Kv7.1/KCNE2 channels resemble those of the K^+ currents in colonic crypt cells of the small intestine and colon and KCNE2 knockout impairs gastric acid secretion in mice (Roepke et al. 2006). KCNE2 interacts with Kv7.2 and Kv7.3 as well as producing a moderate acceleration of activation and inactivation kinetics (Tinel et al. 2000b). Interestingly, KCNE3 and KCNE4 each express in two splice variants: long and short forms in humans (Abbott 2016). Short form of KCNE3 suppresses while the short form of KCNE4 potentiates Kv7.4 currents in *Xenopus* oocytes (Abbott 2016; Strutz-Seebohm et al. 2006) while long forms of either protein have no effect (Abbott 2016). KCNE4 and KCNE5 both strongly suppress Kv7.1 currents (Abbott 2016; Angelo et al. 2002; Bendahhou et al. 2005; Grunnet et al. 2002, 2003, 2005)) with KCNE5 producing a strong positive shift of the voltage dependence of activation (Angelo et al. 2002). Physiological significance of these effects of KCNE2–5 is poorly understood at present.

Apart from the effects on the Kv7 channel gating, KCNE subunits are reported to affect channel trafficking. Thus, disruption of KCNE1 glycosylation (Chandrasekhar et al. 2011) or LQTS-associated KCNE1 mutation L51H (Krumerman et al. 2004) disrupt trafficking and/or insertion of the Kv7.1 to the plasma membrane (reviewed in (Kanda and Abbott 2012)). KCNE1 was also reported to be necessary for the dynamin-dependent internalization of the I_{Ks} complex (Xu et al. 2009). You can read more about KCNE subunits in Chapter "Control of Biophysical and Pharmacological Properties of Potassium Channels by Ancillary Subunits".

4.2 A-Kinase Anchoring Proteins

Kv7 channels associate with and are modulated by A-kinase anchoring proteins (AKAPs), a group of scaffolding proteins binding to the regulatory subunit of protein kinase A (PKA) (Langeberg and Scott 2005). AKAP called *yotiao* (in addition to KCNE1) is necessary for proper functioning of the native I_{Ks} channel complex (Haitin and Attali 2008; Kurokawa et al. 2004; Marx 2003; Marx et al. 2002; Nicolas et al. 2008). Yotiao binds to helix D at Kv7.1 C-terminus and acts as a scaffolding protein to recruit PKA and the protein phosphatase 1 (PP1) controlling the phosphorylation/dephosphorylation state of Kv7.1 at N-terminal Ser-27 and possibly Ser-92 (Dvir et al. 2014; Kurokawa et al. 2004; Marx et al. 2002; Nicolas et al. 2008; Pongs and Schwarz 2010). The PKA-dependent phosphorylation of Kv7.1 leads to an increase in I_{Ks} amplitude and acceleration of action potential repolarization, which is at the core of β-adrenergic modulation of the I_{Ks} complex activity (Kurokawa et al. 2004; Marx et al. 2002; Pongs and Schwarz 2010). AKAP79/150 (numbers denote human and murine orthologs, respectively) interacts with A-B helix region of Kv7.2-Kv7.5 but not Kv7.1 (Hoshi et al. 2003); it recruits PKC to the channel complex, leading to phosphorylation at two threonine residues within helix B (Hoshi et al. 2003). This phosphorylation decreases channel affinity to PIP_2 and sensitizes it towards muscarinic (but not bradykinin) inhibition (Bal et al. 2010; Higashida et al. 2005; Hoshi et al. 2003, 2005; Tunquist et al. 2008; Zhang et al. 2011; Zhang and Shapiro 2012).

4.3 Other Kv7-Interacting Proteins

As discussed above, CaM is considered to be an obligatory resident of the Kv7 channel complex, which is necessary for channel assembly, trafficking and modulation by Ca^{2+}. Several other proteins interact with Kv7 channels. Kv7.2 through its C-terminal A helix was shown to bind syntaxin 1A, which targets the channels to presynaptic sites in neurons (Etzioni et al. 2011; Regev et al. 2009). Kv7.2 and Kv7.3 interact with the ankyrin-G, which results in accumulation of these channels at the axon initial segment (AIS) and nodes of Ranvier (Chung et al. 2006; Cooper 2011; Pan et al. 2006). Kv7s were reported to interact with the ubiquitin-protein ligase Nedd4–2, which controls degradation of the channel proteins (Ekberg et al. 2007; Jespersen et al. 2005; Kurakami et al. 2019).

5 Kv7 Pharmacology

A more detailed account on the pharmacology of the Kv7 channels is given in Chapter "Pharmacological Activation of Neuronal Voltage-Gated Kv7/KCNQ/M-Channels for Potential Therapy of Epilepsy and Pain", therefore here we will only mention some major compounds relevant to the content of this chapter. Linopirdine, and its analogue, XE991, are potent inhibitors of Kv7 channels (Aiken et al. 1995; Lamas et al. 1997; Zaczek et al. 1998). XE991 is commonly used for M current validation in primary cells, it blocks Kv7.1-Kv7.4 channels with the IC_{50}s in the range of 1–5 μM (Sogaard et al. 2001; Wang et al. 1998, 2000), however, Kv7.5 is much less sensitive with $IC_{50} \sim 60$ μM (Jensen et al. 2005; Schroeder et al. 2000a). Chromanol 293B is a 'classical' inhibitor of Kv7.1 and I_{Ks} (Busch et al. 1997; Suessbrich et al. 1996). Since Kv7 channel activity has the potential to silence neurons, there is a big drive to develop Kv7 activators ('openers') as potential medicines for excitability disorders, such as epilepsy and pain and several hundreds of compounds have been identified to date (see some recent summaries in Du and Gamper (2013); Du et al. (2018) and in Chapter "Pharmacological Activation of Neuronal Voltage-Gated Kv7/KCNQ/M-Channels for Potential Therapy of Epilepsy and Pain"). The prototypic M channel openers, retigabine and flupirtine, are potent but relatively non-selective Kv7 openers that activate all Kv7 subunits except Kv7.1, with Kv7.3 being the most sensitive subunit (Schenzer et al. 2005; Tatulian and Brown 2003; Tatulian et al. 2001; Wuttke et al. 2005). Both compounds have been used clinically as an anticonvulsant (retigabine) and an analgesic (flupirtine) but both are currently discontinued for a range of side effects. Nevertheless, retigabine remains a very useful research tool and a benchmark for the development of novel openers.

6 Tissue Distribution of Kv7 Channels

6.1 Nervous System

Distribution of Kv7 subunits throughout the mammalian peripheral and central nervous systems is discussed in Chapter "Pharmacological Activation of Neuronal Voltage-Gated Kv7/KCNQ/M-Channels for Potential Therapy of Epilepsy and Pain", therefore we only briefly outline some key points here. M-type channels (Kv7.2-Kv7.5 and their heteromers) are abundantly expressed throughout the CNS. Kv7.2 and Kv7.3 are particularly widely expressed and heteromeric Kv7.2/Kv7.3 channel is considered to be a 'classical' M channel (Brown and Passmore 2009). These subunits have been found throughout almost all major brain structures with particularly high expression at key sites controlling rhythmic neuronal activity and synchronization (Cooper et al. 2001; Geiger et al. 2006; Kanaumi et al. 2008; Klinger et al. 2011). Kv7.2 and Kv7.3 are also abundantly expressed in the peripheral somatosensory neurons (reviewed in (Du et al. 2018)) and in sympathetic neurons (where the M current was first discovered; reviewed in Brown and Passmore (2009)).

Kv7.4 has a much more restricted expression pattern: it is highly expressed in the auditory and vestibular pathways (Hurley et al. 2006; Kharkovets et al. 2000, 2006), including the cochlea (basal pole of outer hair cells) (Kharkovets et al. 2006), post-synaptic calyx terminals innervating vestibular type I hair cells (Sousa et al. 2009; Spitzmaul et al. 2013) and in auditory brainstem nuclei and tracts (Kharkovets et al. 2000). Accordingly, several *KCNQ4* mutations result in a dominant form of deafness, DFNA2 (Kharkovets et al. 2000, 2006; Kubisch et al. 1999) (see below). Kv7.4 channels are selectively expressed in dopaminergic neurons in the ventral tegmental area of the mesolimbic dopamine system and their downregulation is associated with the depression-like behaviour in mice (Li et al. 2017; Su et al. 2019). Kv7.4 is also expressed in the peripheral nerve endings of cutaneous rapidly adapting hair follicle and Meissner corpuscle mechanoreceptors. In these fibres, Kv7.4 shortens the duration of stimulus-induced bursts (Heidenreich et al. 2012). Accordingly, human DFNA2 carriers are sensitized to low-frequency tactile vibrations (Heidenreich et al. 2012).

Kv7.5 is expressed throughout the CNS and PNS alongside KCNQ2 and KCNQ3. Specific expression was reported in giant glutamatergic terminals (calyx of Held) in the medial nucleus of the trapezoid body (MNTB), where this subunit is likely to be the main constituent of M channels (Huang and Trussell 2011). It is also expressed in the auditory system and in the hippocampus (Tzingounis et al. 2010).

6.2 Cardiovascular System

As mentioned above, Kv7.1/KCNE1 channel complex is one of the main K^+ channels in the heart, forming the I_{Ks} (Barhanin et al. 1996; Sanguinetti et al. 1996). Kv7.1, Kv7.4 and Kv7.5 channels are expressed in smooth muscle layer of mammalian vasculature (Fosmo and Skraastad 2017; Hedegaard et al. 2014; Jepps et al. 2011; Khanamiri et al. 2013; Ng et al. 2011), including human visceral arteries (Fosmo and Skraastad 2017; Ng et al. 2011), where these channels control the excitability and contractility of vascular smooth muscle cells (VSMC) (Chadha et al. 2012, 2014; Jepps et al. 2011; Khanamiri et al. 2013; Mani et al. 2011; Ng et al. 2011; Stott et al. 2015; Tsai et al. 2020). Reduced activity of Kv7s has been suggested to result in increased vascular resistance and hypertension (Chadha et al. 2012; Jepps et al. 2011; Khanamiri et al. 2013; Stott et al. 2015). Kv7 channels in coronary arteries are thought to contribute to the vasodilatory effect of adenosine (Hedegaard et al. 2014; Khanamiri et al. 2013) and to hypoxia-induced vasodilation (Hedegaard et al. 2014) and reactive hyperaemia (Khanamiri et al. 2013).

6.3 Non-vascular Musculature

Kcnq gene transcripts were reported in non-vascular smooth muscles as well. *Kcnq1*, *Kcnq3*, *Kcne1* and *Kcne2* were found in rat gastric antral smooth muscle (Ohya et al. 2002) while *Kcnq4*, *Kcnq5* and *Kcne4* (as well as M-like currents) were reported in murine gastrointestinal tract (Jepps et al. 2009). M-like currents and *Kcnq* transcripts have been reported in rodent and human airway smooth muscle cells, ASMC (Brueggemann et al. 2012, 2014a, 2018), where these are suggested to control excitability and contractility. Accordingly, Kv7 channel activators could have therapeutic value as bronchodilators (Evseev et al. 2013; Mondejar-Parreno et al. 2020). *Kcnq1*, *Kcnq5* and *Kcne4* transcripts were reported in murine myometrial smooth muscle throughout the oestrus cycle (McCallum et al. 2009). Expression of all five *Kcnq* genes has been detected in skeletal muscle (Iannotti et al. 2010; Lerche et al. 2000; Roura-Ferrer et al. 2008; Schroeder et al. 2000a). Kv7 channels were suggested to play a role in skeletal muscle proliferation, differentiation and survival (Iannotti et al. 2010; Roura-Ferrer et al. 2008) as well as in their contractility (Su et al. 2012).

6.4 Epithelia

As mentioned earlier, Kv7.1 contributes to the basolateral K^+ current in epithelia. Together with several KCNE subunits, Kv7.1 is expressed in many epithelial cells including kidney, stomach, colon and small intestine (Dedek and Waldegger 2001; Demolombe et al. 2001; Horikawa et al. 2005; Kunzelmann et al. 2001b; Schroeder et al. 2000b; Vallon et al. 2001). Kv7.1 channel complexes (together with other epithelial K^+ channels) are essential for maintaining a negative membrane potential, providing a driving force for trans-epithelial transport, and for K^+ absorption or secretion. In many epithelia, voltage-gated Kv7.1 channels are converted into voltage-independent, constitutively-open K^+ channels by co-assembly with KCNE2 (Tinel et al. 2000a) or KCNE3 (Schroeder et al. 2000b) (see (Hamilton and Devor 2012; Jespersen et al. 2005; Soldovieri et al. 2011) for review).

6.5 Other Tissues and Cell Types

Adrenal chromaffin cells (Wallace et al. 2002) and some other neuroendocrine cells such as the prolactin releasing cells of the pituitary gland (Sankaranarayanan and Simasko 1996) express functional M current. As in neurons and muscle cells, M-like current is an important mechanism of excitability control in neuroendocrine cells and M current inhibition (e.g. via histamine H1 receptor stimulation) was proposed to stimulate hormone release from these cells (Sankaranarayanan and Simasko 1996; Wallace et al. 2002). M-like currents were also recorded from rod photoreceptors (Wollmuth 1994) and retinal pigment epithelial cells (Takahira and Hughes 1997). Functional expression of Kv7.2 has been reported in keratinocytes (Reilly et al. 2013).

7 Regulation and Modulation

Kv7 channel modulation via GPCR signalling cascades is extensively studied and covered in numerous reviews (Brown et al. 2007; Brown and Passmore 2009; Delmas and Brown 2005; Gamper and Shapiro 2007, 2015; Greene and Hoshi 2017; Hernandez et al. 2008b; Hille et al. 2015; van der Horst et al. 2020; Zhang and Shapiro 2016), as well as in Chapter "Pharmacological Activation of Neuronal Voltage-Gated Kv7/KCNQ/M-Channels for Potential Therapy of Epilepsy and Pain", therefore here we will only outline key themes. As mentioned, M current was discovered as a K^+ conductance inhibited by mAChRs. Subsequent research identified that M_1, M_3 and M_5 mAChRs, as well as many other GPCR coupled to $G_{q/11}$ signalling cascades (e.g. bradykinin B_2, purinergic P_2Y, protease-activated receptor 2, angiotensin II A_1, etc.) inhibit Kv7 channels (Cruzblanca et al. 1998; Filippov et al. 2006; Gamper and Shapiro 2015; Linley et al. 2008; Shapiro et al. 1994; Zaika et al. 2006, 2007) and cause excitation of neurons (Crozier et al. 2007; Linley et al. 2008; Liu et al. 2010). The signalling cascade linking $G_{q/11}$-type GPCR with Kv7 inhibition involves activation of Phospholipase C (PLC) and the depletion of PIP_2 (Li et al. 2005; Suh and Hille 2002; Zhang et al. 2003). Ca^{2+} release from the IP_3-sensitive Ca^{2+} stores (via CaM) (Cruzblanca et al. 1998; Gamper and Shapiro 2003; Kosenko and Hoshi 2013) and PKC-mediated M channel phosphorylation (orchestrated by AKAP79/150) (Hoshi et al. 2003; Zhang et al. 2011) also contribute to Kv7 suppression; the contribution of the above mechanisms to the overall Kv7 channel suppression varies substantially between the individual receptor types, as well as in different types of cells (reviewed in Gamper and Shapiro (2015);

Hernandez et al. (2008b). As mentioned above, another physiologically important endogenous signalling cascade targeting Kv7.1 and I_{Ks} channels is β-adrenergic modulation via the PKA-dependent phosphorylation of Kv7.1 (Kurokawa et al. 2004; Marx et al. 2002; Pongs and Schwarz 2010), reviewed in (van der Horst et al. 2020). There is evidence that Kv7.2 and Kv7.5 are also potentiated by PKA phosphorylation (reviewed in van der Horst et al. (2020). Kv7 channels are suppressed by Src kinase phosphorylation (Gamper et al. 2003; Li et al. 2004b) and potentiated by reactive oxygen species (Gamper et al. 2006), arginine methylation (Kim et al. 2016) and intracellular zinc (Gao et al. 2017) although physiological processes where these pathways may be involved remain to be elucidated.

8 Transcriptional Control of Kv7 Gene Expression

KCNQ genes expression is regulated by a number of transcription factors, including positive regulation by Sp1 (Mucha et al. 2010) and NFAT (Zhang and Shapiro 2012) and negative regulation by the repressor element 1-silencing transcription factor (REST, NRSF) (Iannotti et al. 2013; Mucha et al. 2010). Regulation by NFAT is activity-dependent: increased firing of sympathetic and hippocampal neurons results in Ca^{2+} accumulation, calcineurin-dependent phosphorylation and nuclear translocation of NFAT, which interacts with NFAT-binding regulatory elements of *Kcnq2* and *Kcnq3* genes, resulting in augmented mRNA levels and increased I_M amplitudes (Zhang and Shapiro 2012). This pathway was suggested to protect neurons from overexcitability and seizures. In contrast, REST suppresses the expression of *Kcnq* genes and this was reported as one of the mechanisms resulting in overexcitable phenotype of peripheral afferents in neuropathic and inflammatory pain models (Rose et al. 2011; Uchida et al. 2010; Zhang et al. 2019). Tonic expression of REST in neurons is believed to be low but it was shown to increase greatly following inflammation (Mucha et al. 2010) or after neuropathic injury (Rose et al. 2011; Uchida et al. 2010; Zhang et al. 2019). Interestingly, dramatic reduction in *Kcnq2* (Rose et al. 2011; Zheng et al. 2013) and *Kcnq3* (Zheng et al. 2013) transcripts and proteins following experimentally-induced chronic pain development has been reported. Genetic deletion of REST in sensory neurons alleviated neuropathic pain (Zhang et al. 2018, 2019) and prevented downregulation of some of its target genes, including *Kcnq2* (Zhang et al. 2019). On the other hand, viral overexpression of REST in sensory neurons recapitulated neuropathic-like hyperalgesia in mice (Zhang et al. 2019). Thus, M channel downregulation by REST may represent a feature of chronic pain-associated remodelling of peripheral afferents.

9 Kv7 Channels and Excitability Disorders

The role of Kv7 channels in epilepsy and pain is considered in Chapter "Pharmacological Activation of Neuronal Voltage-Gated Kv7/KCNQ/M-Channels for Potential Therapy of Epilepsy and Pain", below we will focus on the other disorders and pathologies where contribution of Kv7 channels has been documented or suggested.

9.1 Cardiac Arrhythmias

Cardiac arrhythmia refers to an array of changes to contractions of the heart causing it to beat irregularly (Antzelevitch and Burashnikov 2011). Long QT syndrome (LQTS) is an arrhythmia where ventricular repolarization is prolonged, resulting in episodic ventricular tachyarrhythmia. This is essentially a prolonged depolarization that can lead to ventricular fibrillation and sudden death in otherwise healthy people (Vohra 2007). LQTS often occurs due to a loss of function in the Kv7.1 channel complex. Here we focus on the role of Kv7.1 but some KCNE1 mutations also lead to LQT syndrome (Schulze-Bahr et al. 1997).

Over 500 arrhythmia-associated *KCNQ1* mutations have been identified thus far, most are loss-of-function. Inherited LQTS most commonly manifests as Romano-Ward syndrome (RW), the autosomal dominant form of LQTS, where a dominant-negative effect is imposed by the mutant on the unmutated copies. The autosomal-recessive form of LQTS, Jervell and Lange-Nielsen syndrome (JLN), whereby the mutants do not have an impact on the unmutated subunits, is less common and also associated with bilateral deafness (see below). The *KCNQ1* mutations that result in these forms of LQTS separate broadly into those that impair ion selectivity, voltage dependence, trafficking or multimerization of the channel complex. The lack of dominant-negative effect of mutations in JLN was originally thought to be due to the loss of the mutant subunits ability to form tetramers. This would exert no effect on the wild-type channels as the wild-type copies could still tetramerize. This is evidenced by a significant number of JLN causing mutations resulting in truncations of the C-terminus and missense mutations in the A-domain (Loussouarn et al. 2006; Schmitt et al. 2000), areas that are crucial for Kv7 tetramerization (Howard et al. 2007; Maljevic et al. 2003; Schwake et al. 2003, 2006). Though, importantly there are mutations resulting in JLN that do not impair tetramerization (Xu and Minor Jr. 2009) but do hamper trafficking/insertion of the mutant channel to the plasma membrane (Marx et al. 2002).

In autosomal dominant RW syndrome, mutations in the N terminus are often related to limitations in trafficking. A number of N-terminal mutations lead to the channel not escaping the endoplasmic reticulum (Dahimene et al. 2006; Peroz et al. 2009) or impaired recycling via Rab-dependent endocytosis and exocytosis (Seebohm et al. 2007; Seebohm et al. 2008). Mutations are continuing to be discovered, some of which lead to the expected impaired trafficking through ER retention and some leading to novel impairments such as reductions in current amplitude only at positive voltages (Hammami Bomholtz et al. 2020).

Structural X-ray crystallography has revealed that some mutations believed to be preventing tetramerization or trafficking in the C-terminus A domain were actually likely disrupting the channel binding to yotiao, thus disrupting protein kinase A (PKA) binding (Howard et al. 2007). This, in turn, disrupts β-adrenergic potentiation of the I_{Ks} and disturbs regularity of cardiac action potential when heart rate is increased by catecholamine action. This has been shown to be the case in patients with PKA insensitive *KCNQ1* mutations, who often only exhibiting arrhythmia after stress and not at rest (Bartos et al. 2014).

At least three mutations in the PIP_2 binding sites of Kv7.1 have resulted in LQTS through reduced channel activity (Park et al. 2005). Similarly, some Kv7.1 mutations that result in LQTS impair the interaction between Kv7.1 and CaM, this was shown by two independent groups to be due to impaired gating of the channel and preventing the correct assembly of channel subunits, resulting in a dominant-negative loss of function (Ghosh et al. 2006; Shamgar et al. 2006). Indeed, a mutation in a PIP_2 and CaM binding sites in helix B leads to LQTS seemingly by interfering with CaM binding under PIP_2 depletion (Tobelaim et al. 2017). Mutations in the S4 voltage sensing domain have also been recently discovered that cause a shift in voltage dependence to more positive potentials, likely impairing action potential adaptation to increased heart rate (Moreno et al. 2017). Additionally, a number of high-risk mutations associated with sudden death were localized to the selectivity filter of Kv7.1 and were shown to disrupt the carbonyl oxygen atoms, preventing the selective permeation of potassium through the channel (Burgess et al. 2012).

Although the clinical manifestation of many of these mutations is not drastically different except for perhaps severity, understanding the molecular basis of these mutations aids in the treatment of this disorder and was instrumental in learning about the functioning of this highly modulated and complex channel.

A few gain-of-function mutations have also been discovered in Kv7.1 that lead to other types of cardiac arrhythmia, namely short QT syndrome (SQTS) and familial atrial fibrillation. These mutations result in a much earlier repolarization phase of the cardiac action potential. This is significant as the prolonged action potential refractory periods in the heart allow for ventricular filling to occur between contractions. Gain of function mutations manifest as increased activation rate, decreased deactivation rate, removing inactivation or shifting of voltage dependence or various combinations of these, resulting in a constitutively active channel (Bellocq et al. 2004; Chen et al. 2003; Restier et al. 2008; Seebohm et al. 2001). Additionally, a novel mutation in Kv7.1 reduced co-assembly with KCNE1 that led to accelerated activation kinetics and ultimately SQTS (Moreno et al. 2015).

Importance of the Kv7.1 complex for the heartbeat infers a possibility to target these channels for the treatment of the above heart conditions. In fact, the compound Rottlerin has been promising in treating LQTS in part through activation of Kv7.1 channel (Lubke et al. 2020). Rottlerin has been shown to increase Kv7.1/KCNE1 deactivation time; some related analogues have similar effects. This suggests a potential treatment for LQTS, though the efficacy of Rottlerin and related compounds to augment disease-associated Kv7.1 mutants remains to be established.

9.2 Other Cardiovascular Disorders

The Kv7 channels expressed in VSMC are increasingly recognized as strong regulators of vascular function and key to the adaptability of the system. Generally, Kv7 activation would lead to vasodilation while Kv7 inhibition results in vasoconstriction; the latter could cause hypertension, a disease that is increasingly problematic to overall health and a risk factor for stroke and cardiovascular disease (Fosmo and Skraastad 2017; Stott et al. 2014). Vasopressin, one of the main vasoconstricting hormones, acts on $G_{q/11}$ receptor V1A in the vasculature (Holmes et al. 2003) resulting in Kv7.4/Kv7.5 and Kv7.5 channel inhibition (largely mediated by PKC in this case) and activation of L-type calcium channels, to allow the influx of calcium and resulting contraction of smooth muscle (Brueggemann et al. 2007; Mani et al. 2013; Tsai et al. 2020). Opposingly, β-adrenergic receptor activation (acting via PKA) potentiates Kv7 channel activity and produces a vasodilatory effect (Chadha et al. 2012). In rat and mouse models of hypertension, the condition was accompanied by a significant reduction in Kv7.4 channel expression, suggesting that reduced Kv7 activity may contribute to the maintenance of chronic hypertension (Jepps et al. 2011). Recent work has expanded these findings suggesting that angiotensin II inhibits the interaction between Kv7.4 and the heatshock protein HSP90, enhancing its interaction with CHIP (HSP70 interacting protein C terminus), an E3 ubiquitin ligase that leads to channel degradation (Barrese et al. 2018). Furthermore, expression of Kv7.4 is regulating the susceptibility of vascular Kv7 channels to modulation through controlling heteromerization with Kv7.5. Since Kv7.4 homomers are largely insensitive to PKC-dependent phosphorylation (Brueggemann et al. 2014c) and the β-adrenergic receptor agonist, isoproterenol (Mani et al. 2016), inclusion of Kv7.4 into Kv7.5-containing tetramers would reduce sensitivity of the resulting channel complex to β-adrenergic modulation.

Downregulation of Kv7.4 was seen in the renal artery, which may be indicative of renal hypoperfusion, resulting in the release of renin and subsequent activation of the renin-angiotensin-aldosterone cascade that results in hypertension (Fosmo and Skraastad 2017). Kv7 channels are also partly responsible for vasodilation in response to hypoxia, as XE991 significantly prevents vasodilation during hypoxia (Hedegaard et al. 2014).

Aside from the role of Kv7 in the muscle, Kv7 channels are also expressed in the autonomic nervous system innervating the cardiovascular system. Indeed Kv7.2, Kv7.3 and Kv7.5 are expressed in nodose ganglia and baroreceptor terminals (Wladyka et al. 2008). Seizure patients with loss of function mutations in *KCNQ2* sometimes present with bradycardia (Weckhuysen et al. 2013), potentially as a result of parasympathetic disinhibition. Thus, the role of autonomic Kv7 dysfunction should not be ignored when discussing the role of Kv7 in vascular disease.

9.3 Deafness

The role of Kv7 channels in the sensory nervous system is widely understood and reviewed, however, some Kv7 subunits are expressed in the sensory organs themselves. Specifically, Kv7.4 and Kv7.1 have been implicated in certain forms of deafness, both due to expression in the organ cells.

Initially, it was observed that some JLN patients also presented with bilateral deafness and a mutated version of Kv7.1 was localized to the stria vascularis in the inner ear (Neyroud et al. 1997). Since then a number of mutations in *KCNQ1* were shown to result in both arrhythmia (JLN) and deafness (Wang et al. 2002; Wei et al. 2000), and both Kv7.1 and KCNE1 have been identified in vestibular dark cells in the stria vascularis (Nicolas et al. 2001). Interestingly, KCNE1 knockout mice exhibit epithelial degeneration here, which collapses the lymph space, as is the case with some JLN mutations in *KCNQ1* (Nicolas et al. 2001), suggesting a lack of Kv7.1/KCNE1 co-assembly to be one cause of JLN-associated deafness. Overall, *KCNQ1* mutations impact the endolymph through secretion or causing collapse of the space for the endolymph in syndromic (presents with symptoms elsewhere in the body) forms of deafness.

Non-syndromic autosomal dominant deafness (DNFA2) is characterized by high-frequency hearing loss that progresses to full hearing loss over time. DNFA2 has been attributed to mutations in *KCNQ4*, which was found to be expressed in the sensory outer hair cells (OHC) (Kharkovets et al. 2000; Kubisch et al. 1999). Mutations in *KCNQ4* lead to deafness through preventing the OHC functioning correctly, often through degeneration (Kubisch et al. 1999; Nouvian et al. 2003). As OHC are largely responsible for amplifying auditory signals, it is rare that a loss of OHC function leads to complete hearing loss. In fact, it was determined that *KCNQ4* is also expressed in the inner hair cells (IHC) (Beisel et al. 2000) that relay sensory information through the afferent sensory neurons in the spiral ganglia, which have also had Kv7-like currents identified (Lv et al. 2010). Indeed, loss of *KCNQ4* in spiral ganglia sensory neurons leads to neuronal degeneration and progressive hearing loss (Carignano et al. 2019; Lv et al. 2010). In addition, loss of OHC only leads to loss of hearing at higher frequencies (Ryan and Dallos 1975) and IHC have the same sensory input as OHC and are the cells that synapse onto spiral ganglia neurons. This supports a suggestion that DNFA2 is not solely due to OHC dysfunction. Yet, in *KCNQ4* knockout mice, only the OHC degenerated with otherwise largely intact sensory systems (Kharkovets et al. 2006). Clearly, more work is needed to discern the relative contribution of various cell types (and Kv7.4 expressed therein) in this complex system.

Many more *KCNQ4* mutations linked to DFNA2 were identified since the original discovery. Many of these are loss of function mutations in *KCNQ4* associated with the pore region, some of which interfere with the GYG potassium selectivity filter or stabilize the closed state of the channel (Coucke et al. 1999; Kubisch et al. 1999; Talebizadeh et al. 1999; Van Camp et al. 2002; Van Hauwe et al. 2000; Xia et al. 2020). These pore-related mutations seem to be the most common and interfere with channel function through loss of conduction. Although, a pore mutation has been identified that exerts a dominant-negative effect on wild-type channels through reducing the surface expression (Mencia et al. 2008) and other mutations that interfere with channel trafficking have been identified as well (Gao et al. 2013; Xia et al. 2020). Mutations altering the voltage sensor have also been shown to lead to DNFA2 (Su et al. 2007) as have the cytoplasmic region mutations (Mehregan et al. 2019) that may be linked to changes in channel modulation.

The mechanism for DNFA2 associated with defects in Kv7.4 is not entirely resolved but the OHC/IHC degeneration may be either due to potassium overload and chronic depolarization itself or due to the chronic influx of calcium due to the depolarization (Holt et al. 2007; Xu et al. 2007). It is suggested that OHC extrude K^+ through Kv7.4 channels in order to keep the concentration in the cells low to allow influx of potassium to occur upon deflection of cilia on the cell by vibration. Indeed, linopirdine and XE991 selectively block K^+ conductance ($G_{K,n}$) in OHC, as do some of the *KCNQ4* mutations (Kharkovets et al. 2006; Marcotti and Kros 1999; Oliver et al. 2003). However, the $G_{K,n}$ activates at more negative potentials than heterologously-expressed Kv7.4. Recent reports suggest that clustering of Kv7.4 channels is sufficient to shift voltage activation to more negative potentials and that these clusters form in OHC (Perez-Flores et al. 2020). Interestingly, these clusters may convey some mechanosensitivity in the OHC, which opens up further avenues of research into the role of Kv7.4 in the auditory system (Perez-Flores et al. 2020).

The potential for treating DNFA2 by rescuing Kv7.4 current is ongoing, but activation of Kv7 channels with selective agonists is only effective in rescuing current from non-pore-effecting mutants. Though, promisingly, Kv7 agonists do reverse the dominant-negative effect of both pore and non-pore localized mutants when these are expressed with wild-type channels (Leitner et al. 2012; Xia et al. 2020).

9.4 Behavioural Disorders

9.4.1 Depression

Growing evidence suggests a role for M channels in depression. In 2007 Krishnan and colleagues performed a gene expression analysis in a mouse model of chronic depression caused by stress of social defeat. This is a well-accepted animal model of depression that identifies susceptible to depression and resilient (unsusceptible) mice and allows comparison between those. These authors found that *Kcnq3* expression in the ventral tegmental area (VTA) of the midbrain was upregulated in resilient mice and hypothesized that this may contribute to resilience due to a decrease in dopaminergic VTA neuron excitability (Krishnan et al. 2007). A later study supported this hypothesis by showing that overexpression of *Kcnq3* in the VTA dopaminergic neurons normalized neuronal hyperactivity and alleviated depressive behaviour (Friedman et al. 2016). Recently, Hailin Zhang's group identified Kv7.4 as a crucial M channel subunit in VTA DA neurons (Fig. 3; (Li et al. 2017; Su et al. 2019)). These authors demonstrated that Kv7.4 is a dominant modulator of VTA DA neuron excitability in vitro and in vivo and suggested that downregulation of Kv7.4 could be

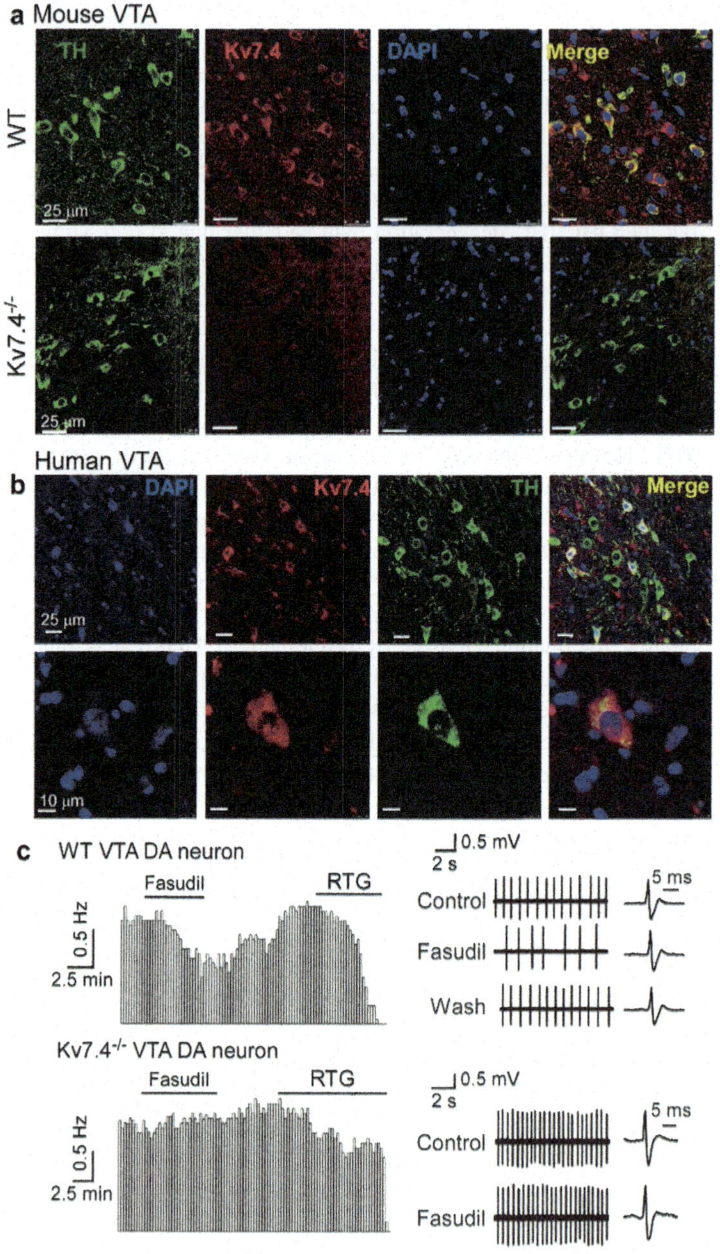

Fig. 3 Functional expression of Kv7.4 in dopaminergic neurons of ventral tegmental area. (**a**) Immunoreactivity of Kv7.4 (red) and a dopaminergic neuron marker, tyrosine hydroxylase (TH, green), in the VTA of WT and Kv7.4 knockout (Kv7.4$^{-/-}$) mice. (**b**) Kv7.4 (red) and TH (green) immunoreactivity in a VTA section from a human brain. (**c**) Cell-attached recording of spontaneous firing of dopaminergic neurons from VTA of WT (top) or Kv7.4$^{-/-}$ mice . Shown are the effects of Kv7.4-selective potentiator, fasudil (10 μM) and a pan-Kv7 opener, retigabine (RTG; 10 μM) on the spontaneous firing frequency of VTA dopaminergic neurons. Typical recordings and action potential waveforms are shown on the right. Panels are reproduced from Li et al. (2017) with permission

a causal factor of the increased excitability of VTA DA neurons and depression-like behaviour (Li et al. 2017). The anti-depressant effects of pan-Kv7 openers retigabine (Feng et al. 2019; Friedman et al. 2016), Lu AA41178 (Grupe et al. 2020) or Kv7.4-selective opener, fasudil (Li et al. 2017) add weight to the idea of augmentation of the Kv7 channel in VTA as a potential treatment of depression.

9.4.2 Anxiety

Kv7 openers BMS-204352 and retigabine were shown to reduce unconditioned anxiety-like behaviours in the mouse zero maze and marble burying models of anxiety; the effects were blocked by the R-enantiomer of BMS-204352 and XE991, respectively (Korsgaard et al. 2005). It was also demonstrated that retigabine attenuated anxiety induced by *status epilepticus* in open-field test, capture and handling test and elevated plus maze (Slomko et al. 2014). A clinical test involving 124 patients also showed that flupirtine-treated patients showed significantly reduced VAS scores for preoperative anxiety (fear of surgical harm), as compared to placebo-treated patients (Yadav et al. 2017). These studies suggest that the neuronal Kv7 channels can be targeted for an anxiolytic effect. However, retigabine did not display any anxiolytic actions in the rat conditional emotional response model of anxiety (Munro et al. 2007) and Frankel and colleagues even observed that neonatal exposure to retigabine could induce increased anxiety-like behaviour in adult animals (Frankel et al. 2016). Thus, systemic pan-Kv7 openers may not be appropriate for specifically targeting brain circuits of anxiety.

9.4.3 Schizophrenia

As discussed above, all the Kv7 channels require PIP_2 for activity. Posphatidylinositol-4-phosphate 5-kinase II alpha (PIP5K2A) is an enzyme that produces PIP_2 by phosphorylation of its precursor, phosphatidylinositol-4-phosphate (PIP) (Gamper and Rohacs 2012). The association between loss-of-function PIP5K2A mutation N251S with schizophrenia was found in German, Dutch and Russian populations (Fedorenko et al. 2013; He et al. 2007; Schwab et al. 2006). Fedorenko and colleagues observed that this mutation abolished the activating effect of WT PIP5K2A on Kv7.2/Kv7.3 and Kv7.3/Kv7.5 channels and hypothesized that loss of PIP5K2A function may result in reduced tonic M channel activity in the brain which, in turn, may contribute to the clinical phenotype of schizophrenia (Fedorenko et al. 2008). Additionally, pretreatment with retigabine or Kv7.2/Kv7.3 opener ICA-27243 was shown to attenuate hyperactivity produced by administration of amphetamine and chlordiazepoxide in a mouse model of mania (Redrobe and Nielsen 2009). There is also a series of studies suggesting Kv7 as a new therapeutic target for treating schizophrenia (Grunnet et al. 2014; Grupe et al. 2020; Peng et al. 2017; Zhao et al. 2017), yet current literature is still inconsistent with respect to whether increased Kv7 channel activity is helpful or detrimental (see below).

9.4.4 Learning, Memory and Neurodevelopmental Disorders

There is a body of evidence suggesting a critical role of Kv7 channels in memory. Yet, the consensus on the exact mechanisms and circuits where these channels may be important is yet to be reached as both positive and negative role of M channel activity on memory formation are being reported. Thus, M current suppression with conditionally expressed dominant-negative Kv7.2 subunits in the brain was shown to change hippocampal morphology and resulted in increased neuronal excitability and deficits in hippocampus-dependent spatial memory (Peters et al. 2005). Similarly, it was shown that acute stress transiently decreased the expression of *Kcnq2* and *Kcnq3* in rat hippocampus and impaired the spatial memory retrieval and hippocampal LTP; flupirtine prevented these impairments, and the protective effects of flupirtine were blocked by XE991 (Li et al. 2014). In drosophila, dKCNQ gene expression declines with age, which parallels cognitive and memory decline commonly associated with ageing (Cavaliere et al. 2013). Loss-of-function mutations in dKCNQ impaired associative short- and long-term memory, suggesting that Kv7 function in the mushroom body a/b-neurons is required for short-term memory (Cavaliere et al. 2013). It was also suggested that β-amyloid peptide can reduce the expression of Kv7.2 and Kv7.3 which, in turn, affects the hippocampal activity balance underlying learning and memory processes impaired in Alzheimer's disease (Mayordomo-Cava et al. 2015). All these studies reported positive effects of neuronal M channel activity on memory.

Yet, there is another line of observation reporting detrimental effects of M channel activity on memory. In aged monkeys, the memory-related firing of aged DELAY neurons was partially restored to more youthful levels by inhibiting cAMP signalling which reportedly decreases the Kv7 current, or by blocking HCN or Kv7 channels (Wang et al. 2011). Inhibition of M/Kv7 channels by M1 muscarinic receptors in the basolateral amygdala was shown to promote fear memory consolidation (Young and Thomas 2014). Another study demonstrated that mice with a knock-in substitution of WT Kv7.2 with Kv7.2(S559A), a mutation which significantly reduces sensitivity of M current to muscarinic inhibition, had impaired long-term object recognition memory which was restored by the administration of XE991 (Kosenko et al. 2020). Additionally, Kv7 blockers infused into rat medial prefrontal cortex (mPFC) improved cognitive function and prevented acute pharmacological stress-induced memory deficits. (Arnsten et al. 2019; Zwierzyńska et al. 2017). Genetic or pharmacological (XE991) suppression of Kv7.2 channel function in the forebrain alleviated NMDA antagonist (MK-801)-induced cognitive decline in mice (Wang et al. 2020a). Neuroprotective effects of intracerebroventricular administration of XE991 in rat models of Parkinson's disease have also been demonstrated (Liu et al. 2018).

Thus, current literature reports both memory-promoting and memory-impairing effects of M channel activity. The key to the apparent contention is likely to lie in the specific circuits and neuron types affected by the manipulations with the M channel activity implemented in different studies. For example, global heterozygous knockout of *Kcnq2* in mice is associated with reduced prepulse inhibition of the acoustic startle response (PPI) (Kapfhamer et al. 2010) (PPI is a neurobehavioral phenomenon related to the sensorimotor gating, which is reduced in schizophrenia (Mena et al. 2016)). Yet, a forebrain-specific (αCaMKII promoter) suppression of Kv7 function by overexpression of a dominant-negative Kv7.2 G279S pore mutant resulted in enhanced PPI (Wang et al. 2020a), suggesting that inhibition of Kv7 channels in this area may have therapeutic potential for schizophrenia. Thus, future studies with circuit-specific manipulations of Kv7 activity are required to unambiguously characterize roles of M channels in cognition.

Given the importance of M current for memory and other cognitive functions, the possibility of M channel involvement in neurodevelopmental disorders has also been investigated. Neurodevelopmental disorders (NDD) include autism spectrum disorder, intellectual disability developmental delay and other forms of cognitive impairments. A number of missense mutations and truncations in *KCNQ2* and *KCNQ3* were reported in NDD cohorts with or without epilepsy (Geisheker et al. 2017; Li et al. 2020). A missense mutation in *KCNQ5* was also reported in four probands with intellectual disability, abnormal neurological findings, and treatment-resistant epilepsy (Lehman et al. 2017). In mice, heterozygous loss of *Kcnq2* induced autism-associated behaviours such as reduced sociability and enhanced repetitive behaviours (Kim et al. 2020). Interestingly, some gain-of-function mutations (R144Q, R201C, and R201H in *KCNQ2* and R230C in *KCNQ3)* were found in patients with epileptic encephalopathies and/or intellectual disability (Miceli et al. 2015), adding to the notion that both reduced and enhanced M channel activity may result in cognitive dysfunction.

9.4.5 Addiction

A series of studies revealed the association of Kv7 channels with alcohol-related quantitative trait loci (QTLs) and alcohol consumption and withdrawal (McGuier et al. 2016; Metten et al. 2014; Rinker et al. 2017; Rinker and Mulholland 2017; reviewed in Cannady et al. 2018; Vigil et al. 2020). An early report found a reduction of M current amplitude by ethanol (Moore et al. 1990). More recent studies revealed that Kv7.2 and Kv7.3 expression in the lateral habenula (LHb), a key brain region involved in anxiety (Pobbe and Zangrossi Jr. 2008), is markedly reduced after withdrawal from the systemic alcohol administration in rats (Kang et al. 2017; Shah et al. 2017). The effect was well correlated with the increased excitability of LHb neurons and increased anxiety. Furthermore, retigabine reduced alcohol consumption in rodent models of alcoholism (McGuier et al. 2016; Rinker et al. 2017). Similarly, retigabine was shown to reduce locomotor activity induced by cocaine, methylphenidate and phencyclidine and also reduced the cocaine-seeking in rats (Hansen et al. 2007; Parrilla-Carrero et al. 2018).

10 Other Diseases and Pathologies

10.1 Pulmonary Disorders

Kv7 channels are expressed in smooth muscle cells of pulmonary arteries and bronchial tubes of the lung, airway epithelia and peripheral somatosensory and sympathetic fibres innervating the airways (with distinct expression profiles in cells of different type; see above and Mondejar-Parreno et al. (2020)). In general terms, Kv7 channel activation in the airway smooth muscle cells (i.e. with the Kv7 openers) exerts a bronchorelaxant effect while M channel inhibition (i.e. by bronchoconstrictors such as methacholine, carbachol or histamine) produces bronchoconstriction (Brueggemann et al. 2014a, b; Haick et al. 2017). M channel inhibition in afferent C-fibres innervating the lungs may depolarize the nerve endings causing local release of substance P and calcitonin gene-related peptide (CGRP) and, subsequently, a neurogenic inflammation (Barnes 1986; Sun et al. 2019). Finally, Kv7.1 channel complex activity in lung epithelia is necessary for Cl^- secretion (via K^+ co-secretion) and mucociliary clearance (Mall et al. 2000; Mondejar-Parreno et al. 2020; Schroeder et al. 2000b). All these effects implicate M channels (with their notoriously rich profile of modulation by endogenous hormones, neuro- and vasoactive peptides and inflammatory mediators) in a number of pulmonary diseases such as asthma, chronic obstructive pulmonary disease (COPD), pulmonary hypertension, chronic cough and cystic fibrosis (Mondejar-Parreno et al. 2020). Additionally, some studies implicated Kv7 channels in some types of lung cancer. Thus, increased expression of Kv7.1 was found in tumour samples from patients with lung adenocarcinoma (Girault et al. 2014), while activity of Kv7.3 and Kv7.5 was suggested to be important for offsetting drug resistance in lung adenocarcinoma cell line (Choi et al. 2017).

10.2 Gastrointestinal Tract Disorders

In 2000, Thomas Jentsch's group identified and cloned a new beta-subunit, KCNE3, which colocalized with KCNQ1 in intestine crypt cells (Schroeder et al. 2000b). They also found that this potassium channel complex plays an important role in cAMP-stimulated intestinal Cl^- secretion, which is disturbed in secretory diarrhoea and cystic fibrosis. Several other studies reported functional expression of Kv7.1 in gastrointestinal tract, specifically, in colonic epithelial and stomach parietal cells, which are involved in intestinal Cl^- and gastric acid secretion (Grahammer et al. 2001; Heitzmann et al. 2004; Inagaki et al. 2019; Kunzelmann et al. 2001a, c; Lambrecht et al. 2005; Waldegger 2003).

Kv7 channels are also involved in modulation of gastrointestinal motor activity (Currò 2014). Jepps and colleagues reported that Kv7.4 and Kv7.5 are expressed in smooth muscles throughout the GI tract, including the circular muscle layer of the colon; XE991 and linopirdine caused an increase in spontaneous contractile activity of the distal colon (Jepps et al. 2009). Another study demonstrated functional expression of Kv7 channels in rat gastric fundus; Kv7.4 and Kv7.5 were also found to be the main subtypes. As expected, XE991 stimulated contraction while retigabine and flupirtine caused relaxation of the stomach strips (Ipavec et al. 2011). Similar results were later reported in the human taenia coli by the same group (Adduci et al. 2013). The above findings suggest that selective inhibition of Kv7.4/7.5 might be therapeutic for the range of GIT disorders such as constipation associated with irritable bowel syndrome, secretory diarrhoea and peptic ulcer disease.

10.3 Bladder Dysfunctions

Presence of *KCNQ1*, *KCNQ3-KCNQ5* and *KCNE1–5* mRNA was reported in human urinary bladder; a 3.4-fold up-regulation of *KCNQ1* mRNA was observed in patients with bladder outflow obstruction (Svalø et al. 2015). Kv7 activators reduced the bladder tone while muscarinic agonist carbachol and XE991 produced constriction (Seefeld et al. 2018; Svalø et al. 2015). The Kv7 subunit expression and their importance for bladder function were also observed in guinea pigs and rats (Afeli et al. 2013; Anderson et al. 2013; Provence et al. 2015, 2018; Svalø et al. 2013). There are however studies that did not find significant functional importance of Kv7 channels in the bladder. Thus, Tykocki and colleagues observed that there were no retigabine-sensitive Kv7 channels expressed in mouse urinary bladder smooth muscle cells and suggested that the activation of Kv7 channels in afferent nerves innervating the bladder is responsible for the effect of retigabine on the bladder function (Tykocki et al. 2019). Another study suggested that BK channels are the predominant K^+ channels controlling guinea pig bladder contractility while Kv7 serve a subsidiary role (Lee et al. 2018). Despite this discrepancy, all the studies cited above found that modulators of Kv7 channels (be that the ones expressed in smooth muscle or the bladder afferents) do affect bladder contractility and, thus, should be considered as potential therapeutics for urinary bladder dysfunctions. The flip side of bladder-localized Kv7 channels is that these are likely to underlie urinary retention observed in patients receiving retigabine as an anticonvulsant.

10.4 Other Pathologies

Kv7 channels were implicated in itch (Zhang et al. 2020), erectile dysfunction (Lee et al. 2020), high myopia (Liao et al. 2017) and Parkinson's disease (Liu et al. 2018).

Acknowledgments The work in our laboratories is supported by Wellcome Trust (212302/Z/18/Z to NG), Medical Research Council (MR/P015727/1 to NG & FJ), Biotechnology and Biological Sciences Research Council (BB/R003068/1, BB/R02104X/1 to NG) and National Natural Science Foundation of China (81871027 to HG & NG).

References

Abbott GW (2016) Novel exon 1 protein-coding regions N-terminally extend human KCNE3 and KCNE4. FASEB J 30:2959–2969

Abbott GW (2020) KCNQs: ligand- and voltage-gated potassium channels. Front Physiol 11:583

Adduci A, Martire M, Taglialatela M, Arena V, Rizzo G, Coco C, Currò D (2013) Expression and motor functional roles of voltage-dependent type 7 K(+) channels in the human taenia coli. Eur J Pharmacol 721:12–20

Afeli SA, Malysz J, Petkov GV (2013) Molecular expression and pharmacological evidence for a functional role of kv7 channel subtypes in Guinea pig urinary bladder smooth muscle. PLoS One 8:e75875

Aiken SP, Lampe BJ, Murphy PA, Brown BS (1995) Reduction of spike frequency adaptation and blockade of M-current in rat CA1 pyramidal neurones by linopirdine (DuP 996), a neurotransmitter release enhancer. Br J Pharmacol 115:1163–1168

Anderson UA, Carson C, Johnston L, Joshi S, Gurney AM, McCloskey KD (2013) Functional expression of KCNQ (Kv7) channels in Guinea pig bladder smooth muscle and their contribution to spontaneous activity. Br J Pharmacol 169:1290–1304

Angelo K, Jespersen T, Grunnet M, Nielsen MS, Klaerke DA, Olesen SP (2002) KCNE5 induces time- and voltage-dependent modulation of the KCNQ1 current. Biophys J 83:1997–2006

Antzelevitch C, Burashnikov A (2011) Overview of basic mechanisms of cardiac arrhythmia. Card Electrophysiol Clin 3:23–45

Archer CR, Enslow BT, Taylor AB, De la Rosa V, Bhattacharya A, Shapiro MS (2019) A mutually induced conformational fit underlies ca(2+)-directed interactions between calmodulin and the proximal C terminus of KCNQ4 K(+) channels. J Biol Chem 294:6094–6112

Arnsten AFT, Jin LE, Gamo NJ, Ramos B, Paspalas CD, Morozov YM, Kata A, Bamford NS, Yeckel MF, Kaczmarek LK, El-Hassar L (2019) Role of KCNQ potassium channels in stress-induced deficit of working memory. Neurobiol Stress 11:100187

Bal M, Zhang J, Zaika O, Hernandez CC, Shapiro MS (2008) Homomeric and heteromeric assembly of KCNQ (Kv7) K+ channels assayed by total internal reflection fluorescence/ fluorescence resonance energy transfer and patch clamp analysis. J Biol Chem 283:30668–30676

Bal M, Zhang J, Hernandez CC, Zaika O, Shapiro MS (2010) Ca2+/calmodulin disrupts AKAP79/ 150 interactions with KCNQ (M-type) K+ channels. J Neurosci 30:2311–2323

Barhanin J, Lesage F, Guillemare E, Fink M, Lazdunski M, Romey G (1996) K(V)LQT1 and lsK (minK) proteins associate to form the I(Ks) cardiac potassium current. Nature 384:78–80

Barnes PJ (1986) Asthma as an axon reflex. Lancet 1:242–245

Barrese V, Stott JB, Figueiredo HB, Aubdool AA, Hobbs AJ, Jepps TA, McNeish AJ, Greenwood IA (2018) Angiotensin II promotes KV7.4 channels degradation through reduced interaction with HSP90 (heat shock protein 90). Hypertension 71:1091–1100

Barro-Soria R, Rebolledo S, Liin SI, Perez ME, Sampson KJ, Kass RS, Larsson HP (2014) KCNE1 divides the voltage sensor movement in KCNQ1/KCNE1 channels into two steps. Nat Commun 5:3750

Bartos DC, Giudicessi JR, Tester DJ, Ackerman MJ, Ohno S, Horie M, Gollob MH, Burgess DE, Delisle BP (2014) A KCNQ1 mutation contributes to the concealed type 1 long QT phenotype by limiting the Kv7.1 channel conformational changes associated with protein kinase A phosphorylation. Heart Rhythm 11:459–468

Beisel KW, Nelson NC, Delimont DC, Fritzsch B (2000) Longitudinal gradients of KCNQ4 expression in spiral ganglion and cochlear hair cells correlate with progressive hearing loss in DFNA2. Brain Res Mol Brain Res 82:137–149

Bellocq C, van Ginneken AC, Bezzina CR, Alders M, Escande D, Mannens MM, Baro I, Wilde AA (2004) Mutation in the KCNQ1 gene leading to the short QT-interval syndrome. Circulation 109:2394–2397

Bendahhou S, Marionneau C, Haurogne K, Larroque MM, Derand R, Szuts V, Escande D, Demolombe S, Barhanin J (2005) In vitro molecular interactions and distribution of KCNE family with KCNQ1 in the human heart. Cardiovasc Res 67:529–538

Bernardo-Seisdedos G, Nunez E, Gomis-Perez C, Malo C, Villarroel A, Millet O (2018) Structural basis and energy landscape for the ca(2+) gating and calmodulation of the Kv7.2 K(+) channel. Proc Natl Acad Sci U S A 115:2395–2400

Biervert C, Schroeder BC, Kubisch C, Berkovic SF, Propping P, Jentsch TJ, Steinlein OK (1998) A potassium channel mutation in neonatal human epilepsy. Science 279:403–406

Bleich M, Warth R (2000) The very small-conductance K+ channel KvLQT1 and epithelial function. Pflugers Arch 440:202–206

Brown DA, Adams PR (1980) Muscarinic suppression of a novel voltage-sensitive K+ current in a vertebrate neurone. Nature 283:673–676

Brown DA, Passmore GM (2009) Neural KCNQ (Kv7) channels. Br J Pharmacol 156:1185–1195

Brown DA, Abogadie FC, Allen TG, Buckley NJ, Caulfield MP, Delmas P, Haley JE, Lamas JA, Selyanko AA (1997) Muscarinic mechanisms in nerve cells. Life Sci 60:1137–1144

Brown DA, Hughes SA, Marsh SJ, Tinker A (2007) Regulation of M(Kv7.2/7.3) channels in neurons by PIP(2) and products of PIP(2) hydrolysis: significance for receptor-mediated inhibition. J Physiol 582:917–925

Brueggemann LI, Moran CJ, Barakat JA, Yeh JZ, Cribbs LL, Byron KL (2007) Vasopressin stimulates action potential firing by protein kinase C-dependent inhibition of KCNQ5 in A7r5 rat aortic smooth muscle cells. Am J Physiol Heart Circ Physiol 292:H1352–H1363

Brueggemann LI, Kakad PP, Love RB, Solway J, Dowell ML, Cribbs LL, Byron KL (2012) Kv7 potassium channels in airway smooth muscle cells: signal transduction intermediates and pharmacological targets for bronchodilator therapy. Am J Physiol Lung Cell Mol Physiol 302:L120–L132

Brueggemann LI, Haick JM, Neuburg S, Tate S, Randhawa D, Cribbs LL, Byron KL (2014a) KCNQ (Kv7) potassium channel activators as bronchodilators: combination with a beta2-adrenergic agonist enhances relaxation of rat airways. Am J Physiol Lung Cell Mol Physiol 306:L476–L486

Brueggemann LI, Haick JM, Neuburg S, Tate S, Randhawa D, Cribbs LL, Byron KL (2014b) KCNQ (Kv7) potassium channel activators as bronchodilators: combination with a β2-adrenergic agonist enhances relaxation of rat airways. Am J Physiol Lung Cell Mol Physiol 306:L476–L486

Brueggemann LI, Mackie AR, Cribbs LL, Freda J, Tripathi A, Majetschak M, Byron KL (2014c) Differential protein kinase C-dependent modulation of Kv7.4 and Kv7.5 subunits of vascular Kv7 channels. J Biol Chem 289:2099–2111

Brueggemann LI, Cribbs LL, Schwartz J, Wang M, Kouta A, Byron KL (2018) Mechanisms of PKA-dependent potentiation of Kv7.5 channel activity in human airway smooth muscle cells. Int J Mol Sci:19

Burgess DE, Bartos DC, Reloj AR, Campbell KS, Johnson JN, Tester DJ, Ackerman MJ, Fressart V, Denjoy I, Guicheney P, Moss AJ, Ohno S, Horie M, Delisle BP (2012) High-risk long QT syndrome mutations in the Kv7.1 (KCNQ1) pore disrupt the molecular basis for rapid K(+) permeation. Biochemistry 51:9076–9085

Busch AE, Busch GL, Ford E, Suessbrich H, Lang HJ, Greger R, Kunzelmann K, Attali B, Stuhmer W (1997) The role of the IsK protein in the specific pharmacological properties of the IKs channel complex. Br J Pharmacol 122:187–189

Cannady R, Rinker JA, Nimitvilai S, Woodward JJ, Mulholland PJ (2018) Chronic alcohol, intrinsic excitability, and potassium channels: neuroadaptations and drinking behavior. Handbook Exp Pharmacol 248:311–343

Carignano C, Barila EP, Rias EI, Dionisio L, Aztiria E, Spitzmaul G (2019) Inner hair cell and neuron degeneration contribute to hearing loss in a DFNA2-like mouse model. Neuroscience 410:202–216

Cavaliere S, Malik BR, Hodge JJ (2013) KCNQ channels regulate age-related memory impairment. PLoS One 8:e62445
Chadha PS, Zunke F, Zhu HL, Davis AJ, Jepps TA, Olesen SP, Cole WC, Moffatt JD, Greenwood IA (2012) Reduced KCNQ4-encoded voltage-dependent potassium channel activity underlies impaired beta-adrenoceptor-mediated relaxation of renal arteries in hypertension. Hypertension 59:877–884
Chadha PS, Jepps TA, Carr G, Stott JB, Zhu HL, Cole WC, Greenwood IA (2014) Contribution of kv7.4/kv7.5 heteromers to intrinsic and calcitonin gene-related peptide-induced cerebral reactivity. Arterioscler Thromb Vasc Biol 34:887–893
Chandrasekhar KD, Lvov A, Terrenoire C, Gao GY, Kass RS, Kobertz WR (2011) O-glycosylation of the cardiac I(Ks) complex. J Physiol 589:3721–3730
Chang A, Abderemane-Ali F, Hura GL, Rossen ND, Gate RE, Minor DL Jr (2018) A calmodulin C-lobe ca(2+)-dependent switch governs Kv7 channel function. Neuron 97:836–852.e6
Charlier C, Singh NA, Ryan SG, Lewis TB, Reus BE, Leach RJ, Leppert M (1998) A pore mutation in a novel KQT-like potassium channel gene in an idiopathic epilepsy family. Nat Genet 18:53–55
Chen YH, Xu SJ, Bendahhou S, Wang XL, Wang Y, Xu WY, Jin HW, Sun H, Su XY, Zhuang QN, Yang YQ, Li YB, Liu Y, Xu HJ, Li XF, Ma N, Mou CP, Chen Z, Barhanin J, Huang W (2003) KCNQ1 gain-of-function mutation in familial atrial fibrillation. Science 299:251–254
Choi SY, Kim HR, Ryu PD, Lee SY (2017) Regulation of voltage-gated potassium channels attenuates resistance of side-population cells to gefitinib in the human lung cancer cell line NCI-H460. BMC Pharmacol Toxicol 18:14
Choveau FS, Shapiro MS (2012) Regions of KCNQ K(+) channels controlling functional expression. Front Physiol 3:397
Choveau FS, De la Rosa V, Bierbower SM, Hernandez CC, Shapiro MS (2018) Phosphatidylinositol 4,5-bisphosphate (PIP2) regulates KCNQ3 K(+) channels by interacting with four cytoplasmic channel domains. J Biol Chem 293:19411–19428
Chung HJ, Jan YN, Jan LY (2006) Polarized axonal surface expression of neuronal KCNQ channels is mediated by multiple signals in the KCNQ2 and KCNQ3 C-terminal domains. Proc Natl Acad Sci U S A 103:8870–8875
Constanti A, Brown DA (1981) M-currents in voltage-clamped mammalian sympathetic neurones. Neurosci Lett 24:289–294
Cooper EC (2011) Made for "anchorin": Kv7.2/7.3 (KCNQ2/KCNQ3) channels and the modulation of neuronal excitability in vertebrate axons. Semin Cell Dev Biol 22:185–192
Cooper EC, Harrington E, Jan YN, Jan LY (2001) M channel KCNQ2 subunits are localized to key sites for control of neuronal network oscillations and synchronization in mouse brain. J Neurosci 21:9529–9540
Coucke PJ, Hauwe PV, Kelley PM, Kunst H, Schatteman I, Velzen DV, Meyers J, Ensink RJ, Verstreken M, Declau F, Marres H, Kastury K, Bhasin S, McGuirt WT, Smith RJ, Cremers CW, Heyning PV, Willems PJ, Smith SD, Camp GV (1999) Mutations in the KCNQ4 gene are responsible for autosomal dominant deafness in four DFNA2 families. Hum Mol Genet 8:1321–1328
Crozier RA, Ajit SK, Kaftan EJ, Pausch MH (2007) MrgD activation inhibits KCNQ/M-currents and contributes to enhanced neuronal excitability. J Neurosci 27:4492–4496
Cruzblanca H, Koh DS, Hille B (1998) Bradykinin inhibits M current via phospholipase C and Ca2 + release from IP3-sensitive Ca2+ stores in rat sympathetic neurons. Proc Natl Acad Sci U S A 95:7151–7156
Currò D (2014) K+ channels as potential targets for the treatment of gastrointestinal motor disorders. Eur J Pharmacol 733:97–101
Dahimene S, Alcolea S, Naud P, Jourdon P, Escande D, Brasseur R, Thomas A, Baro I, Merot J (2006) The N-terminal juxtamembranous domain of KCNQ1 is critical for channel surface expression: implications in the Romano-Ward LQT1 syndrome. Circ Res 99:1076–1083

Dedek K, Waldegger S (2001) Colocalization of KCNQ1/KCNE channel subunits in the mouse gastrointestinal tract. Pflugers Arch 442:896–902
Delmas P, Brown DA (2005) Pathways modulating neural KCNQ/M (Kv7) potassium channels. Nat Rev Neurosci 6:850–862
Demolombe S, Franco D, de Boer P, Kuperschmidt S, Roden D, Pereon Y, Jarry A, Moorman AF, Escande D (2001) Differential expression of KvLQT1 and its regulator IsK in mouse epithelia. Am J Physiol Cell Physiol 280:C359–C372
Du X, Gamper N (2013) Potassium channels in peripheral pain pathways: expression, function and therapeutic potential. Curr Neuropharmacol 11:621–640
Du X, Gao H, Jaffe D, Zhang H, Gamper N (2018) M-type K+ channels in peripheral nociceptive pathways. Br J Pharmacol 175:2158–2172
Dvir M, Strulovich R, Sachyani D, Ben-Tal Cohen I, Haitin Y, Dessauer C, Pongs O, Kass R, Hirsch JA, Attali B (2014) Long QT mutations at the interface between KCNQ1 helix C and KCNE1 disrupt I(KS) regulation by PKA and PIP(2). J Cell Sci 127:3943–3955
Ekberg J, Schuetz F, Boase NA, Conroy SJ, Manning J, Kumar S, Poronnik P, Adams DJ (2007) Regulation of the voltage-gated K(+) channels KCNQ2/3 and KCNQ3/5 by ubiquitination. Novel role for Nedd4-2. J Biol Chem 282:12135–12142
Etxeberria A, Aivar P, Rodriguez-Alfaro JA, Alaimo A, Villace P, Gomez-Posada JC, Areso P, Villarroel A (2008) Calmodulin regulates the trafficking of KCNQ2 potassium channels. FASEB J 22:1135–1143
Etzioni A, Siloni S, Chikvashvilli D, Strulovich R, Sachyani D, Regev N, Greitzer-Antes D, Hirsch JA, Lotan I (2011) Regulation of neuronal M-channel gating in an isoform-specific manner: functional interplay between calmodulin and Syntaxin 1A. J Neurosci 31:14158–14171
Evseev AI, Semenov I, Archer CR, Medina JL, Dube PH, Shapiro MS, Brenner R (2013) Functional effects of KCNQ K(+) channels in airway smooth muscle. Front Physiol 4:277
Fedorenko O, Strutz-Seebohm N, Henrion U, Ureche ON, Lang F, Seebohm G, Lang UE (2008) A schizophrenia-linked mutation in PIP5K2A fails to activate neuronal M channels. Psychopharmacology 199:47–54
Fedorenko O, Rudikov EV, Gavrilova VA, Boiarko EG, Semke AV, Ivanova SA (2013) Association of (N251S)-PIP5K2A with schizophrenic disorders: a study of the Russian population of Siberia. Zh Nevrol Psikhiatr Im S S Korsakova 113:58–61
Feng M, Crowley NA, Patel A, Guo Y, Bugni SE, Luscher B (2019) Reversal of a treatment-resistant, depression-related brain state with the Kv7 channel opener Retigabine. Neuroscience 406:109–125
Filippov AK, Choi RC, Simon J, Barnard EA, Brown DA (2006) Activation of P2Y1 nucleotide receptors induces inhibition of the M-type K+ current in rat hippocampal pyramidal neurons. J Neurosci 26:9340–9348
Ford CP, Stemkowski PL, Light PE, Smith PA (2003) Experiments to test the role of phosphatidylinositol 4,5-bisphosphate in neurotransmitter-induced M-channel closure in bullfrog sympathetic neurons. J Neurosci 23:4931–4941
Ford CP, Stemkowski PL, Smith PA (2004) Possible role of phosphatidylinositol 4,5 bisphosphate in luteinizing hormone releasing hormone-mediated M-current inhibition in bullfrog sympathetic neurons. Eur J Neurosci 20:2990–2998
Fosmo AL, Skraastad OB (2017) The Kv7 channel and cardiovascular risk factors. Front Cardiovasc Med 4:75
Frankel S, Medvedeva N, Gutherz S, Kulick C, Kondratyev A, Forcelli PA (2016) Comparison of the long-term behavioral effects of neonatal exposure to retigabine or phenobarbital in rats. Epilepsy Behav E&B 57:34–40
Friedman AK, Juarez B, Ku SM, Zhang H, Calizo RC, Walsh JJ, Chaudhury D, Zhang S, Hawkins A, Dietz DM, Murrough JW, Ribadeneira M, Wong EH, Neve RL, Han MH (2016) KCNQ channel openers reverse depressive symptoms via an active resilience mechanism. Nat Commun 7:11671

Gamper N, Rohacs T (2012) Phosphoinositide sensitivity of ion channels, a functional perspective. Subcell Biochem 59:289–333

Gamper N, Shapiro MS (2003) Calmodulin mediates Ca2+−dependent modulation of M-type K+ channels. J Gen Physiol 122:17–31

Gamper N, Shapiro MS (2007) Regulation of ion transport proteins by membrane phosphoinositides. Nat Rev Neurosci 8:921–934

Gamper N, Shapiro MS (2015) KCNQ channels. In: Zheng J, Trudeau MC (eds) Handbook of ion channels. CRC Press, Boca Raton, pp 275–306

Gamper N, Stockand JD, Shapiro MS (2003) Subunit-specific modulation of KCNQ potassium channels by Src tyrosine kinase. J Neurosci 23:84–95

Gamper N, Li Y, Shapiro MS (2005) Structural requirements for differential sensitivity of KCNQ K + channels to modulation by Ca2+/calmodulin. Mol Biol Cell 16:3538–3551

Gamper N, Zaika O, Li Y, Martin P, Hernandez CC, Perez MR, Wang AY, Jaffe DB, Shapiro MS (2006) Oxidative modification of M-type K(+) channels as a mechanism of cytoprotective neuronal silencing. EMBO J 25:4996–5004

Gao Y, Yechikov S, Vazquez AE, Chen D, Nie L (2013) Impaired surface expression and conductance of the KCNQ4 channel lead to sensorineural hearing loss. J Cell Mol Med 17:889–900

Gao HX, Boillat A, Huang DY, Liang C, Peers C, Gamper N (2017) Intracellular zinc activates KCNQ channels by reducing their dependence on phosphatidylinositol 4,5-bisphosphate. Proc Natl Acad Sci U S A 114:E6410–E6419

Geiger J, Weber YG, Landwehrmeyer B, Sommer C, Lerche H (2006) Immunohistochemical analysis of KCNQ3 potassium channels in mouse brain. Neurosci Lett 400:101–104

Geisheker MR, Heymann G, Wang T, Coe BP, Turner TN, Stessman HAF, Hoekzema K, Kvarnung M, Shaw M, Friend K, Liebelt J, Barnett C, Thompson EM, Haan E, Guo H, Anderlid BM, Nordgren A, Lindstrand A, Vandeweyer G, Alberti A, Avola E, Vinci M, Giusto S, Pramparo T, Pierce K, Nalabolu S, Michaelson JJ, Sedlacek Z, Santen GWE, Peeters H, Hakonarson H, Courchesne E, Romano C, Kooy RF, Bernier RA, Nordenskjöld M, Gecz J, Xia K, Zweifel LS, Eichler EE (2017) Hotspots of missense mutation identify neurodevelopmental disorder genes and functional domains. Nat Neurosci 20:1043–1051

Ghosh S, Nunziato DA, Pitt GS (2006) KCNQ1 assembly and function is blocked by long-QT syndrome mutations that disrupt interaction with calmodulin. Circ Res 98:1048–1054

Girault A, Privé A, Trinh NT, Bardou O, Ferraro P, Joubert P, Bertrand R, Brochiero E (2014) Identification of KvLQT1 K+ channels as new regulators of non-small cell lung cancer cell proliferation and migration. Int J Oncol 44:838–848

Grahammer F, Herling AW, Lang HJ, Schmitt-Graff A, Wittekindt OH, Nitschke R, Bleich M, Barhanin J, Warth R (2001) The cardiac K+ channel KCNQ1 is essential for gastric acid secretion. Gastroenterology 120:1363–1371

Greene DL, Hoshi N (2017) Modulation of Kv7 channels and excitability in the brain. Cell Mol Life Sci 74:495–508

Grunnet M, Jespersen T, Rasmussen HB, Ljungstrom T, Jorgensen NK, Olesen SP, Klaerke DA (2002) KCNE4 is an inhibitory subunit to the KCNQ1 channel. J Physiol 542:119–130

Grunnet M, Jespersen T, MacAulay N, Jorgensen NK, Schmitt N, Pongs O, Olesen SP, Klaerke DA (2003) KCNQ1 channels sense small changes in cell volume. J Physiol 549:419–427

Grunnet M, Olesen SP, Klaerke DA, Jespersen T (2005) hKCNE4 inhibits the hKCNQ1 potassium current without affecting the activation kinetics. Biochem Biophys Res Commun 328:1146–1153

Grunnet M, Strøbæk D, Hougaard C, Christophersen P (2014) Kv7 channels as targets for anti-epileptic and psychiatric drug-development. Eur J Pharmacol 726:133–137

Grupe M, Bentzen BH, Benned-Jensen T, Nielsen V, Frederiksen K, Jensen HS, Jacobsen AM, Skibsbye L, Sams AG, Grunnet M, Rottländer M, Bastlund JF (2020) In vitro and in vivo characterization of Lu AA41178: a novel, brain penetrant, pan-selective Kv7 potassium channel opener with efficacy in preclinical models of epileptic seizures and psychiatric disorders. Eur J Pharmacol:173440

Haick JM, Brueggemann LI, Cribbs LL, Denning MF, Schwartz J, Byron KL (2017) PKC-dependent regulation of Kv7.5 channels by the bronchoconstrictor histamine in human airway smooth muscle cells. Am J Physiol Lung Cell Mol Physiol 312:L822–L834

Haitin Y, Attali B (2008) The C-terminus of Kv7 channels: a multifunctional module. J Physiol 586:1803–1810

Hamilton KL, Devor DC (2012) Basolateral membrane K+ channels in renal epithelial cells. Am J Physiol Renal Physiol 302:F1069–F1081

Hammami Bomholtz S, Refaat M, Buur Steffensen A, David JP, Espinosa K, Nussbaum R, Wojciak J, Hjorth Bentzen B, Scheinman M, Schmitt N (2020) Functional phenotype variations of two novel KV 7.1 mutations identified in patients with long QT syndrome. Pacing Clin Electrophysiol 43:210–216

Hansen HH, Andreasen JT, Weikop P, Mirza N, Scheel-Krüger J, Mikkelsen JD (2007) The neuronal KCNQ channel opener retigabine inhibits locomotor activity and reduces forebrain excitatory responses to the psychostimulants cocaine, methylphenidate and phencyclidine. Eur J Pharmacol 570:77–88

He Z, Li Z, Shi Y, Tang W, Huang K, Ma G, Zhou J, Meng J, Li H, Feng G, He L (2007) The PIP5K2A gene and schizophrenia in the Chinese population--a case-control study. Schizophr Res 94:359–365

Hedegaard ER, Nielsen BD, Kun A, Hughes AD, Kroigaard C, Mogensen S, Matchkov VV, Frobert O, Simonsen U (2014) KV 7 channels are involved in hypoxia-induced vasodilatation of porcine coronary arteries. Br J Pharmacol 171:69–82

Heidenreich M, Lechner SG, Vardanyan V, Wetzel C, Cremers CW, de Leenheer EM, Aranguez G, Moreno-Pelayo MA, Jentsch TJ, Lewin GR (2012) KCNQ4 K(+) channels tune mechanoreceptors for normal touch sensation in mouse and man. Nat Neurosci 15:138–145

Heitzmann D, Grahammer F, von Hahn T, Schmitt-Gräff A, Romeo E, Nitschke R, Gerlach U, Lang HJ, Verrey F, Barhanin J, Warth R (2004) Heteromeric KCNE2/KCNQ1 potassium channels in the luminal membrane of gastric parietal cells. J Physiol 561:547–557

Hernandez CC, Zaika O, Shapiro MS (2008a) A carboxy-terminal inter-helix linker as the site of phosphatidylinositol 4,5-bisphosphate action on Kv7 (M-type) K+ channels. J Gen Physiol 132:361–381

Hernandez CC, Zaika O, Tolstykh GP, Shapiro MS (2008b) Regulation of neural KCNQ channels: signalling pathways, structural motifs and functional implications. J Physiol 586:1811–1821

Higashida H, Hoshi N, Zhang JS, Yokoyama S, Hashii M, Jin D, Noda M, Robbins J (2005) Protein kinase C bound with A-kinase anchoring protein is involved in muscarinic receptor-activated modulation of M-type KCNQ potassium channels. Neurosci Res 51:231–234

Hille B, Dickson EJ, Kruse M, Vivas O, Suh BC (2015) Phosphoinositides regulate ion channels. Biochim Biophys Acta 1851:844–856

Holmes CL, Landry DW, Granton JT (2003) Science review: vasopressin and the cardiovascular system part 1--receptor physiology. Crit Care 7:427–434

Holt JR, Stauffer EA, Abraham D, Geleoc GS (2007) Dominant-negative inhibition of M-like potassium conductances in hair cells of the mouse inner ear. J Neurosci 27:8940–8951

Horikawa N, Suzuki T, Uchiumi T, Minamimura T, Tsukada K, Takeguchi N, Sakai H (2005) Cyclic AMP-dependent Cl- secretion induced by thromboxane A2 in isolated human colon. J Physiol 562:885–897

Hoshi N, Zhang JS, Omaki M, Takeuchi T, Yokoyama S, Wanaverbecq N, Langeberg LK, Yoneda Y, Scott JD, Brown DA, Higashida H (2003) AKAP150 signaling complex promotes suppression of the M-current by muscarinic agonists. Nat Neurosci 6:564–571

Hoshi N, Langeberg LK, Scott JD (2005) Distinct enzyme combinations in AKAP signalling complexes permit functional diversity. Nat Cell Biol

Hou P, Eldstrom J, Shi J, Zhong L, McFarland K, Gao Y, Fedida D, Cui J (2017) Inactivation of KCNQ1 potassium channels reveals dynamic coupling between voltage sensing and pore opening. Nat Commun 8:1730

Howard RJ, Clark KA, Holton JM, Minor DL Jr (2007) Structural insight into KCNQ (Kv7) channel assembly and channelopathy. Neuron 53:663–675

Huang H, Trussell LO (2011) KCNQ5 channels control resting properties and release probability of a synapse. Nat Neurosci 14:840–847

Hurley KM, Gaboyard S, Zhong M, Price SD, Wooltorton JR, Lysakowski A, Eatock RA (2006) M-like K+ currents in type I hair cells and calyx afferent endings of the developing rat utricle. J Neurosci 26:10253–10269

Iannotti FA, Panza E, Barrese V, Viggiano D, Soldovieri MV, Taglialatela M (2010) Expression, localization, and pharmacological role of Kv7 potassium channels in skeletal muscle proliferation, differentiation, and survival after myotoxic insults. J Pharmacol Exp Ther 332:811–820

Iannotti FA, Barrese V, Formisano L, Miceli F, Taglialatela M (2013) Specification of skeletal muscle differentiation by repressor element-1 silencing transcription factor (REST)-regulated Kv7.4 potassium channels. Mol Biol Cell 24:274–284

Inagaki A, Hayashi M, Andharia N, Matsuda H (2019) Involvement of butyrate in electrogenic K(+) secretion in rat rectal colon. Pflugers Arch 471:313–327

Ipavec V, Martire M, Barrese V, Taglialatela M, Currò D (2011) KV7 channels regulate muscle tone and nonadrenergic noncholinergic relaxation of the rat gastric fundus. Pharmacol Res 64:397–409

Jensen HS, Callo K, Jespersen T, Jensen BS, Olesen SP (2005) The KCNQ5 potassium channel from mouse: a broadly expressed M-current like potassium channel modulated by zinc, pH, and volume changes. Brain Res Mol Brain Res 139:52–62

Jepps TA, Greenwood IA, Moffatt JD, Sanders KM, Ohya S (2009) Molecular and functional characterization of Kv7 K+ channel in murine gastrointestinal smooth muscles. Am J Physiol Gastrointest Liver Physiol 297:G107–G115

Jepps TA, Chadha PS, Davis AJ, Harhun MI, Cockerill GW, Olesen SP, Hansen RS, Greenwood IA (2011) Downregulation of Kv7.4 channel activity in primary and secondary hypertension. Circulation 124:602–611

Jespersen T, Grunnet M, Olesen SP (2005) The KCNQ1 potassium channel: from gene to physiological function. Physiology (Bethesda) 20:408–416

Kaczmarek LK, Blumenthal EM (1997) Properties and regulation of the minK potassium channel protein. Physiol Rev 77:627–641

Kanaumi T, Takashima S, Iwasaki H, Itoh M, Mitsudome A, Hirose S (2008) Developmental changes in KCNQ2 and KCNQ3 expression in human brain: possible contribution to the age-dependent etiology of benign familial neonatal convulsions. Brain and Development 30:362–369

Kanda VA, Abbott GW (2012) KCNE regulation of K(+) channel trafficking – a Sisyphean task? Front Physiol 3:231

Kang C, Tian C, Sonnichsen FD, Smith JA, Meiler J, George AL Jr, Vanoye CG, Kim HJ, Sanders CR (2008) Structure of KCNE1 and implications for how it modulates the KCNQ1 potassium channel. Biochemistry 47:7999–8006

Kang S, Li J, Zuo W, Fu R, Gregor D, Krnjevic K, Bekker A, Ye JH (2017) Ethanol withdrawal drives anxiety-related behaviors by reducing M-type potassium channel activity in the lateral Habenula. Neuropsychopharmacology 42:1813–1824

Kapfhamer D, Berger KH, Hopf FW, Seif T, Kharazia V, Bonci A, Heberlein U (2010) Protein phosphatase 2a and glycogen synthase kinase 3 signaling modulate prepulse inhibition of the acoustic startle response by altering cortical M-type potassium channel activity. J Neurosci 30:8830–8840

Khanamiri S, Soltysinska E, Jepps TA, Bentzen BH, Chadha PS, Schmitt N, Greenwood IA, Olesen SP (2013) Contribution of Kv7 channels to basal coronary flow and active response to ischemia. Hypertension 62:1090–1097

Kharkovets T, Hardelin JP, Safieddine S, Schweizer M, El-Amraoui A, Petit C, Jentsch TJ (2000) KCNQ4, a K+ channel mutated in a form of dominant deafness, is expressed in the inner ear and the central auditory pathway. Proc Natl Acad Sci U S A 97:4333–4338

Kharkovets T, Dedek K, Maier H, Schweizer M, Khimich D, Nouvian R, Vardanyan V, Leuwer R, Moser T, Jentsch TJ (2006) Mice with altered KCNQ4 K+ channels implicate sensory outer hair cells in human progressive deafness. EMBO J 25:642–652

Kim HJ, Jeong MH, Kim KR, Jung CY, Lee SY, Kim H, Koh J, Vuong TA, Jung S, Yang H, Park SK, Choi D, Kim SH, Kang K, Sohn JW, Park JM, Jeon D, Koo SH, Ho WK, Kang JS, Kim ST, Cho H (2016) Protein arginine methylation facilitates KCNQ channel-PIP2 interaction leading to seizure suppression. eLife 5

Kim EC, Patel J, Zhang J, Soh H, Rhodes JS, Tzingounis AV, Chung HJ (2020) Heterozygous loss of epilepsy gene KCNQ2 alters social, repetitive and exploratory behaviors. Genes Brain Behav 19:e12599

Klinger F, Gould G, Boehm S, Shapiro MS (2011) Distribution of M-channel subunits KCNQ2 and KCNQ3 in rat hippocampus. NeuroImage 58:761–769

Korsgaard MP, Hartz BP, Brown WD, Ahring PK, Strøbaek D, Mirza NR (2005) Anxiolytic effects of Maxipost (BMS-204352) and retigabine via activation of neuronal Kv7 channels. J Pharmacol Exp Ther 314:282–292

Kosenko A, Hoshi N (2013) A change in configuration of the calmodulin-KCNQ channel complex underlies Ca2+−dependent modulation of KCNQ channel activity. PLoS One 8:e82290

Kosenko A, Moftakhar S, Wood MA, Hoshi N (2020) In vivo attenuation of M-current suppression impairs consolidation of object recognition memory. J Neurosci 40:5847–5856

Krishnan V, Han MH, Graham DL, Berton O, Renthal W, Russo SJ, Laplant Q, Graham A, Lutter M, Lagace DC, Ghose S, Reister R, Tannous P, Green TA, Neve RL, Chakravarty S, Kumar A, Eisch AJ, Self DW, Lee FS, Tamminga CA, Cooper DC, Gershenfeld HK, Nestler EJ (2007) Molecular adaptations underlying susceptibility and resistance to social defeat in brain reward regions. Cell 131:391–404

Krumerman A, Gao X, Bian JS, Melman YF, Kagan A, McDonald TV (2004) An LQT mutant minK alters KvLQT1 trafficking. Am J Physiol Cell Physiol 286:C1453–C1463

Kubisch C, Schroeder BC, Friedrich T, Lutjohann B, El-Amraoui A, Marlin S, Petit C, Jentsch TJ (1999) KCNQ4, a novel potassium channel expressed in sensory outer hair cells, is mutated in dominant deafness. Cell 96:437–446

Kunzelmann K, Bleich M, Warth R, Levy-Holzman R, Garty H, Schreiber R (2001a) Expression and function of colonic epithelial KvLQT1 K+ channels. Clin Exp Pharmacol Physiol 28:79–83

Kunzelmann K, Hubner M, Schreiber R, Levy-Holzman R, Garty H, Bleich M, Warth R, Slavik M, von Hahn T, Greger R (2001b) Cloning and function of the rat colonic epithelial K+ channel KVLQT1. J Membr Biol 179:155–164

Kunzelmann K, Hübner M, Schreiber R, Levy-Holzman R, Garty H, Bleich M, Warth R, Slavik M, von Hahn T, Greger R (2001c) Cloning and function of the rat colonic epithelial K+ channel KVLQT1. J Membr Biol 179:155–164

Kurakami K, Norota I, Nasu F, Ohshima S, Nagasawa Y, Konno Y, Obara Y, Ishii K (2019) KCNQ1 is internalized by activation of alpha1 adrenergic receptors. Biochem Pharmacol 169:113628

Kurokawa J, Motoike HK, Rao J, Kass RS (2004) Regulatory actions of the A-kinase anchoring protein Yotiao on a heart potassium channel downstream of PKA phosphorylation. Proc Natl Acad Sci U S A 101:16374–16378

Lamas JA, Selyanko AA, Brown DA (1997) Effects of a cognition-enhancer, linopirdine (DuP 996), on M-type potassium currents (IK(M)) and some other voltage- and ligand-gated membrane currents in rat sympathetic neurons. Eur J Neurosci 9:605–616

Lambrecht NW, Yakubov I, Scott D, Sachs G (2005) Identification of the K efflux channel coupled to the gastric H-K-ATPase during acid secretion. Physiol Genomics 21:81–91

Langeberg LK, Scott JD (2005) A-kinase-anchoring proteins. J Cell Sci 118:3217–3220

Lee K, Isogai A, Antoh M, Kajioka S, Eto M, Hashitani H (2018) Role of K(+) channels in regulating spontaneous activity in the muscularis mucosae of Guinea pig bladder. Eur J Pharmacol 818:30–37

Lee JH, Chae MR, Kang SJ, Sung HH, Han DH, So I, Park JK, Lee SW (2020) Characterization and functional roles of KCNQ-encoded voltage-gated potassium (Kv7) channels in human corpus cavernosum smooth muscle. Pflugers Arch 472:89–102

Lehman A, Thouta S, Mancini GMS, Naidu S, van Slegtenhorst M, McWalter K, Person R, Mwenifumbo J, Salvarinova R, Guella I, McKenzie MB, Datta A, Connolly MB, Kalkhoran SM, Poburko D, Friedman JM, Farrer MJ, Demos M, Desai S, Claydon T (2017) Loss-of-function and gain-of-function mutations in KCNQ5 cause intellectual disability or epileptic encephalopathy. Am J Hum Genet 101:65–74

Leitner MG, Feuer A, Ebers O, Schreiber DN, Halaszovich CR, Oliver D (2012) Restoration of ion channel function in deafness-causing KCNQ4 mutants by synthetic channel openers. Br J Pharmacol 165:2244–2259

Lerche C, Scherer CR, Seebohm G, Derst C, Wei AD, Busch AE, Steinmeyer K (2000) Molecular cloning and functional expression of KCNQ5, a potassium channel subunit that may contribute to neuronal M-current diversity. J Biol Chem 275:22395–22400

Li Y, Gamper N, Shapiro MS (2004a) Single-channel analysis of KCNQ K+ channels reveals the mechanism of augmentation by a cysteine-modifying reagent. J Neurosci 24:5079–5090

Li Y, Langlais P, Gamper N, Liu F, Shapiro MS (2004b) Dual phosphorylations underlie modulation of unitary KCNQ K(+) channels by Src tyrosine kinase. J Biol Chem 279:45399–45407

Li Y, Gamper N, Hilgemann DW, Shapiro MS (2005) Regulation of Kv7 (KCNQ) K+ channel open probability by phosphatidylinositol (4,5)-bisphosphate. J Neurosci 25:9825–9835

Li C, Huang P, Lu Q, Zhou M, Guo L, Xu X (2014) KCNQ/Kv7 channel activator flupirtine protects against acute stress-induced impairments of spatial memory retrieval and hippocampal LTP in rats. Neuroscience 280:19–30

Li L, Sun H, Ding J, Niu C, Su M, Zhang L, Li Y, Wang C, Gamper N, Du X, Zhang H (2017) Selective targeting of M-type potassium Kv 7.4 channels demonstrates their key role in the regulation of dopaminergic neuronal excitability and depression-like behaviour. Br J Pharmacol 174:4277–4294

Li J, Maghera J, Lamothe SM, Marco EJ, Kurata HT (2020) Heteromeric assembly of truncated neuronal Kv7 channels: implications for neurologic disease and pharmacotherapy. Mol Pharmacol 98:192–202

Li X, Zhang Q, Guo P, Fu J, Mei L, Lv D, Wang J, Lai D, Ye S, Yang H, Guo J (2021) Molecular basis for ligand activation of the human KCNQ2 channel. Cell Res 31:52–61

Liao X, Yap MKH, Leung KH, Kao PYP, Liu LQ, Yip SP (2017) Genetic association study of KCNQ5 polymorphisms with high myopia. Biomed Res Int 2017:3024156

Linley JE, Rose K, Patil M, Robertson B, Akopian AN, Gamper N (2008) Inhibition of M current in sensory neurons by exogenous proteases: a signaling pathway mediating inflammatory nociception. J Neurosci 28:11240–11249

Liu W, Devaux JJ (2014) Calmodulin orchestrates the heteromeric assembly and the trafficking of KCNQ2/3 (Kv7.2/3) channels in neurons. Mol Cell Neurosci 58:40–52

Liu B, Linley JE, Du X, Zhang X, Ooi L, Zhang H, Gamper N (2010) The acute nociceptive signals induced by bradykinin in rat sensory neurons are mediated by inhibition of M-type K+ channels and activation of Ca2+−activated Cl- channels. J Clin Invest 120:1240–1252

Liu H, Jia L, Chen X, Shi L, Xie J (2018) The Kv7/KCNQ channel blocker XE991 protects nigral dopaminergic neurons in the 6-hydroxydopamine rat model of Parkinson's disease. Brain Res Bull 137:132–139

Lohrmann E, Burhoff I, Nitschke RB, Lang HJ, Mania D, Englert HC, Hropot M, Warth R, Rohm W, Bleich M et al (1995) A new class of inhibitors of cAMP-mediated Cl- secretion in rabbit colon, acting by the reduction of cAMP-activated K+ conductance. Pflugers Arch 429:517–530

Loussouarn G, Park KH, Bellocq C, Baro II, Charpentier F, Escande D (2003) Phosphatidylinositol-4,5-bisphosphate, PIP(2), controls KCNQ1/KCNE1 voltage-gated potassium channels: a functional homology between voltage-gated and inward rectifier K(+) channels. EMBO J 22:5412–5421

Loussouarn G, Baro I, Escande D (2006) KCNQ1 K+ channel-mediated cardiac channelopathies. Methods Mol Biol 337:167–183

Lubke M, Schreiber JA, Le Quoc T, Korber F, Muller J, Sivanathan S, Matschke V, Schubert J, Strutz-Seebohm N, Seebohm G, Scherkenbeck J (2020) Rottlerin: structure modifications and KCNQ1/KCNE1 ion channel activity. ChemMedChem 15:1078–1088

Lv P, Wei D, Yamoah EN (2010) Kv7-type channel currents in spiral ganglion neurons: involvement in sensorineural hearing loss. J Biol Chem 285:34699–34707

Mackie AR, Brueggemann LI, Henderson KK, Shiels AJ, Cribbs LL, Scrogin KE, Byron KL (2008) Vascular KCNQ potassium channels as novel targets for the control of mesenteric artery constriction by vasopressin, based on studies in single cells, pressurized arteries, and in vivo measurements of mesenteric vascular resistance. J Pharmacol Exp Ther 325:475–483

Maljevic S, Lerche C, Seebohm G, Alekov AK, Busch AE, Lerche H (2003) C-terminal interaction of KCNQ2 and KCNQ3 K+ channels. J Physiol 548:353–360

Mall M, Wissner A, Schreiber R, Kuehr J, Seydewitz HH, Brandis M, Greger R, Kunzelmann K (2000) Role of K(V)LQT1 in cyclic adenosine monophosphate-mediated Cl(-) secretion in human airway epithelia. Am J Respir Cell Mol Biol 23:283–289

Mani BK, Brueggemann LI, Cribbs LL, Byron KL (2011) Activation of vascular KCNQ (Kv7) potassium channels reverses spasmogen-induced constrictor responses in rat basilar artery. Br J Pharmacol 164:237–249

Mani BK, O'Dowd J, Kumar L, Brueggemann LI, Ross M, Byron KL (2013) Vascular KCNQ (Kv7) potassium channels as common signaling intermediates and therapeutic targets in cerebral vasospasm. J Cardiovasc Pharmacol 61:51–62

Mani BK, Robakowski C, Brueggemann LI, Cribbs LL, Tripathi A, Majetschak M, Byron KL (2016) Kv7.5 potassium channel subunits are the primary targets for PKA-dependent enhancement of vascular smooth muscle Kv7 currents. Mol Pharmacol 89:323–334

Marcotti W, Kros CJ (1999) Developmental expression of the potassium current IK,n contributes to maturation of mouse outer hair cells. J Physiol 520(Pt 3):653–660

Marx S (2003) Ion channel macromolecular complexes in the heart. J Mol Cell Cardiol 35:37–44

Marx SO, Kurokawa J, Reiken S, Motoike H, D'Armiento J, Marks AR, Kass RS (2002) Requirement of a macromolecular signaling complex for beta adrenergic receptor modulation of the KCNQ1-KCNE1 potassium channel. Science 295:496–499

Mayordomo-Cava J, Yajeya J, Navarro-López JD, Jiménez-Díaz L (2015) Amyloid-β(25-35) modulates the expression of GirK and KCNQ Channel genes in the Hippocampus. PLoS One 10:e0134385

McCallum LA, Greenwood IA, Tribe RM (2009) Expression and function of K(v)7 channels in murine myometrium throughout oestrous cycle. Pflugers Arch 457:1111–1120

McCrossan ZA, Abbott GW (2004) The MinK-related peptides. Neuropharmacology 47:787–821

McGuier NS, Griffin WC 3rd, Gass JT, Padula AE, Chesler EJ, Mulholland PJ (2016) Kv7 channels in the nucleus accumbens are altered by chronic drinking and are targets for reducing alcohol consumption. Addict Biol 21:1097–1112

Mehregan H, Mohseni M, Akbari M, Jalalvand K, Arzhangi S, Nikzat N, Kahrizi K, Najmabadi H (2019) Novel mutations in KCNQ4, LHFPL5 and COCH genes in Iranian families with hearing impairment. Arch Iran Med 22:189–197

Meisel E, Tobelaim W, Dvir M, Haitin Y, Peretz A, Attali B (2018) Inactivation gating of Kv7.1 channels does not involve concerted cooperative subunit interactions. Channels (Austin) 12:89–99

Mena A, Ruiz-Salas JC, Puentes A, Dorado I, Ruiz-Veguilla M, De la Casa LG (2016) Reduced prepulse inhibition as a biomarker of schizophrenia. Front Behav Neurosci 10:202

Mencia A, Gonzalez-Nieto D, Modamio-Hoybjor S, Etxeberria A, Aranguez G, Salvador N, Del Castillo I, Villarroel A, Moreno F, Barrio L, Moreno-Pelayo MA (2008) A novel KCNQ4 pore-region mutation (p.G296S) causes deafness by impairing cell-surface channel expression. Hum Genet 123:41–53

Metten P, Iancu OD, Spence SE, Walter NA, Oberbeck D, Harrington CA, Colville A, McWeeney S, Phillips TJ, Buck KJ, Crabbe JC, Belknap JK, Hitzemann RJ (2014) Dual-trait selection for ethanol consumption and withdrawal: genetic and transcriptional network effects. Alcohol Clin Exp Res 38:2915–2924

Miceli F, Soldovieri MV, Ambrosino P, De Maria M, Migliore M, Migliore R, Taglialatela M (2015) Early-onset epileptic encephalopathy caused by gain-of-function mutations in the voltage sensor of Kv7.2 and Kv7.3 potassium channel subunits. J Neurosci 35:3782–3793

Mondejar-Parreno G, Perez-Vizcaino F, Cogolludo A (2020) Kv7 channels in lung diseases. Front Physiol 11:634

Moore SD, Madamba SG, Siggins GR (1990) Ethanol diminishes a voltage-dependent K+ current, the M-current, in CA1 hippocampal pyramidal neurons in vitro. Brain Res 516:222–228

Moreno C, Oliveras A, de la Cruz A, Bartolucci C, Munoz C, Salar E, Gimeno JR, Severi S, Comes N, Felipe A, Gonzalez T, Lambiase P, Valenzuela C (2015) A new KCNQ1 mutation at the S5 segment that impairs its association with KCNE1 is responsible for short QT syndrome. Cardiovasc Res 107:613–623

Moreno C, Oliveras A, Bartolucci C, Munoz C, de la Cruz A, Peraza DA, Gimeno JR, Martin-Martinez M, Severi S, Felipe A, Lambiase PD, Gonzalez T, Valenzuela C (2017) D242N, a KV7.1 LQTS mutation uncovers a key residue for IKs voltage dependence. J Mol Cell Cardiol 110:61–69

Mucha M, Ooi L, Linley JE, Mordaka P, Dalle C, Robertson B, Gamper N, Wood IC (2010) Transcriptional control of KCNQ channel genes and the regulation of neuronal excitability. J Neurosci 30:13235–13245

Munro G, Erichsen HK, Mirza NR (2007) Pharmacological comparison of anticonvulsant drugs in animal models of persistent pain and anxiety. Neuropharmacology 53:609–618

Murray CI, Westhoff M, Eldstrom J, Thompson E, Emes R, Fedida D (2016) Unnatural amino acid photo-crosslinking of the IKs channel complex demonstrates a KCNE1:KCNQ1 stoichiometry of up to 4:4. eLife 5

Nakajo K, Kubo Y (2007) KCNE1 and KCNE3 stabilize and/or slow voltage sensing S4 segment of KCNQ1 channel. J Gen Physiol 130:269–281

Nakajo K, Kubo Y (2014) Steric hindrance between S4 and S5 of the KCNQ1/KCNE1 channel hampers pore opening. Nat Commun 5:4100

Nakajo K, Ulbrich MH, Kubo Y, Isacoff EY (2010) Stoichiometry of the KCNQ1 – KCNE1 ion channel complex. Proc Natl Acad Sci U S A 107:18862–18867

Neyroud N, Tesson F, Denjoy I, Leibovici M, Donger C, Barhanin J, Faure S, Gary F, Coumel P, Petit C, Schwartz K, Guicheney P (1997) A novel mutation in the potassium channel gene KVLQT1 causes the Jervell and Lange-Nielsen cardioauditory syndrome. Nat Genet 15:186–189

Ng FL, Davis AJ, Jepps TA, Harhun MI, Yeung SY, Wan A, Reddy M, Melville D, Nardi A, Khong TK, Greenwood IA (2011) Expression and function of the K+ channel KCNQ genes in human arteries. Br J Pharmacol 162:42–53

Nicolas M, Dememes D, Martin A, Kupershmidt S, Barhanin J (2001) KCNQ1/KCNE1 potassium channels in mammalian vestibular dark cells. Hear Res 153:132–145

Nicolas CS, Park KH, El Harchi A, Camonis J, Kass RS, Escande D, Merot J, Loussouarn G, Le Bouffant F, Baro I (2008) IKs response to protein kinase A-dependent KCNQ1 phosphorylation requires direct interaction with microtubules. Cardiovasc Res 79:427–435

Nouvian R, Ruel J, Wang J, Guitton MJ, Pujol R, Puel JL (2003) Degeneration of sensory outer hair cells following pharmacological blockade of cochlear KCNQ channels in the adult Guinea pig. Eur J Neurosci 17:2553–2562

Nunez E, Muguruza-Montero A, Villarroel A (2020) Atomistic insights of calmodulin gating of complete ion channels. Int J Mol Sci 21
Ohya S, Asakura K, Muraki K, Watanabe M, Imaizumi Y (2002) Molecular and functional characterization of ERG, KCNQ, and KCNE subtypes in rat stomach smooth muscle. Am J Physiol Gastrointest Liver Physiol 282:G277–G287
Ohya S, Sergeant GP, Greenwood IA, Horowitz B (2003) Molecular variants of KCNQ channels expressed in murine portal vein myocytes: a role in delayed rectifier current. Circ Res 92:1016–1023
Oliver D, Knipper M, Derst C, Fakler B (2003) Resting potential and submembrane calcium concentration of inner hair cells in the isolated mouse cochlea are set by KCNQ-type potassium channels. J Neurosci 23:2141–2149
Osteen JD, Gonzalez C, Sampson KJ, Iyer V, Rebolledo S, Larsson HP, Kass RS (2010) KCNE1 alters the voltage sensor movements necessary to open the KCNQ1 channel gate. Proc Natl Acad Sci U S A 107:22710–22715
Osteen JD, Barro-Soria R, Robey S, Sampson KJ, Kass RS, Larsson HP (2012) Allosteric gating mechanism underlies the flexible gating of KCNQ1 potassium channels. Proc Natl Acad Sci U S A 109:7103–7108
Owen DG, Marsh SJ, Brown DA (1990) M-current noise and putative M-channels in cultured rat sympathetic ganglion cells. J Physiol Lond 431:269–290
Pan Z, Kao T, Horvath Z, Lemos J, Sul JY, Cranstoun SD, Bennett V, Scherer SS, Cooper EC (2006) A common ankyrin-G-based mechanism retains KCNQ and NaV channels at electrically active domains of the axon. J Neurosci 26:2599–2613
Park KH, Piron J, Dahimene S, Merot J, Baro I, Escande D, Loussouarn G (2005) Impaired KCNQ1-KCNE1 and phosphatidylinositol-4,5-bisphosphate interaction underlies the long QT syndrome. Circ Res 96:730–739
Parrilla-Carrero J, Buchta WC, Goswamee P, Culver O, McKendrick G, Harlan B, Moutal A, Penrod R, Lauer A, Ramakrishnan V, Khanna R, Kalivas P, Riegel AC (2018) Restoration of Kv7 channel-mediated inhibition reduces cued-reinstatement of cocaine seeking. J Neurosci 38:4212–4229
Passmore GM, Selyanko AA, Mistry M, Al-Qatari M, Marsh SJ, Matthews EA, Dickenson AH, Brown TA, Burbidge SA, Main M, Brown DA (2003) KCNQ/M currents in sensory neurons: significance for pain therapy. J Neurosci 23:7227–7236
Peng H, Bian XL, Ma FC, Wang KW (2017) Pharmacological modulation of the voltage-gated neuronal Kv7/KCNQ/M-channel alters the intrinsic excitability and synaptic responses of pyramidal neurons in rat prefrontal cortex slices. Acta Pharmacol Sin 38:1248–1256
Perez-Flores MC, Lee JH, Park S, Zhang XD, Sihn CR, Ledford HA, Wang W, Kim HJ, Timofeyev V, Yarov-Yarovoy V, Chiamvimonvat N, Rabbitt RD, Yamoah EN (2020) Cooperativity of Kv7.4 channels confers ultrafast electromechanical sensitivity and emergent properties in cochlear outer hair cells. Sci Adv 6:eaba1104
Peroz D, Dahimene S, Baro I, Loussouarn G, Merot J (2009) LQT1-associated mutations increase KCNQ1 proteasomal degradation independently of Derlin-1. J Biol Chem 284:5250–5256
Peters HC, Hu H, Pongs O, Storm JF, Isbrandt D (2005) Conditional transgenic suppression of M channels in mouse brain reveals functions in neuronal excitability, resonance and behavior. Nat Neurosci 8:51–60
Pobbe RL, Zangrossi H Jr (2008) Involvement of the lateral habenula in the regulation of generalized anxiety- and panic-related defensive responses in rats. Life Sci 82:1256–1261
Pongs O, Schwarz JR (2010) Ancillary subunits associated with voltage-dependent K+ channels. Physiol Rev 90:755–796
Provence A, Malysz J, Petkov GV (2015) The novel KV7.2/KV7.3 channel opener ICA-069673 reveals subtype-specific functional roles in Guinea pig detrusor smooth muscle excitability and contractility. J Pharmacol Exp Ther 354:290–301

Provence A, Angoli D, Petkov GV (2018) K(V)7 channel pharmacological activation by the novel activator ML213: role for Heteromeric K(V)7.4/K(V)7.5 channels in Guinea pig detrusor smooth muscle function. J Pharmacol Exp Ther 364:131–144

Pusch M (1998) Increase of the single-channel conductance of KvLQT1 potassium channels induced by the association with minK. Pflugers Arch 437:172–174

Redrobe JP, Nielsen AN (2009) Effects of neuronal Kv7 potassium channel activators on hyperactivity in a rodent model of mania. Behav Brain Res 198:481–485

Regev N, Degani-Katzav N, Korngreen A, Etzioni A, Siloni S, Alaimo A, Chikvashvili D, Villarroel A, Attali B, Lotan I (2009) Selective interaction of syntaxin 1A with KCNQ2: possible implications for specific modulation of presynaptic activity. PLoS One 4:e6586

Reilly JM, Passmore GM, Marsh SJ, Brown DA (2013) Kv7/M-type potassium channels in rat skin keratinocytes. Pflugers Arch 465:1371–1381

Restier L, Cheng L, Sanguinetti MC (2008) Mechanisms by which atrial fibrillation-associated mutations in the S1 domain of KCNQ1 slow deactivation of IKs channels. J Physiol 586:4179–4191

Rinker JA, Mulholland PJ (2017) Promising pharmacogenetic targets for treating alcohol use disorder: evidence from preclinical models. Pharmacogenomics 18:555–570

Rinker JA, Fulmer DB, Trantham-Davidson H, Smith ML, Williams RW, Lopez MF, Randall PK, Chandler LJ, Miles MF, Becker HC, Mulholland PJ (2017) Differential potassium channel gene regulation in BXD mice reveals novel targets for pharmacogenetic therapies to reduce heavy alcohol drinking. Alcohol (Fayetteville, NY) 58:33–45

Rocheleau JM, Kobertz WR (2008) KCNE peptides differently affect voltage sensor equilibrium and equilibration rates in KCNQ1 K+ channels. J Gen Physiol 131:59–68

Roepke TK, Anantharam A, Kirchhoff P, Busque SM, Young JB, Geibel JP, Lerner DJ, Abbott GW (2006) The KCNE2 potassium channel ancillary subunit is essential for gastric acid secretion. J Biol Chem 281:23740–23747

Romey G, Attali B, Chouabe C, Abitbol I, Guillemare E, Barhanin J, Lazdunski M (1997) Molecular mechanism and functional significance of the MinK control of the KvLQT1 channel activity. J Biol Chem 272:16713–16716

Rose K, Ooi L, Dalle C, Robertson B, Wood IC, Gamper N (2011) Transcriptional repression of the M channel subunit Kv7.2 in chronic nerve injury. Pain 152:742–754

Roura-Ferrer M, Sole L, Martinez-Marmol R, Villalonga N, Felipe A (2008) Skeletal muscle Kv7 (KCNQ) channels in myoblast differentiation and proliferation. Biochem Biophys Res Commun 369:1094–1097

Ruscic KJ, Miceli F, Villalba-Galea CA, Dai H, Mishina Y, Bezanilla F, Goldstein SA (2013) IKs channels open slowly because KCNE1 accessory subunits slow the movement of S4 voltage sensors in KCNQ1 pore-forming subunits. Proc Natl Acad Sci U S A 110:E559–E566

Ryan A, Dallos P (1975) Effect of absence of cochlear outer hair cells on behavioural auditory threshold. Nature 253:44–46

Sachyani D, Dvir M, Strulovich R, Tria G, Tobelaim W, Peretz A, Pongs O, Svergun D, Attali B, Hirsch JA (2014) Structural basis of a Kv7.1 potassium channel gating module: studies of the intracellular c-terminal domain in complex with calmodulin. Structure 22:1582–1594

Sanguinetti MC, Curran ME, Zou A, Shen J, Spector PS, Atkinson DL, Keating MT (1996) Coassembly of K(V)LQT1 and minK (IsK) proteins to form cardiac I(Ks) potassium channel. Nature 384:80–83

Sankaranarayanan S, Simasko SM (1996) Characterization of an M-like current modulated by thyrotropin-releasing hormone in normal rat lactotrophs. J Neurosci 16:1668–1678

Schenzer A, Friedrich T, Pusch M, Saftig P, Jentsch TJ, Grotzinger J, Schwake M (2005) Molecular determinants of KCNQ (Kv7) K+ channel sensitivity to the anticonvulsant retigabine. J Neurosci 25:5051–5060

Schmitt N, Schwarz M, Peretz A, Abitbol I, Attali B, Pongs O (2000) A recessive C-terminal Jervell and Lange-Nielsen mutation of the KCNQ1 channel impairs subunit assembly. EMBO J 19:332–340

Schroeder BC, Hechenberger M, Weinreich F, Kubisch C, Jentsch TJ (2000a) KCNQ5, a novel potassium channel broadly expressed in brain, mediates M-type currents. J Biol Chem 275:24089–24095

Schroeder BC, Waldegger S, Fehr S, Bleich M, Warth R, Greger R, Jentsch TJ (2000b) A constitutively open potassium channel formed by KCNQ1 and KCNE3. Nature 403:196–199

Schulze-Bahr E, Wang Q, Wedekind H, Haverkamp W, Chen Q, Sun Y, Rubie C, Hordt M, Towbin JA, Borggrefe M, Assmann G, Qu X, Somberg JC, Breithardt G, Oberti C, Funke H (1997) KCNE1 mutations cause jervell and Lange-Nielsen syndrome. Nat Genet 17:267–268

Schwab SG, Knapp M, Sklar P, Eckstein GN, Sewekow C, Borrmann-Hassenbach M, Albus M, Becker T, Hallmayer JF, Lerer B, Maier W, Wildenauer DB (2006) Evidence for association of DNA sequence variants in the phosphatidylinositol-4-phosphate 5-kinase IIalpha gene (PIP5K2A) with schizophrenia. Mol Psychiatry 11:837–846

Schwake M, Jentsch TJ, Friedrich T (2003) A carboxy-terminal domain determines the subunit specificity of KCNQ K(+) channel assembly. EMBO Rep 4:76–81

Schwake M, Athanasiadu D, Beimgraben C, Blanz J, Beck C, Jentsch TJ, Saftig P, Friedrich T (2006) Structural determinants of M-type KCNQ (Kv7) K+ channel assembly. J Neurosci 26:3757–3766

Seebohm G, Scherer CR, Busch AE, Lerche C (2001) Identification of specific pore residues mediating KCNQ1 inactivation. A novel mechanism for long QT syndrome. J Biol Chem 276:13600–13605

Seebohm G, Strutz-Seebohm N, Birkin R, Dell G, Bucci C, Spinosa MR, Baltaev R, Mack AF, Korniychuk G, Choudhury A, Marks D, Pagano RE, Attali B, Pfeufer A, Kass RS, Sanguinetti MC, Tavare JM, Lang F (2007) Regulation of endocytic recycling of KCNQ1/KCNE1 potassium channels. Circ Res 100:686–692

Seebohm G, Strutz-Seebohm N, Ureche ON, Henrion U, Baltaev R, Mack AF, Korniychuk G, Steinke K, Tapken D, Pfeufer A, Kaab S, Bucci C, Attali B, Merot J, Tavare JM, Hoppe UC, Sanguinetti MC, Lang F (2008) Long QT syndrome-associated mutations in KCNQ1 and KCNE1 subunits disrupt normal endosomal recycling of IKs channels. Circ Res 103:1451–1457

Seefeld MA, Lin H, Holenz J, Downie D, Donovan B, Fu T, Pasikanti K, Zhen W, Cato M, Chaudhary KW, Brady P, Bakshi T, Morrow D, Rajagopal S, Samanta SK, Madhyastha N, Kuppusamy BM, Dougherty RW, Bhamidipati R, Mohd Z, Higgins GA, Chapman M, Rouget C, Lluel P, Matsuoka Y (2018) Novel K(V)7 ion channel openers for the treatment of epilepsy and implications for detrusor tissue contraction. Bioorg Med Chem Lett 28:3793–3797

Selyanko AA, Brown DA (1996) Intracellular calcium directly inhibits potassium M channels in excised membrane patches from rat sympathetic neurons. Neuron 16:151–162

Selyanko AA, Brown DA (1999) M-channel gating and simulation. Biophys J 77:701–713

Selyanko AA, Hadley JK, Wood IC, Abogadie FC, Jentsch TJ, Brown DA (2000) Inhibition of KCNQ1-4 potassium channels expressed in mammalian cells via M1 muscarinic acetylcholine receptors. J Physiol (Lond) 522(Pt 3):349–355

Selyanko AA, Hadley JK, Brown DA (2001) Properties of single M-type KCNQ2/KCNQ3 potassium channels expressed in mammalian cells. J Physiol 534:15–24

Sesti F, Goldstein SA (1998) Single-channel characteristics of wild-type IKs channels and channels formed with two minK mutants that cause long QT syndrome. J Gen Physiol 112:651–663

Shah A, Zuo W, Kang S, Li J, Fu R, Zhang H, Bekker A, Ye JH (2017) The lateral habenula and alcohol: role of glutamate and M-type potassium channels. Pharmacol Biochem Behav 162:94–102

Shalaby FY, Levesque PC, Yang WP, Little WA, Conder ML, Jenkins-West T, Blanar MA (1997) Dominant-negative KvLQT1 mutations underlie the LQT1 form of long QT syndrome. Circulation 96:1733–1736

Shamgar L, Ma L, Schmitt N, Haitin Y, Peretz A, Wiener R, Hirsch J, Pongs O, Attali B (2006) Calmodulin is essential for cardiac IKS channel gating and assembly. Impaired function in long-QT mutations. Circ Res 98:1055–1063

Shapiro MS, Wollmuth LP, Hille B (1994) Angiotensin II inhibits calcium and M current channels in rat sympathetic neurons via G proteins. Neuron 12:1319–1329

Shapiro MS, Roche JP, Kaftan EJ, Cruzblanca H, Mackie K, Hille B (2000) Reconstitution of muscarinic modulation of the KCNQ2/KCNQ3 K(+) channels that underlie the neuronal M current. J Neurosci 20:1710–1721

Sihn CR, Kim HJ, Woltz RL, Yarov-Yarovoy V, Yang PC, Xu J, Clancy CE, Zhang XD, Chiamvimonvat N, Yamoah EN (2016) Mechanisms of calmodulin regulation of different isoforms of Kv7.4 K+ channels. J Biol Chem 291:2499–2509

Singh NA, Charlier C, Stauffer D, DuPont BR, Leach RJ, Melis R, Ronen GM, Bjerre I, Quattlebaum T, Murphy JV, McHarg ML, Gagnon D, Rosales TO, Peiffer A, Anderson VE, Leppert M (1998) A novel potassium channel gene, KCNQ2, is mutated in an inherited epilepsy of newborns. Nat Genet 18:25–29

Slomko AM, Naseer Z, Ali SS, Wongvravit JP, Friedman LK (2014) Retigabine calms seizure-induced behavior following status epilepticus. Epilepsy Behav E&B 37:123–132

Sogaard R, Ljungstrom T, Pedersen KA, Olesen SP, Jensen BS (2001) KCNQ4 channels expressed in mammalian cells: functional characteristics and pharmacology. Am J Physiol Cell Physiol 280:C859–C866

Soldovieri MV, Miceli F, Taglialatela M (2011) Driving with no brakes: molecular pathophysiology of Kv7 potassium channels. Physiology (Bethesda) 26:365–376

Sousa AD, Andrade LR, Salles FT, Pillai AM, Buttermore ED, Bhat MA, Kachar B (2009) The septate junction protein caspr is required for structural support and retention of KCNQ4 at calyceal synapses of vestibular hair cells. J Neurosci 29:3103–3108

Spitzmaul G, Tolosa L, Winkelman BH, Heidenreich M, Frens MA, Chabbert C, de Zeeuw CI, Jentsch TJ (2013) Vestibular role of KCNQ4 and KCNQ5 K+ channels revealed by mouse models. J Biol Chem 288:9334–9344

Stansfeld CE, Marsh SJ, Gibb AJ, Brown DA (1993) Identification of M-channels in outside-out patches excised from sympathetic ganglion cells. Neuron 10:639–654

Stott JB, Jepps TA, Greenwood IA (2014) K(V)7 potassium channels: a new therapeutic target in smooth muscle disorders. Drug Discov Today 19:413–424

Stott JB, Barrese V, Jepps TA, Leighton EV, Greenwood IA (2015) Contribution of Kv7 channels to natriuretic peptide mediated vasodilation in normal and hypertensive rats. Hypertension 65:676–682

Strutz-Seebohm N, Seebohm G, Fedorenko O, Baltaev R, Engel J, Knirsch M, Lang F (2006) Functional coassembly of KCNQ4 with KCNE-beta- subunits in Xenopus oocytes. Cell Physiol Biochem 18:57–66

Strutz-Seebohm N, Pusch M, Wolf S, Stoll R, Tapken D, Gerwert K, Attali B, Seebohm G (2011) Structural basis of slow activation gating in the cardiac I Ks channel complex. Cell Physiol Biochem 27:443–452

Su CC, Yang JJ, Shieh JC, Su MC, Li SY (2007) Identification of novel mutations in the KCNQ4 gene of patients with nonsyndromic deafness from Taiwan. Audiol Neurootol 12:20–26

Su TR, Zei WS, Su CC, Hsiao G, Lin MJ (2012) The effects of the KCNQ openers retigabine and flupirtine on myotonia in mammalian skeletal muscle induced by a chloride channel blocker. Evid Based Complement Alternat Med 2012:803082

Su M, Li L, Wang J, Sun H, Zhang L, Zhao C, Xie Y, Gamper N, Du X, Zhang H (2019) Kv7.4 channel contribute to projection-specific auto-inhibition of dopamine neurons in the ventral tegmental area. Front Cell Neurosci 13:557

Suessbrich H, Bleich M, Ecke D, Rizzo M, Waldegger S, Lang F, Szabo I, Lang HJ, Kunzelmann K, Greger R, Busch AE (1996) Specific blockade of slowly activating I (sK) channels by chromanols -- impact on the role of I(sK) channels in epithelia. FEBS Lett 396:271–275

Suh B, Hille B (2002) Recovery from muscarinic modulation of M current channels requires phosphatidylinositol 4,5-bisphosphate synthesis. Neuron 35:507–520

Sun J, MacKinnon R (2017) Cryo-EM structure of a KCNQ1/CaM complex reveals insights into congenital long QT syndrome. Cell 169:1042–1050.e9

Sun J, MacKinnon R (2020) Structural basis of human KCNQ1 modulation and gating. Cell 180:340–347.e9

Sun H, Lin AH, Ru F, Patil MJ, Meeker S, Lee LY, Undem BJ (2019) KCNQ/M-channels regulate mouse vagal bronchopulmonary C-fiber excitability and cough sensitivity. JCI Insight 4

Svalø J, Bille M, Parameswaran Theepakaran N, Sheykhzade M, Nordling J, Bouchelouche P (2013) Bladder contractility is modulated by Kv7 channels in pig detrusor. Eur J Pharmacol 715:312–320

Svalø J, Sheykhzade M, Nordling J, Matras C, Bouchelouche P (2015) Functional and molecular evidence for Kv7 channel subtypes in human detrusor from patients with and without bladder outflow obstruction. PLoS One 10:e0117350

Takahira M, Hughes BA (1997) Isolated bovine retinal pigment epithelial cells express delayed rectifier type and M-type K+ currents. Am J Phys 273:C790–C803

Talebizadeh Z, Kelley PM, Askew JW, Beisel KW, Smith SD (1999) Novel mutation in the KCNQ4 gene in a large kindred with dominant progressive hearing loss. Hum Mutat 14:493–501

Tatulian L, Brown DA (2003) Effect of the KCNQ potassium channel opener retigabine on single KCNQ2/3 channels expressed in CHO cells. J Physiol 549:57–63

Tatulian L, Delmas P, Abogadie FC, Brown DA (2001) Activation of expressed KCNQ potassium currents and native neuronal M- type potassium currents by the anti-convulsant drug retigabine. J Neurosci 21:5535–5545

Telezhkin V, Brown DA, Gibb AJ (2012) Distinct subunit contributions to the activation of M-type potassium channels by PI(4,5)P2. J Gen Physiol 140:41–53

Tian C, Vanoye CG, Kang C, Welch RC, Kim HJ, George AL Jr, Sanders CR (2007) Preparation, functional characterization, and NMR studies of human KCNE1, a voltage-gated potassium channel accessory subunit associated with deafness and long QT syndrome. Biochemistry 46:11459–11472

Tinel N, Diochot S, Borsotto M, Lazdunski M, Barhanin J (2000a) KCNE2 confers background current characteristics to the cardiac KCNQ1 potassium channel. EMBO J 19:6326–6330

Tinel N, Diochot S, Lauritzen I, Barhanin J, Lazdunski M, Borsotto M (2000b) M-type KCNQ2-KCNQ3 potassium channels are modulated by the KCNE2 subunit. FEBS Lett 480:137–141

Tobelaim WS, Dvir M, Lebel G, Cui M, Buki T, Peretz A, Marom M, Haitin Y, Logothetis DE, Hirsch JA, Attali B (2017) Competition of calcified calmodulin N lobe and PIP2 to an LQT mutation site in Kv7.1 channel. Proc Natl Acad Sci U S A 114:E869–E878

Tristani-Firouzi M, Sanguinetti MC (1998) Voltage-dependent inactivation of the human K+ channel KvLQT1 is eliminated by association with minimal K+ channel (minK) subunits. J Physiol Lond 510:37–45

Tsai YM, Jones F, Mullen P, Porter KE, Steely D, Peers C, Gamper N (2020) Vascular Kv7 channels control intracellular Ca2+ dynamics in smooth muscle. Cell Calcium 92:102283

Tunquist BJ, Hoshi N, Guire ES, Zhang F, Mullendorff K, Langeberg LK, Raber J, Scott JD (2008) Loss of AKAP150 perturbs distinct neuronal processes in mice. Proc Natl Acad Sci U S A 105:12557–12562

Tykocki NR, Heppner TJ, Dalsgaard T, Bonev AD, Nelson MT (2019) The K(V) 7 channel activator retigabine suppresses mouse urinary bladder afferent nerve activity without affecting detrusor smooth muscle K(+) channel currents. J Physiol 597:935–950

Tzingounis AV, Heidenreich M, Kharkovets T, Spitzmaul G, Jensen HS, Nicoll RA, Jentsch TJ (2010) The KCNQ5 potassium channel mediates a component of the afterhyperpolarization current in mouse hippocampus. Proc Natl Acad Sci U S A 107:10232–10237

Uchida H, Ma L, Ueda H (2010) Epigenetic gene silencing underlies C-fiber dysfunctions in neuropathic pain. J Neurosci 30:4806–4814

Vallon V, Grahammer F, Richter K, Bleich M, Lang F, Barhanin J, Volkl H, Warth R (2001) Role of KCNE1-dependent K+ fluxes in mouse proximal tubule. J Am Soc Nephrol 12:2003–2011

Van Camp G, Coucke PJ, Akita J, Fransen E, Abe S, De Leenheer EM, Huygen PL, Cremers CW, Usami S (2002) A mutational hot spot in the KCNQ4 gene responsible for autosomal dominant hearing impairment. Hum Mutat 20:15–19

van der Horst J, Greenwood IA, Jepps TA (2020) Cyclic AMP-dependent regulation of Kv7 voltage-gated potassium channels. Front Physiol 11:727

Van Hauwe P, Coucke PJ, Ensink RJ, Huygen P, Cremers CW, Van Camp G (2000) Mutations in the KCNQ4 K+ channel gene, responsible for autosomal dominant hearing loss, cluster in the channel pore region. Am J Med Genet 93:184–187

Vigil FA, Carver CM, Shapiro MS (2020) Pharmacological manipulation of K (v) 7 channels as a new therapeutic tool for multiple brain disorders. Front Physiol 11:688

Villalba-Galea CA (2020) Modulation of KV7 channel deactivation by PI(4,5)P2. Front Pharmacol 11:895

Vohra J (2007) The long QT syndrome. Heart Lung Circ 16(Suppl 3):S5–S12

Waldegger S (2003) Heartburn: cardiac potassium channels involved in parietal cell acid secretion. Pflugers Arch 446:143–147

Wallace DJ, Chen C, Marley PD (2002) Histamine promotes excitability in bovine adrenal chromaffin cells by inhibiting an M-current. J Physiol 540:921–939

Walsh KB, Kass RS (1988) Regulation of a heart potassium channel by protein kinase A and C. Science 242:67–69

Wang KW, Goldstein SA (1995) Subunit composition of minK potassium channels. Neuron 14:1303–1309

Wang Q, Curran ME, Splawski I, Burn TC, Millholland JM, VanRaay TJ, Shen J, Timothy KW, Vincent GM, de Jager T, Schwartz PJ, Toubin JA, Moss AJ, Atkinson DL, Landes GM, Connors TD, Keating MT (1996) Positional cloning of a novel potassium channel gene: KVLQT1 mutations cause cardiac arrhythmias. Nat Genet 12:17–23

Wang HS, Pan Z, Shi W, Brown BS, Wymore RS, Cohen IS, Dixon JE, McKinnon D (1998) KCNQ2 and KCNQ3 potassium channel subunits: molecular correlates of the M-channel. Science 282:1890–1893

Wang HS, Brown BS, McKinnon D, Cohen IS (2000) Molecular basis for differential sensitivity of KCNQ and I(Ks) channels to the cognitive enhancer XE991. Mol Pharmacol 57:1218–1223

Wang Z, Li H, Moss AJ, Robinson J, Zareba W, Knilans T, Bowles NE, Towbin JA (2002) Compound heterozygous mutations in KvLQT1 cause Jervell and Lange-Nielsen syndrome. Mol Genet Metab 75:308–316

Wang M, Gamo NJ, Yang Y, Jin LE, Wang XJ, Laubach M, Mazer JA, Lee D, Arnsten AF (2011) Neuronal basis of age-related working memory decline. Nature 476:210–213

Wang J, Yu W, Gao Q, Ju C, Wang K (2020a) Prefrontal inhibition of neuronal Kv7 channels enhances prepulse inhibition of acoustic startle reflex and resilience to hypofrontality. Br J Pharmacol 177:4720–4733

Wang Y, Eldstrom J, Fedida D (2020b) Gating and regulation of KCNQ1 and KCNQ1 + KCNE1 channel complexes. Front Physiol 11:504

Wangemann P (2002) K+ cycling and the endocochlear potential. Hear Res 165:1–9

Warth R, Riedemann N, Bleich M, Van Driessche W, Busch AE, Greger R (1996) The cAMP-regulated and 293B-inhibited K+ conductance of rat colonic crypt base cells. Pflugers Arch 432:81–88

Weckhuysen S, Ivanovic V, Hendrickx R, van Coster R, Hjalgrim H, Moller RS, Gronborg S, Schoonjans AS, Ceulemans B, Heavin SB, Eltze C, Horvath R, Casara G, Pisano T, Giordano L, Rostasy K, Haberlandt E, Albrecht B, Bevot A, Benkel I, Syrbe S, Sheidley B, Guerrini R, Poduri A, Lemke JR, Mandelstam S, Scheffer I, Angriman M, Striano P, Marini C, Suls A, De Jonghe P, Group KS (2013) Extending the KCNQ2 encephalopathy spectrum: clinical and neuroimaging findings in 17 patients. Neurology 81:1697–1703

Wei A, Jegla T, Salkoff L (1996) Eight potassium channel families revealed by the C. elegans genome project. Neuropharmacology 35:805–829

Wei J, Fish FA, Myerburg RJ, Roden DM, George AL Jr (2000) Novel KCNQ1 mutations associated with recessive and dominant congenital long QT syndromes: evidence for variable hearing phenotype associated with R518X. Hum Mutat 15:387–388

Wen H, Levitan IB (2002) Calmodulin is an auxiliary subunit of KCNQ2/3 potassium channels. J Neurosci 22:7991–8001

Werry D, Eldstrom J, Wang Z, Fedida D (2013) Single-channel basis for the slow activation of the repolarizing cardiac potassium current, IKs. Proc Natl Acad Sci U S A 110:E996–E1005

Westhoff M, Murray CI, Eldstrom J, Fedida D (2017) Photo-cross-linking of IKs demonstrates state-dependent interactions between KCNE1 and KCNQ1. Biophys J 113:415–425

Westhoff M, Eldstrom J, Murray CI, Thompson E, Fedida D (2019) I Ks ion-channel pore conductance can result from individual voltage sensor movements. Proc Natl Acad Sci U S A 116:7879–7888

Wickenden AD, Zou A, Wagoner PK, Jegla T (2001) Characterization of KCNQ5/Q3 potassium channels expressed in mammalian cells. Br J Pharmacol 132:381–384

Wladyka CL, Feng B, Glazebrook PA, Schild JH, Kunze DL (2008) The KCNQ/M-current modulates arterial baroreceptor function at the sensory terminal in rats. J Physiol 586:795–802

Wollmuth LP (1994) Mechanism of Ba2+ block of M-like K channels of rod photoreceptors of tiger salamanders. J Gen Physiol 103:45–66

Wrobel E, Tapken D, Seebohm G (2012) The KCNE tango – how KCNE1 interacts with Kv7.1. Front Pharmacol 3:142

Wuttke TV, Seebohm G, Bail S, Maljevic S, Lerche H (2005) The new anticonvulsant retigabine favors voltage-dependent opening of the Kv7.2 (KCNQ2) channel by binding to its activation gate. Mol Pharmacol 67:1009–1017

Xia X, Zhang Q, Jia Y, Shu Y, Yang J, Yang H, Yan Z (2020) Molecular basis and restoration of function deficiencies of Kv7.4 variants associated with inherited hearing loss. Hear Res 388:107884

Xu Q, Minor DL Jr (2009) Crystal structure of a trimeric form of the K(V)7.1 (KCNQ1) A-domain tail coiled-coil reveals structural plasticity and context dependent changes in a putative coiled-coil trimerization motif. Protein Sci 18:2100–2114

Xu T, Nie L, Zhang Y, Mo J, Feng W, Wei D, Petrov E, Calisto LE, Kachar B, Beisel KW, Vazquez AE, Yamoah EN (2007) Roles of alternative splicing in the functional properties of inner ear-specific KCNQ4 channels. J Biol Chem 282:23899–23909

Xu X, Kanda VA, Choi E, Panaghie G, Roepke TK, Gaeta SA, Christini DJ, Lerner DJ, Abbott GW (2009) MinK-dependent internalization of the IKs potassium channel. Cardiovasc Res 82:430–438

Xu Q, Chang A, Tolia A, Minor DL Jr (2013) Structure of a Ca(2+)/CaM:Kv7.4 (KCNQ4) B-helix complex provides insight into M current modulation. J Mol Biol 425:378–394

Yadav G, Jain G, Singh M (2017) Role of flupirtine in reducing preoperative anxiety of patients undergoing craniotomy procedure. Saudi J Anaesth 11:158–162

Yang Y, Sigworth FJ (1998) Single-channel properties of IKs potassium channels. J Gen Physiol 112:665–678

Yang WP, Levesque PC, Little WA, Conder ML, Shalaby FY, Blanar MA (1997) KvLQT1, a voltage-gated potassium channel responsible for human cardiac arrhythmias. Proc Natl Acad Sci U S A 94:4017–4021

Yang WP, Levesque PC, Little WA, Conder ML, Ramakrishnan P, Neubauer MG, Blanar MA (1998) Functional expression of two KvLQT1-related potassium channels responsible for an inherited idiopathic epilepsy. J Biol Chem 273:19419–19423

Young MB, Thomas SA (2014) M1-muscarinic receptors promote fear memory consolidation via phospholipase C and the M-current. J Neurosci 34:1570–1578

Yus-Najera E, Santana-Castro I, Villarroel A (2002) The identification and characterization of a non-continuous calmodulin binding site in non-inactivating voltage-dependent KCNQ potassium channels. J Biol Chem 24:24

Zaczek R, Chorvat RJ, Saye JA, Pierdomenico ME, Maciag CM, Logue AR, Fisher BN, Rominger DH, Earl RA (1998) Two new potent neurotransmitter release enhancers, 10,10-bis(4-pyridinylmethyl)-9(10H)-anthracenone and 10,10-bis(2-fluoro-4- pyridinylmethyl)-9(10H)-anthracenone: comparison to linopirdine. J Pharmacol Exp Ther 285:724–730

Zaika O, Lara LS, Gamper N, Hilgemann DW, Jaffe DB, Shapiro MS (2006) Angiotensin II regulates neuronal excitability via phosphatidylinositol 4,5-bisphosphate-dependent modulation of Kv7 (M-type) K+ channels. J Physiol 575:49–67

Zaika O, Tolstykh GP, Jaffe DB, Shapiro MS (2007) Inositol triphosphate-mediated Ca2+ signals direct purinergic P2Y receptor regulation of neuronal ion channels. J Neurosci 27:8914–8926

Zaydman MA, Cui J (2014) PIP2 regulation of KCNQ channels: biophysical and molecular mechanisms for lipid modulation of voltage-dependent gating. Front Physiol 5:195

Zaydman MA, Silva JR, Delaloye K, Li Y, Liang H, Larsson HP, Shi J, Cui J (2013) Kv7.1 ion channels require a lipid to couple voltage sensing to pore opening. Proc Natl Acad Sci U S A 110:13180–13185

Zaydman MA, Kasimova MA, McFarland K, Beller Z, Hou P, Kinser HE, Liang H, Zhang G, Shi J, Tarek M, Cui J (2014) Domain-domain interactions determine the gating, permeation, pharmacology, and subunit modulation of the IKs ion channel. eLife 3:e03606

Zhang J, Shapiro MS (2012) Activity-dependent transcriptional regulation of M-type (Kv7) K(+) channels by AKAP79/150-mediated NFAT actions. Neuron 76:1133–1146

Zhang J, Shapiro MS (2016) Mechanisms and dynamics of AKAP79/150-orchestrated multi-protein signalling complexes in brain and peripheral nerve. J Physiol 594:31–37

Zhang H, Craciun LC, Mirshahi T, Rohacs T, Lopes CM, Jin T, Logothetis DE (2003) PIP_2 activates KCNQ channels, and its hydrolysis underlies receptor-mediated inhibition of M currents. Neuron 37:963–975

Zhang J, Bal M, Bierbower S, Zaika O, Shapiro MS (2011) AKAP79/150 signal complexes in G-protein modulation of neuronal ion channels. J Neurosci 31:7199–7211

Zhang J, Chen SR, Chen H, Pan HL (2018) RE1-silencing transcription factor controls the acute-to-chronic neuropathic pain transition and Chrm2 receptor gene expression in primary sensory neurons. J Biol Chem 293:19078–19091

Zhang F, Gigout S, Liu Y, Wang Y, Hao H, Buckley NJ, Zhang H, Wood IC, Gamper N (2019) Repressor element 1-silencing transcription factor drives the development of chronic pain states. Pain 160:2398–2408

Zhang D, Men H, Zhang L, Gao X, Wang J, Li L, Zhu Q, Zhang H, Jia Z (2020) Inhibition of M/K (v)7 currents contributes to chloroquine-induced itch in mice. Front Mol Neurosci 13:105

Zhao C, Su M, Wang Y, Li X, Zhang Y, Du X, Zhang H (2017) Selective modulation of K(+) channel Kv7.4 significantly affects the excitability of DRN 5-HT neurons. Front Cell Neurosci 11:405

Zheng Q, Fang D, Liu M, Cai J, Wan Y, Han JS, Xing GG (2013) Suppression of KCNQ/M (Kv7) potassium channels in dorsal root ganglion neurons contributes to the development of bone cancer pain in a rat model. Pain 154:434–448

Zwierzyńska E, Krupa-Burtnik A, Pietrzak B (2017) The possibility of adverse effect of Kv7-channel opener retigabine on memory processes in rats. Epilepsy Behav E&B 75:170–175

Pharmacological Activation of Neuronal Voltage-Gated Kv7/KCNQ/M-Channels for Potential Therapy of Epilepsy and Pain

Yani Liu, Xiling Bian, and KeWei Wang

Contents

Abstract

Native M-current is a low-threshold, slowly activating potassium current that exerts an inhibitory control over neuronal excitability. The M-channel is primarily co-assembled by heterotetrameric Kv7.2/KCNQ2 and Kv7.3/KCNQ3 subunits that are specifically expressed in the brain and peripheral nociceptive

Y. Liu
Department of Pharmacology, Qingdao University School of Pharmacy, Qingdao, China

X. Bian
Department of Pharmacology, Peking University School of Pharmaceutical Sciences, Beijing, China

K. Wang (✉)
Department of Pharmacology, Qingdao University School of Pharmacy, Qingdao, China

Institute of Innovative Drugs Qingdao University, Qingdao, China
e-mail: wangkw@qdu.edu.cn

N. Gamper, K. Wang (eds.), *Pharmacology of Potassium Channels*,
Handbook of Experimental Pharmacology 267, https://doi.org/10.1007/164_2021_458

and visceral sensory neurons in the spinal cord. Reduction of M-channel function leads to neuronal hyperexcitability that defines the fundamental mechanism of neurological disorders such as epilepsy and pain, indicating that pharmacological activation of Kv7/KCNQ/M-channels may serve the basis for the therapy. The well-known KCNQ opener retigabine (ezogabine or Potiga) was approved by FDA in 2011 as an anticonvulsant used for an adjunctive treatment of partial epilepsies. Unfortunately, retigabine was discontinued in 2017 due to its side effects of blue-colored appearance of the skin and eyes after prolonged intake. In addition, flupirtine, a structural derivative of retigabine and a centrally acting non-opioid analgesic, was also withdrawn in 2018 for liver toxicity. Fortunately, these side effects are compound-structures related and can be avoided. Thus, further identification and development of novel potent and selective Kv7 channel openers may lead to an effective therapy with improved safety window for anti-epilepsy and anti-nociception.

Keywords

Epilepsy · Flupirtine · KCNQ2 · KCNQ3 · KCNQ5 · Pain · Retigabine · SCR2682

1 Introduction

The human genome encodes a superfamily of voltage-gated K^+ channels with 40 members that are grouped into 12 subfamilies, Kv1-12. Among the voltage-gated K^+ channel superfamily, Kv7 (or KCNQ) subfamily consists of five members: Kv7.1-Kv7.5 or KCNQ1-KCNQ5. The cardiac Kv7.1/KCNQ1 (or KvLQT1) is expressed in the heart and is responsible for slow phase repolarization of action potential. Mutations of Kv7.1/KCNQ1 cause the autosomal dominant long-QT syndrome of cardiac arrhythmia as well as the recessive Jervell and Lange-Nielsen cardio-auditory syndrome (Neyroud et al. 1997). Neuronal Kv7.2/KCNQ2 and Kv7.3/KCNQ3 are abundantly expressed in the brain and their mutations cause benign familial neonatal convulsions (BFNC), a form of neonatal epilepsy (Biervert et al. 1998; Charlier et al. 1998; Singh et al. 1998). The heterotetrameric KCNQ2 and KCNQ3 channels resemble the native M-current ($I_{k(M)}$) that was first described in the peripheral sympathetic neurons and named as M-current because of its inhibition by muscarine, acetylcholine receptor (mAChR) agonist (Brown and Adams 1980; Wang et al. 1998). Kv7/KCNQ/M current is a low-threshold, slowly activating K^+ current that exerts an inhibitory control over neuronal excitability. KCNQ4 channel is predominately the sensory outer hair cells of the cochlea and its loss-of-function mutations cause autosomal dominant deafness primarily by affecting endolymph secretion (Kubisch et al. 1999). KCNQ5 is broadly expressed in the brain and skeletal muscles, and can co-assemble with KCNQ3 to form heteromeric channels, also yielding M-type current (Lerche et al. 2000; Schroeder et al. 2000) (For more detailed descriptions of Kv7/KCNQ subfamily, see chapter “Kv7 Channels and Excitability Disorders”).

2 Distributions of Neuronal Kv7.2/7.3 Channels in the Central Nervous System (CNS) and the Peripheral Nervous System (PNS)

Human *KCNQ2* gene, located in the long arm (q) chromosome 20 at position13.3 (22q13.3), encodes a major KCNQ2 protein isoform (KCNQ2 a) with 872 amino acids (aa), and other KCNQ2 isoforms (KCNQ2 b-e) consisting of 854 aa, 844 aa, 841 aa, and 393 aa, respectively (Biervert et al. 1998). Northern blot analysis indicates that KCNQ2 mRNA is specifically expressed in the brain. In the brain KCNQ2 mRNA is detected in the cerebellum, cerebral cortex, hippocampus, amygdala, and caudate nucleus (Biervert et al. 1998). This distribution pattern of KCNQ2 mRNA overlaps KCNQ3 mRNA expression, which is confirmed by in situ hybridization of rat brain slices with antisense probes against KCNQ2 and KCNQ3 (Biervert et al. 1998; Schroeder et al. 1998).

KCNQ2 and KCNQ3 proteins share 41% of identity, and their overlapping distribution patterns in rat brain indicate either homotetrameric or heterotetrameric assembly (Schroeder et al. 1998). In the human brain, immunohistochemistry reveals that KCNQ2 and KCNQ3 proteins are expressed on the somata and dendrites of many polymorphic and pyramidal neurons in the hippocampus and cerebral cortex (Cooper et al. 2000). In the axons and/or termini of hippocampal excitatory neurons (the mossy cells and granule cells), KCNQ2 is robustly detected, but not KCNQ3, suggesting a critical role of KCNQ2 in regulating action potential propagation and neurotransmitter release (Cooper et al. 2000).

In the peripheral nociceptive sensory and visceral sensory neurons, KCNQ2, KCNQ3, and KCNQ5 channel subunit mRNAs and proteins are detected in the small and large DRG neurons using reverse transcription-PCR and single-cell PCR analysis, and visceral sensory neurons of rat nodose ganglia in immunocytochemical staining assay (Passmore et al. 2003; Rose et al. 2011; Wladyka and Kunze 2006; Zhang et al. 2019b). Patch clamp recordings of DRG neurons or nodose neurons confirm the functional expression of endogenous M-current that is sensitively inhibited by specific blockers XE991 and linopirdine (Du et al. 2014; Linley et al. 2008; Liu et al. 2010; Passmore et al. 2003; Wladyka and Kunze 2006; Zhang et al. 2019b). These observations indicate a role of neuronal Kv7 channels in modulation of peripheral pain and visceral pain.

3 Regulation of Kv7.2/7.3 Channels by GPCR Signaling

Each KCNQ2 or KCNQ3 subunit comprises six transmembrane domains (S1-S6), a pore loop between the S5 and S6 that forms a central ion selectivity pore, a positively charged S4 that primarily acts as a voltage sensor, an intracellular N terminal and a long intracellular carboxy-terminal tails (Delmas and Brown 2005).

In neurons, the KCNQ channels can be inhibited by neurotransmitters and neuropeptides via G-protein (Gq and/or G11) coupled receptor. At least two separate downstream signaling pathways are employed by different G-protein (Gq and/or

G11) coupled receptors to suppress KCNQ current (Suh et al. 2004; Zhang et al. 2011). The first pathway, used by muscarinic M1 receptors in sympathetic neurons, acts by depleting phosphatidylinositol-4,5-bisphosphate (PIP_2) and via protein kinase C (PKC) phosphorylation. A certain level of PIP_2 in the cell membrane is required for KCNQ channel opening (Zhang et al. 2003). Depletion of membrane PIP_2 (as a result of PIP_2 hydrolysis catalyzed by G-protein induced activation of phospholipase-Cβ (PLCβ)) leads to closure of KCNQ channel (Delmas and Brown 2005). The C terminal histidine site (H328) of KCNQ2 channel confers the PIP_2 sensitivity (Zhang et al. 2003), and residues 428–484 in KCNQ2 subunit are involved in PIP_2 binding (Hernandez et al. 2008). In the cryo-EM structure of KCNQ1, the PIP2 binding sites are observed in multiple amino acids with positive-charged side chains located in S0, the S2-S3 loop, and the S4-S5 linker (Sun and MacKinnon 2020).

The phosphorylation of KCNQ subunits by protein kinase C (PKC) also partially contributes to muscarinic agonist induced reduction of channel activity, and A-kinase-anchoring protein (AKAP)79/150 is necessary for this process (Hoshi et al. 2003). Expression of a mutated AKAP that disrupts PKC binding prevents Kv7.2 phosphorylation and reduces the inhibition of native ganglionic M channels by muscarinic receptor stimulation (Hoshi et al. 2003; Zhang et al. 2011). When recruited to the KCNQ channel by AKAP 79/150, PKC can phosphorylate two serine sites of KCNQ subunits (Ser523 and Ser530 of KCNQ2) (Delmas and Brown 2005).

The second mechanism, employed by bradykinin B2 receptor, involves intracellular Ca^{2+} signals that act in concert with CaM (calmodulin) (Gamper and Shapiro 2003; Liu et al. 2010). The binding sites of apo-calmodulin (Ca^{2+} free calmodulin) are also located in the C-terminal tail of KCNQ channels. As predicted from crystal structure of KCNQ4 (Xu et al. 2013), interactions between apo-CaM and KCNQ channels involve two conserved C-terminal tail segments, known as the A (a motif consisting of residues 310–372 for KCNQ2) and B segments (a motif consisting of residues 504–527 for KCNQ2), whereas the interaction between Ca^{2+}/CaM and the KCNQ C-terminal tail requires only the conserved α helix forming B segment (Xu et al. 2013). The structural and biochemical data suggest that there is a conformational switch between the apo-CaM and Ca^{2+}/CaM states (Xu et al. 2013) (For more detailed descriptions of interaction between apo-CaM, Ca^{2+}/CaM, and KCNQ channels, see chapter "Kv7 Channels and Excitability Disorders").

4 Validation of Kv7.2/7.3 Channels as Therapeutic Targets for Epilepsy and Pain

It has long been appreciated that closure of KCNQ2/3 channels leads to neuronal hyperactivity. At the cellular level, the precise modulation of neuronal excitability by KCNQ2/3 channel is determined by the channel activation kinetics. The activation time constant of KCNQ2/3 channels is in an order of tens of milliseconds, thus not contributing significantly to the repolarization phase of fast action potentials with duration of several milliseconds (Delmas and Brown 2005). Yet, KCNQ2/3 channel activation during sustained depolarization can act to prevent repetitive firing and contribute to the phenomenon of spike frequency adaptation. For example, KCNQ channel inhibitor XE991 increases spontaneous firing rate of dopaminergic neurons (Gopalakrishnan and Shieh 2004) and evokes spikes of hippocampal pyramidal neurons (Bi et al. 2011; Peters et al. 2005). In addition to spike frequency adaptation, KCNQ2/3 channels contribute to after-hyperpolarization (AHP) (Peters et al. 2005) following spike trains, and work to reduce spike after-depolarization (ADP) primarily driven by persistent sodium current (I_{NaP}) in hippocampal pyramidal neurons. During the spike ADP, the activated KCNQ/M channels produce an opposing outward current against I_{NaP}, limiting ADP duration and preventing the transition from solitary ("simple") spikes to high-frequency bursts (Yue and Yaari 2004). Thus, down-modulation of KCNQ/M channel activity in CA3 region of hippocampus converts the neuronal firing pattern from simple to complex spiking, whereas up-modulation of these channels exerts the opposite effect (Yue and Yaari 2004).

Loss-of-function mutations of both KCNQ2 subunits and KCNQ3 subunits can lead to human BFNC (Biervert et al. 1998; Charlier et al. 1998; Singh et al. 1998). Up to date, more than 100 mutations in KCNQ2 subunits and 15 mutations from KCNQ3 subunits have been found from different BFNC kindreds (see more details at the website of https://www.rikee.org/) (Miceli et al. 2013; Nappi et al. 2020; Singh et al. 2003). While the three KCNQ3 single amino acid mutations (D305G, W309R, G310V) are concentrated in the pore region, the KCNQ2 mutations, either missense, nonsense, or frame-shift mutations, are scattered throughout the channel including the S2, S4, and S5 transmembrane domains, the S4-S5 intracellular loop, the pore, and the C-terminus. Most of these mutations cause reduction of KCNQ current to a various extent. A generation of protein degradation signal is one of the mechanisms underlying KCNQ current reduction for recurrent seizures resulted from two KCNQ2 frame-shift mutations (Q2-2513del1bp: 1-nucleotide deletion at 2513 bp; Q2-2516ins5bp: 5-nucleotide insertion at 2516 bp) and it is recently found that clinical disease severity may be positively related to the extent of the mutation-induced functional K^+ channel impairment (Miceli et al. 2013).

The relationship of reduced KCNQ channel activity and seizure susceptibility has been shown in transgenic mice (Otto et al. 2004; Peters et al. 2005; Singh et al. 2008; Watanabe et al. 2000). Although KCNQ2$^{-/-}$ mice die within a few hours after birth owing to pulmonary atelectasis but not epilepsy, heterozygous mice (KCNQ2$^{+/-}$) with decreased expression of KCNQ2 show hypersensitivity to seizure inducer pentylenetetrazole (PTZ) (Watanabe et al. 2000). Homozygous knock-in mouse models of involving KCNQ2 (or KCNQ3) human BFNC missense mutations exhibit an early onset of spontaneous generalized tonic-clonic seizures, and heterozygous mice exhibit reduced thresholds to electrically induced seizures (Singh et al. 2008). Transgenic suppression of the neuronal KCNQ current in mice is associated with spontaneous seizures and behavioral hyperactivity (Greene et al. 2018; Peters et al. 2005).

It is of interest that many mutations in either KCNQ2 or KCNQ3 do not exert obvious dominant-negative effect, rather the currents are only reduced by a small portion (~25%) when co-expressed with their wild-type at 1:1 ratio to mimic the situation in a heterozygous patient (Schroeder et al. 1998; Singh et al. 2003). Therefore, in heterozygous epileptic patients if a 25% functional loss of KCNQ2 can be recovered, then it is feasible that this form of epilepsy can be treated (Schroeder et al. 1998).

5 Pharmacological Activation of Neuronal Kv7 for Potential Therapy of Epilepsy and Chronic Pain

Because of specific and robust expressions of KCNQ2/3 channels in neurons and the fact that their loss-of-function can lead to neuronal hyperexcitability, pharmacological activation of KCNQ2/3 channels serves the basis for therapy of epilepsy and pain (Fig. 1). Efforts for developing Kv7/KCNQ/M-channel openers have been devoted to therapy of epilepsy, pain, and other hyperexcitability-related disorders (Table 1 and Fig. 2).

5.1 Kv7/KCNQ/M Channel Opener Retigabine for Treatment of Partial Epilepsy

Retigabine or ezogabine (N-[2-amino-4-(4-fluoroben-zylamino)-phenyl] carbamic acid ethyl ester), as the first and also the best characterized activator of KCNQ

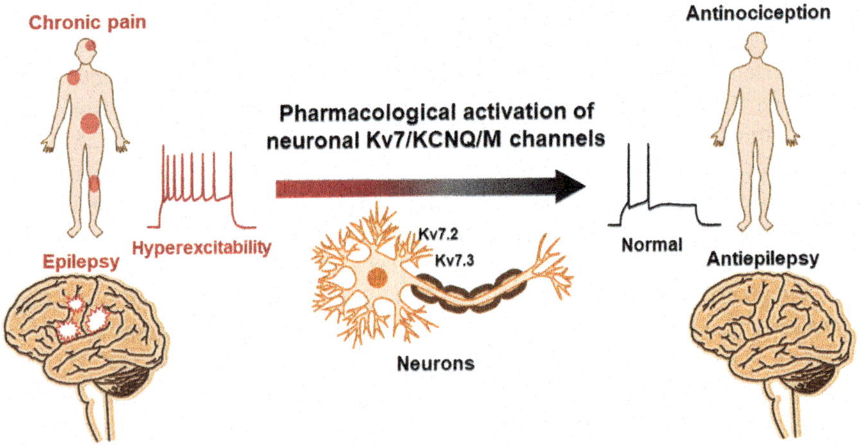

Fig. 1 Pharmacological activation of Kv7/KCNQ/M channels suppresses neuronal hyperexcitability for therapy of epilepsy and chronic pain. Neuronal hyperexcitability defines the fundamental mechanism of neurological disorders such as epilepsy and chronic pain. Voltage-gated Kv7.2 and Kv7.3 channels are specifically expressed in the neurons of peripheral nociceptive and visceral sensory systems and the brain. Neuronal Kv7.2/Kv7.3 channels are a validated therapeutic target for epilepsy and pain. Pharmacological activation of Kv7/KCNQ/M-current by channel openers converts repetitive firing into normal action potential in neurons, serving the basis for therapy of epilepsy and chronic pain or hyperexcitability-related neuropsychiatric disorders

Table 1 Pharmacological landscape of Kv7 channel openers

Name	Target gene	EC_{50} (μM)	Test assay	Clinical condition	Preclinical model	MED[a] (mg/kg)	Ref.
Retigabine (RTG)	Kv7.2 Kv7.3 Kv7.4 Kv7.5 Kv7.2/7.3 Kv7.3/7.5	2.5 0.6 5.2 NR 1.9 1.4	Manual Ephy.	DPN pain	STZ	15 (i.p.)	(Abd-Elsayed et al. 2015; Blackburn-Munro and Jensen 2003; Cai et al. 2015; Djouhri et al. 2019; Rostock et al. 1996; Tatulian et al. 2001; Xu et al. 2010)
				TMD pain	CFA	2.5 (i.p.)	
				Neuropathic pain	ION-CCI	0.56 (i.p.)	
					CCI	5 (p.o.)	
				Inflammatory pain	Formalin	20 (p.o.)	
				Cancer pain	Bone cancer pain	0.85 (i.th.)	
				Seizures	MES	2.87[b] (p.o.)	
					PTZ	13.5[b] (p.o.)	
Flupirtine	Kv7.2 Kv7.2/7.3 Kv7.3/7.5	NR 4.6 1.6	Manual Ephy. Automated Ephy.	Neuropathic pain DPN pain Inflammatory pain Gout arthritis	SNL STZ Orofacial formalin MSU	10 nM 10 (i.p.) 2.5 (i.p.) 7.5 (i.p.)	(Capuano et al. 2011; Friedel and Fitton 1993; Goodchild et al. 2008; Klinger et al. 2012; Rose et al. 2011; Wilenkin et al. 2019; Zhang et al. 2020)
Maxipost (BMS-204352)	Kv7.2 Kv7.3 Kv7.4 Kv7.5 Kv7.3/7.5	10[c] 10[c] 2.4 2.4 10[c]	Manual Ephy.	Anxiety	Mouse zero maze	30 (mouse, i.p.)	(Dupuis et al. 2002; Korsgaard et al. 2005; Schroder et al. 2001)
					Conditioned emotional response	60 (i.p.)	

(continued)

Table 1 (continued)

Name	Target gene	EC_{50} (μM)	Test assay	Clinical condition	Preclinical model	MED[a] (mg/kg)	Ref.
Zinc pyrithione (ZnPy)	Kv7.1 Kv7.2 Kv7.4 Kv7.5 Kv7.2/7.3	3.5 1.5 10[c] 10[c] 10[c]	Manual Ephy.	NR	NR	NR	(Gao et al. 2008; Xiong et al. 2007)
Acrylamide (S)-1	Kv7.2 Kv7.4 Kv7.5	3.3 10.4 30[c]	Manual Ephy. TEVC	Migraine	CSD	10 (i.g.)	(Bentzen et al. 2006; Wu et al. 2003, 2004)
Acrylamide (S)-2	Kv7.2 Kv7.4 Kv7.5	0.06 10[c] 10[c]	Manual Ephy.	NR	NR	NR	(Blom et al. 2009; Wu et al. 2004)
ICA-27243	Kv7.2/7.3 Kv7.4	0.2–0.4 9.7	Manual Ephy.	Seizure	MES	1.5[b] (p.o.)	(Munro and Dalby-Brown 2007; Roeloffs et al. 2008; Wickenden et al. 2008)
					PTZ	2.2[b] (p.o.)	
				Inflammatory pain	Formalin	30 (p.o.)	
					Carrageenan	30 (p.o.)	
				Neuropathic pain	SNL	25[b] (p.o.)	
QO-58	Kv7.1 Kv7.2 Kv7.4 Kv7.2/7.3 Kv7.3/7.5	7.0 1.3 0.6 2.3 5.2	Manual Ephy.	Neuropathic pain	CCI	25 (i.g.)	(Zhang et al. 2013)
Gabapentin	Kv7.3 Kv7.5 Kv7.2/7.3	5.3 nM 1.9 nM 4.3 nM	TEVC	NR	NR	NR	(Manville and Abbott 2018)

RL-81	Kv7.2/7.3 Kv7.4 Kv7.5	0.2 5.1 6.7	Manual Ephy.		NR	NR	(Kumar et al. 2016)
RL-56	Kv7.4 Kv7.2/7.3 Kv7.3/7.5 Kv7.4/7.5	0.38[d] 0.11[d] 0.23[d] 0.28[d]	Automated Ephy.	NR	NR	NR	(Liu et al. 2019)
RL-12	Kv7.2/7.3	1.0[d]	Automated Ephy.	NR	NR	NR	(Liu et al. 2019)
SCR2682	Kv7.2 Kv7.3 Kv7.4 Kv7.2/7.3 Kv7.3/7.5	26.3 nM 11.2 nM 28.8 nM 9.8 nM 11.1 nM	Manual Ephy.	Seizure	MES	1 (i.g.)	(Zhang et al. 2019a)
Quercetin	Kv7.1 Kv7.4 Kv7.2/7.3	100[c] 100[c] 100[c]	Manual Ephy.	NR	NR	NR	(Redford and Abbott 2020)
Diclofenac	Kv7.2/7.3	2.6	Manual Ephy.	Seizure	MES	43[b] (i.p.)	(Peretz et al. 2005)

Notes: *ION-CCI* infraorbital nerve chronic constrictive ligation, *CCI* chronic constriction injury, *DPN* diabetic peripheral neuropathy, *STZ* streptozotocin, *TMD* Temporomandibular disorder, *CFA* complete Freund's adjuvant, *CSD* cortical spreading depression, *TEVC* two electrode voltage clamp, *MED* minimum effective dose, *SNL* spread nerve ligation, *MES* maximal electroshock, *PTZ* pentylenetetrazole, *MSU* monosodium urate, *NR* no report

[a]Data from rats

[b]ED_{50}, 50% antinociceptive/anticonvulsant effect, *i.p.*, intraperitoneal, *i.g.*, intragastric, *p.o.* peros (oral), *i.th.* intrathecal

[c]Tested concentration

[d]15% channel activation (EC_{2x})

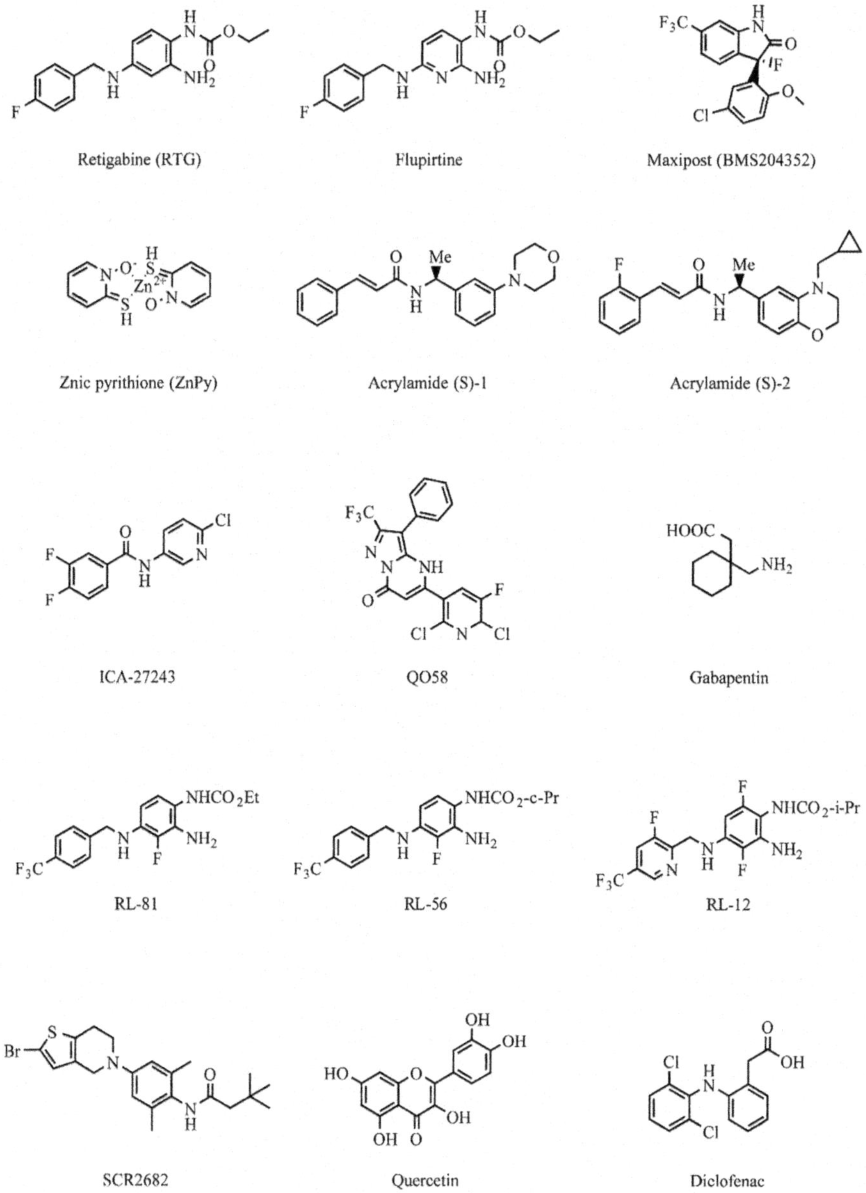

Fig. 2 Current landscape for chemical structures of major Kv7 channel activators

channels, was approved in 2011 by FDA as an anticonvulsant for treatment of partial epilepsies (Deeks 2011). Retigabine (former name D-23129) is a derivative of centrally acting analgesic flupirtine that shows anticonvulsant effects in animal models of epilepsy and in patients with refractory seizures in the mid-1980s. Shortly after its discovery, retigabine was experimentally proved to be a novel anticonvulsant agent at company Asta Medica between 1994 and 1996 (Blackburn-Munro et al. 2005; Rostock et al. 1996). It exhibits potent anticonvulsant activity in nearly all animal models of epilepsy, including genetic models of epilepsy, seizures induced by chemical drugs such as PTZ, kainate, and NMDA, and electrically induced seizures including electrical kindling of the amygdala or maximal electroshock (Blackburn-Munro et al. 2005; Rostock et al. 1996). In rat model of amygdala kindling of the complex partial seizure, which is the most frequent type of conventional drug resistant seizure, peritoneal and oral administrations of retigabine (0.01–5 mg/kg) exhibit a dose-dependent increase of the threshold current for induction of afterdischarges (Tober et al. 1996).

Retigabine was shown to be a specific Kv7/KCNQ/M- channel activator by three independent groups in 2000 (Main et al. 2000; Rundfeldt and Netzer 2000; Wickenden et al. 2000) (Table 1). Their major findings demonstrate that: (1) retigabine enhances KCNQ2/3 current in a dose-dependent manner with an EC_{50} about 1 μM; (2) retigabine elicits a profound hyperpolarizing shift in the voltage dependence of channel activation (−43 mV left shift for KCNQ3 and −24 mV left shift for KCNQ2), which results in an enhancement of KCNQ current and activation of KCNQ current at hyperpolarized potentials (for instance at −80 mV). The increase of channel open probability, but not single channel conductance, leads to an enhancement of whole-cell KCNQ current induced by retigabine (Tatulian and Brown 2003). Subsequent studies also show that retigabine is an opener of KCNQ4 and KCNQ5 (Blackburn-Munro et al. 2005; Schroder et al. 2001).

Retigabine specifically activates KCNQ2-5 but not KCNQ1. A single amino acid, a tryptophan (Trp or W) within the S5 segment, is critical for retigabine sensitivity for KCNQ2-5 channels as the tryptophan mutations in KCNQ2 (W236L), KCNQ3 (W265L), KCNQ4 (W242L), and KCNQ5 (W235L) cause loss of retigabine sensitivity (Schenzer et al. 2005). Retigabine binds to a hydrophobic pocket located between the cytoplasmic parts of the S5 and the S6 transmembrane domains in the open channel configuration (Schenzer et al. 2005; Wuttke et al. 2005). Critical residues including the Trp 236 in the S5 and Gly 301 in S6 of KCNQ2 (also considered as activation gate) are important for the interaction between retigabine and KCNQ2 channel. Using a refined chimeric strategy and structural model guided mutation (Lange et al. 2009), Lange et al. demonstrated that in KCNQ3, Leu-314 in the pore region, the conserved Leu-272 in S5 and Leu-338 extending from S6 of the neighboring subunit can form a hydrophobic binding pocket for retigabine binding. In this pocket, Trp-265 and Leu-314 can represent the upper and lower margins of the putative binding site, Leu-272 contributes to the binding of KCNQ3, and Leu-338 from the neighboring subunit is apparently involved in lining the hydrophobic binding pocket. In addition to residues Trp-265, Leu-272, Leu-314, and Leu-338 of KCNQ3 important for retigabine binding (Lange et al. 2009), the recent cryo-EM structure of KCNQ2 also reveals four residues Trp-236, Ser-303, Leu-299, and Phe-305 critical for retigabine sensitivity and stabilization of the channel open-gate confirmation (Li et al. 2021), which may help explain the hyperpolarizing shift of neuronal Kv7 channels.

However, retigabine was withdrawn from the market in 2017 for side effects such as a blue-colored appearance of the skin or eyes that can be reversible after drug discontinuation (Garin Shkolnik et al. 2014; Mathias and Abou-Khalil 2017). A recent investigation indicates that metabolic oxidation of retigabine and its derivative flupirtine generates the formation of electrophilic quinones that are highly reactive metabolites that might be responsible for the blue tissue discoloration after prolonged intake (retigabine) or induction of liver injury (flupirtine) (Bock et al. 2019). Analysis of structure–activity relationship reveals that sulfide analogs can be devoid of the risk of quinone formation, and possesses potent channel activation and same efficacy with improved toxicity/activity ratio (Bock et al. 2019).

5.2 Other KCNQ Openers for Potential Treatment of Epilepsy

There are several other series of KCNQ/M-channel openers that are shown to potentiate both recombinant KCNQ channels and native M channels through various mechanisms, such as lowering channel activation threshold, increasing channel open probability, and stabilizing the open state of the channel (Table 1 and Fig. 2). Oxindole analogs, represented by BMS-204352 [(3S)-(+)-(5-chloro-2-methoxyphenyl)-1,3-dihydro-3-fluoro-6-(trifluoromethyl)-2H-indol-2-one)], are potent activators of KCNQ channels with an EC_{50} of 2.4 μM for KCNQ2 at −30 mV. BMS-204352 potentiates KCNQ2, KCNQ2/3, and KCNQ3/5 channels to a variable extent (Xiong et al. 2008). Although tested in models of stroke and anxiety, there is no report yet on BMS-204352 in testing its anti-epileptic efficacy.

Zinc pyrithione (ZnPy), widely used for control of dandruff and treatment of psoriasis, is a synthetic KCNQ potentiator that potently activates both heterologous and native M channels by inducing channel opening at the resting potential (Xiong et al. 2007). The EC_{50} of zinc pyrithione for KCNQ2 is 1.5 μM, and for KCNQ2/3 heteromultimers is 2.4 μM (Table 1). Intracellular free zinc at 10 μM is also able to increase the open probability of single KCNQ2/3 channels (Gao et al. 2017). Zinc pyrithione is capable of pharmacological rescue of certain KCNQ genetic mutations found in people who suffer from neonatal epilepsy. However, there are no reports about whether zinc pyrithione is effective in epileptic animal models. The molecular determinants of ZnPy are identified primarily from mutagenesis studies based on KCNQ2 channels (Xiong et al. 2008). The key determinants include a leucine residue in S5 (Leu249), another leucine within the linker (Leu275) between S5 and the pore region, and Ala306 in the S6. Although mutating these two leucines causes larger current increase in response to ZnPy, it completely abolishes the hyperpolarizing shift caused by ZnPy. In contrast, KCNQ2 (A306T) mutation shows little potentiation in overall channel conductance by ZnPy but displays the hyperpolarizing shift (Xiong et al. 2008).

Acrylamide (S)-1, (S)-N-[1-(3-morpholin-4-yl-phenyl)-ethyl]-3-phenyl-acrylamide, has been reported to differentially affect KCNQ channels (Bentzen et al. 2006).

Although it blocks the cardiovascular expressed KCNQ1 current, it is a KCNQ2-5 channel activator at low voltages, with an EC_{50} value of 3.28 μM at −40 mV for KCNQ2 (Bentzen et al. 2006). However, at high voltages, it blocks KCNQ2 and KCNQ2/3, whereas still potentiates KCNQ4 and KCNQ5. A single tryptophan residue (Trp242) in S5 of KCNQ4, known to be crucial for the effect of retigabine, is also shown to be critical for the enhancing effect of (S)-1. (S)-2 is a conformationally restricted molecule and has an EC_{50} of 0.06 μM, making it 55 times more potent than (S)-1 (Xiong et al. 2008).

ICA-27243 [N-(6-chloro-pyridin-3-yl)-3,4-difluoro-benzamide] was synthesized by Icagen Inc. (Wickenden et al. 2008). It is a selective and relatively potent KCNQ2/3 activator with an EC_{50} around 0.2–0.4 μM. ICA-27243 exhibits anticonvulsant activity both in hippocampal slice model of epilepsy and rodent models, including maximal electroshock, pentylenetetrazole-induced seizures, rat amygdala kindling model of partial seizures, and the 6-Hz model of psychomotor seizures in mice(Roeloffs et al. 2008). The binding site of ICA-27243 is distinct to that of retigabine, as ICA-27243 likely binds to novel sites within the S1-S4 voltage-sensor domain of KCNQ channels (Padilla et al. 2009). Although individual amino acids that determine ICA-27243 binding interactions have not identified yet, one motif containing of residues 118–127 is involved in the interaction of ICA-27243 and KCNQ2 subunit (Padilla et al. 2009). This motif consists of a part of extracellular domain of S1-S2 linker and neighboring S2 fragment. Substitution of this motif leads to 17-fold less sensitivity to ICA- 27243. Amino acid sequences of this ICA-27243 binding pocket show high diversity among different KCNQ subunits, which might contribute to the KCNQ2/3 selectivity of ICA-27243 (Table 1).

Series of pyrazolo[1,5-a]pyrimidin-7(4H)-ones (PPOs) were synthesized and identified as Kv7/KCNQ/M channel openers in 2011 (Qi et al. 2011). In atomic Rb^+ efflux assay, a lead PPO-17 compound activates KCNQ2/3 channels with an EC_{50} of 0.06 μM in comparison with EC_{50} of 0.46 μM for retigabine (Qi et al. 2011). In whole-cell patch clamp recordings of CHO cells expressing KCNQ2/3 channels, PPO compound QO-58 shows an EC_{50} of 2.3 μM (Zhang et al. 2013). The EC_{50} values for QO-58 over other Kv7 subtypes are 7.0, 1.3, 0.6, and 5.2 μM for Kv7.1, Kv7.2, Kv7.4, and Kv7.3/Kv7.5 channels, respectively (Zhang et al. 2013). In rodent models of maximal electroshock seizures and PTZ-induced seizures, QO58 (>25 mM/kg) and its hydrophilic lysine salt QO58-lysine (QO58L) exhibit the antiepileptic activity. Other PPO derivatives, such as QO-26, QO-28, QO-40, and QO-41, also can potently augment KCNQ2/3 channels (Jia et al. 2011). In addition, KCNQ opener QO58 shows the effectiveness in rodent models of migraine, trigeminal neuralgia, and inflammation pain induced by complete Freund's adjuvant (CFA) in rats (Zhang et al. 2013).

Gabapentin (or neurontin), originally designed as analogs of the neurotransmitter g-aminobutyric acid (GABA), is recently reported to play an unexpected role in activation of heteromeric KCNQ2/3 channels expressed in *Xenopus* oocytes with EC_{50} of 4.2 nM (Manville and Abbott 2018). Gabapentin also activates homomeric KCNQ3 and KCNQ5 channels with EC_{50}s at 5.3 and 1.9 nM, respectively. Gabapentin activation of heteromeric KCNQ2/3 or homomeric KCNQ3 channels requires KCNQ3-W265, a conserved residue tryptophan in the transmembrane segment. Interestingly, homomeric KCNQ2 or KCNQ4 channels are insensitive to gabapentin, and structure-related pregabalin is unable to activate KCNQ2/3 (Manville and Abbott 2018).

Compound RL-81 and its structure-related analog RL-56 and RL-12, derived from retigabine, can activate Kv7.2/Kv7.3 expressed in HEK293 cells in an automated electrophysiology assay (Liu et al. 2019). They are selective over other subtypes of Kv7.3/Kv7.5, Kv7.4, and Kv7.4/7.5 (Table 1). Compound RL-56 is a potent activator with an EC2x at 0.11 μM, and shows a selectivity index (SI) about 2.5. The SAR analysis reveals the ability of fluorine substituents to substantially alter the potency and selectivity of Kv7 channel activation (Liu et al. 2019).

Compound SCR2682, an N-(4-(2-bromo-6,7-dihydrothieno[3,2-c]pyridin-5 (4H)-yl)-2,6-dimethylphenyl)-3,3-d imethylbutanamide (SCR2682) 2,6-dimethyl-4-(piperidin-yl) phenyl)-amide derivative, is reported to show selective and potent activation of neuronal Kv7/KCNQ/M-channels (Zhang et al. 2019a). In manual patch-clamp recordings of HEK293 cells expressing Kv7.2/7.3 channels, SCR2682 activates the channel with an EC_{50} of 9.8 nM, which is approximately 100-fold more potent than retigabine), and shifts voltage-dependent activation of Kv7.2/7.3 current toward more negative membrane potential about −37 mV ($V_{1/2}$). SCR2682 also activates native M-current in rat hippocampal or cortical neurons, causing marked hyperpolarization and potent inhibition of neuronal firings. Mechanistically, mutating the residue tryptophan 236 located at the fifth transmembrane segment of Kv7.2 is critical for SCR2682-mediated activation of the channel. Intraperitoneal or intragastric administration of SCR2682 results in a dose-dependent inhibition of seizures induced by maximal electroshock (Zhang et al. 2019a).

5.3 Pharmacological Activation of Kv7 by Openers for Therapy of Chronic Pain

Chronic pain is characterized by persistent pain, including somatic pain, visceral pain, and neuropathic pain, which lasts months to years and is an unmet medical need (Dydyk and Conermann 2020). Besides expressed in the central nervous system, KCNQ channel subunits are also expressed in the peripheral nociceptive neurons and visceral sensory system responsible for pain sensation and signaling (Du et al. 2018; Passmore et al. 2003; Wladyka and Kunze 2006). M-currents are identified in both small and large DRG neurons, which can be blocked by XE991 and linopirdine with IC_{50}s of 0.26 μM and 2.1 μM, respectively. KCNQ2-5 mRNA and protein expressions, but not KCNQ1, are detected in DRG neurons (Passmore et al. 2003). Immunocytochemical staining reveals that KCNQ2, KCNQ3, and KCNQ5 are also expressed in the visceral sensory neurons of the nodose ganglia from rats, and both XE991 and linopirdine can inhibit the M-current (Rose et al. 2011; Wladyka and Kunze 2006). Genetic suppression or pharmacological inhibition of KCNQ/M channel function causes the hyperexcitability of central and peripheral sensory neurons and enhances the sensitivity and development of visceral pain, osteoarthritic pain, and bone cancer pain (Bi et al. 2011; Du et al. 2014; Liu et al. 2010; Zhang et al. 2019b; Zheng et al. 2013).

Analgesic drug flupirtine, a retigabine-related small molecule, is a non-opiate centrally acting agent that entered clinical trials for treatment of pain after hysterectomy surgery in 1983 and for patients with migraine in 1984 (Million et al. 1984; Moore et al. 1983), exhibiting analgesic efficacy similar to a semi-synthetic opioid analgesic dihydrocodeine. Flupirtine shows as effective as opiate analgesics codeine, dihydrocodeine and pentazocine, and nonsteroidal anti-inflammatory drugs (NSAIDs) suprofen, diclofenac and ketoprofen for pain relief of surgery, traumatic injury, dental procedures, headache/migraine, arthritis, cancer and abdominal spasms, and neuropathic pain (Friedel and Fitton 1993; Herrmann et al. 1987; Mishra et al. 2013; Scheef 1987). Although an effective analgesic, flupirtine causes serious liver injury and hepatotoxicity (Nicoletti et al. 2016; Puljak 2018) and was withdrawn from market by European Medicines Agency (EMA) in 2018. However, chemical modifications of flupirtine appear to significantly improve toxicity/activity ratio and maintain potent activation of Kv7.2/7.3 channels (Surur et al. 2019).

Retigabine produces analgesic activity in rat models of neuropathic pain, visceral pain induced by an intracolonic injection of capsaicin, and mechanical allodynia of inflammatory temporomandibular joint in rats (Blackburn-Munro and Jensen 2003; Hirano et al. 2007; Xu et al. 2010). In the chronic constriction injury and spared nerve models of neuropathic pain, oral administrations of retigabine (5 and 20 mg/kg) attenuates mechanical hypersensitivity in response to pin prick stimulation, but not von Frey stimulation, of injured hindpaw (Blackburn-Munro and Jensen 2003).

In the formalin test, oral retigabine (20 mg/kg) reduces flinching behavior in the second phase, the effect of which can be completely reversed by XE-991 (3 mg/kg, i.p.). Intraperitoneal administration of retigabine (1, 3 and 10 mg/kg) dose-dependently suppressed visceral pain behavior induced by intracolonic injection of capsaicin and prolonged the latency to first licking (Hirano et al. 2007). In rodent model of paclitaxel-induced peripheral neuropathy (PIPN), retigabine is also effective in attenuating the development of neuropathic pain, suggesting that Kv7 openers may have therapeutic potential for cancer patients receiving paclitaxel treatment and developing sensory peripheral neuropathies (Li et al. 2019).

Retigabine or Kv7 openers without penetration of the CNS can exert analgesic effect through peripheral mechanism for prevention of CNS side effects. The retigabine-induced analgesic effects appear to be mediated by direct activation of peripheral Kv7 channels as intracerebroventricular (i.c.v.) administration of blocker XE991 can reverse the retigabine-induced anticonvulsant effect, but not the analgesic effect (Du et al. 2014; Hayashi et al. 2014; Wu et al. 2017).

Kv7 opener QO58-lysine causes a leftward shift of the voltage-dependent activation of Kv7.2/7.3 to a hyperpolarized potential about −28 mV (Teng et al. 2016). QO58-lysine has a half-life of approximately 3 h in the plasma and absolute bioavailability of 14% for oral dose of 12.5/kg). Oral or intraperitoneal administration of QO58-lysine exhibits a concentration-dependent reduction in the licking times during phase II pain induced by the injection of formalin into mouse hindpaw, and increases the paw withdrawal threshold (Teng et al. 2016).

6 Conclusions/Perspectives

Neuronal Kv7/KCNQ/M-channel is a proven drug target for therapy of neurological diseases including epilepsy and pain. Although the first-generation M-channel openers retigabine and flupirtine have recently been discontinued from the market due to the compound-related side effects of blue skin and liver toxicity, targeting M-channels by pharmacological activation still presents an exciting opportunity for discovery and development of selective and potent channel openers for therapy of epilepsy and pain and beyond.

Acknowledgements This work is supported by research grants to KWW from Natural Science Foundation of China (81573410); and from the Ministry of Science and Technology of China (2018ZX09711001-004-006) and Science and Technology Program of Guangdong (2018B030334001).

References

Abd-Elsayed AA, Ikeda R, Jia Z, Ling J, Zuo X, Li M, Gu JG (2015) KCNQ channels in nociceptive cold-sensing trigeminal ganglion neurons as therapeutic targets for treating orofacial cold hyperalgesia. Mol Pain 11:45

Bentzen BH, Schmitt N, Calloe K, Dalby Brown W, Grunnet M, Olesen SP (2006) The acrylamide (S)-1 differentially affects Kv7 (KCNQ) potassium channels. Neuropharmacology 51:1068–1077

Bi Y, Chen H, Su J, Cao X, Bian X, Wang K (2011) Visceral hyperalgesia induced by forebrain-specific suppression of native Kv7/KCNQ/M-current in mice. Mol Pain 7:84

Biervert C, Schroeder BC, Kubisch C, Berkovic SF, Propping P, Jentsch TJ, Steinlein OK (1998) A potassium channel mutation in neonatal human epilepsy. Science 279:403–406

Blackburn-Munro G, Dalby-Brown W, Mirza NR, Mikkelsen JD, Blackburn-Munro RE (2005) Retigabine: chemical synthesis to clinical application. CNS Drug Rev 11:1–20

Blackburn-Munro G, Jensen BS (2003) The anticonvulsant retigabine attenuates nociceptive behaviours in rat models of persistent and neuropathic pain. Eur J Pharmacol 460:109–116

Blom SM, Schmitt N, Jensen HS (2009) The acrylamide (S)-2 as a positive and negative modulator of Kv7 channels expressed in Xenopus laevis oocytes. PLoS One 4:e8251

Bock C, Surur AS, Beirow K, Kindermann MK, Schulig L, Bodtke A, Bednarski PJ, Link A (2019) Sulfide analogues of flupirtine and retigabine with nanomolar KV 7.2/KV 7.3 channel opening activity. ChemMedChem 14:952–964

Brown DA, Adams PR (1980) Muscarinic suppression of a novel voltage-sensitive K+ current in a vertebrate neurone. Nature 283:673–676

Cai J, Fang D, Liu XD, Li S, Ren J, Xing GG (2015) Suppression of KCNQ/M (Kv7) potassium channels in the spinal cord contributes to the sensitization of dorsal horn WDR neurons and pain hypersensitivity in a rat model of bone cancer pain. Oncol Rep 33:1540–1550

Capuano A, De Corato A, Treglia M, Tringali G, Navarra P (2011) Flupirtine antinociception in the rat orofacial formalin test: an analysis of combination therapies with morphine and tramadol. Pharmacol Biochem Behav 97:544–550

Charlier C, Singh NA, Ryan SG, Lewis TB, Reus BE, Leach RJ, Leppert M (1998) A pore mutation in a novel KQT-like potassium channel gene in an idiopathic epilepsy family. Nat Genet 18:53–55

Cooper EC, Aldape KD, Abosch A, Barbaro NM, Berger MS, Peacock WS, Jan YN, Jan LY (2000) Colocalization and coassembly of two human brain M-type potassium channel subunits that are mutated in epilepsy. Proc Natl Acad Sci U S A 97:4914–4919

Deeks ED (2011) Retigabine (ezogabine): in partial-onset seizures in adults with epilepsy. CNS Drugs 25:887–900

Delmas P, Brown DA (2005) Pathways modulating neural KCNQ/M (Kv7) potassium channels. Nat Rev Neurosci 6:850–862

Djouhri L, Malki MI, Zeidan A, Nagi K, Smith T (2019) Activation of Kv7 channels with the anticonvulsant retigabine alleviates neuropathic pain behaviour in the streptozotocin rat model of diabetic neuropathy. J Drug Target 27:1118–1126

Du X, Gao H, Jaffe D, Zhang H, Gamper N (2018) M-type K(+) channels in peripheral nociceptive pathways. Br J Pharmacol 175:2158–2172

Du X, Hao H, Gigout S, Huang D, Yang Y, Li L, Wang C, Sundt D, Jaffe DB, Zhang H et al (2014) Control of somatic membrane potential in nociceptive neurons and its implications for peripheral nociceptive transmission. Pain 155:2306–2322

Dupuis DS, Schroder RL, Jespersen T, Christensen JK, Christophersen P, Jensen BS, Olesen SP (2002) Activation of KCNQ5 channels stably expressed in HEK293 cells by BMS-204352. Eur J Pharmacol 437:129–137

Dydyk AM, Conermann T (2020) Chronic pain. In: StatPearls, Treasure Island

Friedel HA, Fitton A (1993) Flupirtine. A review of its pharmacological properties, and therapeutic efficacy in pain states. Drugs 45:548–569

Gamper N, Shapiro MS (2003) Calmodulin mediates Ca2+–dependent modulation of M-type K+ channels. J Gen Physiol 122:17–31

Gao H, Boillat A, Huang D, Liang C, Peers C, Gamper N (2017) Intracellular zinc activates KCNQ channels by reducing their dependence on phosphatidylinositol 4,5-bisphosphate. Proc Natl Acad Sci U S A 114:E6410–E6419

Gao Z, Xiong Q, Sun H, Li M (2008) Desensitization of chemical activation by auxiliary subunits: convergence of molecular determinants critical for augmenting KCNQ1 potassium channels. J Biol Chem 283:22649–22658

Garin Shkolnik T, Feuerman H, Didkovsky E, Kaplan I, Bergman R, Pavlovsky L, Hodak E (2014) Blue-gray mucocutaneous discoloration: a new adverse effect of ezogabine. JAMA Dermatol 150:984–989

Goodchild CS, Kolosov A, Tucker AP, Cooke I (2008) Combination therapy with flupirtine and opioid: studies in rat pain models. Pain Med 9:928–938

Gopalakrishnan M, Shieh CC (2004) Potassium channel subtypes as molecular targets for overactive bladder and other urological disorders. Expert Opin Ther Targets 8:437–458

Greene DL, Kosenko A, Hoshi N (2018) Attenuating M-current suppression in vivo by a mutant Kcnq2 gene knock-in reduces seizure burden and prevents status epilepticus-induced neuronal death and epileptogenesis. Epilepsia 59:1908–1918

Hayashi H, Iwata M, Tsuchimori N, Matsumoto T (2014) Activation of peripheral KCNQ channels attenuates inflammatory pain. Mol Pain 10:15

Hernandez CC, Zaika O, Tolstykh GP, Shapiro MS (2008) Regulation of neural KCNQ channels: signalling pathways, structural motifs and functional implications. J Physiol 586:1811–1821

Herrmann WM, Kern U, Aigner M (1987) On the adverse reactions and efficacy of long-term treatment with flupirtine: preliminary results of an ongoing twelve-month study with 200 patients suffering from chronic pain states in arthrosis or arthritis. Postgrad Med J 63 (Suppl 3):87–103

Hirano K, Kuratani K, Fujiyoshi M, Tashiro N, Hayashi E, Kinoshita M (2007) Kv7.2-7.5 voltage-gated potassium channel (KCNQ2-5) opener, retigabine, reduces capsaicin-induced visceral pain in mice. Neurosci Lett 413:159–162

Hoshi N, Zhang JS, Omaki M, Takeuchi T, Yokoyama S, Wanaverbecq N, Langeberg LK, Yoneda Y, Scott JD, Brown DA et al (2003) AKAP150 signaling complex promotes suppression of the M-current by muscarinic agonists. Nat Neurosci 6:564–571

Jia C, Qi J, Zhang F, Mi Y, Zhang X, Chen X, Liu L, Du X, Zhang H (2011) Activation of KCNQ2/3 potassium channels by novel pyrazolo[1,5-a]pyrimidin-7(4H)-one derivatives. Pharmacology 87:297–310

Klinger F, Geier P, Dorostkar MM, Chandaka GK, Yousuf A, Salzer I, Kubista H, Boehm S (2012) Concomitant facilitation of GABAA receptors and KV7 channels by the non-opioid analgesic flupirtine. Br J Pharmacol 166:1631–1642

Korsgaard MP, Hartz BP, Brown WD, Ahring PK, Strobaek D, Mirza NR (2005) Anxiolytic effects of Maxipost (BMS-204352) and retigabine via activation of neuronal Kv7 channels. J Pharmacol Exp Ther 314:282–292

Kubisch C, Schroeder BC, Friedrich T, Lutjohann B, El-Amraoui A, Marlin S, Petit C, Jentsch TJ (1999) KCNQ4, a novel potassium channel expressed in sensory outer hair cells, is mutated in dominant deafness. Cell 96:437–446

Kumar M, Reed N, Liu R, Aizenman E, Wipf P, Tzounopoulos T (2016) Synthesis and evaluation of potent KCNQ2/3-specific channel activators. Mol Pharmacol 89:667–677

Lange W, Geissendorfer J, Schenzer A, Grotzinger J, Seebohm G, Friedrich T, Schwake M (2009) Refinement of the binding site and mode of action of the anticonvulsant Retigabine on KCNQ K + channels. Mol Pharmacol 75:272–280

Lerche C, Scherer CR, Seebohm G, Derst C, Wei AD, Busch AE, Steinmeyer K (2000) Molecular cloning and functional expression of KCNQ5, a potassium channel subunit that may contribute to neuronal M-current diversity. J Biol Chem 275:22395–22400

Li L, Li J, Zuo Y, Dang D, Frost JA, Yang Q (2019) Activation of KCNQ channels prevents paclitaxel-induced peripheral neuropathy and associated neuropathic pain. J Pain 20:528–539

Li X, Zhang Q, Guo P, Fu J, Mei L, Lv D, Wang J, Lai D, Ye S, Yang H et al (2021) Molecular basis for ligand activation of the human KCNQ2 channel. Cell Res 31:52–61

Linley JE, Rose K, Patil M, Robertson B, Akopian AN, Gamper N (2008) Inhibition of M current in sensory neurons by exogenous proteases: a signaling pathway mediating inflammatory nociception. J Neurosci 28:11240–11249

Liu B, Linley JE, Du X, Zhang X, Ooi L, Zhang H, Gamper N (2010) The acute nociceptive signals induced by bradykinin in rat sensory neurons are mediated by inhibition of M-type K+ channels and activation of Ca2+−activated Cl- channels. J Clin Invest 120:1240–1252

Liu R, Tzounopoulos T, Wipf P (2019) Synthesis and optimization of Kv7 (KCNQ) potassium channel agonists: the role of fluorines in potency and selectivity. ACS Med Chem Lett 10:929–935

Main MJ, Cryan JE, Dupere JR, Cox B, Clare JJ, Burbidge SA (2000) Modulation of KCNQ2/3 potassium channels by the novel anticonvulsant retigabine. Mol Pharmacol 58:253–262

Manville RW, Abbott GW (2018) Gabapentin is a potent activator of KCNQ3 and KCNQ5 potassium channels. Mol Pharmacol 94:1155–1163

Mathias SV, Abou-Khalil BW (2017) Ezogabine skin discoloration is reversible after discontinuation. Epilepsy Behav Case Rep 7:61–63

Miceli F, Soldovieri MV, Ambrosino P, Barrese V, Migliore M, Cilio MR, Taglialatela M (2013) Genotype-phenotype correlations in neonatal epilepsies caused by mutations in the voltage sensor of K(v)7.2 potassium channel subunits. Proc Natl Acad Sci U S A 110:4386–4391

Million R, Finlay BR, Whittington JR (1984) Clinical trial of flupirtine maleate in patients with migraine. Curr Med Res Opin 9:204–212

Mishra S, Choudhary P, Joshi S, Bhatnagar S (2013) Successful use of flupirtine in refractory neuropathic pain due to small fiber neuropathy. Am J Hosp Palliat Care 30:91–93

Moore RA, Bullingham RE, Simpson S, O'Sullivan G, Evans PJ, McQuay HJ, Lloyd JW (1983) Comparison of flupirtine maleate and dihydrocodeine in patients following surgery. Br J Anaesth 55:429–432

Munro G, Dalby-Brown W (2007) Kv7 (KCNQ) channel modulators and neuropathic pain. J Med Chem 50:2576–2582

Nappi P, Miceli F, Soldovieri MV, Ambrosino P, Barrese V, Taglialatela M (2020) Epileptic channelopathies caused by neuronal Kv7 (KCNQ) channel dysfunction. Pflugers Arch 472:881–898

Neyroud N, Tesson F, Denjoy I, Leibovici M, Donger C, Barhanin J, Faure S, Gary F, Coumel P, Petit C et al (1997) A novel mutation in the potassium channel gene KVLQT1 causes the Jervell and Lange-Nielsen cardioauditory syndrome. Nat Genet 15:186–189

Nicoletti P, Werk AN, Sawle A, Shen Y, Urban TJ, Coulthard SA, Bjornsson ES, Cascorbi I, Floratos A, Stammschulte T et al (2016) HLA-DRB1*16: 01-DQB1*05: 02 is a novel genetic risk factor for flupirtine-induced liver injury. Pharmacogenet Genomics 26:218–224

Otto JF, Yang Y, Frankel WN, Wilcox KS, White HS (2004) Mice carrying the szt1 mutation exhibit increased seizure susceptibility and altered sensitivity to compounds acting at the m-channel. Epilepsia 45:1009–1016

Padilla K, Wickenden AD, Gerlach AC, McCormack K (2009) The KCNQ2/3 selective channel opener ICA-27243 binds to a novel voltage-sensor domain site. Neurosci Lett 465:138–142

Passmore GM, Selyanko AA, Mistry M, Al-Qatari M, Marsh SJ, Matthews EA, Dickenson AH, Brown TA, Burbidge SA, Main M et al (2003) KCNQ/M currents in sensory neurons: significance for pain therapy. J Neurosci 23:7227–7236

Peretz A, Degani N, Nachman R, Uziyel Y, Gibor G, Shabat D, Attali B (2005) Meclofenamic acid and diclofenac, novel templates of KCNQ2/Q3 potassium channel openers, depress cortical neuron activity and exhibit anticonvulsant properties. Mol Pharmacol 67:1053–1066

Peters HC, Hu H, Pongs O, Storm JF, Isbrandt D (2005) Conditional transgenic suppression of M channels in mouse brain reveals functions in neuronal excitability, resonance and behavior. Nat Neurosci 8:51–60

Puljak L (2018) Flupirtine, an effective analgesic, but hepatotoxicity should limit its use. Anesth Analg 127:309–310

Qi J, Zhang F, Mi Y, Fu Y, Xu W, Zhang D, Wu Y, Du X, Jia Q, Wang K et al (2011) Design, synthesis and biological activity of pyrazolo[1,5-a]pyrimidin-7(4H)-ones as novel Kv7/KCNQ potassium channel activators. Eur J Med Chem 46:934–943

Redford KE, Abbott GW (2020) The ubiquitous flavonoid quercetin is an atypical KCNQ potassium channel activator. Commun Biol 3:356

Roeloffs R, Wickenden AD, Crean C, Werness S, McNaughton-Smith G, Stables J, McNamara JO, Ghodadra N, Rigdon GC (2008) In vivo profile of ICA-27243 [N-(6-chloro-pyridin-3-yl)-3,4-difluoro-benzamide], a potent and selective KCNQ2/Q3 (Kv7.2/Kv7.3) activator in rodent anticonvulsant models. J Pharmacol Exp Ther 326:818–828

Rose K, Ooi L, Dalle C, Robertson B, Wood IC, Gamper N (2011) Transcriptional repression of the M channel subunit Kv7.2 in chronic nerve injury. Pain 152:742–754

Rostock A, Tober C, Rundfeldt C, Bartsch R, Engel J, Polymeropoulos EE, Kutscher B, Loscher W, Honack D, White HS et al (1996) D-23129: a new anticonvulsant with a broad spectrum activity in animal models of epileptic seizures. Epilepsy Res 23:211–223

Rundfeldt C, Netzer R (2000) The novel anticonvulsant retigabine activates M-currents in Chinese hamster ovary-cells tranfected with human KCNQ2/3 subunits. Neurosci Lett 282:73–76

Scheef W (1987) Analgesic efficacy and safety of oral flupirtine in the treatment of cancer pain. Postgrad Med J 63(Suppl 3):67–70

Schenzer A, Friedrich T, Pusch M, Saftig P, Jentsch TJ, Grotzinger J, Schwake M (2005) Molecular determinants of KCNQ (Kv7) K+ channel sensitivity to the anticonvulsant retigabine. J Neurosci 25:5051–5060

Schroder RL, Jespersen T, Christophersen P, Strobaek D, Jensen BS, Olesen SP (2001) KCNQ4 channel activation by BMS-204352 and retigabine. Neuropharmacology 40:888–898

Schroeder BC, Hechenberger M, Weinreich F, Kubisch C, Jentsch TJ (2000) KCNQ5, a novel potassium channel broadly expressed in brain, mediates M-type currents. J Biol Chem 275:24089–24095

Schroeder BC, Kubisch C, Stein V, Jentsch TJ (1998) Moderate loss of function of cyclic-AMP-modulated KCNQ2/KCNQ3 K+ channels causes epilepsy. Nature 396:687–690

Singh NA, Charlier C, Stauffer D, DuPont BR, Leach RJ, Melis R, Ronen GM, Bjerre I, Quattlebaum T, Murphy JV et al (1998) A novel potassium channel gene, KCNQ2, is mutated in an inherited epilepsy of newborns [see comments]. Nat Genet 18:25–29

Singh NA, Otto JF, Dahle EJ, Pappas C, Leslie JD, Vilaythong A, Noebels JL, White HS, Wilcox KS, Leppert MF (2008) Mouse models of human KCNQ2 and KCNQ3 mutations for benign familial neonatal convulsions show seizures and neuronal plasticity without synaptic reorganization. J Physiol 586:3405–3423

Singh NA, Westenskow P, Charlier C, Pappas C, Leslie J, Dillon J, Anderson VE, Sanguinetti MC, Leppert MF (2003) KCNQ2 and KCNQ3 potassium channel genes in benign familial neonatal convulsions: expansion of the functional and mutation spectrum. Brain 126:2726–2737

Suh BC, Horowitz LF, Hirdes W, Mackie K, Hille B (2004) Regulation of KCNQ2/KCNQ3 current by G protein cycling: the kinetics of receptor-mediated signaling by Gq. J Gen Physiol 123:663–683

Sun J, MacKinnon R (2020) Structural basis of human KCNQ1 modulation and gating. Cell 180:340–347 e349

Surur AS, Bock C, Beirow K, Wurm K, Schulig L, Kindermann MK, Siegmund W, Bednarski PJ, Link A (2019) Flupirtine and retigabine as templates for ligand-based drug design of KV7.2/3 activators. Org Biomol Chem 17:4512–4522

Tatulian L, Brown DA (2003) Effect of the KCNQ potassium channel opener retigabine on single KCNQ2/3 channels expressed in CHO cells. J Physiol 549:57–63

Tatulian L, Delmas P, Abogadie FC, Brown DA (2001) Activation of expressed KCNQ potassium currents and native neuronal M-type potassium currents by the anti-convulsant drug retigabine. J Neurosci 21:5535–5545

Teng BC, Song Y, Zhang F, Ma TY, Qi JL, Zhang HL, Li G, Wang K (2016) Activation of neuronal Kv7/KCNQ/M-channels by the opener QO58-lysine and its anti-nociceptive effects on inflammatory pain in rodents. Acta Pharmacol Sin 37:1054–1062

Tober C, Rostock A, Rundfeldt C, Bartsch R (1996) D-23129: a potent anticonvulsant in the amygdala kindling model of complex partial seizures. Eur J Pharmacol 303:163–169

Wang HS, Pan Z, Shi W, Brown BS, Wymore RS, Cohen IS, Dixon JE, McKinnon D (1998) KCNQ2 and KCNQ3 potassium channel subunits: molecular correlates of the M-channel. Science 282:1890–1893

Watanabe H, Nagata E, Kosakai A, Nakamura M, Yokoyama M, Tanaka K, Sasai H (2000) Disruption of the epilepsy KCNQ2 gene results in neural hyperexcitability. J Neurochem 75:28–33

Wickenden AD, Krajewski JL, London B, Wagoner PK, Wilson WA, Clark S, Roeloffs R, McNaughton-Smith G, Rigdon GC (2008) N-(6-chloro-pyridin-3-yl)-3,4-difluoro-benzamide (ICA-27243): a novel, selective KCNQ2/Q3 potassium channel activator. Mol Pharmacol 73:977–986

Wickenden AD, Yu W, Zou A, Jegla T, Wagoner PK (2000) Retigabine, a novel anti-convulsant, enhances activation of KCNQ2/Q3 potassium channels. Mol Pharmacol 58:591–600

Wilenkin B, Burris KD, Eastwood BJ, Sher E, Williams AC, Priest BT (2019) Development of an electrophysiological assay for Kv7 modulators on IonWorks barracuda. Assay Drug Dev Technol 17:310–321

Wladyka CL, Kunze DL (2006) KCNQ/M-currents contribute to the resting membrane potential in rat visceral sensory neurons. J Physiol 575:175–189

Wu YJ, Boissard CG, Chen J, Fitzpatrick W, Gao Q, Gribkoff VK, Harden DG, He H, Knox RJ, Natale J et al (2004) (S)-N-[1-(4-cyclopropylmethyl-3,4-dihydro-2H-benzo[1,4]oxazin-6-yl)-ethyl]-3-(2-fluoro-phenyl)-acrylamide is a potent and efficacious KCNQ2 opener which inhibits induced hyperexcitability of rat hippocampal neurons. Bioorg Med Chem Lett 14:1991–1995

Wu YJ, Boissard CG, Greco C, Gribkoff VK, Harden DG, He H, L'Heureux A, Kang SH, Kinney GG, Knox RJ et al (2003) (S)-N-[1-(3-morpholin-4-ylphenyl)ethyl]- 3-phenylacrylamide: an orally bioavailable KCNQ2 opener with significant activity in a cortical spreading depression model of migraine. J Med Chem 46:3197–3200

Wu Z, Li L, Xie F, Du J, Zuo Y, Frost JA, Carlton SM, Walters ET, Yang Q (2017) Activation of KCNQ channels suppresses spontaneous activity in dorsal root ganglion neurons and reduces chronic pain after spinal cord injury. J Neurotrauma 34:1260–1270

Wuttke TV, Seebohm G, Bail S, Maljevic S, Lerche H (2005) The new anticonvulsant retigabine favors voltage-dependent opening of the Kv7.2 (KCNQ2) channel by binding to its activation gate. Mol Pharmacol 67:1009–1017

Xiong Q, Gao Z, Wang W, Li M (2008) Activation of Kv7 (KCNQ) voltage-gated potassium channels by synthetic compounds. Trends Pharmacol Sci 29:99–107

Xiong Q, Sun H, Li M (2007) Zinc pyrithione-mediated activation of voltage-gated KCNQ potassium channels rescues epileptogenic mutants. Nat Chem Biol 3:287–296

Xu Q, Chang A, Tolia A, Minor DL Jr (2013) Structure of a Ca(2+)/CaM:Kv7.4 (KCNQ4) B-helix complex provides insight into M current modulation. J Mol Biol 425:378–394

Xu W, Wu Y, Bi Y, Tan L, Gan Y, Wang K (2010) Activation of voltage-gated KCNQ/Kv7 channels by anticonvulsant retigabine attenuates mechanical allodynia of inflammatory temporomandibular joint in rats. Mol Pain 6:49

Yue C, Yaari Y (2004) KCNQ/M channels control spike afterdepolarization and burst generation in hippocampal neurons. J Neurosci 24:4614–4624

Zhang F, Liu S, Jin L, Tang L, Zhao X, Yang T, Wang Y, Huo B, Liu R, Li H (2020) Antinociceptive efficacy of retigabine and flupirtine for gout arthritis pain. Pharmacology 105:471–476

Zhang F, Liu Y, Tang F, Liang B, Chen H, Zhang H, Wang K (2019a) Electrophysiological and pharmacological characterization of a novel and potent neuronal Kv7 channel opener SCR2682 for antiepilepsy. FASEB J 33:9154–9166

Zhang F, Liu Y, Zhang D, Fan X, Shao D, Li H (2019b) Suppression of KCNQ/M potassium channel in dorsal root ganglia neurons contributes to the development of osteoarthritic pain. Pharmacology 103:257–262

Zhang F, Mi Y, Qi JL, Li JW, Si M, Guan BC, Du XN, An HL, Zhang HL (2013) Modulation of K (v)7 potassium channels by a novel opener pyrazolo[1,5-a]pyrimidin-7(4H)-one compound QO-58. Br J Pharmacol 168:1030–1042

Zhang H, Craciun LC, Mirshahi T, Rohacs T, Lopes CM, Jin T, Logothetis DE (2003) PIP (2) activates KCNQ channels, and its hydrolysis underlies receptor-mediated inhibition of M currents. Neuron 37:963–975

Zhang J, Bal M, Bierbower S, Zaika O, Shapiro MS (2011) AKAP79/150 signal complexes in G-protein modulation of neuronal ion channels. J Neurosci 31:7199–7211

Zheng Q, Fang D, Liu M, Cai J, Wan Y, Han JS, Xing GG (2013) Suppression of KCNQ/M (Kv7) potassium channels in dorsal root ganglion neurons contributes to the development of bone cancer pain in a rat model. Pain 154:434–448

Potassium Channels in Cancer

Katrin Ganser, Lukas Klumpp, Helmut Bischof, Robert Lukowski, Franziska Eckert, and Stephan M. Huber

Contents

Abstract

Neoplastic transformation is reportedly associated with alterations of the potassium transport across plasma and intracellular membranes. These alterations have been identified as crucial elements of the tumourigenic reprogramming of cells. Potassium channels may contribute to cancer initiation, malignant progression and therapy resistance of tumour cells. The book chapter focusses on (oncogenic) potassium channels frequently upregulated in different tumour entities, upstream and downstream signalling of these channels, their contribution to the maintenance of cancer stemness and the formation of an immunosuppressive tumour

K. Ganser · L. Klumpp · F. Eckert · S. M. Huber (✉)
Department of Radiation Oncology, University of Tübingen, Tübingen, Germany
e-mail: stephan.huber@uni-tuebingen.de

H. Bischof · R. Lukowski
Department of Pharmacology, Toxicology and Clinical Pharmacy, Institute of Pharmacy, University of Tübingen, Tübingen, Germany

N. Gamper, K. Wang (eds.), *Pharmacology of Potassium Channels*,
Handbook of Experimental Pharmacology 267, https://doi.org/10.1007/164_2021_465

microenvironment. In addition, their role in adaptation to tumour hypoxia, metabolic reprogramming, as well as tumour spreading and metastasis is discussed. Finally, we discuss how (oncogenic) potassium channels may confer treatment resistance of tumours against radiation and chemotherapy and thus might be harnessed for new therapy strategies, for instance, by repurposing approved drugs known to target potassium channels.

Keywords

Cancer stem cells · Chemotherapy · Metabolic reprogramming · Metastasis · Oncochannels · Oncoimmunology · Radiation therapy · Tumour biology · Tumour hypoxia

1 Introduction

As became evident in early 1990s, cancer cells express a toolkit of ion channels that differs tremendously from the parental non-transformed cells. Soon, it turned out that this aberrant ion channel expression exerts pivotal functions in tumour biology (Nilius and Wohlrab 1992). In particular, tumour onset is frequently associated with alterations in the expression patterns of specific K^+ channels or with recurring somatic mutations in discrete K^+ channel encoding genes (Liu et al. 2002). It has been demonstrated recently in different gene-targeted mouse models that these K^+ channels fundamentally contribute to the malignancy of tumours and that their systemic deletion or pharmacologic targeting represents a suitable strategy to delay cancer onset and/or progression (Steudel et al. 2017; Mohr et al. 2020).

Mechanistically, cell proliferation depends on interconnected oscillations of membrane potential and cell volume that are regulated by K^+ channel activity (Wonderlin and Strobl 1996). Clamping the potential to high negative values has been proposed to lock terminally differentiated cells in G_0 phase of cell cycle. Vice versa, membrane depolarization seems to promote cell proliferation. Fast cycling tumour cells exhibit membrane potentials that oscillate in a depolarized voltage range which requires K^+ channels that operate at these depolarized voltages. To this end, tumour cells may repurpose their K^+ channels for cell cycle regulation; e.g. K^+ channels that repolarize an action potential in non-transformed excitable cells may become involved in cell cycle control upon transformation (Arcangeli et al. 1995). Beyond cell cycle regulation, aberrant K^+ channel expression in tumour cells contributes to oncogenic signalling that renders transformed cells independent of survival and growth factors. In addition, it triggers metabolic reprogramming, maintains a stem cell phenotype, induces tissue infiltration and metastasis, or promotes therapy resistance (for review, see Huber 2013; Pardo and Stuhmer 2014).

This brief chapter aims to introduce some mechanisms that underlie the "onco-physiology" of K^+ channels with an emphasis on clinically relevant aspects. Specifically, we will focus on the putative involvement of K^+ channels in therapy and anti-cancer immune responses. We discuss how this knowledge might be translated into K^+ channel-targeting therapies, for instance, by repurposing approved drugs. Attention will be given to a few K^+ channel types with high oncogenic potential, i.e. those which serve multiple functions in different tumour entities.

2 Oncogenic Potassium Channels

The voltage-gated ether-à-go-go-1 (hEAG1, $K_v10.1$, KCHN1) and the ether-à-go-go-related (hERG1, $K_v11.1$, KCNH2) K^+ channels that are physiologically expressed in brain (hEAG1, hERG1), and other organs such as heart, endocrine and lymphoid tissue and gastrointestinal tract (hERG1) are upregulated in many tumours (for review, see Lastraioli et al. 2015; Cázares-Ordoñez and Pardo 2017). Several pieces of evidence suggest that a further voltage-gated K^+ channel, $K_v1.3$ (KCNA3), which operates physiologically in bone marrow, lymphatic tissue and immune cells, brain and other organs, is also aberrantly expressed in a variety of tumour entities (Teisseyre et al. 2019). In detail, all of these channels have been found in tumours of mesenchymal (leukaemia, sarcoma), epithelial (e.g. gastrointestinal carcinomas), and glial (glioblastoma) origin. Similarly, deregulation of Ca^{2+}-activated intermediate conductance IK_{Ca} K^+ channels (hSK4, $K_{Ca}3.1$, Gardos, KCNN4) which physiologically are expressed in the vasculature, lymphatic tissue, bone marrow and several blood cell types, as well as in the gastrointestinal tract, has been identified in several tumour entities of mesenchymal (sarcoma, leukaemia), epithelial (e.g. prostate carcinoma) and glial origin (for review, see Mohr et al. 2019b). Moreover, aberrant expression of a second Ca^{2+}-activated K^+ channel, the almost ubiquitously expressed large conductance voltage-dependent BK_{Ca} (KCa1.1, Maxi-K, slo1, KCNMA1), has been shown for several carcinomas including breast cancer (Mohr et al. 2020) as well as for glioblastoma (Liu et al. 2002).

Although the picture might be distorted by the scientific attention selectively paid to hEAG1, hERG1, $K_v1.3$, IK_{Ca} and BK_{Ca}, our current knowledge indeed suggests that only few K^+ channel types are particularly associated with cancer. Functional in vitro studies corroborated by preclinical analyses of animal models have accumulated overwhelming evidence for the oncogenic properties of these K^+ channels (for review, see Huber 2013). In addition, several retrospective associations between the expression levels of hEAG1, hERG1, $K_v1.3$, IK_{Ca}, or BK_{Ca} in various tumours and patient outcome have been reported (e.g. Pillozzi et al. 2007; Brevet et al. 2009; Stegen et al. 2015; Mohr et al. 2020). In most cases these associations do not show statistical significance upon multivariate testing (considering patient subgroups, treatment protocols and established predictive/prognostic parameters), which is a prerequisite for defining K^+ channel abundance as an independent prognostic/predictive marker. Nevertheless, these retrospective associations underpin the preclinical evidence for hEAG1, hERG1, $K_v1.3$, IK_{Ca} and BK_{Ca} in modulating cancer development. Notably, the abundance of these channels in tumours may be associated with worse or better patient outcome, depending on tumour entity or even within an entity depending on tumour subgroup and treatment protocol (e.g. Mohr et al. 2019b). Thus, these K^+ channels may be implemented in both promoting and supressing tumour development. How these K^+ channels may drive oncogenic reprogramming of transformed cells is addressed in the next paragraphs.

3 Upstream and Downstream Signalling

In the following, we will give few examples demonstrating that K^+ channels act as integrated modules of biochemical signalling networks. K^+ channel-mediated electrosignalling includes (but is not limited to) the following: (1) conductive and thereby electrogenic channel activity (with setting of the membrane potentials, and hence, the activity of voltage-dependent transmembrane transports; (2) adjustments of K^+ concentrations; (3) regulation of electrochemical driving forces; (4) osmotic regulation (with associated iso-osmotically obliged H_2O fluxes); (5) cell volume regulation; (6) regulation of mitochondrial function and formation of reactive oxygen species (ROS). This bestows K^+ channels with a degree of control over the redox, pH, and Ca^{2+} signalling leading to regulation of downstream effector proteins of those pathways (for review, see Huber 2013; Huber et al. 2013, 2015; Klumpp et al. 2016b; Roth and Huber 2020).

Beyond that, K^+ channels reportedly form molecular complexes with membrane receptors and kinases, thereby transducing signals in the conducting and non-conducting mode. This has been elucidated by the seminal work of Annarosa Arcangeli's group (Cherubini et al. 2002), who showed that hERG1 accomplishes outside-in signalling in the so-called adhesomes that are macromolecular signalling complexes at focal adhesion sites in cancer cells. Within these adhesions, hERG1 channels physically interact with the β1 integrin-, the VEGF-, or the chemokine CXCR4 receptors in the plasma membrane. Notably, hERG channels bridge β1 integrin binding to extracellular matrix proteins to downstream kinases. Among those are extracellular signal-regulated kinase-1/2 (ERK1/2), the phosphoinositide-3-kinases (PI3K)/Akt-, or the Ras-related C3 botulinum toxin substrate-1 (Rac1)/ focal adhesion kinase-1 (FAK) pathways (for review, see Pillozzi and Arcangeli 2010; Becchetti et al. 2019).

An example for oncogenic processes regulated by K^+ channel-mediated electrosignalling is the phenotypical switch from a proliferating "grow" into a migrating/tissue invading "go" phenotype by the stromal cell-derived factor-1 (SDF1, CXCL12)/CXCR4 pathway. This pathway is upregulated in many tumour entities, including glioblastoma, and helps tumour cells to adjust to an adverse environment such as hypoxia or ionizing radiation (for review, see Eckert et al. 2018). Both stressors reportedly stabilize hypoxia-inducible factor-1α (HIF1α) (Li et al. 2007) which induces expression of several target genes including SDF1 and CXCR4. Auto- and paracrine SDF1/CXCR4 signalling, in turn, results in Ca^{2+} release and subsequent activation of Ca^{2+}-entry pathways, as well as of Ca^{2+}-sensitive K^+ channels (Edalat et al. 2016). Most importantly, K^+ channel activity is required to fine-tune Ca^{2+} signals that mediate activation of downstream effector proteins such as Ca^{2+}/calmodulin-dependent kinase-II isoforms (CaMKIIs) (Steinle

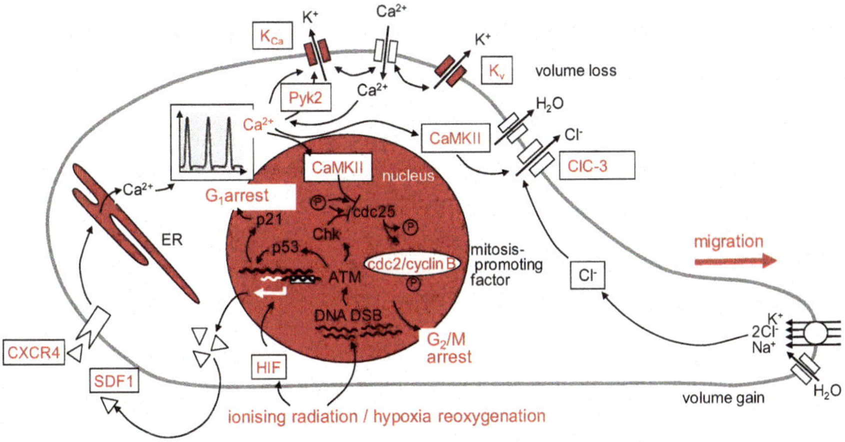

Fig. 1 K^+ channel electrosignalling is required for cellular stress response. Ionizing radiation and hypoxia stabilize hypoxia-inducible factor (HIF) resulting in auto- and paracrine signalling of the stromal cell-derived factor-1 (SDF1) via its C-X-C-motif chemokine receptor-4 (CXCR4) and generation of Ca^{2+} signals. Importantly, Ca^{2+}-activated (K_{Ca}) and voltage-activated (K_v) K^+ channels act downstream and upstream of Ca^{2+} and generate, in concert with Ca^{2+} entry pathway and the Ca^{2+}-dependent tyrosine kinase Pyk2 complex, Ca^{2+} signals needed for activation of downstream Ca^{2+} effector proteins such as Ca^{2+}/calmodulin kinase-II isoforms (CaMKIIs). Directed migration is motorized by cell volume increase at the lamellipodium and volume decrease at the cell rear. Volume changes are accomplished by ion and iso-osmotically obliged H_2O fluxes. In glioblastoma, K^+ channel-regulated cytosolic CaMKII isoforms trigger local cell volume loss by activating the ClC-3 Cl^- efflux pathway. Moreover, K^+ channel-regulated nuclear translocated CaMKII isoforms are required for G_2/M cell cycle arrest in glioblastoma and leukaemia cells. DNA double-strand breaks (DNA DSB) caused by ionizing radiation or reoxygenation of hypoxic cells may induce G_1 cell cycle arrest via activation of the protein kinase ataxia-telangiectasia mutated (ATM), subsequent stabilization of p53, and induction of p21 as well as G_2/M arrest via activation of checkpoint kinases (Chk), and inhibitory phosphorylation of the cdc25 phosphatases. The latter results in inactivity of the cdc2/cyclin B mitosis-promoting factor. Importantly, inhibition of cdc25 also requires its phosphorylation by CaMKII. Activation of the latter crucially depends on K^+ channel signalling that is probably triggered by the HIF/SDF1/CXCR4 axis (for details see main text)

et al. 2011; Stegen et al. 2016). K^+ channel-dependent activation of cytosolic CaMKII and further Ca^{2+}-dependent kinases such as the proline-rich-tyrosine-kinase-2 (Pyk2), in turn, have been demonstrated to programme stress-stimulated migration (Steinle et al. 2011; Edalat et al. 2016; Stegen et al. 2016) (Fig. 1).

Similarly, irradiation triggers Ca^{2+} signalling and activation of nuclear translated CaMKII isoforms in a K^+ channel-dependent manner. Intriguingly, BK_{Ca} and IK_{Ca} in glioblastoma (Stegen et al. 2016), as well as $K_v3.4$ and hERG1 K^+ channel in leukaemia (Palme et al. 2020, 2013) seem to exert antagonizing effects on cytosolic free Ca^{2+} levels hinting to a complex crosstalk between different K^+ channel types with Ca^{2+} entry, release and/or sequestration and extrusion. Radiation-induced CaMKII activation halts cell cycle progression by triggering G_2/M cell cycle arrest via inactivating phosphorylation of cdc25 isoforms. Active cdc25 is required to de-phosphorylate and thereby activate cdc2 (cyclin-dependent kinase-1, CDK1), a subunit of the mitosis-promoting factor which initiates entry into mitosis (Heise et al. 2010; Steinle et al. 2011; Palme et al. 2013; Klumpp et al. 2017) (Fig. 1).

Besides electrosignalling at the plasma membrane, K^+ channels residing in intracellular membranes of, e.g. mitochondria and the nucleus, also have been reported to exert signalling functions (for review, see Leanza et al. 2013). For instance, BK_{Ca} channel activity in the nuclear envelope has been shown to regulate gene expression in neurons via modulating nuclear Ca^{2+} signalling (Li et al. 2014a). In cardiomyocytes, mitochondrial BK_{Ca} channels ($_{mt}BK_{Ca}$) affect ischaemia/perfusion-induced ROS production (Frankenreiter et al. 2017). K^+ channels may also signal via modulating intracellular K^+ concentrations as described in the next paragraph.

4 Metabolic Reprogramming and Adaptation to Tumour Hypoxia

Already in the 1920, Otto Heinrich Warburg hypothesized that cancer cells reprogramme their metabolisms towards aerobic glycolysis at the expense of the much more energy-efficient mitochondrial oxidative phosphorylation. In aerobic glycolysis, pyruvate is reduced to and extruded as lactic acid even at high oxygen pressure (Warburg et al. 1927). Among enzymes that directly utilize K^+ as cofactor (or depend via other mechanisms on K^+ concentration) are pyruvate kinase (Kachmar and Boyer 1953), pyruvate dehydrogenase kinase (Kato et al. 2005), ribokinase (Quiroga-Roger et al. 2015) and hexokinase II (Rose and Warms 1982). In particular, recent research demonstrated that cellular K^+ depletion as well as the presence and activation of specific K^+ channels in the plasma membrane alters the glucose metabolism and energy homeostasis of hexokinase II positive cancer cells in culture (Bischof et al. 2021). Whether such K^+ dependence is sufficient to alter the enzyme activities under (patho-)physiological alterations has not been defined.

All solid tumours at a certain size experience hypoxia due to insufficient tumour perfusion. Under hypoxia the O_2-sensitive hypoxia-inducible transcription factors (HIFs) and the cellular energy-sensing AMP kinase trigger downregulation of anabolic metabolism, upregulation of nutrient import and anaerobic glycolysis, an adaptation of glutamine metabolisms to maintain fuelling of the citrate pool, alteration of lipid metabolism, as well as upregulation of oxidative defence and attenuation of mitochondrial superoxide anion formation (for review, see Xie and Simon 2017; Eckert et al. 2019). The latter is promoted, e.g. during reoxygenation of intermittently hypoxic volumes of solid tumours upon mitochondrial Ca^{2+} overflow and $\Delta\Psi_m$ hyperpolarization (Fig. 2; for review, see Huber et al. 2013).

Beyond their expression in the plasma membrane, some K^+ channels are found in the inner mitochondrial membrane; these include ATP-sensitive mitochondrial (mt) $_{mt}K_{ir}6.x$ (Fu et al. 2003; Paggio et al. 2019), Ca^{2+}-activated $_{mt}BK_{Ca}$ (Singh et al. 2013; Gu et al. 2014; Frankenreiter et al. 2017), $_{mt}IK_{Ca}$ (De Marchi et al. 2009) and $_{mt}SK_{Ca}$ (Stowe et al. 2013), as well as voltage-gated (e.g. $_{mt}K_v1.3$, KCNA3)

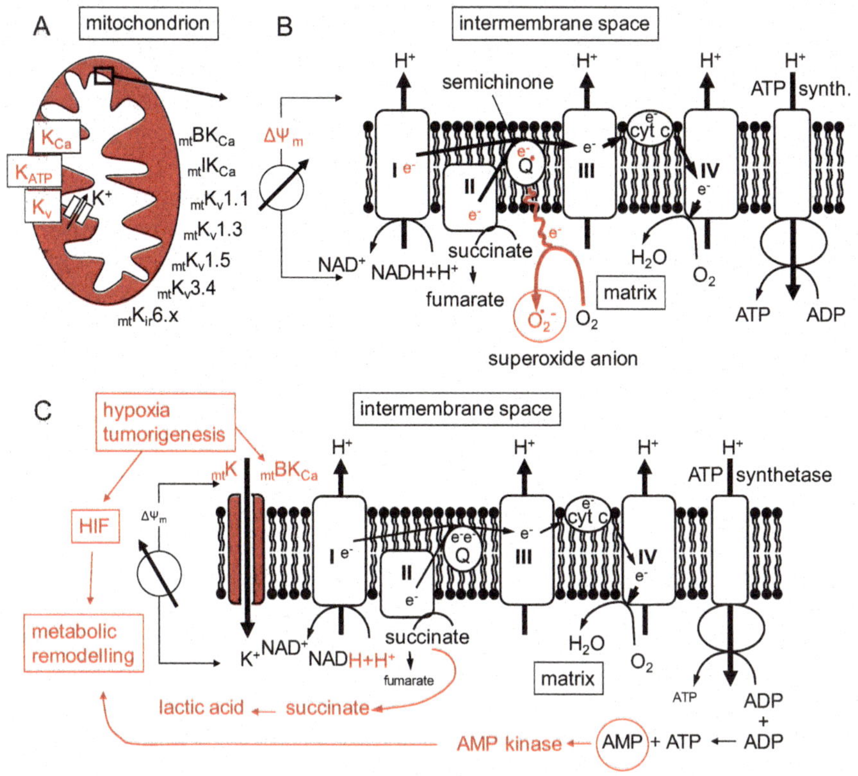

Fig. 2 Ca^{2+}-activated ($_{mt}K_{Ca}$), voltage-activated ($_{mt}K_V$) and ATP-sensitive ($_{mt}K_{ATP}$, $_{mt}K_{ir}6.x$) K^+ channels in the inner mitochondrial membrane may be involved in oxygen sensing, formation of superoxide anion radicals, metabolic reprogramming and adaptation to a radioresistance-conferring hypoxic tumour microenvironment (a hypothetical model). (**a**) K^+ channel expressed in the inner mitochondrial membrane. (**b**) Inner mitochondrial membrane ($\Delta\Psi_m$) hyperpolarization-associated superoxide anion ($O^{\bullet}{}_2{}^-$) formation by the mitochondrial electron transport chain that particularly occurs during reoxygenation of intermittently hypoxic cells. Complexes I to IV and the ATP synthase in the inner mitochondrial membrane are shown (Q: ubiquinone, cyt C: cytochrome C, e^-: electron). (**c**) Partial dissipation of $\Delta\Psi_m$ by upregulation of $_{mt}K^+$ channels in cancer cells lowers $O^{\bullet}{}_2{}^-$ formation, oxidative phosphorylation and metabolic ATP yield. The latter might contribute to metabolic reprogramming towards aerobic glycolysis and lactic acid formation via activation of AMP kinase. $_{mt}BK_{Ca}$ channels are reportedly activated by hypoxia and have been proposed to act as oxygen sensors (for details see main text)

channel (Roberta et al. 2020). These and other K^+ channels may dissipate the mitochondrial inner membrane potential ($\Delta\Psi_m$). Mechanistically, $_{mt}BK_{Ca}$ is reportedly involved in oxygen sensing by increasing its open probability with decreasing oxygen pressure (Gu et al. 2014). $_{mt}K^+$ channels-mediated $\Delta\Psi_m$ dissipation lowers oxidative phosphorylation capacity and thereby ATP production and superoxide anion formation, which are directly associated with $\Delta\Psi_m$ (Kulawiak et al. 2008). By these mechanisms, $_{mt}K^+$ channels may contribute to adaptation of solid tumours to chronic or intermittent hypoxia as deduced from the phenomenon of cardiac ischaemic preconditioning whereby the upregulation of $_{mt}K^+$ channel protects the heart from ischaemia-reperfusion injury (Fig. 2; for review, see Madreiter-Sokolowski et al. 2019). In contrast to this proposed mechanism, $_{mt}K_V1.3$ channels have been demonstrated to directly interact with complex I of the respiratory chain in leukaemia and melanoma cells where inhibition of $_{mt}K_V1.3$ results in superoxide anion formation and triggering of cell death (Roberta et al. 2020).

Hypoxia-stimulated metabolic reprogramming may be accompanied by an increase in cellular plasticity of tumour and stroma cells that drives tumour heterogeneity. Well documented is a hypoxia-driven epithelial-mesenchymal transition (EMT) (or neural/glial-mesenchymal transition in brain tumours) or induction of cancer stem(-like)/tumour initiating cells (CSCs). Importantly, EMT and upregulation of CSC properties are accompanied by a change from a "grow" to a "go" phenotype increasing the risk of tissue infiltration and metastasis as discussed in the next paragraphs (for review, see Eckert et al. 2019).

5 Maintenance of Cancer Stemness

A small population of tumour cells, the CSCs, has been suggested to exhibit higher therapy resistance and propensity to invade healthy tissue as compared to the bulk of "differentiated" cancer cells. CSCs, therefore, are held responsible for local therapy failure and metastases. K^+ channels contribute to maintenance of the CSC phenotype. This can be deduced from in vitro and animal studies demonstrating that salinomycin, an antibiotic with probable K^+ ionophore function, specifically targets CSCs in breast-, lung- and colorectal cancer as well as in leukaemia (Fuchs et al. 2009; Gupta et al. 2009; Dong et al. 2011; Wang 2011; Alama et al. 2012). Notably, CSCs of ovarian cancer which exhibit ABC transporter-mediated salinomycin resistance are eradicated by a synthetic K^+ ionophore (Shen et al. 2020). Together, this might suggest that the CSC phenotype is more sensitive to disruption of K^+ channel function and/or K^+ homeostasis as compared to "differentiated" bulk tumour cells. Along those lines, tetraethylammonium (TEA), a largely unselective K^+ channel antagonist, reportedly decreases clonogenicity of CSCs in endometrial cancer (Schickling et al. 2011). Likewise, clonogenicity of $CD34^+$ haematopoietic progenitor cells and the proliferative subpopulation of blasts in acute myeloid leukaemia (AML) reportedly requires hERG channel activity (Pillozzi et al. 2002).

In breast cancer cells, overexpression of G protein-activated inward rectifier K^+ channel-1 (Kir3.1, KCNJ3) has been shown to boost mammosphere formation suggesting that the self-renewal capacity depends on Kir3.1 (Kammerer et al. 2016). Moreover, K_v channel blockers or a K_v7 opener has been shown to sensitize CSCs of a lung cancer cell line to the tyrosine kinase inhibitor gefitinib (Choi et al. 2017). Furthermore, CSC subpopulations of neuroblastoma and glioblastoma cell lines have been demonstrated to exhibit elevated expression of BK_{Ca} as compared to non-CSCs (Park et al. 2010; Rosa et al. 2017). Importantly, high BK_{Ca}-expressing glioblastoma CSCs (GSCs) tend to favour neuronal above astrocyte lineage differentiation suggesting that BK_{Ca} contributes to lineage priming (Rosa et al. 2017). In addition, migration of GSCs requires BK_{Ca} (Rosa et al. 2017) and IK_{Ca} (Ruggieri et al. 2012) activity. Notably, IK_{Ca} upregulation is associated with the migration/invasion-prone mesenchymal subpopulation of GSCs (Klumpp et al. 2018), suggesting that tumour spread and metastasis are tightly controlled by K^+ channel pathways, a topic which is introduced in the next paragraph.

6 Tumour Spreading and Metastasis

To adopt a migratory, invasive and infiltrating phenotype, tumour cells undergo epithelial-mesenchymal (EMT) or – in the case of gliomas – glial-mesenchymal transition (GMT) that reportedly is accompanied by remodelling of the Ca^{2+} signalosome and associated with altered K^+ channel expression and activity (for review, see Klumpp et al. 2016b). Directed cell migration is orchestrated by polarized and local cell volume changes that are paralleled by reorganization of the cytoskeleton and the plasma membrane. The mechanics of cell migration is propelled by H_2O influx-driven protrusion of the lamellipodium and H_2O efflux-mediated retraction of the cell rear. To maintain iso-osmotic balance these H_2O fluxes require transmembrane transport of ions, in particular Cl^- and K^+. In addition, osmolyte transport is a key process for the motorization of migration (for review, see Schwab et al. 2012).

In glioblastoma, a highly infiltrating primary brain tumour, the involved transport processes have been identified by the pioneering work of Harald Sontheimer's group. These colleagues reported that at the lamellipodium, a $Na^+/K^+/2Cl^-$-cotansport not only acquires cell volume for protrusion of the leading edge but also accumulates Cl^- in the cytosol highly above its electrochemical equilibrium. Cl^-, which reaches cytosolic concentrations of up to 100 mM, in turn, can be used as osmolyte for highly effective and fast local cell volume decrease upon efflux of Cl^- together with the counterion K^+ and iso-osmotically obliged H_2O (for review, see Sontheimer 2008). Glioblastoma cells invade the brain parenchyma by squeezing through interstitial spaces alongside vessels or axon bundles which are used as tracks. Therefore, effective cell volume regulation is a prerequisite for brain infiltration, contributing to the high malignancy of glioblastoma cells (Watkins and Sontheimer 2011) (Fig. 1).

In glioblastoma (Catacuzzeno et al. 2011) as well as in endometric, hepatocellular, pancreatic, prostate and lung carcinoma mouse and cell models (Liu et al. 2015b; Bonito et al. 2016; Storck et al. 2017; Du et al. 2019; Zhang et al. 2019), IK_{Ca} channels have been identified as essential drivers of basal or agonist-stimulated tumour cell migration. This suggests that IK_{Ca} is not only associated with a pro-migratory mesenchymal cancer stem cell phenotype (see above), but also essentially required for the execution and/or programming of cell migration of various tumour entities from different origins. BK_{Ca} activity, on the other hand, is required for ionizing radiation-induced hypermigration of glioblastoma cells as demonstrated in vitro (Steinle et al. 2011) and in an orthotopic mouse model (Edalat et al. 2016).

Besides boosting migration, K^+ channels may promote therapy resistance as described in more detail in the next paragraphs.

7 Therapy Resistance

About half of all cancer patients undergo radiotherapy in order to induce lethal DNA damages such as DNA double-strand breaks (DSBs) in the targeted cancer cells. Beyond the dose of ionizing radiation, the extent of DNA damages depends on the O_2 tension of the tumour microenvironment, the activity of mitochondrial metabolism and the cellular anti-oxidative defence. Passing through mitosis with unrepaired DSBs introduces chromosome aberrations in proliferating cells, finally leading to mitotic catastrophe and, eventually, programmed cell death (for review, see Roth and Huber 2020). On the other hand, DSBs trigger the cellular DNA damage response which comprises deceleration of the cell cycle, activation of DNA repair pathways and metabolic reprogramming. Radiogenic K^+ channel activity has been suggested to contribute to all these DSB-triggered responses.

Within 1–2 h after ionizing radiation, BK_{Ca} (Steinle et al. 2011) and IK_{Ca} (Stegen et al. 2015) are activated in the plasma membrane of glioblastoma cells. Radiogenic IK_{Ca} activity has also been shown in human lung adenocarcinoma (Gibhardt et al. 2015; Roth et al. 2015), in murine breast cancer (Mohr et al. 2019a) and in human T cell leukaemia cells (Klumpp et al. 2016a; Voos et al. 2018). Moreover, in chronic myeloid leukaemia (CML) cells, ionizing radiation stimulates the activity of $K_v3.4$ (KCNC4) (Palme et al. 2013) and hERG1 (Palme et al. 2020). Radiogenic K^+ channel activity is preceded and/or accompanied by cytoplasmic Ca^{2+} signalling. In particular, in T-cell leukaemia (Klumpp et al. 2016a), CML (Heise et al. 2010) and glioblastoma cell lines (Klumpp et al. 2017) radiogenic activation of transient receptor potential (TRP) Ca^{2+} permeable nonselective cation channels has been proposed to trigger K^+ channel activity by membrane depolarization and/or a rise of the cytosolic free Ca^{2+} concentration. By modulating driving force and/or activity of Ca^{2+} entry pathways, radiogenic electrosignalling by K^+ channels, in turn, contributes to complex Ca^{2+} signals that regulate downstream Ca^{2+} effector proteins involved in DNA damage response (Fig. 1). Importantly, pharmacological blockage or knockdown of these channels radiosensitizes the cells in vitro and in preclinical mouse models (Palme et al. 2020, 2013; Stegen et al. 2015; Mohr et al. 2019a), indicating the pivotal function of K^+ channels in DNA damage response of irradiated cancer cells.

A significant fraction of radiogenic DNA damages occurs late (hours after irradiation) through Ca^{2+}-mediated mitochondrial formation of ROS. It is hardly surprising that adaptation to hypoxia, for example by upregulation of mitochondrial K^+ ($_{mt}K^+$) channels (see above) and uncoupling proteins (Braun et al. 2015) in the inner mitochondrial membrane at the same time, confers radioresistance (Braun et al. 2015). Moreover, hypoxic tumour volumes are per se more radioresistant than normoxic tumours due to the oxygen-dependent efficacy of ionizing radiation (for review, see Eckert et al. 2019; Roth and Huber 2020). Consequently, tumour cells automatically acquire radioresistance when residing in hypoxic areas.

Along those lines, metabolic reprogramming of tumour cells from mitochondrial oxidative phosphorylation towards aerobic glycolysis and lactic acid formation is associated with a gain in radioresistance (Cruz-Gregorio et al. 2019). This reprogramming also involves altered membrane transports including upregulation of glucose uptake across the plasma membrane. Glucose uptake via electrogenic Na^+-coupled cotransport (SGLT) is reportedly upregulated during DNA damage response of irradiated lung adenocarcinoma and head and neck squamous cell carcinoma cell lines (Huber et al. 2012). This increased glucose uptake is likely to counteract the elevated ATP consumption and provides the overhang of carbohydrates required as a source of metabolites for chromatin remodelling by histone acetylation and DNA repair (Dittmann et al. 2013). Radiogenic co-activation of IK_{Ca} in the plasma membrane (Roth et al. 2015) most probably counteracts SGLT-mediated membrane depolarization (Huber et al. 2012) and maintains the inwardly directed driving force for Na^+-coupled glucose uptake. Hence, besides $_{mt}K^+$ channel activity, plasmalemmal K^+ channels may also contribute to metabolic reprogramming of tumour cells towards a radioresistant phenotype.

In addition to radioresistance, K^+ channels have been demonstrated to modulate the resistance against chemotherapeutics in cancer cells. To give few examples, doxorubicin resistance induced in acute lymphoblastic leukaemia cells by co-culture with bone marrow mesenchymal cells reportedly depends on the formation of a hERG1/β_1-integrin/CXCR4 signalling complex that triggers pro-survival signals via ERK1/2 and PI3K/Akt (Pillozzi et al. 2011). Likewise, hERG1 expression or activity has been demonstrated to confer cisplatin and 5-fluorouracil resistance to colorectal cancer cells (Fortunato 2017; Pillozzi et al. 2018; Petroni 2020), as well as the resistance to vincristine, paclitaxel and hydroxy-camptothecin to lung adenocarcinoma cells (Chen et al. 2005). Additionally hERG1 is associated with chemoresistance of retinoblastoma cells (Fortunato et al. 2010). A chemoresistance-promoting action has been documented for further K^+ channel types such as hEAG1 (Hui et al. 2015; Sales et al. 2016; García-Quiroz et al. 2019), BK_{Ca} (Samuel et al. 2016) and Kir2.1 (KCNJ2) (Liu et al. 2015a).

On the contrary, in liver carcinoma and HeLa-derived KB cells, hERG1 channel activity does not alter cisplatin sensitivity (Liang et al. 2005) while in gastric cancer cells, hERG1 expression has been shown to be required for cisplatin-induced apoptotic cell death (Zhang et al. 2012). Similarly, BK_{Ca} (Samuel et al. 2016) and IK_{Ca} (Pillozzi et al. 2018) activity reportedly promotes cisplatin sensitivity in ovarian and colorectal cancer, respectively. Notably, in colorectal cancer pharmacological IK_{Ca} activation and hERG1 blockage exert additive effects on cisplatin sensitivity, suggesting an antagonizing electrosignalling by both channel types during stress response to cisplatin-mediated DNA damage (Pillozzi et al. 2018).

Finally, BK_{Ca} has been demonstrated to modulate hormone treatment in breast cancer in vitro and in orthotopic mouse models (Mohr et al. 2020). Together, these data indicate that K^+ channels interfere with therapy response in cancer cells, highlighting the potential of their modulation as new anti-cancer treatment concepts.

Nonetheless, such targeting of K^+ channels may also interfere with immunotherapies.

8 Immune Surveillance, Immunosuppression and Anti-cancer Immune Response

Lifelong our immune system eradicates tumour cells which emerge permanently by neoplastic transformation. To form a primary tumour and micro-metastasis, nascent and spread tumour cells, respectively, must evade the local immune surveillance. To this end, transformed or metastasized cells may pursue perfidious strategies such as dormancy or normal tissue mimicry to hide themselves from the immune system (for review, see Klumpp et al. 2016b). Once established, solid tumours build up an immunosuppressive microenvironment with the help of recruited immune cells such as tumour-associated macrophages (TAM), regulatory T cells (T_{reg}) or, in the case of brain tumours, microglia cells. Currently developed and applied immunotherapies aim to attenuate immunosuppression in the tumour and to boost the anti-tumour immune response (for review, see Eckert et al. 2019). Innate and adaptive immunity in oncology, however, critically depend on ion channel function. Hence, any strategy of targeting K^+ channels for anti-cancer therapy must also consider the effects on anti-cancer immune responses, especially if K^+ channel targeting should be combined with immunotherapy.

As comprehensively summarized by Feske and colleagues, ion channels play major roles in all stages of immune responses (Feske et al. 2015). Thus, T cell activation following antigen recognition and activation is dependent on Ca^{2+} influx with subsequent calcineurin signalling (Trebak and Kinet 2019). This signalling is tightly regulated by several ion channels including $K_v1.3$ and IK_{Ca} (Chiang et al. 2017). The role of ion channels in immunology is also highlighted by the fact that immunosuppression is a key feature of genetic channelopathies such as XMEN (X-linked immunodeficiency with magnesium defect, Epstein-Barr virus infection and neoplasia) disease (Li et al. 2014b) and CRAC channelopathies (disorders caused by mutations in ORAI1 or STIM1, major constituents of the Ca^{2+} release activated Ca^{2+} channel, CRAC) (Lacruz and Feske 2015; Feske 2019).

$K_v1.3$ and IK_{Ca} seem to have different roles in different T cell subsets. While naïve T cells mostly express $K_v1.3$, after activation the abundance of IK_{Ca} increases. $K_v1.3$ has been described to play a central role in $CD4^+$ Th17 cells, as well as effector memory T cells. T_{reg} cells show IK_{Ca} expression. The role in central memory T cells led to the development of $K_v1.3$ channel blockers for the treatment of Psoriasis with Shk-186 (dalazatide) (Tarcha et al. 2017).

In the cancer microenvironment, the function of several tumour infiltrating cell types is influenced by K^+ channel activity. In vitro and in vivo data show that human natural killer (NK) cells express both $K_v1.3$ and IK_{Ca}. Blocking of those channels results in enhanced proliferation of NK cells and reduced growth of *xeno*grafted tumours (Koshy et al. 2013). In an orthotopic glioblastoma mouse model, IK_{Ca} inhibition led to a macrophage/microglia polarization towards M1 and, thus, to a tumour inhibiting phenotype (Grimaldi et al. 2016). In addition, IK_{Ca} inhibition has been demonstrated to mitigate microglia activation in an orthotopic mouse glioblastoma model and to reduce phagocytosis and chemotactic activity of primary microglia cells exposed to glioblastoma-conditioned medium (D'Alessandro et al. 2013).

Moreover, in tumour and blood samples acquired from patients with head and neck squamous cell carcinomas, $K_v1.3$ expression correlated with several markers of $CD8^+$ T cell activation (Chimote et al. 2017). In addition, IK_{Ca} activity has been demonstrated to contribute to chemotaxis of $CD8^+$ T cells in the presence of immunosuppressive agent, adenosine, in the TME (Chimote et al. 2018). These effects may suggest an immunosuppressing effect of IK_{Ca} or $K_v1.3$ targeting. In addition, tumour-associated macrophages seem to need the activity of IK_{Ca} to release pro-inflammatory IL-6 and IL-8 cytokines that promote tumour invasiveness in colorectal cancer (Xu et al. 2014).

Besides the abundance and activity of potassium channels in different immune cell subsets, the ionic tumour microenvironment also plays a role in cancer immunology (Eil et al. 2016). High interstitial K^+ concentrations $[K^+]_e$ suppress the function of tumour infiltrating lymphocytes (Chandy and Norton 2016; Eil et al. 2016; Conforti 2017) by impairing the proliferation of central memory and effector memory T cells, T cell cytokine production and anti-tumour cytotoxicity. In addition, high $[K^+]_e$ stimulates upregulation of the immune checkpoint protein PD-1 in activated T cells. Importantly, pharmacological IK_{Ca} activation rescues T cell function in high $[K^+]_e$ environment (Ong et al. 2019).

Combined, the limited data available suggest that targeting of K^+ channels such as $K_v1.3$ and IK_{Ca} may both stimulate and suppress the anti-tumour immune response. These opposing effects may differ between the various tumour entities. Many approved drugs that are directed against K^+ channels or modulate them by off-target effects are available. These drugs may be repurposed for anti-cancer therapy as discussed in the next paragraphs.

9 Repurposing Drugs for Potassium Channel-Targeting Therapies

Approved drugs targeting directly or indirectly K^+ channels comprise anaesthetics, neuroleptics, antidepressants, antihistamines, fluoroquinolone antibiotics, analgesics, anticonvulsants, antiarrhythmics, antihypertensives, antidiabetics and others (for review, see Kale et al. 2015). The observed side effects of pharmacological or molecular K^+ channel targeting in preclinical models range from mild hypertension to life-threatening cardiac arrhythmias. As an example, for the latter, "torsadogenic" hERG blockers may delay cardiac repolarization and cause *Torsade de Pointes* and sudden cardiac death, due to the pivotal function of hERG1 in the repolarization of the heart action potential (see also chapter "Cardiac hERG K^+ channel as safety and pharmacological target"). Nevertheless, pharmacological modulation of oncogenic hERG1 channels may remain an option. Drugs such as some antiepileptics exert off-target inhibition of hERG1 partially with submicromolar IC_{50}s without being "torsadogenic". One reason might be that these "non-torsadogenic" hERG1 inhibitors do not achieve steady-state free plasma concentrations required for clinically relevant hERG1 inhibition (Lehmann et al. 2018). Another reason for their "non-torsadogenicity" might be that the potency of a certain drug to induce fatal arrhythmia relies not only on hERG1 inhibition but also on the whole spectrum of other ion channels modulated by this specific drug (for commentary, see Arcangeli and Becchetti 2017). Accordingly, "non-torsadogenic" hERG1 inhibitors might impact tumour biology without exerting severe side effects. Along those lines, multivariate analysis of glioblastoma defined hERG1 protein abundance in the tumour specimen as an independent prognostic parameter for shorter patient survival. Importantly, overall survival of patients with highly hERG1-expressing glioblastoma (but not of those with low hERG1 abundance in the tumour) seemed to benefit from "non-torsadogenic" hERG1-inhibiting drugs (Pointer et al. 2017). Consistently, several mouse tumour models have proven the anti-cancer efficacy of hERG1 targeting strategies (Pillozzi et al. 2011).

Beyond that, approved drugs for targeting of oncogenic EAG1 such as the neuroleptic chlorpromazine (Wang et al. 2020), the tricycle antidepressant imipramine (García-Ferreiro et al. 2004), or the second-generation antihistamine astemizole (García-Ferreiro et al. 2004) are available. In preclinical in vitro and mouse tumour models (García-Quiroz et al. 2014), astemizole exhibits tumouricidal and/or chemo- or hormone therapy-sensitizing activity. Likewise, EAG2 can be targeted by the antipsychotic drug thioridazine. EAG2 expression has been reported at the trailing edge of migrating medulloblastoma cells where it regulates local cell volume dynamics and cell motility. Blockage of EAG2 by thioridazine has been shown to reduce growth and metastasis of medulloblastoma cell *xeno*grafts (Huang et al. 2015).

Classical neuroleptics such as haloperidol or chlorpromazine inhibit BK_{Ca} channels with IC_{50}s in the low micromolar range (Lee et al. 1997). As well as in a bunch of other cell types, BK_{Ca} channels are expressed in neurons of the central nervous system (Sausbier et al. 2006). Haloperidol and chlorpromazine may accumulate in the human brain up to micromolar concentrations (Huang and Ruskin 1964; Korpi et al. 1984). Reported side effects of BK_{Ca} channel blockage in the central nervous systems are ataxia and uncontrollable tremors as impressively illustrated by ovine "ryegrass staggers" caused by intoxication of paxilline, a very potent BK_{Ca} channel blocker produced by the endophytic fungus *Penicillium paxilli* (Imlach et al. 2008).

The topically administered antifungal imidazole derivative clotrimazole inhibits IK_{Ca} channels with submicromolar IC_{50}s (Jensen et al. 1998). Systemically, clotrimazole is no longer in clinical use due to the inhibition of cytochrome P450-dependent enzymes and associated liver toxicity. By replacing the imidazole ring of clotrimazole, responsible for the inhibition of cytochrome P450-dependent enzymes, further IK_{Ca} channel blockers were developed. Among those, the orally bioavailable ICA-17043 (Senicapoc®; Icagen) inhibits IK_{Ca} with an IC_{50} of 11 nM (for review, see Wulff and Castle 2010), and was well tolerated in clinical trials (Ataga et al. 2008).

Besides ERG, EAG, BK_{Ca} and IK_{Ca}, further K_v-, Ca^{2+}-activated-, background (K2P)- or inwardly rectifying K^+ channel types are druggable with pharmaceutics in clinical use. As an example for the latter channel types, K_{ATP} channels (Kir6.1,-2, KCNJ8, -11) can be targeted by sulphonylurea receptor antagonists such as glibenclamide used in type-2 diabetes. Additionally, $K_v1.3$ channels can be targeted by the anti-psoriasis drug dalazatide which is well tolerated in clinical trials (for review, see Kale et al. 2015). Combined, this suggests that clinical trials on K^+ channel-targeted therapies in cancer might be feasible. By repurposing approved drugs that are in clinical use with well-described pharmacokinetics, the achievable concentrations in the organ of interest are known. Thus, efficacy of pharmaceutical K^+ channel modulation should become predictable. The same holds true for the associated side effects.

10 Concluding Remarks

Theoretically, prescription of certain drugs (for whatever reason) which modulate oncogenic K^+ channels "off-target" should lower the incidence of tumours or improve the survival of cancer patients in that specific population. Such epidemiological evidence, however, is rare (Pointer et al. 2017) or missing. In addition, to the best of our knowledge, not a single clinical trial on K^+ channel-targeted therapies in cancer has been announced up to date (https://clinicaltrials.gov). Thus, robust clinical proof-of-concepts for the oncogenic function of K^+ channels are lacking. This might be a reason why K^+ channel-targeting seems not to be considered in currently developing concepts of anti-cancer therapy. Given the pivotal functions of K^+ channels in physiology, the fear of unmanageable side effects might further hinder the development of K^+ channel-targeting therapies, particularly if dose escalation is required to reach therapeutic concentrations in the tumour.

A better tumour specificity of K^+ channel-targeting might be achieved by tailoring K^+ channel-targeting drugs to tumour-associated/specific variants or molecular complexes of K^+ channels. For instance, glioblastoma upregulates a variant of BK_{Ca}, gBK_{Ca}, that contains an additional 34-amino-acid exon at splice site 2 in the C-terminal tail of BK_{Ca} channels (Liu et al. 2002). Additional approaches to translate our current knowledge on oncogenic K^+ channels into clinical applications include the use of K^+ channels as tumour biomarkers (Erdem et al. 2015) or delivery of targeting molecules to tumour cells by coupling them to antibodies directed against tumour-associated K^+ channels (Hartung and Pardo 2016). Beyond that, pharmacological modulation of K^+ channel activity in normal tissue may be used to mitigate the normal tissue toxicity of anti-cancer therapy (Li et al. 2019).

The "executive" functions of K^+ channels in the plasma membrane for generating ionic currents and potentials, transmembrane transports and cell volume regulation during, e.g. cell cycling of cancer cells, are fairly well understood. There are also a very few examples of specific oncogenic pathways, where K^+ channels, either within molecular complexes with receptors and kinases or by reciprocal interaction with the Ca^{2+} signalosome and downstream Ca^{2+} effector proteins, function as integral modules of biochemical signalling cascades. However, for designing of treatment protocols and time schedules of K^+ channel-targeting therapies (in addition to standard radio/chemotherapy) a more comprehensive understanding of the specific oncogenic functions of the relevant K^+ channel types in a given tumour entity is urgently needed. We are just at the very beginning of understanding how K^+ channels in the inner mitochondrial membrane might contribute to redox signalling, metabolic reprogramming, or adaptation to intermittent or chronic hypoxia of cancer cells. The functions of other intracellular K^+ channels such as those in the inner and outer membrane of the nuclear envelope, by contrast, are almost completely undefined. Likewise, it cannot be foreseen at present how pharmacological targeting of K^+ channels in the tumour alters the highly complex crosstalk between tumour and stroma cells that may also express the targeted channel type. Indeed, IK_{Ca} and $K_v1.3$ channels upregulated in many solid tumours are also expressed in immune cells and required for, e.g. T cell priming and clonal expansion. Targeting these channels might, therefore, be a double-edged sword leading to tumouricidal, but also unintended immunosuppressive effects.

Acknowledgement FE was partly funded by the Else-Kroener-Fresenius Research Foundation (Grant 2015_Kolleg.14) and the Gesellschaft für Kinderkrebsforschung, SMH by a grant from the German Cancer Aid (70112872, 70113144). HB is a fellow of the Alexander von Humboldt Foundation. RL received funding from the ICEPHA Graduate Program "Membrane-associated Drug Targets in Personalized Cancer Medicine".

Conflict of Interest All authors declare no competing interests.

References

Alama A, Orengo AM, Ferrini S, Gangemi R (2012) Targeting cancer-initiating cell drug-resistance: a roadmap to a new-generation of cancer therapies? Drug Discov Today 17:435–442. https://doi.org/10.1016/j.drudis.2011.02.005

Arcangeli A, Becchetti A (2017) hERG channels: from antitargets to novel targets for cancer therapy. Clin Cancer Res 23:3–5. https://doi.org/10.1158/1078-0432.Ccr-16-2322

Arcangeli A, Bianchi L, Becchetti A et al (1995) A novel inward-rectifying K+ current with a cell-cycle dependence governs the resting potential of mammalian neuroblastoma cells. J Physiol 489(Pt 2):455–471

Ataga KI, Smith WR, De Castro LM et al (2008) Efficacy and safety of the Gardos channel blocker, senicapoc (ICA-17043), in patients with sickle cell anemia. Blood 111:3991–3997

Becchetti A, Petroni G, Arcangeli A (2019) Ion channel conformations regulate integrin-dependent signaling. Trends Cell Biol 29:298–307. https://doi.org/10.1016/j.tcb.2018.12.005

Bischof H, Burgstaller S, Springer A et al (2021) Potassium ions promote hexokinase-II dependent glycolysis. eScience 24:102346. https://doi.org/10.1016/j.isci.2021.102346

Bonito B, Sauter DR, Schwab A, Djamgoz MB, Novak I (2016) K(Ca)3.1 (IK) modulates pancreatic cancer cell migration, invasion and proliferation: anomalous effects on TRAM-34. Pflugers Arch 468:1865–1875. https://doi.org/10.1007/s00424-016-1891-9

Braun N, Klumpp D, Hennenlotter J, Bedke J, Duranton C, Bleif M, Huber SM (2015) UCP-3 uncoupling protein confers hypoxia resistance to renal epithelial cells and is upregulated in renal cell carcinoma. Sci Rep 5:13450

Brevet M, Fucks D, Chatelain D, Regimbeau JM, Delcenserie R, Sevestre H, Ouadid-Ahidouch H (2009) Deregulation of 2 potassium channels in pancreas adenocarcinomas: implication of KV1.3 gene promoter methylation. Pancreas 38:649–654. https://doi.org/10.1097/MPA.0b013e3181a56ebf

Catacuzzeno L, Aiello F, Fioretti B et al (2011) Serum-activated K and Cl currents underlay U87-MG glioblastoma cell migration. J Cell Physiol 226:1926–1933

Cázares-Ordoñez V, Pardo LA (2017) Kv10.1 potassium channel: from the brain to the tumors. Biochem Cell Biol 95:531–536. https://doi.org/10.1139/bcb-2017-0062

Chandy KG, Norton RS (2016) Immunology: channelling potassium to fight cancer. Nature 537:497–499. https://doi.org/10.1038/nature19467

Chen SZ, Jiang M, Zhen YS (2005) HERG K+ channel expression-related chemosensitivity in cancer cells and its modulation by erythromycin. Cancer Chemother Pharmacol 56:212–220. https://doi.org/10.1007/s00280-004-0960-5

Cherubini A, Pillozzi S, Hofmann G et al (2002) HERG K+ channels and beta1 integrins interact through the assembly of a macromolecular complex. Ann N Y Acad Sci 973:559–561

Chiang EY, Li T, Jeet S et al (2017) Potassium channels Kv1.3 and KCa3.1 cooperatively and compensatorily regulate antigen-specific memory T cell functions. Nat Commun 8:14644. https://doi.org/10.1038/ncomms14644

Chimote AA, Hajdu P, Sfyris AM, Gleich BN, Wise-Draper T, Casper KA, Conforti L (2017) Kv1.3 channels mark functionally competent CD8+ tumor-infiltrating lymphocytes in head and neck cancer. Cancer Res 77:53–61. https://doi.org/10.1158/0008-5472.Can-16-2372

Chimote AA, Balajthy A, Arnold MJ et al (2018) A defect in KCa3.1 channel activity limits the ability of CD8(+) T cells from cancer patients to infiltrate an adenosine-rich microenvironment. Sci Signal 11. https://doi.org/10.1126/scisignal.aaq1616

Choi SY, Kim HR, Ryu PD, Lee SY (2017) Regulation of voltage-gated potassium channels attenuates resistance of side-population cells to gefitinib in the human lung cancer cell line NCI-H460. BMC Pharmacol Toxicol 18:14. https://doi.org/10.1186/s40360-017-0118-9

Conforti L (2017) Potassium channels of T lymphocytes take center stage in the fight against cancer. J Immunother Cancer 5:2. https://doi.org/10.1186/s40425-016-0202-5

Cruz-Gregorio A, Martínez-Ramírez I, Pedraza-Chaverri J, Lizano M (2019) Reprogramming of energy metabolism in response to radiotherapy in head and neck squamous cell carcinoma. Cancers (Basel) 11. https://doi.org/10.3390/cancers11020182

D'Alessandro G, Catalano M, Sciaccaluga M et al (2013) KCa3.1 channels are involved in the infiltrative behavior of glioblastoma in vivo. Cell Death Dis 4:e773

De Marchi U, Sassi N, Fioretti B, Catacuzzeno L, Cereghetti GM, Szabò I, Zoratti M (2009) Intermediate conductance Ca2+-activated potassium channel (KCa3.1) in the inner mitochondrial membrane of human colon cancer cells. Cell Calcium 45:509–516. https://doi.org/10.1016/j.ceca.2009.03.014

Dittmann K, Mayer C, Rodemann HP, Huber SM (2013) EGFR cooperates with glucose transporter SGLT1 to enable chromatin remodeling in response to ionizing radiation. Radiother Oncol 107:247–251. https://doi.org/10.1016/j.radonc.2013.03.016

Dong TT, Zhou HM, Wang LL, Feng B, Lv B, Zheng MH (2011) Salinomycin selectively targets 'CD133+' cell subpopulations and decreases malignant traits in colorectal cancer lines. Ann Surg Oncol 18:1797–1804. https://doi.org/10.1245/s10434-011-1561-2

Du Y, Song W, Chen J et al (2019) The potassium channel KCa3.1 promotes cell proliferation by activating SKP2 and metastasis through the EMT pathway in hepatocellular carcinoma. Int J Cancer 145:503–516. https://doi.org/10.1002/ijc.32121

Eckert F, Schilbach K, Klumpp L et al (2018) Potential role of CXCR4 targeting in the context of radiotherapy and immunotherapy of Cancer. Front Immunol 9:3018. https://doi.org/10.3389/fimmu.2018.03018

Eckert F, Zwirner K, Boeke S, Thorwarth D, Zips D, Huber SM (2019) Rationale for combining radiotherapy and immune checkpoint inhibition for patients with hypoxic tumors. Front Immunol 10:407. https://doi.org/10.3389/fimmu.2019.00407

Edalat L, Stegen B, Klumpp L et al (2016) BK K+ channel blockade inhibits radiation-induced migration/brain infiltration of glioblastoma cells. Oncotarget 7:14259–14278. https://doi.org/10.18632/oncotarget.7423

Eil R, Vodnala SK, Clever D et al (2016) Ionic immune suppression within the tumour microenvironment limits T cell effector function. Nature 537:539–543

Erdem M, Tekiner TA, Fejzullahu A et al (2015) herg1b expression as a potential specific marker in pediatric acute myeloid leukemia patients with HERG 897K/K genotype. Pediatr Hematol Oncol 32:182–192. https://doi.org/10.3109/08880018.2014.949941

Feske S (2019) CRAC channels and disease – from human CRAC channelopathies and animal models to novel drugs. Cell Calcium 80:112–116. https://doi.org/10.1016/j.ceca.2019.03.004

Feske S, Wulff H, Skolnik EY (2015) Ion channels in innate and adaptive immunity. Annu Rev Immunol 33:291–353. https://doi.org/10.1146/annurev-immunol-032414-112212

Fortunato A (2017) The role of hERG1 ion channels in epithelial-mesenchymal transition and the capacity of riluzole to reduce cisplatin resistance in colorectal cancer cells. Cell Oncol (Dordr) 40:367–378. https://doi.org/10.1007/s13402-017-0328-6

Fortunato P, Pillozzi S, Tamburini A, Pollazzi L, Franchi A, La Torre A, Arcangeli A (2010) Irresponsiveness of two retinoblastoma cases to conservative therapy correlates with up-regulation of hERG1 channels and of the VEGF-A pathway. BMC Cancer 10:504

Frankenreiter S, Bednarczyk P, Kniess A et al (2017) cGMP-elevating compounds and ischemic conditioning provide cardioprotection against ischemia and reperfusion injury via cardiomyocyte-specific BK channels. Circulation 136:2337–2355. https://doi.org/10.1161/circulationaha.117.028723

Fu C, Cao CM, Xia Q, Yang J, Lu Y (2003) Reactive oxygen species and mitochondrial KATP-sensitive channels mediated cardioprotection induced by TNF-alpha during hypoxia and reoxygenation. Sheng Li Xue Bao 55:284–289

Fuchs D, Heinold A, Opelz G, Daniel V, Naujokat C (2009) Salinomycin induces apoptosis and overcomes apoptosis resistance in human cancer cells. Biochem Biophys Res Commun 390:743–749. https://doi.org/10.1016/j.bbrc.2009.10.042

García-Ferreiro RE, Kerschensteiner D, Major F, Monje F, Stühmer W, Pardo LA (2004) Mechanism of block of hEag1 K+ channels by imipramine and astemizole. J Gen Physiol 124:301–317. https://doi.org/10.1085/jgp.200409041

García-Quiroz J, García-Becerra R, Santos-Martínez N et al (2014) In vivo dual targeting of the oncogenic Ether-à-go-go-1 potassium channel by calcitriol and astemizole results in enhanced antineoplastic effects in breast tumors. BMC Cancer 14:745. https://doi.org/10.1186/1471-2407-14-745

García-Quiroz J, González-González ME, Díaz L et al (2019) Astemizole, an inhibitor of ether-à-go-go-1 potassium channel, increases the activity of the tyrosine kinase inhibitor gefitinib in breast cancer cells. Rev Invest Clin 71:186–194. https://doi.org/10.24875/ric.18002840

Gibhardt CS, Roth B, Schroeder I et al (2015) X-ray irradiation activates K+ channels via H2O2 signaling. Sci Rep 5:13861

Grimaldi A, D'Alessandro G, Golia MT et al (2016) KCa3.1 inhibition switches the phenotype of glioma-infiltrating microglia/macrophages. Cell Death Dis 7:e2174. https://doi.org/10.1038/cddis.2016.73

Gu XQ, Pamenter ME, Siemen D, Sun X, Haddad GG (2014) Mitochondrial but not plasmalemmal BK channels are hypoxia-sensitive in human glioma. Glia 62:504–513. https://doi.org/10.1002/glia.22620

Gupta PB, Onder TT, Jiang G, Tao K, Kuperwasser C, Weinberg RA, Lander ES (2009) Identification of selective inhibitors of cancer stem cells by high-throughput screening. Cell 138:645–659. https://doi.org/10.1016/j.cell.2009.06.034

Hartung F, Pardo LA (2016) Guiding TRAIL to cancer cells through Kv10.1 potassium channel overcomes resistance to doxorubicin. Eur Biophys J 45:709–719. https://doi.org/10.1007/s00249-016-1149-7

Heise N, Palme D, Misovic M et al (2010) Non-selective cation channel-mediated Ca2+-entry and activation of Ca2+/calmodulin-dependent kinase II contribute to G2/M cell cycle arrest and survival of irradiated leukemia cells. Cell Physiol Biochem 26:597–608

Huang CL, Ruskin BH (1964) Determination of serum chlorpromazine metabolites in psychotic patients. J Nerv Ment Dis 139:381–386

Huang X, He Y, Dubuc AM et al (2015) EAG2 potassium channel with evolutionarily conserved function as a brain tumor target. Nat Neurosci 18:1236–1246. https://doi.org/10.1038/nn.4088

Huber SM (2013) Oncochannels. Cell Calcium 53:241–255

Huber SM, Misovic M, Mayer C, Rodemann HP, Dittmann K (2012) EGFR-mediated stimulation of sodium/glucose cotransport promotes survival of irradiated human A549 lung adenocarcinoma cells. Radiother Oncol 103:373–379

Huber SM, Butz L, Stegen B, Klumpp D, Braun N, Ruth P, Eckert F (2013) Ionizing radiation, ion transports, and radioresistance of cancer cells. Front Physiol 4:212

Huber SM, Butz L, Stegen B, Klumpp L, Klumpp D, Eckert F (2015) Role of ion channels in ionizing radiation-induced cell death. Biochim Biophys Acta 1848:2657–2664. https://doi.org/10.1016/j.bbamem.2014.11.004

Hui C, Lan Z, Yue-li L, Li-lin H, Li-lin H (2015) Knockdown of Eag1 expression by RNA interference increases chemosensitivity to cisplatin in ovarian cancer cells. Reprod Sci 22:1618–1626. https://doi.org/10.1177/1933719115590665

Imlach WL, Finch SC, Dunlop J, Meredith AL, Aldrich RW, Dalziel JE (2008) The molecular mechanism of "ryegrass staggers," a neurological disorder of K+ channels. J Pharmacol Exp Ther 327:657–664

Jensen BS, Strobaek D, Christophersen P et al (1998) Characterization of the cloned human intermediate-conductance Ca2+-activated K+ channel. Am J Phys 275:C848–C856. https://doi.org/10.1152/ajpcell.1998.275.3.C848

Kachmar JF, Boyer PD (1953) Kinetic analysis of enzyme reactions. II. The potassium activation and calcium inhibition of pyruvic phosphoferase. J Biol Chem 200:669–682

Kale VP, Amin SG, Pandey MK (2015) Targeting ion channels for cancer therapy by repurposing the approved drugs. Biochim Biophys Acta 1848:2747–2755. https://doi.org/10.1016/j.bbamem.2015.03.034

Kammerer S, Sokolowski A, Hackl H et al (2016) KCNJ3 is a new independent prognostic marker for estrogen receptor positive breast cancer patients. Oncotarget 7:84705–84717. https://doi.org/10.18632/oncotarget.13224

Kato M, Chuang JL, Tso SC, Wynn RM, Chuang DT (2005) Crystal structure of pyruvate dehydrogenase kinase 3 bound to lipoyl domain 2 of human pyruvate dehydrogenase complex. EMBO J 24:1763–1774. https://doi.org/10.1038/sj.emboj.7600663

Klumpp D, Misovic M, Szteyn K, Shumilina E, Rudner J, Huber SM (2016a) Targeting TRPM2 channels impairs radiation-induced cell cycle arrest and fosters cell death of T cell leukemia cells in a Bcl-2-dependent manner. Oxidative Med Cell Longev 2016:8026702. https://doi.org/10.1155/2016/8026702

Klumpp L, Sezgin EC, Eckert F, Huber SM (2016b) Ion channels in brain metastasis. Int J Mol Sci 17

Klumpp D, Frank SC, Klumpp L et al (2017) TRPM8 is required for survival and radioresistance of glioblastoma cells. Oncotarget 8:95896–95913. https://doi.org/10.18632/oncotarget.21436

Klumpp L, Sezgin EC, Skardelly M, Eckert F, Huber SM (2018) KCa3.1 channels and glioblastoma: in vitro studies. Curr Neuropharmacol 16:627–635. https://doi.org/10.2174/1570159x15666170808115821

Korpi ER, Kleinman JE, Costakos DT, Linnoila M, Wyatt RJ (1984) Reduced haloperidol in the post-mortem brains of haloperidol-treated patients. Psychiatry Res 11:259–269

Koshy S, Wu D, Hu X et al (2013) Blocking KCa3.1 channels increases tumor cell killing by a subpopulation of human natural killer lymphocytes. PLoS One 8:e76740. https://doi.org/10.1371/journal.pone.0076740

Kulawiak B, Kudin AP, Szewczyk A, Kunz WS (2008) BK channel openers inhibit ROS production of isolated rat brain mitochondria. Exp Neurol 212:543–547

Lacruz RS, Feske S (2015) Diseases caused by mutations in ORAI1 and STIM1. Ann N Y Acad Sci 1356:45–79. https://doi.org/10.1111/nyas.12938

Lastraioli E, Lottini T, Bencini L, Bernini M, Arcangeli A (2015) hERG1 potassium channels: novel biomarkers in human solid cancers. Biomed Res Int 2015:896432. https://doi.org/10.1155/2015/896432

Leanza L, Biasutto L, Managò A, Gulbins E, Zoratti M, Szabo I (2013) Intracellular ion channels and cancer. Front Physiol 4:227. https://doi.org/10.3389/fphys.2013.00227

Lee K, McKenna F, Rowe IC, Ashford ML (1997) The effects of neuroleptic and tricyclic compounds on BKCa channel activity in rat isolated cortical neurones. Br J Pharmacol 121:1810–1816

Lehmann DF, Eggleston WD, Wang D (2018) Validation and clinical utility of the hERG IC50: Cmax ratio to determine the risk of drug-induced Torsades de pointes: a meta-analysis. Pharmacotherapy 38:341–348. https://doi.org/10.1002/phar.2087

Li F, Sonveaux P, Rabbani ZN et al (2007) Regulation of HIF-1alpha stability through S-nitrosylation. Mol Cell 26:63–74. https://doi.org/10.1016/j.molcel.2007.02.024

Li B, Jie W, Huang L et al (2014a) Nuclear BK channels regulate gene expression via the control of nuclear calcium signaling. Nat Neurosci 17:1055–1063. https://doi.org/10.1038/nn.3744

Li FY, Chaigne-Delalande B, Su H, Uzel G, Matthews H, Lenardo MJ (2014b) XMEN disease: a new primary immunodeficiency affecting Mg2+ regulation of immunity against Epstein-Barr virus. Blood 123:2148–2152. https://doi.org/10.1182/blood-2013-11-538686

Li L, Li J, Zuo Y, Dang D, Frost JA, Yang Q (2019) Activation of KCNQ channels prevents paclitaxel-induced peripheral neuropathy and associated neuropathic pain. J Pain 20:528–539. https://doi.org/10.1016/j.jpain.2018.11.001

Liang XJ, Taylor B, Cardarelli C et al (2005) Different roles for K+ channels in cisplatin-resistant cell lines argue against a critical role for these channels in cisplatin resistance. Anticancer Res 25:4113–4122

Liu X, Chang Y, Reinhart PH, Sontheimer H, Chang Y (2002) Cloning and characterization of glioma BK, a novel BK channel isoform highly expressed in human glioma cells. J Neurosci 22:1840–1849. https://doi.org/10.1523/jneurosci.22-05-01840.2002

Liu H, Huang J, Peng J, Wu X, Zhang Y, Zhu W, Guo L (2015a) Upregulation of the inwardly rectifying potassium channel Kir2.1 (KCNJ2) modulates multidrug resistance of small-cell lung cancer under the regulation of miR-7 and the Ras/MAPK pathway. Mol Cancer 14:59. https://doi.org/10.1186/s12943-015-0298-0

Liu Y, Zhao L, Ma W et al (2015b) The blockage of KCa3.1 channel inhibited proliferation, migration and promoted apoptosis of human hepatocellular carcinoma cells. J Cancer 6:643–651. https://doi.org/10.7150/jca.11913

Madreiter-Sokolowski CT, Ramadani-Muja J, Ziomek G et al (2019) Tracking intra- and inter-organelle signaling of mitochondria. FEBS J 286:4378–4401. https://doi.org/10.1111/febs.15103

Mohr CJ, Gross D, Sezgin EC, Steudel FA, Ruth P, Huber SM, Lukowski R (2019a) KCa3.1 channels confer radioresistance to breast cancer cells. Cancers (Basel) 11. https://doi.org/10.3390/cancers11091285

Mohr CJ, Steudel FA, Gross D et al (2019b) Cancer-associated intermediate conductance Ca(2+)-activated K^+ channel K(Ca)3.1. Cancers (Basel) 11. https://doi.org/10.3390/cancers11010109

Mohr CJ, Schroth W, Murdter TE et al (2020) Subunits of BK channels promote breast cancer development and modulate responses to endocrine treatment in preclinical models. Br J Pharmacol. https://doi.org/10.1111/bph.15147

Nilius B, Wohlrab W (1992) Potassium channels and regulation of proliferation of human melanoma cells. J Physiol 445:537–548. https://doi.org/10.1113/jphysiol.1992.sp018938

Ong ST, Ng AS, Ng XR et al (2019) Extracellular K+ dampens T cell functions: implications for immune suppression in the tumor microenvironment. Bioelectricity 1:169–179

Paggio A, Checchetto V, Campo A et al (2019) Identification of an ATP-sensitive potassium channel in mitochondria. Nature 572:609–613. https://doi.org/10.1038/s41586-019-1498-3
Palme D, Misovic M, Schmid E, Klumpp D, Salih HR, Rudner J, Huber SM (2013) Kv3.4 potassium channel-mediated electrosignaling controls cell cycle and survival of irradiated leukemia cells. Pflugers Arch 465:1209–1221. https://doi.org/10.1007/s00424-013-1249-5
Palme D, Misovic M, Ganser K, Klumpp L, Salih HR, Zips D, Huber SM (2020) hERG K(+) channels promote survival of irradiated leukemia cells. Front Pharmacol 11:489. https://doi.org/10.3389/fphar.2020.00489
Pardo LA, Stuhmer W (2014) The roles of K(+) channels in cancer. Nat Rev Cancer 14:39–48. https://doi.org/10.1038/nrc3635
Park JH, Park SJ, Chung MK et al (2010) High expression of large-conductance Ca2+-activated K+ channel in the CD133+ subpopulation of SH-SY5Y neuroblastoma cells. Biochem Biophys Res Commun 396:637–642. https://doi.org/10.1016/j.bbrc.2010.04.142
Petroni G (2020) Targeting potassium channels and autophagy to defeat chemoresistance. Mol Cell Oncol 7:1745038. https://doi.org/10.1080/23723556.2020.1745038
Pillozzi S, Arcangeli A (2010) Physical and functional interaction between integrins and hERG1 channels in cancer cells. Adv Exp Med Biol 674:55–67
Pillozzi S, Brizzi MF, Balzi M et al (2002) HERG potassium channels are constitutively expressed in primary human acute myeloid leukemias and regulate cell proliferation of normal and leukemic hemopoietic progenitors. Leukemia 16:1791–1798
Pillozzi S, Brizzi MF, Bernabei PA et al (2007) VEGFR-1 (FLT-1), beta1 integrin, and hERG K+ channel for a macromolecular signaling complex in acute myeloid leukemia: role in cell migration and clinical outcome. Blood 110:1238–1250
Pillozzi S, Masselli M, De Lorenzo E et al (2011) Chemotherapy resistance in acute lymphoblastic leukemia requires hERG1 channels and is overcome by hERG1 blockers. Blood 117:902–914
Pillozzi S, D'Amico M, Bartoli G et al (2018) The combined activation of KCa3.1 and inhibition of Kv11.1/hERG1 currents contribute to overcome cisplatin resistance in colorectal cancer cells. Br J Cancer 118:200–212. https://doi.org/10.1038/bjc.2017.392
Pointer KB, Clark PA, Eliceiri KW, Salamat MS, Robertson GA, Kuo JS (2017) Administration of non-torsadogenic human ether-a-go-go-related gene inhibitors is associated with better survival for high hERG-expressing glioblastoma patients. Clin Cancer Res 23:73–80. https://doi.org/10.1158/1078-0432.Ccr-15-3169
Quiroga-Roger D, Babul J, Guixe V (2015) Role of monovalent and divalent metal cations in human ribokinase catalysis and regulation. Biometals 28:401–413. https://doi.org/10.1007/s10534-015-9844-x
Roberta P, Andrea M, Michele A et al (2020) Insight into the mechanism of cytotoxicity of membrane-permeant psoralenic Kv1.3 channel inhibitors by chemical dissection of a novel member of the family. Redox Biology 37:101705. https://doi.org/10.1016/j.redox.2020.101705
Rosa P, Sforna L, Carlomagno S et al (2017) Overexpression of large-conductance calcium-activated potassium channels in human glioblastoma stem-like cells and their role in cell migration. J Cell Physiol 232:2478–2488. https://doi.org/10.1002/jcp.25592
Rose IA, Warms JVB (1982) Stability of hexokinase II in vitro and in ascites tumor cells. Arch Biochem Biophys 213:625–634. https://doi.org/10.1016/0003-9861(82)90592-6
Roth B, Huber SM (2020) Ion transport and radioresistance. Rev Physiol Biochem Pharmacol. https://doi.org/10.1007/112_2020_33
Roth B, Gibhardt CS, Becker P et al (2015) Low-dose photon irradiation alters cell differentiation via activation of hIK channels. Pflugers Arch 467:1835–1849. https://doi.org/10.1007/s00424-014-1601-4
Ruggieri P, Mangino G, Fioretti B et al (2012) The inhibition of KCa3.1 channels activity reduces cell motility in glioblastoma derived cancer stem cells. PLoS One 7:e47825
Sales TT, Resende FF, Chaves NL, Titze-De-Almeida SS, Báo SN, Brettas ML, Titze-De-Almeida R (2016) Suppression of the Eag1 potassium channel sensitizes glioblastoma cells to injury caused by temozolomide. Oncol Lett 12:2581–2589. https://doi.org/10.3892/ol.2016.4992

Samuel P, Pink RC, Caley DP, Currie JM, Brooks SA, Carter DR (2016) Over-expression of miR-31 or loss of KCNMA1 leads to increased cisplatin resistance in ovarian cancer cells. Tumour Biol 37:2565–2573. https://doi.org/10.1007/s13277-015-4081-z

Sausbier U, Sausbier M, Sailer CA, Arntz C, Knaus HG, Neuhuber W, Ruth P (2006) Ca2+ -activated K+ channels of the BK-type in the mouse brain. Histochem Cell Biol 125:725–741

Schickling BM, Aykin-Burns N, Leslie KK, Spitz DR, Korovkina VP (2011) An inhibitor of K+ channels modulates human endometrial tumor-initiating cells. Cancer Cell Int 11:25. https://doi.org/10.1186/1475-2867-11-25

Schwab A, Fabian A, Hanley PJ, Stock C (2012) Role of ion channels and transporters in cell migration. Physiol Rev 92:1865–1913. https://doi.org/10.1152/physrev.00018.2011

Shen F-F, Dai S-Y, Wong N-K, Deng S, Wong AS-T, Yang D (2020) Mediating K+/H+ transport on organelle membranes to selectively eradicate cancer stem cells by a small molecule. J Am Chem Soc 142(24):10769–10779

Singh H, Lu R, Bopassa JC, Meredith AL, Stefani E, Toro L (2013) MitoBK(Ca) is encoded by the Kcnma1 gene, and a splicing sequence defines its mitochondrial location. Proc Natl Acad Sci U S A 110:10836–10841. https://doi.org/10.1073/pnas.1302028110

Sontheimer H (2008) An unexpected role for ion channels in brain tumor metastasis. Exp Biol Med (Maywood) 233:779–791

Stegen B, Butz L, Klumpp L, Zips D, Dittmann K, Ruth P, Huber SM (2015) Ca2+-activated IK K+ channel blockade radiosensitizes glioblastoma cells. Mol Cancer Res 13:1283–1295. https://doi.org/10.1158/1541-7786.Mcr-15-0075

Stegen B, Klumpp L, Misovic M et al (2016) K(+) channel signaling in irradiated tumor cells. Eur Biophys J 45:585–598. https://doi.org/10.1007/s00249-016-1136-z

Steinle M, Palme D, Misovic M et al (2011) Ionizing radiation induces migration of glioblastoma cells by activating BK K(+) channels. Radiother Oncol 101:122–126. https://doi.org/10.1016/j.radonc.2011.05.069

Steudel FA, Mohr CJ, Stegen B et al (2017) SK4 channels modulate Ca(2+) signalling and cell cycle progression in murine breast cancer. Mol Oncol 11:1172–1188

Storck H, Hild B, Schimmelpfennig S et al (2017) Ion channels in control of pancreatic stellate cell migration. Oncotarget 8:769–784. https://doi.org/10.18632/oncotarget.13647

Stowe DF, Gadicherla AK, Zhou Y et al (2013) Protection against cardiac injury by small Ca(2+)-sensitive K(+) channels identified in guinea pig cardiac inner mitochondrial membrane. Biochim Biophys Acta 1828:427–442. https://doi.org/10.1016/j.bbamem.2012.08.031

Tarcha EJ, Olsen CM, Probst P, Peckham D, Muñoz-Elías EJ, Kruger JG, Iadonato SP (2017) Safety and pharmacodynamics of dalazatide, a Kv1.3 channel inhibitor, in the treatment of plaque psoriasis: a randomized phase 1b trial. PLoS One 12:e0180762. https://doi.org/10.1371/journal.pone.0180762

Teisseyre A, Palko-Labuz A, Sroda-Pomianek K, Michalak K (2019) Voltage-gated potassium channel Kv1.3 as a target in therapy of cancer. Front Oncol 9:933. https://doi.org/10.3389/fonc.2019.00933

Trebak M, Kinet JP (2019) Calcium signalling in T cells. Nat Rev Immunol 19:154–169. https://doi.org/10.1038/s41577-018-0110-7

Voos P, Fuck S, Weipert F et al (2018) Ionizing radiation induces morphological changes and immunological modulation of Jurkat cells. Front Immunol 9:922. https://doi.org/10.3389/fimmu.2018.00922

Wang Y (2011) Effects of salinomycin on cancer stem cell in human lung adenocarcinoma A549 cells. Med Chem 7:106–111. https://doi.org/10.2174/157340611794859307

Wang ZJ, Soohoo SM, Tiwari PB, Piszczek G, Brelidze TI (2020) Chlorpromazine binding to the PAS domains uncovers the effect of ligand modulation on EAG channel activity. J Biol Chem 295:4114–4123. https://doi.org/10.1074/jbc.RA119.012377

Warburg O, Wind F, Negelein E (1927) The metabolism of tumors in the body. J Gen Physiol 8:519–530. https://doi.org/10.1085/jgp.8.6.519

Watkins S, Sontheimer H (2011) Hydrodynamic cellular volume changes enable glioma cell invasion. J Neurosci 31:17250–17259. https://doi.org/10.1523/jneurosci.3938-11.2011

Wonderlin WF, Strobl JS (1996) Potassium channels, proliferation and G1 progression. J Membr Biol 154:91–107. https://doi.org/10.1007/s002329900135

Wulff H, Castle NA (2010) Therapeutic potential of KCa3.1 blockers: recent advances and promising trends. Expert Rev Clin Pharmacol 3:385–396. https://doi.org/10.1586/ecp.10.11

Xie H, Simon MC (2017) Oxygen availability and metabolic reprogramming in cancer. J Biol Chem 292:16825–16832. https://doi.org/10.1074/jbc.R117.799973

Xu H, Lai W, Zhang Y et al (2014) Tumor-associated macrophage-derived IL-6 and IL-8 enhance invasive activity of LoVo cells induced by PRL-3 in a KCNN4 channel-dependent manner. BMC Cancer 14:330. https://doi.org/10.1186/1471-2407-14-330

Zhang R, Tian P, Chi Q et al (2012) Human ether-à-go-go-related gene expression is essential for cisplatin to induce apoptosis in human gastric cancer. Oncol Rep 27:433–440. https://doi.org/10.3892/or.2011.1515

Zhang Y, Zhang P, Chen L, Zhao L, Zhu J, Zhu T (2019) The long non-coding RNA-14327.1 promotes migration and invasion potential of endometrial carcinoma cells by stabilizing the potassium channel Kca3.1. Onco Targets Ther 12:10287–10297. https://doi.org/10.2147/ott.S226737

Kir Channel Molecular Physiology, Pharmacology, and Therapeutic Implications

Meng Cui, Lucas Cantwell, Andrew Zorn, and Diomedes E. Logothetis

Contents

M. Cui · L. Cantwell · A. Zorn · D. E. Logothetis (✉)
Department of Pharmaceutical Sciences, School of Pharmacy and Pharmaceutical Sciences, Bouvé College of Health Sciences, Northeastern University, Boston, MA, USA

Department of Chemistry and Chemical Biology, College of Science, Northeastern University, Boston, MA, USA
e-mail: d.logothetis@northeastern.edu

N. Gamper, K. Wang (eds.), *Pharmacology of Potassium Channels*,
Handbook of Experimental Pharmacology 267, https://doi.org/10.1007/164_2021_501

Abstract

For the past two decades several scholarly reviews have appeared on the inwardly rectifying potassium (Kir) channels. We would like to highlight two efforts in particular, which have provided comprehensive reviews of the literature up to 2010 (Hibino et al., Physiol Rev 90(1):291–366, 2010; Stanfield et al., Rev Physiol Biochem Pharmacol 145:47–179, 2002). In the past decade, great insights into the 3-D atomic resolution structures of Kir channels have begun to provide the molecular basis for their functional properties. More recently, computational studies are beginning to close the time domain gap between in silico dynamic and patch-clamp functional studies. The pharmacology of these channels has also been expanding and the dynamic structural studies provide hope that we are heading toward successful structure-based drug design for this family of K^+ channels. In the present review we focus on placing the physiology and pharmacology of this K^+ channel family in the context of atomic resolution structures and in providing a glimpse of the promising future of therapeutic opportunities.

Keywords

Cytosolic G-loop gate · GIRK · Helix bundle crossing gate · K^+ transport channel · K_{ATP} · Phosphoinositides · Potassium inward rectifiers · Resting potential

1 Historical Perspective: The Pre-structure Era

Inward Rectification and Dependence of Conductance on $[K^+]_o$: Kir currents were first described in skeletal muscle fibers, where Bernard Katz observed (Katz 1949a, b) that when these cells were immersed in solutions containing high potassium the membrane conductance was larger at hyperpolarizing potentials. This behavior was the opposite from what Cole and Curtis had reported in squid giant axon (Cole and Curtis 1941), where the depolarization-induced non-linear conductance had been described as rectification, borrowing from electrical engineering terminology used for diodes. The fact that the first example of outward rectification, demonstrated by the delayed rectifier K_v current, was also in Katz's laboratory with Hodgkin and Huxley (Hodgkin et al. 1952; Katz 1949a, b), led Katz to describe the skeletal muscle conductance in high K^+ as "anomalous rectification" ("propriétés détectrices anormales," from the French description). This anomalous rectification was later (Adrian et al. 1970) renamed "inward rectification," the term most used at the present. Another difference between the Kir and the outwardly rectifying K_v currents was their respective voltage dependence. Kir currents were voltage-independent but instead their activation depended on the driving force (V_m-E_K, i.e. the distance of the membrane potential V_m from the equilibrium potential E_K for K^+ ions). Activation was shown to depend on (V_m-E_K) if $[K^+]_o$, but not $[K^+]_i$, was changed (Hagiwara and Yoshii 1979; Leech and Stanfield 1981; Stanfield et al. 2002). The dependence of Kir current rectification and conductance on $[K^+]_o$ is shown in Fig. 1a (Hagiwara et al. 1976). Early on, the inward rectification was attributed to block of the channel by particles from the cytosol (Armstrong 1969; Hille and Schwarz 1978; Standen and Stanfield 1978), and later the intracellular blocking substances were identified to be the millimolar levels of Mg^{2+} as well as polyamines that exist in sub-millimolar concentrations inside of cells (Lopatin et al. 1994; Matsuda et al. 1987). Upon hyperpolarization, the inward Kir current showed a time-independent increase due to fast Mg^{2+} unblock, followed by a time-dependent increase due to slow polyamine unblock (Lopatin et al. 1995).

1.1 Tissue Distribution of Kir Family Members

Kir channels have been found in a wide variety of cells, including: cardiac myocytes (Beeler and Reuter 1970; Kurachi 1985; McAllister and Noble 1966; Rougier et al. 1967), neurons (Brown and Carpentier 1990; Gahwiler and Brown 1985; Lacey et al. 1988; North et al. 1987; Takahashi 1990; Williams et al. 1988), blood cells (Lewis et al. 1991; McKinney and Gallin 1988), osteoclasts (Sims and Dixon 1989), endothelial cells (Silver and DeCoursey 1990), glial cells (Kuffler and Nicholls 1966; Newman 1984), epithelial cells (Greger et al. 1990; Hebert et al. 2005; Lorenz et al. 2002; Lu et al. 2002), and oocytes (Hagiwara et al. 1978; Hagiwara et al. 1976; Hagiwara and Takahashi 1974). Figure 2 shows the tissue/organ expression of each Kir subfamily throughout the body. The Kir tissue expression has been comprehensively reviewed (de Boer et al. 2010; Hibino et al. 2010). The role of Kir channels lies in contributing to the cellular resting potential and keeping V_m near E_K. This serves to reduce action potential firing in excitable cells, to control K^+ transport in non-excitable cells or to transduce extracellular (by external stimuli) to intracellular (by internal metabolites) communication and vice versa. In 1993 the first 3 of the 16 members of the Kir channel family were cloned (Dascal et al. 1993;

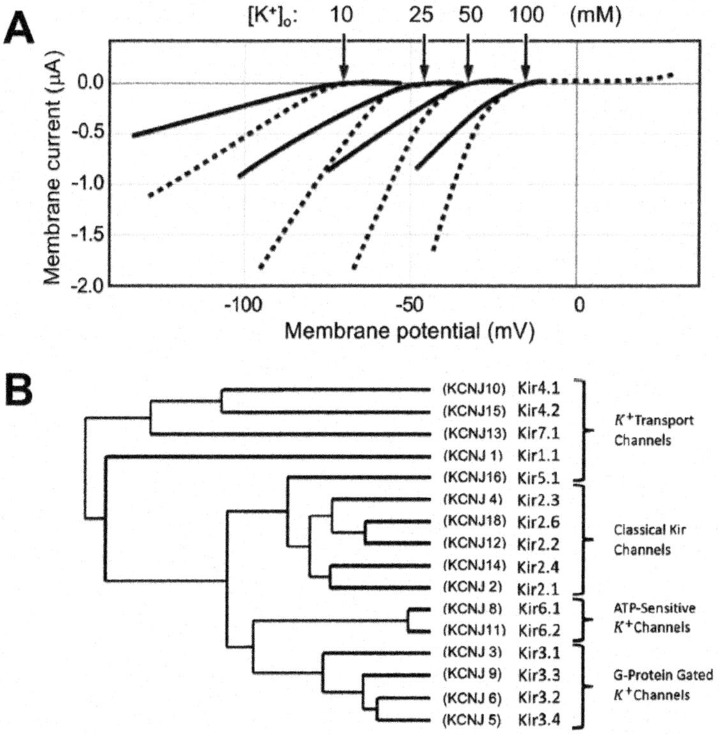

Fig. 1 Kir channel inward rectification and phylogenetic/functional classification. (**a**) I-V relationship of the starfish Kir at four different $[K^+]_o$. Continuous and broken lines indicate instantaneous and steady-state currents, respectively (from Hagiwara et al. 1976) (**b**) Phylogenetic analysis of the 16 known subunits of human Kir channels (for identity/similarity see Fig. 10). These subunits have been classified into four functional groups

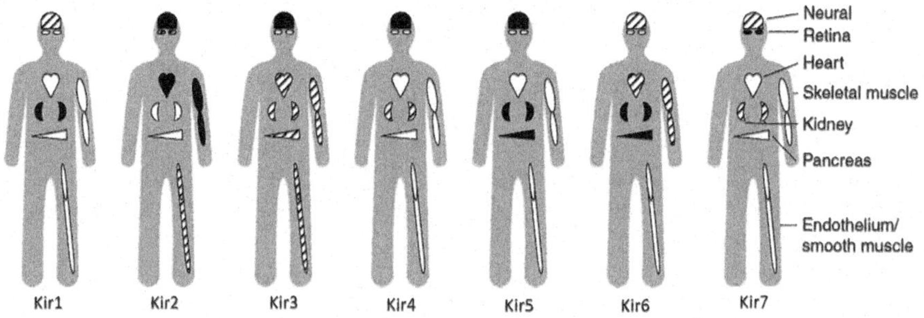

Fig. 2 Expression in indicated organs and tissues of Kir channel members of each of the seven subfamilies. Blank indicates no, striations intermediate, and black prominent levels of expression. Adapted (permission obtained) from de Boer et al. (2010)

Ho et al. 1993; Kubo et al. 1993a; Kubo et al. 1993b). Based on sequence alignment (see Fig. 9) and phylogenetic analysis, the 16 human isoforms have been classified into seven subfamilies (Kir1-7) and four functional groups (a. the K^+ transport, b. the classical, c. the ATP-sensitive, and d. the G protein-gated K^+ channels) (Fig. 1b). The strong sequence similarity among family members (30–99%) (Fig. 10) allows for formation of heteromeric assemblies along with homomeric ones, in one case across subfamilies (Kir4.1 and Kir5.1) and in most cases within subfamilies (e.g., Kir3.1/3.2 in neuronal tissues and Kir3.1/3.4 in atrial cells of the heart). The resulting heteromers display unique functional properties compared to the homomers that will be discussed individually for each subfamily.

1.2 Kir Channel Gating

The signaling phospholipid, phosphatidylinositol 4,5-bisphosphate [PI(4,5)P_2 or PIP_2] found in the inner leaflet of the plasma membrane plays an essential role in supporting Kir channel gating (Hilgemann and Ball 1996; Huang et al. 1998; Logothetis et al. 2015; Petit-Jacques et al. 1999; Sui et al. 1998). Excision into inside-out patches in ATP-free solutions showed a typical gradual decline in channel activity, referred to as "run-down" (Hilgemann and Ball 1996). Run-down activity could be restored by hydrolyzable forms of ATP or by PIP_2 (Hilgemann and Ball 1996; Huang et al. 1998; Petit-Jacques et al. 1999; Sui et al. 1998). The seven Kir subfamily members could be classified into four groups depending on the degree of stereospecificity and affinity to PI(4,5)P_2 (over PI(3,4,5)P_3 and PI(3,4)P_2) (Rohacs et al. 2003): *Highest* or Group 1 (Kir2.1, Kir2.4, Kir4.1), *Moderate* or Group 2 (Kir1.1, Kir2.2, Kir2.3, Kir4.2, Kir7.1), *Weak* or Group 3 (Kir3.4, Kir3.1/3.4), *Lowest* or Group 4 (Kir6.2, Kir6.2/SURs). Mutagenesis studies identified channel regions critical for the effects of PIP_2 and linked basic and non-basic residues to sensitivity to PIP_2 and Kir channelopathies (Lopes et al. 2002).

1.3 Kir Channel Trafficking

Kir channel mutations that lead to channelopathies, especially those that affect trafficking, have been recently reviewed (Zangerl-Plessl et al. 2019). As other transmembrane proteins, Kir channels are translated in the endoplasmic reticulum (ER) and transported through the Golgi apparatus and the trans-Golgi network to the plasma membrane through the process known as forward trafficking. Upon removal from the plasma membrane, channel proteins enter the degradation pathway through trafficking to the early endosome and the multivesicular body to the lysosome, through a process known as backward trafficking. Kir channels could instead of becoming degraded be recycled back to the plasma membrane from the early endosome to the trans-Golgi network. Kir channel mutations resulting in loss (LOF) or gain (GOF) of function are associated with a variety of human diseases, including Bartter syndrome type II (Kir1.1 LOF), Andersen-Tawil syndrome (Kir2.1

LOF and GOF), thyrotoxic hypokalemic periodic paralysis (Kir2.6 LOF), Keppen-Lubinsky syndrome (Kir3.2 LOF), familial hyperaldosteronism type III and long QT syndrome 12 (Kir3.4 LOF), EAST/SeSAME syndrome (Kir4.1 LOF), Cantú syndrome (Kir6.1 or SUR2 GOF) and hyperinsulinism and hypoglycemia (Kir6.2 LOF) and diabetes (Kir6.2 GOF), Leber congenital amaurosis type 16 and Snowflake vitreoretinal degeneration (Kir7.1 LOF). In many of the diseases mentioned above LOF has been associated with defective forward trafficking, while GOF mutations or loss of specificity mutations (e.g., some Kir3.4 mutations) are not likely to be related to trafficking abnormalities. The likely causes of trafficking defects are either (1) defects in trafficking motifs compromising interactions with the proteins involved in the trafficking machinery, or (2) protein structure defects leading to channel misfolding, destabilization and ER-associated degradation (Zangerl-Plessl et al. 2019).

1.4 The Structural Era

1.4.1 Kir Structures by 2020

The first Kir channel high-resolution crystal structure to be solved in 2002 was the cytoplasmic domain (CTD) of Kir3.1 (Nishida and MacKinnon 2002). In this crystal structure, the transmembrane domain of the channel was removed, and the N-terminus was fused to the CTD and expressed in bacteria (PDBID: 1N9P). Soon after this structure, the CTDs of Kir2.1 (PDBID: 1U4F, 2005), (Pegan et al. 2005) pointing to the potential of a cytosolic constriction, coined as the G-loop gate, and Kir3.2 (PDBID: 2E4F, 2007) (Inanobe et al. 2007) were solved. Subsequently, in 2007, a crystal structure of a Kir3.1 prokaryotic Kir channel chimera was solved (PDBID: 2QKS) (Table 1). In this structure, two thirds of the transmembrane domains (TMDs) were replaced with the corresponding region of a homologous prokaryotic Kir channel, revealing the continuity of the permeation pathway from the membrane (2/3 prokaryotic, 1/3 mammalian) to the cytosolic (mammalian) domains of the Kir3.1 channel. Two distinct conformations of the G-loop gate were captured, in one structure the apex of the CTD was open (or dilated), and in the other it was closed (or constricted) (Nishida et al. 2007). The transmembrane gate, referred to as the helix bundle crossing (HBC) gate was captured in the closed conformation in both structures. Even though Nishida and colleagues were not able to show that the chimera they constructed was functional (Nishida et al. 2007), a subsequent paper showed that the purified chimera once reconstituted in planar lipid bilayers with PIP2 was indeed functional (Leal-Pinto et al. 2010). In 2009, a crystal structure of the chicken Kir2.2, 90% identical to the human Kir2.2, was solved (PDBID:3JYC), where large structured turrets and an unusual selectivity filter entryway were seen, which may explain the relative insensitivity of eukaryotic Kir channels to toxins (Tao et al. 2009). In 2011, the crystal structure of Kir2.2 channel in complex with a short-chain derivative of PIP_2 was solved [PDBID:3SPI] (see Fig. 3b). This structure showed that PIP_2 binds at an interface between the TMD and CTD of the Kir2.2 channel. Upon PIP_2 binding, a flexible linker between the TMD and CTD transforms to a helical structure, which causes a 6 Å translation of the CTD

Table 1 Selected functional Kir channel structures solved by X-ray crystallography (X-ray) or cryo-electron microscopy (Cryo-EM)

PDBID	Channel	Resolution	Method	Year	References
2QKS	Kir3.1-prokaryotic Kir channel chimera	2.2 Å	X-ray	2007	Nishida et al. (2007), Leal-Pinto et al. (2010)
3JYC	Kir2.2	3.11 Å	X-ray	2009	Tao et al. (2009)
3SPI	Kir2.2/PIP_2	3.31 Å	X-ray	2011	Hansen et al. (2011)
3SYO	Kir3.2/sodium	3.54 Å	X-ray	2011	Whorton and MacKinnon (2011)
3SYA	Kir3.2 in complex with sodium and PIP_2	2.98 Å	X-ray	2011	Whorton and MacKinnon (2011)
4KFM	Kir3.2/βγ G protein subunits	3.45 Å	X-ray	2013	Whorton and MacKinnon (2013)
5WUA	ATP-sensitive K^+ channel	5.6 Å	Cryo-EM	2017	Li et al. (2017)
5YW8/ 5YWC/ 6YKE	Pancreatic ATP-sensitive potassium channel bound with ATPγS	4.4 Å	Cryo-EM	2017	Wu et al. (2018)
6BAA	Kir6.2/SUR1/Glibenclamide, ATP	3.63 Å	Cryo-EM	2017	Martin et al. (2017)
6C3O/ 6C3P	Kir6.2/SUR1/ATP(ADP), Mg^{2+}	3.9 Å/ 5.6 Å	Cryo-EM	2017	Lee et al. (2017)
6JB1	Pancreatic ATP-sensitive K^+ channel bound with repaglinide and ATPγS	3.3 Å	Cryo-EM	2019	Ding et al. (2019)

toward the TMD, as the HBC gate begins to open (Hansen et al. 2011). In the same year (2011), the crystal structure of Kir3.2 in complex with sodium and PIP_2 was solved (PDBID:3SYA). The structures suggested that the presence of PIP_2 couples the G loop and HBC gates to open in a coordinated manner. The intracellular Na^+ binding site was also confirmed in this structure (Whorton and MacKinnon 2011). In 2013, the crystal structure of Kir3.2-βγ G-protein (Gβγ) complex was solved (PDBID:4KFM). In this structure, the Gβγ subunit bound to the interfaces between four pairs of adjacent Kir3.2 channel CTD subunits that were also bound to sodium and a PIP_2 analog. The structure was thought to represent a "pre-open" state, an intermediate between the closed and open but still a non-conducting conformation of the channel (Whorton and MacKinnon 2013) (shown later in Fig. 7b). In 2017, a hetero-octameric pancreatic K_{ATP} channel in complex with the non-competitive inhibitor glibenclamide was solved by cryo-electron microscopy (EM) at a 5.6-Å resolution [PBDID:5WUA]. This structure showed four SUR1 regulatory subunits located in the periphery of a centrally located Kir6.2 channel tetramer (Li et al.

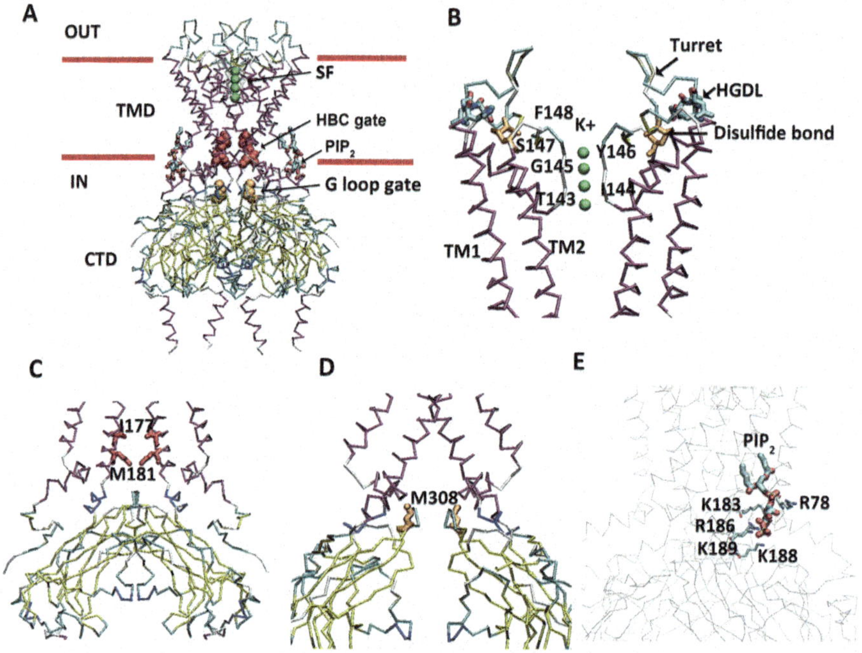

Fig. 3 (**a**) Crystal structure of Kir2.2 (PDBID: 3SPI) in complex with PIP$_2$. (**b**) Selectivity Filter. (**c**) Helix bundle crossing gate. (**d**) G-loop gate. (**e**) PIP$_2$ binding site

2019). In 2017, Wu and colleagues solved complexes of the K_{ATP} channel (KIR6.2/SUR1) with ATP, Mg-ADP, and ATP/glibenclamide, respectively, by using cryo-EM (PDBID: 5YW8, 5YWC, 5YKE). These structures depict the binding site of glibenclamide, ATP, and suggested a mechanism of how Mg-ADP binding on nucleotide-binding domains drives a conformational change of the SUR1 subunit (Wu et al. 2018). Martin and colleagues solved a cryo-EM structure of SUR1/Kir6.2 channel bound to glibenclamide and ATP at 3.63 Å resolution (PDBID: 6BAA). The structure showed that the drug bonded to the transmembrane bundle of SUR1, and mutation of the interacting residues in the binding site reduced the channel sensitivity to the drug (Martin et al. 2017). Lee and colleagues solved two cryo-EM structures of the human K_{ATP} channel (Kir6.2/SUR1) in complex with Mg^{2+} and nucleotides, referred to as the quatrefoil (PDBID: 6C3O) and propeller (PDBID: 6C3P) forms. In both forms, ATP binds to the inhibitory site in Kir6.2. Mg^{2+}-ATP and Mg^{2+}-ADP bind to the degenerate and consensus sites, respectively, in the nucleotide-binding domains of SUR1. A lasso extension interface between Kir6.2 and SUR1 formed in the propeller form but was disrupted in the quatrefoil form (Lee et al. 2017). In 2019, a higher resolution (3.3 Å) cryo-EM structure of the same pancreatic K_{ATP} channel was achieved in complex with the short-acting insulin secretagogue repaglinide and adenosine-5′-(γ-thio)-triphosphate (ATPγS) (PDBID: 6JB1) (Ding et al. 2019) (shown later in Fig. 8). Table 1 summarizes crystal and cryo-EM Kir channel structures for which functional expression has been demonstrated.

1.4.2 Structural Features of Kir Channels

The Kir channel structures have provided new insights into the mechanism of channel function, such as channel gating, selectivity, rectification, and modulation by PIP_2, sodium ions, alcohol, G proteins, ATP, etc. Figure 3 shows the classical Kir2.2 crystal structure with which we will illustrate some key features of this channel family. Kir channels are either homotetramers or heterotetramers formed by four subunits. Each subunit contains a transmembrane region, the TM1 (or outer) and TM2 (or inner) helices (see Fig. 3b), and a cytoplasmic domain (Fig. 3d).

Selectivity Filter (SF)

The selectivity filter (SF) of the Kir channel is similar to that of other potassium channels (Fig. 3b), but instead of the canonical filter sequence TXGYGDX (X: aliphatic amino acid), the corresponding sequence in Kir channels is TIGYGXR (X: Y/F/H, V/L, T, G), with very few exceptions (Kir2.4-S147, Kir6.1/6.2-F143/133, Kir7.1-M125) (see Fig. 9). The conserved Kir2.2(R149) forms an ionized hydrogen bond with E139 in the pore helix. Kir channels also contain a highly conserved disulfide bond flanking the pore region (C123-C155) (see Figs. 3b and 9). Between the outer helix and pore region "turret," there is the conserved 3–10 helical HGDL signature sequence for Kir channels (Tao et al. 2009), with very few exceptions: Kir3.1/3.3/3.4 (RGDL), Kir3.2 (RGDM), Kir1.1 (HKDL), and Kir7.1 (NGDL) (see Figs. 3b and 9).

Gates

The residues I177 and M181 on the TM2 helix in Kir2.2 form two hydrophobic seals that close the path from the pore to the CTD, which is referred to as the helix bundle crossing (HBC) gate (Fig. 3c). In other K^+ channels, such as KcsA and Kv channels, a small or polar amino acid exists at the position corresponding to I177. Alternate residues can be utilized in other Kirs at these positions [at I177: V/L/T/C and at M181: Kir2 (M), Kir3/6 (F), Kir1/4/5 (L), Kir7.1 (V)] (see Figs. 3c and 9). In addition to the HBC gate, Kir channels have a second unique cytoplasmic gate (G loop) at the apex of the cytoplasmic domain (see Figs. 3d and 9) (Tao et al. 2009).

PIP_2 Binding Site

PIP_2 binds to the interface between TMD and CTD and is thought to cause a 6 Å translation of the CTD toward the TMD (Fig. 3a, e). The negatively charged phosphate groups interact with the Kir channel through highly conserved salt bridge interactions. The 1′ phosphate interacts with the R78 and R80 of the RWR highly conserved sequence motif with a few exceptions (Kir2.2 numbers, unless otherwise specified): at R78 – Kir2/7 (R), Kir1/3.1/3.2/3.4/4.2/5/6 (K), Kir3.3/4 (Q) and at R80 – Kir6.2(P69) (see Figs. 3e and 9) at the N-terminus of the outer helix. The 4′ and 5′ phosphates interact with the TM2 residue K183 (absolutely conserved) and the highly conserved residues of the B-loop region sequence motif – 186-RPKKR-190 – with the following exceptions: R186 (Kir3/6 (Q), Kir5.1(T174)); K188 (Kir6 (H), Kir3.3 (N)); and K189 (Kir6 (R); Kir7.1(N165)) (see Figs. 3e and 9). These basic amino acids are critical for PIP_2 activation of Kir channels (Hansen et al. 2011; Tao et al. 2009).

Cholesterol Regulation

Cholesterol enrichment or depletion was shown to decrease or increase, respectively, Kir2.1 currents in endothelial cells (Romanenko et al. 2002). Through mutagenesis structural determinants could be identified, suggesting strongly that specific interactions of cholesterol with the Kir proteins were responsible for the observed effects (Epshtein et al. 2009; Rosenhouse-Dantsker 2019). Cholesterol enrichment caused significant inhibition to several active Kir channels (Kir1.1, Kir2.1, Kir3.1 (F137S), Kir6.2Δ36) and activation (rather than inhibition) to others (Kir3.2, Kir3.4 (S143T)). Modeling studies and site-directed mutagenesis suggested that the interaction sites were located between α-helices of two adjacent channel subunits and involved hydrophobic and aromatic residues (Rosenhouse-Dantsker 2019). Even though some key determinant residues for cholesterol sensitivity overlapped with those for PIP_2 sensitivity, others did not, making the interrelationship between cholesterol and PIP_2 influences on Kir channels unclear. A recent cryo-EM structure of Kir3.2 with a cholesterol analog (in the presence and absence of PIP_2) suggests that cholesterol binds near PIP_2 potentiating its effects (Mathiharan et al. 2020).

Rectification Determinants

Kir channels can be classified as strong, intermediate, and weak rectifiers based on their rectification properties (Table 2; Yellow – Kir2, Blue – Kir3, Green – Transport, Orange – Kir6). The rectification features of Kir channel are generally believed to occur through blocking of the channels by Mg^{2+} and polyamines (spermine, spermidine, and putrescine) (Nichols and Lee 2018). Key acidic residues involved in the rectification of Kir channels have been identified through mutagenesis. A TM2 aspartate residue (corresponding to D173 in the Kir2.2 channel) was the first residue identified as critical for inward rectification. Mutation of this residue affected both polyamine-blocking affinity and voltage dependence (Fakler et al. 1996; Lopatin et al. 1994; Wible et al. 1994; Yang et al. 1995). The corresponding residue in intermediate/weak rectifiers such as Kir1.1 and Kir6.2 is an asparagine. Mutation of this asparagine to an aspartate residue converted these channels to strong rectifier channels (Lopatin et al. 1994; Lu and MacKinnon 1994; Shyng et al. 1997). Thus, this residue has been termed as the "rectification controller." Several other negatively charged residues in the pore-lining region of the CTD in Kir channels are important for rectification. For example, mutation of E224, E229, D259, and E299 of Kir2.1 (numbers are one less than Kir2.2) to a neutral residue reduced the intensity of inward rectification (Fujiwara and Kubo 2006; Guo and Lu 2003; Kubo and Murata 2001; Taglialatela et al. 1995; Xie et al. 2002; Yang et al. 1995). Figure 4 shows a comparison of these residues between the strong rectifier Kir2.2 and the weak rectifier Kir6.2. Table 2 shows that the number of acidic residues per Kir channel subunit correlates well with the functional rectification properties of each subfamily member, such that 4–5 negatively charged residues per subunit (Q/s) result in strong, 3 Q/s in intermediate and 1 Q/s in weak rectification.

Table 2 Kir channel critical residues for inward rectification

R	Channel	173	225	256	260	300	Q/s
S	Kir2.1	D	E	D	D	E	5
S	Kir2.2	D	E	D	D	E	5
S	Kir2.4	D	E	D	D	E	5
S	Kir2.3	D	E	D	D	E	5
S	Kir2.6	D	E	D	D	E	5
NF	Kir3.1	D	S	S	D	E	
S	Kir3.2	N	E	Y	D	E	3
S	3.1/3.2	D/N	S/E	S/Y	D	E	4
NF	Kir3.3	N	E	D	D	E	
S	Kir3.4	N	E	D	D	E	4
S	3.1/3.4	D/N	S/E	S/D	D	E	5
I	Kir4.1	E	G	D	D	S	3
I	Kir4.2	E	Q	D	E	N	3
NF	Kir5.1	N	E	K	D	I	
S	4.1/5.1	E/N	G/E	D/K	D	S/I	4
S	4.2/5.1	E/N	Q/E	D/K	E/D	N/I	4
I	Kir1.1	N	G	D	E	D	3
I	Kir7.1	E	S	D	E	S	3
W	Kir6.1	N	S	P	N	E	1
W	Kir6.2	N	S	G	N	E	1

Yellow – Kir2 family, blue – Kir3 Family, green – Transport family, orange – Kir6 family
R Rectification, *S* Strong, *I* Intermediate, *W* Weak, *NF* Non-functional, *Q/s* number of charges per subunit for each Kir channel
Residue numbers from Kir2.2 channel

Recent microsecond MD simulation studies on polyamine blocking of the mKir3.2 channel were conducted by Chen et al. (2020). Two binding sites for putrescine were identified, one located close to the E236 (hE234) in the CTD, and another located close to the T154 (hT152) from the SF. These observations were consistent with previous experimentally identified residues. By applying a cross-membrane electric field, putrescine was transferred from the CTD binding site to the SF site. In contrast to putrescine (+2 charges), spermine (+4 charges) did not transfer to the SF site, but remained bound to the CTD site. An additional force was needed to be applied to make spermine transfer from the CTD to the SF binding site (Chen et al. 2020). These results contribute toward a dynamic molecular insight of the rectification mechanism through polyamines. In future studies, additional dynamic simulations will be needed to provide further molecular insight into the dynamic rectification mechanisms of Kir channels, such as how polyamines limit potassium ion permeation.

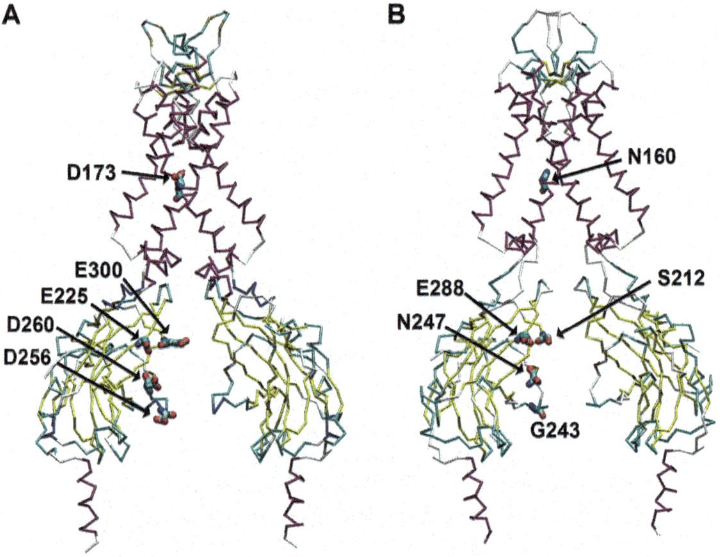

Fig. 4 Critical residues for Kir channel rectification. (**a**) Kir2.2 (PDBID: 3SPI); (**b**) Kir6.2 (PDBID: 6C3O)

2 Classical Kir2 Channels

2.1 Historical Perspective

The classical (also sometimes referred to as "canonical") rectifiers display strong inward rectification, with little outward current at depolarizing potentials. As we mentioned already in Sect. 1, they were differentiated early on from voltage-gated K^+ (Kv) channels (outward rectifiers) in that at membrane potentials around the K^+ equilibrium potential (E_K), where K_v channels are closed, they are constitutively open, albeit conducting small but physiologically relevant outward currents. They conduct very little to not at all at depolarized membrane potentials, where K_v channels are open and conduct the most. This property of strong inward rectifiers is critical for setting a background K^+ conductance in excitable tissues, such as in muscle (both skeletal and cardiac, where Kir2 channels are expressed – see below), driving the resting potential near E_K and enabling rapid conduction of excitation and coordinated contraction (Nichols et al. 1996). As already mentioned, the molecular mechanisms underlying Kir rectification, such as the blockade by polyamines, have been studied extensively (Fujiwara and Kubo 2006; Guo and Lu 2003; Ishihara and Ehara 2004; Kubo and Murata 2001; Nishida et al. 2007; Taglialatela et al. 1995; Xie et al. 2003; Yang et al. 1995). Yet, several aspects of the mechanisms involved remain unclear (Liu et al. 2012; Xu et al. 2009). One of these is the voltage dependence of inward rectification (i.e., polyamine block) displaying shifts that occur with changes in E_K when $[K^+]_o$ but not $[K^+]_i$ is varied (see Fig. 1a) (Hagiwara and Takahashi 1974; Hagiwara and Yoshii 1979; Hestrin 1981; Kubo 1996; Kubo et al. 1993a; Lopatin and Nichols 1996; Nishida et al. 2007). Thus, the probability of channels being open (P_{open}) shifts along the voltage axis by 25 mV for an e-fold change in $[K^+]_o$ and the open channel conductance (inward or outward) is proportional to $[K^+]_o$ with a power of 0.2–0.6, referred to as a "square root" proportionality (Hagiwara and Takahashi 1974; Hille and Schwarz 1978; Kubo et al. 1993a; Matsuda 1988; Ohmori 1978; Sackmann et al. 1984). It has been argued that this dependence of the Kir conductance on $[K^+]_o$, particularly the outward conductance, plays a significant role in regulating the myocardial action potential duration (Matsuoka et al. 2003; Nichols et al. 1996; Shimoni et al. 1992), cardiac contractility (Bouchard et al. 2004), and arrhythmogenicity (Asakura et al. 2014; Ishihara et al. 2009; Maruyama et al. 2011). Ishihara has recently shown (2018) that in the absence of Na^+ in the external solutions, the open conductance of Kir2.1 does not change over a wide $[K^+]_o$ range but in its presence, Na^+ competitively inhibits K^+ conductance in a voltage-dependent manner. Thus, it is an impermeant physiological cation that mediates the apparent $[K^+]_o$ dependence of the open Kir channel conductance (Ishihara 2018). The structural determinants of the Na^+ binding site have not yet been elucidated. A guiding clue comes from Kir7.1, that differs from other Kir channels in lacking to a great extent in the property of being activated by $[K^+]_o$. We will revisit this clue of the structural determinants for the dependence of the K^+ conductance on $[K^+]_o$ and how Na^+ may be involved when we discuss Kir7.1 in the "K^+ transport Kir channels" in Sect. 5.3.

2.2 Subfamily Members and Tissue Distribution

The first subfamily member Kir2.1 (also referred to as IRK1) was cloned in 1993 from a mouse macrophage cell line (Kubo et al. 1993a). In the following 2 years three additional members were cloned, Kir2.2, Kir2.3, and Kir2.4 (Bond et al. 1994; Bredt et al. 1995; Morishige et al. 1994). Kir2.5 was cloned from fish in 2008 (Crucian carp; *Carassius carassius*) (Hassinen et al. 2008), while Kir2.6 was revealed as the fifth mammalian isoform in 2010 (Ryan et al. 2010).

In excitable cells, Kir2 channels are expressed in skeletal muscle (Kir2.1, Kir2.2, Kir2.6), brain (Kir2.1, Kir2.2, Kir2.3, Kir2.4), and heart (Kir2.1, Kir2.2, Kir2.3), while an intermediate level of expression is seen in smooth muscle tissue and the retina (Fig. 2) (de Boer et al. 2010; Hibino et al. 2010). Their role in non-excitable cells remains to be elucidated (e.g., in macrophages where Kir2.1 was first cloned from) (Kubo et al. 1993a). Not only these channel subunits assemble as homotetramers but also as heterotetramers, both in heterologous expression systems and in native tissues where they are expressed, endowing tissues with greater versatility in the physiological roles they play (Hibino et al. 2010). Only a small fraction of Kir2.6 manages to leave the endoplasmic reticulum to make it to the cell surface, whilst the almost identical subunit Kir2.2 (6 differences in the human clones Kir2.6-Kir2.2: L15S, A56E, V100I, H118R, L156P, G430E) is trafficked robustly. Dassau and colleagues showed that the L156P and P156L accounted for the majority of the endoplasmic reticulum rescue and retention of Kir2.6 and Kir2.2, respectively.

In addition, they showed that the wild-type Kir2.6 co-assembles with Kir2.1 and Kir2.2 in vitro and in skeletal muscle acting as a dominant negative subunit, limiting Kir2.1 and Kir2.2 localization to the cell surface (Dassau et al. 2011).

2.3 Physiology/Pathophysiology

The physiology and pathophysiology of Kir2 channels has been previously reviewed (de Boer et al. 2010; Hibino et al. 2010). Here, we will highlight the major roles of this channel subfamily members in the various tissues they are expressed.

Heart: Kir2 channels are critically involved in determining the shape of the cardiac action potential by (1) setting the resting potential, (2) permitting the plateau phase (through their Mg^{2+} and polyamine block at depolarized potentials), and (3) inducing a rapid final stage of repolarization. Contributions of different Kir2 subunits in comprising cardiac Kir currents have revealed that Kir2.1 knockout in mice abolishes ventricular Kir2 currents, while Kir2.2 knockout reduces the currents by 50% (Zaritsky et al. 2001). This suggests that Kir2.1 is the dominant expressing subunit and Kir2.2 contributes to the current by assembling with Kir2.1. Kir2.1 dysfunction due to 21 mutations identified in 30 families have been identified to cause Andersen-Tawil syndrome (ATS) by multiple mechanisms, including allosteric decreases in channel-PIP_2 interactions (Donaldson et al. 2004; Lopes et al. 2002). ATS is an autosomal-dominant disorder resulting in cardiac arrhythmias (long Q-T syndrome 7, LQT7), periodic paralysis, and dysmorphic bone structure in the face and fingers (Tawil et al. 1994). The cardiac symptoms entail a depolarized V_{REST} and this loss of the stabilization of V_m can trigger arrhythmias. The inability to contribute to the late phase of the repolarization of the action potential results in a prolonged action potential duration, hence the LQT7 arrhythmia. Differential expression is seen within tissues of the same organ, as in the heart for example, where dominant expression in the atria is seen for Kir2.3, while in the ventricles for Kir2.1 (Anumonwo and Lopatin 2010). Besides ATS, Kir2.1 has been implicated in atrial fibrillation (AF), the most prevalent arrhythmia that ranges from 1–2% in the general population to 9–10% in the elderly population (Dobrev et al. 2005). Kir3.1 and Kir3.4 have also been implicated in AF but will be discussed in the next section devoted to Kir3 channels. Enhanced expression of the Kir2.1 channel leads to AF and this has been linked with downregulation for two microRNAs: miR-1 (Girmatsion et al. 2009; Yang et al. 2007) and miR-26 (Luo et al. 2013). Additionally, the Kir2.1(V93I) is a gain-of-function mutation that has been linked to hereditary AF (Xia et al. 2005).

Skeletal muscle: The periodic paralysis symptom of ATS is due to the effects of the $I_{Kir2.1}$ reduction in skeletal muscle. There, the depolarized V_{REST} inactivates Na_v channels making them unavailable for initiation and propagation of the action potential, leading to paralysis. Kir2.1 has been reported to be essential for myoblast differentiation (Konig et al. 2004) and also for fusion of mononucleated myoblasts to form multinucleated skeletal muscle fibers (Fischer-Lougheed et al. 2001). Kir2.1 underlies a 60 mV hyperpolarization in the V_{REST} of myoblasts during development, which drives Ca^{2+} entry through Ca^{2+}-permeant ion channels that promotes differentiation and fusion of myoblasts. Even though there are no gross skeletal muscle developmental defects in ATS patients, their "slender" built could be related to the mild developmental defects caused by the Kir2.1 dysfunction. Kir2.6 is also expressed in skeletal muscle and as already mentioned is nearly identical to Kir2.2 (Ryan et al. 2010). It was discovered in a screen for candidate genes responsible for thyrotoxic hypokalemic periodic paralysis, a serious complication of hyperthyroidism characterized by skeletal muscle paralysis and hypokalemia, affecting young adult male patients of Asian descent (Kung et al. 2006). The gene coding for Kir2.6 (*KCNJ18*) is transcriptionally regulated by thyroid hormone via a thyroid-responsive element in its promoter region. Thirty three percent of the patients afflicted by this condition bear mutations mainly localized in the C-terminus causing decreased current densities (de Boer et al. 2010). Destabilizing V_{REST} would inactivate Na_v channels yielding paralysis.

Bone: The dysmorphic facial bone structure defects point out the role of Kir2.1 channels in bone development. Indeed, it has been recently reported that Kir2.1 is important for efficient signaling by the bone morphogenic proteins in mammalian face development (Belus et al. 2018).

Blood vessels – Endothelial cells: Both the endothelial and smooth muscle cells that comprise the vasculature express Kir2 channels (Adams and Hill 2004; Nilius and Droogmans 2001). In fact, for vascular endothelial cells these are considered the most prominent channels expressed (Nilius and Droogmans 2001; Nilius et al. 1993; von Beckerath et al. 1996). By setting V_{REST} near E_K, they drive Ca^{2+} entry into endothelial cells (Kwan et al. 2003; Wellman and Bevan 1995) that triggers NO-mediated vasodilation. More recently, Kir2.1 channels were shown to boost the endothelial cell-dependent vasodilation generated by Ca^{2+}-dependent activation of small and intermediate conductance Ca^{2+}-activated K^+ channels in resistance-sized arteries (Sonkusare et al. 2016). In aortic endothelial cells evidence has suggested that Kir2.2 channels are the dominant Kir2 conductance (Fang et al. 2005).

Smooth muscle cells: A mild increase in $[K^+]_o$ (from 6 to 15 mM) hyperpolarizes the V_m of smooth muscle cells by 15 mV (from −45 to −60 mV) by increasing Kir conductance and vasodilating cerebral and coronary arteries (Knot et al. 1996; McCarron and Halpern 1990; Nelson et al. 1995). The hyperpolarization closes Ca_v channels reducing $[Ca^{2+}]_i$ leading to vasodilation (Knot and Nelson 1998). In cerebral arteries, evidence has been presented arguing for astrocyte secretion of K^+ upon neuronal stimulation, suggesting a way to couple neuronal activity to local blood flow in the brain (Filosa et al. 2006). Kir2.1, rather than Kir2.2 or Kir2.3, is thought to underlie these hyperpolarization effects in vascular smooth muscle cells (Bradley et al. 1999; Zaritsky et al. 2000). This capillary-to-arteriole coupling, whereby neuronal activity results in a small increase in extracellular K^+, is sensed by the Kir2.1 channels of capillaries to increase Kir2.1 outward current and hyperpolarization that in turn vasodilates the arterioles to cause an increase in local cerebral blood flow (Longden et al. 2017). Interestingly, this is a regulated process, as signaling through Gq-protein coupled receptors (GqPCRs) that hydrolyzes PIP_2 prevents activation of Kir2.1 by the $[K^+]_o$ increase and uncouples this intricate sensing mechanism (Harraz et al. 2018).

Neurons: Kir2 channels are abundantly and differentially expressed in somata and dendrites of neurons in the brain: diffusely and weakly in the whole brain (Kir2.1), moderately throughout the forebrain and strongly in the cerebellum (Kir2.2), mainly in the forebrain and olfactory bulb (Kir2.3), and in the cranial nerve motor nuclei in the midbrain, pons (Kir2.4) (Hibino et al. 2010). PIP_2 depletion, once again, modulates activity, this time of striatopallidal neurons, as stimulation of their dendritic spines that contain Kir2.3 channels results in enhancement of dendritic excitability (Shen et al. 2007). The microvilli of Schwann cells at the node of Ranvier express Kir2.1 and Kir2.3 which may be serving a buffering role to maintain $[K^+]_o$ by absorbing excess K^+ released by excited neurons as done by astroglia (Mi et al. 1996).

2.4 Pharmacology

Kir2 channel inhibitors: A number of small molecule inhibitors for Kir2 channels have been reported. *Tamoxifen*, an estrogen receptor antagonist for breast cancer treatment, inhibits Kir2.1, Kir2.2, and Kir2.3. The experimental results suggested the compound inhibits the channels by interfering with channel–PIP_2 interactions (Ponce-Balbuena et al. 2009). Chloroquine, an important anti-malaria drug, inhibits the Kir2.1 channel and can induce lethal ventricular arrhythmias. Molecular modeling and mutagenesis suggested that chloroquine blocks the Kir2.1 channel by plugging the cytoplasmic conduction pathway, interacting with the negatively charged and aromatic residues within the central pocket (Rodriguez-Menchaca et al. 2008).

Gambogic acid (GA) was discovered as a Kir2.1 inhibitor through a library screening of 720 naturally occurring compounds. GA acts slowly at nanomolar concentrations to abolish Kir2.1 but not $K_v2.1$, HERG or Kir1.1 channel activity. GA could interfere with Kir2.1 channel trafficking to the cell surface (Zaks-Makhina et al. 2009). VU573 was discovered through a thallium (Tl^+) flux-based high-throughput screen of a Kir1.1 inhibitor library. The compound inhibits Kir3, Kir2.3, and Kir7.1 over Kir1.1 and Kir2.1 channels (Raphemot et al. 2011). ML133 was discovered through a high-throughput screen (HTS) of a library (>300,000 small molecules). It inhibits Kir2.1 with little selectivity against other Kir2 family channels. However, ML133 has no effect on Kir1.1, and a weak activating effect on Kir4.1 and Kir7.1 channels. Using a chimera between Kir2.1 and Kir1.1, the molecular determinants were identified to be residues D172 (the "rectification controller" residue) and I176 (one of the two HBC gates, I177 in Kir2.2) of the TM2 segment in Kir2.1 (Wang et al. 2011). Chloroethylclonidine (CEC) has been reported as an agonist for α_2-adrenergic receptors and an antagonist for α_{1x}-receptors. It inhibits Kir2.1 by directly blocking the channel pore, possibly at an intracellular polyamine binding site. Mutation of E172 in Kir2.1 to asparagine abolished CEC inhibition (Barrett-Jolley et al. 1999). Celastrol has been discovered as an inhibitor for the Kir2.1 and hERG channels and causes QT prolongation. It also alters the rate of channel transport and causes a reduction of channel density at the cell surface (Sun et al. 2006a). 3-bicyclo [2.2.1] hept-2-yl-benzene-1,2-diol was discovered as an inhibitor of the Kir2.1 and $K_v2.1$ channels through screening of

10,000 small molecules from a combinatorial chemical library. It is a weak inhibitor of Kir2.1 compared to Kv2.1 channels (Zaks-Makhina et al. 2004).

Kir2 channel activators: Pregnenolone sulfate (PREGS) belongs to the neurosteroid family and has been shown to be a Kir2.3 activator from the extracellular side of the membrane with no effect from the intracellular side. The activation was not affected by changes in the external pH. Other Kir channels, such as, Kir1.1, Kir2.1, Kir2.2, and Kir3 channels, were insensitive to PREGS (Kobayashi et al. 2009). Tenidap was discovered as a potent activator of Kir2.3 channels using an $^{86}Rb^{+}$ efflux assay. The action of tenidap was from the extracellular side of the membrane. It had little or no effect on Kir2.1, Kv1.5, and Na_v channels. Tenidap could serve as a pharmacological tool, an opener of Kir2.3 channels (Liu et al. 2002) (Table 3).

Table 3 Small molecule inhibitors and activators of Kir2 channels

Drug name	Species system	Potency (IC_{50}/ EC_{50})	References
Inhibitor			
ML133	HEK-293 cells	Kir2.1 (IC_{50} = 1.8 μM) Kir2.6 (IC_{50} = 2.8 μM) Kir2.2 (IC_{50} = 2.9 μM) Kir2.3 (IC_{50} = 4.0 μM)	Wang et al. (2011)
VU573	HEK-293 cells	Kir2.3 (IC_{50} = 4.7 μM)	Raphemot et al. (2011)
Tamoxifen (antiestrogens)	HEK-293 cells	Kir2.1 (IC_{50} = 0.93 μM) Kir2.2 (IC_{50} = 0.87 μM) Kir2.3 (IC_{50} = 0.31 μM)	Ponce-Balbuena et al. (2009)
Gambogic acid (xanthonoid)	HEK-293 cells	Kir2.1 (IC_{50} = 27 nM)	Zaks-Makhina et al. (2009)
Chloroquine (4-aminoquinoline)	HEK-293 cells	Kir2.1 (IC_{50} = 8.7 μM	Rodriguez-Menchaca et al. (2008)
Chloroethylclonidine (imidazoline)	Skeletal muscle	Kir2.1 (IC_{50} = 37 μM)	Barrett-Jolley et al. (1999)
Celastrol (dienonephenolic triterpene)	HEK-293 cells	Kir2.1 (IC_{50} = 20 μM)	Sun et al. (2006a)
3-Bicyclo[2.2.1] hept-2-yl-benzene-1,2-diol	HEK-293 cells	Kir2.1 (IC_{50} = 60 μM)	Zaks-Makhina et al. (2004)
Activator			
Pregnenolone sulfate (neurosteroids)	Xenopus oocytes	Kir2.3 (EC_{50} = 1.43 μM)	Kobayashi et al. (2009)
Tenidap (indoles)	CHO cells	Kir2.3 (EC_{50} = 402 nM)	Liu et al. (2002)

2.5 Structural Studies

Channel activation of all Kir channels requires PIP_2, but Kir2 channels have been shown to require an additional secondary non-specific phospholipid (PL-) (i.e., PA, phosphatidic acid; PG, phosphatidylglycerol; PS, phosphatidylserine; PI, phosphatidylinositol) for high PIP_2 sensitivity. Lee et al. (2013) using molecular docking simulations and experimental validation identified a putative anionic phospholipid [PL(−)] binding site, which is located adjacent to the PIP_2 binding site formed by two lysine residues, K64 and K219. Neutralization of either of these residues to Cys reduces channel activity, but lipid-tethering K64C with decyl-MTS (methanethiosulfonate) induces high-affinity PIP_2 activation even in the absence of PL(−). These results suggest a molecular mechanism for a PL(−) synergistic effect on PIP_2-dependent activation of Kir2 channels. Interestingly, the residues K64 and K219 in Kir2.1 are not highly conserved in the Kir channel family. For example, the corresponding residues in Kir 3.2 are E71 and N229 (Fig. 9). Thus, for Kir3 channels, instead of PL(−), it is the regulators $G_{\beta\gamma}$ and Na^+ that play a critical role in stimulating channel activity by enhancing channel-PIP_2 interactions (Lee et al. 2013). Lee and colleagues further characterized the PL(−) site by introducing a Trp mutation at the K62 position in Kir2.2 channel (corresponding to the K64 in Kir2.1), which enhances PIP_2 sensitivity even in absence of PL(−). High-resolution crystal structures of Kir2.2(K62W) were solved in the presence and absence of PIP_2. The mutation caused a tight tethering of the CTD to the TMD of the channel regardless of the presence of PIP_2 and mimicked the PL(−) binding to the second site, inducing formation of the high affinity primary PIP_2 site. MD simulations have revealed a more extensive hydrogen bonding to basic residues at the PIP_2 binding site in the Kir2.2 (K62W) mutant compared to the wild-type channel (Lee et al. 2016). To elucidate the open conformation and mechanism of K^+ ion permeation, Zangerl-Plessl et al. (2020) introduced an additional G178D mutation in the Kir2.2 (K62W) mutant channel to generate the "forced open" mutant channel (KW/GD). Crystal structures of the KW/GD mutant channel were solved in the presence and absence of PIP_2. The PIP_2 bound form of the KW/GD structure showed a slight widening of the HBC gate (~1.5 Å) compared to the PIP_2 bound K62W structure. However, microsecond MD simulations for KW/GD in lipid bilayer showed opening of the HBC gate and K^+ permeation. Simulation results showed K^+ ion permeation through the SF via a strict ion-ion knock-on mechanism with rates comparable to the experimentally measured ion conductance (Zangerl-Plessl et al. 2020).

3 G Protein Kir3 Channels

3.1 Historical Perspective

In the mid-1950s after the advent of intracellular microelectrodes and voltage-clamp techniques Burgen and Terroux (1953) revisited the vagal induced hyperpolarization of heart muscle reported by Gaskell in the mid-1880s (Gaskell 1886; Gaskell 1887).

They first confirmed that hyperpolarization is induced by ACh application (Burgen and Terroux 1953) and then by measuring the effect of external K^+ concentration on resting potential in the absence and presence of ACh, they also demonstrated that an increased cell permeability to potassium may underlie hyperpolarization. Del Castillo and Katz used microelectrodes to directly show hyperpolarization of the sinus node upon vagal stimulation (Del Castillo and Katz 1955). Voltage clamp allowed Trautwein and Dudel to directly confirm changes in cell potassium permeability by measuring K^+ reversal potentials (Trautwein and Dudel 1958). Noma and Trautwein studied activation kinetics of ACh-induced K^+ currents and concluded that ACh binding activates a specific ion channel, K_{ACh} (Noma and Trautwein 1978). The introduction of the patch-clamp technique (Hamill et al. 1981) led to the first single-channel recordings of K_{ACh} currents ($I_{K\text{-}ACh}$) (Sakmann et al. 1983), which clearly demonstrated kinetic properties distinct from other background potassium channels. In the mid-1980s five studies shed light as to how ACh activated $I_{K\text{-}ACh}$. First, following the advent of the patch-clamp technique, Soejima and Noma demonstrated the membrane-delimited nature of $I_{K\text{-}ACh}$ activation, when in cell-attached recordings they showed that ACh could only activate the channels if it was perfused (or loaded) directly in the pipette but not in the bath, arguing that no second messenger was involved in activating $I_{K\text{-}ACh}$ (Soejima and Noma 1984). The following year two reports published back-to-back provided evidence that pertussis toxin- (PTX-) sensitive $G_{i/o}$ proteins coupled muscarinic receptors to $I_{K\text{-}ACh}$, either by showing that PTX treatment disabled ACh stimulation of channel activity (Pfaffinger et al. 1985) or showing that non-hydrolyzable GTP analogs activated $I_{K\text{-}ACh}$ constitutively (Breitwieser and Szabo 1985). The following year using inside-out patches from atrial cells Kurachi and colleagues showed that in the presence of adenosine (or ACh) in the pipette GTP perfusion in the bath supported activation of $I_{K\text{-}ACh}$ (Kurachi et al. 1986). In 1987, Logothetis and colleagues showed that purified $G\beta\gamma$ (but not $G\alpha$ subunits) activated $I_{K\text{-}ACh}$ in a manner similar to non-hydrolyzable GTP analogs, identifying $I_{K\text{-}ACh}$ as the first direct effector of the $G\beta\gamma$ subunits (Fig. 5a – top). A controversy arose as Birnbaumer and Brown argued that the activated $G\alpha$ (i.e., $G\alpha$-GTPγS) and not the $G\beta\gamma$ subunits were the activators of $I_{K\text{-}ACh}$ (Birnbaumer and Brown 1987). The controversy was resolved 7 years later when Reuveny and colleagues confirmed with recombinant channels in *Xenopus* oocytes that $G\beta\gamma$ subunits were the G protein activators of $I_{K\text{-}ACh}$ (Reuveny et al. 1994). Intracellular Na^+ ions were also shown to activate Kir3.4 and Kir3.2 homo- and heterotetramers in a G protein-independent manner (Ho and Murrell-Lagnado 1999; Sui et al. 1996). Ethanol was also shown to activate Kir3 currents (Kobayashi et al. 1999; Lewohl et al. 1999). In 1998, Kir3 currents activated by $G\beta\gamma$ or Na^+ were shown to require PIP_2 in the plasma membrane for activation (Huang et al. 1998; Petit-Jacques et al. 1999; Sui et al. 1998). Although for classical Kir channels, PIP_2 was sufficient to stimulate activity, Kir3 channels required a gating molecule (e.g., Na^+, $G\beta\gamma$, alcohol, etc.) together with PIP_2 to stimulate activity (Huang et al. 1998; Sui et al. 1998). These gating molecules seemed to strengthen the interactions of Kir3 channels with PIP_2, as a single point mutant (Kir3.4-I229L) removed the

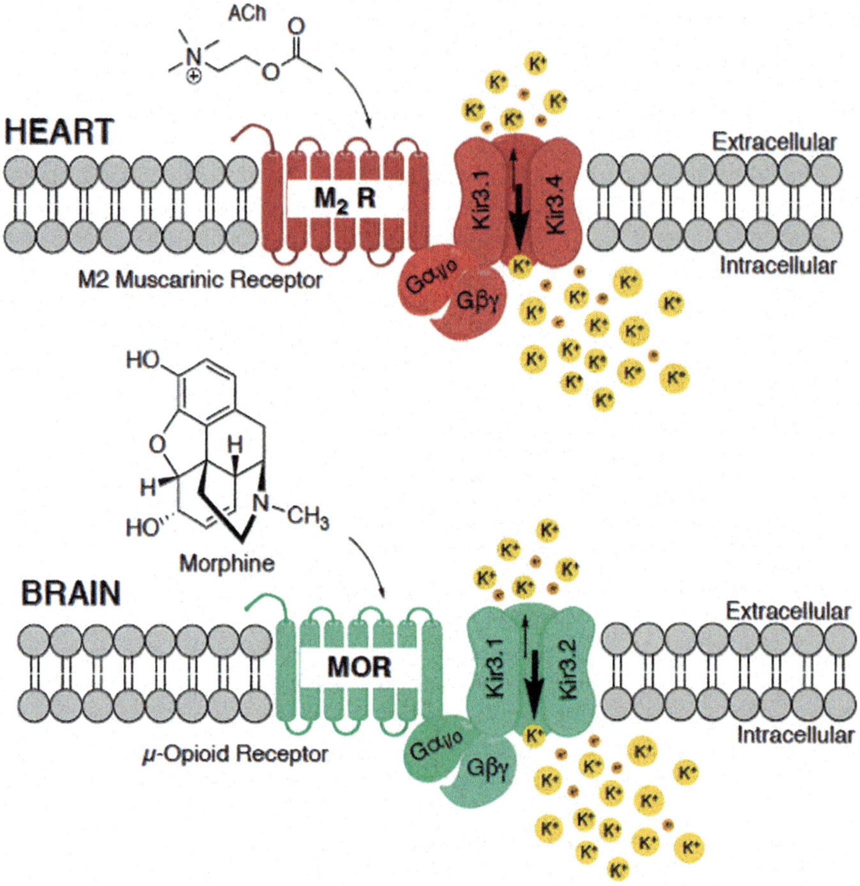

Fig. 5 The Kir3 GEMMA ($G_{i/o}$pcr-Effector-Macromolecular-Membrane-Assembly) before activation (name/concept adapted from Ferré et al. 2021)

requirement for gating molecules, allowing PIP_2 to activate on its own (Zhang et al. 1999).

3.2 Subfamily Members and Tissue Distribution

It wasn't until 1993 that the first member Kir3.1 of this subfamily was cloned (Dascal et al. 1993; Kubo et al. 1993b). The following year the neuronal Kir3.2 and Kir3.3 subunits were identified (Lesage et al. 1994). Finally, in 1995 the Kir3.4 subunit was cloned and when co-expressed together with Kir3.1 was shown to produce the biophysical properties of $I_{K\text{-}ACh}$ expressed in heterologous cells (Chan et al. 1996; Krapivinsky et al. 1995). Kir3.2 and Kir3.4 are capable of conducting currents through homotetramers, while Kir3.1 and Kir3.3 need to heteromerize with other subunits to produce functional channels. In the brain, multiple signals including ACh, adenosine, dopamine, opioids, GABA, etc. bind their respective $G_{i/o}$ protein-coupled receptors ($G_{i/o}$PCRs) to activate Kir3 channels using the Gβγ subunits of PTX-sensitive G proteins (Fig. 5b – bottom). Expression of different subunits within distinct neuronal populations defines their responsiveness to neurotransmitters. The ventral tegmental area (VTA) in the reward centers of the brain illustrates well the relevance of the Kir3 subunit composition (Lujan et al. 2014). Dopaminergic neurons within the VTA do not express Kir3.1 but do express Kir3.2 and Kir3.3, producing Kir3.2 homotetramers as well as Kir3.2/3.3 heterotetramers (Cruz et al. 2004; Jelacic et al. 2000; Kotecki et al. 2015). GABAergic neurons within the VTA also express Kir3.2 and Kir3.3 along with Kir3.1 producing also the dominating Kir3.1/3.2 heterotetramers. VTA dopaminergic neurons show less sensitivity to GABA than GABAergic neurons do (Cruz et al. 2004; Labouebe et al. 2007). Genetic ablation of Kir3.3 in the dopaminergic neurons enhances their sensitivity to GABA, suggesting an inhibitory role by Kir3.3. Thus, low concentrations of agonists preferentially inhibit GABAergic neurons and thereby disinhibit dopaminergic neurons. This disinhibition might confer reinforcing properties on addictive $GABA_B$ receptor agonists (Labouebe et al. 2007).

Three major alternatively spliced isoforms of Kir3.2 are differentially expressed in the brain Kir3.2a-c. Kir3.2c is longer than Kir3.2a, which is longer than Kir3.2b and also shows differences in 8 C-terminal aa (Hibino et al. 2010). In the periphery, Kir3 channels show moderate expression in the heart and pancreas (Fig. 2) as well as in the testis and the pituitary and adrenal glands.

Heart: Kir3.1 and Kir3.4 subunits are predominantly expressed in the atria, in nodal cells and pulmonary vein myocardial sleeves (Ehrlich et al. 2004; Greener et al. 2011; Krapivinsky et al. 1995). Expression in mouse ventricles has also been reported but it is not thought to impact cardiac physiology or ventricular arrhythmogenesis (Anderson et al. 2018).

Pancreas: Four Kir3 isoforms have been reported to be expressed and play important roles in the physiology of the pancreas. Kir3.4 subunits exist in α, β, and δ cells of pancreatic islets and in the exocrine pancreas, whereas the Kir3.2c isoform is expressed in α and δ cells (Ferrer et al. 1995; Iwanir and Reuveny 2008; Vaughn et al. 2000; Yoshimoto et al. 1999).

Testis: Kir3.1 and Kir3.2d (another splice isoform, 18 aa shorter than Kir3.2c, that has been described only in the testis) have been reported to be expressed in the spermatids and in spermatogonia and spermatocytes (Inanobe et al. 1999).

Anterior pituitary lobe: Kir3.1, Kir3.2, and Kir3.4 have all been reported to be expressed in the anterior pituitary (Gregerson et al. 2001; Morishige et al. 1999).

Adrenal gland: Since 2011 it was reported that in patients with severe hereditary hypertension Kir3.4 mutations were found in adrenal aldosterone-producing adenomas to cause hyperaldosteronism. These findings brought a greater appreciation of the role of Kir3.4 in aldosterone production and disease (Choi et al. 2011; Gomez-Sanchez and Oki 2014). Other tissues including the eye, skin, and the colon have also been reported to express Kir3 channels (Huang et al. 2018a; Yamada et al. 1998).

Central nervous system (CNS): Kir3.1-Kir3.3 are expressed throughout, with highest levels seen in the olfactory bulb, neocortex, hippocampus, and the granule cell layer of the cerebellum but notably also in the amygdala, thalamus, substantia nigra, ventral tegmental area, locus coeruleus, some nuclei of the brainstem, and the spinal cord (Kobayashi and Ikeda 2006; Lujan and Aguado 2015). In contrast, Kir3.4 is expressed in relatively fewer brain regions with highest levels in deep cortical pyramidal neurons, the endopiriform nucleus and claustrum of the insular cortex, the globus pallidus, the ventromedial hypothalamic nucleus, parafascicular and paraventricular thalamic nuclei, and a few brainstem nuclei, such as the inferior olive and vestibular nuclei (Wickman et al. 2000).

3.3 Physiology/Pathophysiology

Physiology: The Kir3 channels and the adenylyl cyclase enzymes have been studied extensively as prototypical effectors coupled to G proteins and their corresponding GPCRs. There is a vast body of literature supporting the existence of pre-coupled macromolecular complexes that have been coined as GEMMAs, GPCR-Effector-Macromolecular-Membrane-Assemblies (Ferré et al. 2021). Thus, the various $G_{i/o}$PCRs-$G\alpha_{i/o}\beta\gamma$-Kir3 channel tetramers constitute unique GEMMAs (Fig. 5). The GEMMA concept is significant in that it extends the collision coupling model by which receptors, G proteins, and effectors communicate and transduce signals. In a GEMMA, agonists set a series of conformational changes in the pre-coupled elements that communicate with each other like clock gears rather than billiard balls. Clearly, the best evidence for GEMMAs is yet to come from high-resolution structures capturing them in their pre-coupled inactive and active states and elucidating their interactions at atomic resolution. With cryogenic electron microscopy (cryo-EM) having achieved technological breakthroughs in the past decade, structures of complexes are becoming a reality and it is only a matter of time before we obtain high-resolution images of GEMMAs. Computational power to perform dynamic simulations of large systems is also in the midst of technological breakthroughs, such that the workings of GEMMAs can be captured in the course of a computer simulation. This unprecedented progress in our ability to make reliable dynamic structure-based models of macromolecular membrane assemblies of proteins promises to illuminate our understanding of the molecular physiology and pharmacology of GEMMAs.

GEMMAs use a core assembly, as shown in Fig. 5 with the examples of the K_{ACh} and K_{MOR} GEMMAs, to carry out their basic function but can accommodate in a dynamic manner the coming and going of multiple other partners to fine-tune effector function in response to external and internal signals. Each member of the core GEMMA can recruit additional regulatory partners or scaffold proteins that orchestrate a symphony of regulatory partners with the GEMMA core proteins. RGS (regulator of G protein signaling) proteins, for example, enhance the GTPase activity of G proteins and as such they accelerate receptor-induced Kir3 current deactivation kinetics and inhibit GPCR-Kir3 signaling (Doupnik 2015; Doupnik et al. 1997;

Sjogren 2011). Molecular platforms to achieve subcellular localization of membrane proteins or post-translational modifications needed to serve the physiology of the GEMMA have been reviewed by Ferré and colleagues (2021).

Utilization of motifs in the C-termini of any of the core proteins of the GEMMA can recruit specific modulatory proteins. As an example, let us consider the Kir3 channels using their PDZ domains to recruit scaffold proteins. The PDZ domain is a common structural domain of 80–90 amino acids found in signaling proteins from bacteria to animals. Proteins containing PDZ domains play a key role in anchoring receptor proteins in the membrane to cytoskeletal components. Proteins with these domains help hold together and organize signaling complexes at cellular membranes. PDZ got its name by combining the first letters of the first three proteins discovered to share the domain – post synaptic density protein (PSD95), Drosophila disc large tumor suppressor (Dlg1), and zonula occludens-1 protein (zo-1). Scaffold proteins of the PSD-95 and Shank families recognize class I PDZ motifs (x-S/T-x-F: F is a hydrophobic aa and x is any aa) and can bind the C-termini of these channel subunits. Kir3.2c and Kir3.3 possess a C-terminal motif for class I PDZ domain-containing proteins (ESKV) that is absent in Kir3.2a. PDZ-scaffold proteins are considered to be one of the major determinants of postsynaptic localization of various proteins (Sheng and Sala 2001). Thus, Kir3.2c and Kir3.3 channels could be localized post-synaptically through direct or indirect interactions with PDZ-scaffold proteins. Classical Kir2 channels, for which there is no evidence that they are part of a GEMMA, also contain class I PDZ motifs. Interestingly, Kir2, but not Kir3 subunits, has been found to interact directly with PSD-95 (Nehring et al. 2000), indicating that the mere presence of the binding motif does not necessarily imply a direct binding of the scaffold protein. Instead, the scaffold protein sorting nexin 27 (SNX27) binds the Kir3.2c and Kir3.3 channels but not the Kir2 (or a Kir4 channel that also contains the motif (Balana et al. 2011; Lunn et al. 2007)). SNX27 association with the Kir3 subunits leads to a reduction of signaling at the plasma membrane, likely by promoting internalization of the channel (Lunn et al. 2007). This example points to how subunit composition (in this case Kir3.2c and Kir3.3) makes the channel and potentially its GEMMA prone to another regulatory control of channel density at the plasma membrane to control cellular function. Interactions of Kir3.1/3.2 with NCAM (neural cell adhesion molecule) in lipid raft microdomains of hippocampal neurons have been reported to decrease Kir3.1/3.2 cell surface density (Delling et al. 2002).

Finally, let us consider post-translational modifications of Kir3 channels, such as phosphorylation. Kir3 (as well as several other Kir) channels have been reported to be modified in a functionally meaningful manner by phosphorylation, e.g. by protein kinases A and C (PKA, PKC): (Keselman et al. 2007; Leaney et al. 2001; Lei et al. 2001; Lopes et al. 2007; Mao et al. 2004; Medina et al. 2000; Mullner et al. 2000; Sharon et al. 1997; Sohn et al. 2007; Stevens et al. 1999; Witkowski et al. 2008; Gada and Logothetis 2021). Another large (more than 50-member) class of scaffolding proteins called AKAPs (A kinase anchoring proteins) binds to a common motif on effectors or GPCRs to organize recruitment of specific kinases and phosphatases (e.g., AKAP5 anchors PKA, PKC, calcineurin to proteins it binds) to modify the protein(s) of interest (Ferré et al. 2021). Is phosphorylation of Kir3 channels involving the use of scaffolding partner proteins, such as AKAPs? Although direct evidence is not yet available for Kir channels, GPCRs have been studied extensively and as parts of a GEMMA they could provide the organization needed for phosphorylation/dephosphorylation of Kir channels. *Pathophysiology* Studies on mutant mice lacking specific Kir3 subunits have provided evidence for physiological/pathophysiological effects suggesting these channels as therapeutic targets and highlighting the usefulness of potential Kir3 activators/inhibitors (Fig. 6).

Cardiac arrhythmias: As we have discussed above, stimulation of M_2R in supraventricular tissues by ACh released by the vagus nerve activates $I_{K\text{-}ACh}$ slowing down heart rate and endowing the heart with the adaptability needed to respond to rate adjustments (heart rate variability or HRV) and meet the demands of the body,

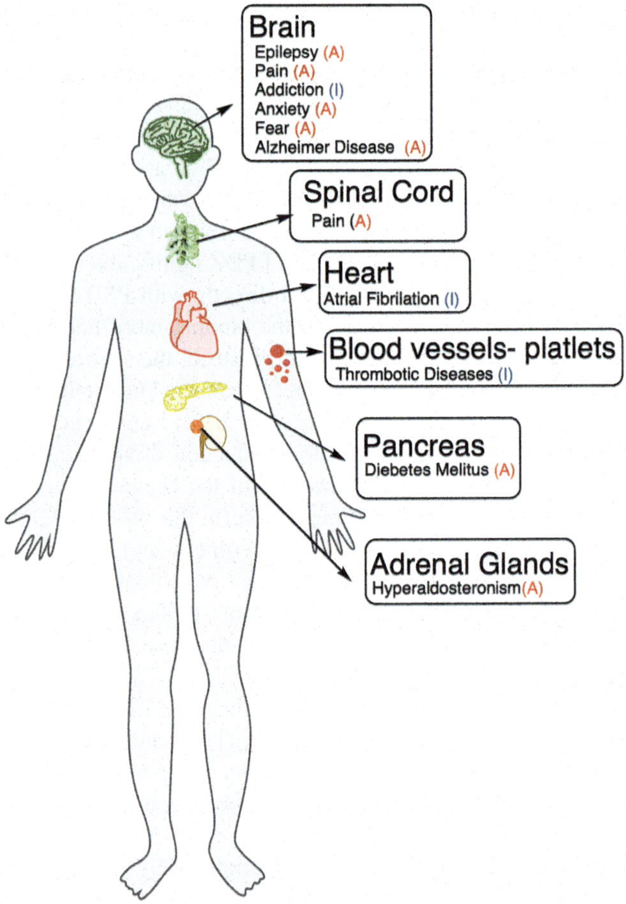

Fig. 6 Pathophysiological Kir3 conditions and drug regulators needed as potential therapeutics (A) Activators (I) Inhibitors. *CB*, cannabinoids, *ETOH* ethanol, *MOR* μ-opioid receptor, *PSVT* paroxysmal supraventricular tachycardia

as in exercise. Adenosine has similar effects on $I_{K\text{-}ACh}$ (Kurachi et al. 1986), acting through adenosine 1 receptors (A1Rs). Although Kir3.4 or Kir3.1 knockout in mice resulted at best in mild resting tachycardia, these deletions strongly decreased chronotropic response and HRV (Bettahi et al. 2002; Wickman et al. 1998). Kir3.4 loss of function mutations in people have been described resulting in Long QT (LQT) syndrome (Yang et al. 2010). Similarly, in cases of paroxysmal supraventricular tachycardia (PSVT), adenosine and other A1R receptor agonists have been used successfully for its acute treatment (Prystowsky et al. 2003). Vagal denervation (or knockout of the Kir3.4 channel) prevents the induction of atrial fibrillation (AF), the most common arrhythmia in clinical practice (Kovoor et al. 2001). On the other hand, gain of function conditions, such as vagal stimulation, predispose to AF. In AF, a constitutive increase in I_{KACh}, in the absence of channel stimulation by acetylcholine, promotes the shortening of action potential duration (Carlsson et al. 2010; Dobrev et al. 2005; Kovoor et al. 2001). This increase in basal activity of the channel is ATP dependent and is enhanced in the presence of phosphatase inhibitors (Makary et al. 2011), suggesting kinase-mediated activity. Conventional PKC isoforms inhibit, while novel PKC isoforms stimulate $I_{K\text{-}ACh}$ (Makary et al. 2011). A decrease in PKCα and a concomitant increase in PKCε expression have been reported from patients with AF (Voigt et al. 2014; Gada and Logothetis 2021).

Hyperaldosteronism: As mentioned above, Kir3.4 mutations were found in adrenal aldosterone-producing adenomas causing hyperaldosteronism in patients with severe hereditary hypertension (Choi et al. 2011).

Hyperalgesia and decreased analgesia: Kir3.1 and/or Kir3.2 knockout mice exhibit hyperalgesia (Marker et al. 2004). Moreover, $G_{i/o}$PCR ligands for opioid, M2 muscarinic, α_2 adrenergic, $GABA_B$, and cannabinoid receptors show reduced analgesic effects in Kir3.2 and/or Kir3.3 knockout mice (Blednov et al. 2003; Cruz et al. 2008; Lotsch et al. 2010; Lujan et al. 2014; Luscher and Slesinger 2010; Mitrovic et al. 2003; Nishizawa et al. 2009; Smith et al. 2008; Tipps and Buck 2015). Gene polymorphisms in *KCNJ6* (coding of Kir3.2) in humans have been associated with analgesia and pain sensitivity (Bruehl et al. 2013; Nishizawa et al. 2014). Thus, a plethora of studies converge on the mechanism that Kir3 channel activation induces analgesia accounting for the action of many drugs that reduce pain.

Drug addiction: As discussed earlier, Kir3 channels are thought to play an important role in the communication between the midbrain GABAergic and VTA dopaminergic neurons in the mesolimbic reward system, resulting in disinhibition of the VTA dopaminergic neurons which have been connected to reward behaviors (Kotecki et al. 2015; Marron Fernandez de Velasco et al. 2015; Tipps and Buck 2015). Kir3 channels play a role in mediating the rewarding effects of ethanol and addictive drugs that target GPCRS, such as the MOR and cannabinoid type 1 receptor (Blednov et al. 2003; Blednov et al. 2001a; Blednov et al. 2001b; Marron Fernandez de Velasco et al. 2015; Nagi and Pineyro 2014; Nassirpour et al. 2010; Rifkin et al. 2017; Tipps and Buck 2015). Kir3.2 and Kir3.3 knockout mice showed reduced self-administration of cocaine compared to wild-type mice indicating that in the absence of Kir3 channels the behavioral response to drugs of abuse is altered (Morgan et al. 2003). It has been shown in multiple cell lines that exposure to these addictive drugs reduces Kir3 currents (Marron Fernandez de Velasco et al. 2015; Rifkin et al. 2017; Tipps and Buck 2015). Recent studies have attempted to generate cell-specific Kir3 channel knockouts with cre-lox methodologies (Kotecki et al. 2015). Whether specific Kir3 subunit composition selective modulators could become useful pharmacotherapies for managing addiction remains to be tested.

Epilepsy: Unlike Kir3.4 (Wickman et al. 2000), Kir3.2 knockout mice exhibit spontaneous seizures and increased susceptibility to a convulsant agent (Signorini et al. 1997). Adenosine-induced hyperpolarization in hippocampal brain slice recordings could be blocked by a Kir3 blocker, which induced seizure activity (Hill et al. 2020). In temporal lobe epilepsy, the most common epilepsy in adults (Tellez-Zenteno and Hernandez-Ronquillo 2012) that progresses to drug-resistant seizures, N-methyl-D-aspartate receptor (NMDAR) hyperactivity triggers apoptotic mechanisms, including activation of cysteinyl aspartate-specific proteases (caspases) that characterize the severity of epileptic seizures. It has been recently shown that caspase-3 targets the Kir3.1 and Kir3.2 proteins, cleaving their C-terminal ends to compromise Gβγ activation and cell surface localization of these channels, implicating them strongly in the progression of this disease (Baculis et al. 2017).

Down syndrome: The KCNJ6 that encodes for Kir3.2 is located in human chromosome 21. In Down syndrome there is a trisomy of chromosome 21, resulting in a duplication of the KCNJ6 gene that alters synaptic transmission (Reeves et al. 1995; Sago et al. 1998). Full or partial duplication of the corresponding mouse gene that contains KCNJ6 produced hallmark characteristics of the disease (Reeves et al. 1995; Sago et al. 1998). Kir3.2 protein upregulation resulted in altered reward mechanisms, cognitive functions, and synaptic plasticity (Reeves et al. 1995; Sago et al. 1998; Siarey et al. 1999). To control for the fact that there are several gene duplications that make the pathophysiology more complex in this disorder, a single trisomy of the KCNJ6 gene was produced in mice recapitulating deficits characteristic of the syndrome, suggesting that triplication of only this gene is key in some of the abnormal neurological phenotypes seen in Down syndrome (Cooper et al. 2012).

Alzheimer's disease (AD): AD is characterized by progressive cognitive decline accompanied by formation of amyloid-beta (Aβ) plaques, neurofibrillary tangles, and aggregates of hyperphosphorylated tau protein. The main brain regions affected by AD are the hippocampus and amygdala, where early onset AD pathogenesis is seen (Arriagada et al. 1992; Haass and Selkoe 2007; Hardy and Selkoe 2002; Huang and Mucke 2012; Swanberg et al. 2004; Zald 2003). Imbalances of excitatory/inhibitory synaptic transmission occur early in the pathogenesis of AD, leading to hippocampal hyperexcitability and causing synaptic, network, and cognitive dysfunction. Neuronal Kir3 channels control neuronal excitability contributing to the inhibitory signaling in the hippocampus. The tonic Kir3 currents related to $GABA_B$ receptors contribute critically to the balancing of membrane excitability and synaptic transmission (Lujan et al. 2009; Nava-Mesa et al. 2013; Sanchez-Rodriguez et al. 2020). Intracerebroventricular injections of Aβ impaired inhibitory signaling and disrupted synaptic plasticity (Sanchez-Rodriguez et al. 2017). $A\beta_{25-35}$ significantly decreased Kir3 channel conductance in the hippocampus (Mayordomo-Cava et al. 2015). RT-qPCR revealed that in CA3-CA1 hippocampal neurons application of Aβ decreased mRNA transcripts for Kir3.2, Kir3.3, and Kir3.4 subunits (Mayordomo-Cava et al. 2015). Contradictory results have also been published suggesting that a Kir3-mediated potassium efflux triggers apoptosis in hippocampal neurons (May et al. 2017). These conflicting results will need to be resolved in future studies.

Fear: Kir3.2 knockout mice showed impairments in both hippocampal-dependent contextual fear conditioning and hippocampal-independent cue fear conditioning (Victoria et al. 2016). Mice subjected to auditory cue-induced fear extinguish the fear by activating pathways between basolateral amygdala (BLA) and the nucleus accumbens (NAc) in the ventral striatum. BLA expresses all Kir3 isoforms, except Kir3.4, and displays baclofen-induced Kir3 currents through $GABA_B$ receptors (Xu et al. 2020).

3.4 Pharmacology

Kir3 channel inhibitors: Table 4 lists a number of inhibitors of Kir3 currents that have been studied in supraventricular cardiac arrhythmias and AF in particular.

3.4.1 Inhibitors in Disease

Atrial fibrillation (AF): The discovery that potent low nanomolar inhibition of Kir3.1/3.4 could be achieved by tertiapin (TPN), the 21 aa peptide extracted from the honey bee venom that also inhibits similarly Kir1.1 channels (Jin and Lu 1998), provided a useful tool to explore the role of this channel in AF. A non-oxidizable variant of TPN(M13Q) (or TPNQ) shows a similar Ki value to TPN and is often used instead (Jin and Lu 1999). Its specificity limitation, however, encouraged exploration for small molecule inhibitors.

Three compounds in the benzopyran class emerged, NIP-142, NIP-151, and NTC-801 with outstanding potency but not high enough specificity, especially against the neuronal Kir3.1/3.2. NTC-801 in particular inhibited Kir3.1/3.4 channels with a sub-nanomolar IC_{50} but inhibited Kir3.1/3.2 with an IC_{50} of 24 nM (Machida et al. 2011).

Although NTC-801 was able to convert AF to normal sinus rhythm in canine tachypacing models (Yamamoto et al. 2014), in clinical trials it failed to reduce AF burden in 20 patients with paroxysmal AF (Podd et al. 2016). The reason for this failure was likely that due to lack of specificity between Kir3.1/3.2 and Kir3.1/3.4 it was not possible to reach concentrations required for antiarrhythmic effect without producing CNS side effects. Cui and colleagues have revealed the mechanism of action of this class of compounds by the prototypic Benzopyran-G1 (likely to be NTC-801) revealing its binding site in Kir3.1 and showing that it acts non-specifically in Kir3.1 heteromeric channels (Cui et al. 2021). Recently, XAF-1407 was reported to inhibit Kir3.1/3.4 with an IC_{50} of 1 nM and to decrease the AF rate in an equine model of persistent AF (Fenner et al. 2020) but whether due to specificity issues it will have a similar fate as NTC-801 remains to be seen.

Pain: TPNQ (tertiapin-Q) was shown to reduce MOR agonist activity in the immersion tail flick test (Marker et al. 2004). Additionally, intrathecal administration

Table 4 Small molecule inhibitors and activators of Kir3 channels

Drug name	Species system	Kir3.1/3.2 ~IC_{50}/EC_{50} or Tc (Tc: typical concentration) *NR: not reported*	Kir3.1/3.4 ~IC_{50}/EC_{50} Tc (Tc: typical concentration) *N/A: not applicable*	References
Inhibitor				
Tertiapin-Q (peptide)	Xenopus oocyte	NR	0.0082 μM	Jin and Lu (1998)
SCH3390 (D_1R antagonist)	CHO cells	7.78 μM	NR	Kuzhikandathil and Oxford (2002)
NIP-151 (tricyclic benzopyran)	HEK-293 cells	NR	0.0016 μM	Hashimoto et al. (2008)
NIP-142 (tricyclic benzopyran)	Xenopus oocyte HEK293 cells	NR	10.0 μM 0.64 μM	Matsuda et al. (2005, 2006)
NTC-801 (BMS914392) BP-G1 (tricyclic benzopyran)	Xenopus oocytes Guinea pig atrial cells	0.024	0.0007 μM	Machida et al. (2011) Cui et al. (2021)
Halothane (general anesthetic)	Xenopus oocyte	Tc = 2MAC	NR	Yamakura et al. (2001)
Haloperidol (antipsychotic)	Xenopus oocyte	75.5 μM	40.9 μM	Kobayashi et al. (2000)
Desipramine (tricyclic antidepressant)	Xenopus oocyte	36.4 μM	53.9 μM	Kobayashi et al. (2004)
Methadone (MOR agonist)	Xenopus oocyte	53.3 μM	NR	Ulens et al. (1999)
Ifenprodil (α-1R antagonist NMDAR inhibitor)	Xenopus oocyte	7.01 μM	2.83 μM	Kobayashi et al. (2006)
Fluoxetine (prozac -SSRI)	Xenopus oocyte	16.9 μM	18.4 μM	Kobayashi et al. (2003)
Falcatin-A (terpenoid)	HEK-293 cells	NR	2.5 μM	Vasas et al. (2016)
VU573	HEK-293 cells	1.3 μM	1.3 μM	Raphemot et al. (2011)
XAF-1407	HEK-293 cells	NR	0.0011 μM	Fenner et al. (2020)

(continued)

Table 4 (continued)

Drug name	Species system	Kir3.1/3.2 ~IC_{50}/ EC_{50} or Tc (Tc: typical concentration) *NR: not reported*	Kir3.1/3.4 ~IC_{50}/ EC_{50} Tc (Tc: typical concentration) *N/A: not applicable*	References
Activator				
Ethanol (n-alcohol)	Xenopus oocyte	Tc = 100 mM	Tc = 100 mM	Kobayashi et al. (1999), Lewohl et al. (1999)
Naringin (flavonoid)	Xenopus oocyte	111.0 μM	120.9 μM	Yow et al. (2011)
ML297 (N-phenyl pyrazole urea)	HEK-293 cells	0.16 μM	1.8 μM	Kaufmann et al. (2013)
VU0466551 (N-benzyl pyrazole urea)	HEK-293 cells	0.07 μM	0.11 μM	Wen et al. (2013)
VU464 (N-cyclohexyl pyrazole acetamide)	HEK-293 cells	0.165 μM	0.72 μM	Wieting et al. (2017)
GAT1508 (N-phenyl pyrazole urea)	Xenopus oocyte HEK-293 cells	0.37 μM 0.075 μM	N/A	Xu et al. (2020)
VU331	HEK-293 cells	5.2 μM (Kir3.2 = 5.1 μM)	N/A	Kozek et al. (2019)
Ivermectin (antiparasitic)	Xenopus oocyte	3.5 μM	7.5 μM	Chen et al. (2017)
GiGA1 (urea)	HEK-293 cells	31 μM	NR	Zhao et al. (2020)

of TPNQ attenuated oxycodone-induced antinociceptive effects in mice (Nakamura et al. 2014).

Epilepsy: TPNQ administered intrathecally induced seizures (Mazarati et al. 2006). Several drugs used clinically as antipsychotics and antidepressants in the early 2000s (e.g., haloperidol, clozapine, desipramine, fluoxetine) were identified to inhibit Kir3 channels at higher concentrations and to induce seizures as a side effect (Kobayashi et al. 2000; Kobayashi et al. 2003; Kobayashi et al. 2004; Kobayashi et al. 2006; Luscher and Slesinger 2010; Ulens et al. 1999; Yamakura et al. 2001).

Kir3 channel activators: Table 4 also lists a number of Kir3 activators. Specifically, there is a highly potent urea scaffold class that gave rise to a completely neuronal specific Kir3.1/3.2 activator with no effect on the cardiac Kir3.1/3.4 channels (i.e., GAT1508 – see below – Xu et al. 2020).

Ethanol was the first direct Kir3 channel drug activator, as a series of N-alkyl alcohols were shown to activate Kir3.1/3.2 and Kir3.1/3.4 channels with 1-propanol being the most effective and potent activator (Kobayashi et al. 1999; Lewohl et al. 1999). In 2011 it was identified that at high micromolar concentrations the flavonoid Naringin from grapefruit could activate Kir3.1/3.2 channels (Yow et al. 2011). In 2013, data mining and development of the thallium flux fluorescent assay for Kir channels allowed for early identification of CID736191 (nF-ML297), revealing an urea core scaffold that was connecting a left side benzene ring, to the N-phenyl pyrazole on the right side (see Fig. 11, Kir3 Channel Activators, ML297) (Kaufmann et al. 2013; Weaver et al. 2004; Wen et al. 2013). Structure-activity relationship studies produced compound ML297 [1-(3,4-difluorophenyl)-3-(3-methyl-1-phenyl-1*H*-pyrazol-5-yl) urea]. ML297 has a Kir3.1/3.2 EC_{50} of 0.160 μM and a Kir3.1/3.4 EC_{50} of 1.8 μM and did not show any local motor impairments at efficacious doses in the rotor rod behavioral test in mice. ML297 was effective in two models of epilepsy in mice (Kaufmann et al. 2013). A slight chemical modification of the ML297 scaffold changing a phenyl group to a benzyl group produced compound VU0466551, 1-(1-benzyl-3-methyl-1*H*-pyrazol-5-yl)-3-(3,4-difluorophenyl) urea. VU0466551 has a Kir3.1/3.2 EC_{50} of 0.07 μM and Kir3.1/3.4 EC_{50} of 0.11 μM and was efficacious in two rodent models of pain (Abney et al. 2019; Wen et al. 2013). A second scaffold was data mined and screened for Kir3.1-containing channel activity by Weaver and colleagues, VU0259369, (2-(2-chlorophenyl)-*N*-(3-(*N*,*N*-dimethylsulfamoyl)-4-methylphenyl) acetamide) with an acetamide core, similar to the Urea-pyrazole class of compounds. However, this compound lacked selectivity and potency and was not pursued (Ramos-Hunter et al. 2013). The following SAR reported in 2017 was a scaffold merging of ML297's pyrazole core and replacing the urea with the acetamide core from VU0259369 to yield the hybrid compound VU0810464, 2-(3-chloro-4-fluorophenyl)-*N*-(1-cyclohexyl-3-methyl-1*H*-pyrazol-5-yl) acetamide. VU0810464 was synthesized to improve blood brain barrier (BBB) penetration (Wieting et al. 2017). In 2018, VU0810464 was evaluated in cultured hippocampal neurons and a stress induced hyperthermia model for anxiety and was found to be effective at reducing induced hyperthermia and activated Kir3 currents in hippocampal neurons (Vo et al. 2019). However, VU0810464 was not efficacious in the elevated plus maze model of anxiety in mice, similar to its parent compound ML297 (Vo et al. 2019). Recent studies identified GAT1508, a highly selective potent and efficacious Kir3.1/3.2 activator, with an EC_{50} of 0.075 μM, which is devoid of cardiac Kir3.1/3.4 channel activity (Xu et al. 2020). GAT1508 was evaluated in a rat auditory cue-induced fear extinction model of post-traumatic stress disorder (PTSD) that was found to be effective at extinguishing conditioned fear (Xu et al. 2020). GAT1508 also was found to induce Kir3 currents in basolateral amygdala (BLA) brain slice recordings where at low ineffective concentrations it potentiated baclofen-induced current showing synergism (Xu et al. 2020). Additionally, in 2017 a larger macrocyclic antiparasitic drug, Ivermectin was also shown to activate Kir3.1/3.2 channels with an EC_{50} of 3.5 μM and Kir3.1/3.4 channel with an EC_{50} of 7.5 μM (Chen et al. 2017).

3.4.2 Activators in Disease

Pain: Flupirtine, a non-opioid analgesic that non-selectively activates potassium channels and indirectly antagonizes NMDA receptors, activated Kir3 channels in rat hippocampal neurons and in cultured retinal ganglion cells (Jakob and Krieglstein 1997; Sattler et al. 2008). When given in combination with morphine it potentiated antinociception in two rat models of pain, suggesting a role for Kir3 in flupirtine-induced analgesia (Devulder 2010; Jakob and Krieglstein 1997).

Epilepsy: $G_{i/o}$PCR agonists, like a selective A1 adenosine receptor ligand, could be effective in suppressing seizures in a model of pharmaco-resistant epilepsy in mice (Gouder et al. 2003). *Urea-based activators* (e.g., ML297) of Kir3.1/3.2 heteromers have been shown to exert antiepileptic efficacy both in maximal electric shock-induced and in chemically-induced models of epilepsy in rodents (Huang et al. 2018b; Kaufmann et al. 2013; Zhao et al. 2020).

Alzheimer's disease (AD): Activation of Kir3 channels by ML297 was able to rescue all hippocampal deficits induced by intracerebroventricular injection of $A\beta_{1-42}$, restoring proper excitability in CA3-CA1 synapses (Sanchez-Rodriguez et al. 2017). ML297 prevented increased excitability, restored long-term potentiation (LTP) hindered by Aβ, and recovered hippocampal oscillatory activity (Sanchez-Rodriguez et al. 2017). ML297 was also able to restore long-term potentiation and novel object recognition deficits (Sanchez-Rodriguez et al. 2017). These findings suggest that Kir3 channel activation may represent a novel therapeutic strategy to recover excitation imbalances conferred by Aβ in early onset AD.

3.5 Structural Studies

Numerous MD simulation studies have been performed with Kir3 channels. Pioneering MD simulations were conducted based on homology models of Kir1.1, Kir3.1, and Kir6.2 (Haider et al. 2007; Stansfeld et al. 2009). The results provided insights into the nature of Kir channel–lipid interactions, in particular the importance of the slide helix and linker of the TMD and CTD in forming interactions with the headgroups of lipids. However, the timescale of the simulations at these early studies was limited to 10 ns that is a relatively short time period for an adequate sampling of protein motions (Haider et al. 2007). Using a combination of coarse-grained (1.5 μs) and atomistic MD simulations (20 ns), the interactions of PIP_2 and KirBac1.1, Kir3.1-KirBac1.3 chimera, and Kir 6.2 channels were explored. The PIP_2-binding site was identified at the N-terminal end of the slide helix and interface between adjacent subunits of Kir channels (Stansfeld et al. 2009). Meng et al. (2012) provided the first dynamic molecular view of PIP_2-induced channel gating by conducting MD simulations on the Kir3.1-KirBac1.3 chimera in the presence and absence of PIP_2 (100 ns). Simulations of the closed state with PIP_2 revealed an intermediate state between the closed and open conformations of the channel. A PIP_2-driven movement of the N-terminus and C-linker led to CD-loop stabilization of the G-loop gate in the open state (Meng et al. 2012). Due to timescale limitations of the MD simulations, the opening of the HBC gate of Kir channels had not been observed in any of these early studies. A highly conserved glycine residue in the middle of TM2 plays an important pivoting role in channel gating. Replacement of the residue immediately following this glycine by a proline in Kir3 channels leads to constitutively active channels (Jin et al. 2002; Sadja et al. 2001). Meng et al. (2016) conducted MD simulations on the Kir3.1 chimera (M170P) mutant (2QKS) in the presence of PIP_2 (100 ns). Interestingly, the HBC gate of the channel mutant was opened within 30 ns of the simulations. The open HBC gate reached 6 Å in diameter, which allowed partial hydrated K^+ ions to pass through. During the gating process, a cooperative rotation of TM1 and TM2 in counterclockwise direction (viewed from the extracellular side) was observed. Three K^+ ions passed through the HBC gate during a 100 ns simulation. Introduction of the proline mutation decreased the free energy barrier for opening the channel by 1.4 kcal/mol (Meng et al. 2016).

Lacin et al. (2017) studied the role of a conserved basic residue, Kir3.2(K200) at the tether helix region, combining functional studies with MD simulations (400 ns). Both experiments and simulations demonstrated that K200 in Kir3.2 supports a dynamic interaction with PIP_2. When K200 was mutated to a Tyr, it activated the channel by enhancing the interaction with PIP_2. The mutant K200Y opened the HBC gate during a 400 ns simulation (Lacin et al. 2017). Bernsteiner et al. (2019) conducted multi-microsecond–timescale MD simulations based on the crystal structures of Kir3.2 bound to PIP_2. The simulations provided detailed insights into the channel's gating dynamics, as well as the movement of K^+ ions through the channel under an electric field across the membrane (Bernsteiner et al. 2019). Li et al. (2019) also conducted microsecond-scale MD simulations based on the Kir3.2 crystal structure (PDB code 4KFM) in complex with PIP_2, Na^+, and $G_{\beta\gamma}$ to understand which gates are controlled by Na^+ and $G_{\beta\gamma}$ and how each regulator uses the channel domain movements to control gate transitions. The simulation results suggested that Na^+ ions control the cytosolic gate of the channel through an anticlockwise rotation, whereas $G_{\beta\gamma}$ stabilizes the transmembrane gate in the open state through a rocking movement of the cytosolic domain. Both effects alter the way in which the channel interacts with PIP_2 and thereby stabilize the open states of the respective gates (Fig. 7) (Li et al. 2019). Li et al. (2016a, b) predicted Tertiapin and Kir3.2 channel interactions using MD simulations and PMF (potentials of mean force) calculations. The residue K17 of TPN was predicted to protrude into the pore of the channel and form a hydrogen bond with carbonyl group of T157, and R7 of TPN to form a salt bridge with the E127 from the channel turret (Li et al. 2016b). As discussed under the "Kir3 Channel Activators" (Sect. 3.4), ML297 and particularly its derivative GAT1508 proved to be potent and specific activators of Kir3.1/3.2 channels. Molecular docking and MD simulations predicted a GAT1508 binding site validated by mutagenesis experiments, providing molecular insights into how GAT1508 interacts with the channel and allosterically modulates channel–PIP_2 interactions (Xu et al. 2020).

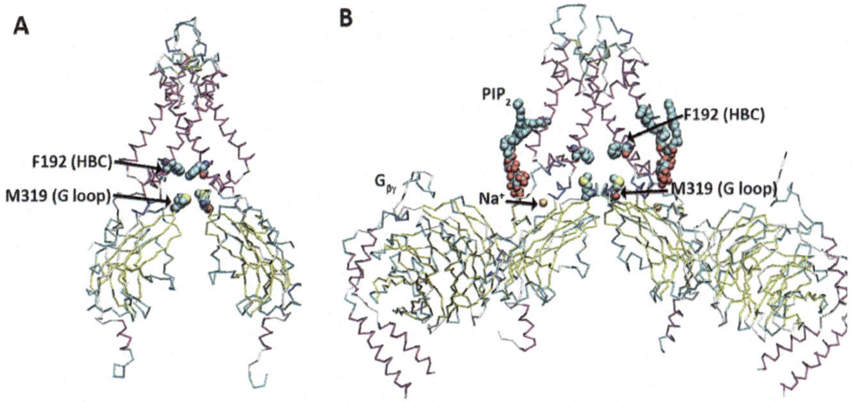

Fig. 7 Modulators, PIP_2, Na^+, and $G_{\beta\gamma}$ stabilize Kir3.2 channel in open state conformation. (**a**) Kir3.2 (Apo) is in a close conformation. (**b**) Kir3.2 (Holo, with PIP_2, Na^+ and $G_{\beta\gamma}$) is in an open conformation. Residues F192 form HBC gate, and M319 form G loop gate. The structural coordinates were obtained from MD simulations (Li et al. 2019)

4 ATP-Sensitive Kir6 Channels

4.1 Introduction

The pharmacology of ATP-sensitive K^+ channels (K_{ATP}) is in part addressed by Li and colleagues (2021, in this issue chapter). K_{ATP} channels are comprised of four pore-forming subunits (Kir6.1 or Kir6.2) and four sulfonylurea receptors subunits (SUR1, SUR2A or SUR2B) creating an octomeric structure (eight total subunits). Here we complement (Li et al. 2021) with emphasis on the Kir6 subunits of the K_{ATP} channels, as members of the Kir channel family at large. K_{ATP} channels (different combinations of SUR/Kir6 subunits) are found in cardiomyocytes, pancreatic β cells, neurons, skeletal muscle, smooth muscle, endothelial cells, and in membranes of mitochondria (Rodrigo and Standen 2005; Li et al. 2021). They couple metabolic changes (ATP to ADP ratio) to cellular excitability. They are regulated by a variety of physiological regulators including H_2S, long chain CoA esters, phosphatidylinositol phosphates (PIPs), carbon monoxide (CO), and changes in intracellular pH. As the intracellular pH acidifies, the channel activates, increasing rectification (Baukrowitz et al. 1999; Li et al. 2016a; Tinker et al. 2018; Rohacs et al. 2003; Shumilina et al. 2006). Like most other ion channels, K_{ATP} is regulated by post-translational modifications, such as S-palmitoylation (Yang et al. 2020) and phosphorylation affecting open probability and functional channel cell surface expression (Beguin et al. 1999). Analogous to all other Kir channels, K_{ATP} channels are ultimately regulated by phosphatidylinositol (4,5) bisphosphate (PIP_2). The sensitivity of the channel to ATP is regulated by PIP_2. PIP_2 decreases the ATP sensitivity resulting in a reversal of channel closure by ATP and stabilizing the open state (Shyng et al. 2000). However, these channels are the least stereospecific to regulation by $PI(4,5)P_2$ relative to other PIPs. The electrostatic interactions, dictated by the number of negative charges in the head group of PIPs, and the stereoselective nature, length of acyl chain, is less important for K_{ATP} channels, allowing $PI(4,5)P_2$, $PI(3,4,5)P_3$, $PI(3,4)P_2$, and PI(4)P to regulate the channel equally (Rohacs et al. 2003). Furthermore, long chain CoA esters activate and regulate the K_{ATP} channel by binding residues 311–332 on the C-terminus, a site distinct to both PIP_2 and ATP binding sites (Branstrom et al. 2007; Shyng et al. 2000). The metabolic sensing capability and its coupling to the electrical activity of K_{ATP} channels along with its distinct tissue localization has put much of the clinical emphasis on insulin regulation, cardiac rhythm, and neurological function.

4.2 Channel Regulation and Trafficking

ATP binds directly to Kir6 while MgADP binds to SUR. As ATP levels increase, ATP binds to the Kir6 subunits causing closure of the channel gates leading to a depolarization of the membrane and thus an increase in electrical activity. Conversely, as ATP levels fall and the ADP levels increase, MgADP binds to the SUR subunits causing the channels to open leading to hyperpolarization of the membrane and reduction in electrical activity (Ashcroft 2005).

Trafficking: Both SUR and Kir have a putative ER retention signal, RKR, in the distal portion of the C-termini. The SUR and Kir subunits co-assemble and in doing so mask each other's ER retention signals. This allows for the octameric complex to exit the ER and localize to the plasma membrane (Zerangue et al. 1999). The truncated Kir6.2 channel (Kir6.2ΔC36) is a functional channel in the absence of any SUR subunits due to its lack of ER retention motif and is an experimentally useful tool (Reimann et al. 2001; Zerangue et al. 1999). K_{ATP} channels may also be post-translationally trafficked to specific excitable membrane domains by the cytoskeletal adapter ankyrin-B (AnkB) (Kline et al. 2009). SUR and Kir6 are regulated in the ER by proteasome-mediated degradation as well as through protein recognition and translocation by Derlin-1 and folding by the chaperone protein Hsp90 (Wang et al. 2012; Yan et al. 2010).

4.3 Physiology/Pathophysiology and Tissue Distribution

We will briefly summarize major points of channel characteristics, disease implications, and tissue distribution. For a more detailed discussion of channel characteristics and disease implications, refer to (Li et al. 2021) as well as to more in-depth reviews (e.g., Tinker et al. 2018).

Pancreatic: The K_{ATP} channels of pancreatic β cells are comprised of SUR1/Kir6.2. These channels couple the metabolic state of the cell to the release of insulin. In the presence of elevated blood glucose levels, ATP levels rise, bind directly to the Kir6 subunit, and cause channel closure. In β cells this closure depolarizes the cell leading to activation of voltage-gated Ca^{2+} channels, thus increasing levels of intracellular Ca^{2+}, which results in insulin vesicle fusion to the plasma membrane and secretion (Hibino et al. 2010). There are a range of polymorphisms in the SUR1 and Kir6 subunit that leads to insulin secretory disorders. Hyperinsulinemia, neonatal diabetes, and DEND syndrome have all been linked to loss of function or gain of function mutations in the Kir6.2 subunit (Ashcroft 2005).

Cardiac: Cardiac K_{ATP} channels were initially thought to be an assembly of SUR2A and Kir6.2 due to most studies being conducted with ventricular cells. However, there has been discovery of other subunit assemblies within various tissues in the heart (for a list see Table 3 in Li et al. 2021). Cardiac K_{ATP} channels play an important cardioprotective role. Depending on their location in the heart, conductance and Ki of ATP do vary. K_{ATP} channels appear to be predominantly cardiac adaptive/protective in the presence of stressors (Tinker et al. 2018). Activation of cardiac K_{ATP} channels plays a significant role in ischemic preconditioning, and by shortening the action potential duration during myocardial ischemia increases cell survival (Tamargo et al. 2004). There is also evidence of K_{ATP} channels having altered expression patterns due to exercise leading to physiological adaptation (Tinker et al. 2018; Zingman et al. 2011).

Skeletal muscle: The classical skeletal muscle K_{ATP} channel assembly is SUR2A/Kir6.2, however skeletal muscles have also been shown to contain SUR1/Kir6.2. This has been found in fast twitch muscle, where the current density is larger (Tinker et al. 2018). Like cardiac K_{ATP} channels, in the skeletal muscle these channels contribute little to the membrane potential at rest. Their roles in skeletal muscle are still being investigated but it is evident they prevent muscle fatigue and muscle damage as well as regulate glucose uptake and cellular metabolism (Tinker et al. 2018). These findings suggest potential roles in muscle disorders, weight gain, and glucose regulation.

Vasculature and smooth muscle: Vascular smooth muscle, smooth muscle, and endothelial K_{ATP} channels are all comprised of Kir6.1/SUR2B (Suzuki et al. 2001; Aziz et al. 2017). Regulation of vascular smooth muscle through K_{ATP} channels is due to the ability of channel activity to regulate membrane potential. When hyperpolarization occurs through activation, intracellular Ca^{2+} is reduced causing dilation. This can occur through the G_s/PKA phosphorylation pathway as well as indirectly through hypoxic conditions via the mitochondria. Contraction occurs through the G_i/PLC$\beta_{2/3}$/PKC pathway that inhibits channel activity, causes depolarization and therefore contraction due to increased intracellular Ca^{2+} (Tinker et al. 2018). Smooth muscle K_{ATP}, like cardiac and neuronal K_{ATP}, plays a protective role in hypoxia and ischemic reperfusion (Aziz et al. 2017). However, there may be variation in expression depending on the location of the smooth muscle. K_{ATP} channels have been found in smooth muscle in the GI tract, bladder, uterus, urethra, and respiratory tract.

Neuronal: There is extensive expression of K_{ATP} channels in neuronal populations including glial cells (SUR1/Kir6.1 or Kir6.2) and neurons in the cortex, hippocampus, dorsal root ganglion (Kir6.2/SUR1 or SUR2), hypothalamus (Kir6.2/SUR1), and substantia nigra pars reticulata (Kir6.2/SUR1) (Liss et al. 1999; Sun et al. 2006b). The differential expression of Kir6 and SUR in neuronal populations leads to distinct roles in metabolic and non-metabolic sensing (Sun et al. 2006b;

Tinker et al. 2018). K_{ATP} channels have been implicated in pain, locomotion and behavior, nerve function, immune activity of glial cells, and most notably neuroprotection against ischemia in hypoxic conditions (Sun and Feng 2013; Tinker et al. 2018). In hypoxic conditions, K_{ATP} channels open leading to hyperpolarization which dampens neuronal firing and reduces metabolic demands. Depolarization increases Ca^{2+} entry which can lead to neuronal death. The neuroprotective nature of K_{ATP} appears to be linked directly to the Kir6 subunit (Sun et al. 2006b; Heron-Milhavet et al. 2004; Tinker et al. 2018).

4.4 Pharmacology

Therapeutic compounds that regulate K_{ATP} channels belong to a few main classes, the sulfonylureas and the general class of potassium channel openers (KCOs) or potassium channel blockers (KBOs). These drug classes exert their effects on K_{ATP} channels by binding to the sulfonylurea receptors (SURs) and allosterically inhibiting or activating the channel or by directly engaging with the channel pore.

Small molecule drugs may exert their regulatory properties affecting gating by allosteric modulation or direct interaction with the binding pockets of PIP_2, ATP, or the channel pore. Experimental and MD studies have uncovered other specific chemical scaffolds that bind directly to the Kir6 channels to exert regulatory effects (see Fig. 11). The exact mechanism of action by these chemicals varies and, in some cases, remains elusive. Therapeutic compounds that bind directly to the Kir6 channels, which have been experimentally confirmed, include the imidazoline compounds, thiazolidinedione ("glitazone") compounds, morpholinoguanidines, quinine, the antiarrhythmic, cibenzoline, biguanides and various anesthetics such as propofol and bupivacaine (Cui et al. 2003; Grosse-Lackmann et al. 2003; Kawahito et al. 2011; Kawano et al. 2004a; Kawano et al. 2004b; Kovalev et al. 2004; Mukai et al. 1998; Proks and Ashcroft 1997; Rui Zhang et al. 2010; Yu et al. 2012).

Computational simulations, including MD, have suggested travoprost, betaxolol, and ritodrine as other drugs that bind directly to the Kir6 subunit to exert their effects (Chen et al. 2019). These latter drugs, along with others uncovered through in silico screens and dynamic studies, will need to be experimentally validated for their binding affinity for the Kir6 subunit and the mechanistic impact on channel gating. However, they provide important insights into new therapeutic chemical scaffolds for Kir6 subunits. It remains an open question whether Kir6 targeting drugs control activity either directly by altering channel-PIP_2 interactions or indirectly by altering channel-ATP interactions. With the differential expression of both SUR and Kir6 subunits, drugs that bind each with varying affinities offer substantial advantage to subtype specific targeting and reduced off target effects.

Another mechanism of K_{ATP} channel regulation is the rescue of surface expression impairments causing biogenesis or trafficking defects, through pharmacological correctors (Shyng et al. 2000). There are well-established mutations in SUR domains that cause trafficking defects of the K_{ATP} complex to the cell surface (Zhou et al.

2014). Carbamazepine is a clinically approved anticonvulsant drug, approved for the treatment of epilepsy and bipolar disorder, known to bind to Na_V channels, Ca_V channels, and $GABA_A$ receptors (Chen et al. 2013). Carbamazepine rescued trafficking of K_{ATP} with trafficking mutations occurring in the TMD0 of SUR1 (Zhou et al. 2014). The TMD0 is part of the N-terminal region of the SUR which binds and couples to the Kir6 channel to form the K_{ATP} complex (Chan et al. 2003; Chen et al. 2013). Interestingly, carbamazepine also appears to affect MgADP binding to the NBD domains of SUR, so while it rescues trafficking and expression can inhibit the channel by abolishing the MgADP activation effect (Zhou et al. 2014). It has also been established that sulfonylurea drugs may also be able to rescue trafficking and folding in specific SUR mutants (Yan et al. 2004). Again, they also have an impact on K_{ATP} activity and conductance once folding and cell surface expression is rescued, increasing the complexity of the pharmacology of such drugs.

The following listed drugs (Table 5 and Fig. 11) were validated with the functional channels Kir6.2ΔC26 or Kir6.2ΔC36 that do not require the presence of SUR subunits, therefore suggesting they act directly on the Kir6 subunit. While they all provide valuable insights into structure-based drug discovery, they all require further computational and experimental investigation to better understand the binding sites and mechanisms of gating.

4.4.1 Drugs Acting Through the Kir6 Subunits

Imidazolines: Imidazolines, including clinically relevant clonidine and phentolamine, and others not on the clinical market, such as idazoxan, efaroxan, and RX871024, bind directly to Kir6 channels (Kawahito et al. 2011; Proks and Ashcroft 1997; Rui Zhang et al. 2010). Clonidine and phentolamine made it to the market based on their ability to bind α-adrenergic receptors but later discovered to bind endogenous imidazoline receptors (Bousquet et al. 2020). Computational studies have elucidated various putative binding sites on the Kir6 channels including the PIP_2 binding site (Idazoxan) and near the slide helix (clonidine and efaroxan).

Despite such differences, they all bind close to the interface between the transmembrane domain and cytosolic domain of the Kir6 subunit (Rui Zhang et al. 2010). Functional studies of clonidine and phentolamine (Kir6.2ΔC36 channel in HEK293 cells and *Xenopus* oocyte macropatches, respectively) suggest they do not act through SUR (Grosse-Lackmann et al. 2003; Kawahito et al. 2011; Proks and Ashcroft 1997). This finding has been further validated (Kawahito et al. 2011). RX871024 binds in the channel pore forming two hydrogen bonds with interior pore-lining residues. This causes direct block of the channel (Rui Zhang et al. 2010). Aside from RX871024, imidazoline drugs are likely candidates to be working through direct or allosteric alteration of channel–PIP_2 interactions. The ATP-insensitive mutant (Kir6.2ΔC36-K185Q) showed no reduction in current in the presence of ATP but both clonidine (1 mM) and phentolamine (1 μM) were still able to block the current as effectively as in the Kir6.2ΔC36 channel, suggesting the binding site of these two imidazolines is not directly at the ATP binding site or involving key ATP binding residues (Kawahito et al. 2011; Proks and Ashcroft 1997).

Table 5 Small molecule inhibitors directly binding Kir6

Phentolamine (imidazoline)	Xenopus oocytes	KirΔC36: 0.77 μM	Kir6.2/SUR1: 1.22 μM	Proks and Ashcroft (1997)
Clonidine (imidazoline)	HEK293 cells Native HIT-T15 cells	KirΔC26: 40.1 μM	Kir6.1/SUR2B: 1.21 μM Native: 44.2 μM	Kawahito et al. (2011), Grosse-Lackmann et al. (2003)
PNU-37883A/ (morpholinoguanidine)	HEK293 cells Xenopus oocytes	KirΔC26: 4.6 μM	Kir6.1/SUR2B: 4.88 μM	Kovalev et al. (2004), Cui et al. (2003)
Rosiglitazone/ (thiazolidinedione)	HEK293 cells	KIR6.2ΔC36: 45 μM	KIR6.1/SUR2B: 10 μM KIR6.2/SUR1: 45 μM KIR6.2/SUR2A: 37 μM KIR6.2/SUR2B: 50 μM	Yu et al. (2012), Chen et al. (2019)
Cibenzoline/ (diphenylmethane)	COS-7 cells	Kir6.2ΔC36: 22.2 μM	Kir6.2/SUR1: 30.9 μM	Mukai et al. (1998)
Propofol (cumene)	COS-7 cells	Kir6.2ΔC36: 78 μM	Kir6.2/SUR1: 77 μM Kir6.2/SUR2A: 72 μM Kir6.2SUR2B: 71 μM	Kawano et al. (2004b)
Quinine	HEK293 cells	Kir6.2ΔC26: 14.6 μM	Kir6.2Δ26C/SUR1:29.8 μM	Grosse-Lackmann et al. (2003)
Bupivacaine/ (piperidinecarboxamides)	COS-7 cells	Kir6.2ΔC36: 366 μM	Kir6.2/SUR2A: 52 μM Kir6.2/SUR2B: 396 μM Kir6.2/SUR2B: 379 μM	Kawano et al. (2004a)
Phenformin (biguanide)	HEK293 cells	Kir6.2ΔC26: 1,780 μM	Kir6.1/SUR2B: 550 μM Kir6.2/SUR2B: 1,096 μM Kir6.2/SUR1: 9,330 μM	Aziz et al. (2010)
SpTx-1	Xenopus oocytes	Kir6.2ΔC26: 8.5 nM (Kd)	Kir6.2/SUR1: 8.42 nM (Kd)	Ramu et al. (2018)
Mitiglinide (benzylsuccinic acid derivative)	HEK293	KIR6.2ΔC36 > 1 mM	Kir6.2/SJUR1: (1) 3.8 nM, (2) 4.1 mM Kir6.2/ SUR2A: (1) 3.2 nM, (2) 2.5 mM, Kir6.2/SUR2B: (1) 5 μM, (2) 2.9 mM	Reimann et al. (2001)

Thiazolidinediones: The thiazolidinediones (TZDs), also known as the glitazones, are a drug class whose members exert their clinical anti-diabetic effect by binding to the nuclear transcription factor, PPAR and as such they do have increased cardiovascular risks (Chen et al. 2020; Nanjan et al. 2018). Rosiglitazone (RSG) and pioglitazone are two important clinically approved anti-diabetic drugs of this class. TZDs are reported to have a direct and indirect mechanism of action in relation to diabetes. It is also well characterized that TZDs can directly bind to the Kir6 subunits of the K_{ATP} channel through functional studies (Yu et al. 2012). More interestingly, it has been shown computationally that RSG binds in close proximity to the helix bundle crossing (HBC) gate and PIP_2, without complete PIP_2 binding site occupation (Chen et al. 2020). The main impact of RSG on Kir6 channel current is increased mean closed times (Yu et al. 2012). Based on the putative binding site proposed by Stary-Weinzinger, RSG binds residues in or near the slide helix separating the putative binding site to be distinct from that of the ATP binding site (Chen et al. 2019; Martin et al. 2017).

Morpholinoguanidines: The most historically studied morpholinoguanidine in relation to K_{ATP} channels is PNU-37883A. It acts directly on the Kir6 subunit and has selectivity for Kir6.1 over Kir6.2 making it more selective for smooth muscle with some selectivity for the vascular K_{ATP} channels (Cui et al. 2003). The selectivity of PNU-37883A for K_{ATP} channels is largely impacted by the SUR that complexes with the Kir6.1 (Cui et al. 2003). This likely has to do with either obstruction of the drug binding pocket or an allosteric conformational change of Kir6 upon SUR binding (Humphrey 1999; Kovalev et al. 2004). The binding region for the PNU-38773A has been mapped to amino acids 200–280 on the C-terminus of Kir6 (Kovalev et al. 2004). It is still unclear if binding to the C-terminus allows the molecule to position into the channel pore or it causes a conformational rearrangement of the gating machinery to close the channel. The effect of PNU-37883A on the binding of ATP or PIP_2 to the Kir6 channel has not been adequately explored. Neither the precise binding site of PNU-37883A is determined nor how its binding results in current inhibition. Could it promote ATP binding? Could it disrupt MgADP binding to the SUR, since the SUR NBD interfaces with the Kir6.2 CTD? Could it work by disrupting channel–PIP_2 interactions directly or allosterically? Such questions would need to be further addressed in future studies (Lee et al. 2017). The non-K_{ATP} channel-specific nature of the drug leading to a variety of pharmacological actions does not make it a good drug candidate but its further study will provide a chemical and structural framework for developing more specific K_{ATP} channel inhibitors (Teramoto 2006).

Anesthetics (propofol, bupivacaine): Propofol is a general anesthetic that is used widely throughout the world. Its pharmacological action has been largely attributed to its action on the $GABA_AR$ (Tang and Eckenhoff 2018). Growing functional evidence has uncovered its action at K_{ATP} channels. Propofol is selective for K_{ATP} channels containing Kir6.2 over Kir6.1 and directly binds the Kir6 subunit of the K_{ATP} channel (Kawano et al. 2004b). Utilizing the ATP-insensitive mutant and the R31E mutation, inhibition was abolished indicating propofol has at least some partial overlap with the ATP binding site and with an N-terminal binding site

(Kawano et al. 2004b). There is also evidence to suggest that propofol affects the expression level of Kir6.1 (Zhang et al. 2016).

Bupivacaine is a local anesthetic with a chemical structure different and more complex than that of propofol (see Fig. 11). Bupivacaine was experimentally proven to have a potency on Kir6.2ΔC36 equal to that of Kir6.1/SUR2B. It has a greater affinity for the cardiac channel (Kir6.2/SUR2A). It had higher affinity over levobupivacaine and ropivacaine due to the stereoselective nature of the Kir6 binding pocket (Kawano et al. 2004a). It likely exerts its effects by binding residues at the cytosolic end of TM2 which are important in the channel gating properties likely relinquishing any interference with ATP binding to the channel (Kawano et al. 2004a). Bupivacaine and other similar local anesthetics bind to Na_V channels as well as Ca_V and K_V channels leading to their major cardiotoxic effects. It has been proposed that local anesthetics inhibit Kir3 channels through antagonism of channel–PIP_2 interactions (Zhou et al. 2001). If local anesthetics work at the same conserved site on K_{ATP} as they do on Kir3 channels, PIP_2 binding residues may be important candidates to explore.

Biguanides: Metformin and phenformin are the most well-known, clinically relevant biguanides. They are used to increase insulin sensitivity in type 2 DM (Hansen 2006). Phenformin was withdrawn from the market in the 1970s. Biguanides activate AMPK with several downstream implications of the metabolic state of the cell as well as potentially increasing K_{ATP} cell surface expression with evidence that they directly affect K_{ATP} channels independently of AMPK signaling (Aziz et al. 2010). Aziz et al. (2010) were able to show that there is differential activity of phenformin based on SUR subunit composition but equivalent activity of Kir6 in the absence of SUR subunits. While they appear to bind Kir6 directly, SUR does play a role in modulating the affinity of biguanides to the channel (Aziz et al. 2010).

Drugs acting through the SURs: K_{ATP} channel openers and inhibitors acting through the SURs of the K_{ATP} channel complexes have been discussed in chapter "The pharmacology of ATP-sensitive K^+ channels (K_{ATP})" (Li et al. 2021).

Miscellaneous drugs: Quinine, chloroquine, and quinacrine are all structurally related drugs containing the quinolone bicyclic ring. These drugs have been shown to bind and alter the conductance of Kirs (Lopez-Izquierdo et al. 2011; Grosse-Lackmann et al. 2003). Quinolones are used as one of the most common antibacterial in the world and their mechanism may limit their potential use on K_{ATP} regulation for the current clinically used quinolones (Aldred et al. 2014).

The well-known antiarrhythmic agent, cibenzoline, is a diphenylmethane with inhibitory properties on the Kir6 subunit shown by functional studies. There is evidence that cibenzoline binds the K^+-recognition site on K_{ATP} channels similar to how it binds the H^+, K^+ -ATPase (Tabuchi et al. 2001). This may prove to be a novel mechanism of binding but would decrease its selectivity for Kir6 over other Kirs given the conservation of the putative binding site. In 2018 Ramu and colleagues uncovered a 54-residue protein toxin recovered from the venom of *S. polymorpha* (SpTx) that inhibited Kir6 from the extracellular side (Ramu et al. 2018). SpTx binds the Kir subunit and not the SUR subunit. Further investigation is needed to determine the mechanism of channel closure but nonetheless this discovery opens a new possibility in the pharmacology of Kir6 channel gating. There is evidence to suggest that glinides or the benzylsuccinic acid derivative, mitiglinide, bind both to the SUR at a high affinity site and directly to the Kir6 subunit at a lower affinity site (Aziz et al. 2010; Reimann et al. 2001). Similarly, the repaglinide binding site is on the SUR, as resolved by cryo-EM, but it also has a partial contribution by the N-terminus of the Kir6.2 subunit. This was confirmed through binding studies in which co-expression of Kir6.2 enhanced binding affinity and deletion of the N-terminal peptide of Kir6.2 abolished the binding enhancement (Ding et al. 2019).

4.5 Structural Studies

Before atomic resolution structures of the K_{ATP} were available, in order to identify the PIP_2 binding site, Haider and colleagues conducted molecular docking and MD simulations (10 ns) based on a Kir6.2 homology model. The PIP_2 head group was predicted to interact with K39, N41, and R54 in the N-terminus, K67 in the transmembrane domain and R176, R177, E179, and R301 in the C-terminus. The model predictions are consistent with a large body of functional data, suggesting how PIP_2 binding may lead to an increase in Kir6.2 open probability and a reduction in ATP sensitivity (Haider et al. 2007). In order to understand the mechanism by which disease mutations exert their deleterious effects, in 2017, Cooper and colleagues worked on a Cantu syndrome (CS) mutant, where the Kir6.1(V65M)/Kir6.2(V64M) GoF mutations in the slide helix enhance the channel activity when co-expressed with the SUR1 or SUR2 regulatory subunits (Cooper et al. 2017). The Val to Leu mutation that does not cause the disease was used as a negative control. The disease mutations abolished the sensitivity of K_{ATP} to ATP and to the blocker glibenclamide, which acts on the SUR subunit. This finding suggested that sulfonylurea therapy may not be successful, at least for some CS mutations. Homology modeling and MD simulations (100 ns) were conducted on the Kir6.1(WT), Kir6.1(V65L), and Kir6.1(V65M) systems. The stabilization of the open state of the Val to Met mutations (but not to Leu) that was obtained experimentally could not be shown computationally in the 100 ns simulation time frame. The same year in 2017, near-atomic resolution structures were obtained from three independent labs (Chen in Beijing, Shyng in Oregon and MacKinnon in New York). These structures were solved for the Kir6.2/SUR1 in complex with ADP, ATP, and glibenclamide (Lee et al. 2017; Martin et al. 2017; Wu et al. 2018). In 2019, a higher resolution (3.3 Å) cryo-EM structure of Kir6.2/SUR1 in complex with ATPγS and repaglinide was solved by the Chen lab. Figure 8 shows the Kir6.2 channel surrounded by four SUR1 subunits, the two distinct binding sites for ATPγS (one in Kir6.2 and the other in SUR1), and the binding site for repaglinide in the transmembrane domain of SUR1

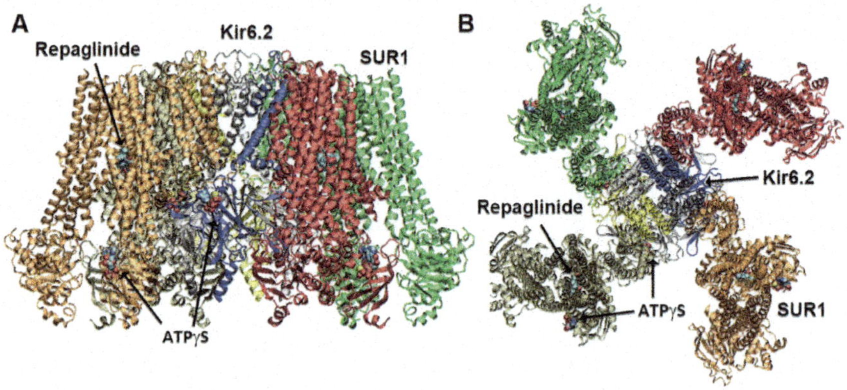

Fig. 8 Cryo-EM structure of K_{ATP} channel in complex with ATPγS and Repaglinide (PDBID:6JB1), (**a**) side view. (**b**) top view. The K_{ATP} channel and SUR1 are drawn in Newcartoon, the ATPγS and Repaglinide are drawn in VDW

(Ding et al. 2019). This study also suggested through cryo-EM and binding studies that the N-terminal portion of Kir6.2 adds to the binding pocket for repaglinide. This is further supported by earlier functional studies done by Reimann et al. (2001) in which a structurally similar drug, mitiglinide, had a biphasic dose-response curve, suggesting a high affinity binding site at SUR and a low affinity binding site at Kir6.2. Chen and colleagues conducted MD simulations (at a microsecond time scale) on Kir6.1 channel to investigate the blocking mechanism by rosiglitazone (RSG). The putative RSG binding site was identified through unbiased MD simulations of Kir6.1 channel with 20 RSG molecules randomly placed in the solvent, and followed by free energy calculations. Based on the predicted RSG binding site, dynamic pharmacophore models were constructed and used for screening of hits in the DrugBank database. Three new high affinity blockers, betaxolol, ritodrine, and travoprost, were identified and subsequently tested functionally. Using the inside-out patch-clamp mode in HEK293T cells expressing Kir6.2/SUR2A, travoprost, betaxolol, and ritodrine had IC_{50}s of 2.46 μM, 22.06 μM, and 7.09 μM, respectively (Chen et al. 2019). Given the computational identification of direct binding to Kir6, the expectation would be that these three drugs should have similar potency when tested on Kir6.2ΔC channels. Computational simulations to understand the dynamics of K_{ATP} channel structure and function are still lacking. The recent cryo-EM structures have provided us with the opportunity to characterize the channel dynamic function using long timescale MD simulations, and a molecular basis for structure-based drug discovery for these important drug targets.

5 K^+ Transport Kir (1, 4, 4/5, 7) Channels

5.1 Kir1.1: Historical Perspective

The first member of the Kir family to be cloned (see below) was initially referred to as the "rat outer medullary K^+" (ROMK1) channel for its prominent role in the kidney.

The task of renal epithelial cells is to maintain ionic homeostatic control of K^+, Na^+, and Cl^- in the urine and blood and Kir1 channels play a key role in achieving this task. The channels are specifically localized in the apical rather than the basolateral membrane of these cells. To appreciate the strategic importance of Kir1 channels let us consider their role in the thick ascending loop of Henle (TAL), the segment of the nephron responsible for 25–30% of the total Na^+ reabsorbed by the kidney and the site of action of loop diuretics. Here, K^+ efflux through Kir1 channels also maintains the Na^+-K^+-$2Cl^-$ symporter active by supplying K^+ to the extracellular site of the transporter and enabling the uptake of NaCl along with K^+ into the TAL cells. The Na^+ entering the cytoplasm from the apical (lumen or urine side) through the symporter fuels the basolateral Na^+-K^+ ATPase to continue transporting Na^+ to the blood side. K^+ channels (e.g., Kir1, Kir4, Kir7) co-expressed with the Na^+-K^+-ATPase in basolateral membranes supply K^+ to the extracellular side of the pump to maintain its activity. This is referred to as "K^+ recycling" (Hibino et al. 2010). Furthermore, the hyperpolarization caused by Kir1 channels accelerates Cl^- exit from basolateral Cl^- channels, establishing "the lumen positive transepithelial potential," which serves as the main driving force for paracellular Na^+, Ca^{2+}, and Mg^{2+} transport from the lumen to blood side. Deletion of Kir1 channels impairs renal NaCl absorption yielding "Barter's syndrome"-like phenotypes, as we will discuss below (Bleich et al. 1990; Greger et al. 1990). Moreover, Kir1 channels on the apical membrane of TAL epithelial cells are functionally coupled to Cl^- channels (the cystic fibrosis transmembrane regulator – CFTR) in that CFTR decreases Kir1 activity. This effect can be reversed by PKA phosphorylation and may underlie the actions of the antidiuretic hormone arginine vasopressin that increases Kir1 activity, limiting K^+ secretion and urinary K^+ loss (Field et al. 1984).

5.1.1 Subfamily Members and Tissue Distribution

In early 1993 the first two Kir channel cDNAs were reported, Kir1.1 and Kir2.1, followed by Kir3.1 later the same year (Dascal et al. 1993; Ho et al. 1993; Kubo et al. 1993a; Kubo et al. 1993b). Six alternatively spliced isoforms of Kir1.1(a-f) have been identified but only two of these (a, c) give rise to distinct proteins from the other four isoforms (b, d-f) (Boim et al. 1995; Ho et al. 1993; Kondo et al. 1996; Shuck et al. 1994; Zhou et al. 1994). The splice isoforms differ in their N-termini with Kir1.1b being the shortest, Kir1.1a 19 aa and Kir1.1c 26 aa longer than Kir1.1b (the human isoform shown in Fig. 9 is the Kir1.1a isoform with 391 aa). No Kir1.1 heteromers with members of other Kir subfamilies have been reported. As mentioned above, Kir1.1 is expressed in the kidney (TAL, distal convoluted tubule, and cortical collecting duct) and, as shown by in situ hybridization, in cortical and hippocampal neurons (Kenna et al. 1994).

5.1.2 Physiology/Pathophysiology

Kir1.1 function is regulated by internal pH, phosphorylation by a number of different kinases, miRNAs, and ubiquitination. Acidification closes Kir1.1 channels with a pK_a of ~6.5 (Choe et al. 1997; Doi et al. 1996; Fakler et al. 1996; McNicholas et al. 1998; Tsai et al. 1995). Multiple studies have implicated the involvement of several residues in pH sensitivity but structural insights are still lacking (Hibino et al. 2010). Kir1.1 has ER retention signals (368R-X/A-R370 and K370-R371) (Ma et al. 2001; Yoo et al. 2005). It is regulated by phosphorylation of S44 (by both PKA, SGK – Ser/Thr protein kinases), increasing trafficking to the plasma membrane (McNicholas et al. 1994; Xu et al. 1996; Yoo et al. 2005). Additionally, PKA-mediated phosphorylation of S219 and S313 increases the channel open probability, likely by enhancing channel–PIP_2 interactions (Liou et al. 1999; MacGregor et al. 1998). Another class of Ser/Thr kinases, WNKs (With-no-K – K = Lys) decreases Kir1.1 cell surface expression by unique mechanisms (Hibino et al. 2010). A number of microRNAs have been shown to be upregulated in high-K^+ diet and, in turn, they regulate Kir1.1: miR-802 increased Kir1.1 surface expression and channel activity by reducing caveolin-1 that limits Kir1.1 expression at the plasma membrane (Lin et al. 2014); miR-194 also enhanced Kir1.1 surface expression by counteracting the opposite regulation by WNK and intersectin 1 (He et al. 2007). Finally, monoubiquitination of K22 decreased cell surface Kir1.1 serving as a signal to process the channel for lysosomal degradation (Lin et al. 2005). The Kir1.1 knockout mouse resulted in mice afflicted with Barter's syndrome (BS) (Lorenz et al. 2002; Lu et al. 2002). BS is an autosomal recessive renal tubulopathy characterized by hypokalemic metabolic alkalosis, renal salt wasting, hyperreninemia, and hyperaldosteronism (Asteria 1997; Guay-Woodford 1995; Hebert 2003; Karolyi et al. 1998; Peters et al. 2002; Rodriguez-Soriano 1998). Loss-of-function mutations of Kir1.1 (type II BS) affect many aspects of the functional integrity of the channel and its regulation, including interactions with PIP_2 (Hibino et al. 2010; Lopes et al. 2002). BS patients and Kir1.1 mice are characterized by hypokalemia and excess K^+ in the urine, which is contrary to the role of Kir1.1 as providing the main secretory pathway of K^+ in renal tubules. Bailey and colleagues addressed the conundrum showing that in the absence of Kir1.1 in the TAL cells K^+ secretion is sustained in the late distal tubule by maxi-K^+ channels (Bailey et al. 2006).

5.2 Kir4/Kir5: Historical Perspective and Tissue Distribution

Kir4 are expressed as homotetramers as well as heterotetramers with Kir5.1, while when Kir5.1 is expressed alone in a heterologous expression system it shows no function.

These channels are the molecular identities of neuro- and retinal glial cell K^+ channels. Their cloning provided a boost to their long-recognized importance (Kuffler and Nicholls 1966; Newman 1984). Kir4channel cDNAs were first identified in the mid-90s and Kir4.1 has been referred to under a variety of names: BIR10 (Bond et al. 1994), K_{AB}-2 (Takumi et al. 1995), BIRK-1 (Bredt et al. 1995), and Kir1.2 (Shuck et al. 1997). In situ hybridization showed Kir4.1 predominantly expressed in glial cells in the CNS (brain and spinal cord) and retina (Müller cells) (Hibino et al. 2010; Takumi et al. 1995). Kir4.1 has also been reported to be expressed in the kidney (reviewed in Manis et al. 2020), stomach (Fujita et al. 2002), and the cochlea of the inner ear (Hibino and Kurachi 2006). Kir4.2 was first isolated from a kidney library, where it is highly expressed, along with liver, embryonic fibrocytes, and microvascular endothelial cells (Lachheb et al. 2008; Pearson et al. 1999; Shuck et al. 1997). A single member of the Kir5 subfamily, Kir5.1, was identified in 1994 (Bond et al. 1994). It is highly expressed in kidney, spleen, adrenal glands, liver, in several brain regions and in spermatozoa and spermatogenic tissue (Bond et al. 1994; Salvatore et al. 1999). Kir5.1 is only functional as a heteromer with Kir4.1 and Kir4.2. The nephron serves as a great example of where expression and heteromerization with one, the other, or both of the Kir4 isoforms best serves overall renal salt handling (see below in Sect. 5.2.1). Its distribution and co-expression with the Kir4 isoforms vary along the nephron to fulfill specific functional characteristics of each tubule segment (full validation of the expression and activity of Kir4.2 is still pending). Thus, in the proximal convoluted tubule Kir5.1 is co-expressed with Kir4.2, in the cortical collecting duct and also in the connecting tubule (connecting the distal convoluted tubule to the cortical collecting duct) it is co-expressed with Kir4.1, while in the thick ascending limb of Henle it is not expressed at all (only Kir4.1 is expressed), and in the distal convoluted tubule it is co-expressed with both Kir4 isoforms (Manis et al. 2020).

5.2.1 Physiology/Pathophysiology

Kir5.1 confers several unique functional characteristics to Kir4 isoforms: (1) greater pH sensitivity (Tucker et al. 2000) (2) maintenance of pH homeostasis (Puissant et al. 2019), (3) inhibition of the dependence of channel conductance on $[K^+]_o$ (Edvinsson et al. 2011a; Edvinsson et al. 2011b) (shown only for Kir4.2 thus far), and (4) activation by $[Na^+]_i$ (Rosenhouse-Dantsker et al. 2008). Kir4.1/Kir5.1 channels have been shown to act as K^+ sensors in the distal nephron (Cuevas et al. 2017) regulating expression of apical Na^+ transporting proteins and renal salt handling in response to dietary salt intake (Manis et al. 2020). Dysfunction of these channels has been proposed to contribute to the pathogenesis of salt-sensitive hypertension (Palygin et al. 2017; Palygin et al. 2018). Knockouts of either Kir4.1 (Cuevas et al. 2017) or Kir5.1 (Wu et al. 2019) compromise $[K^+]_o$ sensing and K^+ excretion. A unifying link for action of various Kir4/5 modulators could be the allosteric changes in channel–PIP_2 interactions. The modulators proposed to work in this way include $[Na^+]_i$ (entering through the apical side; Rosenhouse-Dantsker et al. 2008)), $[K^+]_o$ at the basolateral side (Harraz et al. 2018), changes in intracellular pH or phosphorylation by kinases, such as WNK-4, (Du et al. 2004). Differences in

Kir4.1 versus Kir4.2 and their relative heteromers with Kir5.1, along with their characteristic expression may underlie their distinct physiological roles. Thus, Kir4.2 is more pH sensitive than Kir4.1. Heteromerization of Kir4.2 with Kir5.1, unlike Kir4.1, converts Kir4.2 from a strong to a weak rectifier, rendering it more sensitive to intracellular pH. Knockout mice have shown that Kir4.2 plays a distinct role in acid-base homeostasis compared to Kir4.1, resulting in metabolic acidosis with intracellular alkanization and membrane depolarization in the proximal convoluted tubules of the kidney (Manis et al. 2020). In the CNS, the Kir4 isoforms and their regulatory Kir5.1 subunits exhibit very interesting physiology by allowing for control of $[K^+]_o$ in nearby neurons to regulate neuronal excitability (Hibino et al. 2010).

5.3 Kir7.1: Historical Perspective

Kir7.1 is the only member identified in this subfamily and the most distant relative of all other Kir channels, showing the highest homology/identity to the Kir4 isoforms (~50% similarity and 40% identity; Fig. 10). Three groups reported cloning of Kir7.1 from CNS neurons in 1998 (Doring et al. 1998; Krapivinsky et al. 1998; Partiseti et al. 1998). A splice variant in native human retinal pigment epithelium has been described to be missing a region from exon 2 (amino acids 244–479), yielding a non-functional truncated Kir7.1 that is expressed at fourfold lower levels than the wild-type protein and does not interfere with the activity of the full-length protein (Yang et al. 2007). Another two splice variants of Kir7.1 were described recently in several mouse tissues. These lacked most of the C-terminal domain, Kir7.1-R166X (in the B-loop, Fig. 9) and Kir7.1-Q219X (in βE, Fig. 9). These truncation variants would be comparable to mutations associated with Leber congenital amaurosis, a rare recessive hereditary retinal disease that results in vision loss at early age. Simultaneous expression with the full-length Kir7.1, however, led to a reduction in activity of the wild-type channel (possibly due to partial proteasome degradation of WT-mutant channel heteromers) (Vera et al. 2019).

5.3.1 Tissue Distribution

The Kir7.1 protein has been reported to be expressed along with the Na^+/K^+-ATPase for "K^+ recycling" either in the apical membranes of epithelial cells of the choroid plexus (Doring et al. 1998; Hasselblatt et al. 2006; Nakamura et al. 1999) and retinal pigment epithelium (Kusaka et al. 2001; Shimura et al. 2001) or in basolateral membranes of thyroid follicular cells (Nakamura et al. 1999), in the distal convoluted tubule, proximal tubule, and collecting duct of renal epithelia (Derst et al. 2001; Ookata et al. 2000) as well as in the small intestine and stomach (Partiseti et al. 1998). Kir7.1 has also been reported to be expressed in hypothalamic arcuate neurons involved in the appetite control circuit, where they are coupled to the MC4R G Protein Coupled Receptor in a G protein-independent manner (Ghamari-Langroudi et al. 2015) as well as in the myometrial smooth muscle, where it controls uterine excitability throughout pregnancy (McCloskey et al. 2014).

5.3.2 Physiology/Pathophysiology

A hallmark of Kir channels is a dependence of their conductance on $[K^+]_o$ (Fig. 1a). Kir7.1 is also characterized by very weak rectification and a very low unitary conductance (Krapivinsky et al. 1998). The crucial difference appears to lie in the structure of the outer mouth of the selectivity filter, where Kir7.1 has a Met residue (M125), instead of Arg that is found in every other Kir channel (Fig. 9). M125R point mutation in Kir7.1 not only increased its unitary conductance by nearly 20-fold, strengthened its inward rectification and sensitivity to Ba^{2+} but also reinstated strong dependence on $[K^+]_o$ (Doring et al. 1998; Krapivinsky et al. 1998). The corresponding residue R148 in Kir2.1 had been shown to have the opposite effect, namely the Kir2.1(R148Y) mutant showed a much reduced $[K^+]_o$ dependence (Kubo 1996) while the R148H mutation increased sensitivity to pH_o (but not pH_i) (Shieh et al. 1999). Taken together, this evidence suggests that a positively charged residue is needed at the outer mouth of the selectivity filter to confer the dependence on (V-E_K) shown by Kir channels. Several questions remain for a better understanding of this property. How do extracellular Na^+ ions serve as blocking particles to regulate this dependence (Ishihara 2018; Soe et al. 2009) and how such interactions couple allosterically to control channel–PIP_2 interactions (Harraz et al. 2018)? Dynamic computational studies will be instrumental in answering these questions. We will next focus discussion of the role of Kir7.1 channels in health and disease in the best studied system, the retinal pigment epithelium (RPE).

Retinal pigment epithelium: The RPE is a hexagonally packed, tight-junction connected monolayer of post-mitotic, pigmented cells between the neuroretina and the choroids. The apical membrane of the polarized RPE cells faces the photoreceptor outer segment (POS), while the basolateral membrane faces the fenestrated endothelium of the choriocapillaris. Interactions between the RPE and the POS are essential for visual function. The RPE's primary functions are: a) to transport nutrients, ions, and water between the blood supply and the POS, b) to absorb light and protect against photo-oxidation of proteins and phospholipids of the POS, c) the isomerization of retinal which serves as the agonist of the rhodopsin GPCR to signal to the phosphodiesterase to cleave cGMP to 5'-GMP, d) phagocytosis of shed photoreceptor membranes, and e) secretion of essential elements for the morphological integrity of the retina (Kumar and Pattnaik 2014). In the dark, the cGMP levels are maintained to signal to the cGMP-gated (CNG) cation channels in the POS and allow the influx of Na^+ and Ca^{2+} to the photoreceptor cell (rod or cone). This is countered by a net K^+ efflux in the photoreceptor inner segment (PIS) mediated by various K^+ channels (e.g., KCNQ, and 8 Kir currents, with Kir7.1 being the predominant one) giving rise to the so-called dark current. The efflux at the PIS pulls the resting V_m toward E_K but the cationic influx at the POS counters the K^+ efflux, maintaining a rather depolarized level in the dark. Upon light illumination, the resulting 5'-GMP no longer keeps the CNG channels open, causing the K^+ efflux to dominate and an ensuing hyperpolarization of the photoreceptor cell. This signal will decrease glutamate release that once integrated to the retinal ganglion cells, it will be communicated to the brain through the optic nerve and be interpreted as "light." The new balance of Na^+ and K^+ ions decreases the PIS Na^+/K^+ ATPase activity and the $[K^+]_o$ falls from 5 mM in the dark to 2 mM in the light, making E_K more negative increasing Kir currents toward re-establishing $[K^+]_o$ to the 5 mM levels. The predominant Kir7.1 resting conductance, unlike other Kir currents, does not decrease its conductance with the drop of $[K^+]_o$ as we discussed above, carrying the weight of re-establishing the "normal" higher $[K^+]_o$ levels. This suggests another case of K^+ recycling in conjunction with the Na^+/K^+ ATPase (Kumar and Pattnaik 2014). Loss of function mutations in Kir7.1 result in the autosomal-dominant disease called snowflake vitreoretinal degeneration, which progressively causes fibrillar degeneration of the vitreous humor, early-onset cataract, minute crystalline deposits, and retinal detachment. The R162W mutation, characteristic of this disease has been suggested to affect channel–PIP_2 interactions (Pattnaik et al. 2013; Zhang et al. 2013).

5.4 Transport Channel Pharmacology

Kir1, Kir4, and Kir7 channel inhibitors: Dozens of small molecule inhibitors for Kir1, Kir4, and Kir7 channels have been reported (Table 6). VU590 was discovered through HTS of 126,009 small molecules as Kir1.1 modulators. It inhibits Kir1.1 with submicromolar affinity, but has no effect on Kir2.1 or Kir4.1. It also inhibits Kir7.1 at low micromolar concentrations. Electrophysiological studies indicated VU590 is an intracellular pore blocker (Lewis et al. 2009). In an effort to achieve more selective inhibitors for Kir1.1, VU591 was developed based on VU590. VU591 is a potent inhibitor like VU590 for Kir1.1, but it is selective for Kir1.1 over Kir7.1. It blocks the intracellular pore of the channel through interactions with V168 and N171 (Bhave et al. 2011; Swale et al. 2015). VU573 was discovered through a thallium (Tl^+) flux-based high-throughput screen of a Kir1.1 inhibitor library for modulators of Kir3. It inhibits selectively Kir2.3, Kir3.X, and Kir7.1 with similar potency over Kir1.1 and Kir2.1 (Raphemot et al. 2011). ML418 was discovered based on a lead compound VU714, which was identified by a HTS screen as a novel Kir7.1 channel inhibitor. It selectively inhibits Kir7.1 and Kir6.2/SUR1 over Kir1.1, Kir2.1, Kir2.3, Kir3.1/Kir3.2, and Kir4.1 channels (Swale et al. 2016).

VU0134992 was discovered through HTS of 76,575 compounds from a Vanderbilt University library for small molecule modulators of Kir4.1. It inhibits selectively Kir4.1 and Kir3.1/3.2, Kir3.1/3.4 and Kir4.2 with similar potency over (30-fold selective) Kir1.1, Kir2.1, and Kir2.2. It also weakly inhibits Kir2.3, Kir6.2/SUR1, and Kir7.1 (Kharade et al. 2018). Tricyclic antidepressants (TCA), nortriptyline, amitriptyline, desipramine, and imipramine were found to act as inhibitors for Kir4.1 channel expressed in HEK293T cells, using the whole-cell patch-clamp technique. These compounds inhibited Kir4.1 currents in a voltage-dependent fashion, and marginally affected neuronal Kir2.1 currents (Su et al. 2007). Antidepressant, selective serotonin reuptake inhibitor (SSRI), fluoxetine, was found to inhibit the Kir4.1 channel expressed in HEK293T cells using whole-cell patch clamp. It inhibited the Kir4.1 channel in a concentration-dependent and voltage-independent manner. Fluoxetine had little or no effect on Kir1.1 and Kir2.1 channels. Other SSRIs, sertraline and fluvoxamine also inhibited the Kir4.1 channel (Ohno et al.

Table 6 Kir1, Kir4, and Kir7 channel inhibitors

Drug name	Species system	Potency (IC_{50}/ EC_{50})	References
Inhibitor			
VU 590	HEK-293 cells	Kir1.1 ($IC_{50} = 294$ nM) Kir7.1 ($IC_{50} = 8$ μM)	Lewis et al. (2009), Kharade et al. (2018)
VU 591	HEK-293 cells	Kir1.1 ($IC_{50} = 300$ nM)	Bhave et al. (2011)
VU 0134992	HEK-293 cells	Kir4.1 ($IC_{50} = 0.97$ μM)	Kharade et al. (2018)
Pentamidine (aromatic diamine)	HEK-293 cells	Kir4.1 ($IC_{50} = 97$ nM)	Arechiga-Figueroa et al. (2017)
Quinacrine (cationic amphiphilic drug)	HEK-293 cells	Kir4.1 ($IC_{50} = 1.8$ μM)	Marmolejo-Murillo et al. (2017a)
Nortriptyline (tricyclic antidepressant)	HEK293T cells	Kir4.1 ($IC_{50} = 16$ μM)	Su et al. (2007)
Fluoxetine (SSRI)	HEK293T cells	Kir4.1 ($IC_{50} = 15.2$ μM)	Ohno et al. (2007)
Chloroquine (4-aminoquinoline)	HEK-293 cells	Kir4.1 ($IC_{50} = 0.5$ μM)	Marmolejo-Murillo et al. (2017b)
Gentamicin (aminoglycoside antibiotic)	HEK293T cells	Kir4.1 ($IC_{50} = 6.2$ μM)	Moran-Zendejas et al. (2020)
VU573	HEK-293 cells	Kir7.1 ($IC_{50} = 4.9$ μM)	Raphemot et al. (2011)
ML 418	HEK-293 cells	Kir7.1 ($IC_{50} = 0.31$ μM)	Swale et al. (2016)

2007). Aréchiga-Figueroa and colleagues discovered pentamidine, a drug for treatment of protozoan infections, inhibited Kir4.1 channels. This drug potently inhibited the Kir4.1 channel from the cytoplasmic side in the inside-out patch-clamp configuration. The inhibition was voltage dependent. Molecular modeling predicted the binding of pentamidine to the transmembrane pore region interacting with residues T127, T128, and E158 (Arechiga-Figueroa et al. 2017). Quinacrine, an old antimalarial drug, was discovered as an inhibitor of the Kir4.1 channel expressed on HEK293 cells, using patch clamp. It inhibited Kir4.1 channel with an IC_{50} of 1.8 μM and in a voltage-dependent manner. Molecular modeling and mutagenesis studies suggested quinacrine blocks Kir4.1 through the central cavity interacting with residues E158 and T128 (Marmolejo-Murillo et al. 2017a). Similarly, chloroquine, an aminoquinoline derivative anti-malarial drug, was discovered as an inhibitor for Kir4.1 channel (Marmolejo-Murillo et al. 2017b). More recently, Morán-Zendejas and colleagues discovered that aminoglycoside antibiotics (AGAs), such as gentamicin, neomycin, kanamycin, inhibited the Kir4.1 channel. Using patch-clamp, mutagenesis, and molecular modeling, the AGAs were characterized as pore blockers plugging the central cavity of the channel. The residues T128 and E158 were identified as critical determinants for AGA inhibition activity (Moran-Zendejas et al. 2020). From these studies, we can conclude that residues E158 and T128 of the central pore in Kir4.1 are critical for the activity of several channel blockers. Insight can be gained by comparing conservation of the key residue determinants of intracellular pore blockers. The residue E158 (Kir4.1) is conserved in Kir4.2 (E157) and in Kir7.1(E149), but not in Kir1.1(N171) (Fig. 9). Kir1.1-specific intracellular pore blockers of Kir1.1, e.g. VU590 or VU591, utilize position 171 to exert their specific Kir1.1 effect. Indeed, VU591 interacts with Kir1.1 through this residue as a hydrogen bond donor with the nitro group acceptor of the compound. While the negatively charged residues, such as Glu and Asp, form unfavorable, repulsive interactions with the nitro group of the compound (see Fig. 11). A double mutation, Kir1.1(N171E/K80M), abolished VU591 activity (Swale et al. 2015). This explains that VU591 selectively inhibits Kir1.1 over Kir7.1, and UV590 has no effect on Kir4.1 and Kir2.1 (D172 is the corresponding residue). On the other hand, antibiotics, chloroquine, quinacrine, pentamidine utilize their amine groups to interact with Kir4.1 through E158 as a hydrogen bond acceptor to block the channel. Although discovery of selective channel blockers is challenging, the structural information of determinant residues could be useful for future structure-based drug design for selective channel inhibitors. In addition, identification of new allosteric binding sites could be valuable for discovery of novel selective modulators.

5.4.1 Transport Channel Structural Determinants

Kir1, 4, and 7 Channel Structure and Dynamics

Linder et al. (2015) conducted MD simulations (200 ns) on wild-type KirBac1.1 and a stimulatory mutant (G143E) to investigate activation gating of KirBac channels. The simulations revealed that a Glu mutation at G143 position caused significant widening at the HBC gate, enabling water flux in the cavity. Both global and local rearrangements were observed. Opening of the HBC gate could trigger twisting of the CTD, which was mediated by electrostatic interactions between the TMD and CTD. In addition, the channel's slide-helix and C-linker interactions with lipids were strengthened during gating the channel open (Linder et al. 2015). Hu and colleagues predicted the TPNQ (Tertiapin-Q) and Kir1.1 channel interactions using homology modeling, protein docking, and MD simulations. The results suggested TPNQ toxin interacts with Kir1.1 through its helical domain as its interacting surface along with residue H12 as a pore-blocking residue. Residues F146 and F148 in Kir1.1 formed dominant nonpolar interaction with the toxin (Hu et al. 2013). To understand the mechanism for the antidepressant–Kir4.1 channel interaction, Furutani and colleagues conducted homology modeling and molecular docking simulations on Kir4.1 channel with fluoxetine and nortriptyline. Chimeric and site-directed mutagenesis studies suggested the residues T128 and E158 on the TM2 were critical for the drug inhibitory activity. In addition, a 3D quantitative structure-activity relationship (3D-QSAR) model of antidepressants was generated, which suggested common features of a hydrogen bond acceptor and positively charged moiety from the drugs interacting with the channel (Furutani et al. 2009). Using similar homology modeling and molecular docking approaches, in combination with mutagenesis and electrophysiology experiments, several Kir4.1 channel inhibitors were characterized: Quinacrine (Marmolejo-Murillo et al. 2017a), Pentamidine (Arechiga-Figueroa et al. 2017), Chloroquine (Marmolejo-Murillo et al. 2017b), and aminoglycoside antibiotics (Moran-Zendejas et al. 2020; Arechiga-Figueroa et al. 2017; Marmolejo-Murillo et al. 2017a, b; Moran-Zendejas et al. 2020; Swale et al. 2016).

Swale et al. (2016) discovered a novel Kir7.1 channel inhibitor, VU714, through fluorescence-based high-throughput screening. Homology model and docking predicted the binding site of VU714 in Kir7.1 central pore cavity. Site-directed mutagenesis suggested that residues E149 and A150 are essential determinants of VU714 activity. Lead optimization based on VU14 generated ML418, which exhibited high potency ($IC_{50} = 310$ nM) and superior selectivity over other channels (Kir1.1, Kir2.1, Kir2.2, Kir2.3, Kir3.1/3.2, and Kir4.1) except for Kir6.2/SUR1 (Swale et al. 2016).

6 Conclusions

Kir channels are mostly active at negative potentials influencing cells to rest near E_K. They are either constitutively active or opened by external signals (e.g., neurotransmitters and G protein signaling) or closed by internal signals (e.g., ATP/ADP ratio). They are fundamental to the proper functioning of many cells and systems in our body and their malfunction yields a multitude of diseases, making them desirable targets for therapeutic drug discovery and development. Great insights from 3-D atomic resolution structures of these K^+ channels have begun to provide insights for the molecular basis of their fundamental functional properties, such as inward rectification, dependence on $[K^+]_o$, and interactions with PIP_2 to control gating. More specialized functional attributes are also being deciphered, such as $G\beta\gamma$ binding and allosteric control of gating or ATP binding in relation to the gates and the interplay with PIP_2 gating. Computational studies have begun to offer dynamic views of ion permeation and gating by physiological regulators. The pharmacology of these channels has also been expanding providing insights as to how to best synthesize specific and effective drugs. Dynamic structural studies are paving the way toward successful structure-based drug design for this family of K^+ channels. Thus, the physiology and pharmacology of Kir channels in the context of dynamic atomic resolution structures have provided the groundwork for numerous therapeutic opportunities. The next decade promises great advances in Kir pharmacology with therapeutic potential (Figs. 9, 10, and 11).

Acknowledgements We dedicate this review to Dr. Yoshihisa Kurachi who by devoting his career to Kir channel structure, function, and pharmacology has inspired many of us to follow his example. We are grateful to Professor Leigh Plant for reading this review and providing us with critical feedback. We thank the members of the Logothetis and Plant groups for pursuing and freely sharing their mechanistic Kir channel results. Last but not least, we thank the National Institutes of Health (NIH) in the United States of America (USA) for funding our research and thus allowing us to continue working in this exciting field (supported by NIH R01-HL059949-23 to D.E.L.)

Appendix

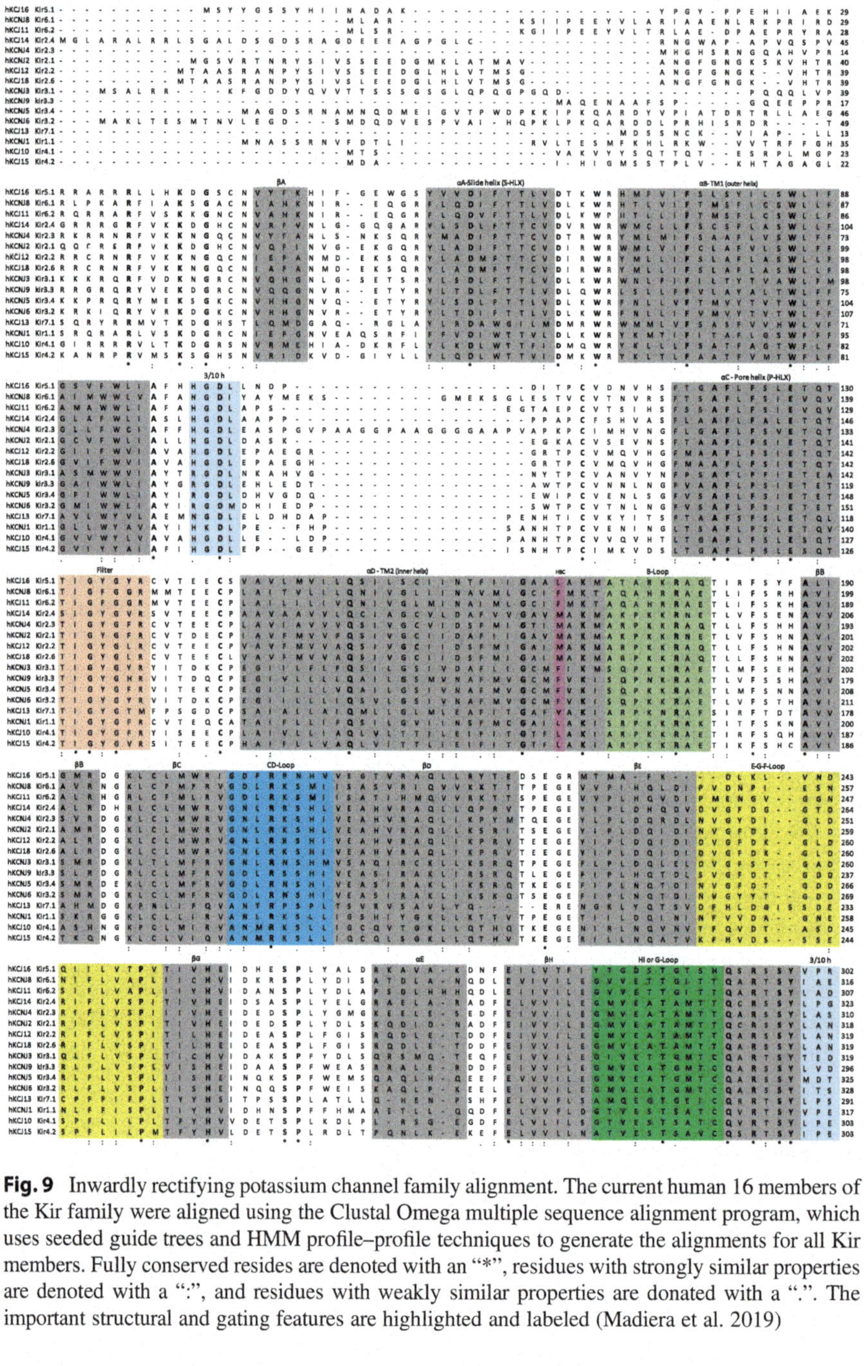

Fig. 9 Inwardly rectifying potassium channel family alignment. The current human 16 members of the Kir family were aligned using the Clustal Omega multiple sequence alignment program, which uses seeded guide trees and HMM profile–profile techniques to generate the alignments for all Kir members. Fully conserved resides are denoted with an "*", residues with strongly similar properties are denoted with a ":", and residues with weakly similar properties are donated with a ".". The important structural and gating features are highlighted and labeled (Madiera et al. 2019)

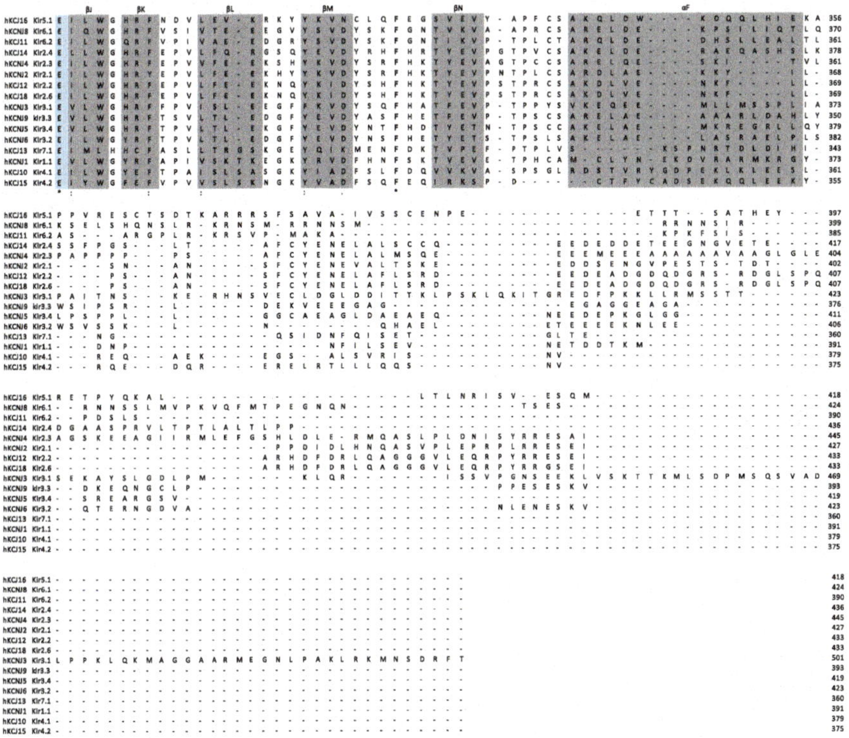

Fig. 9 (continued)

	Kir Family - % Similarity/ % Identity															
Kir1.1	100/100															
Kir2.1	54/40	100/100														
Kir2.2	54/42	80/71	100/100													
Kir2.3	53/41	70/61	71/64	100/100												
Kir2.4	49/37	67/58	64/57	62/54	100/100											
Kir2.6	54/42	79/70	99/99	70/63	64/56	100/100										
Kir3.1	51/40	57/45	55/43	51/40	50/40	54/43	100/100									
Kir3.2	52/40	57/45	55/43	56/44	55/41	54/43	62/53	100/100								
Kir3.3	49/38	60/49	60/51	62/50	57/49	59/50	64/55	76/68	100/100							
Kir3.4	51/41	57/44	55/45	57/45	55/44	55/45	62/54	76/72	71/65	100/100						
Kir4.1	56/45	50/37	50/38	49/36	47/36	50/38	49/36	48/35	48/36	48/36	100/100					
Kir4.2	57/45	49/36	49/37	47/36	47/35	49/37	46/33	47/33	45/31	48/35	71/61	100/100				
Kir5.1	48/34	52/42	51/41	51/40	50/40	51/40	47/34	46/33	49/36	47/32	47/34	46/32	100/100			
Kir6.1	49/37	53/42	53/41	51/40	50/36	53/41	52/39	52/41	55/42	50/39	49/35	46/32	47/35	100/100		
Kir6.2	49/37	56/43	55/44	54/43	54/41	55/44	53/40	56/42	54/43	53/42	50/35	47/33	49/36	79/69	100/100	
Kir7.1	46/35	44/32	44/32	42/30	44/33	43/32	41/31	41/30	40/31	41/30	51/40	50/37	41/28	41/28	39/27	100/100
	Kir1.1	Kir2.1	Kir2.2	Kir2.3	Kir2.4	Kir2.6	Kir3.1	Kir3.2	Kir3.3	Kir3.4	Kir4.1	Kir4.2	Kir5.1	Kir6.1	Kir6.2	Kir7.1

Fig. 10 Inwardly rectifying potassium channel family similarity and identity. The 16 members of the Kir family were analyzed for percent similarity and percent identity using the Sequence Identity And Similarity (SIAS) tool which uses a pairwise analyses utilizing several methods. Amino acid similarity and sequence length is used to calculate the values (% = 100*(identicall similar residues/sequence length)). Parameters were set to the standard settings relative to amino acid similarities. Sequences were input from the Clustal Omega alignment. The values are listed as percent similarity/% identity

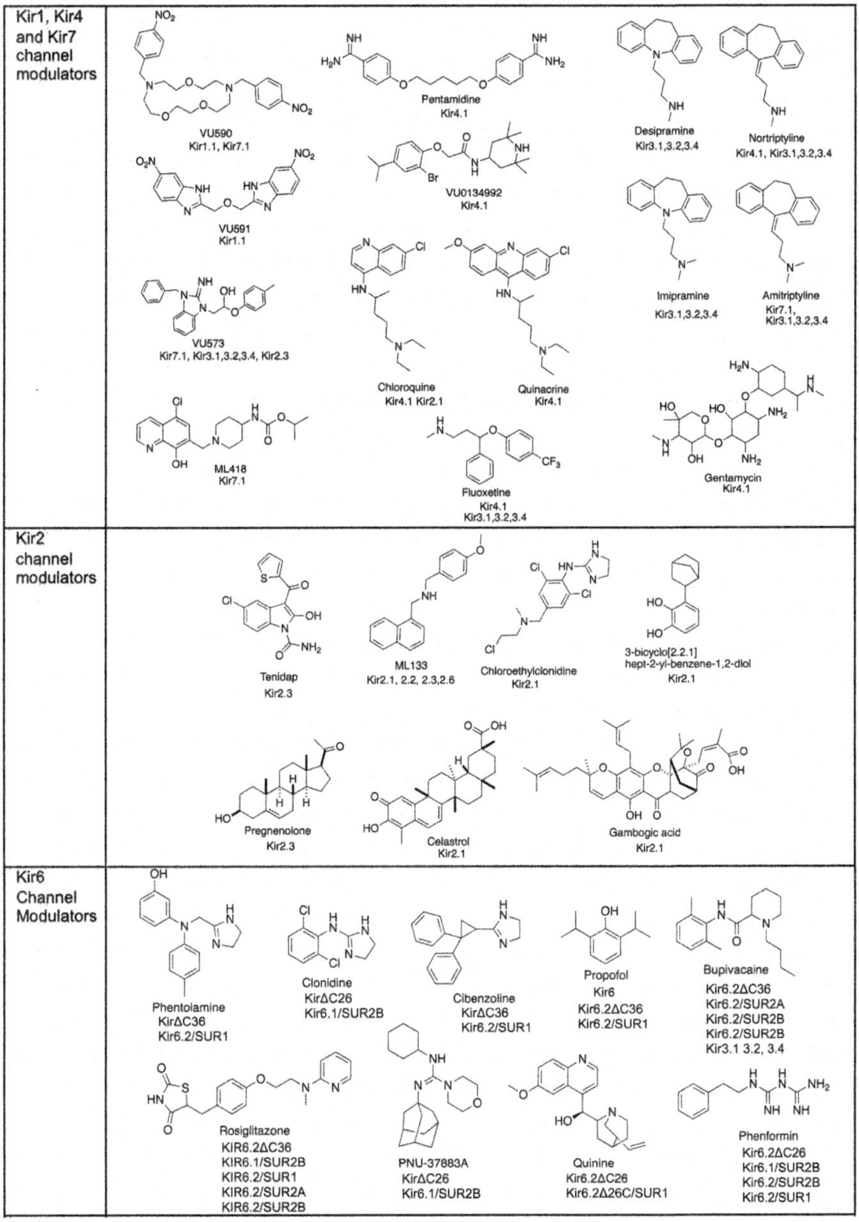

Fig. 11 Chemical structures of drugs targeting Kir channels

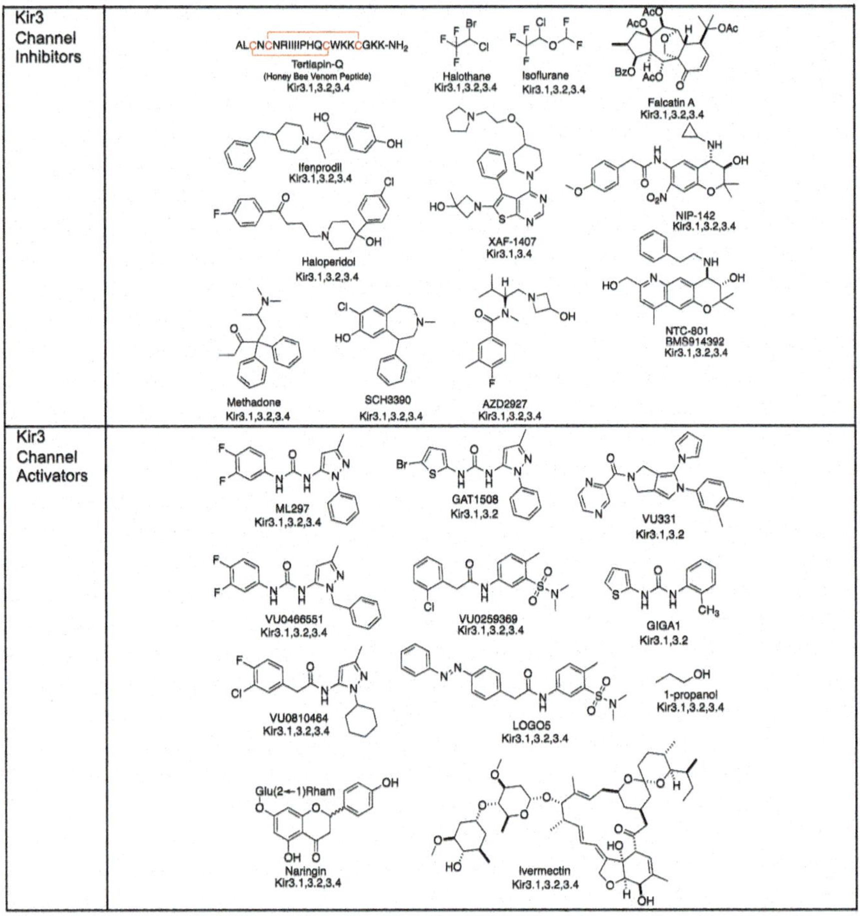

Fig. 11 (continued)

References

Abney KK, Bubser M, Du Y et al (2019) Correction to analgesic effects of the GIRK activator, VU0466551, alone and in combination with morphine in acute and persistent pain models. ACS Chem Nerosci 10(5):2621. https://doi.org/10.1021/acschemneuro.9b00165

Adams DJ, Hill MA (2004) Potassium channels and membrane potential in the modulation of intracellular calcium in vascular endothelial cells. J Cardiovasc Electrophysiol 15(5):598–610. https://doi.org/10.1046/j.1540-8167.2004.03277.x

Adrian RH, Chandler WK, Hodgkin AL (1970) Slow changes in potassium permeability in skeletal muscle. J Physiol 208(3):645–668. https://doi.org/10.1113/jphysiol.1970.sp009140

Aldred KJ, Kerns RJ, Osheroff N (2014) Mechanism of quinolone action and resistance. Biochemistry 53(10):1565–1574. https://doi.org/10.1021/bi5000564

Anderson A, Kulkarni K, Marron Fernandez de Velasco E et al (2018) Expression and relevance of the G protein-gated K(+) channel in the mouse ventricle. Sci Rep 8(1):1192. https://doi.org/10.1038/s41598-018-19719-x

Anumonwo JM, Lopatin AN (2010) Cardiac strong inward rectifier potassium channels. J Mol Cell Cardiol 48(1):45–54. https://doi.org/10.1016/j.yjmcc.2009.08.013

Arechiga-Figueroa IA, Marmolejo-Murillo LG, Cui M et al (2017) High-potency block of Kir4.1 channels by pentamidine: molecular basis. Eur J Pharmacol 815:56–63. https://doi.org/10.1016/j.ejphar.2017.10.009

Armstrong WM (1969) Effect of external potassium and ouabain on sodium efflux from frog sartorius muscle. Proc Soc Exp Biol Med 130(4):1264–1270. https://doi.org/10.3181/00379727-130-33769

Arriagada PV, Growdon JH, Hedley-Whyte ET, Hyman BT (1992) Neurofibrillary tangles but not senile plaques parallel duration and severity of Alzheimer's disease. Neurology 42(3 Pt 1):631–639. https://doi.org/10.1212/wnl.42.3.631

Asakura K, Cha CY, Yamaoka H et al (2014) EAD and DAD mechanisms analyzed by developing a new human ventricular cell model. Prog Biophys Mol Biol 116(1):11–24. https://doi.org/10.1016/j.pbiomolbio.2014.08.008

Ashcroft FM (2005) ATP-sensitive potassium channelopathies: focus on insulin secretion. J Clin Invest 115(8):2047–2058. https://doi.org/10.1172/JCI25495

Asteria C (1997) Molecular basis of Bartter's syndrome: new insights into the correlation between genotype and phenotype. Eur J Endocrinol 137(6):613–615. https://doi.org/10.1530/eje.0.1370613

Aziz Q, Thomas A, Khambra T, Tinker A (2010) Phenformin has a direct inhibitory effect on the ATP-sensitive potassium channel. Eur J Pharmacol 634(1–3):26–32. https://doi.org/10.1016/j.ejphar.2010.02.023

Aziz Q, Li Y, Anderson N, Ojake L, Tsisanova E, Tinker A (2017) Molecular and functional characterization of the endothelial ATP-sensitive potassium channel. J Biol Chem 292(43):17587–17597. https://doi.org/10.1074/jbc.M117.810325

Baculis BC, Weiss AC, Pang W et al (2017) Prolonged seizure activity causes caspase dependent cleavage and dysfunction of G-protein activated inwardly rectifying potassium channels. Sci Rep 7(1):12313. https://doi.org/10.1038/s41598-017-12508-y

Bailey MA, Cantone A, Yan Q et al (2006) Maxi-K channels contribute to urinary potassium excretion in the ROMK-deficient mouse model of type II Bartter's syndrome and in adaptation to a high-K diet. Kidney Int 70(1):51–59. https://doi.org/10.1038/sj.ki.5000388

Balana B, Maslennikov I, Kwiatkowski W et al (2011) Mechanism underlying selective regulation of G protein-gated inwardly rectifying potassium channels by the psychostimulant-sensitive sorting nexin 27. Proc Natl Acad Sci U S A 108(14):5831–5836. https://doi.org/10.1073/pnas.1018645108

Barrett-Jolley R, Dart C, Standen NB (1999) Direct block of native and cloned (Kir2.1) inward rectifier K+ channels by chloroethylclonidine. Br J Pharmacol 128(3):760–766. https://doi.org/10.1038/sj.bjp.0702819

Baukrowitz T, Tucker SJ, Schulte U, Benndorf K, Ruppersberg JP, Fakler B (1999) Inward rectification in KATP channels: a pH switch in the pore. EMBO J 18(4):847–853. https://doi.org/10.1093/emboj/18.4.847

Beeler GW Jr, Reuter H (1970) The relation between membrane potential, membrane currents and activation of contraction in ventricular myocardial fibres. J Physiol 207(1):211–229. https://doi.org/10.1113/jphysiol.1970.sp009057

Beguin P, Nagashima K, Nishimura M, Gonoi T, Seino S (1999) PKA-mediated phosphorylation of the human K(ATP) channel: separate roles of Kir6.2 and SUR1 subunit phosphorylation. EMBO J 18(17):4722–4732. https://doi.org/10.1093/emboj/18.17.4722

Belus MT, Rogers MA, Elzubeir A et al (2018) Kir2.1 is important for efficient BMP signaling in mammalian face development. Dev Biol 444(Suppl 1):S297–S307. https://doi.org/10.1016/j.ydbio.2018.02.012

Bernsteiner H, Zangerl-Plessl EM, Chen X, Stary-Weinzinger A (2019) Conduction through a narrow inward-rectifier K(+) channel pore. J Gen Physiol 151(10):1231–1246. https://doi.org/10.1085/jgp.201912359

Bettahi I, Marker CL, Roman MI, Wickman K (2002) Contribution of the Kir3.1 subunit to the muscarinic-gated atrial potassium channel IKACh. J Biol Chem 277(50):48282–48288. https://doi.org/10.1074/jbc.M209599200

Bhave G, Chauder BA, Liu W et al (2011) Development of a selective small-molecule inhibitor of Kir1.1, the renal outer medullary potassium channel. Mol Pharmacol 79(1):42–50. https://doi.org/10.1124/mol.110.066928

Birnbaumer L, Brown AM (1987) G protein opening of K+ channels. Nature 327(6117):21–22. https://doi.org/10.1038/327021a0

Blednov YA, Stoffel M, Chang SR, Harris RA (2001a) GIRK2 deficient mice. Evidence for hyperactivity and reduced anxiety. Physiol Behav 74(1–2):109–117. https://doi.org/10.1016/s0031-9384(01)00555-8

Blednov YA, Stoffel M, Chang SR, Harris RA (2001b) Potassium channels as targets for ethanol: studies of G-protein-coupled inwardly rectifying potassium channel 2 (GIRK2) null mutant mice. J Pharmacol Exp Ther 298(2):521–530

Blednov YA, Stoffel M, Alva H, Harris RA (2003) A pervasive mechanism for analgesia: activation of GIRK2 channels. Proc Natl Acad Sci U S A 100(1):277–282. https://doi.org/10.1073/pnas.012682399

Bleich M, Schlatter E, Greger R (1990) The luminal K+ channel of the thick ascending limb of Henle's loop. Pflugers Arch 415(4):449–460. https://doi.org/10.1007/BF00373623

Boim MA, Ho K, Shuck ME et al (1995) ROMK inwardly rectifying ATP-sensitive K+ channel. II. Cloning and distribution of alternative forms. Am J Physiol 268(6 Pt 2):F1132–F1140. https://doi.org/10.1152/ajprenal.1995.268.6.F1132

Bond CT, Pessia M, Xia XM, Lagrutta A, Kavanaugh MP, Adelman JP (1994) Cloning and expression of a family of inward rectifier potassium channels. Recept Channels 2(3):183–191

Bouchard R, Clark RB, Juhasz AE, Giles WR (2004) Changes in extracellular K+ concentration modulate contractility of rat and rabbit cardiac myocytes via the inward rectifier K+ current IK1. J Physiol 556(Pt 3):773–790. https://doi.org/10.1113/jphysiol.2003.058248

Bousquet P, Hudson A, Garcia-Sevilla JA, Li JX (2020) Imidazoline receptor system: the past, the present, and the future. Pharmacol Rev 72(1):50–79. https://doi.org/10.1124/pr.118.016311

Bradley KK, Jaggar JH, Bonev AD et al (1999) Kir2.1 encodes the inward rectifier potassium channel in rat arterial smooth muscle cells. J Physiol 515(Pt 3):639–651. https://doi.org/10.1111/j.1469-7793.1999.639ab.x

Branstrom R, Leibiger IB, Leibiger B et al (2007) Single residue (K332A) substitution in Kir6.2 abolishes the stimulatory effect of long-chain acyl-CoA esters: indications for a long-chain acyl-CoA ester binding motif. Diabetologia 50(8):1670–1677. https://doi.org/10.1007/s00125-007-0697-x

Bredt DS, Wang TL, Cohen NA, Guggino WB, Snyder SH (1995) Cloning and expression of two brain-specific inwardly rectifying potassium channels. Proc Natl Acad Sci U S A 92(15):6753–6757. https://doi.org/10.1073/pnas.92.15.6753

Breitwieser GE, Szabo G (1985) Uncoupling of cardiac muscarinic and beta-adrenergic receptors from ion channels by a guanine nucleotide analogue. Nature 317(6037):538–540. https://doi.org/10.1038/317538a0

Brown RA, Carpentier RG (1990) Effects of acetaldehyde on membrane potentials of sinus node pacemaker fibers. Alcohol 7(1):33–36. https://doi.org/10.1016/0741-8329(90)90057-j

Bruehl S, Denton JS, Lonergan D et al (2013) Associations between KCNJ6 (GIRK2) gene polymorphisms and pain-related phenotypes. Pain 154(12):2853–2859. https://doi.org/10.1016/j.pain.2013.08.026

Burgen AS, Terroux KG (1953) On the negative inotropic effect in the cat's auricle. J Physiol 120(4):449–464. https://doi.org/10.1113/jphysiol.1953.sp004910

Carlsson L, Duker G, Jacobson I (2010) New pharmacological targets and treatments for atrial fibrillation. Trends Pharmacol Sci 31(8):364–371. https://doi.org/10.1016/j.tips.2010.05.001
Chan KW, Langan MN, Sui JL et al (1996) A recombinant inwardly rectifying potassium channel coupled to GTP-binding proteins. J Gen Physiol 107(3):381–397. https://doi.org/10.1085/jgp.107.3.381
Chan KW, Zhang H, Logothetis DE (2003) N-terminal transmembrane domain of the SUR controls trafficking and gating of Kir6 channel subunits. EMBO J 22(15):3833–3843. https://doi.org/10.1093/emboj/cdg376
Chen PC, Olson EM, Zhou Q et al (2013) Carbamazepine as a novel small molecule corrector of trafficking-impaired ATP-sensitive potassium channels identified in congenital hyperinsulinism. J Biol Chem 288(29):20942–20954. https://doi.org/10.1074/jbc.M113.470948
Chen IS, Tateyama M, Fukata Y, Uesugi M, Kubo Y (2017) Ivermectin activates GIRK channels in a PIP2 -dependent, Gbetagamma -independent manner and an amino acid residue at the slide helix governs the activation. J Physiol 595(17):5895–5912. https://doi.org/10.1113/JP274871
Chen X, Garon A, Wieder M et al (2019) Computational identification of novel Kir6 channel inhibitors. Front Pharmacol 10:549. https://doi.org/10.3389/fphar.2019.00549
Chen X, Brundl M, Friesacher T, Stary-Weinzinger A (2020) Computational insights into voltage dependence of polyamine block in a strong inwardly rectifying K(+) channel. Front Pharmacol 11:721. https://doi.org/10.3389/fphar.2020.00721
Choe H, Zhou H, Palmer LG, Sackin H (1997) A conserved cytoplasmic region of ROMK modulates pH sensitivity, conductance, and gating. Am J Physiol 273(4):F516–F529. https://doi.org/10.1152/ajprenal.1997.273.4.F516
Choi M, Scholl UI, Yue P et al (2011) K+ channel mutations in adrenal aldosterone-producing adenomas and hereditary hypertension. Science 331(6018):768–772. https://doi.org/10.1126/science.1198785
Cole KS, Curtis HJ (1941) Membrane potential of the squid giant axon during current flow. J Gen Physiol 24(4):551–563. https://doi.org/10.1085/jgp.24.4.551
Cooper A, Grigoryan G, Guy-David L, Tsoory MM, Chen A, Reuveny E (2012) Trisomy of the G protein-coupled K+ channel gene, Kcnj6, affects reward mechanisms, cognitive functions, and synaptic plasticity in mice. Proc Natl Acad Sci U S A 109(7):2642–2647. https://doi.org/10.1073/pnas.1109099109
Cooper PE, McClenaghan C, Chen X, Stary-Weinzinger A, Nichols CG (2017) Conserved functional consequences of disease-associated mutations in the slide helix of Kir6.1 and Kir6.2 subunits of the ATP-sensitive potassium channel. J Biol Chem 292(42):17387–17398. https://doi.org/10.1074/jbc.M117.804971
Cruz HG, Ivanova T, Lunn ML, Stoffel M, Slesinger PA, Luscher C (2004) Bi-directional effects of GABA(B) receptor agonists on the mesolimbic dopamine system. Nat Neurosci 7(2):153–159. https://doi.org/10.1038/nn1181
Cruz HG, Berton F, Sollini M et al (2008) Absence and rescue of morphine withdrawal in GIRK/Kir3 knock-out mice. J Neurosci 28(15):4069–4077. https://doi.org/10.1523/JNEUROSCI.0267-08.2008
Cuevas CA, Su XT, Wang MX et al (2017) Potassium sensing by renal distal tubules requires Kir4.1. J Am Soc Nephrol 28(6):1814–1825. https://doi.org/10.1681/ASN.2016090935
Cui Y, Tinker A, Clapp LH (2003) Different molecular sites of action for the KATP channel inhibitors, PNU-99963 and PNU-37883A. Br J Pharmacol 139(1):122–128. https://doi.org/10.1038/sj.bjp.0705228
Cui M, Alhamshari Y, Cantwell L et al (2021) A benzopyran with anti-arrhythmic activity is an inhibitor of Kir3.1-containing potassium channels. J Biol Chem 296:100535. https://doi.org/10.1016/j.jbc.2021.100535
Dascal N, Schreibmayer W, Lim NF et al (1993) Atrial G protein-activated K+ channel: expression cloning and molecular properties. Proc Natl Acad Sci U S A 90(21):10235–10239. https://doi.org/10.1073/pnas.90.21.10235

Dassau L, Conti LR, Radeke CM, Ptacek LJ, Vandenberg CA (2011) Kir2.6 regulates the surface expression of Kir2.x inward rectifier potassium channels. J Biol Chem 286(11):9526–9541. https://doi.org/10.1074/jbc.M110.170597

de Boer TP, Houtman MJ, Compier M, van der Heyden MA (2010) The mammalian K(IR)2.x inward rectifier ion channel family: expression pattern and pathophysiology. Acta Physiol (Oxf) 199(3):243–256. https://doi.org/10.1111/j.1748-1716.2010.02108.x

Del Castillo J, Katz B (1955) Production of membrane potential changes in the frog's heart by inhibitory nerve impulses. Nature 175(4467):1035. https://doi.org/10.1038/1751035a0

Delling M, Wischmeyer E, Dityatev A et al (2002) The neural cell adhesion molecule regulates cell-surface delivery of G-protein-activated inwardly rectifying potassium channels via lipid rafts. J Neurosci 22(16):7154–7164. https://doi.org/10.1523/JNEUROSCI.22-16-07154.2002

Derst C, Hirsch JR, Preisig-Muller R et al (2001) Cellular localization of the potassium channel Kir7.1 in guinea pig and human kidney. Kidney Int 59(6):2197–2205. https://doi.org/10.1046/j.1523-1755.2001.00735.x

Devulder J (2010) Flupirtine in pain management: pharmacological properties and clinical use. CNS Drugs 24(10):867–881. https://doi.org/10.2165/11536230-000000000-00000

Ding D, Wang M, Wu JX, Kang Y, Chen L (2019) The structural basis for the binding of repaglinide to the pancreatic KATP Channel. Cell Rep 27(6):1848–1857.e4. https://doi.org/10.1016/j.celrep.2019.04.050

Dobrev D, Friedrich A, Voigt N et al (2005) The G protein-gated potassium current I(K,ACh) is constitutively active in patients with chronic atrial fibrillation. Circulation 112(24):3697–3706. https://doi.org/10.1161/CIRCULATIONAHA.105.575332

Doi T, Fakler B, Schultz JH et al (1996) Extracellular K+ and intracellular pH allosterically regulate renal Kir1.1 channels. J Biol Chem 271(29):17261–17266. https://doi.org/10.1074/jbc.271.29.17261

Donaldson MR, Yoon G, Fu YH, Ptacek LJ (2004) Andersen-Tawil syndrome: a model of clinical variability, pleiotropy, and genetic heterogeneity. Ann Med 36(Suppl 1):92–97. https://doi.org/10.1080/17431380410032490

Doring F, Derst C, Wischmeyer E et al (1998) The epithelial inward rectifier channel Kir7.1 displays unusual K+ permeation properties. J Neurosci 18(21):8625–8636

Doupnik CA (2015) RGS redundancy and implications in GPCR-GIRK signaling. Int Rev Neurobiol 123:87–116. https://doi.org/10.1016/bs.irn.2015.05.010

Doupnik CA, Davidson N, Lester HA, Kofuji P (1997) RGS proteins reconstitute the rapid gating kinetics of gbetagamma-activated inwardly rectifying K+ channels. Proc Natl Acad Sci U S A 94(19):10461–10466. https://doi.org/10.1073/pnas.94.19.10461

Du X, Zhang H, Lopes C, Mirshahi T, Rohacs T, Logothetis DE (2004) Characteristic interactions with phosphatidylinositol 4,5-bisphosphate determine regulation of Kir channels by diverse modulators. J Biol Chem 279(36):37271–81

Edvinsson JM, Shah AJ, Palmer LG (2011a) Kir4.1 K+ channels are regulated by external cations. Channels (Austin) 5(3):269–279. https://doi.org/10.4161/chan.5.3.15827

Edvinsson JM, Shah AJ, Palmer LG (2011b) Potassium-dependent activation of Kir4.2 K(+) channels. J Physiol 589(Pt 24):5949–5963. https://doi.org/10.1113/jphysiol.2011.220731

Ehrlich JR, Cha TJ, Zhang L et al (2004) Characterization of a hyperpolarization-activated time-dependent potassium current in canine cardiomyocytes from pulmonary vein myocardial sleeves and left atrium. J Physiol 557(Pt 2):583–597. https://doi.org/10.1113/jphysiol.2004.061119

Epshtein Y, Chopra AP, Rosenhouse-Dantsker A, Kowalsky GB, Logothetis DE, Levitan I (2009) Identification of a C-terminus domain critical for the sensitivity of Kir2.1 to cholesterol. Proc Natl Acad Sci U S A 106(19):8055–8060. https://doi.org/10.1073/pnas.0809847106

Fakler B, Schultz JH, Yang J et al (1996) Identification of a titratable lysine residue that determines sensitivity of kidney potassium channels (ROMK) to intracellular pH. EMBO J 15 (16):4093–4099

Fang Y, Schram G, Romanenko VG et al (2005) Functional expression of Kir2.x in human aortic endothelial cells: the dominant role of Kir2.2. Am J Physiol Cell Physiol 289(5):C1134–C1144. https://doi.org/10.1152/ajpcell.00077.2005

Fenner MF, Carstensen H, Dalgas Nissen S et al (2020) Effect of selective IK,ACh inhibition by XAF-1407 in an equine model of tachypacing-induced persistent atrial fibrillation. Br J Pharmacol 177(16):3778–3794. https://doi.org/10.1111/bph.15100

Ferré SCF, Dessauer CW, González-Maeso J, Hébert TE, Jockers R, Logothetis DE, Pardo L (2021) G protein-coupled receptor-effector macromolecular membrane assemblies (GEMMA). Pharmacol Ther. Under revision

Ferrer J, Nichols CG, Makhina EN et al (1995) Pancreatic islet cells express a family of inwardly rectifying K+ channel subunits which interact to form G-protein-activated channels. J Biol Chem 270(44):26086–26091. https://doi.org/10.1074/jbc.270.44.26086

Field MJ, Stanton BA, Giebisch GH (1984) Influence of ADH on renal potassium handling: a micropuncture and microperfusion study. Kidney Int 25(3):502–511. https://doi.org/10.1038/ki.1984.46

Filosa JA, Bonev AD, Straub SV et al (2006) Local potassium signaling couples neuronal activity to vasodilation in the brain. Nat Neurosci 9(11):1397–1403. https://doi.org/10.1038/nn1779

Fischer-Lougheed J, Liu JH, Espinos E et al (2001) Human myoblast fusion requires expression of functional inward rectifier Kir2.1 channels. J Cell Biol 153(4):677–686. https://doi.org/10.1083/jcb.153.4.677

Fujita A, Horio Y, Higashi K et al (2002) Specific localization of an inwardly rectifying K(+) channel, Kir4.1, at the apical membrane of rat gastric parietal cells; its possible involvement in K(+) recycling for the H(+)-K(+)-pump. J Physiol 540(Pt 1):85–92. https://doi.org/10.1113/jphysiol.2001.013439

Fujiwara Y, Kubo Y (2006) Functional roles of charged amino acid residues on the wall of the cytoplasmic pore of Kir2.1. J Gen Physiol 127(4):401–419. https://doi.org/10.1085/jgp.200509434

Furutani K, Ohno Y, Inanobe A, Hibino H, Kurachi Y (2009) Mutational and in silico analyses for antidepressant block of astroglial inward-rectifier Kir4.1 channel. Mol Pharmacol 75(6):1287–1295. https://doi.org/10.1124/mol.108.052936

Gada K, Logothetis DE (2021) Regulation of ion channels by protein kinase C isoforms. Commissioned Review, JBC. In Revision

Gahwiler BH, Brown DA (1985) GABAB-receptor-activated K+ current in voltage-clamped CA3 pyramidal cells in hippocampal cultures. Proc Natl Acad Sci U S A 82(5):1558–1562. https://doi.org/10.1073/pnas.82.5.1558

Gaskell WH (1886) The electrical changes in the quiescent cardiac muscle which accompany stimulation of the Vagus nerve. J Physiol 7(5–6):451–452. https://doi.org/10.1113/jphysiol.1886.sp000235

Gaskell WH (1887) On the action of Muscarin upon the heart, and on the electrical changes in the non-beating cardiac muscle brought about by stimulation of the inhibitory and Augmentor nerves. J Physiol 8(6):404–4i8. https://doi.org/10.1113/jphysiol.1887.sp000269

Ghamari-Langroudi M, Digby GJ, Sebag JA et al (2015) G-protein-independent coupling of MC4R to Kir7.1 in hypothalamic neurons. Nature 520(7545):94–98. https://doi.org/10.1038/nature14051

Girmatsion Z, Biliczki P, Bonauer A et al (2009) Changes in microRNA-1 expression and IK1 up-regulation in human atrial fibrillation. Heart Rhythm 6(12):1802–1809. https://doi.org/10.1016/j.hrthm.2009.08.035

Gomez-Sanchez CE, Oki K (2014) Minireview: potassium channels and aldosterone dysregulation: is primary aldosteronism a potassium channelopathy? Endocrinology 155(1):47–55. https://doi.org/10.1210/en.2013-1733

Gouder N, Fritschy JM, Boison D (2003) Seizure suppression by adenosine A1 receptor activation in a mouse model of pharmacoresistant epilepsy. Epilepsia 44(7):877–885. https://doi.org/10.1046/j.1528-1157.2003.03603.x

Greener ID, Monfredi O, Inada S et al (2011) Molecular architecture of the human specialised atrioventricular conduction axis. J Mol Cell Cardiol 50(4):642–651. https://doi.org/10.1016/j.yjmcc.2010.12.017

Greger R, Bleich M, Schlatter E (1990) Ion channels in the thick ascending limb of Henle's loop. Ren Physiol Biochem 13(1–2):37–50. https://doi.org/10.1159/000173346

Gregerson KA, Flagg TP, O'Neill TJ et al (2001) Identification of G protein-coupled, inward rectifier potassium channel gene products from the rat anterior pituitary gland. Endocrinology 142(7):2820–2832. https://doi.org/10.1210/endo.142.7.8236

Grosse-Lackmann T, Zunkler BJ, Rustenbeck I (2003) Specificity of nonadrenergic imidazoline binding sites in insulin-secreting cells and relation to the block of ATP-sensitive K(+) channels. Ann N Y Acad Sci 1009:371–377. https://doi.org/10.1196/annals.1304.050

Guay-Woodford LM (1995) Molecular insights into the pathogenesis of inherited renal tubular disorders. Curr Opin Nephrol Hypertens 4(2):121–129. https://doi.org/10.1097/00041552-199503000-00004

Guo D, Lu Z (2003) Interaction mechanisms between polyamines and IRK1 inward rectifier K+ channels. J Gen Physiol 122(5):485–500. https://doi.org/10.1085/jgp.200308890

Haass C, Selkoe DJ (2007) Soluble protein oligomers in neurodegeneration: lessons from the Alzheimer's amyloid beta-peptide. Nat Rev Mol Cell Biol 8(2):101–112. https://doi.org/10.1038/nrm2101

Hagiwara S, Takahashi K (1974) The anomalous rectification and cation selectivity of the membrane of a starfish egg cell. J Membr Biol 18(1):61–80. https://doi.org/10.1007/BF01870103

Hagiwara S, Yoshii M (1979) Effects of internal potassium and sodium on the anomalous rectification of the starfish egg as examined by internal perfusion. J Physiol 292:251–265. https://doi.org/10.1113/jphysiol.1979.sp012849

Hagiwara S, Miyazaki S, Rosenthal NP (1976) Potassium current and the effect of cesium on this current during anomalous rectification of the egg cell membrane of a starfish. J Gen Physiol 67 (6):621–638. https://doi.org/10.1085/jgp.67.6.621

Hagiwara S, Miyazaki S, Moody W, Patlak J (1978) Blocking effects of barium and hydrogen ions on the potassium current during anomalous rectification in the starfish egg. J Physiol 279:167–185. https://doi.org/10.1113/jphysiol.1978.sp012338

Haider S, Khalid S, Tucker SJ, Ashcroft FM, Sansom MS (2007) Molecular dynamics simulations of inwardly rectifying (Kir) potassium channels: a comparative study. Biochemistry 46 (12):3643–3652. https://doi.org/10.1021/bi062210f

Hamill OP, Marty A, Neher E, Sakmann B, Sigworth FJ (1981) Improved patch-clamp techniques for high-resolution current recording from cells and cell-free membrane patches. Pflugers Arch 391(2):85–100. https://doi.org/10.1007/BF00656997

Hansen JB (2006) Towards selective Kir6.2/SUR1 potassium channel openers, medicinal chemistry and therapeutic perspectives. Curr Med Chem 13(4):361–376. https://doi.org/10.2174/092986706775527947

Hansen SB, Tao X, MacKinnon R (2011) Structural basis of PIP2 activation of the classical inward rectifier K+ channel Kir2.2. Nature 477(7365):495–498. https://doi.org/10.1038/nature10370

Hardy J, Selkoe DJ (2002) The amyloid hypothesis of Alzheimer's disease: progress and problems on the road to therapeutics. Science 297(5580):353–356. https://doi.org/10.1126/science.1072994

Harraz OF, Longden TA, Dabertrand F, Hill-Eubanks D, Nelson MT (2018) Endothelial GqPCR activity controls capillary electrical signaling and brain blood flow through PIP2 depletion. Proc Natl Acad Sci U S A 115(15):E3569–E3577. https://doi.org/10.1073/pnas.1800201115

Hashimoto N, Yamashita T, Tsuruzoe N (2008) Characterization of in vivo and in vitro electrophysiological and antiarrhythmic effects of a novel IKACh blocker, NIP-151: a comparison with an IKr-blocker dofetilide. J Cardiovasc Pharmacol 51(2):162–169. https://doi.org/10.1097/FJC.0b013e31815e854c

Hasselblatt M, Bohm C, Tatenhorst L et al (2006) Identification of novel diagnostic markers for choroid plexus tumors: a microarray-based approach. Am J Surg Pathol 30(1):66–74. https://doi.org/10.1097/01.pas.0000176430.88702.e0
Hassinen M, Paajanen V, Vornanen M (2008) A novel inwardly rectifying K+ channel, Kir2.5, is upregulated under chronic cold stress in fish cardiac myocytes. J Exp Biol 211 (Pt 13):2162–2171. https://doi.org/10.1242/jeb.016121
He G, Wang HR, Huang SK, Huang CL (2007) Intersectin links WNK kinases to endocytosis of ROMK1. J Clin Invest 117(4):1078–1087. https://doi.org/10.1172/JCI30087
Hebert SC (2003) Bartter syndrome. Curr Opin Nephrol Hypertens 12(5):527–532. https://doi.org/10.1097/00041552-200309000-00008
Hebert SC, Desir G, Giebisch G, Wang W (2005) Molecular diversity and regulation of renal potassium channels. Physiol Rev 85(1):319–371. https://doi.org/10.1152/physrev.00051.2003
Heron-Milhavet L, Xue-Jun Y, Vannucci SJ et al (2004) Protection against hypoxic-ischemic injury in transgenic mice overexpressing Kir6.2 channel pore in forebrain. Mol Cell Neurosci 25 (4):585–593. https://doi.org/10.1016/j.mcn.2003.10.012
Hestrin S (1981) The interaction of potassium with the activation of anomalous rectification in frog muscle membrane. J Physiol 317:497–508. https://doi.org/10.1113/jphysiol.1981.sp013839
Hibino H, Kurachi Y (2006) Molecular and physiological bases of the K+ circulation in the mammalian inner ear. Physiology (Bethesda) 21:336–345. https://doi.org/10.1152/physiol.00023.2006
Hibino H, Inanobe A, Furutani K, Murakami S, Findlay I, Kurachi Y (2010) Inwardly rectifying potassium channels: their structure, function, and physiological roles. Physiol Rev 90 (1):291–366. https://doi.org/10.1152/physrev.00021.2009
Hilgemann DW, Ball R (1996) Regulation of cardiac Na+,Ca2+ exchange and KATP potassium channels by PIP2. Science 273(5277):956–959. https://doi.org/10.1126/science.273.5277.956
Hill E, Hickman C, Diez R, Wall M (2020) Role of A1 receptor-activated GIRK channels in the suppression of hippocampal seizure activity. Neuropharmacology 164:107904. https://doi.org/10.1016/j.neuropharm.2019.107904
Hille B, Schwarz W (1978) Potassium channels as multi-ion single-file pores. J Gen Physiol 72 (4):409–442. https://doi.org/10.1085/jgp.72.4.409
Ho IH, Murrell-Lagnado RD (1999) Molecular determinants for sodium-dependent activation of G protein-gated K+ channels. J Biol Chem 274(13):8639–8648. https://doi.org/10.1074/jbc.274.13.8639
Ho K, Nichols CG, Lederer WJ et al (1993) Cloning and expression of an inwardly rectifying ATP-regulated potassium channel. Nature 362(6415):31–38. https://doi.org/10.1038/362031a0
Hodgkin AL, Huxley AF, Katz B (1952) Measurement of current-voltage relations in the membrane of the giant axon of Loligo. J Physiol 116(4):424–448. https://doi.org/10.1113/jphysiol.1952.sp004716
Hu J, Qiu S, Yang F, Cao Z, Li W, Wu Y (2013) Unique mechanism of the interaction between honey bee toxin TPNQ and rKir1.1 potassium channel explored by computational simulations: insights into the relative insensitivity of channel towards animal toxins. PLoS One 8(7):e67213. https://doi.org/10.1371/journal.pone.0067213
Huang Y, Mucke L (2012) Alzheimer mechanisms and therapeutic strategies. Cell 148 (6):1204–1222. https://doi.org/10.1016/j.cell.2012.02.040
Huang CL, Feng S, Hilgemann DW (1998) Direct activation of inward rectifier potassium channels by PIP2 and its stabilization by Gbetagamma. Nature 391(6669):803–806. https://doi.org/10.1038/35882
Huang X, Lee SH, Lu H, Sanders KM, Koh SD (2018a) Molecular and functional characterization of inwardly rectifying K(+) currents in murine proximal colon. J Physiol 596(3):379–391. https://doi.org/10.1113/JP275234
Huang Y, Zhang Y, Kong S et al (2018b) GIRK1-mediated inwardly rectifying potassium current suppresses the epileptiform burst activities and the potential antiepileptic effect of ML297. Biomed Pharmacother 101:362–370. https://doi.org/10.1016/j.biopha.2018.02.114

Humphrey SJ (1999) Pharmacology of the K-ATP channel blocking morpholinoguanidine PNU-37883A. Cardiovasc Drug Rev 17(4):295–328

Inanobe A, Horio Y, Fujita A et al (1999) Molecular cloning and characterization of a novel splicing variant of the Kir3.2 subunit predominantly expressed in mouse testis. J Physiol 521 (Pt 1):19–30. https://doi.org/10.1111/j.1469-7793.1999.00019.x

Inanobe A, Matsuura T, Nakagawa A, Kurachi Y (2007) Structural diversity in the cytoplasmic region of G protein-gated inward rectifier K+ channels. Channels (Austin) 1(1):39–45

Ishihara K (2018) External K(+) dependence of strong inward rectifier K(+) channel conductance is caused not by K(+) but by competitive pore blockade by external Na(.). J Gen Physiol 150 (7):977–989. https://doi.org/10.1085/jgp.201711936

Ishihara K, Ehara T (2004) Two modes of polyamine block regulating the cardiac inward rectifier K + current IK1 as revealed by a study of the Kir2.1 channel expressed in a human cell line. J Physiol 556(Pt 1):61–78. https://doi.org/10.1113/jphysiol.2003.055434

Ishihara K, Sarai N, Asakura K, Noma A, Matsuoka S (2009) Role of mg(2+) block of the inward rectifier K(+) current in cardiac repolarization reserve: a quantitative simulation. J Mol Cell Cardiol 47(1):76–84. https://doi.org/10.1016/j.yjmcc.2009.03.008

Iwanir S, Reuveny E (2008) Adrenaline-induced hyperpolarization of mouse pancreatic islet cells is mediated by G protein-gated inwardly rectifying potassium (GIRK) channels. Pflugers Arch 456 (6):1097–1108. https://doi.org/10.1007/s00424-008-0479-4

Jakob R, Krieglstein J (1997) Influence of flupirtine on a G-protein coupled inwardly rectifying potassium current in hippocampal neurones. Br J Pharmacol 122(7):1333–1338. https://doi.org/10.1038/sj.bjp.0701519

Jelacic TM, Kennedy ME, Wickman K, Clapham DE (2000) Functional and biochemical evidence for G-protein-gated inwardly rectifying K+ (GIRK) channels composed of GIRK2 and GIRK3. J Biol Chem 275(46):36211–36216. https://doi.org/10.1074/jbc.M007087200

Jin W, Lu Z (1998) A novel high-affinity inhibitor for inward-rectifier K+ channels. Biochemistry 37(38):13291–13299. https://doi.org/10.1021/bi981178p

Jin W, Lu Z (1999) Synthesis of a stable form of tertiapin: a high-affinity inhibitor for inward-rectifier K+ channels. Biochemistry 38(43):14286–14293. https://doi.org/10.1021/bi991205r

Jin T, Peng L, Mirshahi T et al (2002) The (beta)gamma subunits of G proteins gate a K(+) channel by pivoted bending of a transmembrane segment. Mol Cell 10(3):469–481. https://doi.org/10.1016/s1097-2765(02)00659-7

Karolyi L, Koch MC, Grzeschik KH, Seyberth HW (1998) The molecular genetic approach to "Bartter's syndrome". J Mol Med (Berl) 76(5):317–325. https://doi.org/10.1007/s001090050223

Katz B (1949a) Les constants electriques de la membrane du muscle. Arch Sci Physiol 3:285–299

Katz B (1949b) The efferent regulation of the muscle spindle in the frog. J Exp Biol 26(2):201–217

Kaufmann K, Romaine I, Days E et al (2013) ML297 (VU0456810), the first potent and selective activator of the GIRK potassium channel, displays antiepileptic properties in mice. ACS Chem Nerosci 4(9):1278–1286. https://doi.org/10.1021/cn400062a

Kawahito S, Kawano T, Kitahata H et al (2011) Molecular mechanisms of the inhibitory effects of clonidine on vascular adenosine triphosphate-sensitive potassium channels. Anesth Analg 113 (6):1374–1380. https://doi.org/10.1213/ANE.0b013e3182321142

Kawano T, Oshita S, Takahashi A et al (2004a) Molecular mechanisms of the inhibitory effects of bupivacaine, levobupivacaine, and ropivacaine on sarcolemmal adenosine triphosphate-sensitive potassium channels in the cardiovascular system. Anesthesiology 101(2):390–398. https://doi.org/10.1097/00000542-200408000-00020

Kawano T, Oshita S, Takahashi A et al (2004b) Molecular mechanisms of the inhibitory effects of propofol and thiamylal on sarcolemmal adenosine triphosphate-sensitive potassium channels. Anesthesiology 100(2):338–346. https://doi.org/10.1097/00000542-200402000-00024

Kenna S, Roper J, Ho K, Hebert S, Ashcroft SJ, Ashcroft FM (1994) Differential expression of the inwardly-rectifying K-channel ROMK1 in rat brain. Brain Res Mol Brain Res 24 (1–4):353–356. https://doi.org/10.1016/0169-328x(94)90150-3

Keselman I, Fribourg M, Felsenfeld DP, Logothetis DE (2007) Mechanism of PLC-mediated Kir3 current inhibition. Channels (Austin) 1(2):113–123. https://doi.org/10.4161/chan.4321

Kharade SV, Kurata H, Bender AM et al (2018) Discovery, characterization, and effects on renal fluid and electrolyte excretion of the Kir4.1 potassium channel pore blocker, VU0134992. Mol Pharmacol 94(2):926–937. https://doi.org/10.1124/mol.118.112359

Kline CF, Kurata HT, Hund TJ et al (2009) Dual role of K ATP channel C-terminal motif in membrane targeting and metabolic regulation. Proc Natl Acad Sci U S A 106(39):16669–16674. https://doi.org/10.1073/pnas.0907138106

Knot HJ, Nelson MT (1998) Regulation of arterial diameter and wall [Ca2+] in cerebral arteries of rat by membrane potential and intravascular pressure. J Physiol 508(Pt 1):199–209. https://doi.org/10.1111/j.1469-7793.1998.199br.x

Knot HJ, Zimmermann PA, Nelson MT (1996) Extracellular K(+)-induced hyperpolarizations and dilatations of rat coronary and cerebral arteries involve inward rectifier K(+) channels. J Physiol 492(Pt 2):419–430. https://doi.org/10.1113/jphysiol.1996.sp021318

Kobayashi T, Ikeda K (2006) G protein-activated inwardly rectifying potassium channels as potential therapeutic targets. Curr Pharm Des 12(34):4513–4523. https://doi.org/10.2174/138161206779010468

Kobayashi T, Ikeda K, Kojima H et al (1999) Ethanol opens G-protein-activated inwardly rectifying K+ channels. Nat Neurosci 2(12):1091–1097. https://doi.org/10.1038/16019

Kobayashi T, Ikeda K, Kumanishi T (2000) Inhibition by various antipsychotic drugs of the G-protein-activated inwardly rectifying K(+) (GIRK) channels expressed in xenopus oocytes. Br J Pharmacol 129(8):1716–1722. https://doi.org/10.1038/sj.bjp.0703224

Kobayashi T, Washiyama K, Ikeda K (2003) Inhibition of G protein-activated inwardly rectifying K + channels by fluoxetine (Prozac). Br J Pharmacol 138(6):1119–1128. https://doi.org/10.1038/sj.bjp.0705172

Kobayashi T, Washiyama K, Ikeda K (2004) Inhibition of G protein-activated inwardly rectifying K + channels by various antidepressant drugs. Neuropsychopharmacology 29(10):1841–1851. https://doi.org/10.1038/sj.npp.1300484

Kobayashi T, Washiyama K, Ikeda K (2006) Inhibition of G protein-activated inwardly rectifying K + channels by ifenprodil. Neuropsychopharmacology 31(3):516–524. https://doi.org/10.1038/sj.npp.1300844

Kobayashi T, Washiyama K, Ikeda K (2009) Pregnenolone sulfate potentiates the inwardly rectifying K channel Kir2.3. PLoS One 4(7):e6311. https://doi.org/10.1371/journal.pone.0006311

Kondo C, Isomoto S, Matsumoto S et al (1996) Cloning and functional expression of a novel isoform of ROMK inwardly rectifying ATP-dependent K+ channel, ROMK6 (Kir1.1f). FEBS Lett 399(1–2):122–126. https://doi.org/10.1016/s0014-5793(96)01302-6

Konig S, Hinard V, Arnaudeau S et al (2004) Membrane hyperpolarization triggers myogenin and myocyte enhancer factor-2 expression during human myoblast differentiation. J Biol Chem 279(27):28187–28196. https://doi.org/10.1074/jbc.M313932200

Kotecki L, Hearing M, McCall NM et al (2015) GIRK channels modulate opioid-induced motor activity in a cell type- and subunit-dependent manner. J Neurosci 35(18):7131–7142. https://doi.org/10.1523/JNEUROSCI.5051-14.2015

Kovalev H, Quayle JM, Kamishima T, Lodwick D (2004) Molecular analysis of the subtype-selective inhibition of cloned KATP channels by PNU-37883A. Br J Pharmacol 141(5):867–873. https://doi.org/10.1038/sj.bjp.0705670

Kovoor P, Wickman K, Maguire CT et al (2001) Evaluation of the role of I(KACh) in atrial fibrillation using a mouse knockout model. J Am Coll Cardiol 37(8):2136–2143. https://doi.org/10.1016/s0735-1097(01)01304-3

Kozek KA, Du Y, Sharma S et al (2019) Discovery and characterization of VU0529331, a synthetic small-molecule activator of Homomeric G protein-gated, inwardly rectifying, potassium (GIRK) channels. ACS Chem Nerosci 10(1):358–370. https://doi.org/10.1021/acschemneuro.8b00287

Krapivinsky G, Gordon EA, Wickman K, Velimirovic B, Krapivinsky L, Clapham DE (1995) The G-protein-gated atrial K+ channel IKACh is a heteromultimer of two inwardly rectifying K(+)-channel proteins. Nature 374(6518):135–141. https://doi.org/10.1038/374135a0

Krapivinsky G, Medina I, Eng L, Krapivinsky L, Yang Y, Clapham DE (1998) A novel inward rectifier K+ channel with unique pore properties. Neuron 20(5):995–1005. https://doi.org/10.1016/s0896-6273(00)80480-8

Kubo Y (1996) Effects of extracellular cations and mutations in the pore region on the inward rectifier K+ channel IRK1. Recept Channels 4(2):73–83

Kubo Y, Murata Y (2001) Control of rectification and permeation by two distinct sites after the second transmembrane region in Kir2.1 K+ channel. J Physiol 531(Pt 3):645–660. https://doi.org/10.1111/j.1469-7793.2001.0645h.x

Kubo Y, Baldwin TJ, Jan YN, Jan LY (1993a) Primary structure and functional expression of a mouse inward rectifier potassium channel. Nature 362(6416):127–133. https://doi.org/10.1038/362127a0

Kubo Y, Reuveny E, Slesinger PA, Jan YN, Jan LY (1993b) Primary structure and functional expression of a rat G-protein-coupled muscarinic potassium channel. Nature 364(6440):802–806. https://doi.org/10.1038/364802a0

Kuffler SW, Nicholls JG (1966) The physiology of neuroglial cells. Ergeb Physiol 57:1–90

Kumar M, Pattnaik BR (2014) Focus on Kir7.1: physiology and channelopathy. Channels (Austin) 8(6):488–495. https://doi.org/10.4161/19336950.2014.959809

Kung AW, Lau KS, Cheung WM, Chan V (2006) Thyrotoxic periodic paralysis and polymorphisms of sodium-potassium ATPase genes. Clin Endocrinol (Oxf) 64(2):158–161. https://doi.org/10.1111/j.1365-2265.2005.02442.x

Kurachi Y (1985) Voltage-dependent activation of the inward-rectifier potassium channel in the ventricular cell membrane of guinea-pig heart. J Physiol 366:365–385. https://doi.org/10.1113/jphysiol.1985.sp015803

Kurachi Y, Nakajima T, Sugimoto T (1986) On the mechanism of activation of muscarinic K+ channels by adenosine in isolated atrial cells: involvement of GTP-binding proteins. Pflugers Arch 407(3):264–274. https://doi.org/10.1007/BF00585301

Kusaka S, Inanobe A, Fujita A et al (2001) Functional Kir7.1 channels localized at the root of apical processes in rat retinal pigment epithelium. J Physiol 531(Pt 1):27–36. https://doi.org/10.1111/j.1469-7793.2001.0027j.x

Kuzhikandathil EV, Oxford GS (2002) Classic D1 dopamine receptor antagonist R-(+)-7-chloro-8-hydroxy-3-methyl-1-phenyl-2,3,4,5-tetrahydro-1H-3-benzazepine hydrochloride (SCH23390) directly inhibits G protein-coupled inwardly rectifying potassium channels. Mol Pharmacol 62(1):119–126. https://doi.org/10.1124/mol.62.1.119

Kwan HY, Leung PC, Huang Y, Yao X (2003) Depletion of intracellular Ca2+ stores sensitizes the flow-induced Ca2+ influx in rat endothelial cells. Circ Res 92(3):286–292. https://doi.org/10.1161/01.res.0000054625.24468.08

Labouebe G, Lomazzi M, Cruz HG et al (2007) RGS2 modulates coupling between GABAB receptors and GIRK channels in dopamine neurons of the ventral tegmental area. Nat Neurosci 10(12):1559–1568. https://doi.org/10.1038/nn2006

Lacey MG, Mercuri NB, North RA (1988) On the potassium conductance increase activated by GABAB and dopamine D2 receptors in rat substantia nigra neurones. J Physiol 401:437–453. https://doi.org/10.1113/jphysiol.1988.sp017171

Lachheb S, Cluzeaud F, Bens M et al (2008) Kir4.1/Kir5.1 channel forms the major K+ channel in the basolateral membrane of mouse renal collecting duct principal cells. Am J Physiol Renal Physiol 294(6):F1398–F1407. https://doi.org/10.1152/ajprenal.00288.2007

Lacin E, Aryal P, Glaaser IW et al (2017) Dynamic role of the tether helix in PIP2-dependent gating of a G protein-gated potassium channel. J Gen Physiol 149(8):799–811. https://doi.org/10.1085/jgp.201711801

Leal-Pinto E, Gomez-Llorente Y, Sundaram S et al (2010) Gating of a G protein-sensitive mammalian Kir3.1 prokaryotic Kir channel chimera in planar lipid bilayers. J Biol Chem 285 (51):39790–39800. https://doi.org/10.1074/jbc.M110.151373

Leaney JL, Dekker LV, Tinker A (2001) Regulation of a G protein-gated inwardly rectifying K+ channel by a Ca(2+)-independent protein kinase C. J Physiol 534(Pt. 2):367–379. https://doi.org/10.1111/j.1469-7793.2001.00367.x

Lee SJ, Wang S, Borschel W, Heyman S, Gyore J, Nichols CG (2013) Secondary anionic phospholipid binding site and gating mechanism in Kir2.1 inward rectifier channels. Nat Commun 4:2786. https://doi.org/10.1038/ncomms3786

Lee SJ, Ren F, Zangerl-Plessl EM et al (2016) Structural basis of control of inward rectifier Kir2 channel gating by bulk anionic phospholipids. J Gen Physiol 148(3):227–237. https://doi.org/10.1085/jgp.201611616

Lee KPK, Chen J, MacKinnon R (2017) Molecular structure of human KATP in complex with ATP and ADP. eLife 6. https://doi.org/10.7554/eLife.32481

Leech CA, Stanfield PR (1981) Inward rectification in frog skeletal muscle fibres and its dependence on membrane potential and external potassium. J Physiol 319:295–309. https://doi.org/10.1113/jphysiol.1981.sp013909

Lei Q, Talley EM, Bayliss DA (2001) Receptor-mediated inhibition of G protein-coupled inwardly rectifying potassium channels involves G(alpha)q family subunits, phospholipase C, and a readily diffusible messenger. J Biol Chem 276(20):16720–16730. https://doi.org/10.1074/jbc.M100207200

Lesage F, Duprat F, Fink M et al (1994) Cloning provides evidence for a family of inward rectifier and G-protein coupled K+ channels in the brain. FEBS Lett 353(1):37–42. https://doi.org/10.1016/0014-5793(94)01007-2

Lewis DL, Ikeda SR, Aryee D, Joho RH (1991) Expression of an inwardly rectifying K+ channel from rat basophilic leukemia cell mRNA in Xenopus oocytes. FEBS Lett 290(1–2):17–21. https://doi.org/10.1016/0014-5793(91)81215-t

Lewis LM, Bhave G, Chauder BA et al (2009) High-throughput screening reveals a small-molecule inhibitor of the renal outer medullary potassium channel and Kir7.1. Mol Pharmacol 76 (5):1094–1103. https://doi.org/10.1124/mol.109.059840

Lewohl JM, Wilson WR, Mayfield RD, Brozowski SJ, Morrisett RA, Harris RA (1999) G-protein-coupled inwardly rectifying potassium channels are targets of alcohol action. Nat Neurosci 2 (12):1084–1090. https://doi.org/10.1038/16012

Li CG, Cui WY, Wang H (2016a) Sensitivity of KATP channels to cellular metabolic disorders and the underlying structural basis. Acta Pharmacol Sin 37(1):134–142. https://doi.org/10.1038/aps.2015.134

Li D, Chen R, Chung SH (2016b) Molecular dynamics of the honey bee toxin tertiapin binding to Kir3.2. Biophys Chem 219:43–48. https://doi.org/10.1016/j.bpc.2016.09.010

Li N, Wu JX, Ding D, Cheng J, Gao N, Chen L (2017) Structure of a pancreatic ATP-sensitive potassium channel. Cell 168(1–2):101–110.e10. https://doi.org/10.1016/j.cell.2016.12.028

Li D, Jin T, Gazgalis D, Cui M, Logothetis DE (2019) On the mechanism of GIRK2 channel gating by phosphatidylinositol bisphosphate, sodium, and the Gbetagamma dimer. J Biol Chem 294 (49):18934–18948. https://doi.org/10.1074/jbc.RA119.010047

Li Y, Aziz Q, Tinker A (2021) The pharmacology of ATP-sensitive K^+ channels (K_{ATP}). In: Handbook of experimental pharmacology. Springer, Cham

Lin DH, Sterling H, Wang Z et al (2005) ROMK1 channel activity is regulated by monoubiquitination. Proc Natl Acad Sci U S A 102(12):4306–4311. https://doi.org/10.1073/pnas.0409767102

Lin DH, Yue P, Zhang C, Wang WH (2014) MicroRNA-194 (miR-194) regulates ROMK channel activity by targeting intersectin 1. Am J Physiol Renal Physiol 306(1):F53–F60. https://doi.org/10.1152/ajprenal.00349.2013

Linder T, Wang S, Zangerl-Plessl EM, Nichols CG, Stary-Weinzinger A (2015) Molecular dynamics simulations of KirBac1.1 mutants reveal global gating changes of Kir channels. J Chem Inf Model 55(4):814–822. https://doi.org/10.1021/acs.jcim.5b00010

Liou HH, Zhou SS, Huang CL (1999) Regulation of ROMK1 channel by protein kinase A via a phosphatidylinositol 4,5-bisphosphate-dependent mechanism. Proc Natl Acad Sci U S A 96 (10):5820–5825. https://doi.org/10.1073/pnas.96.10.5820

Liss B, Bruns R, Roeper J (1999) Alternative sulfonylurea receptor expression defines metabolic sensitivity of K-ATP channels in dopaminergic midbrain neurons. EMBO J 18(4):833–846. https://doi.org/10.1093/emboj/18.4.833

Liu Y, Liu D, Printzenhoff D, Coghlan MJ, Harris R, Krafte DS (2002) Tenidap, a novel anti-inflammatory agent, is an opener of the inwardly rectifying K+ channel hKir2.3. Eur J Pharmacol 435(2–3):153–160. https://doi.org/10.1016/s0014-2999(01)01590-4

Liu TA, Chang HK, Shieh RC (2012) Revisiting inward rectification: K ions permeate through Kir2.1 channels during high-affinity block by spermidine. J Gen Physiol 139(3):245–259. https://doi.org/10.1085/jgp.201110736

Logothetis DE, Mahajan R, Adney SK et al (2015) Unifying mechanism of controlling Kir3 channel activity by G proteins and phosphoinositides. Int Rev Neurobiol 123:1–26. https://doi.org/10.1016/bs.irn.2015.05.013

Longden TA, Dabertrand F, Koide M et al (2017) Capillary K(+)-sensing initiates retrograde hyperpolarization to increase local cerebral blood flow. Nat Neurosci 20(5):717–726. https://doi.org/10.1038/nn.4533

Lopatin AN, Nichols CG (1996) [K+] dependence of polyamine-induced rectification in inward rectifier potassium channels (IRK1, Kir2.1). J Gen Physiol 108(2):105–113. https://doi.org/10.1085/jgp.108.2.105

Lopatin AN, Makhina EN, Nichols CG (1994) Potassium channel block by cytoplasmic polyamines as the mechanism of intrinsic rectification. Nature 372(6504):366–369. https://doi.org/10.1038/372366a0

Lopatin AN, Makhina EN, Nichols CG (1995) The mechanism of inward rectification of potassium channels: "long-pore plugging" by cytoplasmic polyamines. J Gen Physiol 106(5):923–955. https://doi.org/10.1085/jgp.106.5.923

Lopes CM, Zhang H, Rohacs T, Jin T, Yang J, Logothetis DE (2002) Alterations in conserved Kir channel-PIP2 interactions underlie channelopathies. Neuron 34(6):933–944. https://doi.org/10.1016/s0896-6273(02)00725-0

Lopes CM, Remon JI, Matavel A et al (2007) Protein kinase A modulates PLC-dependent regulation and PIP2-sensitivity of K+ channels. Channels (Austin) 1(2):124–134. https://doi.org/10.4161/chan.4322

Lopez-Izquierdo A, Arechiga-Figueroa IA, Moreno-Galindo EG et al (2011) Mechanisms for Kir channel inhibition by quinacrine: acute pore block of Kir2.x channels and interference in PIP2 interaction with Kir2.x and Kir6.2 channels. Pflugers Arch 462(4):505–517. https://doi.org/10.1007/s00424-011-0995-5

Lorenz JN, Baird NR, Judd LM et al (2002) Impaired renal NaCl absorption in mice lacking the ROMK potassium channel, a model for type II Bartter's syndrome. J Biol Chem 277 (40):37871–37880. https://doi.org/10.1074/jbc.M205627200

Lotsch J, Pruss H, Veh RW, Doehring A (2010) A KCNJ6 (Kir3.2, GIRK2) gene polymorphism modulates opioid effects on analgesia and addiction but not on pupil size. Pharmacogenet Genomics 20(5):291–297. https://doi.org/10.1097/FPC.0b013e3283386bda

Lu Z, MacKinnon R (1994) Electrostatic tuning of Mg2+ affinity in an inward-rectifier K+ channel. Nature 371(6494):243–246. https://doi.org/10.1038/371243a0

Lu M, Wang T, Yan Q et al (2002) Absence of small conductance K+ channel (SK) activity in apical membranes of thick ascending limb and cortical collecting duct in ROMK (Bartter's) knockout mice. J Biol Chem 277(40):37881–37887. https://doi.org/10.1074/jbc.M206644200

Lujan R, Aguado C (2015) Localization and targeting of GIRK channels in mammalian central neurons. Int Rev Neurobiol 123:161–200. https://doi.org/10.1016/bs.irn.2015.05.009

Lujan R, Maylie J, Adelman JP (2009) New sites of action for GIRK and SK channels. Nat Rev Neurosci 10(7):475–480. https://doi.org/10.1038/nrn2668

Lujan R, Marron Fernandez de Velasco E, Aguado C, Wickman K (2014) New insights into the therapeutic potential of Girk channels. Trends Neurosci 37(1):20–29. https://doi.org/10.1016/j.tins.2013.10.006

Lunn ML, Nassirpour R, Arrabit C et al (2007) A unique sorting nexin regulates trafficking of potassium channels via a PDZ domain interaction. Nat Neurosci 10(10):1249–1259. https://doi.org/10.1038/nn1953

Luo X, Pan Z, Shan H et al (2013) MicroRNA-26 governs profibrillatory inward-rectifier potassium current changes in atrial fibrillation. J Clin Invest 123(5):1939–1951. https://doi.org/10.1172/JCI62185

Luscher C, Slesinger PA (2010) Emerging roles for G protein-gated inwardly rectifying potassium (GIRK) channels in health and disease. Nat Rev Neurosci 11(5):301–315. https://doi.org/10.1038/nrn2834

Ma D, Zerangue N, Lin YF et al (2001) Role of ER export signals in controlling surface potassium channel numbers. Science 291(5502):316–319. https://doi.org/10.1126/science.291.5502.316

MacGregor GG, Xu JZ, McNicholas CM, Giebisch G, Hebert SC (1998) Partially active channels produced by PKA site mutation of the cloned renal K+ channel, ROMK2 (kir1.2). Am J Physiol 275(3):F415–F422. https://doi.org/10.1152/ajprenal.1998.275.3.F415

Machida T, Hashimoto N, Kuwahara I et al (2011) Effects of a highly selective acetylcholine-activated K+ channel blocker on experimental atrial fibrillation. Circ Arrhythm Electrophysiol 4 (1):94–102. https://doi.org/10.1161/CIRCEP.110.951608

Madiera F, Park YM, Lee J, Buso N, Gur T, Madhusoodanan N, Basutka P, Tivey ARN, Potter SC, Finn RD, Lopez R (2019) The EMBL-EBI search and sequence analysis tools APIs in 2019. Nucleic Acids Research. W1:W636–641. https://doi.org/10.1093/nar/gkz268

Makary S, Voigt N, Maguy A et al (2011) Differential protein kinase C isoform regulation and increased constitutive activity of acetylcholine-regulated potassium channels in atrial remodeling. Circ Res 109(9):1031–1043. https://doi.org/10.1161/CIRCRESAHA.111.253120

Manis AD, Hodges MR, Staruschenko A, Palygin O (2020) Expression, localization, and functional properties of inwardly rectifying K(+) channels in the kidney. Am J Physiol Renal Physiol 318 (2):F332–F337. https://doi.org/10.1152/ajprenal.00523.2019

Mao J, Wang X, Chen F et al (2004) Molecular basis for the inhibition of G protein-coupled inward rectifier K(+) channels by protein kinase C. Proc Natl Acad Sci U S A 101(4):1087–1092. https://doi.org/10.1073/pnas.0304827101

Marker CL, Stoffel M, Wickman K (2004) Spinal G-protein-gated K+ channels formed by GIRK1 and GIRK2 subunits modulate thermal nociception and contribute to morphine analgesia. J Neurosci 24(11):2806–2812. https://doi.org/10.1523/JNEUROSCI.5251-03.2004

Marmolejo-Murillo LG, Arechiga-Figueroa IA, Cui M et al (2017a) Inhibition of Kir4.1 potassium channels by quinacrine. Brain Res 1663:87–94. https://doi.org/10.1016/j.brainres.2017.03.009

Marmolejo-Murillo LG, Arechiga-Figueroa IA, Moreno-Galindo EG et al (2017b) Chloroquine blocks the Kir4.1 channels by an open-pore blocking mechanism. Eur J Pharmacol 800:40–47. https://doi.org/10.1016/j.ejphar.2017.02.024

Marron Fernandez de Velasco E, McCall N, Wickman K (2015) GIRK channel plasticity and implications for drug addiction. Int Rev Neurobiol 123:201–238. https://doi.org/10.1016/bs.irn.2015.05.011

Martin GM, Kandasamy B, DiMaio F, Yoshioka C, Shyng SL (2017) Anti-diabetic drug binding site in a mammalian KATP channel revealed by Cryo-EM. eLife 6. https://doi.org/10.7554/eLife.31054

Maruyama M, Lin SF, Xie Y et al (2011) Genesis of phase 3 early afterdepolarizations and triggered activity in acquired long-QT syndrome. Circ Arrhythm Electrophysiol 4(1):103–111. https://doi.org/10.1161/CIRCEP.110.959064

Mathiharan YKGI, Zhao Y, Robertson MJ, Skiniotis G, Slesinger PA (2020) Structural basis of GIRK2 channel modulation by cholesterol and PIP_2. bioRxiv. https://doi.org/10.1101/2020.06.04.134544

Matsuda H (1988) Open-state substructure of inwardly rectifying potassium channels revealed by magnesium block in guinea-pig heart cells. J Physiol 397:237–258. https://doi.org/10.1113/jphysiol.1988.sp016998

Matsuda H, Saigusa A, Irisawa H (1987) Ohmic conductance through the inwardly rectifying K channel and blocking by internal Mg2+. Nature 325(7000):156–159. https://doi.org/10.1038/325156a0

Matsuda T, Takeda K, Ito M et al (2005) Atria selective prolongation by NIP-142, an antiarrhythmic agent, of refractory period and action potential duration in guinea pig myocardium. J Pharmacol Sci 98(1):33–40. https://doi.org/10.1254/jphs.fpj04045x

Matsuda T, Ito M, Ishimaru S et al (2006) Blockade by NIP-142, an antiarrhythmic agent, of carbachol-induced atrial action potential shortening and GIRK1/4 channel. J Pharmacol Sci 101 (4):303–310. https://doi.org/10.1254/jphs.fp0060324

Matsuoka S, Sarai N, Kuratomi S, Ono K, Noma A (2003) Role of individual ionic current systems in ventricular cells hypothesized by a model study. Jpn J Physiol 53(2):105–123. https://doi.org/10.2170/jjphysiol.53.105

May LM, Anggono V, Gooch HM et al (2017) G-protein-coupled inwardly rectifying potassium (GIRK) channel activation by the p75 Neurotrophin receptor is required for amyloid beta toxicity. Front Neurosci 11:455. https://doi.org/10.3389/fnins.2017.00455

Mayordomo-Cava J, Yajeya J, Navarro-Lopez JD, Jimenez-Diaz L (2015) Amyloid-beta(25-35) modulates the expression of GirK and KCNQ channel genes in the Hippocampus. PLoS One 10 (7):e0134385. https://doi.org/10.1371/journal.pone.0134385

Mazarati A, Lundstrom L, Sollenberg U, Shin D, Langel U, Sankar R (2006) Regulation of kindling epileptogenesis by hippocampal galanin type 1 and type 2 receptors: the effects of subtype-selective agonists and the role of G-protein-mediated signaling. J Pharmacol Exp Ther 318 (2):700–708. https://doi.org/10.1124/jpet.106.104703

McAllister RE, Noble D (1966) The time and voltage dependence of the slow outward current in cardiac Purkinje fibres. J Physiol 186(3):632–662. https://doi.org/10.1113/jphysiol.1966.sp008060

McCarron JG, Halpern W (1990) Potassium dilates rat cerebral arteries by two independent mechanisms. Am J Physiol 259(3 Pt 2):H902–H908. https://doi.org/10.1152/ajpheart.1990.259.3.H902

McCloskey C, Rada C, Bailey E et al (2014) The inwardly rectifying K+ channel KIR7.1 controls uterine excitability throughout pregnancy. EMBO Mol Med 6(9):1161–1174. https://doi.org/10.15252/emmm.201403944

McKinney LC, Gallin EK (1988) Inwardly rectifying whole-cell and single-channel K currents in the murine macrophage cell line J774.1. J Membr Biol 103(1):41–53. https://doi.org/10.1007/BF01871931

McNicholas CM, Wang W, Ho K, Hebert SC, Giebisch G (1994) Regulation of ROMK1 K+ channel activity involves phosphorylation processes. Proc Natl Acad Sci U S A 91 (17):8077–8081. https://doi.org/10.1073/pnas.91.17.8077

McNicholas CM, MacGregor GG, Islas LD, Yang Y, Hebert SC, Giebisch G (1998) pH-dependent modulation of the cloned renal K+ channel, ROMK. Am J Physiol 275(6):F972–F981. https://doi.org/10.1152/ajprenal.1998.275.6.F972

Medina I, Krapivinsky G, Arnold S, Kovoor P, Krapivinsky L, Clapham DE (2000) A switch mechanism for G beta gamma activation of I(KACh). J Biol Chem 275(38):29709–29716. https://doi.org/10.1074/jbc.M004989200

Meng XY, Zhang HX, Logothetis DE, Cui M (2012) The molecular mechanism by which PIP (2) opens the intracellular G-loop gate of a Kir3.1 channel. Biophys J 102(9):2049–2059. https://doi.org/10.1016/j.bpj.2012.03.050

Meng XY, Liu S, Cui M, Zhou R, Logothetis DE (2016) The molecular mechanism of opening the Helix bundle crossing (HBC) gate of a Kir Channel. Sci Rep 6:29399. https://doi.org/10.1038/srep29399

Mi H, Deerinck TJ, Jones M, Ellisman MH, Schwarz TL (1996) Inwardly rectifying K+ channels that may participate in K+ buffering are localized in microvilli of Schwann cells. J Neurosci 16 (8):2421–2429

Mitrovic I, Margeta-Mitrovic M, Bader S, Stoffel M, Jan LY, Basbaum AI (2003) Contribution of GIRK2-mediated postsynaptic signaling to opiate and alpha 2-adrenergic analgesia and analgesic sex differences. Proc Natl Acad Sci U S A 100(1):271–276. https://doi.org/10.1073/pnas.0136822100

Moran-Zendejas R, Delgado-Ramirez M, Xu J et al (2020) In vitro and in silico characterization of the inhibition of Kir4.1 channels by aminoglycoside antibiotics. Br J Pharmacol. https://doi.org/10.1111/bph.15214

Morgan AD, Carroll ME, Loth AK, Stoffel M, Wickman K (2003) Decreased cocaine self-administration in Kir3 potassium channel subunit knockout mice. Neuropsychopharmacology 28(5):932–938. https://doi.org/10.1038/sj.npp.1300100

Morishige K, Takahashi N, Jahangir A et al (1994) Molecular cloning and functional expression of a novel brain-specific inward rectifier potassium channel. FEBS Lett 346(2–3):251–256. https://doi.org/10.1016/0014-5793(94)00483-8

Morishige K, Inanobe A, Yoshimoto Y et al (1999) Secretagogue-induced exocytosis recruits G protein-gated K+ channels to plasma membrane in endocrine cells. J Biol Chem 274 (12):7969–7974. https://doi.org/10.1074/jbc.274.12.7969

Mukai E, Ishida H, Horie M, Noma A, Seino Y, Takano M (1998) The antiarrhythmic agent cibenzoline inhibits KATP channels by binding to Kir6.2. Biochem Biophys Res Commun 251 (2):477–481. https://doi.org/10.1006/bbrc.1998.9492

Mullner C, Vorobiov D, Bera AK et al (2000) Heterologous facilitation of G protein-activated K(+) channels by beta-adrenergic stimulation via cAMP-dependent protein kinase. J Gen Physiol 115 (5):547–558. https://doi.org/10.1085/jgp.115.5.547

Nagi K, Pineyro G (2014) Kir3 channel signaling complexes: focus on opioid receptor signaling. Front Cell Neurosci 8:186. https://doi.org/10.3389/fncel.2014.00186

Nakamura N, Suzuki Y, Sakuta H, Ookata K, Kawahara K, Hirose S (1999) Inwardly rectifying K+ channel Kir7.1 is highly expressed in thyroid follicular cells, intestinal epithelial cells and choroid plexus epithelial cells: implication for a functional coupling with Na+,K+-ATPase. Biochem J 342(Pt 2):329–336

Nakamura A, Fujita M, Ono H et al (2014) G protein-gated inwardly rectifying potassium (KIR3) channels play a primary role in the antinociceptive effect of oxycodone, but not morphine, at supraspinal sites. Br J Pharmacol 171(1):253–264. https://doi.org/10.1111/bph.12441

Nanjan MJ, Mohammed M, Prashantha Kumar BR, Chandrasekar MJN (2018) Thiazolidinediones as antidiabetic agents: a critical review. Bioorg Chem 77:548–567. https://doi.org/10.1016/j.bioorg.2018.02.009

Nassirpour R, Bahima L, Lalive AL, Luscher C, Lujan R, Slesinger PA (2010) Morphine- and CaMKII-dependent enhancement of GIRK channel signaling in hippocampal neurons. J Neurosci 30(40):13419–13430. https://doi.org/10.1523/JNEUROSCI.2966-10.2010

Nava-Mesa MO, Jimenez-Diaz L, Yajeya J, Navarro-Lopez JD (2013) Amyloid-beta induces synaptic dysfunction through G protein-gated inwardly rectifying potassium channels in the fimbria-CA3 hippocampal synapse. Front Cell Neurosci 7:117. https://doi.org/10.3389/fncel.2013.00117

Nehring RB, Wischmeyer E, Doring F, Veh RW, Sheng M, Karschin A (2000) Neuronal inwardly rectifying K(+) channels differentially couple to PDZ proteins of the PSD-95/SAP90 family. J Neurosci 20(1):156–162

Nelson MT, Cheng H, Rubart M et al (1995) Relaxation of arterial smooth muscle by calcium sparks. Science 270(5236):633–637. https://doi.org/10.1126/science.270.5236.633

Newman EA (1984) Regional specialization of retinal glial cell membrane. Nature 309 (5964):155–157. https://doi.org/10.1038/309155a0
Nichols CG, Lee SJ (2018) Polyamines and potassium channels: a 25-year romance. J Biol Chem 293(48):18779–18788. https://doi.org/10.1074/jbc.TM118.003344
Nichols CG, Makhina EN, Pearson WL, Sha Q, Lopatin AN (1996) Inward rectification and implications for cardiac excitability. Circ Res 78(1):1–7. https://doi.org/10.1161/01.res.78.1.1
Nilius B, Droogmans G (2001) Ion channels and their functional role in vascular endothelium. Physiol Rev 81(4):1415–1459. https://doi.org/10.1152/physrev.2001.81.4.1415
Nilius B, Schwarz G, Droogmans G (1993) Modulation by histamine of an inwardly rectifying potassium channel in human endothelial cells. J Physiol 472:359–371. https://doi.org/10.1113/jphysiol.1993.sp019951
Nishida M, MacKinnon R (2002) Structural basis of inward rectification: cytoplasmic pore of the G protein-gated inward rectifier GIRK1 at 1.8 a resolution. Cell 111(7):957–965. https://doi.org/10.1016/s0092-8674(02)01227-8
Nishida M, Cadene M, Chait BT, MacKinnon R (2007) Crystal structure of a Kir3.1-prokaryotic Kir channel chimera. EMBO J 26(17):4005–4015. https://doi.org/10.1038/sj.emboj.7601828
Nishizawa D, Nagashima M, Katoh R et al (2009) Association between KCNJ6 (GIRK2) gene polymorphisms and postoperative analgesic requirements after major abdominal surgery. PLoS One 4(9):e7060. https://doi.org/10.1371/journal.pone.0007060
Nishizawa D, Fukuda K, Kasai S et al (2014) Association between KCNJ6 (GIRK2) gene polymorphism rs2835859 and post-operative analgesia, pain sensitivity, and nicotine dependence. J Pharmacol Sci 126(3):253–263. https://doi.org/10.1254/jphs.14189fp
Noma A, Trautwein W (1978) Relaxation of the ACh-induced potassium current in the rabbit sinoatrial node cell. Pflugers Arch 377(3):193–200. https://doi.org/10.1007/BF00584272
North RA, Williams JT, Surprenant A, Christie MJ (1987) Mu and delta receptors belong to a family of receptors that are coupled to potassium channels. Proc Natl Acad Sci U S A 84 (15):5487–5491. https://doi.org/10.1073/pnas.84.15.5487
Ohmori H (1978) Inactivation kinetics and steady-state current noise in the anomalous rectifier of tunicate egg cell membranes. J Physiol 281:77–99. https://doi.org/10.1113/jphysiol.1978.sp012410
Ohno Y, Hibino H, Lossin C, Inanobe A, Kurachi Y (2007) Inhibition of astroglial Kir4.1 channels by selective serotonin reuptake inhibitors. Brain Res 1178:44–51. https://doi.org/10.1016/j.brainres.2007.08.018
Ookata K, Tojo A, Suzuki Y et al (2000) Localization of inward rectifier potassium channel Kir7.1 in the basolateral membrane of distal nephron and collecting duct. J Am Soc Nephrol 11 (11):1987–1994
Palygin O, Pochynyuk O, Staruschenko A (2017) Role and mechanisms of regulation of the basolateral Kir 4.1/Kir 5.1K(+) channels in the distal tubules. Acta Physiol (Oxf) 219 (1):260–273. https://doi.org/10.1111/apha.12703
Palygin O, Pochynyuk O, Staruschenko A (2018) Distal tubule basolateral potassium channels: cellular and molecular mechanisms of regulation. Curr Opin Nephrol Hypertens 27(5):373–378. https://doi.org/10.1097/MNH.0000000000000437
Partiseti M, Collura V, Agnel M, Culouscou JM, Graham D (1998) Cloning and characterization of a novel human inwardly rectifying potassium channel predominantly expressed in small intestine. FEBS Lett 434(1–2):171–176. https://doi.org/10.1016/s0014-5793(98)00972-7
Pattnaik BR, Tokarz S, Asuma MP et al (2013) Snowflake vitreoretinal degeneration (SVD) mutation R162W provides new insights into Kir7.1 ion channel structure and function. PLoS One 8(8):e71744. https://doi.org/10.1371/journal.pone.0071744
Pearson WL, Dourado M, Schreiber M, Salkoff L, Nichols CG (1999) Expression of a functional Kir4 family inward rectifier K+ channel from a gene cloned from mouse liver. J Physiol 514 (Pt 3):639–653. https://doi.org/10.1111/j.1469-7793.1999.639ad.x

Pegan S, Arrabit C, Zhou W et al (2005) Cytoplasmic domain structures of Kir2.1 and Kir3.1 show sites for modulating gating and rectification. Nat Neurosci 8(3):279–287. https://doi.org/10.1038/nn1411
Peters M, Jeck N, Reinalter S et al (2002) Clinical presentation of genetically defined patients with hypokalemic salt-losing tubulopathies. Am J Med 112(3):183–190. https://doi.org/10.1016/s0002-9343(01)01086-5
Petit-Jacques J, Sui JL, Logothetis DE (1999) Synergistic activation of G protein-gated inwardly rectifying potassium channels by the betagamma subunits of G proteins and Na(+) and Mg(2+) ions. J Gen Physiol 114(5):673–684. https://doi.org/10.1085/jgp.114.5.673
Pfaffinger PJ, Martin JM, Hunter DD, Nathanson NM, Hille B (1985) GTP-binding proteins couple cardiac muscarinic receptors to a K channel. Nature 317(6037):536–538. https://doi.org/10.1038/317536a0
Podd SJ, Freemantle N, Furniss SS, Sulke N (2016) First clinical trial of specific IKACh blocker shows no reduction in atrial fibrillation burden in patients with paroxysmal atrial fibrillation: pacemaker assessment of BMS 914392 in patients with paroxysmal atrial fibrillation. Europace 18(3):340–346. https://doi.org/10.1093/europace/euv263
Ponce-Balbuena D, Lopez-Izquierdo A, Ferrer T, Rodriguez-Menchaca AA, Arechiga-Figueroa IA, Sanchez-Chapula JA (2009) Tamoxifen inhibits inward rectifier K+ 2.x family of inward rectifier channels by interfering with phosphatidylinositol 4,5-bisphosphate-channel interactions. J Pharmacol Exp Ther 331(2):563–573. https://doi.org/10.1124/jpet.109.156075
Proks P, Ashcroft FM (1997) Phentolamine block of KATP channels is mediated by Kir6.2. Proc Natl Acad Sci U S A 94(21):11716–11720. https://doi.org/10.1073/pnas.94.21.11716
Prystowsky EN, Niazi I, Curtis AB et al (2003) Termination of paroxysmal supraventricular tachycardia by tecadenoson (CVT-510), a novel A1-adenosine receptor agonist. J Am Coll Cardiol 42(6):1098–1102. https://doi.org/10.1016/s0735-1097(03)00987-2
Puissant MM, Muere C, Levchenko V et al (2019) Genetic mutation of Kcnj16 identifies Kir5.1-containing channels as key regulators of acute and chronic pH homeostasis. FASEB J 33 (4):5067–5075. https://doi.org/10.1096/fj.201802257R
Ramos-Hunter SJ, Engers DW, Kaufmann K et al (2013) Discovery and SAR of a novel series of GIRK1/2 and GIRK1/4 activators. Bioorg Med Chem Lett 23(18):5195–5198. https://doi.org/10.1016/j.bmcl.2013.07.002
Ramu Y, Xu Y, Lu Z (2018) A novel high-affinity inhibitor against the human ATP-sensitive Kir6.2 channel. J Gen Physiol 150(7):969–976. https://doi.org/10.1085/jgp.201812017
Raphemot R, Lonergan DF, Nguyen TT et al (2011) Discovery, characterization, and structure-activity relationships of an inhibitor of inward rectifier potassium (Kir) channels with preference for Kir2.3, Kir3.x, and Kir7.1. Front Pharmacol 2:75. https://doi.org/10.3389/fphar.2011.00075
Reeves RH, Irving NG, Moran TH et al (1995) A mouse model for down syndrome exhibits learning and behaviour deficits. Nat Genet 11(2):177–184. https://doi.org/10.1038/ng1095-177
Reimann F, Proks P, Ashcroft FM (2001) Effects of mitiglinide (S 21403) on Kir6.2/SUR1, Kir6.2/SUR2A and Kir6.2/SUR2B types of ATP-sensitive potassium channel. Br J Pharmacol 132 (7):1542–1548. https://doi.org/10.1038/sj.bjp.0703962
Reuveny E, Slesinger PA, Inglese J et al (1994) Activation of the cloned muscarinic potassium channel by G protein beta gamma subunits. Nature 370(6485):143–146. https://doi.org/10.1038/370143a0
Rifkin RA, Moss SJ, Slesinger PA (2017) G protein-gated potassium channels: a link to drug addiction. Trends Pharmacol Sci 38(4):378–392. https://doi.org/10.1016/j.tips.2017.01.007
Rodrigo GC, Standen NB (2005) ATP-sensitive potassium channels. Curr Pharm Des 11 (15):1915–1940. https://doi.org/10.2174/1381612054021015
Rodriguez-Menchaca AA, Navarro-Polanco RA, Ferrer-Villada T et al (2008) The molecular basis of chloroquine block of the inward rectifier Kir2.1 channel. Proc Natl Acad Sci U S A 105 (4):1364–1368. https://doi.org/10.1073/pnas.0708153105
Rodriguez-Soriano J (1998) Bartter and related syndromes: the puzzle is almost solved. Pediatr Nephrol 12(4):315–327. https://doi.org/10.1007/s004670050461

Rohacs T, Lopes CM, Jin T, Ramdya PP, Molnar Z, Logothetis DE (2003) Specificity of activation by phosphoinositides determines lipid regulation of Kir channels. Proc Natl Acad Sci U S A 100 (2):745–750. https://doi.org/10.1073/pnas.0236364100

Romanenko VG, Rothblat GH, Levitan I (2002) Modulation of endothelial inward-rectifier K+ current by optical isomers of cholesterol. Biophys J 83(6):3211–3222. https://doi.org/10.1016/S0006-3495(02)75323-X

Rosenhouse-Dantsker A (2019) Cholesterol binding sites in inwardly rectifying potassium channels. Adv Exp Med Biol 1135:119–138. https://doi.org/10.1007/978-3-030-14265-0_7

Rosenhouse-Dantsker A, Sui JL, Zhao Q et al (2008) A sodium-mediated structural switch that controls the sensitivity of Kir channels to PtdIns(4,5)P(2). Nat Chem Biol 4(10):624–631. https://doi.org/10.1038/nchembio.112

Rougier O, Vassort G, Stampfli R (1967) Voltage clamp experiments on cardiac muscle fibers with the aid of the sucrose gap technic. J Physiol Paris 59(4 Suppl):490

Rui Zhang ZW, Ling B, Liu Y, Liu C (2010) Docking and molecular dynamics studies on the interaction of four imidazoline derivatives with potassium ion channel (Kir6.2). Mol Simul 36 (2):166–174. https://doi.org/10.1080/08927020903141035

Ryan DP, da Silva MR, Soong TW et al (2010) Mutations in potassium channel Kir2.6 cause susceptibility to thyrotoxic hypokalemic periodic paralysis. Cell 140(1):88–98. https://doi.org/10.1016/j.cell.2009.12.024

Sackmann E, Kotulla R, Heiszler FJ (1984) On the role of lipid-bilayer elasticity for the lipid-protein interaction and the indirect protein-protein coupling. Can J Biochem Cell Biol 62 (8):778–788. https://doi.org/10.1139/o84-099

Sadja R, Smadja K, Alagem N, Reuveny E (2001) Coupling Gbetagamma-dependent activation to channel opening via pore elements in inwardly rectifying potassium channels. Neuron 29 (3):669–680. https://doi.org/10.1016/s0896-6273(01)00242-2

Sago H, Carlson EJ, Smith DJ et al (1998) Ts1Cje, a partial trisomy 16 mouse model for down syndrome, exhibits learning and behavioral abnormalities. Proc Natl Acad Sci U S A 95 (11):6256–6261. https://doi.org/10.1073/pnas.95.11.6256

Sakmann B, Noma A, Trautwein W (1983) Acetylcholine activation of single muscarinic K+ channels in isolated pacemaker cells of the mammalian heart. Nature 303(5914):250–253. https://doi.org/10.1038/303250a0

Salvatore L, D'Adamo MC, Polishchuk R, Salmona M, Pessia M (1999) Localization and age-dependent expression of the inward rectifier K+ channel subunit Kir 5.1 in a mammalian reproductive system. FEBS Lett 449(2–3):146–152. https://doi.org/10.1016/s0014-5793(99)00420-2

Sanchez-Rodriguez I, Temprano-Carazo S, Najera A et al (2017) Activation of G-protein-gated inwardly rectifying potassium (Kir3/GirK) channels rescues hippocampal functions in a mouse model of early amyloid-beta pathology. Sci Rep 7(1):14658. https://doi.org/10.1038/s41598-017-15306-8

Sanchez-Rodriguez I, Djebari S, Temprano-Carazo S et al (2020) Hippocampal long-term synaptic depression and memory deficits induced in early amyloidopathy are prevented by enhancing G-protein-gated inwardly rectifying potassium channel activity. J Neurochem 153(3):362–376. https://doi.org/10.1111/jnc.14946

Sattler MB, Williams SK, Neusch C et al (2008) Flupirtine as neuroprotective add-on therapy in autoimmune optic neuritis. Am J Pathol 173(5):1496–1507. https://doi.org/10.2353/ajpath.2008.080491

Sharon D, Vorobiov D, Dascal N (1997) Positive and negative coupling of the metabotropic glutamate receptors to a G protein-activated K+ channel, GIRK, in Xenopus oocytes. J Gen Physiol 109(4):477–490. https://doi.org/10.1085/jgp.109.4.477

Shen W, Tian X, Day M et al (2007) Cholinergic modulation of Kir2 channels selectively elevates dendritic excitability in striatopallidal neurons. Nat Neurosci 10(11):1458–1466. https://doi.org/10.1038/nn1972

Sheng M, Sala C (2001) PDZ domains and the organization of supramolecular complexes. Annu Rev Neurosci 24:1–29. https://doi.org/10.1146/annurev.neuro.24.1.1
Shieh RC, Chang JC, Kuo CC (1999) K+ binding sites and interactions between permeating K+ ions at the external pore mouth of an inward rectifier K+ channel (Kir2.1). J Biol Chem 274 (25):17424–17430. https://doi.org/10.1074/jbc.274.25.17424
Shimoni Y, Clark RB, Giles WR (1992) Role of an inwardly rectifying potassium current in rabbit ventricular action potential. J Physiol 448:709–727. https://doi.org/10.1113/jphysiol.1992.sp019066
Shimura M, Yuan Y, Chang JT et al (2001) Expression and permeation properties of the K(+) channel Kir7.1 in the retinal pigment epithelium. J Physiol 531(Pt 2):329–346. https://doi.org/10.1111/j.1469-7793.2001.0329i.x
Shuck ME, Bock JH, Benjamin CW et al (1994) Cloning and characterization of multiple forms of the human kidney ROM-K potassium channel. J Biol Chem 269(39):24261–24270
Shuck ME, Piser TM, Bock JH, Slightom JL, Lee KS, Bienkowski MJ (1997) Cloning and characterization of two K+ inward rectifier (Kir) 1.1 potassium channel homologs from human kidney (Kir1.2 and Kir1.3). J Biol Chem 272(1):586–593. https://doi.org/10.1074/jbc.272.1.586
Shumilina E, Klocker N, Korniychuk G, Rapedius M, Lang F, Baukrowitz T (2006) Cytoplasmic accumulation of long-chain coenzyme A esters activates KATP and inhibits Kir2.1 channels. J Physiol 575(Pt 2):433–442. https://doi.org/10.1113/jphysiol.2006.111161
Shyng S, Ferrigni T, Nichols CG (1997) Control of rectification and gating of cloned KATP channels by the Kir6.2 subunit. J Gen Physiol 110(2):141–153. https://doi.org/10.1085/jgp.110.2.141
Shyng SL, Barbieri A, Gumusboga A et al (2000) Modulation of nucleotide sensitivity of ATP-sensitive potassium channels by phosphatidylinositol-4-phosphate 5-kinase. Proc Natl Acad Sci U S A 97(2):937–941. https://doi.org/10.1073/pnas.97.2.937
Siarey RJ, Carlson EJ, Epstein CJ, Balbo A, Rapoport SI, Galdzicki Z (1999) Increased synaptic depression in the Ts65Dn mouse, a model for mental retardation in down syndrome. Neuropharmacology 38(12):1917–1920. https://doi.org/10.1016/s0028-3908(99)00083-0
Signorini S, Liao YJ, Duncan SA, Jan LY, Stoffel M (1997) Normal cerebellar development but susceptibility to seizures in mice lacking G protein-coupled, inwardly rectifying K+ channel GIRK2. Proc Natl Acad Sci U S A 94(3):923–927. https://doi.org/10.1073/pnas.94.3.923
Silver MR, DeCoursey TE (1990) Intrinsic gating of inward rectifier in bovine pulmonary artery endothelial cells in the presence or absence of internal Mg2+. J Gen Physiol 96(1):109–133. https://doi.org/10.1085/jgp.96.1.109
Sims SM, Dixon SJ (1989) Inwardly rectifying K+ current in osteoclasts. Am J Physiol 256(6 Pt 1): C1277–C1282. https://doi.org/10.1152/ajpcell.1989.256.6.C1277
Sjogren B (2011) Regulator of G protein signaling proteins as drug targets: current state and future possibilities. Adv Pharmacol 62:315–347. https://doi.org/10.1016/B978-0-12-385952-5.00002-6
Smith SB, Marker CL, Perry C et al (2008) Quantitative trait locus and computational mapping identifies Kcnj9 (GIRK3) as a candidate gene affecting analgesia from multiple drug classes. Pharmacogenet Genomics 18(3):231–241. https://doi.org/10.1097/FPC.0b013e3282f55ab2
Soe R, Andreasen M, Klaerke DA (2009) Modulation of Kir4.1 and Kir4.1-Kir5.1 channels by extracellular cations. Biochim Biophys Acta 1788(9):1706–1713. https://doi.org/10.1016/j.bbamem.2009.07.002
Soejima M, Noma A (1984) Mode of regulation of the ACh-sensitive K-channel by the muscarinic receptor in rabbit atrial cells. Pflugers Arch 400(4):424–431. https://doi.org/10.1007/BF00587544
Sohn JW, Lim A, Lee SH, Ho WK (2007) Decrease in PIP(2) channel interactions is the final common mechanism involved in PKC- and arachidonic acid-mediated inhibitions of GABA(B)-activated K+ current. J Physiol 582(Pt 3):1037–1046. https://doi.org/10.1113/jphysiol.2007.137265

Sonkusare SK, Dalsgaard T, Bonev AD, Nelson MT (2016) Inward rectifier potassium (Kir2.1) channels as end-stage boosters of endothelium-dependent vasodilators. J Physiol 594 (12):3271–3285. https://doi.org/10.1113/JP271652

Standen NB, Stanfield PR (1978) Inward rectification in skeletal muscle: a blocking particle model. Pflugers Arch 378(2):173–176. https://doi.org/10.1007/BF00584452

Stanfield PR, Nakajima S, Nakajima Y (2002) Constitutively active and G-protein coupled inward rectifier K+ channels: Kir2.0 and Kir3.0. Rev Physiol Biochem Pharmacol 145:47–179. https://doi.org/10.1007/BFb0116431

Stansfeld PJ, Hopkinson R, Ashcroft FM, Sansom MS (2009) PIP(2)-binding site in Kir channels: definition by multiscale biomolecular simulations. Biochemistry 48(46):10926–10933. https://doi.org/10.1021/bi9013193

Stevens EB, Shah BS, Pinnock RD, Lee K (1999) Bombesin receptors inhibit G protein-coupled inwardly rectifying K+ channels expressed in Xenopus oocytes through a protein kinase C-dependent pathway. Mol Pharmacol 55(6):1020–1027

Su S, Ohno Y, Lossin C, Hibino H, Inanobe A, Kurachi Y (2007) Inhibition of astroglial inwardly rectifying Kir4.1 channels by a tricyclic antidepressant, nortriptyline. J Pharmacol Exp Ther 320 (2):573–580. https://doi.org/10.1124/jpet.106.112094

Sui JL, Chan KW, Logothetis DE (1996) Na+ activation of the muscarinic K+ channel by a G-protein-independent mechanism. J Gen Physiol 108(5):381–391. https://doi.org/10.1085/jgp.108.5.381

Sui JL, Petit-Jacques J, Logothetis DE (1998) Activation of the atrial KACh channel by the betagamma subunits of G proteins or intracellular Na+ ions depends on the presence of phosphatidylinositol phosphates. Proc Natl Acad Sci U S A 95(3):1307–1312. https://doi.org/10.1073/pnas.95.3.1307

Sun HS, Feng ZP (2013) Neuroprotective role of ATP-sensitive potassium channels in cerebral ischemia. Acta Pharmacol Sin 34(1):24–32. https://doi.org/10.1038/aps.2012.138

Sun H, Liu X, Xiong Q, Shikano S, Li M (2006a) Chronic inhibition of cardiac Kir2.1 and HERG potassium channels by celastrol with dual effects on both ion conductivity and protein trafficking. J Biol Chem 281(9):5877–5884. https://doi.org/10.1074/jbc.M600072200

Sun HS, Feng ZP, Miki T, Seino S, French RJ (2006b) Enhanced neuronal damage after ischemic insults in mice lacking Kir6.2-containing ATP-sensitive K+ channels. J Neurophysiol 95 (4):2590–2601. https://doi.org/10.1152/jn.00970.2005

Suzuki M, Li RA, Miki T et al (2001) Functional roles of cardiac and vascular ATP-sensitive potassium channels clarified by Kir6.2-knockout mice. Circ Res 88(6):570–577. https://doi.org/10.1161/01.res.88.6.570

Swale DR, Sheehan JH, Banerjee S et al (2015) Computational and functional analyses of a small-molecule binding site in ROMK. Biophys J 108(5):1094–1103. https://doi.org/10.1016/j.bpj.2015.01.022

Swale DR, Kurata H, Kharade SV et al (2016) ML418: the first selective, sub-micromolar pore blocker of Kir7.1 potassium channels. ACS Chem Nerosci 7(7):1013–1023. https://doi.org/10.1021/acschemneuro.6b00111

Swanberg MM, Tractenberg RE, Mohs R, Thal LJ, Cummings JL (2004) Executive dysfunction in Alzheimer disease. Arch Neurol 61(4):556–560. https://doi.org/10.1001/archneur.61.4.556

Tabuchi Y, Yashiro H, Hoshina S, Asano S, Takeguchi N (2001) Cibenzoline, an ATP-sensitive K (+) channel blocker, binds to the K(+)-binding site from the cytoplasmic side of gastric H(+),K (+)-ATPase. Br J Pharmacol 134(8):1655–1662. https://doi.org/10.1038/sj.bjp.0704422

Taglialatela M, Ficker E, Wible BA, Brown AM (1995) C-terminus determinants for Mg2+ and polyamine block of the inward rectifier K+ channel IRK1. EMBO J 14(22):5532–5541

Takahashi T (1990) Inward rectification in neonatal rat spinal motoneurones. J Physiol 423:47–62. https://doi.org/10.1113/jphysiol.1990.sp018010

Takumi T, Ishii T, Horio Y et al (1995) A novel ATP-dependent inward rectifier potassium channel expressed predominantly in glial cells. J Biol Chem 270(27):16339–16346. https://doi.org/10.1074/jbc.270.27.16339

Tamargo J, Caballero R, Gomez R, Valenzuela C, Delpon E (2004) Pharmacology of cardiac potassium channels. Cardiovasc Res 62(1):9–33. https://doi.org/10.1016/j.cardiores.2003.12.026

Tang P, Eckenhoff R (2018) Recent progress on the molecular pharmacology of propofol. F1000Res 7:123. https://doi.org/10.12688/f1000research.12502.1

Tao X, Avalos JL, Chen J, MacKinnon R (2009) Crystal structure of the eukaryotic strong inward-rectifier K+ channel Kir2.2 at 3.1 a resolution. Science 326(5960):1668–1674. https://doi.org/10.1126/science.1180310

Tawil R, Ptacek LJ, Pavlakis SG et al (1994) Andersen's syndrome: potassium-sensitive periodic paralysis, ventricular ectopy, and dysmorphic features. Ann Neurol 35(3):326–330. https://doi.org/10.1002/ana.410350313

Tellez-Zenteno JF, Hernandez-Ronquillo L (2012) A review of the epidemiology of temporal lobe epilepsy. Epilepsy Res Treat 2012:630853. https://doi.org/10.1155/2012/630853

Teramoto N (2006) Pharmacological profile of U-37883A, a channel blocker of smooth muscle-type ATP-sensitive K channels. Cardiovasc Drug Rev 24(1):25–32. https://doi.org/10.1111/j.1527-3466.2006.00025.x

Tinker A, Aziz Q, Li Y, Specterman M (2018) ATP-sensitive potassium channels and their physiological and pathophysiological roles. Compr Physiol 8(4):1463–1511. https://doi.org/10.1002/cphy.c170048

Tipps ME, Buck KJ (2015) GIRK channels: a potential link between learning and addiction. Int Rev Neurobiol 123:239–277. https://doi.org/10.1016/bs.irn.2015.05.012

Trautwein W, Dudel J (1958) Mechanism of membrane effect of acetylcholine on myocardial fibers. Pflugers Arch Gesamte Physiol Menschen Tiere 266(3):324–334. https://doi.org/10.1007/BF00416781

Tsai TD, Shuck ME, Thompson DP, Bienkowski MJ, Lee KS (1995) Intracellular H+ inhibits a cloned rat kidney outer medulla K+ channel expressed in Xenopus oocytes. Am J Physiol 268 (5 Pt 1):C1173–C1178. https://doi.org/10.1152/ajpcell.1995.268.5.C1173

Tucker SJ, Imbrici P, Salvatore L, D'Adamo MC, Pessia M (2000) pH dependence of the inwardly rectifying potassium channel, Kir5.1, and localization in renal tubular epithelia. J Biol Chem 275(22):16404–16407. https://doi.org/10.1074/jbc.C000127200

Ulens C, Daenens P, Tytgat J (1999) The dual modulation of GIRK1/GIRK2 channels by opioid receptor ligands. Eur J Pharmacol 385(2–3):239–245. https://doi.org/10.1016/s0014-2999(99)00736-0

Vasas A, Forgo P, Orvos P et al (2016) Myrsinane, premyrsinane, and cyclomyrsinane diterpenes from *Euphorbia falcata* as potassium ion channel inhibitors with selective G protein-activated inwardly rectifying ion channel (GIRK) blocking effects. J Nat Prod 79(8):1990–2004. https://doi.org/10.1021/acs.jnatprod.6b00260

Vaughn J, Wolford JK, Prochazka M, Permana PA (2000) Genomic structure and expression of human KCNJ9 (Kir3.3/GIRK3). Biochem Biophys Res Commun 274(2):302–309. https://doi.org/10.1006/bbrc.2000.3136

Vera E, Cornejo I, Burgos J, Niemeyer MI, Sepulveda FV, Cid LP (2019) A novel Kir7.1 splice variant expressed in various mouse tissues shares organisational and functional properties with human Leber amaurosis-causing mutations of this K(+) channel. Biochem Biophys Res Commun 514(3):574–579. https://doi.org/10.1016/j.bbrc.2019.04.169

Victoria NC, Marron Fernandez de Velasco E, Ostrovskaya O et al (2016) G protein-gated K(+) channel ablation in forebrain pyramidal neurons selectively impairs fear learning. Biol Psychiatry 80(10):796–806. https://doi.org/10.1016/j.biopsych.2015.10.004

Vo BN, Abney KK, Anderson A et al (2019) VU0810464, a non-urea G protein-gated inwardly rectifying K(+) (Kir 3/GIRK) channel activator, exhibits enhanced selectivity for neuronal Kir 3 channels and reduces stress-induced hyperthermia in mice. Br J Pharmacol 176 (13):2238–2249. https://doi.org/10.1111/bph.14671

Voigt N, Abu-Taha I, Heijman J, Dobrev D (2014) Constitutive activity of the acetylcholine-activated potassium current IK,ACh in cardiomyocytes. Adv Pharmacol 70:393–409. https://doi.org/10.1016/B978-0-12-417197-8.00013-4

von Beckerath N, Dittrich M, Klieber HG, Daut J (1996) Inwardly rectifying K+ channels in freshly dissociated coronary endothelial cells from guinea-pig heart. J Physiol 491(Pt 2):357–365. https://doi.org/10.1113/jphysiol.1996.sp021221

Wang HR, Wu M, Yu H et al (2011) Selective inhibition of the K(ir)2 family of inward rectifier potassium channels by a small molecule probe: the discovery, SAR, and pharmacological characterization of ML133. ACS Chem Biol 6(8):845–856. https://doi.org/10.1021/cb200146a

Wang F, Olson EM, Shyng SL (2012) Role of Derlin-1 protein in proteostasis regulation of ATP-sensitive potassium channels. J Biol Chem 287(13):10482–10493. https://doi.org/10.1074/jbc.M111.312223

Weaver CD, Harden D, Dworetzky SI, Robertson B, Knox RJ (2004) A thallium-sensitive, fluorescence-based assay for detecting and characterizing potassium channel modulators in mammalian cells. J Biomol Screen 9(8):671–677. https://doi.org/10.1177/1087057104268749

Wellman GC, Bevan JA (1995) Barium inhibits the endothelium-dependent component of flow but not acetylcholine-induced relaxation in isolated rabbit cerebral arteries. J Pharmacol Exp Ther 274(1):47–53

Wen W, Wu W, Romaine IM et al (2013) Discovery of 'molecular switches' within a GIRK activator scaffold that afford selective GIRK inhibitors. Bioorg Med Chem Lett 23 (16):4562–4566. https://doi.org/10.1016/j.bmcl.2013.06.023

Whorton MR, MacKinnon R (2011) Crystal structure of the mammalian GIRK2 K+ channel and gating regulation by G proteins, PIP2, and sodium. Cell 147(1):199–208. https://doi.org/10.1016/j.cell.2011.07.046

Whorton MR, MacKinnon R (2013) X-ray structure of the mammalian GIRK2-betagamma G-protein complex. Nature 498(7453):190–197. https://doi.org/10.1038/nature12241

Wible BA, Taglialatela M, Ficker E, Brown AM (1994) Gating of inwardly rectifying K+ channels localized to a single negatively charged residue. Nature 371(6494):246–249. https://doi.org/10.1038/371246a0

Wickman K, Nemec J, Gendler SJ, Clapham DE (1998) Abnormal heart rate regulation in GIRK4 knockout mice. Neuron 20(1):103–114. https://doi.org/10.1016/s0896-6273(00)80438-9

Wickman K, Karschin C, Karschin A, Picciotto MR, Clapham DE (2000) Brain localization and behavioral impact of the G-protein-gated K+ channel subunit GIRK4. J Neurosci 20 (15):5608–5615

Wieting JM, Vadukoot AK, Sharma S et al (2017) Discovery and characterization of 1H-pyrazol-5-yl-2-phenylacetamides as novel, non-urea-containing GIRK1/2 potassium channel activators. ACS Chem Nerosci 8(9):1873–1879. https://doi.org/10.1021/acschemneuro.7b00217

Williams JT, Colmers WF, Pan ZZ (1988) Voltage- and ligand-activated inwardly rectifying currents in dorsal raphe neurons in vitro. J Neurosci 8(9):3499–3506

Witkowski G, Szulczyk B, Rola R, Szulczyk P (2008) D(1) dopaminergic control of G protein-dependent inward rectifier K(+) (GIRK)-like channel current in pyramidal neurons of the medial prefrontal cortex. Neuroscience 155(1):53–63. https://doi.org/10.1016/j.neuroscience.2008.05.021

Wu JX, Ding D, Wang M, Kang Y, Zeng X, Chen L (2018) Ligand binding and conformational changes of SUR1 subunit in pancreatic ATP-sensitive potassium channels. Protein Cell 9 (6):553–567. https://doi.org/10.1007/s13238-018-0530-y

Wu P, Gao ZX, Zhang DD, Su XT, Wang WH, Lin DH (2019) Deletion of Kir5.1 impairs renal ability to excrete potassium during increased dietary potassium intake. J Am Soc Nephrol 30 (8):1425–1438. https://doi.org/10.1681/ASN.2019010025

Xia M, Jin Q, Bendahhou S, He Y, Larroque MM, Chen Y, Zhou Q, Yang Y, Liu Y, Liu B, Zhu Q, Zhou Y, Lin J, Liang B, Li L, Dong X, Pan Z, Wang R, Wan H, Qiu W, Xu W, Eurlings P, Barhanin J, Chen Y (2005) A Kir2.1 gain-of-function mutation underlies familial atrial fibrillation. Biochem Biophys Res Commun 332(4):1012–9

Xie LH, John SA, Weiss JN (2002) Spermine block of the strong inward rectifier potassium channel Kir2.1: dual roles of surface charge screening and pore block. J Gen Physiol 120(1):53–66. https://doi.org/10.1085/jgp.20028576
Xie LH, John SA, Weiss JN (2003) Inward rectification by polyamines in mouse Kir2.1 channels: synergy between blocking components. J Physiol 550(Pt 1):67–82. https://doi.org/10.1113/jphysiol.2003.043117
Xu ZC, Yang Y, Hebert SC (1996) Phosphorylation of the ATP-sensitive, inwardly rectifying K+ channel, ROMK, by cyclic AMP-dependent protein kinase. J Biol Chem 271(16):9313–9319. https://doi.org/10.1074/jbc.271.16.9313
Xu Y, Shin HG, Szep S, Lu Z (2009) Physical determinants of strong voltage sensitivity of K(+) channel block. Nat Struct Mol Biol 16(12):1252–1258. https://doi.org/10.1038/nsmb.1717
Xu Y, Cantwell L, Molosh AI et al (2020) The small molecule GAT1508 activates brain-specific GIRK1/2 channel heteromers and facilitates conditioned fear extinction in rodents. J Biol Chem 295(11):3614–3634. https://doi.org/10.1074/jbc.RA119.011527
Yamada M, Inanobe A, Kurachi Y (1998) G protein regulation of potassium ion channels. Pharmacol Rev 50(4):723–760
Yamakura T, Lewohl JM, Harris RA (2001) Differential effects of general anesthetics on G protein-coupled inwardly rectifying and other potassium channels. Anesthesiology 95(1):144–153. https://doi.org/10.1097/00000542-200107000-00025
Yamamoto W, Hashimoto N, Matsuura J et al (2014) Effects of the selective KACh channel blocker NTC-801 on atrial fibrillation in a canine model of atrial tachypacing: comparison with class Ic and III drugs. J Cardiovasc Pharmacol 63(5):421–427. https://doi.org/10.1097/FJC.0000000000000065
Yan F, Lin CW, Weisiger E, Cartier EA, Taschenberger G, Shyng SL (2004) Sulfonylureas correct trafficking defects of ATP-sensitive potassium channels caused by mutations in the sulfonylurea receptor. J Biol Chem 279(12):11096–11105. https://doi.org/10.1074/jbc.M312810200
Yan FF, Pratt EB, Chen PC et al (2010) Role of Hsp90 in biogenesis of the beta-cell ATP-sensitive potassium channel complex. Mol Biol Cell 21(12):1945–1954. https://doi.org/10.1091/mbc.E10-02-0116
Yang J, Jan YN, Jan LY (1995) Control of rectification and permeation by residues in two distinct domains in an inward rectifier K+ channel. Neuron 14(5):1047–1054. https://doi.org/10.1016/0896-6273(95)90343-7
Yang B, Lin H, Xiao J et al (2007) The muscle-specific microRNA miR-1 regulates cardiac arrhythmogenic potential by targeting GJA1 and KCNJ2. Nat Med 13(4):486–491. https://doi.org/10.1038/nm1569
Yang Y, Yang Y, Liang B et al (2010) Identification of a Kir3.4 mutation in congenital long QT syndrome. Am J Hum Genet 86(6):872–880. https://doi.org/10.1016/j.ajhg.2010.04.017
Yang HQ, Martinez-Ortiz W, Hwang J, Fan X, Cardozo TJ, Coetzee WA (2020) Palmitoylation of the KATP channel Kir6.2 subunit promotes channel opening by regulating PIP2 sensitivity. Proc Natl Acad Sci U S A 117(19):10593–10602. https://doi.org/10.1073/pnas.1918088117
Yoo D, Fang L, Mason A, Kim BY, Welling PA (2005) A phosphorylation-dependent export structure in ROMK (Kir 1.1) channel overrides an endoplasmic reticulum localization signal. J Biol Chem 280(42):35281–35289. https://doi.org/10.1074/jbc.M504836200
Yoshimoto Y, Fukuyama Y, Horio Y, Inanobe A, Gotoh M, Kurachi Y (1999) Somatostatin induces hyperpolarization in pancreatic islet alpha cells by activating a G protein-gated K+ channel. FEBS Lett 444(2–3):265–269. https://doi.org/10.1016/s0014-5793(99)00076-9
Yow TT, Pera E, Absalom N et al (2011) Naringin directly activates inwardly rectifying potassium channels at an overlapping binding site to tertiapin-Q. Br J Pharmacol 163(5):1017–1033. https://doi.org/10.1111/j.1476-5381.2011.01315.x
Yu L, Jin X, Cui N et al (2012) Rosiglitazone selectively inhibits K(ATP) channels by acting on the K(IR) 6 subunit. Br J Pharmacol 167(1):26–36. https://doi.org/10.1111/j.1476-5381.2012.01934.x

Zaks-Makhina E, Kim Y, Aizenman E, Levitan ES (2004) Novel neuroprotective K+ channel inhibitor identified by high-throughput screening in yeast. Mol Pharmacol 65(1):214–219. https://doi.org/10.1124/mol.65.1.214

Zaks-Makhina E, Li H, Grishin A, Salvador-Recatala V, Levitan ES (2009) Specific and slow inhibition of the kir2.1 K+ channel by gambogic acid. J Biol Chem 284(23):15432–15438. https://doi.org/10.1074/jbc.M901586200

Zald DH (2003) The human amygdala and the emotional evaluation of sensory stimuli. Brain Res Brain Res Rev 41(1):88–123. https://doi.org/10.1016/s0165-0173(02)00248-5

Zangerl-Plessl EM, Qile M, Bloothooft M, Stary-Weinzinger A, van der Heyden MAG (2019) Disease associated mutations in KIR proteins linked to aberrant inward rectifier channel trafficking. Biomol Ther 9(11). https://doi.org/10.3390/biom9110650

Zangerl-Plessl EM, Lee SJ, Maksaev G et al (2020) Atomistic basis of opening and conduction in mammalian inward rectifier potassium (Kir2.2) channels. J Gen Physiol 152(1). https://doi.org/10.1085/jgp.201912422

Zaritsky JJ, Eckman DM, Wellman GC, Nelson MT, Schwarz TL (2000) Targeted disruption of Kir2.1 and Kir2.2 genes reveals the essential role of the inwardly rectifying K(+) current in K(+)-mediated vasodilation. Circ Res 87(2):160–166. https://doi.org/10.1161/01.res.87.2.160

Zaritsky JJ, Redell JB, Tempel BL, Schwarz TL (2001) The consequences of disrupting cardiac inwardly rectifying K(+) current (I(K1)) as revealed by the targeted deletion of the murine Kir2.1 and Kir2.2 genes. J Physiol 533(Pt 3):697–710. https://doi.org/10.1111/j.1469-7793.2001.t01-1-00697.x

Zerangue N, Schwappach B, Jan YN, Jan LY (1999) A new ER trafficking signal regulates the subunit stoichiometry of plasma membrane K(ATP) channels. Neuron 22(3):537–548. https://doi.org/10.1016/s0896-6273(00)80708-4

Zhang H, He C, Yan X, Mirshahi T, Logothetis DE (1999) Activation of inwardly rectifying K+ channels by distinct PtdIns(4,5)P2 interactions. Nat Cell Biol 1(3):183–188. https://doi.org/10.1038/11103

Zhang W, Zhang X, Wang H, Sharma AK, Edwards AO, Hughes BA (2013) Characterization of the R162W Kir7.1 mutation associated with snowflake vitreoretinopathy. Am J Physiol Cell Physiol 304(5):C440–C449. https://doi.org/10.1152/ajpcell.00363.2012

Zhang J, Xia Y, Xu Z, Deng X (2016) Propofol suppressed hypoxia/reoxygenation-induced apoptosis in HBVSMC by regulation of the expression of Bcl-2, Bax, Caspase3, Kir6.1, and p-JNK. Oxid Med Cell Longev 2016:1518738. https://doi.org/10.1155/2016/1518738

Zhao Y, Ung PM, Zahoranszky-Kohalmi G et al (2020) Identification of a G-protein-independent activator of GIRK channels. Cell Rep 31(11):107770. https://doi.org/10.1016/j.celrep.2020.107770

Zhou H, Tate SS, Palmer LG (1994) Primary structure and functional properties of an epithelial K channel. Am J Physiol 266(3 Pt 1):C809–C824. https://doi.org/10.1152/ajpcell.1994.266.3.C809

Zhou W, Arrabit C, Choe S, Slesinger PA (2001) Mechanism underlying bupivacaine inhibition of G protein-gated inwardly rectifying K+ channels. Proc Natl Acad Sci U S A 98(11):6482–6487. https://doi.org/10.1073/pnas.111447798

Zhou Q, Chen PC, Devaraneni PK, Martin GM, Olson EM, Shyng SL (2014) Carbamazepine inhibits ATP-sensitive potassium channel activity by disrupting channel response to MgADP. Channels (Austin) 8(4):376–382. https://doi.org/10.4161/chan.29117

Zingman LV, Zhu Z, Sierra A et al (2011) Exercise-induced expression of cardiac ATP-sensitive potassium channels promotes action potential shortening and energy conservation. J Mol Cell Cardiol 51(1):72–81. https://doi.org/10.1016/j.yjmcc.2011.03.010

The Pharmacology of ATP-Sensitive K^+ Channels (K_{ATP})

Yiwen Li, Qadeer Aziz, and Andrew Tinker

Contents

Abstract

ATP-sensitive K^+ channels (K_{ATP}) are inwardly-rectifying potassium channels, broadly expressed throughout the body. K_{ATP} is regulated by adenine nucleotides, characteristically being activated by falling ATP and rising ADP levels thus playing an important physiological role by coupling cellular metabolism with membrane excitability. The hetero-octameric channel complex is

Y. Li · Q. Aziz · A. Tinker (✉)
The Heart Centre, Centre for Clinical Pharmacology, William Harvey Research Centre, Queen Mary University of London, London, UK
e-mail: a.tinker@qmul.a.uk

N. Gamper, K. Wang (eds.), *Pharmacology of Potassium Channels*,
Handbook of Experimental Pharmacology 267, https://doi.org/10.1007/164_2021_466

formed of 4 pore-forming inward rectifier Kir6.x subunits (Kir6.1 or Kir6.2) and 4 regulatory sulfonylurea receptor subunits (SUR1, SUR2A, or SUR2B). These subunits can associate in various tissue-specific combinations to form functional K_{ATP} channels with distinct electrophysiological and pharmacological properties. K_{ATP} channels play many important physiological roles and mutations in channel subunits can result in diseases such as disorders of insulin handling, cardiac arrhythmia, cardiomyopathy, and neurological abnormalities. The tissue-specific expression of K_{ATP} channel subunits coupled with their rich and diverse pharmacology makes K_{ATP} channels attractive therapeutic targets in the treatment of endocrine and cardiovascular diseases.

Keywords

Cardiac arrhythmia · Diabetes · K_{ATP} · Sulfonylureas · SUR

1 Introduction

ATP-sensitive K^+ channels (K_{ATP}) were first described in heart muscle in the early 1980s, where treatment with metabolic poisons or hypoxia evoked an outward K^+ current that was inhibited by ATP (Noma 1983). Since then, they have been described in a variety of other tissues, most prominently in the cardiovascular, endocrine, and nervous systems including in pancreatic β-cells (Ashcroft et al. 1984), neurones (Ashford et al. 1988), skeletal muscle (Spruce et al. 1985), smooth muscle (Aziz et al. 2014; Standen et al. 1989), and endothelium (Aziz et al. 2017; Li et al. 2020).

K_{ATP} channels open in response to changes in cellular metabolism, activated by a decline in intracellular ATP and/or an increase in ADP levels, and thus play an important functional role by linking cellular metabolism to membrane excitability. In addition to regulation by changes in the ATP/ADP ratio, K_{ATP} channels are also modulated by a number of cell signalling pathways. They have an established pharmacological profile and some compounds are in routine clinical use. K_{ATP} channels have a number of important physiological functions, especially in the cardiovascular and endocrine systems such as regulation of insulin release, cardioprotection, and control of blood pressure. Extensive reviews of K_{ATP} channel function have been already published (Tinker et al. 2014, 2018). In this chapter, we will briefly discuss the structure and regulation of K_{ATP} channels, their physiological roles and pathophysiology in human diseases, and their pharmacology with a focus on their therapeutic use.

2 The Structure and Regulation of K_{ATP} Channels

The K_{ATP} channel complex is constituted of four pore-forming subunits (Kir6.1 or Kir6.2) and four sulphonylurea receptor subunits (members of the ATP binding cassette family of proteins; SUR1, SUR2A, and SUR2B) to form an octameric

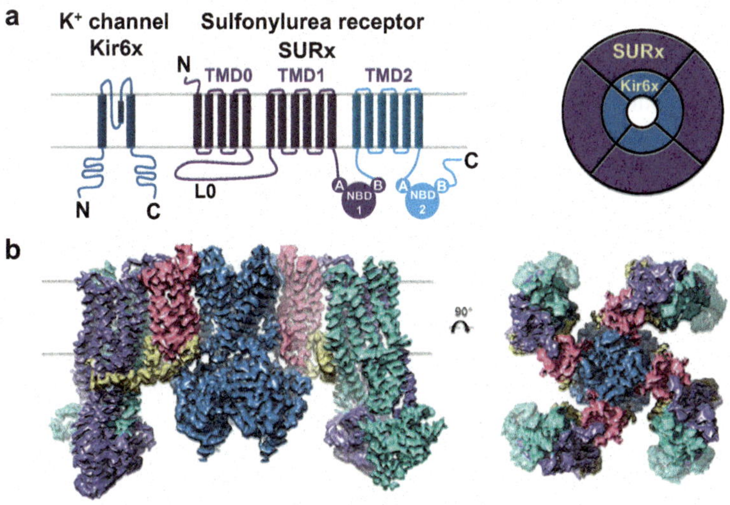

Fig. 1 Molecular composition of a K_{ATP} channel. (**a**) Four pore-forming Kir6.x subunits (belonging to the inward-rectifying K^+ channel family (Kir)) and four regulatory sulphonylurea receptor subunits (belonging to the ATP binding cassette (ABC) family of proteins) form a functional K_{ATP} channel. Kir6x consists of 2 transmembrane domains (M1 and M2), a pore-forming region (H5) with the K^+ selectivity sequence and intracellular N and C termini. SUR consists of 3 transmembrane domains (TMDs) composed of 5, 6, and 6 transmembrane segments, respectively. L0, the intracellular loop between TMD0 and TMD1, provides the physical interaction with Kir6x. Two nucleotide binding domains (NBD1 and NDB2) comprised of Walker A and B nucleotide binding motifs provide the binding sites for magnesium-complexed adenine nucleotides. (**b**) A side view (left) of the cryo-EM density map of the pancreatic K_{ATP} channel (3.63 Å resolution) and the extracellular view (right) of the channel. The position of the membrane is indicated by the grey bars. (This figure is reproduced with permission from Martin et al. 2017)

channel complex (Fig. 1). The association of a particular SUR with a specific Kir6.x subunit constitutes the K_{ATP} current in a specific tissue. The Kir6.x subunits are targets for inhibition by ATP and the SUR proteins for activation by MgADP.

The Kir6.x subunits have two transmembrane domains (M1 and M2), a pore-forming region (H5) with the K^+ selectivity sequence (GYG or GFG) and intracellular N and C termini (Tinker et al. 1996). SUR subunits consist of three transmembrane domains (TMD 0, 1, and 2) comprised of five, six, and six membrane spanning helices, respectively (Fig. 1).Each of these domains is connected by cytosolic linkers; N-terminus is extracellular while C-terminus resides intracellularly (Conti et al. 2001). Physical interaction with Kir6.1.x subunits is via the intracellular loop between TMD0 and TMD1, L0. Two nucleotide binding domains (NBD1 and NDB2), comprised of Walker A and B nucleotide binding motifs in the TMD1-TMD2 linker, and C-terminus provide the binding sites for magnesium-complexed adenine nucleotides.

K_{ATP} channels are highly selective for potassium (P_{Na}/P_K ~ 0.01) and display diverse unitary conductances in different tissues, for example, 70–90 pS in cardiac muscle, 55–75 pS in skeletal muscle, and 50–90 pS in pancreatic β-cells (Ashcroft

1988; Hibino et al. 2010; Quayle et al. 1997). However, smooth muscle K_{ATP} channels have unique properties; they have a lower (~35 pS) single-channel conductance and they absolutely require cytosolic nucleotide diphosphates being present in the solution to be active and this has led to the moniker "K_{NDP}" current in some of the literature (Beech et al. 1993). In addition, these channels are generally less sensitive to ATP inhibition than Kir6.2-containing channels (Beech et al. 1993; Cui et al. 2002). Furthermore, activation by ADP is dependent on the presence of magnesium, without which ADP inhibits the channels (Findlay 1987).

As well as direct regulation of K_{ATP} channels by ATP/ADP, other cell signalling pathways can also modulate channel activity. Membrane phosphoinositides, notably phosphatidylinositol 4,5-bisphosphate have been shown to antagonise ATP inhibition leading to opening of the K_{ATP} channels (Shyng and Nichols 1998). In vascular smooth muscle cells, K_{ATP} channels can be regulated by a number vasodilating (for example, adenosine and CGRP) and vasoconstricting (for example, angiotensin and endothelin) hormones. The binding of a vasodilator to a G_s-protein-coupled receptor leads to downstream activation of protein kinase A (PKA), direct phosphorylation of the K_{ATP} channel complex, leading to the opening of K_{ATP} channels, hyperpolarisation of the cell membrane and vasodilatation (Quinn et al. 2004; Shi et al. 2007). In contrast, vasoconstrictors act via $G_{q/11}$-protein-coupled receptors, leading to activation of protein kinase C (PKC), inhibition of K_{ATP} channels and membrane depolarisation, increased calcium entry and vasoconstriction (Aziz et al. 2012; Shi et al. 2008). The depletion of phosphatidylinositol (4,5) bisphosphate is another potential mediator after phospholipase C activation though this may not be critical with the vascular KATP channel (Quinn et al. 2003). In addition, PKC has been shown to modulate cardiac K_{ATP} channels (Light et al. 1996), and there is evidence of PKA-dependent modulation of the pancreatic K_{ATP} channel (Light et al. 2002).

3 The Pharmacological Properties of K_{ATP} Channels

K_{ATP} channels have a rich and well-developed pharmacology, with both activators and inhibitors existing. K_{ATP} channel openers (KCOs) and blockers (KCBs) have diverse chemical and structural properties (Tables 1 and 2). Importantly, from both research and therapeutic perspectives, there is a degree of channel subtype-specificity for some of these compounds allowing for some tissue-specific targeting.

3.1 K_{ATP} Channel Openers

Pharmacological compounds of diverse structures are able to potentiate K_{ATP} channel activity. These include benzothiadiazines (diazoxide), pyrimidine sulphates (minoxidil), pyridyl nitrates (nicorandil), benzopyrans (cromakalim), carbothiamides (aprikalim), and cyanoguanidines (pinacidil). Many of the KCOs show selectivity to different SUR subunits, for example, pinacidil, cromakalim, and

Table 1 Structure and pharmacology of KCOs acting on K_{ATP} channel

KCO (chemical class)	Chemical structure	Location of K_{ATP} channels	Same class drugs
First generation			
Pinacidil (cyanoguanidines)	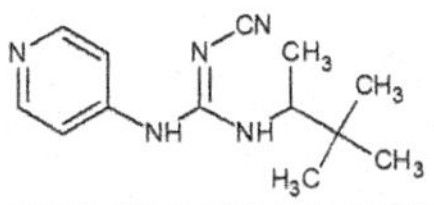	Cardiomyocytes Smooth muscle	P-1075
Diazoxide (benzothiadiazines)	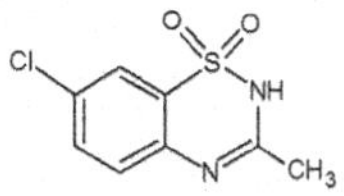	Cardiomyocytes Smooth muscle Pancreas Mitochondria	LN-5330
Cromakalim (benzopyrans)	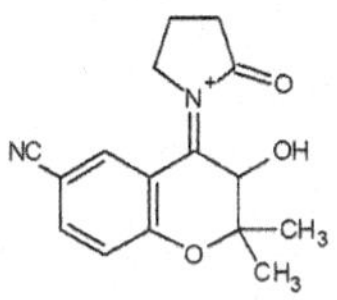	Cardiomyocytes Smooth muscle	Levcromakallm Blmakallm Cellkallm Rilmakalim Y-27152
Nicorandil (pyridyl nitrates)	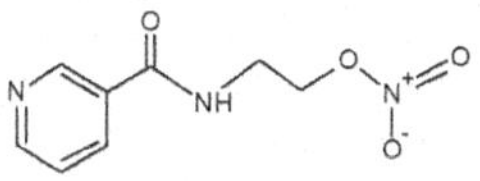	Cardiomyocytes	KRN-2391
Minoxidil (pyrimidine sulphate)	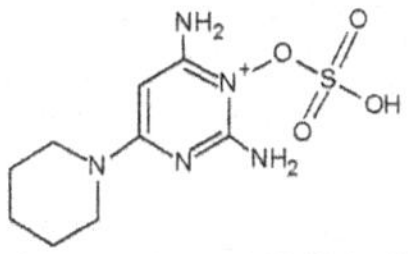	Cardiomyocytes	LP-805
Aprikalim (carbothiamides)	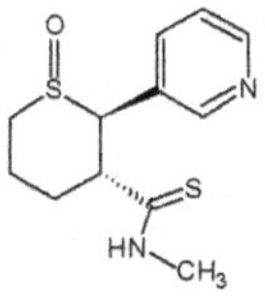	Cardiomyocytes Smooth muscle	MCC-134
Second generation			
WAY-151616 (cyclobutenediones)	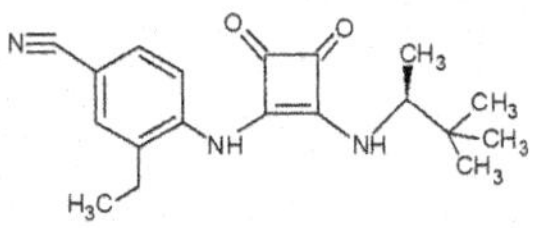	Smooth muscle	WAY-133537
ZM-244085 (dihydropyridine)	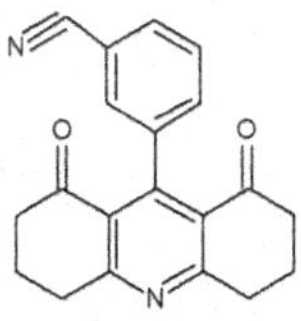	Smooth muscle	ZD-0947
ZD-6169 (tertiary carbinols)	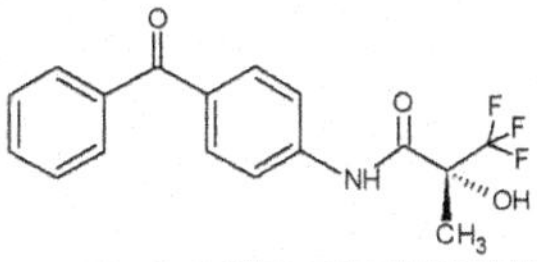	Smooth muscle	A-151892

Table 2 Structure and pharmacology of blockers acting on K_{ATP} channel

KCBs (chemical class)	Chemical structure	Location of K_{ATP} channels	Same class drugs
First generation			
Tolbutamide (sulfonylureas)	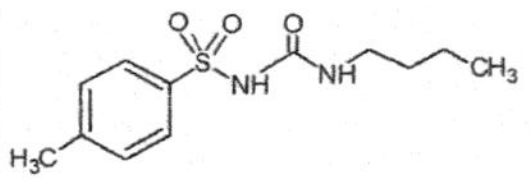	Pancreas	Chlorpropamide Acetohexamide Tolazamide
Second generation			
Glibenclamide (sulfonylureas)	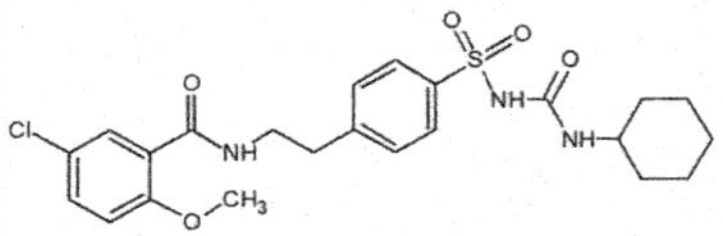	Cardiomyocytes Smooth muscle Mitochondria Pancreas	Gliclazide glimepiride Glipizide
Third generation			
Meglitinide (benzoic acid derivatives)	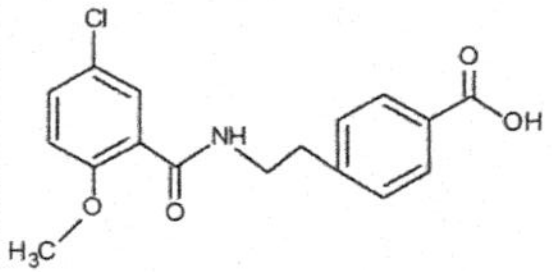	Pancreas	Repaglinide Nateglinide Mitiglinide
HMR-1098 (sulfonylureas)	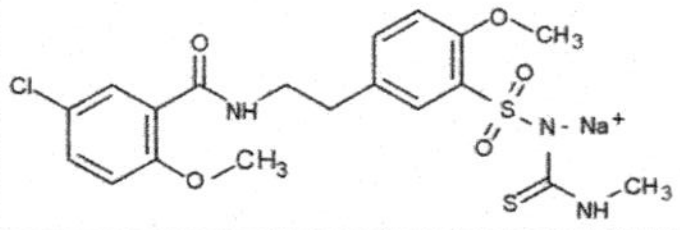	Cardiomyocytes	HMR-1883

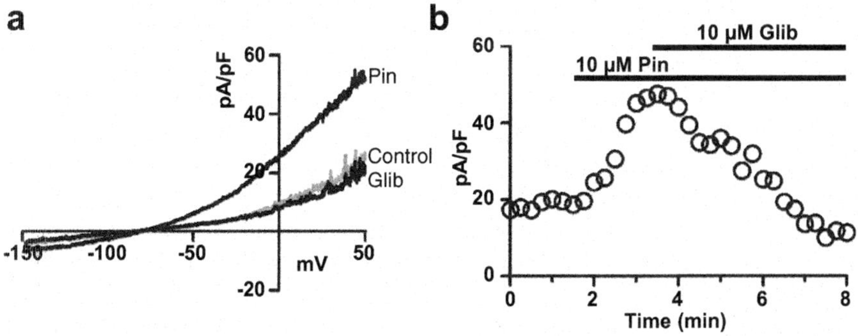

Fig. 2 The effects of pinacidil and glibenclamide on the K_{ATP} current in a murine aortic vascular smooth muscle cell. (**a**) Current Density-Voltage relationship in the presence of 10 μM pinacidil (Pin) and 10 μM glibenclamide (Glib). (**b**) Time course of the effects of pinacidil and glibenclamide at +40 mV. The data are from our own studies and these recordings are unpublished

nicorandil are selective for SUR2-containing channels, whereas diazoxide is more specific for SUR1-containing subunits but also activates SUR2B-containing channels (Giblin et al. 2002; Mannhold 2004); diazoxide has also been shown to activate SUR2A-containing channels under specific circumstances with high MgADP levels (D'hahan et al. 1999). Figure 2 shows activation of a K_{ATP} current by pinacidil in vascular smooth muscle cells.

The use of chimaeras between SURs and radioligand binding experiments has identified regions within TMD2, in particular the cytoplasmic linker between TM13 and TM14 and the last TM helices, TM16 and TM17, as important for pinacidil and cromakalim binding (Babenko et al. 2000; Moreau et al. 2000; Uhde et al. 1999). The binding site for diazoxide is less well-mapped, though it is known that binding is nucleotide-dependent and occurs between TM6 to TM11 and NBD1 (Babenko et al. 2000). The presence of more than one binding site on SUR for KCOs helps explain the structural diversity of KCOs.

KCOs were initially developed based on their ability to relax smooth muscle. However, studies using in vivo models showed differences in their physiological actions. For example, diazoxide was found to have both hypotensive and hyperglycaemic effects (RUBIN et al. 1962; Wolff 1964). Nicorandil, used in the treatment of angina, also acts on cardiac K_{ATP} channels thus potentially conferring cardioprotection (Horinaka 2011). Pinacidil, on the other hand, failed to reverse glibenclamide-induced hypoglycaemia in rats (Clapham et al. 1994), but shows a potent hypotensive effect in man (Carlsen et al. 1983; Ward et al. 1984).

KCOs have many potential therapeutic roles, including treatment of insulinomas with insulin hypersecretion, congenital hyperinsulinism, hypertension, myocardial ischaemia, congestive heart failure, bronchial asthma, urinary incontinence, and certain skeletal muscle myopathies (Hibino et al. 2010). Despite this, they have not been widely adopted in clinical practice because of side effects including profound hypotension, fluid retention, and others such as headache and flushing.

3.2 K_{ATP} Channel Blockers

3.2.1 Sulfonylureas

The hypoglycaemic action of sulfonylureas was inadvertently discovered in studies looking at their potential use as a treatment for typhoid fever. They work by inhibiting K_{ATP} channels in the pancreatic β-cell by binding to the SUR1 subunit, thus preventing K^+ efflux leading to the depolarisation of the β-cell membrane, opening of voltage-dependent calcium channels, increased Ca^{2+} influx, and subsequently insulin release. The first group of sulfonylureas such as tolbutamide and chlorpropamide has relatively low affinity for K_{ATP} channels. The early generations of sulfonylureas were initially used exclusively for treatment of type 2 diabetes mellitus; however, these compounds were also found to act on cardiac K_{ATP} channels with potential undesired cardiovascular side effects (Garratt et al. 1999). Subsequently, a more potent second generation (e.g. glibenclamide, gliclazide, and glipizide) was developed. The second generation drugs are relatively selective for the pancreatic channel and despite the potential for weight gain are still used in the treatment of type II diabetes mellitus. Advances in sulfonylurea chemistry led to the synthesis of a third generation of derivatives that show greater tissue selectivity. For example, HMR-1098 (Table 2) has a 400–800-fold selectivity for the cardiac K_{ATP} over the pancreatic K_{ATP} channel (Manning Fox et al. 2002).

There are differential effects of these agents between SUR1 and SUR2 containing channels. For example, glibenclamide and glimepiride show high-affinity block in both SUR1 and SUR2 containing channels whilst tolbutamide, gliclazide, and chlorpropamide have higher affinity for SUR1 (Gribble and Reimann 2003). In order to find the binding sites for these agents on SUR1 and SUR2, a chimeric approach was used and high-affinity inhibition was assayed (Ashfield et al. 1999). The last group of transmembrane domains, specifically the cytoplasmic loop between helices 15 and 16, was found to be important for binding (Ashfield et al. 1999). S1237 has been identified as a key amino acid residue for the binding of glibenclamide (Hansen et al. 2002) and introduction of serine at an equivalent residue in SUR2B led to an increase in the affinity of glibenclamide binding (Hambrock et al. 2001). Recent elucidation of the structure of SUR1 and Kir6.2 in complex using cryo-EM indicates that the binding site for glibenclamide might lie close to residues S1237, R1246, and R1300 but also that it might closely interact with residue Y230 in the linker between TMD0 and TMD1 (L0) (Ding et al. 2019; Martin et al. 2017) and R306 in TMD1 (Ding et al. 2019). Additional support for this model comes from biochemical studies. For example, glibenclamide binding is abolished with the deletion of TMD0 and L0 but not TMD0 alone. The L0 domain interlinks with Kir6.2 and, thus, is perfectly placed to regulate Kir6.2 gating. Finally, glibenclamide might prevent channel activation by altering the interaction between the NBDs preventing their alignment and dimerisation (Li et al. 2017; Martin et al. 2017). The cryo-EM structure showing drug–SUR interaction is shown in Fig. 3.

Physiologically, the actions of sulphonylureas can be affected by endogenous modulators such as MgADP. For example, in intact whole cell or in the presence of MgADP the action of tolbutamide is much more complete than in inside-out patches

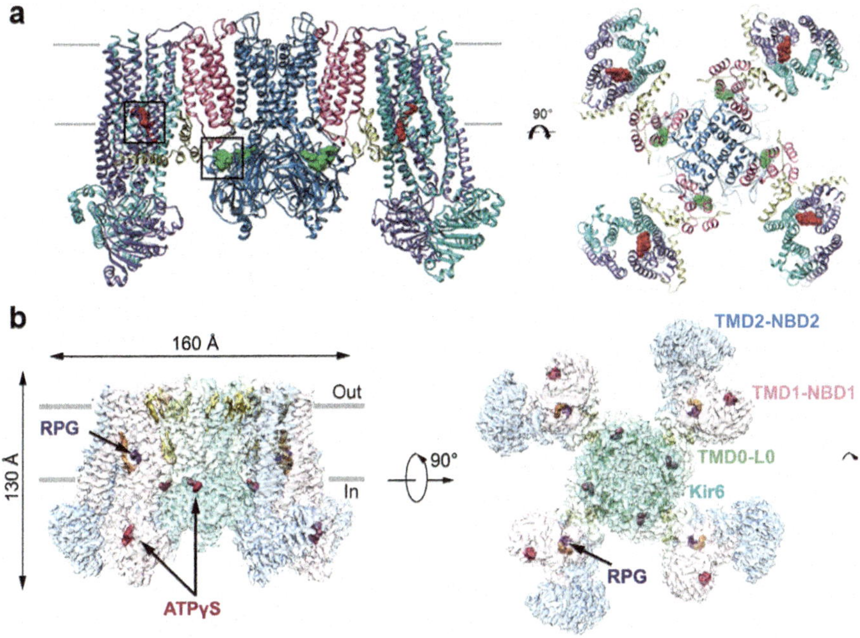

Fig. 3 The structure of the pancreatic K_{ATP} channel in complex with the K_{ATP} channel blockers glibenclamide and repaglinide. (**a**) A side view (left) of the model of the K_{ATP} channel in complex with ATP (green) and glibenclamide (red) and the model viewed from the extracellular side of membrane (right). (This figure is reproduced with permission from Martin et al. 2017). (**b**) Cryo-EM density map of the K_{ATP} channel in complex with repaglinide (RPG, purple) and ATPγS (red) viewed from the side (left) and intracellularly (right). (This figure was reproduced with permission from Ding et al. 2019)

where MgADP is absent (Gribble et al. 1997). This interaction with MgADP is not a feature of SUR2-containing channels and in fact sulphonylureas are less effective when MgADP concentrations are high (Reimann et al. 2003).

3.2.2 Nonsulphonylurea Drugs

Amongst the third generation of K_{ATP} blockers, a new chemical class of *nonsulphonylurea* drugs of benzoic acid derivatives was developed for the treatment of type 2 diabetes mellitus (Table 2). These include meglitinide, repaglinide, nateglinide, and mitiglinide. Benzoic acid derivatives are insulin secretagogues that bind primarily to pancreatic K_{ATP} channels, for example mitiglinide is highly tissue-specific and has a 1,000-fold greater affinity for pancreatic K_{ATP} channels over the cardiac and smooth muscle K_{ATP} channels (Reimann et al. 2001). Meglitinide and repaglinide show high-affinity block in both SUR1 and SUR2 containing channels whilst nateglinide does not exhibit high-affinity block with SUR2 (Gribble and Reimann 2003). A recent cryo-EM study has resolved the structure of the pancreatic K_{ATP} channel in complex with repaglinide (Ding et al. 2019). The images reveal that repaglinide shares a region of the glibenclamide binding site in TMD2 (R1246 and R1300) for its carboxyl group but other parts of the structure bind to distinct regions of SUR. Specifically, residues in TMD1 including M441, L592, V596, F433, W430, L434, Y377, and I381 provide a suitable pocket for the hydrophobic portion of repaglinide. The positioning of the binding site between TMD1 and TMD2 inhibits NBD dimerisation preventing K_{ATP} channel activation (Fig. 3).

Other agents used clinically may exhibit their actions through K_{ATP} channels. For example, baclofen may exhibit its antidepressant-like effect through inhibition of K_{ATP} channels (Nazari et al. 2016). The anti-epilepsy drug carbamazepine can inhibit K_{ATP} channel activity by disrupting the response to MgADP (Zhou et al. 2014). Some drugs that display anticonvulsant properties such as the inotropic calcium sensitiser levosimendan (Gooshe et al. 2017), glycolytic inhibitor 2-deoxy-D-glucose (Yang et al. 2013), K^+-sparing diuretic triamterene (Shafaroodi et al. 2016), hypnotic agent zolpidem (Sheikhi et al. 2016), fatty acid caprylic acid (Socała et al. 2015), and gabapentin (Ortiz et al. 2010) could also exert their action through K_{ATP} channels.

3.2.3 K_{ATP} Channel Pore Blockers

Generic agents such as barium, tetraethylammonium, and 4-aminopyridine can also block K_{ATP} channels by directly occluding the pore (Ashcroft and Ashcroft 1990; Takano and Ashcroft 1996). In addition, derivatives from the cyanoguanidine K_{ATP} opener P1075 such as PNU-37883A, PNU-89692, PNU-97025E, PNU-99963, and PNU-9470 are also K_{ATP} channel blockers.

Of these, PNU-37883A has been extensively investigated and early studies suggested a potential for future therapeutic use. PNU-37883A is a morpholinoguanidine drug that has been shown to be selective for vascular smooth muscle K_{ATP} channels (Meisheri et al. 1993). Studies using different recombinantly expressed K_{ATP} channel subunit combinations showed a preference for Kir6.1-containing K_{ATP} channels (Kovalev et al. 2001; Surah-Narwal et al. 1999). Further investigations revealed a higher affinity for Kir6.1 over Kir6.2 and a chimeric approach found that the C-terminus of Kir6.1 was important for PNU binding (Kovalev et al. 2004). Specifically, residues 200–280 of Kir6.1 are critical for the inhibitory effect. Interestingly, the choice of SUR subunit complexing with Kir6.1 is also important for PNU potency. Thus, PNU is being more potent when Kir6.1 is partnered with SUR2B, as compared to Kir6.1-SUR1 complex, explaining the relative specificity of PNU-37883A for vascular smooth muscle channels (thought to be constituted of Kir6.1/SUR2B).

The early promise of PNU-37883A for therapeutic use has subsided, although it is not clear whether this is a result of off-target effects or lack of potency. Nevertheless, it is routinely used in the research environment to distinguish between Kir6.1 and Kir6.2-containing K_{ATP} channels.

4 The Physiology and Pathophysiology of K_{ATP} Channels

Defective K_{ATP} channel function because of mutations (both loss-of-function [LoF] mutations and gain-of- function [GoF] mutations) can lead to diseases in neurological, cardiac, and endocrine systems.

4.1 Pancreas

K_{ATP} channels in the pancreas, particularly in the β-cells, have been extensively studied. A combination of Kir6.2/SUR1 makes up the K_{ATP} channel population in the insulin-regulating β-cells and glucagon-secreting α-cells (Table 3). In pancreatic β-cells, K_{ATP} channels couple cellular metabolism to electrical activity in response to changes in blood glucose. When blood glucose is low, ATP production is reduced allowing K_{ATP} channels to open thus hyperpolarising the membrane and preventing an increase in intracellular Ca^{2+} and subsequent insulin release. At high blood glucose concentrations, ATP production increases leading to channel inhibition, an increase in intracellular Ca^{2+} and insulin release. In pancreatic α-cells, glucagon is released to promote the mobilisation of glucose, and this process is inhibited by increased blood glucose levels.

Congenital hyperinsulinism is a genetic disorder in which there are abnormally high levels of insulin secretion from pancreatic β-cells leading to hypoglycaemia. It typically occurs in infants and young children in approximately 1/25–50,000 births where, if left untreated, persistent hypoglycaemia increases the risk for serious complications such as breathing difficulties, seizures, intellectual disability, vision loss, brain damage, and coma. Congenital hyperinsulinism is caused by mutations that lead to an overall LoF of K_{ATP} channels and have been identified in both the *KCNJ11* (Kir6.2) and *ABCC8* (SUR1) genes. Loss of channel activity arises from loss of K_{ATP} channels at the membrane due to ER retention, production of non-functional proteins, impaired pore-opening, loss of MgADP sensitivity, and reduced sensitivity to metabolic inhibition and drug activation (Tinker et al. 2018).

Neonatal diabetes mellitus (NMD) is a rare form of diabetes that occurs within the first 6 months of life in approximately in 1/100,000 births (Rubio-Cabezas and Ellard 2013). Neonatal diabetes can be caused by mutations in *KCNJ11 and ABCC8 that lead to* ATP insensitivity and GoF (Babenko et al. 2006).

Type 2 diabetes mellitus is a common and lifelong condition where the body doesn't produce or respond to insulin, it is usually considered as a disease of peripheral insulin resistance, however, there is evidence that pancreatic β-cell mass is reduced leading to impaired insulin secretion. Type 2 diabetes is associated with variants in *KCNJ11* (Gloyn et al. 2003) and *ABCC8* (Hamming et al. 2009).

Table 3 Tissue-specific subunit composition and properties of K_{ATP} channels

Location		K_{ATP} subunits	Physiological function
Pancreas	α-cells	Kir6.2/SUR1	Regulation of glucagon secretion in response to changes in blood glucose
	β-cells	Kir6.2/SUR1	Regulation of insulin release in response to changes in metabolism
Enteroendocrine cells		Kir6.2/SUR1	Involved in the stimulus-secretion coupling of gut hormones such as GIP, GLP-1, and PYY
Skeletal muscle		Kir6.2/SUR2A/SUR1	Adaptation to strenuous exercise, regulation of glucose uptake and metabolism
Heart	Atria	Kir6.2/SUR1	Action potential repolarisation, adaptation to cell swelling
	Ventricle	Kir6.2/SUR2A	Protection against Ca^{2+} overload during hypoxia, adaptation response to stress
	Conduction system	Kir6.1/Kir6.2/SUR2B	Adaptation to stress, regulation of pacemaker activity
Smooth muscle	Endothelium	Kir6.1/Kir6.2/SUR2B	Vasodilation, blood pressure regulation
	Vascular smooth muscle	Kir6.2/SUR2B	Relaxation, contraction
	Non-vascular smooth muscle	Kir6.1/Kir6.2/SUR2B	Vasodilation, protective during ischaemia
Nervous system	Hypothalamus	Kir6.2/SUR1	Expressed in AgRP/NPY- and POMC-positive neurons, regulation of neuronal excitability in response to glucose
	Pituitary	Kir6.2/SUR2B/SUR1	Regulation of hormone secretion
	Substantia nigra	Kir6.2/SUR1	Neuroprotection from stress and against seizures, regulation of excitability, release of neurotransmitters such as dopamine, GABA, and glutamate in response to changes in metabolism, play a role in memory, locomotion and behaviour
	Dorsal root ganglion	Kir6.2/SUR1/SUR2	Suppression of hyperalgesia
	Glial cells	Kir6.1/Kir6.2/SUR1	Neuroprotective, potassium siphoning

4.2 Heart

A combination of SUR2A\Kir6.2 subunits underlie the cardiac K_{ATP} channel in ventricular myocytes, but other subunit combinations have been reported in atria and conduction tissues (Table 3). Under basal conditions, the K_{ATP} channels in cardiomyocytes are closed and contribute little to resting membrane potential or action potential repolarisation (Noma 1983). In pathological conditions associated with hypoxia and ischaemia, cardiac K_{ATP} channels are activated leading to shortened action potential duration and attenuated/abolished contraction in myocytes (Lederer et al. 1989; Venkatesh et al. 1991). Similarly, K_{ATP} channels contribute to action potential duration and QT interval shortening in response to high-intensity exercise (Zingman et al. 2002).

Multiple mutations have been identified within K_{ATP} channel subunits that confer susceptibility to cardiac arrhythmia, cardiomyopathy, hypertrophy, and heart failure. Atrial fibrillation is the most common cardiac arrhythmia and can become persistent due to electrophysiological remodelling of the atria. A LoF missense mutation (T1547I) in *ABCC9* (SUR2) has been implicated in atrial fibrillation (Olson et al. 2007). Multiple mutations in *ABCC9* (SUR2), that impair nucleotide hydrolysis at NBD2, causing reduced function are associated with dilated cardiomyopathy (Bienengraeber et al. 2004). In addition, increased left ventricle size and heart failure are associated with the E23K variant in *KCNJ11* (Reyes et al. 2008, 2009). Whereas S422L GoF mutation in *KCNJ8* has been associated with Brugada syndrome, early repolarisation "J-wave" syndrome, atrial and ventricular fibrillation (Barajas-Martínez et al. 2012; Delaney et al. 2012; Haïssaguerre et al. 2009; Medeiros-Domingo et al. 2010). GoF (V734I and S1402C) mutations in *ABCC9* are thought to underlie Brugada and early repolarisation syndromes (Hu et al. 2014).

4.3 Cantu Syndrome

Cantu syndrome (CS) is a relatively new and rare syndrome, the hallmarks of which are hypertrichosis, abnormal facial features, and cardiomegaly (Cantú et al. 1982; Nichols et al. 2013). The features of CS vary among affected individuals and some patients also display other clinical features such as pericardial effusion, patent ductus arteriosus, conduction system abnormalities, pulmonary hypertension, and coarse lax skin (Scurr et al. 2011). Recently, the genetic basis of CS has been revealed showing the involvement of K_{ATP} channels. Specifically, missense mutations in *KCNJ8* and *ABCC9* have been identified. Using standard heterologous expression techniques, these mutations were shown to be GoF mutations leading to increased K_{ATP} channel activity (Cooper et al. 2015; Harakalova et al. 2012) as a result of reduced ATP-sensitivity and increased activation by MgADP (Cooper et al. 2015). The features of CS, particularly the cardiac abnormalities, have been replicated in murine models where both Kir6.1 and SUR2 GoF mutations have been transgenically introduced into a number of cardiovascular tissues (Levin et al. 2016).

4.4 Sudden Infant Death Syndrome

Sudden infant death syndrome (SIDS) is the sudden, unexpected, and unexplained death of an otherwise healthy baby, with most deaths occurring in the first 6 months of life. Although the exact cause of SIDS is unknown, LoF mutations in *KCNJ8* such as the in-frame deletion E332del and missense mutation V346I have been associated (Tester et al. 2011).

4.5 Nervous System

K_{ATP} channels and currents are widely distributed in the nervous system (Table 3). In central and peripheral neurones, the channels are largely thought to be constituted of Kir6.2 (Sun et al. 2007), with the exception of glial cells, where the current is made up of Kir6.1/SUR1 (Eaton et al. 2002). Neuronal K_{ATP} channels exhibit various physiological functions, including modulation of neuronal excitability (Allen and Brown 2004), suppression of hyperalgesia (Zoga et al. 2010), control of locomotion and behaviour (Deacon et al. 2006), influencing nutrient sensing and satiety (Rother et al. 2008) and control of autonomic function and thus modulation of heart rate (Almond and Paterson 2000; Mohan and Paterson 2000).

Mutations in K_{ATP} channels that underlie congenital hyperinsulinism, neonatal diabetes, and Cantu syndrome all display varying degrees of neurological pathology, it is likely that the abnormalities in K_{ATP} channel expression in neurons are the main reason for this.

4.6 Pulmonary Circulation

Pulmonary arterial hypertension (PAH) is a rare but serious condition and can affect people of all ages. It is characterised by raised pulmonary artery pressure and increased pulmonary vascular resistance and can lead to right heart failure and death. Mutations in multiple genes have been implicated in the development of PAH. Alongside mutations in genes such as *BMPR2* (bone morphogenic protein receptor type 2), LoF mutations in K_{ATP} channels (as well as other K^+ channels) have also been identified (McClenaghan et al. 2019). Reduced K^+ channel activity causes vasoconstriction leading to an increase in blood pressure. Interestingly, patients with Cantu syndrome with GoF mutations in *ABCC9* and *KCNJ8* can also develop PAH, possibly due to systemic feedback involving the renin-angiotensin-aldosterone system (McClenaghan et al. 2019). KCOs such as iptakalim and diazoxide may have therapeutic potential as a treatment for pulmonary hypertension.

4.7 Current Therapeutic Uses of KCOs and Inhibitors of K_{ATP} Channel

K_{ATP} channel openers are only used as second line agents for the treatment of diseases such as stable angina and hypertension. Three KCOs are used in clinical practice; nicorandil (stable angina), diazoxide (hypertension, congenital hyperinsulinism in some patients), and minoxidil sulphate (hypertension and male pattern baldness). Iptakalim, a relatively new KCO thought to be specific for vascular smooth muscle K_{ATP} channels, has shown promise in the treatment of mild to moderate essential hypertension and may also have therapeutic potential in the treatment of pulmonary hypertension (Sikka et al. 2012).

K_{ATP} channel inhibitors, the sulphonylureas, are still used in the treatment of type 2 diabetes in patients who are intolerant of metformin and in combination therapy, although the earlier generation of sulphonylureas are not recommended due to their possible inhibition of cardiac K_{ATP} channels. Sulphonylureas can lead to unwanted side effects including weight gain and also there is a risk of hypoglycaemia particularly in elderly patients (O'Hare et al. 2015).

The recent unmasking of the genetic basis of neonatal diabetes mellitus has revolutionised the management of the disease (Pearson et al. 2006). These patients were traditionally treated with insulin but the discovery that disease pathogenesis was due to over-activity of pancreatic K_{ATP} channels led to the use of sulphonylureas. The use of sulfonylureas normalised glucose homeostasis in many patients with normal responses of insulin release following a meal (Pearson et al. 2006). The presence of neurological symptoms requires higher doses and where the disease is due to mutations that do not affect ATP sensitivity the use of sulfonylureas may fail (Ashcroft et al. 2017; Babiker et al. 2016). Sulphonylurea treatment of the neurological deficits is not as effective as these deficits may have a developmental component (Koster et al. 2008; Shah et al. 2012; Slingerland et al. 2008). It is important that therapy with sulphonylureas is initiated as early as possible, as with time there is a decline in treatment efficacy (Babiker et al. 2016).

The characterisation of the genetic basis (GoF of K_{ATP} channels) of Cantu syndrome has given rise to the possibility of pharmacological intervention with sulphonylureas. Recent studies on murine models suggest that glibenclamide shows promising effects on the cardiovascular abnormalities that occur in CS such as reversing cardiac hypertrophy and increasing blood pressure (McClenaghan et al. 2020).

5 Conclusions

K_{ATP} channels are ubiquitously expressed in the body and have diverse functions in different tissue types. The physiological role of K_{ATP} channels is best described in the pancreatic β-cell, but recent work has also revealed their important pathophysiological roles in other cell types including cardiac, vascular, and nervous cells. K_{ATP} channels have a rich and diverse pharmacology that has the potential to be exploited to develop novel therapeutic agents for the treatment of various human diseases.

Acknowledgement This work was supported by the British Heart Foundation (RG/15/15/31742) and The National Institute for Health Research Barts Cardiovascular Biomedical Research. The authors have no conflicts of interest to declare.

References

Allen TG, Brown DA (2004) Modulation of the excitability of cholinergic basal forebrain neurones by KATP channels. J Physiol 554:353–370. https://doi.org/10.1113/jphysiol.2003.055889

Almond SC, Paterson DJ (2000) Sulphonylurea-sensitive channels and NO-cGMP pathway modulate the heart rate response to vagal nerve stimulation in vitro. J Mol Cell Cardiol 32:2065–2073. https://doi.org/10.1006/jmcc.2000.1237

Ashcroft FM (1988) Adenosine 5′-triphosphate-sensitive potassium channels. Annu Rev Neurosci 11:97–118. https://doi.org/10.1146/annurev.ne.11.030188.000525

Ashcroft SJ, Ashcroft FM (1990) Properties and functions of ATP-sensitive K-channels. Cell Signal 2:197–214. https://doi.org/10.1016/0898-6568(90)90048-f

Ashcroft FM, Harrison DE, Ashcroft SJ (1984) Glucose induces closure of single potassium channels in isolated rat pancreatic beta-cells. Nature 312:446–448

Ashcroft FM, Puljung MC, Vedovato N (2017) Neonatal diabetes and the K. Trends Endocrinol Metab 28:377–387. https://doi.org/10.1016/j.tem.2017.02.003

Ashfield R, Gribble FM, Ashcroft SJ, Ashcroft FM (1999) Identification of the high-affinity tolbutamide site on the SUR1 subunit of the K(ATP) channel. Diabetes 48:1341–1347

Ashford ML, Sturgess NC, Trout NJ, Gardner NJ, Hales CN (1988) Adenosine-5′-triphosphate-sensitive ion channels in neonatal rat cultured central neurones. Pflugers Arch 412:297–304. https://doi.org/10.1007/BF00582512

Aziz Q, Thomas AM, Khambra T, Tinker A (2012) Regulation of the ATP-sensitive potassium channel subunit, Kir6.2, by a Ca2+-dependent protein kinase C. J Biol Chem 287:6196–6207. https://doi.org/10.1074/jbc.M111.243923

Aziz Q, Thomas AM, Gomes J, Ang R, Sones WR, Li Y, Ng KE, Gee L, Tinker A (2014) The ATP-sensitive potassium channel subunit, Kir6.1, in vascular smooth muscle plays a major role in blood pressure control. Hypertension 64:523–529. https://doi.org/10.1161/HYPERTENSIONAHA.114.03116

Aziz Q, Li Y, Anderson N, Ojake L, Tsisanova E, Tinker A (2017) Molecular and functional characterization of the endothelial ATP-sensitive potassium channel. J Biol Chem 292:17587–17597. https://doi.org/10.1074/jbc.M117.810325

Babenko AP, Gonzalez G, Bryan J (2000) Pharmaco-topology of sulfonylurea receptors. Separate domains of the regulatory subunits of K(ATP) channel isoforms are required for selective interaction with K(+) channel openers. J Biol Chem 275:717–720. https://doi.org/10.1074/jbc.275.2.717

Babenko AP, Polak M, Cavé H, Busiah K, Czernichow P, Scharfmann R, Bryan J, Aguilar-Bryan L, Vaxillaire M, Froguel P (2006) Activating mutations in the ABCC8 gene in neonatal diabetes mellitus. N Engl J Med 355:456–466. https://doi.org/10.1056/NEJMoa055068

Babiker T, Vedovato N, Patel K, Thomas N, Finn R, Mannikko R, Chakera AJ, Flanagan SE, Shepherd MH, Ellard S, Ashcroft FM, Hattersley AT (2016) Successful transfer to sulfonylureas in KCNJ11 neonatal diabetes is determined by the mutation and duration of diabetes. Diabetologia 59:1162–1166

Barajas-Martínez H, Hu D, Ferrer T, Onetti CG, Wu Y, Burashnikov E, Boyle M, Surman T, Urrutia J, Veltmann C, Schimpf R, Borggrefe M, Wolpert C, Ibrahim BB, Sánchez-Chapula JA, Winters S, Haïssaguerre M, Antzelevitch C (2012) Molecular genetic and functional association of Brugada and early repolarization syndromes with S422L missense mutation in KCNJ8. Heart Rhythm 9:548–555. https://doi.org/10.1016/j.hrthm.2011.10.035

Beech DJ, Zhang H, Nakao K, Bolton TB (1993) K channel activation by nucleotide diphosphates and its inhibition by glibenclamide in vascular smooth muscle cells. Br J Pharmacol 110:573–582. https://doi.org/10.1111/j.1476-5381.1993.tb13849.x

Bienengraeber M, Olson TM, Selivanov VA, Kathmann EC, O'Cochlain F, Gao F, Karger AB, Ballew JD, Hodgson DM, Zingman LV, Pang YP, Alekseev AE, Terzic A (2004) ABCC9 mutations identified in human dilated cardiomyopathy disrupt catalytic KATP channel gating. Nat Genet 36:382–387. https://doi.org/10.1038/ng1329

Cantú JM, Sánchez-Corona J, Hernándes A, Nazará Z, García-Cruz D (1982) Individualization of a syndrome with mental deficiency, macrocranium, peculiar facies, and cardiac and skeletal anomalies. Clin Genet 22:172–179. https://doi.org/10.1111/j.1399-0004.1982.tb01431.x
Carlsen JE, Kardel T, Jensen HA, Tangø M, Trap-Jensen J (1983) Pinacidil, a new vasodilator: pharmacokinetics and pharmacodynamics of a new retarded release tablet in essential hypertension. Eur J Clin Pharmacol 25:557–561. https://doi.org/10.1007/BF00542128
Clapham JC, Trail BK, Hamilton TC (1994) K+ channel activators, acute glucose tolerance and glibenclamide-induced hypoglycaemia in the hypertensive rat. Eur J Pharmacol 257:79–85. https://doi.org/10.1016/0014-2999(94)90697-1
Conti LR, Radeke CM, Shyng SL, Vandenberg CA (2001) Transmembrane topology of the sulfonylurea receptor SUR1. J Biol Chem 276:41270–41278. https://doi.org/10.1074/jbc.M106555200
Cooper PE, Sala-Rabanal M, Lee SJ, Nichols CG (2015) Differential mechanisms of Cantú syndrome-associated gain of function mutations in the ABCC9 (SUR2) subunit of the KATP channel. J Gen Physiol 146:527–540. https://doi.org/10.1085/jgp.201511495
Cui Y, Tran S, Tinker A, Clapp LH (2002) The molecular composition of K(ATP) channels in human pulmonary artery smooth muscle cells and their modulation by growth. Am J Respir Cell Mol Biol 26:135–143. https://doi.org/10.1165/ajrcmb.26.1.4622
D'hahan N, Moreau C, Prost AL, Jacquet H, Alekseev AE, Terzic A, Vivaudou M (1999) Pharmacological plasticity of cardiac ATP-sensitive potassium channels toward diazoxide revealed by ADP. Proc Natl Acad Sci U S A 96:12162–12167. https://doi.org/10.1073/pnas.96.21.12162
Deacon RM, Brook RC, Meyer D, Haeckel O, Ashcroft FM, Miki T, Seino S, Liss B (2006) Behavioral phenotyping of mice lacking the K ATP channel subunit Kir6.2. Physiol Behav 87:723–733. https://doi.org/10.1016/j.physbeh.2006.01.013
Delaney JT, Muhammad R, Blair MA, Kor K, Fish FA, Roden DM, Darbar D (2012) A KCNJ8 mutation associated with early repolarization and atrial fibrillation. Europace 14:1428–1432. https://doi.org/10.1093/europace/eus150
Ding D, Wang M, Wu JX, Kang Y, Chen L (2019) The structural basis for the binding of repaglinide to the pancreatic K. Cell Rep 27:1848–1857.e4. https://doi.org/10.1016/j.celrep.2019.04.050
Eaton MJ, Skatchkov SN, Brune A, Biedermann B, Veh RW, Reichenbach A (2002) SURI and Kir6.1 subunits of K(ATP)-channels are co-localized in retinal glial (Muller) cells. Neuroreport 13:57–60
Findlay I (1987) The effects of magnesium upon adenosine triphosphate-sensitive potassium channels in a rat insulin-secreting cell line. J Physiol 391:611–629. https://doi.org/10.1113/jphysiol.1987.sp016759
Garratt KN, Brady PA, Hassinger NL, Grill DE, Terzic A, Holmes DR (1999) Sulfonylurea drugs increase early mortality in patients with diabetes mellitus after direct angioplasty for acute myocardial infarction. J Am Coll Cardiol 33:119–124. https://doi.org/10.1016/s0735-1097(98)00557-9
Giblin JP, Cui Y, Clapp LH, Tinker A (2002) Assembly limits the pharmacological complexity of ATP-sensitive potassium channels. J Biol Chem 277:13717–13723. https://doi.org/10.1074/jbc.M112209200
Gloyn AL, Weedon MN, Owen KR, Turner MJ, Knight BA, Hitman G, Walker M, Levy JC, Sampson M, Halford S, McCarthy MI, Hattersley AT, Frayling TM (2003) Large-scale association studies of variants in genes encoding the pancreatic beta-cell KATP channel subunits Kir6.2 (KCNJ11) and SUR1 (ABCC8) confirm that the KCNJ11 E23K variant is associated with type 2 diabetes. Diabetes 52:568–572
Gooshe M, Tabaeizadeh M, Aleyasin AR, Mojahedi P, Ghasemi K, Yousefi F, Vafaei A, Amini-Khoei H, Amiri S, Dehpour AR (2017) Levosimendan exerts anticonvulsant properties against PTZ-induced seizures in mice through activation of nNOS/NO pathway: role for K. Life Sci 168:38–46. https://doi.org/10.1016/j.lfs.2016.11.006

Gribble FM, Reimann F (2003) Sulphonylurea action revisited: the post-cloning era. Diabetologia 46:875–891

Gribble FM, Tucker SJ, Ashcroft FM (1997) The interaction of nucleotides with the tolbutamide block of cloned ATP-sensitive K+ channel currents expressed in Xenopus oocytes: a reinterpretation. J Physiol 504(Pt 1):35–45. https://doi.org/10.1111/j.1469-7793.1997.00035.x

Haïssaguerre M, Chatel S, Sacher F, Weerasooriya R, Probst V, Loussouarn G, Horlitz M, Liersch R, Schulze-Bahr E, Wilde A, Kääb S, Koster J, Rudy Y, Le Marec H, Schott JJ (2009) Ventricular fibrillation with prominent early repolarization associated with a rare variant of KCNJ8/KATP channel. J Cardiovasc Electrophysiol 20:93–98. https://doi.org/10.1111/j.1540-8167.2008.01326.x

Hambrock A, Löffler-Walz C, Russ U, Lange U, Quast U (2001) Characterization of a mutant sulfonylurea receptor SUR2B with high affinity for sulfonylureas and openers: differences in the coupling to Kir6.x subtypes. Mol Pharmacol 60:190–199. https://doi.org/10.1124/mol.60.1.190

Hamming KS, Soliman D, Matemisz LC, Niazi O, Lang Y, Gloyn AL, Light PE (2009) Coexpression of the type 2 diabetes susceptibility gene variants KCNJ11 E23K and ABCC8 S1369A alter the ATP and sulfonylurea sensitivities of the ATP-sensitive K(+) channel. Diabetes 58:2419–2424

Hansen AM, Christensen IT, Hansen JB, Carr RD, Ashcroft FM, Wahl P (2002) Differential interactions of nateglinide and repaglinide on the human beta-cell sulphonylurea receptor 1. Diabetes 51:2789–2795

Harakalova M, van Harssel JJ, Terhal PA, van Lieshout S, Duran K, Renkens I, Amor DJ, Wilson LC, Kirk EP, Turner CL, Shears D, Garcia-Minaur S, Lees MM, Ross A, Venselaar H, Vriend G, Takanari H, Rook MB, van der Heyden MA, Asselbergs FW, Breur HM, Swinkels ME, Scurr IJ, Smithson SF, Knoers NV, van der Smagt JJ, Nijman IJ, Kloosterman WP, van Haelst MM, van Haaften G, Cuppen E (2012) Dominant missense mutations in ABCC9 cause Cantú syndrome. Nat Genet 44:793–796. https://doi.org/10.1038/ng.2324

Hibino H, Inanobe A, Furutani K, Murakami S, Findlay I, Kurachi Y (2010) Inwardly rectifying potassium channels: their structure, function, and physiological roles. Physiol Rev 90:291–366. https://doi.org/10.1152/physrev.00021.2009

Horinaka S (2011) Use of nicorandil in cardiovascular disease and its optimization. Drugs 71:1105–1119

Hu D, Barajas-Martínez H, Terzic A, Park S, Pfeiffer R, Burashnikov E, Wu Y, Borggrefe M, Veltmann C, Schimpf R, Cai JJ, Nam GB, Deshmukh P, Scheinman M, Preminger M, Steinberg J, López-Izquierdo A, Ponce-Balbuena D, Wolpert C, Haïssaguerre M, Sánchez-Chapula JA, Antzelevitch C (2014) ABCC9 is a novel Brugada and early repolarization syndrome susceptibility gene. Int J Cardiol 171:431–442. https://doi.org/10.1016/j.ijcard.2013.12.084

Koster JC, Cadario F, Peruzzi C, Colombo C, Nichols CG, Barbetti F (2008) The G53D mutation in Kir6.2 (KCNJ11) is associated with neonatal diabetes and motor dysfunction in adulthood that is improved with sulfonylurea therapy. J Clin Endocrinol Metab 93:1054–1061. https://doi.org/10.1210/jc.2007-1826

Kovalev H, Lodwick D, Quayle JM (2001) Inhibition of cloned KATP channels by the morpholinoguanidine PNU-37883A. J Physiol (Lond) 531P:S173

Kovalev H, Quayle JM, Kamishima T, Lodwick D (2004) Molecular analysis of the subtype-selective inhibition of cloned KATP channels by PNU-37883A. Br J Pharmacol 141:867–873. https://doi.org/10.1038/sj.bjp.0705670

Lederer WJ, Nichols CG, Smith GL (1989) The mechanism of early contractile failure of isolated rat ventricular myocytes subjected to complete metabolic inhibition. J Physiol 413:329–349. https://doi.org/10.1113/jphysiol.1989.sp017657

Levin MD, Singh GK, Zhang HX, Uchida K, Kozel BA, Stein PK, Kovacs A, Westenbroek RE, Catterall WA, Grange DK, Nichols CG (2016) K(ATP) channel gain-of-function leads to increased myocardial L-type Ca(2+) current and contractility in Cantu syndrome. Proc Natl Acad Sci U S A 113:6773–6778. https://doi.org/10.1073/pnas.1606465113

Li N, Wu JX, Ding D, Cheng J, Gao N, Chen L (2017) Structure of a pancreatic ATP-sensitive potassium channel. Cell 168:101–110
Li Y, Aziz Q, Anderson N, Ojake L, Tinker A (2020) Endothelial ATP-sensitive potassium channel protects against the development of hypertension and atherosclerosis. Hypertension: HYPERTENSIONAHA12015355. https://doi.org/10.1161/HYPERTENSIONAHA.120.15355
Light PE, Sabir AA, Allen BG, Walsh MP, French RJ (1996) Protein kinase C-induced changes in the stoichiometry of ATP binding activate cardiac ATP-sensitive K+ channels. A possible mechanistic link to ischemic preconditioning. Circ Res 79:399–406. https://doi.org/10.1161/01.res.79.3.399
Light PE, Manning Fox JE, Riedel MJ, Wheeler MB (2002) Glucagon-like peptide-1 inhibits pancreatic ATP-sensitive potassium channels via a protein kinase A- and ADP-dependent mechanism. Mol Endocrinol 16:2135–2144. https://doi.org/10.1210/me.2002-0084
Mannhold R (2004) KATP channel openers: structure-activity relationships and therapeutic potential. Med Res Rev 24:213–266. https://doi.org/10.1002/med.10060
Manning Fox JE, Kanji HD, French RJ, Light PE (2002) Cardioselectivity of the sulphonylurea HMR 1098: studies on native and recombinant cardiac and pancreatic K(ATP) channels. Br J Pharmacol 135:480–488. https://doi.org/10.1038/sj.bjp.0704455
Martin GM, Yoshioka C, Rex EA, Fay JF, Xie Q, Whorton MR, Chen JZ, Shyng SL (2017) Cryo-EM structure of the ATP-sensitive potassium channel illuminates mechanisms of assembly and gating. eLife 6. https://doi.org/10.7554/eLife.24149
McClenaghan C, Woo KV, Nichols CG (2019) Pulmonary hypertension and ATP-sensitive potassium channels. Hypertension 74:14–22
McClenaghan C, Huang Y, Yan Z, Harter TM, Halabi CM, Chalk R, Kovacs A, van Haaften G, Remedi MS, Nichols CG (2020) Glibenclamide reverses cardiovascular abnormalities of Cantu syndrome driven by KATP channel overactivity. J Clin Invest 130:1116–1121. https://doi.org/10.1172/JCI130571
Medeiros-Domingo A, Tan BH, Crotti L, Tester DJ, Eckhardt L, Cuoretti A, Kroboth SL, Song C, Zhou Q, Kopp D, Schwartz PJ, Makielski JC, Ackerman MJ (2010) Gain-of-function mutation S422L in the KCNJ8-encoded cardiac K(ATP) channel Kir6.1 as a pathogenic substrate for J-wave syndromes. Heart Rhythm 7:1466–1471. https://doi.org/10.1016/j.hrthm.2010.06.016
Meisheri KD, Humphrey SJ, Khan SA, Cipkus-Dubray LA, Smith MP, Jones AW (1993) 4-morpholinecarboximidine-N-1-adamantyl-N'-cyclohexylhydrochloride (U-37883A): pharmacological characterization of a novel antagonist of vascular ATP-sensitive K+ channel openers. J Pharmacol Exp Ther 266:655–665
Mohan RM, Paterson DJ (2000) Activation of sulphonylurea-sensitive channels and the NO-cGMP pathway decreases the heart rate response to sympathetic nerve stimulation. Cardiovasc Res 47:81–89. https://doi.org/10.1016/s0008-6363(00)00057-2
Moreau C, Jacquet H, Prost AL, D'hahan N, Vivaudou M (2000) The molecular basis of the specificity of action of K(ATP) channel openers. EMBO J 19:6644–6651. https://doi.org/10.1093/emboj/19.24.6644
Nazari SK, Nikoui V, Ostadhadi S, Chegini ZH, Oryan S, Bakhtiarian A (2016) Possible involvement of ATP-sensitive potassium channels in the antidepressant-like effect of baclofen in mouse forced swimming test. Pharmacol Rep 68:1214–1220. https://doi.org/10.1016/j.pharep.2016.07.006
Nichols CG, Singh GK, Grange DK (2013) KATP channels and cardiovascular disease: suddenly a syndrome. Circ Res 112:1059–1072. https://doi.org/10.1161/CIRCRESAHA.112.300514
Noma A (1983) ATP-regulated K+ channels in cardiac muscle. Nature 305:147–148
O'Hare JP, Hanif W, Millar-Jones D, Bain S, Hicks D, Leslie RD, Barnett AH (2015) NICE guidelines for type 2 diabetes: revised but still not fit for purpose. Diabet Med 32:1398–1403. https://doi.org/10.1111/dme.12952
Olson TM, Alekseev AE, Moreau C, Liu XK, Zingman LV, Miki T, Seino S, Asirvatham SJ, Jahangir A, Terzic A (2007) KATP channel mutation confers risk for vein of Marshall

adrenergic atrial fibrillation. Nat Clin Pract Cardiovasc Med 4:110–116. https://doi.org/10.1038/ncpcardio0792

Ortiz MI, Ponce-Monter HA, Fernández-Martínez E, Macías A, Rangel-Flores E, Izquierdo-Vega JA, Sánchez-Gutiérrez M (2010) Pharmacological interaction between gabapentin and glibenclamide in the formalin test in the diabetic rat. Proc West Pharmacol Soc 53:49–51

Pearson ER, Flechtner I, Njølstad PR, Malecki MT, Flanagan SE, Larkin B, Ashcroft FM, Klimes I, Codner E, Iotova V, Slingerland AS, Shield J, Robert JJ, Holst JJ, Clark PM, Ellard S, Søvik O, Polak M, Hattersley AT, Neonatal Diabetes International Collaborative Group (2006) Switching from insulin to oral sulfonylureas in patients with diabetes due to Kir6.2 mutations. N Engl J Med 355:467–477. https://doi.org/10.1056/NEJMoa061759

Quayle JM, Nelson MT, Standen NB (1997) ATP-sensitive and inwardly rectifying potassium channels in smooth muscle. Physiol Rev 77:1165–1232. https://doi.org/10.1152/physrev.1997.77.4.1165

Quinn KV, Cui Y, Giblin JP, Clapp LH, Tinker A (2003) Do anionic phospholipids serve as cofactors or second messengers for the regulation of activity of cloned ATP-sensitive K+ channels? Circ Res 93:646–655. https://doi.org/10.1161/01.RES.0000095247.81449.8E

Quinn KV, Giblin JP, Tinker A (2004) Multisite phosphorylation mechanism for protein kinase A activation of the smooth muscle ATP-sensitive K+ channel. Circ Res 94:1359–1366. https://doi.org/10.1161/01.RES.0000128513.34817.c4

Reimann F, Proks P, Ashcroft FM (2001) Effects of mitiglinide (S 21403) on Kir6.2/SUR1, Kir6.2/SUR2A and Kir6.2/SUR2B types of ATP-sensitive potassium channel. Br J Pharmacol 132:1542–1548. https://doi.org/10.1038/sj.bjp.0703962

Reimann F, Dabrowski M, Jones P, Gribble FM, Ashcroft FM (2003) Analysis of the differential modulation of sulphonylurea block of beta-cell and cardiac ATP-sensitive K+ (K(ATP)) channels by Mg-nucleotides. J Physiol 547:159–168. https://doi.org/10.1113/jphysiol.2002.031625

Reyes S, Terzic A, Mahoney DW, Redfield MM, Rodeheffer RJ, Olson TM (2008) K(ATP) channel polymorphism is associated with left ventricular size in hypertensive individuals: a large-scale community-based study. Hum Genet 123:665–667. https://doi.org/10.1007/s00439-008-0519-3

Reyes S, Park S, Johnson BD, Terzic A, Olson TM (2009) KATP channel Kir6.2 E23K variant overrepresented in human heart failure is associated with impaired exercise stress response. Hum Genet 126:779–789. https://doi.org/10.1007/s00439-009-0731-9

Rother E, Könner AC, Brüning JC (2008) Neurocircuits integrating hormone and nutrient signaling in control of glucose metabolism. Am J Physiol Endocrinol Metab 294:E810–E816. https://doi.org/10.1152/ajpendo.00685.2007

RUBIN AA, ROTH FE, TAYLOR RM, ROSENKILDE H (1962) Pharmacology of diazoxide, an antihypertensive, nondiuretic benzothiadiazine. J Pharmacol Exp Ther 136:344–352

Rubio-Cabezas O, Ellard S (2013) Diabetes mellitus in neonates and infants: genetic heterogeneity, clinical approach to diagnosis, and therapeutic options. Horm Res Paediatr 80:137–146. https://doi.org/10.1159/000354219

Scurr I, Wilson L, Lees M, Robertson S, Kirk E, Turner A, Morton J, Kidd A, Shashi V, Stanley C, Berry M, Irvine AD, Goudie D, Turner C, Brewer C, Smithson S (2011) Cantú syndrome: report of nine new cases and expansion of the clinical phenotype. Am J Med Genet A 155A:508–518. https://doi.org/10.1002/ajmg.a.33885

Shafaroodi H, Barati S, Ghasemi M, Almasirad A, Moezi L (2016) A role for ATP-sensitive potassium channels in the anticonvulsant effects of triamterene in mice. Epilepsy Res 121:8–13. https://doi.org/10.1016/j.eplepsyres.2016.01.003

Shah RP, Spruyt K, Kragie BC, Greeley SA, Msall ME (2012) Visuomotor performance in KCNJ11-related neonatal diabetes is impaired in children with DEND-associated mutations and may be improved by early treatment with sulfonylureas. Diabetes Care 35:2086–2088

Sheikhi M, Shirzadian A, Dehdashtian A, Amiri S, Ostadhadi S, Ghasemi M, Dehpour AR (2016) Involvement of ATP-sensitive potassium channels and the opioid system in the anticonvulsive effect of zolpidem in mice. Epilepsy Behav 62:291–296. https://doi.org/10.1016/j.yebeh.2016.07.014

Shi Y, Wu Z, Cui N, Shi W, Yang Y, Zhang X, Rojas A, Ha BT, Jiang C (2007) PKA phosphorylation of SUR2B subunit underscores vascular KATP channel activation by beta-adrenergic receptors. Am J Physiol Regul Integr Comp Physiol 293:R1205–R1214. https://doi.org/10.1152/ajpregu.00337.2007

Shi Y, Cui N, Shi W, Jiang C (2008) A short motif in Kir6.1 consisting of four phosphorylation repeats underlies the vascular KATP channel inhibition by protein kinase C. J Biol Chem 283:2488–2494. https://doi.org/10.1074/jbc.M708769200

Shyng SL, Nichols CG (1998) Membrane phospholipid control of nucleotide sensitivity of KATP channels. Science 282:1138–1141. https://doi.org/10.1126/science.282.5391.1138

Sikka P, Kapoor S, Bindra VK, Saini M, Saxena KK (2012) Iptakalim: a novel multi-utility potassium channel opener. J Pharmacol Pharmacother 3:12–14. https://doi.org/10.4103/0976-500X.92495

Slingerland AS, Hurkx W, Noordam K, Flanagan SE, Jukema JW, Meiners LC, Bruining GJ, Hattersley AT, Hadders-Algra M (2008) Sulphonylurea therapy improves cognition in a patient with the V59M KCNJ11 mutation. Diabet Med 25:277–281. https://doi.org/10.1111/j.1464-5491.2007.02373.x

Socała K, Nieoczym D, Pieróg M, Wlaź P (2015) Role of the adenosine system and glucose restriction in the acute anticonvulsant effect of caprylic acid in the 6 Hz psychomotor seizure test in mice. Prog Neuropsychopharmacol Biol Psychiatry 57:44–51. https://doi.org/10.1016/j.pnpbp.2014.10.006

Spruce AE, Standen NB, Stanfield PR (1985) Voltage-dependent ATP-sensitive potassium channels of skeletal muscle membrane. Nature 316:736–738

Standen NB, Quayle JM, Davies NW, Brayden JE, Huang Y, Nelson MT (1989) Hyperpolarizing vasodilators activate ATP-sensitive K+ channels in arterial smooth muscle. Science 245:177–180

Sun HS, Feng ZP, Barber PA, Buchan AM, French RJ (2007) Kir6.2-containing ATP-sensitive potassium channels protect cortical neurons from ischemic/anoxic injury in vitro and in vivo. Neuroscience 144:1509–1515

Surah-Narwal S, Xu SZ, McHugh D, McDonald RL, Hough E, Cheong A, Partridge C, Sivaprasadarao A, Beech DJ (1999) Block of human aorta Kir6.1 by the vascular KATP channel inhibitor U37883A. Br J Pharmacol 128:667–672. https://doi.org/10.1038/sj.bjp.0702862

Takano M, Ashcroft FM (1996) The Ba2+ block of the ATP-sensitive K+ current of mouse pancreatic beta-cells. Pflugers Arch 431:625–631. https://doi.org/10.1007/BF02191912

Tester DJ, Tan BH, Medeiros-Domingo A, Song C, Makielski JC, Ackerman MJ (2011) Loss-of-function mutations in the KCNJ8-encoded Kir6.1 K(ATP) channel and sudden infant death syndrome. Circ Cardiovasc Genet 4:510–515. https://doi.org/10.1161/CIRCGENETICS.111.960195

Tinker A, Jan YN, Jan LY (1996) Regions responsible for the assembly of inwardly rectifying potassium channels. Cell 87:857–868. https://doi.org/10.1016/s0092-8674(00)81993-5

Tinker A, Aziz Q, Thomas A (2014) The role of ATP-sensitive potassium channels in cellular function and protection in the cardiovascular system. Br J Pharmacol 171:12–23. https://doi.org/10.1111/bph.12407

Tinker A, Aziz Q, Li Y, Specterman M (2018) ATP-sensitive potassium channels and their physiological and pathophysiological roles. Compr Physiol 8:1463–1511. https://doi.org/10.1002/cphy.c170048

Uhde I, Toman A, Gross I, Schwanstecher C, Schwanstecher M (1999) Identification of the potassium channel opener site on sulfonylurea receptors. J Biol Chem 274:28079–28082. https://doi.org/10.1074/jbc.274.40.28079

Venkatesh N, Lamp ST, Weiss JN (1991) Sulfonylureas, ATP-sensitive K+ channels, and cellular K + loss during hypoxia, ischemia, and metabolic inhibition in mammalian ventricle. Circ Res 69:623–637. https://doi.org/10.1161/01.res.69.3.623

Ward JW, McBurney A, Farrow PR, Sharp P (1984) Pharmacokinetics and hypotensive effect in healthy volunteers of pinacidil, a new potent vasodilator. Eur J Clin Pharmacol 26:603–608. https://doi.org/10.1007/BF00543493

Wolff F (1964) Diazoxide hyperglycaemia and its continued relief by tolbutamide. Lancet 1:309–310

Yang H, Guo R, Wu J, Peng Y, Xie D, Zheng W, Huang X, Liu D, Liu W, Huang L, Song Z (2013) The antiepileptic effect of the glycolytic inhibitor 2-deoxy-D-glucose is mediated by upregulation of K(ATP) channel subunits Kir6.1 and Kir6.2. Neurochem Res 38:677–685. https://doi.org/10.1007/s11064-012-0958-z

Zhou Q, Chen PC, Devaraneni PK, Martin GM, Olson EM, Shyng SL (2014) Carbamazepine inhibits ATP-sensitive potassium channel activity by disrupting channel response to MgADP. Channels (Austin) 8:376–382. https://doi.org/10.4161/chan.29117

Zingman LV, Hodgson DM, Bast PH, Kane GC, Perez-Terzic C, Gumina RJ, Pucar D, Bienengraeber M, Dzeja PP, Miki T, Seino S, Alekseev AE, Terzic A (2002) Kir6.2 is required for adaptation to stress. Proc Natl Acad Sci U S A 99:13278–13283. https://doi.org/10.1073/pnas.212315199

Zoga V, Kawano T, Liang MY, Bienengraeber M, Weihrauch D, McCallum B, Gemes G, Hogan Q, Sarantopoulos C (2010) KATP channel subunits in rat dorsal root ganglia: alterations by painful axotomy. Mol Pain 6:6. https://doi.org/10.1186/1744-8069-6-6

Calcium-Activated K^+ Channels (K_{Ca}) and Therapeutic Implications

Srikanth Dudem, Gerard P. Sergeant, Keith D. Thornbury, and Mark A. Hollywood

Contents

S. Dudem · G. P. Sergeant · K. D. Thornbury · M. A. Hollywood (✉)
Smooth Muscle Research Centre, Dundalk Institute of Technology, Dundalk, Ireland
e-mail: Mark.Hollywood@dkit.ie

N. Gamper, K. Wang (eds.), *Pharmacology of Potassium Channels*,
Handbook of Experimental Pharmacology 267, https://doi.org/10.1007/164_2021_459

Abstract

Potassium channels are the most diverse and ubiquitous family of ion channels found in cells. The Ca^{2+} and voltage gated members form a subfamily that play a variety of roles in both excitable and non-excitable cells and are further classified on the basis of their single channel conductance to form the small conductance (SK), intermediate conductance (IK) and big conductance (BK) K^+ channels.

In this chapter, we will focus on the mechanisms underlying the gating of BK channels, whose function is modified in different tissues by different splice variants as well as the expanding array of regulatory accessory subunits including β, γ and LINGO subunits. We will examine how BK channels are modified by these regulatory subunits and describe how the channel gating is altered by voltage and Ca^{2+} whilst setting this in context with the recently published structures of the BK channel. Finally, we will discuss how BK and other calcium-activated channels are modulated by novel ion channel modulators and describe some of the challenges associated with trying to develop compounds with sufficient efficacy, potency and selectivity to be of therapeutic benefit.

Keywords

BK channels · BK channel modulators · BK channel structure · Pharmacology · Regulatory subunits

1 K^+ Channels

Potassium channels are the most diverse and ubiquitous family of ion channels found in cells. To date, more than 50 individual potassium channels have been cloned and these can be classified, based on the number of transmembrane (TM) domains, into 3 broad families to form the 2TM, 4TM and 6TM families (Ptacek and Fu 2004). The 6TM K^+ channels are further subclassified into 16 subfamilies on the basis of their amino acid sequence homology and include voltage activated (K_v) channels and Slo channels that are activated by both voltage and/or intracellular ions (Sandhiya and Dkhar 2009). The K_v channel family is comprised of pore-forming subunits K_v1 to K_v4, K_v7 and K_v10 to K_v12 in addition to the non-pore forming, regulatory subunits K_v5, K_v6, K_v8 and K_v9 ((Grizel et al. 2014; Bocksteins 2016); see also chapter "Comparison of K^+ Channel Families"). Similarly, the Slo channel family can be further divided into 3 sub-categories based on sequence homology. Thus, Slo1 channels are activated by voltage and elevations in intracellular Ca^{2+} ($[Ca^{2+}]_i$) and Mg^{2+} concentrations ($[Mg^{2+}]_i$). In contrast, the Slo2 paralogues Slick (Slo 2.1) and Slack (Slo 2.2) channels appear to lack significant voltage dependent activation, but are stimulated by intracellular Na^+ and Cl^- (Yuan et al. 2000; Bhattacharjee et al. 2003). The final member of this family is called Slo3 and it is an evolutionary duplication of the Slo1 gene in mammals. These channels are activated by voltage and also by alkaline intracellular pH (Leonetti et al. 2012).

Given the huge number of K^+ channels present and their many roles, they have been targeted in numerous drug discovery programmes in the pharmaceutical industry. Although these studies have generated a large number of 'candidate molecules', concerns over 'off-target' effects and ion channel selectivity have so far made it difficult to bring specific ion channel agonists to market.

1.1 Ca^{2+} Activated K^+ Channels

The Ca^{2+} activated K^+ channels have been further divided on the basis of their conductance into three types. The channels with the largest conductance (250–300 pS) are called BK (Big K^+) channels, whereas those with intermediate (30–80 pS) and small conductances (10 pS) are known as IK and SK channels, respectively (Wei et al. 2005; Adelman et al. 2012). The pharmacology of SK and IK channels has recently been covered in detail by Cui et al. (2014) and Kshatri et al. (2018), so readers are directed to these excellent reviews for further information. In this chapter, we will briefly cover the pharmacology of SK and IK channels, but focus on BK channels and their modulation by voltage, Ca^{2+} and ion channel modulators.

1.2 BK Channel Physiology

The expression profile of BK channels is widespread across most of the major tissues in the body where they play crucial roles in modulating different physiological functions including muscle contraction (Brayden and Nelson 1992), neurotransmitter release (Robitaille et al. 1993), neuronal excitability (Adams et al. 1982) and endothelial function (Nilius and Droogmans 2001). Their function in different tissues is modified not only by the presence of different complements of regulatory subunits, but also by splice variants, which can finely tune the BK channel biophysical properties in each tissue. The large single channel conductance (250–300 pS) of BK channels allows huge K^+ efflux and can result in significant hyperpolarisation.

In excitable cells, BK channels can help limit Ca^{2+} influx by repolarising the cells, whereas in non-excitable cells such as epithelial and glial cells, BK channels can also modulate Ca^{2+} entry via voltage independent Ca^{2+} channels, such as P2X purinoreceptors and TRP channels (Nilius and Droogmans 2001) by regulating the driving force for Ca^{2+}.Thus, in non-excitable cells the BK channels help contribute to Ca^{2+} homeostasis and are an important regulator of physiological functions including osmoregulation, K^+ secretion, cell migration and cell proliferation (Nilius and Droogmans 2001).

1.3 BK Channel Pathophysiology

BK channels are implicated in the pathogenesis of several diseases including epilepsy (Du et al. 2005), diabetes (Rajan et al. 1990), hypertension (Brenner et al. 2000b), autism and mental retardation (Laumonnier et al. 2006), urinary incontinence (Meredith et al. 2004) and cerebellar ataxia (Sausbier et al. 2004).

A missense mutation in the RCK1 domain of the BK channel's α subunit has been linked to a condition in which epilepsy and paroxysmal dyskinesia are the main symptoms (Du et al. 2005). The D434G mutation is responsible for increasing the apparent calcium sensitivity of the channel (Yang et al. 2010). This gain of function mutation ultimately leads to neuronal excitability resulting in generalised epilepsy and paroxysmal dyskinesia (GEPD) syndrome (Du et al. 2005). Abnormal gait and deficient sensorimotor coordination are also observed in genetically modified mice lacking BKα channels and thus highlights their role in movement disorders (Sausbier et al. 2004).

BK channels are therapeutic targets in urinary incontinence and overactive bladder, since they are involved in smooth muscle function of the urinary bladder. Spontaneous and nerve evoked contractions of the urinary bladder are increased in the BK knock out mice tissues (Meredith et al. 2004). In the same knock out mice, frequent urination is also observed, which is consistent with the enhanced contractility of the urinary bladder. These results revealed the notable role of BK channel in urinary incontinence and overactive bladder (Meredith et al. 2004).

Increases in arterial tone and blood pressure are also observed in genetically modified mice lacking the BK regulatory β1 subunit (Brenner et al. 2000b), although the severity of the hypertension appears less when blood pressure is measured by telemetry (Xu et al. 2011). Nevertheless, it is thought that the targeted deletion of the β1 subunit reduces the calcium sensitivity of the BK channels, leading to reduced coupling between the calcium sparks and BK channel activation, and manifests itself as an increase in arterial tone and blood pressure. However, it has also been shown that the observed hypertension in these animals may also be caused by perturbations in K^+ secretion, resulting in renal K^+ retention and hyperkalaemia (Grimm et al. 2009).

2 BK Channel Auxiliary Subunits

Although BK channels are encoded by a single gene KCNMA1, the channels display quite different phenotypes in different cells and tissues. This phenotypical variation can be explained by a combination of alternative splicing, the presence of auxiliary subunits, and metabolic regulation (Orio et al. 2002). To date two main groups of regulatory subunits (β_{1-4} and γ_{1-4}) have been identified (Uebele et al. 2000; Yan and Aldrich 2010, 2012), but recently, another group of BK regulatory subunits called the LINGO family has been discovered (Dudem et al. 2020).

2.1 BK Channel β Subunits

Four β subunits (β1, β2, β3 and β4) have been cloned from mammals (Knaus et al. 1994; Uebele et al. 2000; Brenner et al. 2000a) and each one shows different tissue-specific expression. These subunits are ~191–235 amino acids long and consist of two membrane spanning domains (TM1 and TM2) connected by a large extracellular loop (~148 aa) and two short N and C intracellular termini (Knaus et al. 1994; Orio et al. 2002). β1–β2 and β2–β3 subunits share ~65% conserved sequence similarity between these subunits. Sequence similarity between β1–β4 subunits is less pronounced (Fig. 1; Brenner et al. 2000a). The calcium sensitivity, kinetic and pharmacological properties of BK channels are significantly altered with the co-expression of β subunits.

2.2 β1 Subunit

The BK channel β1 auxiliary subunit was the first BK regulatory subunit discovered and was identified in tracheal smooth muscle cells (Garcia-Calvo et al. 1994; Knaus et al. 1994). It is abundantly expressed in smooth muscle cells of the bladder, uterus, trachea and blood vessels. Co-expression of the β1 subunit with BKα subunits in heterologous systems such as HEK293 cells and Xenopus oocytes results in BK channels with increased apparent calcium sensitivity and altered gating behaviour. The G-V relationships of BKαβ1 channels at different calcium concentrations suggest that β1 subunits cause a −50 to −70 mV leftward shift in $V_{1/2}$ at 1–10 μM $[Ca^{2+}]_i$ (McManus et al. 1995; Wallner et al. 1995; Meera et al. 1996; Tanaka et al. 1997; Brenner et al. 2000a; Bao and Cox 2005).

2.3 β1 Subunit Specific Effects on BK Channel Activation

It is well established that β1 enhances the apparent Ca^{2+} sensitivity of the BK channel. Based on open probability (P_o) measurements of BK channels with and without β1 subunits in Ca^{2+} free conditions, it was revealed that β1 subunits increased the BK channel burst duration ~20-fold and it increased the gaps between the bursts ~threefold. Nimigean and Magleby (2000) first suggested that the apparent increase in the Ca^{2+} sensitivity of BK channels in the presence of β1 was through a Ca^{2+} independent mechanism. In 2005, Bao and Cox used gating current and macroscopic current recordings to demonstrate that β1 subunits did not alter the gating charge of the channel, but instead, stabilised voltage sensor activation. Based on these results, they suggested that in the presence of β1, voltage sensor activation occurs at more negative membrane potentials. Also, they suggested that β1 decreased the true Ca^{2+} affinity of the closed channel by increasing its Ca^{2+} dissociation constant from ~3.7 μM to between 4.7 and 7.1 μM (Bao and Cox 2005).

To date, many biophysical interpretations have been proposed to explain how β1 subunits shift voltage sensor activation to negative membrane potentials. In 2008,

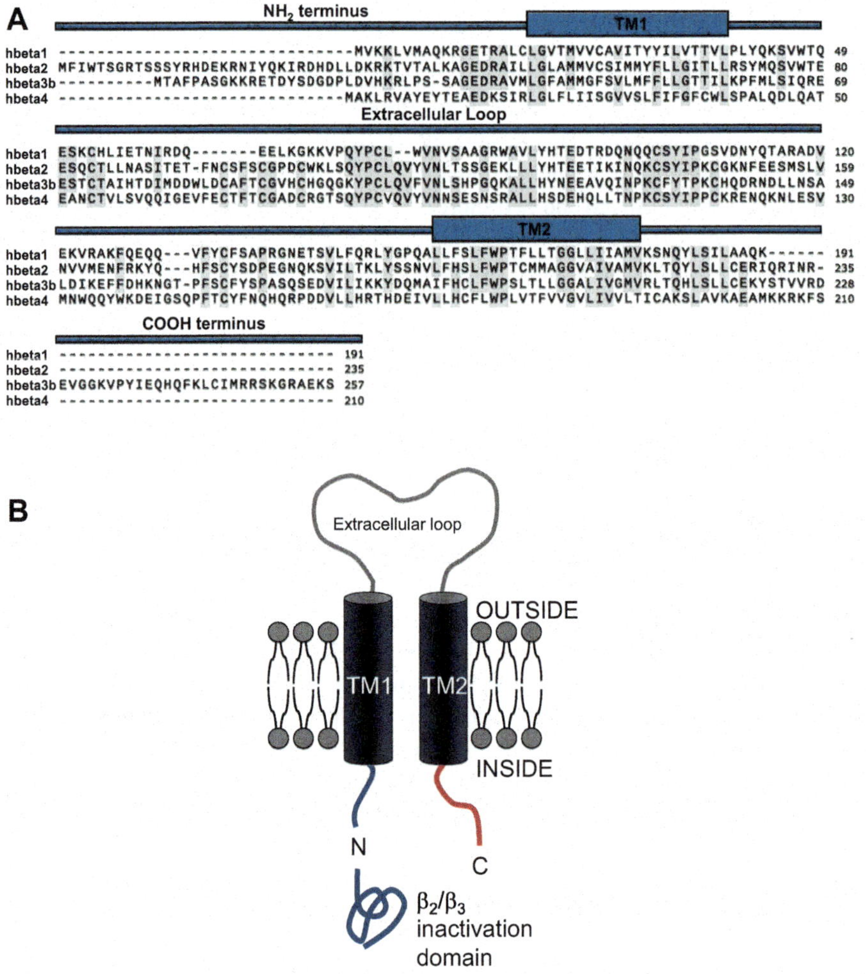

Fig. 1 Sequence alignment and topology of β subunits. (**a**) Multiple sequence alignment of the four regulatory human β subunits (β1–β4). The main domains are highlighted from their predicted structure. The highlighted sequences are two transmembrane domains (TM1 and TM2), extracellular loop, cytosolic NH_2 and COOH termini; conserved residues are shaded in grey. (**b**) Schematic representation shared by β1–β4 subunits. The β subunits have two transmembrane domains, a large extracellular loop and short cytosolic N and C termini. The β2 and β3 subunits contain longer NH_2 termini that constitute the inactivation particle

Yang et al. investigated the effect of BKα voltage sensor mutations on β1 subunit specific effects. They demonstrated that the R167A BKα voltage sensor mutation completely abolished the β1-induced leftward GV shift of the BK channel in 0–100 μM $[Ca^{2+}]_i$. They proposed that β1- specific effects were solely due to the conformational changes in voltage sensor movement (Yang et al. 2008). In 2015, Castillo et al. used a chimeric approach with the β1–3 subunits and convincingly demonstrated that the N-terminus of the β1 subunit, in particular the two positively charged residues K3 & K4, were essential for retaining the effect of β1 on BK channels. They proposed that these two residues in β1 were required to stabilise the activated state of the voltage sensors in the BKα subunit and thus were responsible for the observed apparent enhancement in Ca^{2+} sensitivity of BKβ1 channels.

2.4 β2 Subunit

Although BK channels were thought to produce non-inactivating, voltage and Ca^{2+} dependent currents in most tissue types, a number of studies in the 1990s demonstrated inactivating BK currents in chromaffin cells, pancreatic β cell lines and hippocampal neurones (Solaro and Lingle 1992; Hicks and Marrion 1998). Different groups identified the accessory β2 and β3 subunits which produced the inactivating currents when co-expressed with BKα subunit (Xia et al. 1999; Uebele et al. 2000). The inactivation was removed by the intracellular application of trypsin, suggesting a similar inactivation mechanism to Shaker or K_v1 channels was present (Ding et al. 1998). The major difference in the inactivation of BK channels, in comparison with other potassium channels, was that cytosolic blockers like quaternary ammonium ions did not alter the rate of inactivation of BK channels in rat chromaffin cells (Solaro et al. 1997).

β2 subunits are mainly expressed in chromaffin cells and the brain (Xia et al. 1999). Like the β1 subunit, β2 also increases the apparent calcium sensitivity of BKα and slows down channel activation and deactivation. When β2 is co-expressed with BKα the resultant currents were very similar to those observed in chromaffin cells (Xia et al. 1999). The N-terminus of the β2 subunit has a region of 31 amino acids consisting of a hydrophobic region followed by a series of charged residues. This pattern of residues is characteristic of channels showing N-type inactivation. (Wallner et al. 1999; Brenner et al. 2000a; Xia et al. 2003; Orio and Latorre 2005). Cysteine cross-linking experiments suggest that the position of the β2 subunit TM2 domain is near to BKα S0 whereas TM1 is closer to the S1 and S2 domains of the BKα subunit (Wu et al. 2013).

The β2 mediated inactivation of BK channels possessed five major characteristics. First, the time constant of inactivation (τ_i) was 250 ms in 100 nM Ca^{2+} at +100 mV, became faster with increased $[Ca^{2+}]_i$ and at more depolarising membrane potentials. Second, the intracellular application of trypsin gradually removed the inactivation process in patches containing either single or multiple channels. Trypsin application led to a gradual slowing of the time constants of inactivation, consistent with earlier reports that more than one β2 subunit was associated with a tetrameric channel. Third, the cytosolic blocker of BK channels, QX-314, did not slow the inactivation process, demonstrating that the inactivation mechanism did not interfere with the site occupied by cytosolic blockers. Fourth, the co-expression of β2 with the BKα channels altered the pharmacology of BK channels since BKαβ2 channels showed reduced sensitivity to charybdotoxin (CTX) and increased sensitivity to DHS-I compared to BKα alone (Wallner et al. 1999). Fifth, like β1, co-expression of β2 subunits also shifted activation curves in

10 μM Ca^{2+} to more negative membrane potentials compared to BKα alone (Xia et al. 1999; Wallner et al. 1999).

2.5 β3 Subunit

The β3 subunit is preferentially expressed in pancreas, spleen and testes (Xia et al. 2000). Protein sequence alignment of β2 and β3 subunits suggests that these two subunits have ~60% conserved residues. There are four splice variants of the β3 subunit (β3a–d) and they originate from a single gene (KCNMB3). The co-expression of β3a–c splice variants with BKα channels produced incomplete inactivation of BK channels (Uebele et al. 2000; Xia et al. 2000). The inactivation time constant (τ_i) of β3a and β3c splice variants at +80 mV was 45 ± 15 ms and 60 ± 6 ms, respectively. In contrast, co-expression of β3b splice variant with BKα channels produced very rapidly inactivating currents ($\tau_i = 1.5 \pm 0.2$ ms). The splice variant β3d did not show any inactivation properties (Brenner et al. 2000a; Uebele et al. 2000).

There are also five major characteristics of β3b mediated inactivation of BK channels. Firstly, the BKβ3b currents are voltage and Ca^{2+} dependent, since increases in $[Ca^{2+}]_i$ shift the activation $V_{1/2}$ to more negative membrane potentials. Compared to β1 and β2 subunits, the β3b subunit produced a greater shift in the $V_{1/2}$ of activation at $[Ca^{2+}]_i$ below 10 μM, but was less effective at doing this in higher $[Ca^{2+}]_i$. Secondly, the currents were very rapid to inactivate (τ_i = 1–2 ms), but the process of inactivation was incomplete compared to that observed with β2. The time constant of inactivation was moderately dependent on voltage and practically calcium independent at more positive membrane potentials. Thirdly, the pharmacological properties of BKβ3b currents were similar to BKα channels alone, suggesting that co-expression of β3b subunits with BKα did not alter the pharmacology of BK channels. Fourthly, the cytosolic blockers of BK channels like TEA did not compete with the β3b mediated inactivation. Finally, removal of 21 amino acids of N-terminus of β3b subunit abolished the β3b mediated inactivation (Xia et al. 2000).

2.6 β4 Subunit

The β4 subunit is predominantly expressed in the brain but is also present in proximal convoluted tubules and bladder smooth muscle (Weiger et al. 2000). The conductance-voltage curves from co-expression of β4 with BKα subunits revealed that with $[Ca^{2+}]_i < 1$ μM, the $V_{1/2}$ shifted towards more positive potentials compared to BKα alone. However, when the $[Ca^{2+}]_i$ was >1 μM, the G-V curves shift towards more negative membrane potentials compared to BKα subunit alone (Brenner et al. 2000a; Wang et al. 2006). Although the β4 subunit slowed down the activation kinetics of the channel, the deactivation kinetics were similar to BKα currents.

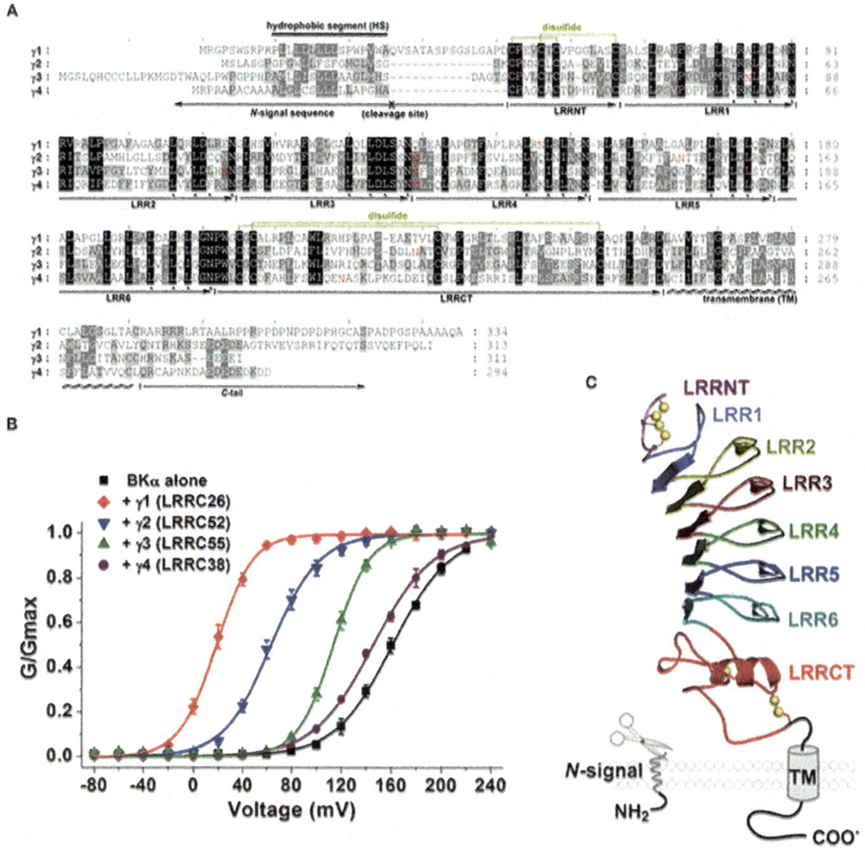

Fig. 2 Sequence alignment and topology of the γ subunits of BK channel. (**a**) The sequence alignment of the four regulatory γ-subunits (γ1–γ4). The LRR domain, transmembrane domain and cytosolic C-tail are highlighted with black bars. (**b**) Modulatory effects of γ subunits on the voltage dependence of BK channel activation in the absence of $[Ca^{2+}]_i$, upon heterologous expression of the γ subunit in HEK-293 cells. (**c**) Prediction of leucine-rich repeat domain structure and membrane topology of the γ subunit (Adapted from Zhang and Yan 2014)

2.7 BK Channel γ Subunits

Gessner et al. (2006) first identified unusual K^+ currents in a prostate cancer cell line (LNCaP) which shared many characteristics of BK channels. Hence, they had a large single channel conductance (250 ps) and showed Ca^{2+} dependent activation in the μM range and Mg^{2+} dependent activation in the mM range. However, under Ca^{2+} free conditions, they activated at very negative potentials (Fig. 2b). They were unable to ascertain why these channels activated so negatively and initially called them BK-like channels (BK_L) (Gessner et al. 2006). However in 2010, Yan and Aldrich used a proteomic approach to identify a novel interacting partner of BK channels, which was responsible for modifying the gating properties of BK channels in LNCaP cells. Based on pull-down experiments, they identified a 35 kDa leucine-rich repeat containing protein called LRRC26. Knockdown of LRRC26 in LnCaP cells abolished the negatively activating component of BK current in 0 $[Ca^{2+}]_i$. Furthermore, co-expression of LRRC26 with BKα subunits in HEK 293 cells shifted the half maximal activation of BK channels by ~ −140 mV (Yan and Aldrich 2010). They also demonstrated that three other leucine-rich repeat containing proteins LRRC52, LRRC55 and LRRC38 were present in different tissues. When co-expressed with BKα subunits, they shifted the $V_{1/2}$ by −100 mV, −50 mV and − 20 mV, respectively (Yan and Aldrich 2012). The four LRRC proteins were classified as auxiliary γ subunits of the BK channel and reported to display a tissue-specific distribution. Thus, LRRC26 (γ1) was predominantly expressed in salivary glands, prostate and trachea, whereas LRRC52 (γ2) was mainly expressed in testes. LRRC55 (γ3) was abundantly expressed in the medial habenular nucleus, cerebellum and LRRC38 (γ4) was highly expressed in skeletal muscle, thymus and adrenal glands (Yan and Aldrich 2012; Zhang et al. 2018).

All four γ subunits share the same structural similarities and there is 35–40% sequence similarity between them (Fig. 2a). The crystal structure of these subunits has not yet been solved, but their predicted structure is represented in Fig. 2c. The mature protein of the γ subunits contains a single transmembrane domain, an extracellular N-terminal LRR domain and an intracellular short C-terminal tail. The LRR domain of the γ subunits are comprised of 6 LRR units, containing a classic consensus sequence of LxxLxLxxN and two cysteine-rich repeat regions called LRRNT and LRRCT. The transmembrane domain of the γ subunits was predicted from the presence of charged residues at both sides of the proposed membrane spanning helix. The cytosolic C-terminal is followed by single TM domain and comprises polyproline residues in the γ1 subunit and polyacidic residues in γ2, γ3 and γ4 subunits (Zhang and Yan 2014).

Recently, Dudem et al. (2020) demonstrated that another LRRC protein, called LINGO-1, was a novel regulatory subunit of BK channels, which shifted the voltage dependence of activation, induced inactivation and significantly downregulated BK channel plasmalemmal expression. This protein is one of the 36 members of the LRRIG family, so called because they possess extracellular LRR domains and, in contrast to the BKγ subunits, also has an Ig domain (Homma et al. 2009). As shown in Fig. 3a, the Ig domain is connected via a short extracellular linker to a single transmembrane helix and the ~40 residue long C-terminus is intracellular.

There are four members of this family (LINGO1–4) and LINGO-2 shares 61% sequence similarity with LINGO-1, whereas LINGO-3 and LINGO-4 share 56% and 44% sequence similarity with LINGO-1, respectively (Mi et al. 2004, 2013). LINGO-2 and LINGO-3 also produce inactivating currents when co-expressed with BKα subunits (Dudem, Sergeant, Thornbury & Hollywood, unpublished observations). Intriguingly, we have been unable to demonstrate that LINGO-4 induces inactivating BK currents. This may be due to the fact that it has a shorter intracellular tail and lacks the C-terminus 'MKMI' residues responsible for LINGO1-induced inactivation (Dudem et al. 2020).

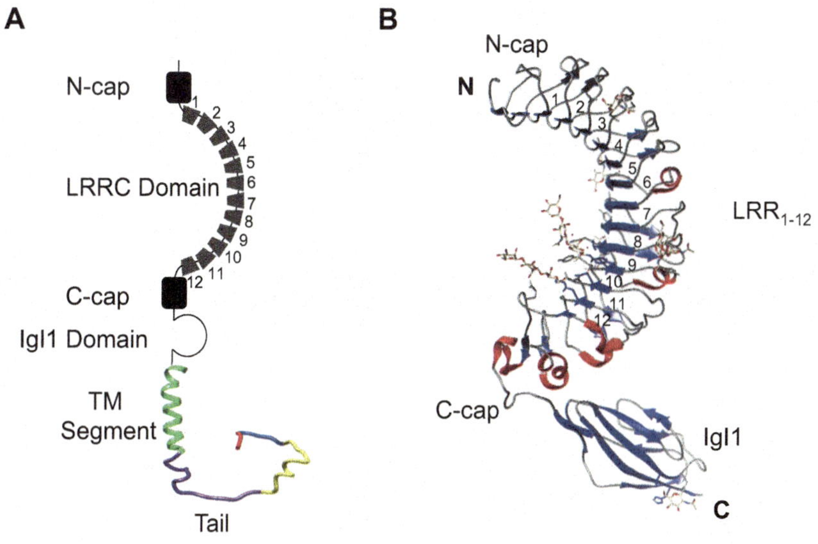

Fig. 3 LINGO-1 structure and architecture. (**a**) shows a cartoon of the main features of LINGO-1 extracellular, transmembrane and intracellular domains. Panel (**b**) shows the crystal structure of the ectodomain obtained by Mosyak et al. (2006). The colours represent different secondary structures: beige-coil; blue-strand; red-helix. Disulphide bonds are represented in green and the yellow shows the N-linked carbohydrates. The 12 leucine-rich repeats are numbered in the figure and the Ig domain is also shown in the bottom right

LINGO-1 appears to be expressed predominantly in the central nervous system (Mi et al. 2007) and has highest expression noted in the cortex of the brain and lowest in the spinal cord (Mi et al. 2007, 2013). The expression of this protein seems to be specific to the CNS, as Northern and Western blot analysis of a variety of different tissues including muscle, heart, kidney, lung and liver failed to detect LINGO-1 mRNA or protein, respectively (Mi et al. 2013).

LINGO-1 was originally discovered as the third component of the Nogo receptor (NgR1)/p75 signalling complex by Mi et al. (2004). When these three (NgR1/p75/LINGO-1) receptors associate they activate Rho A kinase, which then acts as a negative regulator of remyelination and thus inhibits axon regeneration.

Although the structure of the ectodomain of LINGO-1 has been resolved at a resolution of 2.7 Å, the protein used lacked both the transmembrane and cytosolic tail (Mosyak et al. 2006), the latter of which has been demonstrated to induce inactivation when co-expressed with BK channels (Dudem et al. 2020). Nevertheless, it is clear from the crystal structure (PBD:2ID5) of the ectodomain (shown in Fig. 3b) that each LINGO-1 monomer contains the LRR and Ig modules which fold into a bent question mark shaped structure. The N-terminal LRR module is very similar to that of the NgR ectodomain with 15 parallel β strands and only slight differences in curvature and arc length. Interestingly LINGO proteins have double the number of LRRs (12) found in BKγ subunits (Fig. 2c).

The main function of LINGO-1 appears to be the inhibition of axonal regeneration following spinal injury. LINGO-1 also suppresses oligodendrocyte precursor cell maturation and the production of the myelin sheath (Foale et al. 2017). A possible role of LINGO-1 in neurodegenerative diseases like multiple sclerosis (MS) and Parkinson's disease (PD) has also been highlighted. In support of this, increased expression of LINGO-1 was demonstrated in oligodendrocyte progenitor cells of postmortem brain in MS patients and in the dopaminergic neurons of PD patients (Mi et al. 2013). Similarly, Dudem et al. (2020) confirmed the earlier finding of Delay et al. (2014) that LINGO-1 levels were increased in Parkinson's disease and demonstrated that LINGO-1 and BK channels could be co-immunoprecipitated from human cerebellum. Given that BK:LINGO-1 co-expression downregulated plasmalemmal BK channel expression in HEK cells and leads to 'functional knock-out' of these channels (Dudem et al. 2020), it will be of interest to examine if this contributes to the aetiology of Parkinson's disease.

3 BK Channel Structure

BK channels are often called Slo1, MaxiK or KCa1.1 channels (Yang et al. 2015). The basic functional unit of this channel is the pore-forming BKα subunit, encoded by the KCNMA1 gene, which when expressed tetramerises to form a functional BK channel (Quirk and Reinhart 2001). BK channels are unique in the K^+ channel family because they possess 7 transmembrane domains, as shown in Fig. 4a. Initial hydropathy analyses were interpreted to suggest that the BKα subunit contained 10 hydrophobic segments (Pallanck and Ganetzky 1994), of which 6 were transmembranous. However, subsequent electrophysiological and epitope tagged immunocytochemistry studies (Wallner et al. 1996; Meera et al. 1997) demonstrated that BKα subunits actually had 7 membrane spanning domains (S0–S6) in addition to the 4 cytosolic tail domains (S7–S10).

The *Aplysia californica* BK channel cryo-EM structure was first published in the presence of Ca^{2+} and Mg^{2+} at a resolution of 3.5 Å (Tao et al. 2017) and this was followed, in 2019, with the cryo-EM structure of the human BK channel in complex with the β4 regulatory subunit (Tao and MacKinnon 2019). In both structures, the gating ring contributes to almost half the length of the BK channel tetramer which has the dimensions 110 Å × 110 Å × 130 Å (Fig. 4b, Tao et al. 2017). Each BKα subunit can be divided into three functional modules called the voltage-sensing domain (VSD), the pore gate domain (PGD) and the cytosolic tail domain (CTD) show in red, blue and grey/cyan, respectively in Fig. 4.

The transmembrane helices of the BK channel contain an extra TM helix called S0, which is linked to the extracellular N-terminal tail (Jiang et al. 2002a; Fodor and Aldrich 2006). This S0 domain is essential for interactions between the α and regulatory β subunits as well as voltage sensor modulation (Morrow et al. 2006). Close interactions between the S0 domain and the β4 subunit were recently observed in the cryo-EM structure of human BKαβ4 (Tao and MacKinnon 2019). The main residues involved in voltage sensing in BKα are distributed throughout S1–S4,

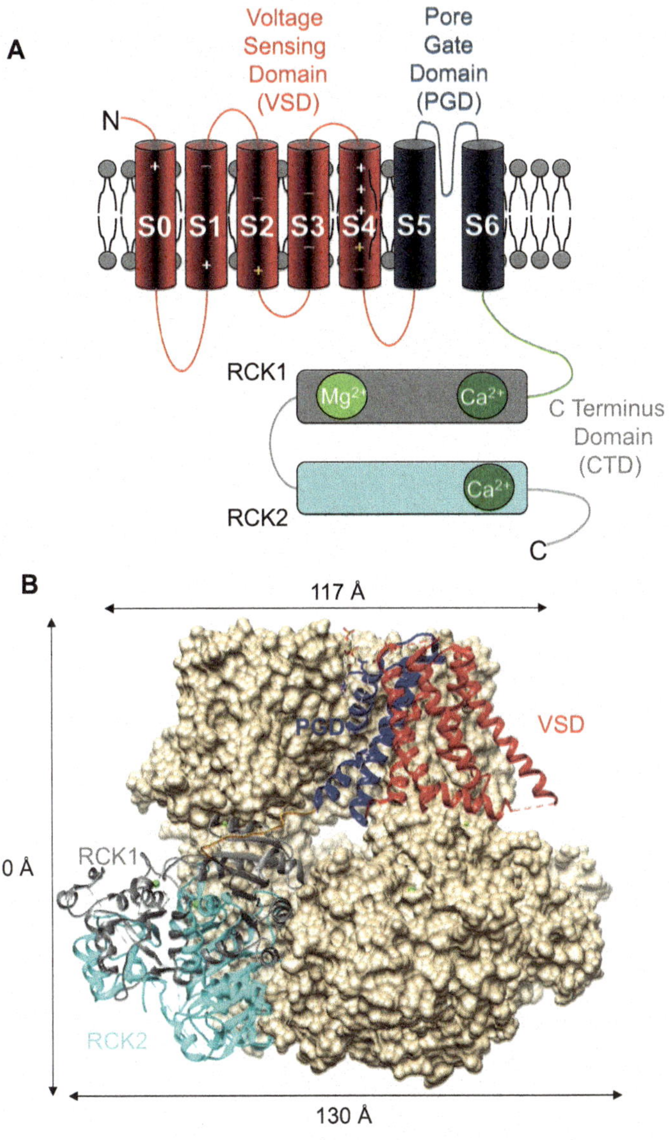

Fig. 4 Schematic diagram and the cryo-EM structure of the BK channel. (**a**) The BKα subunit which forms the pore consists of the VSD (red), PGD (blue) and CTD (grey & cyan). The binding sites for Mg^{2+} and calcium present in the CTD are also highlighted in light and dark green, respectively. The S6-RCK1 linker (orange) connects S6 segment of the PGD and the CTD. (**b**) Aplysia BK tetramer side view (PDB:5TJ6), showing the VSD domain (orange), pore domain (blue), RCK1 (grey) and RCK2 (cyan) domains of one subunit shown in ribbons and other three subunits shown in volume representation

whereas the S5–S6 helices constitute the Pore Gate Domain (PGD), which controls the permeation of K^+ ions through the channel. The large CTD senses Ca^{2+} and Mg^{2+} and is connected to the S6 segment via an S6-RCK1 linker (Fig. 4b orange line, Tao et al. 2017; Tao and MacKinnon 2019).

3.1 Voltage Sensor Domain (VSD)

As Fig. 5a shows, the BKα VSD is not domain swapped, in contrast to the VSD of the $K_v1.2–K_v2.1$ paddle chimaera and voltage gated K^+ (K_v) channels (Long et al. 2005a, b). Consequently, this structure results in the VSD of each BKα interacting with the pore domain of the same subunit. This appears to occur because the S4–S5 linker of the BKα is a short ordered loop rather than a longer helical structure demonstrated in other published K^+ channel structures (Sun and MacKinnon 2017).

An important point to note is that the VSD of one BKα subunit interacts with the CTD of a neighbouring subunit (see Fig. 4b, Tao et al. 2017). Other features of the BKα structure are that the S4 helix is tightly packed against the S5 helix of the BK channel and runs antiparallel to it (Fig. 5b), whilst the pore domain is situated adjacent to the VSD of the same BKα subunit (Fig. 5a).

3.2 Voltage Dependent Activation

Prior to the publication of the cryo-EM structures of BKα, many studies assumed that BK channels shared a number of features with K_v channels. A number of studies (Atkinson et al. 1991; Adelman et al. 1992; Butler et al. 1993) attempted to identify if the S4 transmembrane helix served as a distinct voltage-sensing domain in BK channels. The first direct evidence to support the idea that BK channels were voltage activated came from the gating charge-voltage (QV) relationship obtained from gating currents of BK channels in the absence of Ca^{2+} (Stefani et al. 1997; Horrigan and Aldrich 1999). Also, the conductance-voltage (GV) relationship of macroscopic BKα currents, in the virtual absence of calcium, strongly supported the idea that BK channels could be activated by voltage alone (Cui et al. 1997), although they were less sensitive to changes in voltage compared to K_v channels. Thus, when the BK effective gating charge was determined, it was only ~2.3e, compared to 12–13e observed in Shaker channels (Bezanilla 2000).

In 2006, Ma et al. carried out a rigorous electrophysiology and mutagenesis-based study to examine which charged residues in the TM segments contributed to the gating charge. Intriguingly, they found that although 4 charged residues (D153, R167, D186 and R213) contributed to gating charge in BK channels, these were distributed across different transmembrane helices, rather than concentrated in the S4 segment. In fact, of the 3 arginines in S4 (R207, R210 and R213), only neutralisation of R213 significantly reduced the effective gating charge (by 1.2e), again supporting the idea that the voltage dependence of channel opening must also involve charged residues outside S4. Through their methodical work, Ma et al.

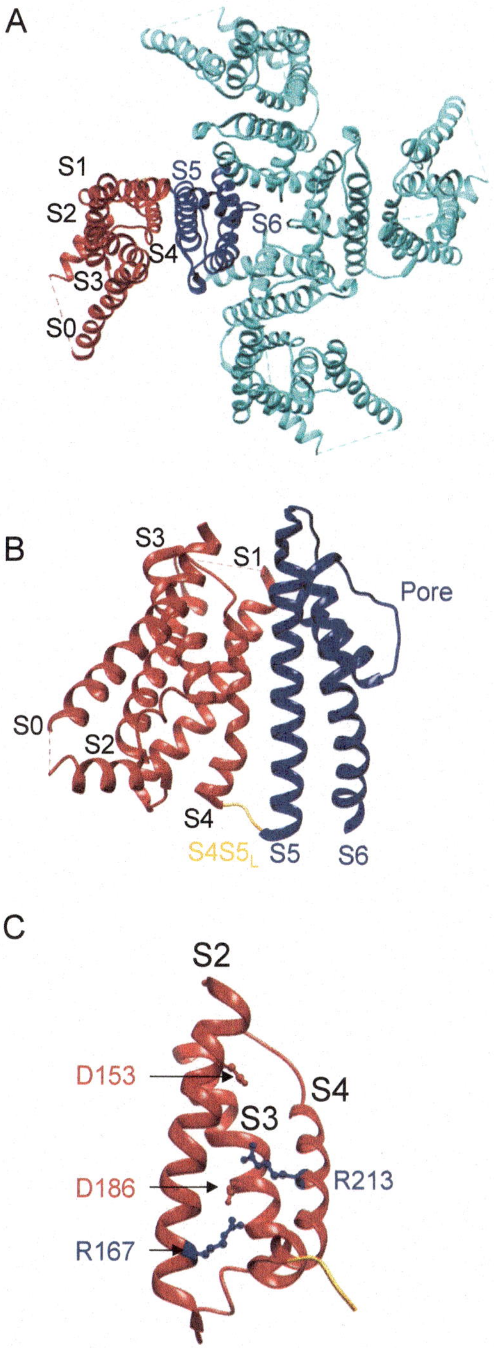

Fig. 5 The cryo-EM structure and topology of the BK channel VSD. (**a**) The transmembrane regions of the BK channel are shown as a top view with the VSD of one subunit shown in red and the PD in dark blue. (**b**) The ribbon representation of the transmembrane region of one subunit of a BK channel is also shown as a side view. In (**c**) the VSD of the BK channel is shown; the residues shown by mutagenesis to be important for voltage-sensing are highlighted in red (D153 in S2 and D186 in S3) and blue (R167 in S2 and R213 in S4)

(2006) established that residues D153 and R167 in S2 and D186 in S3, when neutralised, reduced the effective gating charge by 0.92e, 0.48e and 0.88e, respectively, consistent with the idea that S2 and S3 were also involved in BK channel voltage sensing.

3.3 Cytosolic Tail Domain (CTD)

The CTD of BKα is large, consisting of approximately 800 amino acids (Lingle 2007) and has two RCK (regulator of conductance for potassium) domains: RCK1 and RCK2 (Fig. 4). Wu et al. (2010) solved the crystal structure of this CTD from the human BK channel in the absence of Ca^{2+}. The published crystal structure of the entire cytoplasmic region of the human BK channel shows some similarity to the prokaryotic Ca^{2+} gated K^+ channel (MthK). Interestingly, although the RCK1 domain of BK channels shares a high sequence similarity with the MthK channel, this is not the case with the second RCK domain (Jiang et al. 2002b; Wu et al. 2010). The more recent cryo-EM structure of the *Aplysia californica* BK channel gating ring, in the open conformation, is very similar to the X-ray crystal structure of the BK channel Ca^{2+} bound gating ring (Yuan et al. 2011; Tao et al. 2017). The eight RCK domains from the four α subunits of the BK channel assemble as a 'Gating Ring' as shown in Fig. 6d. Each RCK domain consists of 3 subdomains. First, the N-terminal Rossman folded subdomain (βA-βF) forms the central core of the gating ring. Second, the C-terminal subdomain, which interacts with the other RCK domain within the same subunit. Third, the intermediate helix-crossover domain (αF turn αG) which connects two RCK domains. (Fig. 6e, Wu et al. 2010; Yuan et al. 2010).

3.4 Calcium Dependent Activation

Intracellular Ca^{2+} binds to the CTD of the BK channel and results in channel opening. Electrophysiology and mutation studies identified that the cytosolic tail domain of the BK channel comprised two main calcium binding sites. The first site called the Ca^{2+} bowl, was located in the RCK2 domain and contained a series of aspartic acid residues (Schreiber and Salkoff 1997). This was confirmed when the stuctures of human BK channel CTD and the full length *Aplysia* BK channel were published, showing that the side chain carboxylates of D905 and D907 (D895, D897 in human BK sequence) and the main chain carbonyl groups of Q899 and D902 (Q889 and D892 in human BK sequence) directly bound Ca^{2+} ions (Yuan et al. 2010; Tao et al. 2017). Interestingly, as shown in Fig. 7b, the cryo-EM structures also showed that residue N438 in an adjacent subunit also contributed to Ca^{2+} coordination in the Ca^{2+} bowl (Tao et al. 2017).

Although mutations in the Ca^{2+} bowl reduced the Ca^{2+} sensitivity of BK channels, they did not eliminate it, suggesting that a second Ca^{2+} binding site was present in the CTD (Schreiber and Salkoff 1997). Mutational studies in the RCK1 domain utilised a D362A/D367A mutant to demonstrate that these residues

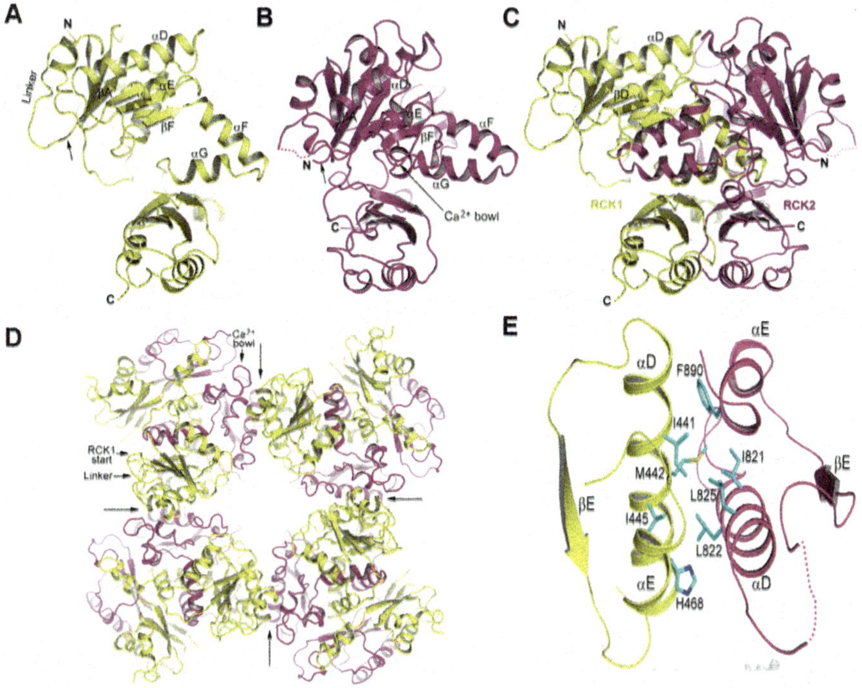

Fig. 6 RCK1, RCK2 and gating ring structures. (**a** and **b**) RCK1 is denoted in yellow and RCK2 in magenta. The starting point of each RCK is indicated by the arrow and the linker connects RCK1 to the channel pore. (**c**) RCK1 and RCK2 form the BK intracellular subunit, and the orientation of RCK1 shown in the subunit assembly is the same as in (A). The long loop connecting the C-terminus of RCK1 and N-terminus of RCK2 is disordered. (**d**) View of the gating ring structure viewed down the fourfold axis from the extracellular side. The position of the inter-subunit assembly interfaces is indicated by the four long arrows and they define the boundary of each subunit. The linker between RCK1 and the channel pore, the starting point of RCK1, and the position of the Ca^{2+} bowl are labelled on one subunit (upper left corner). (**e**) An enlarged view of the assembly interface formed by helices αD and αE from both RCK1 and RCK2. The side chains of those hydrophobic residues important for protein–protein contacts are also shown. The dotted line represents the disordered loop between αD and βE (Adapted from Wu et al. 2010)

participated in binding Ca^{2+} in the RCK1 site (Xia et al. 2002; Zeng et al. 2005). Interestingly an M513I mutant also significantly reduced the Ca^{2+} sensitivity, suggesting that this residue contributed to Ca^{2+} binding. Unfortunately, the Yang et al. (2010) study was unable to resolve any Ca^{2+} binding sites in the crystal structure of the BK channel CTD, even in the presence of 50 mM Ca^{2+}. However, as shown in Fig. 7c, the cryo-EM structure of the *Aplysia* BK channel (Tao et al. 2017) provided a clear answer and showed that Ca^{2+} is coordinated by the main chain carbonyl oxygen atoms of R503 (R514 of human BK), G523 and G591 residues and the side chain carboxylates of the D356 and E525 (corresponding to D367 and E535 in human BK) residues (Fig. 7c, d; Yuan et al. 2010; Tao et al. 2017). The other residues (M513 and D362), which had been previously proposed to

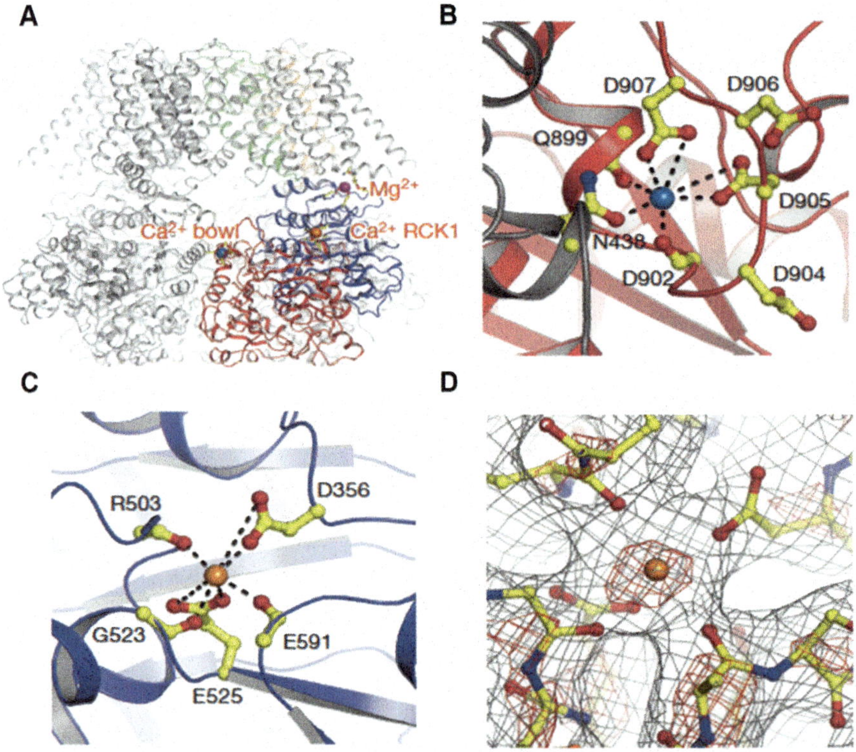

Fig. 7 Ca^{2+} binding sites in the cryo-EM structure of Aplysia BK. (**a**) Ca^{2+} bowl and Ca^{2+} RCK1 sites are shown in the one subunit of the BK tetramer. The remaining three subunits coloured grey. Divalent cations are represented as spheres. (**b**) Ca^{2+} bowl site. Channel is shown as ribbon with the RCK2 domain coloured red and neighbouring RCK1 domain grey. (**c**) Ca^{2+} RCK1 site. Channel is shown as blue ribbon. (**d**) Density at the Ca^{2+} RCK1 site (Adapted from Tao et al. 2017 with permission)

be involved in Ca^{2+} binding do not seem to be part of the Ca^{2+} binding pocket and it appears likely that these mutations produced their effects by altering the structure at the RCK1 Ca^{2+} binding site.

3.5 Pore Gate Domain (PGD)

The PGD is formed by the S5-pore loop-S6 segments and permit K^+ permeation in response to membrane depolarisation or increase in $[Ca^{2+}]_i$ (Magleby 2003; Salkoff et al. 2006). The BK channel PGD is homologous to those of prokaryotic and eukaryotic 2TM and 6TM K^+ channels (Yang et al. 2015). The conduction pathway for K^+ channels is composed of a large intracellular inner vestibule and a shallower extracellular outer vestibule connected with the selectivity filter (Doyle et al. 1998;

Zhou et al. 2001). The inner vestibule of BK channels allows cations such as Na^+, Tl^+, NH_4^+, K^+, Ca^{2+}, Rb^+, Mg^{2+}, Sr^+ and Ba^{2+} to enter the pore, although only NH_4^+, K^+, Tl^+ and Rb^+ cross through the selectivity filter. The other cations act as a fast or flickery blockers in the inner vestibule of the BK channel (Brelidze and Magleby 2005). BK channels contain a signature sequence (TVGYG) in their selectivity filter similar to other K^+ channels. The five-carbonyl oxygen atoms of these amino acids provide four potential binding sites to coordinate the movement of K^+ ions through the pore (Heginbotham et al. 1994).

3.6 Activation Gate

Voltage sensor activation and Ca^{2+} binding ultimately lead to conformational changes in the PGD, which results in the BK channel gate opening. Despite similarities with other K^+ channels, the BK channel activation gate possesses specific structural and functional features which distinguish it from other K^+ channels.

Firstly, a large conductance (250–300 pS) is a signature feature of BK channels compared with other K^+ channels. Eight negatively charged residues (E321 and E324 from each subunit) at the inner vestibule region of the BK channel are responsible for this large conductance. They increase the K^+ concentration at the inner vestibule region, through an electrostatic mechanism (Brelidze et al. 2003). Also, residue D292, located at the external vestibule and close to the selectivity filter region, helps to concentrate K^+ ions at the external surface of the channel and thus helps increase the conductance (Haug et al. 2004). These residues, combined with the large diameter of the inner vestibule region, ensure the large single channel conductance of the BK channel.

Secondly, the large inner vestibule region and wider entrance at the intracellular region of the pore also differentiate BK channels from other K^+ channels. Various sizes and functional properties of chemicals have been used to ascertain the dimension of the inner vestibule region. Irrespective of the closed or open state of the BK channel, quaternary ammonium (QA) ions like tetrabutylammonium (TBA) have free access to the inner vestibule region. Compared to K_v channels, these QAs show faster blocking and unblocking kinetics in BK channels, consistent with a large diameter inner vestibule region (Li and Aldrich 2004; Wilkens and Aldrich 2006). The diameter of the cytosolic pore mouth was estimated ~20 Å, based on changes in K^+ diffusion rates from the intracellular region to inner vestibule region of BK channel, by sucrose interference (Brelidze and Magleby 2005).

Thirdly, the BK channel S6 transmembrane pore-lining residues orientation appears different from K_v channels. Cysteine scanning mutagenesis and state dependent cysteine scanning accessibility studies first suggested that residues A313, A316 and S317 are likely to face towards the inner pore region. In contrast, in the other K^+ channels the same corresponding residues face away from aqueous environment (Zhou et al. 2011), suggesting that the movement of the pore-lining S6 residues differ in BK and K_v channels. Site directed mutagenesis and pharmacological studies of the BK channel S6 segment pore residues (L312, A313 and A316) demonstrated that side chain reorientation occurs during channel gating (Chen et al. 2014). Mutagenesis of either one of these residues to 18 different amino acids revealed that polar or charged side chain substitution of L312 or A313 or A316 constitutively open the channels, irrespective of voltage or Ca^{2+} dependency. This occurs presumably by exposing the hydrophilic side chains to the aqueous environment of the pore to reduce their side chain solvation energy (Chen et al. 2014).

3.7 Allosteric Gating of BK Channels

BK channels can open, albeit with a very low open probability (Horrigan and Aldrich 2002), in the absence of voltage sensor activation or Ca^{2+} binding, suggesting that neither is essential for channel opening. The ability of voltage sensor activation or Ca^{2+} binding to influence channel opening, in different subunits, in a non-obligatory fashion, can be explained by the Monod-Wyman-Changeux (MWC) model. The basic idea of this model is that ligand binding to any of the ligand binding sites promotes conformational changes that alter the ligand binding site and its affinity for calcium (Monod et al. 1965). Horrigan and Aldrich (2002) proposed a model, shown in Fig. 8, to help explain the allosteric gating mechanism of BK channels based on steady state data recorded across a wide range of voltages and Ca^{2+} concentrations (Horrigan and Aldrich 1999, 2002). The Horrigan Aldrich model of BK channel gating suggests that the channel gate can undergo conformational change from closed (C) to open (O), which is allosterically coupled to four independent and identical voltage sensors and four Ca^{2+} sensors. The Ca^{2+} sensors can exist in an unbound Ca^{2+} (X) or a Ca^{2+} bound state (X.Ca^{2+}), whereas the voltage sensors can be in either the resting (R) or activated (A) state. Each conformational change is represented by equilibrium constants for gate opening (L), voltage sensor activation (J) and Ca^{2+} binding (K). The coupling or energy transfer

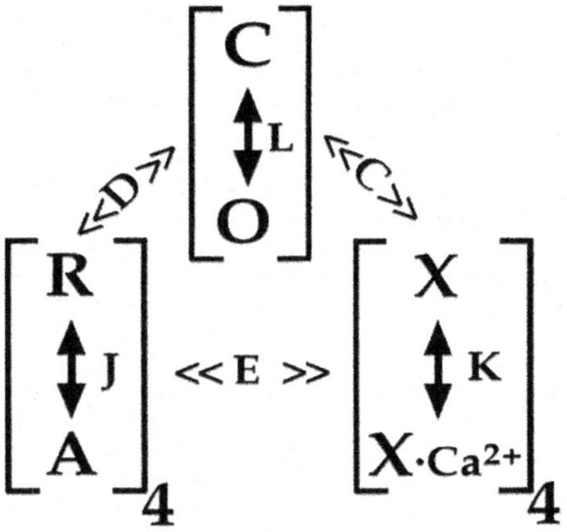

Fig. 8 The Horrigan and Aldrich Model. The allosteric model suggests the possible conformations of the gate (C and O), voltage sensors (R and A) and Ca^{2+} sensors (X and X.Ca^{2+}) in each of the 4 subunits of the BK channel. The allosteric factors (C, D and E) describe the energetic coupling between these three parts of the channel. Each conformational state is represented by equilibrium constants L, J and K

between domains is represented by allosteric factors C, D and E which represents the coupling between (1) Ca^{2+} binding and pore (C), (2) Voltage sensor activation and the pore (D), (3) Ca^{2+} binding and voltage sensor activation (Horrigan and Aldrich 2002), respectively.

3.8 Calcium Sensor/Gate Coupling

The coupling between the Ca^{2+} sensors and the gate is the easiest to measure and consequently is the best understood at a molecular level. In MthK channels, Ca^{2+} binding to the CTD causes conformational changes in the gating ring which pull the S6/RCK1 linker to promote channel opening (Jiang et al. 2002b). In 2004, Niu et al. proposed that the linker and gating ring formed a passive spring which could pull on the gate to open the channel. Their study demonstrated that shortening the S6/RCK1 linker increased channel activity and lengthening the linker length decreased it, suggesting that the linker remains under tension, even in the absence of Ca^{2+} (Niu et al. 2004). Crystal structures of the MthK and BK channels suggest that in the presence of Ca^{2+}, the gating ring diameter expands by up to 12 Å (Ye et al. 2006; Wu et al. 2010; Yuan et al. 2011) and this expansion would increase the tension on the S6/RCK1 linker. Ca^{2+} sensor/gate coupling and Ca^{2+} sensitivity mutations in the AC region of the BK channel RCK1 domain suggest that the flexibility of RCK1/S6 linker is essential for transmitting CTD conformational changes to the gate (Yang et al. 2010). Recent results suggest that the N-terminal half of the RCK1 undergoes extensive reorientation compared with other gating ring regions upon Ca^{2+} binding. This generates the possibility that conformational changes in RCK1 AC region are coupled to the gate through direct contact with the PGD (Yuan et al. 2011). Mutations of RCK1 and RCK2 Ca^{2+} binding sites and measurement of P_o at -80 mV, in a wide range of Ca^{2+} concentrations, performed by Sweet and Cox (2008), attempted to determine the contribution of each Ca^{2+} sensing domain to coupling energy. They demonstrated that the Ca^{2+} sensor/gate coupling energy for the RCK1 site was 3.74 kcal mol^{-1} and the coupling energy for the RCK2 site was 3.04 kcal mol^{-1}. Cumulatively, the coupling of these two Ca^{2+} binding sites was higher than the WT BK channel Ca^{2+} sensor/gate coupling (5.0 kcal mol^{-1}), suggesting that negative co-operativity may exist between these two sites (Sweet and Cox 2008). However, Hite et al. (2017) demonstrated in their cryo-EM structures that positive co-operativity between the two sites is likely to occur during Ca^{2+} binding, as a result of the tilting of the N-lobe of RCK1. This not only leads to its expansion, but the resultant conformational changes simultaneously 'complete' Ca^{2+} coordination. Consequently, the coordination of Ca^{2+} at both Ca^{2+} binding sites shifts the position of the RCK1 N-lobe and favours its open conformation.

3.9 Voltage Sensor/Gate Coupling

The interactions between the S4–S5 linker and the S6 helix are considered to account for voltage sensor/gate coupling, which is responsible for electromechanical coupling in K_V channels (Lu et al. 2002; Tristani-Firouzi et al. 2002; Long et al. 2005b) and it is likely to be important in BK channels (Sun et al. 2012). What is clear from the recently published cryo-EM structures, however, is that the S4-S5 linker is a short loop in BK channels (Hite et al. 2017; Tao et al. 2017; Tao and MacKinnon 2019), in contrast to the longer helix observed in other K^+ channels. This difference in structure may help inform the design of more selective BK channel agonists in the future that are targeted to this region of the channel.

Although a number of studies have demonstrated that the voltage sensor is distributed across S2, S3 and S4 in BK (Fig. 5, Bao and Cox 2005), the precise mechanism of how a change in voltage leads to channel opening remains unclear. Intriguingly in the cryo-EM studies of Hite et al. 2017 and Tao et al. 2017, it is clear that the S4 helix is displaced ~2 Å towards the cytoplasmic surface in the closed configuration (Ca^{2+} free), compared to the open (10 mM Ca^{2+}) state. However, this displacement of the voltage sensors appears to be caused by the relative positions of the N-lobes of RCK1 associated with Ca^{2+} binding. Intriugingly, there was no major change in the rotamer conformation of the Arginine residues in S4 between the Ca^{2+} free and Ca^{2+} bound structures, suggesting that either the small reorientation of the Arginine sidechains was sufficient to account for the observed displacement of the voltage sensors or that the voltage-sensing mechanism is very different to that observed in other K^+ channels. Indeed Hite et al. (2017) hypothesised that the BK channel voltage sensor may have evolved to be a modifier of the Ca^{2+} sensor, which 'bias' the RCK1 N-lobe conformation.

4 BK Channel Activators

Over the last 30 years a large number of BK channel openers have been designed, in the hope of developing treatments for a variety of diseases including stroke, asthma, erectile dysfunction, bladder overactivity, hypertension, ischaemic heart disease and hypermotility (Nardi and Olesen 2008). However, none have made it past Phase III clinical trials, to date, presumably due to a lack of efficacy and/or poor selectivity. The search continues for more selective BK channel openers, which can mediate their effects at physiological membrane potentials. With the publication of the human BK channel structure (Tao and MacKinnon 2019), we now have a valuable tool to help generate a 3D pharmacophore which may guide rational drug design. However, caution is called for, as the cryo-EM structures published represent the channel in two extreme states, one in the presumed closed state in Ca^{2+} free conditions and the other in saturating Ca^{2+} and Mg^{2+} concentrations and neither of these states may be encountered physiologically. Nevertheless, they represent an excellent starting position and will undoubtedly help in the design of more selective BK channel openers.

The BK channel activators discovered to date can be broadly classified into endogenous, synthetic and natural activators and act either on BKα subunits alone or require the presence of BK regulatory β subunits (see Table 1). Interestingly, a

Table 1 Modulators of BK channels

Molecule	BK subunits required	Shift in $V_{1/2}$ in mV (concentration used)	EC_{50}	Reference
NS1619	BKα	~−40 (30 μM)	–	Olesen et al. (1994)
17β-estradiol (E2)	BKα + β1	−21 ± 3 (3 μM)	2.6 μM	Valverde et al. (1999)
Tamoxifen	BKα + β1	−10 ± 3 (1 μM)	650 nM	Dick et al. (2001)
Pimaric acid	BKα	~−20 (10 μM)	–	Imaizumi et al. (2002)
Chlorzoxazone	BKα + β1	~−20 (30 μM)	–	Liu et al. (2003)
NS11021	BKα	~−60 (10 μM)	400 nM	Bentzen et al. (2007)
Zonisamide	BKα + β1	–	34 μM	Huang et al. (2007)
Human β-defensin	BKα + β1	~−25 mV (100 nM)	1.4 nM	Liu et al. (2013)
Docosahexaenoic acid (DHA)	BKα + β1, BKα + β4	−59 ± 4 (3 μM) −61 ± 3 (3 μM)	480 nM –	Hoshi et al. (2013a, b)
GoSlo-SR-5-6	BKα + β1	−107 ± 7 mV (10 μM)	2 μM	Roy et al. (2012)
GoSlo-SR-5-69	BKα + β1	>−104 ± 9 mV (1 μM)	251 nM	Roy et al. (2014)
GoSlo-SR-5-6	BKα	>−120 (10 μM)	3.2 μM	Webb et al. (2015)
GoSlo-SR-5-130	BKα + β1 BKα + β4	−92 ± 9 (10 μM) −84 ± 4 (10 μM)	2.8 μM –	Large et al. (2015)

number of modulators possess different calcium dependencies and the putative drug binding sites vary (Bentzen et al. 2014).

4.1 Endogenous BK Channel Activators

Endogenous chemicals such as metabolites of cytochrome 450, like epoxygenase, lipoxygenase and arachidonic acid, regulate vascular tone by increasing BK channel activity (Hou et al. 2009). The omega-3 fatty acid DHA is a β1 and β4 subunit dependent BK channel activator (Hoshi et al. 2013a) and the effects of DHA appear to depend on residues R11 and C18 of β1 and both E12 and R19 of the β4 residues (Hoshi et al. 2013b). Some steroidal hormones such as 17β-estradiol (E2) also activate BK channels and require the presence of β1 subunits (Valverde et al. 1999). E2 can activate BK channels when applied extracellularly, suggesting that the binding site might be located at an extracellular region of the β1 subunit (Morrow et al. 2006). Interestingly, Morrow et al. (2006) also found that deletion of the first

20 residues in the BKα subunits (αΔN20 mutant), co-expressed with β1 subunits, also abolished the effects of E2. The xenoestrogen, tamoxifen which is commonly used in the treatment of breast cancer, also stimulates BK channels. It increases the open probability of BK channels and also requires the β1 subunit to mediate its effects (Dick et al. 2001).

The antimicrobial peptide human β-defensin 2, which plays an important role in the innate immune system, also increased the open probability of BK channels and shifted their activation $V_{1/2}$ by ~ − 25 mV (Liu et al. 2013). The effects of this cysteine-rich peptide were also dependent on the presence of β1 subunits and were significantly reduced when two residues in the extracellular loop of the β1 subunit (L41 and Q43) were mutated (Liu et al. 2013).

4.2 Natural BK Channel Activators

One of the earliest BK channel activators discovered, DHS-1 (McManus et al. 1993) was initally extracted from the medicinal herb *Desmodium adscendes* and was used in Ghana as a 'folk remedy' for the treatment of asthma (Addy and Awumey 1984). When purified, the extracts contained three glycosylated triterpenes, namely dehydrosoyasaponin-1 (DHS-1) and soyasaponins I and III. DHS-1 appeared to be the most potent BK channel opener of the three, since when applied intracellularly at a concentration of 10 nM, it increased the P_o of smooth muscle BK channels (McManus et al. 1993). DHS1 also required the presence of regulatory β1 subunit to mediate its effects.

Pimaric acid, isolated from the resin of *Pinus genus*, has also been shown to open BK channels when applied to either side of the membrane at concentrations >1 μM (Imaizumi et al. 2002). The polyphenol trans-resveratrol, present in the weed *Polygonum cuspidatum*, has been shown to increase BK channel activity in vascular endothelial cells (Li et al. 2000). However, these effects were rather small given that the $V_{1/2}$ was only shifted by ~ − 15 mV in the presence of 30 μM of this compound. Furthermore, like so many BK channel openers, its selectivity appears poor and it was shown to activate SK channels in the same study (Li et al. 2000). Nordihydroguaiaretic acid, a natural lignin, has also been shown to be a BK channel activator in smooth muscle cells from the porcine coronary artery (Nagano et al. 1996), which increased P_o in a concentration dependent manner. Flavonoids like 5-hydroxyflavone, 5-methoxyflavone and 7-hydroxyflavone also appear to induce vasorelaxation via increased BK channel activity and the hydroxyl group at position 5 in the flavonoid structure appears to be critically important for the observed effects (Calderone et al. 2004).

4.3 Synthetic BK Channel Activators

Some of the first synthetic BK channel openers to be discovered were the benzimidazoles NS004 and NS1619 (Olesen et al. 1994) and these have served as reference structures for a number of other synthetic molecules. Olesen et al. (1994) demonstrated that NS1619 shifted the voltage required for half maximal activation ($V_{1/2}$) of BK channels by −40 mV, albeit at a relatively high concentration (~30 μM). However, like many BK channel modulators, NS1619 lacked selectivity and has a number of off-target effects, including blockade of Ca^{2+} channels (Edwards et al. 1994; Holland et al. 1996).

Another Neurosearch compound, the biarylthiourea NS11021, has been shown to be a more efficacious and selective BK channel opener since it shifted the activation $V_{1/2}$ of BK channels by ~ − 60 mV at a concentration of 10 μM (Bentzen et al. 2007). Furthermore, NS11021 seemed to be more selective and had little effect on T or L-type Ca^{2+} channels, Na^+ channels, inward rectifiers or a range of K_v channels. However, it did significantly activate Kv7.4 channels when applied at a concentration of 30 μM. NS11021 has been shown to improve erectile responses in rats (Kun et al. 2009) and cause relaxation in small penile arteries via an effect on BK channels (Király et al. 2013). In addition, it has been shown to reduce contractility in the guinea pig bladder (Layne et al. 2010).

Other compounds appear to be more potent at activating BK channels. For example, 300 nM of the voltage-sensitive fluorescent dye, bis-(1, 3-dibutylbarbituric acid) trimethine oxonol [DiBAC(4)(3)] shifted the $V_{1/2}$ of BKα co-expressed with β1 and β4 subunits (Morimoto et al. 2007) by about −20 mV, yet appeared to have no effect on BKα channels alone. However, Scornik et al. (2013) later demonstrated that DiBAC(4)(3) could activate BK channels in the absence of regulatory subunits.

In 1996, Cotton et al. demonstrated that the dye, Cibacron blue, could activate bladder BK channels. In 2012, Roy et al. utilised the anilinoanthraquinone core common to Cibacron blue and Acid Blue 25, to synthesise a novel group of BK channel activators called the GoSlo family, for the potential treatment of overactive bladder (OAB). These compounds were effective on rabbit bladder smooth muscle cells and shifted the voltage dependent half maximal activation ($V_{1/2}$) of BK channels by >-100 mV at 10 μM. GoSlo-SR-5-6 and GoSlo-SR-5-44 were the most potent and efficacious molecules in the GoSlo family, shifted the $V_{1/2}$ by −107 ± 7 mV and − 142 ± 8 mV, respectively, with an EC_{50} of ~2 μM (Roy et al. 2012). Webb et al. (2015) identified the molecular mechanism of action of GoSlo-SR-5-6 and revealed that its effects were practically abolished when three residues, L227 in the S4–S5 linker and both S317 and I326 in S6 were mutated (Webb et al. 2015). For an excellent review on the mechanism of action of BK channel openers, readers are referred to Hoshi and Heinemann (2016). In 2014, Roy et al. synthesised additional derivatives and identified a much more potent member of the GoSlo family, the tetrahydro-2-naphthalene derivative GoSlo-SR-5-69. This compound shifted the activation $V_{1/2}$ by ~-100 mV at 1 μM concentration and the EC_{50} of of this compound was 251 nM. The same group (Large et al. 2015) also found that although GoSlo-SR-5-6 could mediate its full effects in the absence of regulatory subunits, the analogue GoSlo-SR-5-130 was much more effective on BKαβ1 and BKαβ4 channels.

Interestingly, the efficacy of some of the GoSlo-SR family was also altered by regulatory γ1 subunits. For example, Kshatri et al. (2016) demonstrated a 70%

reduction in the ability of GoSlo-SR-5-44 to shift activation $V_{1/2}$ in BKα channels co-expressing regulatory γ1 subunits. Almassy and Begenisich (2012) demonstrated a similar reduction in the efficacy of the BK channel opener mallotoxin (Zakharov et al. 2005), in native and heterologously expressed parotid BK channels, providing further evidence that γ1 subunits can markedly alter the pharmacological properties of BK channels.

Different compounds have been studied in animal models, but no modulator has cleared phase III clinical trials. For example, a study by Cheney et al. (2001) showed that BMS204352 reduced regional cerebral edema and motor impairment after brain surgery in mice. Although these data supported the idea that it had neuroprotective effects and may be beneficial in the treatment for acute ischaemic stroke, it failed in phase III clinical trials (Bozik et al. 2000). Andolast is the only BK channel opener which shows fewer side effects and desirable therapeutic effects with respect to treatment of asthma. However, this tetrazolyl-benzamide derivative appears rather non-selective, since in addition to opening BK channels, it also inhibits AMP production, but does reduce bronchoconstriction in asthma patients (Arshad et al. 2007). It has undergone randomised, controlled, double blind multicentre tests (Malerba et al. 2015) and produced a dose dependent improvement in forced expiratory volume (FEV1), particularly in patients with moderate airways obstruction. However, the status of its commercial development is currently unclear.

4.4 Lack of Selectivity of BK Channel Openers

The commercialisation of BK channel openers has been plagued by their lack of selectivity and consequent 'off-target' effects against other ion channel families. For example, some of the early NS compounds such as NS1619 were notoriously unselective and not only blocked Ca^{2+} channels and Kv channels (Edwards et al. 1994; Holland et al. 1996) but also activated CFTR channels (Al-Nakkash et al. 2001). A lack of selectivity has been observed with other compounds including BMS-204352, NS11021 and GoSlo-SR-5-6, which have all been shown to be efficacious activators of Kv7 channels (Schrøder et al. 2001; Bentzen et al. 2007; Zavaritskaya et al. 2020). Recently, Schewe et al. (2019) identified a common feature amongst some efficacious K^+ channel agonists which may help explain the problematic lack of selectivity against other K^+ channels. Their data suggest that the polypharmacology observed with negatively charged activators (NCA) is due to (1) the compounds possessing a common negatively charged tetrazole or carboxylate group and (2) acting at a common site, namely the selectivity filter (SF) of some ion channels. Schewe et al. (2019) convincingly demonstrated that the NCA compounds competed with pore blockers, suggesting overlapping binding sites. Their study utilised a combination of electrophysiology, X-ray crystallography and molecular dynamics simulations to support that idea that NCAs such as NS11021 and GoSlo-SR-5-6 possess a common pharmacophore which increases ion permeation and channel open probability, predominantly by enhancing K^+ occupancy below and within the selectivity filter. These findings have significant implications for drug development and strongly suggest that NCAs are unlikely to be subtype-selective ion channel agonists.

5 SK Channel Modulators

Small conductance Ca^{2+}activated K^+ (SK or KCa2) channels consist of three family members (SK1–3), which are highly expressed in the central nervous system and play role in neuronal excitability and afterhyperpolarisation (Faber and Sah 2007). These channels have been targeted therapeutically for diseases including Parkinson's disease, Alzheimer's, Schizophrenia as well as heart diseases. As their name suggests, the SK channels have a small conductance (10pS). These channels are activated by an increase in intracellular Ca^{2+} (Wei et al. 2005) which binds to calmodulin and induces conformational changes resulting in channel opening. SK channel modulators produced so far can be classified into three main types, namely blockers, positive gating modulators and negative gating modulators (Lam et al. 2013; Christophersen and Wulff 2015). These gating modulators can increase or decrease the Ca^{2+} sensitivity of the channels, whereas the blockers prevent ion channel conduction via SK channels (Christophersen and Wulff 2015).

Of the blockers, the bee venom toxin apamin is the most widely known SK channel blocker, but its ability to block the channels depends on the subtype and indeed the species from which the channel isoform is isolated. Thus, SK2 channels are the most sensitive to apamin ($IC_{50} \sim 70$ pM), SK3 channels have a reduced sensitivity with an IC_{50} of ~0.6–6 nM, the human SK1 channel is least sensitive ($IC_{50} \sim 1$–8 nM) and the rat isoform is resistant to apamin (D'hoedt et al. 2004). Apamin acts via an allosteric mechanism and involves a single amino acid histidine, which is located in the outer vestibule of the SK channel pore (Nolting et al. 2007; Lamy et al. 2010). Scuvée-Moreau et al. (2004) demonstrated that two compounds, Methyl-Laudanosine and Methyl-Noscapine blocked the SK channel mediated hyperpolarisation of dopaminergic neurones, but neither compound shows selectivity for the three different channel isoforms. The antidepressant molecule fluoxetine has also been shown to block the three human SK channel subtypes with similar potencies ($IC_{50} = 9$ μM, 7 μM and 20 μM, for SK1–3, respectively, Terstappen et al. 2003). A few other SK channel blockers have been described including UCL1684 and UCL1848, which are non-peptide bis-quinolinium cyclophanes (Conejo-Garcia and Campos 2008) but show little selectivity amongst the three isoforms. However, the tamapin is a remarkably potent and SK2 selective peptide scorpion toxin, which has an IC_{50} of ~24 pM and this concentration is ~1750 lower than the IC_{50} in SK1 and ~70-fold lower than than required to block SK3 channels (Pedarzani et al. 2002).

A number of SK channel gating modulators have been developed but they vary in their specificity and selectivity. The first reported non-selective positive gating modulator, 1-EBIO (1-ethyl-2-benzimidazolinone) had a relatively high EC_{50} value of ~300 μM (Devor et al. 1996; Wulff et al. 2007). Later NS309, which is structurally related to 1-EBIO was shown to activate SK subtypes at submicromolar concentrations (Strøbæk et al. 2004). SKA-31 is another non-specific positive gating modulator which was shown to reduce excitability and decrease contractility in human and guinea pig detrusor smooth muscle, which may be a potential approach to treat bladder dysfunction disorders (Soder et al. 2013; Parajuli et al. 2011).

The hunt for more selective positive SK modulators led to the development of NS13001, which selectively activates SK2 and SK3 subtypes with an EC_{50} of 5 μM and 0.14 μM, respectively. Interestingly this compound prevented Purkinje cell degeneration and improved motor deficits in spinocerebellar ataxia type 2 mice (Kasumu et al. 2012). Hougaard et al. (2009) described GW542573X as an SK1 selective activator (EC_{50} value of 8 μM) and demonstrated that it activated SK1 channels even in the absence of Ca^{2+} (Hougaard et al. 2009).

The selectivity of negative SK channel modulators is also variable, as exemplified by NS8593 and NS11757. Whereas NS8593 blocked all subtypes of SK channels in submicromolar concentrations and demonstrated atrial antiarrhythmic properties (Strøbæk et al. 2006; Haugaard et al. 2015), NS11757 appeared to block only SK3 channels with a Kd value of 9 nM (Sørensen et al. 2008).

6 IK Channel Modulators

Intermediate conductance Ca^{2+}activated K^+ channels (IK or KCa3) have a single channel conductance of 20–80pS and their expression, at the transcriptional level, has been shown in prostate, colon, lung, placenta, spleen, thymus, bone marrow and lymph nodes (Ishii et al. 1997; Logsdon et al. 1997). In these tissues, these channels are thought to play a role in cell proliferation, volume regulation and secretion (Wulff et al. 2007). Three different groups cloned IK channels at the same time (Joiner et al. 1997; Ishii et al. 1997; Logsdon et al. 1997) and characterised their biophysical properties. Like SK channels, the IK channels are activated via Ca^{2+} binding to calmodulin, which in turn mediates conformational changes in the channel and opens it. IK channels are insensitive to apamin and a series of modulators classified as blockers, positive gating modulators or negative gating modulators have been identified (Cui et al. 2014; Christophersen and Wulff 2015).

The Kv channel scorpion peptide toxins Maurotoxin and Charybdotoxin are potent blockers of IK channels with an IC_{50} of 1 nM and 5 nM, respectively (Miller et al. 1985; Castle et al. 2003). The Kv1 toxin Margatoxin (MgTX) is also an effective blocker of these channels (IC_{50} of 459 nM), as is the sea anemone toxin stichodactyla toxin (ShK) which has an IC_{50} of 291 nM (Jensen et al. 1998). A number of non-peptide small molecules including clotrimazole, cetiedil, nitrendipine and promethazine have been shown to block the IK channels with varied potencies (Roxburgh et al. 1996; Jensen et al. 1998; Wittekindt et al. 2006).

Most of the gating modulators developed to modify SK channels also modulate IK channels, presumably due to the structural similarities between the two channel types (Wulff et al. 2007). The muscle relaxants chlorzoxazone and zoxazolamine have been shown to activate IK channels at micromolar concentrations (Syme et al. 2000). SKA-111 and SKA-121 are reasonably selective IK channel positive gating modulators ($EC_{50} \sim 100$ nM), and showed an ~100-fold selectivity over SK channels

(Coleman et al. 2014). The SK channel positive gating modulator 1-EBIO and its structural derivatives DCEBIO and NS309 were also found to activate IK channels with varying potencies. Thus, 1-EBIO opened IK channels with an EC_{50} of 70 μM, but its derivative DCEBIO was more potent (EC_{50}–750 nM) and the compound NS309 had an EC_{50} of 10 nM, making it one of the most potent activators of IK channels (Singh et al. 2001; Strøbæk et al. 2004). The contribution of IK channels to cell proliferation has been studied using 1-EBIO and riluzole on PC-3 and LNCaP cell lines. Both drugs enhanced cell proliferation and these effects were abolished when IK channels were inhibited by the IK channel blocker clotrimazole (Jensen et al. 1998), suggesting that IK channels may play a role in regulating prostate cancer cell proliferation (Parihar et al. 2003).

Acknowledgements The authors work has been funded by the BREATH project by the EU, under the Interreg VA Programme, managed by the Special EU Programmes Body (to KT, GS & MH). SD is funded by Dundalk Institute of Technology Research Office.

References

Adams PR, Constanti A, Brown DA, Clark RB (1982) Intracellular Ca^{2+} activates a fast voltage-sensitive K^+ current in vertebrate sympathetic neurones. Nature 296(5859):746–749

Addy ME, Awumey EM (1984) Effects of the extracts of desmodium adscendens on anaphylaxis. J Ethnopharmacol 11(3):283–292

Adelman JP, Shen KZ, Kavanaugh MP, Warren RA, Wu YN, Lagrutta A, Bond CT, North RA (1992) Calcium-activated potassium channels expressed from cloned complementary DNAs. Neuron 9(2):209–216

Adelman JP, Maylie J, Sah P (2012) Small-conductance Ca^{2+}-activated K^+ channels: form and function. Annu Rev Physiol 74(1):245–269

Almassy J, Begenisich T (2012) The LRRC26 protein selectively alters the efficacy of BK channel activators. Mol Pharmacol 81:21–23

Al-Nakkash L, Hu S, Li M, Hwang TC (2001) A common mechanism for cysticfibrosis transmembrane conductance regulator protein activation by genistein andbenzimidazolone analogs. J Pharmacol Exp Ther 296:464–472

Arshad SH, Bateman B, Sadeghnejad A, Gant C, Matthews SM (2007) Andolast, a novel calcium-activated potassium-channel opener, inhibits AMP and exercise induced bronchoconstriction in asthma. J Allergy Clin Immunol 119(2):307–313

Atkinson NS, Robertson GA, Ganetzky B (1991) A component of calcium-activated potassium channels encoded by the Drosophila slo locus. Science 253(5019):551–555

Bao L, Cox DH (2005) Gating and ionic currents reveal how the BK_{Ca} channel's Ca^{2+} sensitivity is enhanced by its β1 subunit. J Gen Physiol 126(4):393–412

Bentzen BH, Nardi A, Calloe K, Madsen LS, Olesen SP, Grunnet M (2007) The small molecule NS11021 is a potent and specific activator of Ca^{2+}-activated big-conductance K^+ channels. Mol Pharmacol 72(4):1033–1044

Bentzen BH, Olesen SP, Rønn LCB, Grunnet M (2014) BK channel activators and their therapeutic perspectives. Front Physiol 5:389

Bezanilla F (2000) The voltage sensor in voltage-dependent ion channels. Physiol Rev 80 (2):555–592

Bhattacharjee A, Joiner WJ, Wu M, Yang Y, Sigworth FJ, Kaczmarek LK (2003) Slick (Slo2.1), a rapidly-gating sodium-activated potassium channel inhibited by ATP. J Neurosci 23 (37):11681–11691

Bocksteins E (2016) K_v5, K_v6, K_v8, and K_v9 subunits: no simple silent bystanders. J Gen Physiol 147(2):105–125

Bozik ME, Smith JM, Sullivan MA, Braga JM, Warach S, Luby M (2000) POST: double-blind placebo controlled, safety and efficacy trial of intravenous BMS-204352 in patients with acute stroke. Stroke 31:1–269

Brayden JE, Nelson MT (1992) Regulation of arterial tone by activation of calcium-dependent potassium channels. Science 256(5056):532–535

Brelidze TI, Magleby KL (2005) Probing the geometry of the inner vestibule of BK channels with sugars. J Gen Physiol 126(2):105–121

Brelidze TI, Niu X, Magleby KL (2003) A ring of eight conserved negatively charged amino acids doubles the conductance of BK channels and prevents inward rectification. Proc Natl Acad Sci 100(15):9017–9022

Brenner R, Jegla TJ, Wickenden A, Liu Y, Aldrich RW (2000a) Cloning and functional characterization of novel large conductance calcium-activated potassium channel beta subunits, hKCNMB3 and hKCNMB4. J Biol Chem 275(9):6453–6461

Brenner R, Peréz GJ, Bonev AD, Eckman DM, Kosek JC, Wiler SW, Patterson AJ, Nelson MT, Aldrich RW (2000b) Vasoregulation by the β1 subunit of the calcium-activated potassium channel. Nature 407(6806):870–876

Butler A, Tsunoda S, McCobb DP, Wei A, Salkoff L (1993) mSlo, a complex mouse gene encoding "maxi" calcium-activated potassium channels. Science 261(5118):221–224

Calderone V, Chericoni S, Martinelli C, Testai L, Nardi A, Morelli I, Breschi MC, Martinotti E (2004) Vasorelaxing effects of flavonoids: investigation on the possible involvement of potassium channels. Naunyn Schmiedeberg's Arch Pharmacol 370(4):290–298

Castillo K, Contreras GF, Pupo A, Torres YP, Neely A, González C, Latorre R (2015) Molecular mechanism underlying β1 regulation in voltage- and calcium-activated potassium (BK) channels. Proc Natl Acad Sci 112(15):4809–4814

Castle NA, London DO, Creech C, Fajloun Z, Stocker JW, Sabatier JM (2003) Maurotoxin: a potent inhibitor of intermediate conductance Ca^{2+} activated potassium channels. Mol Pharmacol 63(2):409–418

Chen X, Yan J, Aldrich RW (2014) BK channel opening involves side-chain reorientation of multiple deep-pore residues. Proc Natl Acad Sci 111(1):79–88

Cheney JA, Weisser JD, Bareyre FM, Laurer HL, Saatman KE, Raghupathi R, Gribkoff V, Starrett JE Jr, McIntosh TK (2001) The maxi-K channel opener BMS-204352 attenuates regional cerebral edema and neurologic motor impairment after experimental brain injury. J Cereb Blood Flow Metab 21(4):396–403

Christophersen P, Wulff H (2015) Pharmacological gating modulation of small- and intermediate-conductance Ca^{2+}-activated K^+ channels (KCa2.X and KCa3.1). Channels 9(6):336–343

Coleman N, Brown BM, Oliván-Viguera A, Singh V, Olmstead MM, Valero MS, Köhler R, Wulff H (2014) New positive Ca^{2+} activated K^+ channel gating modulators with selectivity for KCa3.1. Mol Pharmacol 86(3):342–357

Conejo-Garcia A, Campos J (2008) Bis-quinolinium cyclophanes: highly potent and selective non-peptidic blockers of the apamin-sensitive Ca^{2+}-activated K^+ channel. Curr Med Chem 15 (13):1305–1315

Cotton KD, Hollywood MA, Thornbury KD, McHale NG (1996) Effect of purinergic blockers on outward current in isolated smooth muscle cells of the sheep bladder. Am J Phys Cell Phys 270 (3):C969–C973

Cui J, Cox DH, Aldrich RW (1997) Intrinsic voltage dependence and ca^{2+} regulation of mslo large conductance Ca-activated K^+ channels. J Gen Physiol 109(5):647–673

Cui M, Qin G, Yu K, Bowers MS, Zhang M (2014) Targeting the small and intermediate conductance Ca^{2+} activated potassium channels: the drug-binding pocket at the channel/calmodulin interface. Neurosignals 22(2):65–78

D'hoedt D, Hirzel K, Pedarzani P, Stocker M (2004) Domain analysis of the calcium-activated potassium channel SK1 from rat brain. Functional expression and toxin sensitivity. J Biol Chem 279(13):12088–12092

Delay C, Tremblay C, Brochu E, Paris-Robidas S, Emond V, Rajput AH, Rajput A, Calon F (2014) Increased LINGO1 in the cerebellum of essential tremor patients. Mov Disord 29 (13):1637–1647
Devor DC, Singh AK, Frizzell RA, Bridges RJ (1996) Modulation of cl^- secretion by benzimidazolones. I. Direct activation of a Ca^{2+} dependent K^+ channel. Am J Phys Lung Cell Mol Phys 271(5):L775–L784
Dick GM, Rossow CF, Smirnov S, Horowitz B, Sanders KM (2001) Tamoxifen activates smooth muscle BK channels through the regulatory β1 subunit. J Biol Chem 276(37):34594–34599
Ding JP, Li ZW, Lingle CJ (1998) Inactivating BK channels in rat chromaffin cells may arise from heteromultimeric assembly of distinct inactivation-competent and noninactivating subunits. Biophys J 74(1):268–289
Doyle DA, Morais Cabral J, Pfuetzner RA, Kuo A, Gulbis JM, Cohen SL, Chait BT, MacKinnon R (1998) The structure of the potassium channel: molecular basis of K^+ conduction and selectivity. Science 280(5360):69–77
Du W, Bautista JF, Yang H, Diez-Sampedro A, You SA, Wang L, Kotagal P, Lüders HO, Shi J, Cui J, Richerson GB, Wang QK (2005) Calcium-sensitive potassium channelopathy in human epilepsy and paroxysmal movement disorder. Nat Genet 37(7):733–738
Dudem S, Large RJ, Kulkarni S, McClafferty H, Tikhonova IG, Sergeant GP, Thornbury KD, Shipston MJ, Perrino BA, Hollywood MA (2020) LINGO1 is a regulatory subunit of large conductance, Ca2+−activated potassium channels. Proc Natl Acad Sci 117(4):2194–2200
Edwards G, Niederste-Hollenberg A, Schneider J, Noack T, Weston AH (1994) Ion channel modulation by NS 1619, the putative BKCa channel opener, in vascular smooth muscle. Br J Pharmacol 113(4):1538–1547
Faber E, Sah P (2007) Functions of SK channels in central neurons. Clin Exp Pharmacol Physiol 34 (10):1077–1083
Foale S, Berry M, Logan A, Fulton D, Ahmed Z (2017) LINGO-1 and AMIGO3, potential therapeutic targets for neurological and dysmyelinating disorders? Neural Regen Res 12 (8):1247–1251
Fodor AA, Aldrich RW (2006) Statistical limits to the identification of ion channel domains by sequence similarity. J Gen Physiol 127(6):755–766
Garcia-Calvo M, Knaus HG, McManus OB, Giangiacomo KM, Kaczorowski GJ, Garcia ML (1994) Purification and reconstitution of the high-conductance, calcium-activated potassium channel from tracheal smooth muscle. J Biol Chem 269(1):676–682
Gessner G, Schönherr K, Soom M, Hansel A, Asim M, Baniahmad A, Derst C, Hoshi T, Heinemann SH (2006) BK_{Ca} channels activating at resting potential without calcium in LNCaP prostate cancer cells. J Membr Biol 208(3):229–240
Grimm PR, Irsik DL, Settles DC, Holtzclaw JD, Sansom SC (2009) Hypertension of Kcnmb1−/−is linked to deficient K secretion and aldosteronism. Proc Natl Acad Sci 106(28):11800–11805
Grizel AV, Glukhov GS, Sokolova OS (2014) Mechanisms of activation of voltage-gated potassium channels. Acta Nat 6(4):10–26
Haug T, Sigg D, Ciani S, Toro L, Stefani E, Olcese R (2004) Regulation of K^+ flow by a ring of negative charges in the outer pore of BK_{Ca} channels. Part I. J Gen Physiol 124(2):173–184
Haugaard MM, Hesselkilde EZ, Pehrson S, Carstensen H, Flethøj M, Præstegaard KF, Sørensen US, Dines JG, Grunnet M, Buhl R, Jespersen T (2015) Pharmacologic inhibition of small conductance calcium activated potassium (SK) channels by NS8593 reveals atrial antiarrhythmic potential in horses. Heart Rhythm 12(4):825–835
Heginbotham L, Lu Z, Abramson T, MacKinnon R (1994) Mutations in the K^+ channel signature sequence. Biophys J 66(4):1061–1067
Hicks GA, Marrion NV (1998) Ca^{2+}-dependent inactivation of large conductance Ca^{2+}-activated K^+ (BK) channels in rat hippocampal neurones produced by pore block from an associated particle. J Physiol 508(Pt 3):721–734
Hite R, Tao X, MacKinnon R (2017) Structural basis for gating the high-conductance Ca2+−activated K+ channel. Nature 541(7635):52–57

Holland M, Langton PD, Standen NB, Boyle JP (1996) Effects of the BKCa channel activator, NS1619, on rat cerebral artery smooth muscle. Br J Pharmacol 117(1):119–129
Homma S, Shimada T, Hikake T, Yaginuma H (2009) Expression pattern of LRR and Ig domain-containing protein (LRRIG protein) in the early mouse embryo. Gene Expr Patterns 9(1):1–26
Horrigan FT, Aldrich RW (1999) Allosteric voltage gating of potassium channels II. Mslo channel gating charge movement in the absence of Ca^{2+}. J Gen Physiol 114(2):305–336
Horrigan FT, Aldrich RW (2002) Coupling between voltage sensor activation, Ca^{2+} binding and channel opening in large conductance (BK) potassium channels. J Gen Physiol 120(3):267–305
Hoshi T, Heinemann SH (2016) Modulation of BK channels by small endogenous molecules and pharmaceutical channel openers. Int Rev Neurobiol 128:193–237
Hoshi T, Wissuwa B, Tian Y, Tajima N, Xu R, Bauer M, Heinemann SH, Hou S (2013a) Omega-3 fatty acids lower blood pressure by directly activating large-conductance Ca^{2+}-dependent K^+ channels. Proc Natl Acad Sci 110(12):4816–4821
Hoshi T, Tian Y, Xu R, Heinemann SH, Hou S (2013b) Mechanism of the modulation of BK potassium channel complexes with different auxiliary subunit compositions by the omega-3 fatty acid DHA. Proc Natl Acad Sci 110(12):4822–4827
Hou S, Heinemann SH, Hoshi T (2009) Modulation of BK_{Ca} channel gating by endogenous signaling molecules. Physiology 24(1):26–35
Hougaard C, Jensen ML, Dale TJ, Miller DD, Davies DJ, Eriksen BL, Strøbæk D, Trezise DJ, Christophersen P (2009) Selective activation of the SK1 subtype of human small conductance Ca^{2+} activated K^+ channels by 4-(2-methoxyphenylcarbamoyloxymethyl)-piperidine-1-carboxylic acid tert-butyl ester (GW542573X) is dependent on serine 293 in the S5 segment. Mol Pharmacol 76(3):569–578
Huang C, Huang C, Wu S (2007) Activation by zonisamide, a newer antiepileptic drug, of large-conductance calcium-activated potassium channel in differentiated hippocampal neuron-derived H19-7 cells. J Pharmacol Exp Ther 321(1):98–106
Imaizumi Y, Sakamoto K, Yamada A, Hotta A, Ohya S, Muraki K, Uchiyama M, Ohwada T (2002) Molecular basis of pimarane compounds as novel activators of large-conductance Ca^{2+}-activated K^+ channel alpha-subunit. Mol Pharmacol 62(4):836–846
Ishii TM, Silvia C, Hirschberg B, Bond CT, Adelman JP, Maylie J (1997) A human intermediate conductance calcium activated potassium channel. Proc Natl Acad Sci 94(21):11651–11656
Jensen BS, Strobaek D, Christophersen P, Jorgensen TD, Hansen C, Silahtaroglu A et al (1998) Characterization of the cloned human intermediate-conductance Ca^{2+}-activated K^+ channel. Am J Physiol 275(3):C848–C856
Jiang Y, Lee A, Chen J, Cadene M, Chait BT, MacKinnon R (2002a) The open pore conformation of potassium channels. Nature 417(6888):523–526
Jiang Y, Lee A, Chen J, Cadene M, Chait BT, MacKinnon R (2002b) Crystal structure and mechanism of a calcium-gated potassium channel. Nature 417(6888):515–522
Joiner WJ, Wang LY, Tang MD, Kaczmarek LK (1997) hSK4, a member of a novel subfamily of calcium-activated potassium channels. Proc Natl Acad Sci 94(20):11013–11018
Kasumu AW, Hougaard C, Rode F, Jacobsen TA, Sabatier JM, Eriksen BL, Strøbæk D, Linang X, Egorova P, Vorontsova D, Christophersen P, Rønn LB, Bezprozvanny I (2012) Selective positive modulator of calcium activated potassium channels exerts beneficial effects in a mouse model of spinocerebellar ataxia type 2. Chem Biol 19(10):1340–1353
Király I, Pataricza J, Bajory Z, Simonsen U, Varro A, Papp JG, Pajor L, Kun A (2013) Involvement of large-conductance Ca^{2+}-activated K^+ channels in both nitric oxide and endothelium-derived hyperpolarization-type relaxation in human penile small arteries. Basic Clin Pharmacol Toxicol 113(1):19–24
Knaus HG, Folander K, Garcia-Calvo M, Garcia ML, Kaczorowski GJ, Smith M, Swanson R (1994) Primary sequence and immunological characterization of beta-subunit of high conductance Ca^{2+}-activated K^+ channel from smooth muscle. J Biol Chem 269(25):17274–17278
Kshatri A, Li Q, Yan J, Large RJ, Sergeant GP, McHale NG, Thornbury KD, Hollywood MA (2016) Differential efficacy of GoSlo-SR compounds on BKα and BKαγ1–4channels. Channels 11(1):66–78

Kshatri A, Gonzalez-Hernandez A, Giraldez T (2018) Physiological roles and therapeutic potential of Ca2+ activated potassium channels in the nervous system. Front Mol Neurosci 11
Kun A, Matchkov VV, Stankevicius E, Nardi A, Hughes AD, Kirkeby HJ, Demnitz J, Simonsen U (2009) NS11021, a novel opener of large-conductance Ca^{2+}-activated K^+ channels, enhances erectile responses in rats. Br J Pharmacol 158(6):1465–1476
Lam J, Coleman N, Garing AA, Wulff H (2013) The therapeutic potential of small-conductance KCa2 channels in neurodegenerative and psychiatric diseases. Expert Opin Ther Targets 17 (10):1203–1220
Lamy C, Goodchild SJ, Weatherall KL, Jane DE, Liégeois J, Seutin V, Marrion NV (2010) Allosteric block of KCa2 channels by apamin. J Biol Chem 285(35):27067–27077
Large RJ, Kshatri A, Webb TI, Roy S, Akande A, Bradley E, Sergeant GP, Thornbury KD, McHale NG, Hollywood MA (2015) Effects of the novel BK (KCa1.1) channel opener GoSlo-SR-5-130 are dependent on the presence of BKβsubunits. Br J Pharmacol 172(10):2544–2556
Laumonnier F, Roger S, Guérin P, Molinari F, M'Rad R, Cahard D, Belhadj A, Halayem M, Persico AM, Elia M, Romano V, Holbert S, Andres C, Chaabouni H, Colleaux L, Constant J, Le Guennec JY, Briault S (2006) Association of a functional deficit of the BK_{Ca} channel, a synaptic regulator of neuronal excitability, with autism and mental retardation. Am J Psychiatr 163 (9):1622–1629
Layne JJ, Nausch B, Olesen S-P, Nelson MT (2010) BK channel activation by NS11021 decreases excitability and contractility of urinary bladder smooth muscle. Am J Physiol 298(2):R378–R384
Leonetti MD, Yuan P, Hsiung Y, Mackinnon R (2012) Functional and structural analysis of the human SLO3 pH- and voltage-gated K^+ channel. Proc Natl Acad Sci 109(47):19274–19279
Li W, Aldrich RW (2004) Unique inner pore properties of BK channels revealed by quaternary ammonium block. J Gen Physiol 124(1):43–57
Li H, Chen S, Wu S (2000) Evidence for the stimulatory effect of resveratrol on Ca2+−activated K + current in vascular endothelial cells. Cardiovasc Res 45(4):1035–1045
Lingle CJ (2007) Gating rings formed by RCK domains: keys to gate opening. J Gen Physiol 129 (2):101–107
Liu Y, Lo Y, Wu S (2003) Stimulatory effects of chlorzoxazone, a centrally acting muscle relaxant, on large conductance calcium-activated potassium channels in pituitary GH3 cells. Brain Res 959(1):86–97
Liu R, Zhang Z, Liu H, Hou P, Lang J, Wang S, Yan H, Li P, Huang Z, Wu H, Rong M, Huang J, Wang H, Lv L, Qiu M, Ding J, Lai R (2013) Human β-Defensin 2 is a novel opener of Ca^{2+}-activated potassium channels and induces vasodilation and hypotension in monkeys. Hypertension 62(2):415–425
Logsdon NJ, Kang J, Togo JA, Christian EP, Aiyar J (1997) A novel gene, hKCa4, encodes the calcium activated Potassium Channel in human T lymphocytes. J Biol Chem 272 (52):32723–32726
Long SB, Campbell EB, Mackinnon R (2005a) Crystal structure of a mammalian voltage-dependent shaker family K^+ channel. Science 309(5736):897–903
Long SB, Campbell EB, Mackinnon R (2005b) Voltage sensor of $K_V1.2$: structural basis of electromechanical coupling. Science 309(5736):903–908
Long SB, Tao X, Campbell EB, MacKinnon R (2007) Atomic structure of a voltage-dependent K^+ channel in a lipid membrane-like environment. Nature 450(7168):376–382
Lu Z, Klem AM, Ramu Y (2002) Coupling between voltage sensors and activation gate in voltage-gated K^+ channels. J Gen Physiol 120(5):663–676
Ma Z, Lou XJ, Horrigan FT (2006) Role of charged residues in the S1–S4 voltage sensor of BK channels. J Gen Physiol 127(3):309–328
Magleby KL (2003) Gating mechanism of BK (Slo1) channels: so near, yet so far. J Gen Physiol 121(2):81–96
Malerba M, D'Amato M, Radaeli A, Giacovelli G, Rovati L, Arshad S, Holgate S (2015) Efficacy of andolast in mild to moderate asthma: a randomized, controlled, double-blind multicenter study (the Andast trial). Curr Pharm Des 21(26):3835–3843

McManus OB, Harris GH, Giangiacomo KM, Feigenbaum P, Reuben JP, Addy ME, Burka JF, Kaczorowski GJ, Garcia ML (1993) An activator of calcium-dependent potassium channels isolated from a medicinal herb. Biochemistry 32(24):6128–6133

McManus OB, Helms LM, Pallanck L, Ganetzky B, Swanson R, Leonard RJ (1995) Functional role of the beta subunit of high conductance calcium-activated potassium channels. Neuron 14 (3):645–650

Meera P, Wallner M, Jiang Z, Toro L (1996) A calcium switch for the functional coupling between α (hslo) and β subunits ($K_{v,Ca}\beta$) of maxi K channels. FEBS Lett 385(1–2):127–128

Meera P, Wallner M, Song M, Toro L (1997) Large conductance voltage and calcium dependent K^+ channel, a distinct member of voltage-dependent ion channels with seven N-terminal transmembrane segments (S0-S6), an extracellular N terminus, and an intracellular (S9-S10) C terminus. Proc Natl Acad Sci 94(25):14066–14071

Meredith AL, Thorneloe KS, Werner ME, Nelson MT, Aldrich RW (2004) Overactive bladder and incontinence in the absence of the BK large conductance Ca^{2+}-activated K^+ channel. J Biol Chem 279(35):36746–36752

Mi S, Lee X, Shao Z, Thill G, Ji B, Relton J, Levesque M, Allaire N, Perrin S, Sands B, Crowell T, Cate RL, McCoy JM, Pepinsky RB (2004) LINGO-1 is a component of the Nogo-66 receptor/p75 signaling complex. Nat Neurosci 7(3):221–228

Mi S, Hu B, Hahm K, Luo Y, Kam Hui ES, Yuan Q, Wong WM, Wang L, Su H, Chu TH, Guo J, Zhang W, So KF, Pepinsky B, Shao Z, Graff C, Garber E, Jung V, Wu EX, Wu W (2007) LINGO-1 antagonist promotes spinal cord remyelination and axonal integrity in MOG-induced experimental autoimmune encephalomyelitis. Nat Med 13(10):1228–1233

Mi S, Blake Pepinsky R, Cadavid D (2013) Blocking LINGO-1 as a therapy to promote CNS repair: from concept to the clinic. CNS Drugs 27(7):493–503

Miller C, Moczydlowski E, Latorre R, Phillips M (1985) Charybdotoxin, a protein inhibitor of single Ca^{2+} activated K^+ channels from mammalian skeletal muscle. Nature 313(6000):316–318

Monod J, Wyman J, Changeux JP (1965) On the nature of allosteric transitions: a plausible model. J Mol Biol 12:88–118

Morimoto T, Sakamoto K, Sade H, Ohya S, Muraki K, Imaizumi Y (2007) Voltage-sensitive oxonol dyes are novel large-conductance Ca^{2+}-activated K^+ channel activators selective for β_1 and β_4 but not for β_2 subunits. Mol Pharmacol 71(4):1075–1088

Morrow JP, Zakharov SI, Liu G, Yang L, Sok AJ, Marx SO (2006) Defining the BK channel domains required for β1-subunit modulation. Proc Natl Acad Sci 103(13):5096–5101

Mosyak L, Wood A, Dwyer B, Buddha M, Johnson M, Aulabaugh A, Zhong X, Presman E, Benard S, Kelleher K, Wilhelm J, Stahl ML, Kriz R, Gao Y, Cao Z, Ling HP, Pangalos MN, Walsh FS, Somers WS (2006) The structure of the Lingo-1 ectodomain, a module implicated in central nervous system repair inhibition. J Biol Chem 281(47):36378–36390

Nagano N, Imaizumi Y, Hirano M, Watanabe M (1996) Opening of Ca^{2+}-dependent K^+ channels by nordihydroguaiaretic acid in porcine coronary arterial smooth muscle cells. Jpn J Pharmacol 70 (3):281–284

Nardi A, Olesen SP (2008) BK channel modulators: a comprehensive overview. Curr Med Chem 15 (11):1126–1146

Nilius B, Droogmans G (2001) Ion channels and their functional role in vascular endothelium. Physiol Rev 81(4):1415–1459

Nimigean CM, Magleby KL (2000) Functional coupling of the β_1 subunit to the large conductance Ca^{2+}-activated K^+ channel in the absence of Ca^{2+}. Increased Ca^{2+} sensitivity from a Ca^{2+}-independent mechanism. J Gen Physiol 115(6):719–736

Niu X, Qian X, Magleby KL (2004) Linker-gating ring complex as passive spring and Ca^{2+}-dependent machine for a voltage- and Ca^{2+}-activated potassium channel. Neuron 42(5):745–756

Nolting A, Ferraro T, D'hoedt D, Stocker M (2007) An amino acid outside the pore region influences apamin sensitivity in small conductance Ca^{2+}-activated K^+ channels. J Biol Chem 282(6):3478–3486

Olesen SP, Munch E, Moldt P, Drejer J (1994) Selective activation of Ca^{2+}-dependent K^+ channels by novel benzimidazolone. Eur J Pharmacol 251(1):53–59

Orio P, Latorre R (2005) Differential effects of β1 and β2 subunits on BK channel activity. J Gen Physiol 125(4):395–411

Orio P, Rojas P, Ferreira G, Latorre R (2002) New disguises for an old channel: MaxiK channel beta-subunits. News Physiol Sci 17:156–161

Pallanck L, Ganetzky B (1994) Cloning and characterization of human and mouse homologs of the Drosophila calcium-activated potassium channel gene, slowpoke. Hum Mol Genet 3 (8):1239–1243

Parajuli SP, Soder RP, Hristov KL, Petkov GV (2011) Pharmacological activation of small conductance calcium activated potassium channels with Naphtho[1,2-d]thiazol-2-ylamine decreases Guinea pig detrusor smooth muscle excitability and contractility. J Pharmacol Exp Ther 340(1):114–123

Parihar AS, Coghlan MJ, Gopalakrishnan M, Shieh C (2003) Effects of intermediate conductance Ca^{2+} activated K^+ channel modulators on human prostate cancer cell proliferation. Eur J Pharmacol 471(3):157–164

Pedarzani P, D'hoedt D, Doorty KB, Wadsworth JD, Joseph JS, Jeyaseelan K, Kini RM, Gadre SV, Sapatnekar SM, Stocker M, Strong PN (2002) Tamapin, a venom peptide from the Indian red scorpion (Mesobuthus tamulus) that targets small conductance Ca^{2+}-activated K^+ channels and afterhyperpolarization currents in central neurons. J Biol Chem 277(48):46101–46109

Ptacek LJ, Fu YH (2004) Channels and disease. Arch Neurol 61(11):1665

Quirk JC, Reinhart PH (2001) Identification of a novel tetramerization domain in large conductance KCa channels. Neuron 32(1):13–23

Rajan AS, Aguilar-Bryan L, Nelson DA, Yaney GC, Hsu WH, Kunze DL, Boyd AE (1990) Ion channels and insulin secretion. Diabetes Care 13(3):340–363

Robitaille R, Garcia ML, Kaczorowski GJ, Charlton MP (1993) Functional colocalization of calcium and calcium-gated potassium channels in control of transmitter release. Neuron 11 (4):645–655

Roxburgh CJ, Ganellin CR, Shiner MA, Benton DC, Dunn PM, Ayalew Y, Jenkinson DH (1996) The synthesis and some pharmacological actions of the enantiomers of the K^+ channel blocker cetiedil. J Pharm Pharmacol 48(8):851–859

Roy S, Morayo Akande A, Large RJ, Webb TI, Camarasu C, Sergeant GP, McHale NG, Thornbury KD, Hollywood MA (2012) Structure-activity relationships of a novel group of large-conductance Ca^{2+}-activated K^+ (BK) channel modulators: the GoSlo-SR family. ChemMedChem 7(10):1763–1769

Roy S, Large RJ, Akande AM, Kshatri A, Webb TI, Domene C, Sergeant GP, McHale NG, Thornbury KD, Hollywood MA (2014) Development of GoSlo-SR-5-69, a potent activator of large conductance Ca^{2+}-activated K^+ (BK) channels. Eur J Med Chem 75:426–437

Salkoff L, Butler A, Ferreira G, Santi C, Wei A (2006) High-conductance potassium channels of the SLO family. Nat Rev Neurosci 7(12):921–931

Sandhiya S, Dkhar SA (2009) Potassium channels in health, disease & development of channel modulators. Indian J Med Res 129(3):223–232

Sausbier M, Hu H, Arntz C, Feil S, Kamm S, Adelsberger H, Sausbier U, Sailer CA, Feil R, Hofmann F, Korth M, Shipston MJ, Knaus HG, Wolfer DP, Pedroarena CM, Storm JF, Ruth P (2004) Cerebellar ataxia and Purkinje cell dysfunction caused by Ca^{2+}-activated K^+ channel deficiency. Proc Natl Acad Sci 101(25):9474–9478

Schewe M, Sun H, Mackenzie A, Pike A, Schulz F, Constantin C, Kiper AK, Conrad LJ, Gonzalez W, DeGroot BL, Decher N, Fakler B, Carpenter EP, Tucker SJ, Baukrowitz T (2019) A pharmacological masterkey mechanism to unlock the selectivity filter gate in K+ channels. Biophys J 116(3):301a–302a

Schreiber M, Salkoff L (1997) A novel calcium-sensing domain in the BK channel. Biophys J 73 (3):1355–1363

Schrøder R, Jespersen T, Christophersen P, Strøbæk D, Jensen B, Olesen S (2001) KCNQ4 channel activation by BMS-204352 and retigabine. Neuropharmacology 40(7):888–898

Scornik FS, Bucciero RS, Wu Y, Selga E, Bosch Calero C, Brugada R, Pérez GJ (2013) DiBAC4 (3) hits a "sweet spot" for the activation of arterial large-conductance Ca2+−activated potassium channels independently of the β1-subunit. Am J Phys Heart Circ Phys 304(11):H1471–H1482

Scuvée-Moreau J, Boland A, Graulich A, Overmeire LV, D'hoedt D, Graulich-Lorge F, Thomas E, Abras A, Stocker M, Liégeois J, Seutin V (2004) Electrophysiological characterization of the SK channel blockers methyl-laudanosine and methyl-noscapine in cell lines and rat brain slices. Br J Pharmacol 143(6):753–764

Singh S, Syme CA, Singh KA, Devor DC, Bridges RJ (2001) Benzimidazolone activators of chloride secretion: potential therapeutics for cystic fibrosis and chronic obstructive pulmonary disease. J Pharmacol Exp Ther 296(2):600–611

Soder RP, Parajuli SP, Hristov KL, Rovner ES, Petkov GV (2013) SK channel selective opening by SKA-31 induces hyperpolarization and decreases contractility in human urinary bladder smooth muscle. Am J Phys Regul Integr Comp Phys 304(2):R155–R163

Solaro CR, Lingle CJ (1992) Trypsin-sensitive, rapid inactivation of a calcium-activated potassium channel. Science 257(5077):1694–1698

Solaro CR, Ding JP, Li ZW, Lingle CJ (1997) The cytosolic inactivation domains of BK_i channels in rat chromaffin cells do not behave like simple, open-channel blockers. Biophys J 73 (2):819–830

Sørensen US, Strøbæk D, Christophersen P, Hougaard C, Jensen ML, Nielsen E, Peters D, Teuber L (2008) Synthesis and structure activity relationship studies of 2-(N-substituted)-aminobenzimidazoles as potent negative gating modulators of small conductance Ca2+−activated K+channels. J Med Chem 51(23):7625–7634

Stefani E, Ottolia M, Noceti F, Olcese R, Wallner M, Latorre R, Toro L (1997) Voltage-controlled gating in a large conductance Ca^{2+}-sensitive K^+ channel (hslo). Proc Natl Acad Sci 94 (10):5427–5431

Strøbæk D, Teuber L, Jørgensen TD, Ahring PK, Kjær K, Hansen RS, Olesen SP, Christophersen P, Skaaning-jensen B (2004) Activation of human IK and SK Ca^{2+} activated K^+ channels by NS309 (6,7-dichloro-1H-indole-2,3-dione 3-oxime). Biochim Biophys Acta Biomembr 1665 (1–2):1–5

Strøbæk D, Hougaard C, Johansen TH, Sørensen US, Nielsen E, Nielsen KS, Taylor R, Pedarzani P, Christophersen P (2006) Inhibitory gating modulation of small conductance Ca^{2+} activated K^+ channels by the synthetic compound (R)-N-(benzimidazol-2-yl)-1,2,3,4-tetrahydro-1-naphtylamine (NS8593) reduces afterhyperpolarizing current in hippocampal CA1 neurons. Mol Pharmacol 70(5):1771–1782

Sun J, MacKinnon R (2017) Cryo-EM structure of a KCNQ1/CaM complex reveals insights into congenital long QT syndrome. Cell 169:1042–1050

Sun L, Adhikari S, Zou S, Horrigan FT (2012) The interaction of voltage-sensor and gate in BK channels. Biophys J 102(3):684a

Sweet TB, Cox DH (2008) Measurements of the BK_{Ca} channel's high-affinity Ca^{2+} binding constants: effects of membrane voltage. J Gen Physiol 132(5):491–505

Syme CA, Gerlach AC, Singh AK, Devor DC (2000) Pharmacological activation of cloned intermediate- and small-conductance Ca^{2+} activated K^+ channels. Am J Phys Cell Phys 278 (3):C570–C581

Tanaka Y, Meera P, Song M, Knaus HG, Toro L (1997) Molecular constituents of maxi K_{Ca} channels in human coronary smooth muscle: predominant alpha + beta subunit complexes. J Physiol 502(Pt 3):545–557

Tao X, MacKinnon R (2019) Molecular structures of the human Slo1 K+ channel in complex with β4. eLife 8

Tao X, Hite RK, MacKinnon R (2017) Cryo-EM structure of the open high-conductance Ca^{2+}-activated K^+ channel. Nature 541(7635):46–51

Terstappen GC, Pellacani A, Aldegheri L, Graziani F, Carignani C, Pula G, Virginio C (2003) The antidepressant fluoxetine blocks the human small conductance calcium-activated potassium channels SK1, SK2 and SK3. Neurosci Lett 346(1–2):85–88

Tristani-Firouzi M, Chen J, Sanguinetti MC (2002) Interactions between S4-S5 linker and S6 transmembrane domain modulate gating of HERG K^+ channels. J Biol Chem 277 (21):18994–19000

Uebele VN, Lagrutta A, Wade T, Figueroa DJ, Liu Y, McKenna E, Austin CP, Bennett PB, Swanson R (2000) Cloning and functional expression of two families of β-subunits of the large conductance calcium-activated K^+ channel. J Biol Chem 275(30):23211–23218

Valverde MA, Rojas P, Amigo J, Cosmelli D, Orio P, Bahamonde MI, Mann GE, Vergara C, Latorre R (1999) Acute activation of maxi-K channels (hSlo) by estradiol binding to the beta subunit. Science 285(5435):1929–1931

Wallner M, Meera P, Ottolia M, Kaczorowski GJ, Latorre R, Garcia ML, Stefani E, Toro L (1995) Characterization of and modulation by a β-subunit of a human maxi K_{Ca} channel cloned from myometrium. Receptors Channels 3(3):185–199

Wallner M, Meera P, Toro L (1996) Determinant for beta-subunit regulation in high-conductance voltage-activated and Ca^{2+}-sensitive K^+ channels: an additional transmembrane region at the N terminus. Proc Natl Acad Sci 93(25):14922–14927

Wallner M, Meera P, Toro L (1999) Molecular basis of fast inactivation in voltage and Ca^{2+}-activated K^+ channels: a transmembrane beta-subunit homolog. Proc Natl Acad Sci 96 (7):4137–4142

Wang B, Rothberg BS, Brenner R (2006) Mechanism of β4 subunit modulation of BK channels. J Gen Physiol 127(4):449–465

Webb TI, Kshatri AS, Large RJ, Akande AM, Roy S, Sergeant GP, McHale NG, Thornbury KD, Hollywood MA (2015) Molecular mechanisms underlying the effect of the novel BK channel opener GoSlo: involvement of the S4/S5 linker and the S6 segment. Proc Natl Acad Sci 112 (7):2064–2069

Wei AD, Gutman GA, Aldrich R, Chandy KG, Grissmer S, Wulff H (2005) International Union of Pharmacology. LII. Nomenclature and molecular relationships of calcium-activated potassium channels. Pharmacol Rev 57(4):463–472

Weiger TM, Holmqvist MH, Levitan IB, Clark FT, Sprague S, Huang WJ, Ge P, Wang C, Lawson D, Jurman ME, Glucksmann MA, Silos-Santiago I, DiStefano PS, Curtis R (2000) A novel nervous system beta subunit that downregulates human large conductance calcium-dependent potassium channels. J Neurosci 20(10):3563–3570

Wilkens CM, Aldrich RW (2006) State-independent block of BK channels by an intracellular quaternary ammonium. J Gen Physiol 128(3):347–364

Wittekindt O, Schmitz A, Lehmann-Horn F, Hänsel W, Grissmer S (2006) The human Ca^{2+} activated K^+ channel, IK, can be blocked by the tricyclic antihistamine promethazine. Neuropharmacology 50(4):458–467

Wu Y, Yang Y, Ye S, Jiang Y (2010) Structure of the gating ring from the human large-conductance Ca^{2+}-gated K^+ channel. Nature 466(7304):393–397

Wu RS, Liu G, Zakharov SI, Chudasama N, Motoike H, Karlin A, Marx SO (2013) Positions of β2 and β3 subunits in the large-conductance calcium- and voltage-activated BK potassium channel. J Gen Physiol 141(1):105–117

Wulff H, Kolski-Andreaco A, Sankaranarayanan A, Sabatier J, Shakkottai V (2007) Modulators of small and intermediate conductance calcium activated potassium channels and their therapeutic indications. Curr Med Chem 14(13):1437–1457

Xia XM, Ding JP, Lingle CJ (1999) Molecular basis for the inactivation of Ca^{2+}- and voltage-dependent BK channels in adrenal chromaffin cells and rat insulinoma tumor cells. J Neurosci 19(13):5255–5264

Xia XM, Ding JP, Zeng XH, Duan KL, Lingle CJ (2000) Rectification and rapid activation at low Ca^{2+} of Ca^{2+}-activated, voltage-dependent BK currents: consequences of rapid inactivation by a novel beta subunit. J Neurosci 20(13):4890–4903

Xia XM, Zeng X, Lingle CJ (2002) Multiple regulatory sites in large-conductance calcium-activated potassium channels. Nature 418(6900):880–884

Xia XM, Ding JP, Lingle CJ (2003) Inactivation of BK channels by the NH_2 terminus of the beta2 auxiliary subunit: an essential role of a terminal peptide segment of three hydrophobic residues. J Gen Physiol 121(2):125–148

Xu H, Garver H, Galligan JJ, Fink GD (2011) Large-conductance Ca2+−activated K+ channel β1-subunit knockout mice are not hypertensive. Am J Phys Heart Circ Phys 300(2):H476–H485

Yan J, Aldrich RW (2010) LRRC26 auxiliary protein allows BK channel activation at resting voltage without calcium. Nature 466(7305):513–516

Yan J, Aldrich RW (2012) BK potassium channel modulation by leucine-rich repeat-containing proteins. Proc Natl Acad Sci 109(20):7917–7922

Yang H, Zhang G, Shi J, Lee US, Delaloye K, Cui J (2008) Subunit-specific effect of the voltage sensor domain on Ca^{2+} sensitivity of BK channels. Biophys J 94(12):4678–4687

Yang J, Krishnamoorthy G, Saxena A, Zhang G, Shi J, Yang H, Delaloye K, Sept D, Cui J (2010) An epilepsy/dyskinesia-associated mutation enhances BK Channel activation by potentiating Ca^{2+} sensing. Neuron 66(6):871–883

Yang H, Zhang G, Cui J (2015) BK channels: multiple sensors, one activation gate. Front Physiol 6:29

Ye S, Li Y, Chen L, Jiang Y (2006) Crystal structures of a ligand-free MthK gating ring: insights into the ligand gating mechanism of K^+ channels. Cell 126(6):1161–1173

Yuan A, Dourado M, Butler A, Walton N, Wei A, Salkoff L (2000) SLO-2, a K^+ channel with an unusual Cl^- dependence. Nat Neurosci 3(8):771–779

Yuan P, Leonetti MD, Pico AR, Hsiung Y, MacKinnon R (2010) Structure of the human BK channel Ca^{2+}-activation apparatus at 3.0 a resolution. Science 329(5988):182–186

Yuan P, Leonetti MD, Hsiung Y, MacKinnon R (2011) Open structure of the Ca^{2+} gating ring in the high-conductance Ca^{2+}-activated K^+ channel. Nature 481(7379):94–97

Zakharov SI, Morrow JP, Liu G, Yang L, Marx SO (2005) Activation of the BK (SLO1) potassium channel by mallotoxin. J Biol Chem 280:30882–30887

Zavaritskaya O, Dudem S, Ma D, Rabab KE, Albrecht S, Tsvetkov D, Kassmann M, Thornbury KD, Maldenov M, Kammermeier C, Sergeant G, Mullins N, Wouappi O, Wurn H, Kannet A, Gollasch M, Hollywood MA, Schubert R (2020) Vasodilation of rat skeletal muscle arteries by the novel BK channel opener GoSlo is mediated by the simultaneous activation of BK and K_v7 channels. Br J Pharmacol 177(5):1164–1186

Zeng XH, Xia XM, Lingle CJ (2005) Divalent cation sensitivity of BK channel activation supports the existence of three distinct binding sites. J Gen Physiol 125(3):273–286

Zhang J, Yan J (2014) Regulation of BK channels by auxiliary γ subunits. Front Physiol 5:401

Zhang YY, Han X, Liu Y, Chen J, Hua L, Ma Q, Huang YY, Tang QY, Zhang Z (2018) +mRNA expression of LRRC55 protein (leucine-rich repeat-containing protein 55) in the adult mouse brain. Plos One 13(1):e0191749

Zhou Y, Morais-Cabral JH, Kaufman A, MacKinnon R (2001) Chemistry of ion coordination and hydration revealed by a K^+ channel-fab complex at 2.0 a resolution. Nature 414(6859):43–48

Zhou Y, Xia X-M, Lingle CJ (2011) Cysteine scanning and modification reveal major differences between BK channels and K_v channels in the inner pore region. Proc Natl Acad Sci 108 (29):12161–12166

The Pharmacology of Two-Pore Domain Potassium Channels

Jordie M. Kamuene, Yu Xu, and Leigh D. Plant

Contents

Abstract

Two-pore domain potassium channels are formed by subunits that each contain two pore-loops moieties. Whether the channels are expressed in yeast or the human central nervous system, two subunits come together to form a single potassium selective pore. TOK1, the first two-domain channel was cloned from *Saccharomyces cerevisiae* in 1995 and soon thereafter, 15 distinct K_{2P} subunits were identified in the human genome. The human K_{2P} channels are stratified into six K_{2P} subfamilies based on sequence as well as physiological or pharmacological similarities. Functional K_{2P} channels pass background (or "leak") K^+ currents that shape the membrane potential and excitability of cells in a broad range of tissues. In the years since they were first described, classical functional assays, latterly coupled with state-of-the-art structural and computational studies have

J. M. Kamuene · Y. Xu · L. D. Plant (✉)
Department of Pharmaceutical Sciences, Northeastern University, Boston, MA, USA
e-mail: l.plant@northeastern.edu

N. Gamper, K. Wang (eds.), *Pharmacology of Potassium Channels*,
Handbook of Experimental Pharmacology 267, https://doi.org/10.1007/164_2021_462

revealed the mechanistic basis of K_{2P} channel gating in response to specific physicochemical or pharmacological stimuli. The growing appreciation that K_{2P} channels can play a pivotal role in the pathophysiology of a growing spectrum of diseases makes a compelling case for K_{2P} channels as targets for drug discovery. Here, we summarize recent advances in unraveling the structure, function, and pharmacology of the K_2P channels.

Keywords

Background current · K2P channel · KCNK · TALK · TASK · THIK · TRAAK · TREK · TRESK · TWIK

1 An Introduction to Two-Pore Domain Potassium Channels

Potassium (K^+) channels are a superfamily of multi-subunit membrane proteins that are fundamental for physiology throughout the tree of life. K^+ channels are complex protein machines with a simple purpose: they open and close (gate) in a coordinated manner that allows the conduction of K^+ ions down their electrochemical gradient, typically from the intracellular to extracellular space in mammalian tissues. Gating occurs in response to a panoply of stimuli and shapes the resting membrane potential and the dynamics of cellular excitability by regulating the flux of K^+ ions. Thus, K^+ channels are essential for many biological processes including neuronal, muscular, and cardiac function (Enyedi and Czirjak 2010).

The superfamily of K^+ channels is stratified into distinct subfamilies based on structural similarities, namely the number of transmembrane domains and pore forming domains present in each subunit. The largest subfamily includes the voltage (K_V) channels and calcium activated (K_{Ca}) channels which are characterized by one reentrant pore loop (P-loop) and (typically) six transmembrane domains per subunit; holo-channels are tetramers. The inwardly rectifying K^+ channels (K_{IR}) also form as tetramers in which each subunit consists of a single P-loop and two transmembrane domains. The notion that all K^+ channels are tetramers changed in 1995 when TOK1 was cloned from *Saccharomyces cerevisiae* (Ketchum et al. 1995). TOK1 channels have a distinct architecture: Functional channels are dimers of subunits with eight transmembrane domains (M1-M8), intracellular amino- and carboxy-terminal tails, and two reentrant P-loops located between transmembrane domains M5-M6 and M7-M8 (Fig. 1) (Ketchum et al. 1995). Although TOK channels are not found beyond fungi, K^+ channel subunits with two P-loops from higher organisms were described soon after.

Unlike the K_V and K_{IR} channel subfamilies, discovery of the K_{2P} channels was made possible using genome database mining rather than by a molecular cloning strategy (Goldstein et al. 1996; Lesage et al. 1996b; Yang and Jan 2013). In 1996, K2PØ (also called KCNKØ, or dORK) was cloned from *Drosophila melanogaster* and K2P1 (also called KCNK1 or TWIK1) was cloned from human kidney (Lesage et al. 1996b). Holo two-pore domain K^+ (K_{2P}) channels are dimers of subunits, with each subunit contributing two-P loops and four transmembrane domains to the

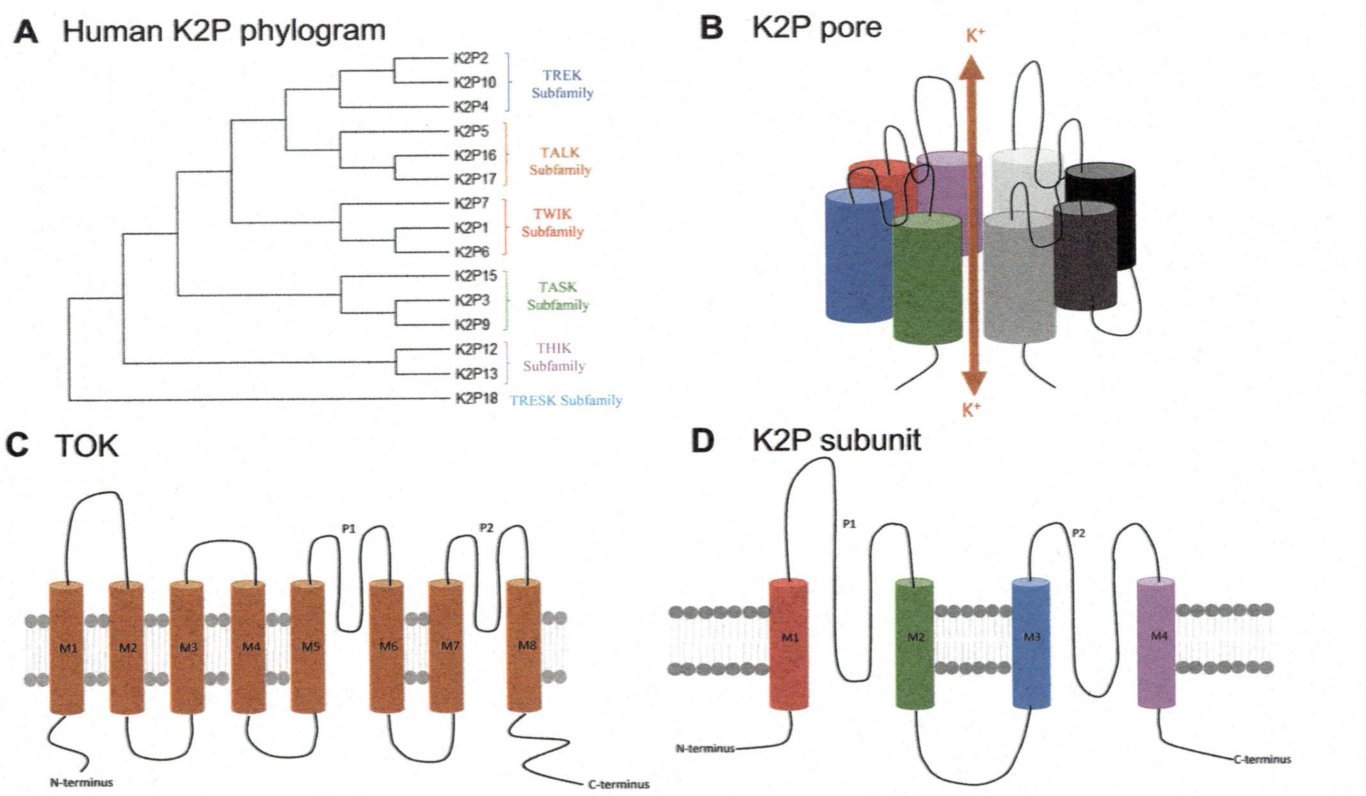

Fig. 1 An overview of the two-pore domain potassium channels. (**a**) A phylogenic tree created using MEGA software to show the relatedness of the K_{2P} subunits expressed in humans. The IUPHAR-standardized names of each subunit are in black. The phylogenic divisions of K_{2P} subunits fall into the different

structure (Feliciangeli et al. 2015; Goldstein et al. 2001; Guyenet et al. 2019; Kollewe et al. 2009; Medhurst et al. 2001; Yang and Jan 2013). In general, heterologous expression of K_{2P} channels produces outward K^+ currents under physiological conditions. The phenomenon of "background'" or "leak" K^+ currents has been appreciated since the 1940s and was ratified in the membrane equations of Hodgkin and Huxley (Enyedi and Czirjak 2010; Goldman 1943; Goldstein et al. 1996, 2001; Hodgkin and Huxley 1952; Lesage et al. 1996b). In the last ~25 years, numerous studies have confirmed the central role that K_{2P} channels play in determining the membrane potential in a broad range of excitable and non-excitable cell types (Goldstein et al. 2001) (Fig. 1).

K^+ channel subunits are identified by a common sequence of amino acid residues that comprise the selectivity filter for K^+ ions within the conduction pore of the channel (see also chapter "Comparison of K^+ Channel Families"). This evolutionarily conserved structural domain is constructed from four P-loops that are held in position in the membrane between two transmembrane helixes that form the channel corpus. The surrounding architecture of the channel is comprised of transmembrane domains that correlate with their unique physiological functions. This architectural arrangement controls when the channels open and for how long (open probability), allowing the conductance of K^+ ions down their electrochemical gradient through the selectivity filter of the pore with high fidelity (Doyle et al. 1998). Gating of K_{2P} channels is regulated by a plethora of physicochemical and mechanical stimuli including stretch, temperature, pH, and various cell signaling, and second-messenger pathways (Chemin et al. 2007; Honore 2007; Lotshaw 2007). Despite significant progress, the mechanistic basis by which each of these stimuli influences the gating machinery, and in turn the activity of K_{2P} channels, remains a matter of ongoing research.

A growing body of work, first using a classical structure-function approach, and more recently via snapshots of channel structures paired with molecular dynamics simulations has revealed that extrinsic regulators typically influence the open probability of K_{2P} channels via allosteric pathways and via c-type gating in particular (Bagriantsev et al. 2011, 2012; Cohen et al. 2008; Lolicato et al. 2014, 2020; Piechotta et al. 2011; Schewe et al. 2016; Zilberberg et al. 2001). Compelling evidence supports that this mode of gating results from constriction of the extracellular region of the channel, occluding the conduction pathway for K^+ ions (Hoshi et al. 1991; Yellen 1998). In common with data from other types of K^+ channel, the c-type gating of K2P channels occurs at the selectivity (SF) (Bagriantsev et al. 2011; Cohen et al. 2008; Piechotta et al. 2011). For example, binding of high affinity quaternary ammonium (QA) deep within the K2P2 channel selectivity filter revealed

Fig. 1 (continued) subfamilies indicated, these are named for their physiological or pharmacological properties. (**b**) A cartoon depicting how the transmembrane domains and P-loops of human K_{2P} channels are organized to create a single channel pore from a dimer of subunits. (**c**) A topological cartoon to show the organization of the 8-transmembrane domains (M1-M8) that comprise a TOK subunit. Note the two pore-loops (P1 and P2) between M5-M6 and M7-M8. (**d**) A topological cartoon to show the organization of the 4-transmembrane domains (M1–M4) that comprise a K_{2P} subunit. Note the two pore-loops (P1 and P2) between M1-M2 and M3-M4 that the gating process was occurring at the SF (Piechotta et al. 2011; Schewe et al. 2016). Furthermore, c-type gating works in concert with the carboxy-terminal tail of the channel to mediate the response to physicochemical stimuli such as temperature and mechanical force (Bagriantsev et al. 2011, 2012), indicating that allosteric interactions can transcend the channel corpus (Bagriantsev et al. 2011, 2012; Zilberberg et al. 2001).

K_{2P} channels assemble as dimers with each subunit composed of four transmembrane domains and two P-loops, one between the M1 and M2 helices and one between the M3 and M4 helices (Fig. 1b, c) (Brohawn et al. 2012; Goldstein et al. 2001; Kollewe et al. 2009; Lolicato et al. 2017; Miller and Long 2012). In addition, the first extracellular loop of each K_{2P} subunit (linking the M1 to M2 helices) contributes to a "cap-domain" located above the axis of the K^+ selectivity filter. This structure bifurcates the pathway for K^+ ions and is proposed to render K_{2P} channels insensitive to many classical K^+ channel blockers (such as protein toxins) by shielding the extracellular mouth of the pore via steric hindrance (Fig. 2) (Lolicato et al. 2017; Miller and Long 2012; Piechotta et al. 2011; Zuniga and Zuniga 2016). The extracellular cap-domain is formed when the two extracellular helices (E1 and E2) assemble (Şterbuleac 2019). The cap-domain has not been observed in other K^+ channels and was first revealed upon elucidation of the structure of K2P1 (TWIK1) and K2P4 (TRAAK) by X-ray crystallography. The placement and movement of the transmembrane helices allow the channel to adopt the two unique states, "up" and "down" (Brohawn et al. 2012; Miller and Long 2012; Şterbuleac 2019). Transitioning from the "up" to the "down" states reveals fenestrations which allow molecules to interact with the channel's inner pore (Feliciangeli et al. 2015; Şterbuleac 2019). The cap-domain has been observed on all K2P channel structures solved to date, including K2P1, K2P2, K2P3, K2P4, and K2P10 (Brohawn et al. 2012; Dong et al. 2015; Lolicato et al. 2017; Miller and Long 2012; Pope et al. 2020; Rödström et al. 2020).

The unique topology of K_{2P} channels is shared among 15 human genes designated "*KCNK*" by the Human Gene Organization nomenclature (Lesage and Barhanin 2011; Yang and Jan 2013) (Table 1). These genes encode 15 K_{2P} channel subunits that are classified into six subfamilies based on similarities in structural and functional properties: **t**andem of pore domains in a **w**eak **i**nward rectifying **K**$^+$ channel (TWIK); **T**WIK-**re**lated **K**$^+$ channel (TREK); **T**WIK-related **a**cid **s**ensitive **K**$^+$ channel (TASK); **T**WIK-related **al**kaline pH-activated **K**$^+$ channel (TALK); **T**WIK-**re**lated **s**pinal cord **K**$^+$ channel (TRESK); and **t**andem pore domain **h**alothane-**i**nhibited **K**$^+$ channel (THIK) (Table 1). To mitigate the variance in the pharmacological and physiological attributes that were subsequently associated with different members of each subfamily the nomenclature of the K_{2P} channels "K_{2PX}" was designated by the International Union of Basic and Clinical Pharmacology (IUPHAR) (Table 1). However, the descriptive names of these channels have utility and remain in common use.

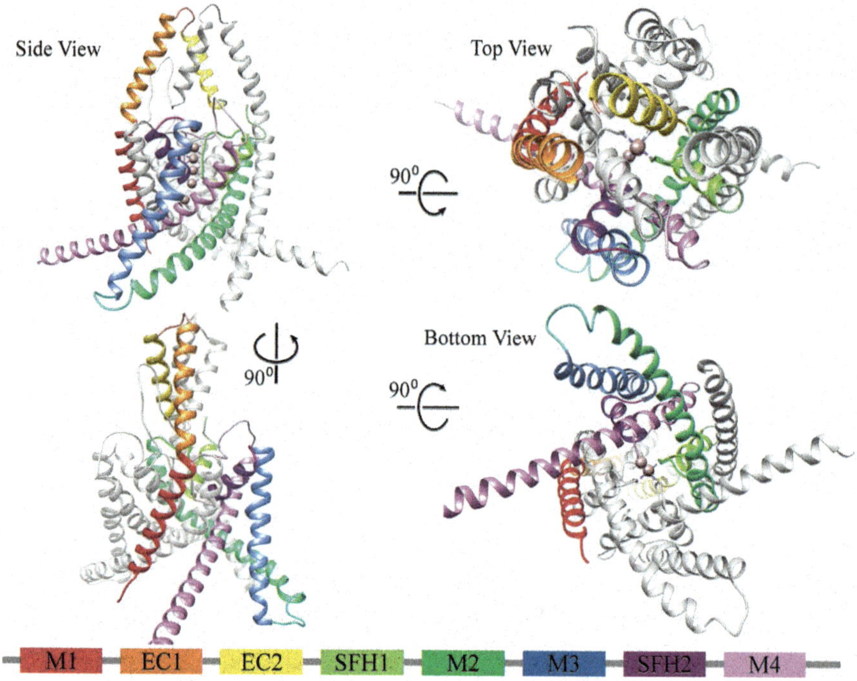

Fig. 2 The architecture of a K_{2P} channel. An overview of the three-dimensional architecture of K2P2 (Crystal structure, PDB ID: 6CQ6) showing views from the side, the top (extracellular), and the bottom (intracellular) of the channel. The helices of one subunit are colored to reflect the segments of a single subunit: four transmembrane domains (M1–M4); two portions of extracellular loop1 that contribute to the cap-domain (EC1 and EC2); two selectivity filter helices (SFH1 and SFH2), one contributes to each P-loops. Images were rendered from the PDB files indicated using UCSF Chimera software (https://www.rbvi.ucsf.edu/chimera)

2 The Role of K_{2P} Channels in Pathology and Pain Signaling

Numerous studies have linked K_{2P} channels to cardiac and neuronal diseases. In this section we highlight examples. K_{2P} channels have also been linked to neurodevelopmental disorders including Birk-Barel syndrome. K_{2P} channels have also been linked to neurodevelopmental disorders including Birk-Barel syndrome. This rare genetic disease is associated with mutation of the glycine residue at position 236 (Gly236) to arginine (a positively charged residue) in the KCNK9 gene (encodes K2P9, also called TASK3) and is characterized by intellectual disability, hypotonia and hyperactivity. Two-electrode voltage-clamp (TEVC) studies of WT and mutant channels expressed in *Xenopus* oocytes revealed that while wild type (WT) channels passed measurable currents, mutant channels had no measurable current. In addition, co-expression of mutant channel with either WT

Table 1 The 15 mammalian K2P channels

K2P subfamily	Channel name	Gene name	Common name
Tandem pore domain halothane-inhibited channel (THIK)	K2P12	KCNK12	THIK2
	K2P13	KCNK13	THIK1
The TWIK-related spinal cord K^+ channel (TRESK)	K2P18	KCNK18	TRESK
TWIK-related alkaline pH-activated K^+ channel (TALK)	K2P5	KCNK5	TASK2
	K2P16	KCNK16	TALK1
	K2P17	KCNK17	TALK2
Tandem of pore domains in a weak inward rectifying K^+ channel (TWIK)	K2P1	KCNK1	TWIK1
	K2P6	KCNK6	TWIK2
	K2P7	KCNK7	kcnk8
TWIK-related K^+ channel (TREK)	K2P2	KCNK2	TREK1
	K2P10	KCNK10	TREK2
	K2P4	KCNK4	TRAAK
TWIK-related acid sensitive K^+ channel (TASK)	K2P3	KCNK3	TASK1
	K2P9	KCNK9	TASK3
	K2P15	KCNK15	TASK5

The 15 unique K2P channels expressed by mammals. Abbreviations: *TWIK* **t**andem of pore domains in a **w**eak **i**nward rectifying **K**$^+$ channel, *TREK* TWIK-**re**lated **K**$^+$ channel, *TASK* TWIK-related **a**cid **s**ensitive **K**$^+$ channel, *TALK* TWIK-related **al**kaline pH-activated **K**$^+$ channel, *TRESK* TWIK-**re**lated **s**pinal cord **K**$^+$ channel, *THIK* **t**andem pore domain **h**alothane-**i**nhibited **K**$^+$ channel, *TRAAK* TWIK-**r**elated **a**rachidonic-**a**cid-stimulated **K**$^+$ channel. See Two P domain potassium channels in the IUPHAR/BPS Guide to Pharmacology Database https://www.guidetopharmacology.org/GRAC/FamilyDisplayForward?familyId=79 for more information about each individual channel

or K2P3 channels (which form functional heterodimers with K2P9) resulted in decreased current (Barel et al. 2008). Using the bacterial K^+ channel KcsA to generate a homology model structure, Barel and colleagues determined that the expected location of the Gly236 residue was in the ion conduction pathway. It was therefore postulated that a mutation to arginine may result in the disruption of physical and electrostatic interactions in the pore that would diminish current by impeding the conduction of K^+ ions.

KCNK18 gene encodes for K2P18 or the TRESK channel and is primarily expressed in trigeminal root ganglion (TRG) and dorsal root ganglion (DRG). Truncations and other mutations in KCNK18 have been associated with familial migraine (Lafrenière et al. 2010). Expression of mutant K2P18 channels resulted in decreased current density when expressed in oocytes. This observation led Lafrenière and colleagues to propose that an increase in the functional expression of WT K2P18 could protect against migraines and that as yet unidentified mutations in KCNK18 could lead to an increase in migraine risk (Lafrenière et al. 2010).

Following whole exome sequencing (WES) studies conducted on patients with arrhythmic disorders, Decher and colleagues identified a heterozygous K2P2 mutation (Ile267Thr) in a patient with right ventricular outflow tract ventricular tachycardia (RVOT-VT). When expressed in *Xenopus* oocytes, K2P2 Ile267Thr channels have decreased current compared to WT channels. Further, co-expression of WT and mutant channel resulted in reduced current density in what is known as a "dominant-negative" behavior (Decher et al. 2017). It was found that the mutant channel was more permeable to sodium (Na^+) ions, unlike WT channels. This change in ion selectivity of the channel was attributed to the mutation of the isoleucine residue in the second pore loop to threonine. A change in the selectivity of K2P channels that permits an increase in the conductance of sodium has previously been observed for development-related alternative-translation initiated truncation variants of K2P2 and for mutation in K2P1 (Thomas et al. 2008). Following their observations, Decher and colleagues sought to reverse this defect in ion selectivity by finding drugs that would "rescue" the channel. Incubation of the channel with the following K2P2 blockers verapamil (62 μM) and fluoxetine (80 μM) and activators 2-APB (50 μM) and riluzole (500 μM) did not alter the selectivity of the channel (Decher et al. 2017). In contrast 5 μM of BL-1249 rescued channel function. Authors hypothesized that BL-1249 may be binding at a unique site that differs from the other compounds.

K_{2P} channels are expressed ubiquitously across excitable and non-excitable tissues (Lesage 2003; Lesage and Lazdunski 2000). Several K_{2P} channels are expressed in the TRG and DRG (Mathie and Veale 2015). The DRG and TRG somatosensory neurons give rise to the peripheral axonal fibers that innervate various tissues including the skin, muscle, and viscera and ascend to the spinal cord (DRG) or brainstem (TRG) (Mathie and Veale 2015; Plant 2012). Damage-sensing (nociceptive) somatosensory neurons detect and respond to noxious stimuli through activation of Aδ fibers which are lightly myelinated neurons that respond to localized pain, and via C-fibers which are unmyelinated neurons that are activated by a range of noxious stimuli (Plant 2012). Aα and Aβ fibers are myelinated fibers that respond to innocuous, mainly mechanical stimuli (Plant 2012). K_{2P} channels expressed in the DRG and TRG modulate neuronal excitability and response to noxious and innocuous mechanical stimuli.

Using a rat neuropathic pain model, Pollema and colleagues demonstrated that following spared nerve injury (SNI) levels of mRNA for KCNK3 and KCNK9 (that encode for K2P3 and K2P9 channels, respectively), were downregulated compared to sham controls. Downregulation of these K2P channels following SNI implicates these channels in neuropathic pain phenotypes. Interestingly, four weeks post SNI, only mRNA for KCNK1 (which encodes for K2P1) remained downregulated hinting at the importance of this channel in maintaining the neuropathic pain phenotype (Pollema-Mays et al. 2013). Contrary to this study, another group found that while still using the SNI model, intrathecal delivery of K2P18 in an adenovirus vector reduced the response of rats to neuropathic pain (Zhou et al. 2013).

2.1 K2P Channel Pharmacology

Although multiple lines of evidence support a role for K_{2P} channels in pain physiology, pharmacological options that target these proteins remain elusive. Given that present pharmacophores lack the ability to selectively inhibit K_{2P} channels, development of selective pharmacological agents is therefore imperative in order to study distinct characteristics of each channel. Intensive efforts to identify selective, potent, and efficacious pharmacophores are in progress. For example, Bagriantsev and colleagues utilized a 384 well plate yeast-based screening assay to identify K_{2P} blockers and activators in a high-throughput fashion. They began by screening a library containing 106,281 small molecules for their ability to inhibit the growth of yeast expressing K2P2. From this screen the library of small molecules was narrowed to 320 compounds that were selected for their ability to inhibit 44–99% of growth (Bagriantsev et al. 2013). A dose-response screen revealed 61 compounds that successfully prevented the growth of yeast expressing K2P2. TEVC experiments conducted in *Xenopus* oocytes revealed that 2 inhibitors ML45, ML58 and 3 activators ML12, ML42, and ML67 altered K2P2 channel activity (Bagriantsev et al. 2013). Bagriantsev et al. selected the activator ML67 which caused an ~11 fold (EC_{50} 213 ± 1.2 μM) increase in K2P2 channel current for further characterization. Through TEVC experiments it was found that the compound activated closely related channels (K2P10, EC_{50} ~ 250 μM) but not the more distantly related K2P3 channel. Substitution of a tricyclic ring to the ML67 compound yielded the compound ML67–33 which was 5 times more potent than the other ML-67 derivatives (Bagriantsev et al. 2013). Mutations at the P1 pore helix (Gly1371) and M4 (Trp275) of K2P2 resulted in decreased channel activity. Conversely, triple glycine mutations at the C-terminal lead to channels that could be activated by the compound. As a result, the authors postulated that ML-67-33 mediates its effects on K2P2 activity by modulating the C-type gate. Compounds such as ML-67-33, a selective and potent activator of K2P2 channels, provide an approach by which similar compounds could be developed and assayed. In this chapter we provide a concise summary of the pharmacology and regulation of K_{2P} channels in that they may be explored further toward the development of novel pharmacophores.

3 The THIK Channels: K2P12 and K2P13

The THIK subfamily is composed of THIK2 (K2P12, *KCNK12*) and THIK1 (K2P13, *KCNK13*) channels (Girard et al. 2001; Rajan et al. 2001). The mammalian K2P12 and K2P13 channels share 64% homology as well as a similar pore region structure (Renigunta et al. 2014). While K2P13 channels are expressed ubiquitously, K2P12 channels are expressed in the lungs, spleen, and brain (Rajan et al. 2001). When expressed heterologously in *Xenopus* oocytes, only K2P13 channel activity can be measured while K2P12 channel activity is largely undetectable. K2P13 currents are activated by arachidonic acid and inhibited by halothane, quinidine, and weakly by hypoxia (a ~ 13% reduction compared to control when Po_2 is decreased to 20 mmHg) (Table 2) (Campanucci et al. 2005; Enyedi and Czirjak 2010; Feliciangeli et al. 2015; Renigunta et al. 2014).

K2P12 channels are one of five channels: K2P1 (TWIK1), K2P6 (TWIK2), K2P7 (kcnk8), and K2P15 (TASK5) that are classified as electrically silent channels because they do not pass measurable K^+ current in either native cells or in

Table 2 Modulators of THIK subfamily of K2P channels[a]

Channel	Activators	Inhibitors
K2P13 (THIK1, KCNK13)	Arachidonic acid (Rajan et al. 2001)	Halothane (Rajan et al. 2001); Quinidine (Chatelain et al. 2013; Rajan et al. 2001); Hypoxia (Campanucci et al. 2005)
K2P12 (THIK2, KCNK12)	No known modulators	Halothane (Rajan et al. 2001)

[a]A regularly updated summary of the activators and inhibitors for all K2P channels is available at https://www.guidetopharmacology.org/GRAC/FamilyDisplayForward?familyId=79

heterologous expression systems (Renigunta et al. 2014). Two groups reported that lack of detectable K2P12 channel activity was a result of the channel possibly being sequestered in the endoplasmic reticulum (ER) and thus resulting in low expression of the channel at the cellular membrane (Blin et al. 2014; Chatelain et al. 2013). However, detection of K2P12 channel activity is possible under specific circumstances. Thus, it was found that substitution of a proline residue within M2 helix or deletion of 18 to 19 AA found in the N-terminus (corresponding to an ER retention/retrieval signaling motif) results in the appearance of macroscopic K2P12 activity that is comparable to K2P13 (* Chatelain et al. 2013; Renigunta et al. 2014). Removal of the AAs from the N-terminus however prevented the channel from being activated by arachidonic acid even though it could still be inhibited by both halothane and quinidine (Renigunta et al. 2014). Of great physiological relevance, heterodimerization of K2Ps 12 and 13 results in functional channels presumably because K2P13 masks the ER retention motif on the K2P12 subunit (Bayliss et al. 2019).

4 The TRESK Subfamily: K2P18

The TRESK subfamily contains only the K2P18 channel, encoded by *KCNK18* (Sano et al. 2003). Discovery of K2P18 in 2003 was made possible following the completion of the human genome project (Sano et al. 2003). Sano and colleagues utilized the human draft sequencing data to clone the K2P18 subunit from the complementary DNA of the spinal cord. Subsequence expression analysis found mRNA transcript for KCNK18 throughout the central and peripheral nervous systems (Bayliss et al. 2019; Enyedi et al. 2012; Enyedi and Czirjak 2015; Gada and Plant 2019; Tulleuda et al. 2011; Weir et al. 2019). In rodents, expression of K2P18 has also been detected in the spleen, thymus, and testis (Enyedi and Czirjak 2010). K2P18 channels contribute to the leak or background K^+ current which plays an important role in the regulation of neuronal excitability (Hwang et al. 2015). When studied using symmetrical K^+ solutions, K2P18 channels displayed outward rectification (Lengyel et al. 2018; Sano et al. 2003). Tulleuda and colleagues reported a decrease in channel activity following neuronal injury, which alters neuronal excitability and thus changes "pain pathways."

K2P18 shares ~19% sequence homology with other members of the K2P family (Lengyel et al. 2018; Sano et al. 2003). Despite this, human K2P18 is predicted to be structurally like the rest of the K2P channels. It however differs in that its intracellular loop found between the second and third transmembrane domains is longer (>120 amino acids (AA) compared to the 20–30 AA in the other K2P channels) and its C-terminal is shorter (30 AA long compared to the ≥120 AA in the other K2P channels) (Enyedi and Czirjak 2015; Sano et al. 2003). In contrast to most other K2P channels, K2P18 has a short C-terminal tail. This structural difference may indicate differential regulation of K2P18, including how regulatory events might allosterically influence the activity of the channel. (Braun et al. 2015).

The activity of K2P18 channels is enhanced by volatile (inhaled) anesthetics (e.g., isoflurane, sevoflurane, halothane, desflurane) but is inhibited by local anesthetics, including bupivacaine, tetracaine, ropivacaine, mepivacaine, lidocaine, as well as unsaturated fatty acids (Table 3) (Czirjak et al. 2004; Liu et al. 2004). Like most K2P channels, K2P18 channels are sensitive to differences in extracellular and intracellular pH, however the degree of sensitivity differs in the human ortholog compared to rodent orthologs (Lotshaw 2007). In contrast to other K2P channels, K2P18 is modulated by the cytosolic concentration of Ca^{2+} ions. Thus, K2P18 channels are regulated by activation of $G\alpha_q$-coupled receptors, which lead to downstream release of Ca^{2+} from intracellular stores (Table 3). However, a series of elegant studies by Czirják et al. showed that the direct application of Ca^{2+} ions to the inside of the membrane was insufficient to stimulate K2P18 in off-cell patches, suggesting that additional cytoplasmic factors are required to activate the channels (Czirjak et al. 2004). Subsequent studies found that the Ca^{2+}-dependent activation of K2P18 is mediated by the calmodulin-dependent protein phosphatase, calcineurin, which interacts with the C-terminal tail of the channel (Czirjak et al. 2004). This regulatory mechanism that activates K2P18 channels can be inhibited by pharmacological inhibitors of calcineurin such as cyclosporine. In addition, mutant channels that lack the calcineurin binding site are still subject to regulation by a novel-type of protein kinase C (Pergel et al. 2019).

5 The TALK Subfamily: K2P5, K2P16, and K2P17

The TALK family includes the K2P5, (TASK2, *KCNK5*), K2P16 (TALK1, *KCNK16*), and K2P17 (TALK2, TASK4, *KCNK17*) channels (Decher et al. 2001; Girard et al. 2001; Reyes et al. 1998). K2P16 and K2P17 channels share 37% homology (Lotshaw 2007). When K2P5 was first cloned from human kidney it was assigned to the TASK subfamily. However, it was later reassigned to the TALK subfamily because it had more sequence similarity (~30%) to K2P16 and K2P17 and, in addition, its pH sensitivity was in the alkaline range, similar to that of K2P16 and K2P17 (Enyedi and Czirjak 2010; Lotshaw 2007; Reyes et al. 1998). In humans, K2P5 expression has been detected in the kidneys, pancreas, and liver. Transcripts for KCNK5 were also detected in DRG and spinal cord (Medhurst et al. 2001) (Enyedi and Czirjak 2010). In humans, mRNA for KCNK17 has been found in the

Table 3 Modulators of the TRESK subfamily of K2P channels[a]

Channel	Activators	Inhibitors
K2P18 (TRESK, KCNK18)	Volatile anesthetics (Liu et al. 2004) Calcium (Czirjak et al. 2004)	Local anesthetics (Czirjak et al. 2004; Liu et al. 2004) Unsaturated fatty acids (Sano et al. 2003)
	Gαq (Czirjak et al. 2004)	Cyclosporin (Czirjak et al. 2004)

[a]A regularly updated summary of the activators and inhibitors for all K2P channels is available at https://www.guidetopharmacology.org/GRAC/FamilyDisplayForward?familyId=79

liver, heart, pancreas, and lungs while K2P16 channels appear to be expressed exclusively in the pancreas (Duprat et al. 2005; Girard et al. 2001; Lotshaw 2007).

All TALK subfamily channels are activated by extracellular and intracellular alkalinization and inhibited by extracellular acidification (Cid et al. 2013) (Table 4). The pH-sensing of K2P5 requires Arg244; substitution of this amino acid with neutral residues abolishes the response of the channel to changes in alkalization of the extracellular pH (pH_o) (Niemeyer et al. 2007). Protonation of Arg244 residue lowers K^+ occupancy of the selectivity filter resulting in pore-blockade (Cid et al. 2013).

TALK channels are also sensitive to changes in the intracellular pH (pH_i) (Niemeyer et al. 2010). It is postulated that lys245, located on the C-terminus of K2P5, acts as a sensor for pH_i (Cid et al. 2013). Given the findings, it may be that the regulation of K2P5 channel activity by pH_o and pH_i occurs via effects on independent gates (Cid et al. 2013; Niemeyer et al. 2010); however, the mechanistic details that subserve this idea are yet to be elucidated.

K2P5 activity can be inhibited by Gβγ subunits of the heterotrimeric G protein (Anazco et al. 2013) (Table 4). Añazco and colleagues suggested that Gβγ modulation plays a role in the channel's ability to react to changes in cell volume (this is a result of neutralization of a lysine residue in the C-terminus that is important for inhibition by Gβγ). Although modulation of K2P5 by Gβγ is possible, it remains an open question in the field. Evidence to support Gβγ-modulation of K2P channel activity can be found in the K2P2 channels (Woo et al. 2012). Finally, Duprat and colleagues demonstrated that both K2P16 and K2P17 channels can be activated by nitric oxide (NO) and reactive oxygen species (ROS) (Table 4) (Duprat et al. 2005).

6 The TWIK Subfamily: K2P1, K2P6, and K2P7

Following its initial description in 1996, K2P1 (TWIK1, *KCNK1*) was observed to have low channel activity in heterologous expression systems (Goldstein et al. 1998; Lesage et al. 1996b; Pountney et al. 1999). However, since mRNA transcripts for KCNK1, the gene that encodes for the K2P1 subunit, are found in the kidney, placenta, lungs, heart, and the brain (Gaborit et al. 2007; Lesage et al. 1996b; Talley et al. 2001), several groups pursued potential cellular and biophysical mechanisms that would limit the activity of K2P1 channels. Data to support three

Table 4 Modulators of TALK subfamily of K2P channels[a]

Channel	Activators	Inhibitors
K2P5 (TASK2, KCNK5)	Alkaline pHo and pHi (Cid et al. 2013; Niemeyer et al. 2007)	Gβγ (Anazco et al. 2013)
K2P16 (TALK1, KCNK16)	Alkaline pHo and pHi (Cid et al. 2013; Niemeyer et al. 2007); Nitric oxide (NO) and Reactive oxygen species (ROS) (Duprat et al. 2005)	No known modulators
K2P17 (TALK2, KCNK17)		No known modulators

[a]A regularly updated summary of the activators and inhibitors for all K2P channels is available at https://www.guidetopharmacology.org/GRAC/FamilyDisplayForward?familyId=79

hypotheses have been presented: SUMOylation of K2P1 channels at the plasma membrane; rapid endocytosis of K2P1 channels from the plasma membrane, and hydrophobic dewetting of the channel pore.

SUMOylation is an enzyme-mediated post-translational modification pathway that links a ~100 amino acid Small Ubiquitin-like MOdifier (SUMO) protein to the epsilon amine-group of lysine residues in specific motifs (Hay 2005). Although SUMOylation was not thought to occur at the plasma membrane, the process was shown to inhibit the activity of K2P1 channels because K^+ selective currents were observed when SUMO was removed from the channel by a SUMO-specific proteases (SENPs), or when the SUMOylation site (K2P1-Lys274) was mutated to prevent SUMO-binding (Plant et al. 2010; Rajan et al. 2005). SUMOylation is now known to regulate the activity of an array of ion channels in multiple tissues. The process is rapid, reversible, and dynamic and is often challenging to capture biochemically. In keeping with observations of numerous soluble SUMO substrates, such as transcriptional regulators, SUMOylation of K2P1 channels is labile and is often not observed when cells and tissues are studied after detergent purification (Feliciangeli et al. 2007; Hay 2005). Therefore, SUMOylation is typically studied in live cells using real-time electrophysiology, spectroscopy, and microscopy (Plant et al. 2010).

Studies in MDCK and HEK293 cells found that the low activity of K2P1 could be attributed to rapid, endocytic recycling of the channel from the plasma membrane (Feliciangeli et al. 2010, 2015). The process is dynamin-dependent based on analysis of a di-isoleucine motif: mutation of Ile293 and Ile294) resulted in measurable currents upon heterologous expression. Further, K2P1 was found to associate with ARF6, a small G protein that modulates endocytosis at the apical surface of epithelial cell (Decressac et al. 2004).

Following the resolution of the crystal structure of human K2P1, molecular dynamic simulations (MDS) of ion permeation identified a "hydrophobic cuff" in the inner vestibule of the channel, below the selectivity filter, comprised of four residues: Leu146 on M2 and Leu261 on M4, from each subunit (Aryal et al. 2014; Miller and Long 2012). MDS revealed that stochastic motion of the cuff restricted the access of water molecules to the internal entrance of the pore, creating an energetic barrier to the permeation of K^+. Based on this model, substitution of Leu146 with hydrophilic residues resulted in a K2P1 channel variant that passed robust currents in *Xenopus* oocytes (Aryal et al. 2014; Chatelain et al. 2012).

Determining how SUMOylation, the hydrophobic gating barrier, and rapid endocytosis contribute individually or together to the regulation of K2P1 in native cells remains an area of active study that is spurred on by the observation that K2P1 knockout mice exhibit altered physiology in several tissues, including pancreatic β cells and the kidney (Chatelain et al. 2012; Nie et al. 2005). Similarly, K2P1 has been shown to play a key physiological and developmental role in the atria of transgenic zebrafish (Christensen et al. 2016). K2P1 has also been shown to mediate arrhythmogenic depolarization of cardiac myocytes exposed to low concentrations of K^+ associated with hypokalemia (Gotter et al. 2011). A part of the enigmatic character of K2P1 can be attributed to heterodimerization with K2P3 and K2P9 subunits in rat neurons and with K2P2 in rat astrocytes (Hwang et al. 2014; Plant et al. 2012). The resultant heteromeric channels have distinct properties. For example, the activity of K2P1-K2P3 and K2P1-K2P9 channels is increased by volatile, halogenated ester-based anesthetics and is subject to regulation by the SUMO pathway (Plant et al. 2012).

The TWIK subfamily is also composed of K2P6 (TWIK2, *KCNK6*) and the K2P7 (Kcnk8, *KCNK7*) channels. K2P6 was described by two independent groups (Chavez et al. 1999; Pountney et al. 1999) and shares 34% sequence identity with K2P1. In contrast, K2P7 is more closely related to K2P6 (94% homology) (Lesage and Lazdunski 2000; Lotshaw 2007). K2P6 and K2P7 are expressed in peripheral tissues and peripheral blood leukocytes, respectively (Lesage and Lazdunski 2000; Medhurst et al. 2001).

In native cells all TWIK channels have low channel activity and as a result they are sometimes considered to be electrically silent (Bockenhauer et al. 2000; Renigunta et al. 2014), limiting functional characterization of the channels as well as the development of selective pharmacological tools (Lotshaw 2007). When active, K2P1 and K2P6 currents are inhibited by barium, quinine, or quinidine (Table 5) (Lesage et al. 1996b). Separately, K2P1 channels can also be inhibited by intracellular (Lesage et al. 1996b) as well as extracellular acidification (Plant et al. 2010). K2P1 is also regulated by PKC activation by phorbol esters such as PMA, which enhances channel activity (Table 5) (Lesage et al. 1996b).

7 The TREK Subfamily: K2P2, K2P10, and K2P4

The TREK subfamily is composed of K2P2 (TREK1, *KCNK2*), K2P10 (TREK2, *KCNK10*), and K2P4 (TWIK-related arachidonic-acid-stimulated K^+ channel or TRAAK, K2P4, *KCNK4*) channels (Bang et al. 2000; Fink et al. 1998). In humans K2P2 and K2P10 tissue expression overlaps in the CNS and periphery tissues while K2P4 expression is most notable in the neurons (Lesage et al. 2000; Meadows et al. 2000; Medhurst et al. 2001). K2P10 channel shares 65% sequence similarity to K2P2 and 45% similarity to K2P4 (Bang et al. 2000; Ozaita and Vega-Saenz de

Table 5 Modulators of TWIK subfamily of K2P channels[a]

Channel	Activators	Inhibitors
K2P1 (TWIK1, KCNK1)	pHo (Rajan et al. 2005) (deSUMOylated channel); PKC (Lesage et al. 1996b)	Barium (Lesage et al. 1996b); Quinine or Quinidine (Lesage et al. 1996b); Acid pHi (Lesage et al. 1996b)
K2P6 (TWIK2, KCNK6)	No known modulators	Barium (Lesage et al. 1996b); Quinidine (Lesage et al. 1996b)
K2P7 (Kcnk8/ KCNK7)	No known modulators	No known modulators

[a]A regularly updated summary of the activators and inhibitors for all K2P channels is available at https://www.guidetopharmacology.org/GRAC/FamilyDisplayForward?familyId=79

Miera 2002). The K2P2 and K2P10 channels exhibit similar outward rectification (Lesage et al. 2000; Maingret et al. 1999; Medhurst et al. 2001). The differences between K2P2 and K2P10 currents can be seen when comparing unitary currents of the two channels under high extracellular concentration of K^+. Under this condition K2P10 exhibits inward rectification (Lesage et al. 2000; Maingret et al. 1999; Medhurst et al. 2001) while both K2P2 and K2P4 exhibit Goldman-Hodgkin-Katz (GHK) rectification (Fink et al. 1998).

The TREK subfamily of K2P channels are noted for their sensitivity to mechanical stimuli. These mechanosensitive channels are modulated by numerous physicochemical stimuli including pH, temperature, mechanical stress (stretch, shear, and swelling), polyunsaturated fatty acids (PUFAs), anesthetics (volatile), and protein phosphorylation (Table 6) (Lotshaw 2007; Maingret et al. 1999). K2P2 channels are also activated by an acidic pH_i (Maingret et al. 2000), likely due to protonation of a glutamic acid residue at position 306 (Glu306). Protonation of this residue is an important regulator of the response of K2P2 channels to mechanical stimulation (Honore et al. 2002).

TREK channels are also activated by heat (Kang et al. 2005; Maingret et al. 2000). Thus, at 37 °C K2P2 channels exhibit outward rectification (Kang et al. 2005; Maingret et al. 2000) that is lost upon cooling (Kang et al. 2005; Maingret et al. 2000). K2P2 and K2P10 are also activated by halogenated volatile anesthetics such as chloroform, ether, halothane, isoflurane (Table 6) (Lesage et al. 2000; Maingret et al. 2000). Halothane is a more effective activator of K2P10 while chloroform is a more efficacious activator of K2P2 (Lesage et al. 2000). All the TREK subfamily channels are activated by riluzole, a neuroprotective drug that transiently activates K2P2 and K2P10 but permanently activates K2P4 (Lesage et al. 2000). The mechanism by which riluzole exerts its effect is believed to be a result of PKA inhibition as a result of cAMP accumulation (Lesage et al. 2000). In 2001, Bockenhauer and colleagues demonstrated that PKA phosphorylation of serine-348 (Ser348) results in an altered voltage-dependence of K2P2 channels, effectively reducing the open probability and thereby the channel activity (Bockenhauer et al. 2001).

Table 6 Modulators of TREK subfamily of K2P channels[a]

Channel	Activators			Inhibitors
K2P2 (TREK1, KCNK2)	NO (Koh et al. 2001)	Acid pHi (Maingret et al. 1999), Volatile anesthetics (Chloroform, ether, halothane, isoflurane) (Lesage et al. 2000; Maingret et al. 2000), Mechanical stress (Lesage et al. 2000; Medhurst et al. 2001),	PUFA (Lesage et al. 2000; Meadows et al. 2000; Medhurst et al. 2001), Riluzole (Lesage and Lazdunski 2000) Heat (Kang et al. 2005; Maingret et al. 2000),	Gαs and Gq (Lesage et al. 2000), Quinidine (Lesage et al. 2000)
K2P10 (TREK2, KCNK10)	Gαi (Lesage et al. 2000)			
K2P4 (TRAAK, KCNK4)	Alkaline pHi (Kim and Gnatenco 2001)	No known modulators		No known modulators

[a]A regularly updated summary of the activators and inhibitors for all K2P channels is available at https://www.guidetopharmacology.org/GRAC/FamilyDisplayForward?familyId=79

Inhibition of K2P2 and K2P10 but not K2P4 was demonstrated to be mediated by activators of protein kinases (Table 6). Lesage and colleagues found that co-expression of K2P10 and $G\alpha_s$-coupled receptor 5HT4 resulted in decreased channel activity when the receptors were activated by 5-hydroxytryptamine. In contrast, co-expression K2P10 and $G\alpha_i$-coupled mGluR2 receptors increased channel activity upon stimulation by glutamate (Lesage et al. 2000). Lastly, co-expression of K2P10 and the $G\alpha_q$-coupled receptor mGluR1 resulted in inhibition of channel activity upon stimulation of mGluR1 by glutamate (Lesage et al. 2000). Signaling through $G\alpha_q$ results in activation of phosholipase C (PLC) which results in the hydrolysis of PIP_2 into diacylglcerol (DAG) and inositol 1,4,5-triphosphate (IP3) production. Lesage et al. postulated that inhibition of the channel may be a result of activation of protein kinase C (PKC) by DAG (Fig. 3).

8 The TASK Subfamily: K2P3, K2P9, and K2P15

The TASK subfamily is composed of K2P3 (TASK1, *KCNK3*), K2P9, (TASK3, *KCNK9*), and K2P15 (TASK5, *KCNK9*) channels (Duprat et al. 1997; Kim and Gnatenco 2001; Kim et al. 2000). The K2P3 channel was first isolated based on its sequence homology to K2P1 and K2P2 (Duprat et al. 1997). In general, the TASK channels share low sequence similarity with other K2P channels (<30%) however, amongst each other TASK channels share relatively high sequence similarity (>50%) (Ashmole et al. 2001; Duprat et al. 1997, 2007; Kim et al. 2000). TASK channels are expressed in most tissues with notable expression in the placenta and pancreas (Ashmole et al. 2001; Duprat et al. 1997; Kim et al. 2000; Rajan et al. 2000). While K2P3 and K2P9 can form functional homodimers or heterodimers,

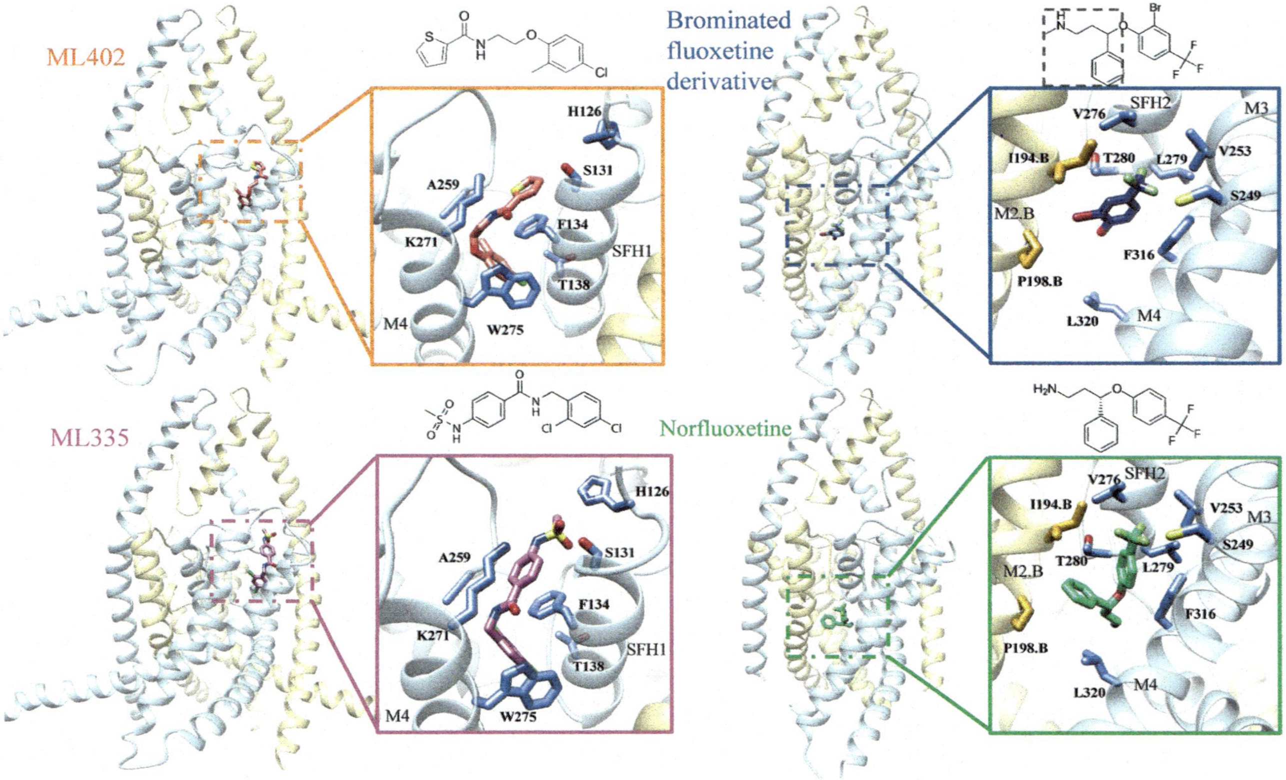

Fig. 3 Interactions between K_{2P} channels and pharmacophores. A comparison of four pharmacophores that have been captured in complex with K2P channels. For each channel, one subunit is shown in light blue and the adjacent subunit is shown in yellow. In each case, the structure of the pharmacophore is given above

Table 7 Modulators of TASK subfamily of K2P channels[a]

<table>
<tr><th>Channel</th><th>Activators</th><th>Inhibitors</th></tr>
<tr><td>K2P3 (TASK1, KCNK3)</td><td rowspan="2">Alkaline pHo (Duprat et al. 1997; Kang et al. 2004), Halothane and Isoflurane (Patel et al. 1999)</td><td rowspan="2">Acidic pHo (Duprat et al. 1997; Kim et al. 2000; Rajan et al. 2000), $G\alpha q$ (Chen et al. 2006)
K2P9: Ruthenium Red (Czirják and Enyedi 2002)</td></tr>
<tr><td>K2P9 (TASK3, KCNK9)</td></tr>
<tr><td>K2P15 (TASK5)</td><td>No known modulators</td><td>No known modulators</td></tr>
</table>

[a]A regularly updated summary of the activators and inhibitors for all K2P channels is available at https://www.guidetopharmacology.org/GRAC/FamilyDisplayForward?familyId=79

K2P15 channels are electrically silent when expressed alone or with other TASK channels (Ashmole et al. 2001; Bayliss and Barrett 2008; Czirják and Enyedi 2002; Duprat et al. 2007). Under physiological conditions activation of TASK1 and TASK3 channels occurs instantaneously and the channels exhibit outward rectification (Duprat et al. 2007; Kim et al. 2000).

The sine qua non of TASK channels is inhibition of the channel activity by extracellular acidification (Table 7) (Czirják and Enyedi 2002; Duprat et al. 1997; Kim et al. 2000; Rajan et al. 2000). In mutational studies of Guinea pig K2P9 (62.3% and 88.3% homology to human K2P3 and K2P9, respectively) Rajan and colleagues found that the histidine at position 98 (His98) conferred pH sensitivity to the channel (Lopes et al. 2000, 2001; Rajan et al. 2000). Similarly, Lopes and colleagues found that protonation of the equivalent residue in K2P3 conferred pH-sensitivity to that channel (Lopes et al. 2000, 2001). Of note, K2P3 and K2P9 heterodimers are also inhibited by by extracellular acidification (Czirják and Enyedi 2002).

Two groups have found that K2P9 homodimers are inhibited by Ruthenium Red (RR) while micromolar concentrations of RR were also unable to inhibit K2P3 homodimers in both *Xenopus* oocytes and COS-7 cells (Table 7). Interestingly, K2P3-K2P9 heterodimers are minimally inhibited by RR (Czirják and Enyedi 2002; Kang et al. 2004). RR appears to inhibit K2P9 homodimers by binding to Glutamate 70 (Glu70) on both subunits (Czirjak and Enyedi 2003). With K2P3-K2P9 heterodimers there is only one subunit with Glu70 for RR to bind which is likely insufficient to cause inhibition (Czirjak and Enyedi 2003).

K2P3 and K2P9 are both inhibited by $G\alpha_q$ (Chen et al. 2006) although whether this result is secondary to hydrolysis of PIP_2 remains an area of active debate. Both K2P3 and K2P9 are activated by volatile anesthetics (halothane and isoflurane) (Kang et al. 2004; Patel et al. 1999).

Fig. 3 (continued) a zoomed-in view of how the molecule interacts with the channel protein. **Right-hand column: ML402** (top) or **ML335** (bottom) interacting with K2P2 channel (PDB ID: 6CQ9, 6CQ8) (Lolicato et al. 2017). **Left-hand column: Brominated fluoxetine derivative** or **Norfluoxetine** binding to K2P10 (PDB ID: 4XDL, 4XDK (Dong et al. 2015)

9 TOK Channels

K^+ channel subunits with two pore domains are not limited to expression in higher order eukaryotes but have also been identified in fungi. The transient outward current (TOK) channels were first cloned and described in *Saccharomyces cerevisiae* following a genome search that identified a P domain peptide sequence homologous to those of other K^+ channels (Ketchum et al. 1995). In contrast to the K2P subunits discussed above, TOK channels are dimers of subunits with eight transmembrane domains with intracellular amino- and carboxy-terminal tails (M1-M8), with two reentrant P-loops located between transmembrane domains composed with two P-loops regions located between M5 and M6 and M7 and M8 (Fig. 3) (Ketchum et al. 1995; Lesage et al. 1996a; Zhou et al. 1995). Expression of the *S. cerevisiae* TOK (ScTOK) channels in *Xenopus* oocytes revealed K^+-selective channels with outward rectification that were activated by depolarizing voltages (Ketchum et al. 1995; Lesage et al. 1996a; Zhou et al. 1995). Activation of ScTOK channels is coupled to the K^+ equilibrium potential (E_K) in that changes in the external concentration of K^+ results in loss of outward rectification (Bertl et al. 1998; Ketchum et al. 1995; Lesage et al. 1996a; Zhou et al. 1995). ScTOK currents are inhibited by barium ions, quinine, or tetraethylammonium (TEA) (Ketchum et al. 1995; Lesage et al. 1996a; Zhou et al. 1995).

TOK channels have now been identified in a range of fungi, including strains that are pathogenic to humans. A comparative study of four pathogenic fungi, *Aspergillus fumigatus* (AfTOK1), *Candida albicans* (CaTOK), and two strains of *Cryptococcus neoformans* (CnTOK and H99TOK), by Lewis and colleagues revealed that the TOK subfamily of K^+ channels share similar biophysical characteristics as ScTOK (Lewis et al. 2020). Their unique distribution in only fungi suggests that these TOK channels could be important therapeutic targets for anti-fungal pharmaceutics. This intriguing proposal is supported by data showing that extracellular K1 killer toxin kills *Saccharomyces* yeast by increasing the open probability of ScTOK and perturbing K^+ homeostasis (Ahmed et al. 1999). In contrast, infection with killer toxin virus protects against the effects of the external toxin, allowing virus-positive cells to propagate (Sesti et al. 2001). Thus, selective, small molecule activators of TOK channels are potential anti-fungal agents.

10 Conclusion and Future Perspectives

The K^+ channels comprise a large, diverse, and ubiquitous superfamily of membrane proteins that regulate various biological processes in both excitable and non-excitable cells (Kuang et al. 2015; Tian et al. 2014). The two-pore domain K^+ channels constitute a subfamily of K^+ channels that are categorized based on structural and sequence similarity. Since the discovery of these channels more than 20 years ago much has been revealed about these channel's physiology and pharmacology. The expression of K_{2P} channels is widespread across various tissues and organ systems. This broad distribution and expression highlight their importance in the biology of many tissues and suggest that K_{2P} channels will continue to emerge as important potential druggable targets for the treatment of diverse diseases. Given the fundamental role that K_{2P} channels play in physiology, it is not surprising that their activity is tightly regulated and modulated by diverse physicochemical and mechanical stimuli including temperature, mechanical stress, pH_i, pH_o, second-messenger pathways, PUFAs, and phosphoinositides.

Despite the growing body of work which has implicated K_{2P} channels in various cardiac and neuronal diseases, there is much that is yet to be learned about K_{2P} physiology and its role in pathophysiology. A present obstacle in attaining this knowledge is the lack of channel selective pharmacophores although this landscape is starting to evolve, particularly for the TREK subfamily of K_{2P} channels. Following the elucidation of several K_{2P} structures we now appreciate that a part of the delay in identifying selective pharmacophores comes from the cap-domain of the K_{2P} channels. This structural feature, seemingly unique amongst K^+ channels, protects the outer mouth of the channel pore from infiltration by classical K^+ channel blockers, particularly protein toxins. However, the same structural revolution that identified the problems has also helped to initiate solutions. Using computational approaches to understand the dynamics of K2P channels, researchers have started to identify druggable pockets and binding sites within the channel corpus. Of note, Lolicato and colleagues identified a cryptic binding pocket behind the pore of the K2P2 channel that can co-ordinate the newly identified channel activators ML335 and ML402 (Lolicato et al. 2017). Bagriantsev and colleagues demonstrated that selective and potent compounds of K_{2P} channels can also be identified using high-throughput screens (Bagriantsev et al. 2013). These powerful approaches promise to break the gridlock in the development of selective new K_{2P} channel modulators in the future.

Conflict of Interest The authors report no competing financial interests.

References

Ahmed A, Sesti F, Ilan N, Shih TM, Sturley SL, Goldstein SA (1999) A molecular target for viral killer toxin: TOK1 potassium channels. Cell 99:283–291. https://doi.org/10.1016/s0092-8674(00)81659-1

Anazco C, Pena-Munzenmayer G, Araya C, Cid LP, Sepulveda FV, Niemeyer MI (2013) G protein modulation of K2P potassium channel TASK-2: a role of basic residues in the C terminus domain. Pflugers Arch 465:1715–1726. https://doi.org/10.1007/s00424-013-1314-0

Aryal P, Abd-Wahab F, Bucci G, Sansom MS, Tucker SJ (2014) A hydrophobic barrier deep within the inner pore of the TWIK-1 K2P potassium channel. Nat Commun 5:1–9

Ashmole I, Goodwin PA, Stanfield PR (2001) TASK-5, a novel member of the tandem pore K+ channel family. Pflugers Arch 442:828–833. https://doi.org/10.1007/s004240100620

Bagriantsev SN, Peyronnet R, Clark KA, Honore E, Minor DL Jr (2011) Multiple modalities converge on a common gate to control K2P channel function. EMBO J 30:3594–3606. https://doi.org/10.1038/emboj.2011.230

Bagriantsev SN, Clark KA, Minor DL Jr (2012) Metabolic and thermal stimuli control K(2P)2.1 (TREK-1) through modular sensory and gating domains. EMBO J 31:3297–3308. https://doi.org/10.1038/emboj.2012.171

Bagriantsev SN, Ang K-H, Gallardo-Godoy A, Clark KA, Arkin MR, Renslo AR, Minor DL Jr (2013) A high-throughput functional screen identifies small molecule regulators of temperature- and mechano-sensitive K2P channels. ACS Chem Biol 8:1841–1851

Bang H, Kim Y, Kim D (2000) TREK-2, a new member of the mechanosensitive tandem-pore K+ channel family. J Biol Chem 275:17412–17419. https://doi.org/10.1074/jbc.M000445200

Barel O, Shalev SA, Ofir R, Cohen A, Zlotogora J, Shorer Z, Mazor G, Finer G, Khateeb S, Zilberberg N (2008) Maternally inherited Birk Barel mental retardation dysmorphism syndrome caused by a mutation in the genomically imprinted potassium channel KCNK9. Am J Hum Genet 83:193–199

Bayliss DA, Barrett PQ (2008) Emerging roles for two-pore-domain potassium channels and their potential therapeutic impact. Trends Pharmacol Sci 29:566–575. https://doi.org/10.1016/j.tips.2008.07.013

Bayliss DA, Czirják G, Enyedi P, Goldstein SA, Lesage F, Minor DL Jr, Plant LD, Sepúlveda F, Winn BT (2019) Two P domain potassium channels (version 2019.4) in the IUPHAR/BPS guide to pharmacology database. IUPHAR/BPS Guide Pharmacol. https://doi.org/10.2218/gtopdb/F79/2019.4

Bertl A, Bihler H, Reid J, Kettner C, Slayman CL (1998) Physiological characterization of the yeast plasma membrane outward rectifying K+ channel, DUK1 (TOK1), in situ. J Membr Biol 162:67–80

Blin S, Chatelain FC, Feliciangeli S, Kang D, Lesage F, Bichet D (2014) Tandem pore domain halothane-inhibited K+ channel subunits THIK1 and THIK2 assemble and form active channels. J Biol Chem 289:28202–28212. https://doi.org/10.1074/jbc.M114.600437

Bockenhauer D, Nimmakayalu MA, Ward DC, Goldstein SA, Gallagher PG (2000) Genomic organization and chromosomal localization of the murine 2 P domain potassium channel gene Kcnk8: conservation of gene structure in 2 P domain potassium channels. Gene 261:365–372

Bockenhauer D, Zilberberg N, Goldstein SA (2001) KCNK2: reversible conversion of a hippocampal potassium leak into a voltage-dependent channel. Nat Neurosci 4:486–491. https://doi.org/10.1038/87434

Braun G, Lengyel M, Enyedi P, Czirjak G (2015) Differential sensitivity of TREK-1, TREK-2 and TRAAK background potassium channels to the polycationic dye ruthenium red. Br J Pharmacol 172:1728–1738. https://doi.org/10.1111/bph.13019

Brohawn SG, del Mármol J, MacKinnon R (2012) Crystal structure of the human K2P TRAAK, a lipid-and mechano-sensitive K+ ion channel. Science 335:436–441

Campanucci V, Brown S, Hudasek K, O'kelly I, Nurse C, Fearon I (2005) O2 sensing by recombinant TWIK-related halothane-inhibitable K+ channel-1 background K+ channels heterologously expressed in human embryonic kidney cells. Neuroscience 135:1087–1094

Chatelain FC, Bichet D, Douguet D, Feliciangeli S, Bendahhou S, Reichold M, Warth R, Barhanin J, Lesage F (2012) TWIK1, a unique background channel with variable ion selectivity. Proc Natl Acad Sci 109:5499–5504

Chatelain FC, Bichet D, Feliciangeli S, Larroque MM, Braud VM, Douguet D, Lesage F (2013) Silencing of the tandem pore domain halothane-inhibited K+ channel 2 (THIK2) relies on combined intracellular retention and low intrinsic activity at the plasma membrane. J Biol Chem 288:35081–35092. https://doi.org/10.1074/jbc.M113.503318

Chavez RA, Gray AT, Zhao BB, Kindler CH, Mazurek MJ, Mehta Y, Forsayeth JR, Yost CS (1999) TWIK-2, a new weak inward rectifying member of the tandem pore domain potassium channel family. J Biol Chem 274:7887–7892. https://doi.org/10.1074/jbc.274.12.7887

Chemin J, Patel AJ, Delmas P, Sachs F, Lazdunski M, Honore E (2007) Regulation of the Mechano-gated K2P channel TREK-1 by membrane phospholipids. Curr Top Membr 59:155–170. https://doi.org/10.1016/S1063-5823(06)59007-6

Chen X, Talley EM, Patel N, Gomis A, McIntire WE, Dong B, Viana F, Garrison JC, Bayliss DA (2006) Inhibition of a background potassium channel by Gq protein alpha-subunits. Proc Natl Acad Sci U S A 103:3422–3427. https://doi.org/10.1073/pnas.0507710103

Christensen AH, Chatelain FC, Huttner IG, Olesen MS, Soka M, Feliciangeli S, Horvat C, Santiago CF, Vandenberg JI, Schmitt N (2016) The two-pore domain potassium channel, TWIK-1, has a role in the regulation of heart rate and atrial size. J Mol Cell Cardiol 97:24–35

Cid LP, Roa-Rojas HA, Niemeyer MI, Gonzalez W, Araki M, Araki K, Sepulveda FV (2013) TASK-2: a K2P K(+) channel with complex regulation and diverse physiological functions. Front Physiol 4:198. https://doi.org/10.3389/fphys.2013.00198

Cohen A, Ben-Abu Y, Hen S, Zilberberg N (2008) A novel mechanism for human K2P2.1 channel gating. Facilitation of C-type gating by protonation of extracellular histidine residues. J Biol Chem 283:19448–19455. https://doi.org/10.1074/jbc.M801273200

Czirják G, Enyedi P (2002) Formation of functional heterodimers between the TASK-1 and TASK-3 two-pore domain potassium channel subunits. J Biol Chem 277:5426–5432

Czirjak G, Enyedi P (2003) Ruthenium red inhibits TASK-3 potassium channel by interconnecting glutamate 70 of the two subunits. Mol Pharmacol 63:646–652. https://doi.org/10.1124/mol.63.3.646

Czirjak G, Toth ZE, Enyedi P (2004) The two-pore domain K+ channel, TRESK, is activated by the cytoplasmic calcium signal through calcineurin. J Biol Chem 279:18550–18558. https://doi.org/10.1074/jbc.M312229200

Decher N, Maier M, Dittrich W, Gassenhuber J, Bruggemann A, Busch AE, Steinmeyer K (2001) Characterization of TASK-4, a novel member of the pH-sensitive, two-pore domain potassium channel family. FEBS Lett 492:84–89. https://doi.org/10.1016/s0014-5793(01)02222-0

Decher N, Ortiz-Bonnin B, Friedrich C, Schewe M, Kiper AK, Rinne S, Seemann G, Peyronnet R, Zumhagen S, Bustos D, Kockskamper J, Kohl P, Just S, Gonzalez W, Baukrowitz T, Stallmeyer B, Schulze-Bahr E (2017) Sodium permeable and "hypersensitive" TREK-1 channels cause ventricular tachycardia. EMBO Mol Med 9:403–414. https://doi.org/10.15252/emmm.201606690

Decressac S, Franco M, Bendahhou S, Warth R, Knauer S, Barhanin J, Lazdunski M, Lesage F (2004) ARF6-dependent interaction of the TWIK1 K+ channel with EFA6, a GDP/GTP exchange factor for ARF6. EMBO Rep 5:1171–1175

Dong YY, Pike AC, Mackenzie A, McClenaghan C, Aryal P, Dong L, Quigley A, Grieben M, Goubin S, Mukhopadhyay S (2015) K2P channel gating mechanisms revealed by structures of TREK-2 and a complex with Prozac. Science 347:1256–1259

Doyle DA, Morais Cabral J, Pfuetzner RA, Kuo A, Gulbis JM, Cohen SL, Chait BT, MacKinnon R (1998) The structure of the potassium channel: molecular basis of K+ conduction and selectivity. Science 280:69–77. https://doi.org/10.1126/science.280.5360.69

Duprat F, Lesage F, Fink M, Reyes R, Heurteaux C, Lazdunski M (1997) TASK, a human background K+ channel to sense external pH variations near physiological pH. EMBO J 16:5464–5471. https://doi.org/10.1093/emboj/16.17.5464

Duprat F, Girard C, Jarretou G, Lazdunski M (2005) Pancreatic two P domain K+ channels TALK-1 and TALK-2 are activated by nitric oxide and reactive oxygen species. J Physiol 562:235–244. https://doi.org/10.1113/jphysiol.2004.071266

Duprat F, Lauritzen I, Patel A, Honore E (2007) The TASK background K2P channels: chemo- and nutrient sensors. Trends Neurosci 30:573–580. https://doi.org/10.1016/j.tins.2007.08.003

Enyedi P, Czirjak G (2010) Molecular background of leak K+ currents: two-pore domain potassium channels. Physiol Rev 90:559–605. https://doi.org/10.1152/physrev.00029.2009

Enyedi P, Czirjak G (2015) Properties, regulation, pharmacology, and functions of the K(2)p channel, TRESK. Pflugers Arch 467:945–958. https://doi.org/10.1007/s00424-014-1634-8

Enyedi P, Braun G, Czirjak G (2012) TRESK: the lone ranger of two-pore domain potassium channels. Mol Cell Endocrinol 353:75–81. https://doi.org/10.1016/j.mce.2011.11.009

Feliciangeli S, Bendahhou S, Sandoz G, Gounon P, Reichold M, Warth R, Lazdunski M, Barhanin J, Lesage F (2007) Does sumoylation control K2P1/TWIK1 background K+ channels? Cell 130:563–569

Feliciangeli S, Tardy MP, Sandoz G, Chatelain FC, Warth R, Barhanin J, Bendahhou S, Lesage F (2010) Potassium channel silencing by constitutive endocytosis and intracellular sequestration. J Biol Chem 285:4798–4805

Feliciangeli S, Chatelain FC, Bichet D, Lesage F (2015) The family of K2P channels: salient structural and functional properties. J Physiol 593:2587–2603

Fink M, Lesage F, Duprat F, Heurteaux C, Reyes R, Fosset M, Lazdunski M (1998) A neuronal two P domain K+ channel stimulated by arachidonic acid and polyunsaturated fatty acids. EMBO J 17:3297–3308. https://doi.org/10.1093/emboj/17.12.3297

Gaborit N, Le Bouter S, Szuts V, Varro A, Escande D, Nattel S, Demolombe S (2007) Regional and tissue specific transcript signatures of ion channel genes in the non-diseased human heart. J Physiol 582:675–693. https://doi.org/10.1113/jphysiol.2006.126714

Gada K, Plant LD (2019) Two-pore domain potassium channels: emerging targets for novel analgesic drugs: IUPHAR review 26. Br J Pharmacol 176:256–266. https://doi.org/10.1111/bph.14518

Girard C, Duprat F, Terrenoire C, Tinel N, Fosset M, Romey G, Lazdunski M, Lesage F (2001) Genomic and functional characteristics of novel human pancreatic 2P domain K(+) channels. Biochem Biophys Res Commun 282:249–256. https://doi.org/10.1006/bbrc.2001.4562

Goldman DE (1943) Potential, impedance, and rectification in membranes. J Gen Physiol 27:37–60. https://doi.org/10.1085/jgp.27.1.37

Goldstein SA, Price LA, Rosenthal DN, Pausch MH (1996) ORK1, a potassium-selective leak channel with two pore domains cloned from Drosophila melanogaster by expression in Saccharomyces cerevisiae. Proc Natl Acad Sci 93:13256–13261

Goldstein SA, Wang KW, Ilan N, Pausch MH (1998) Sequence and function of the two P domain potassium channels: implications of an emerging superfamily. J Mol Med (Berl) 76:13–20. https://doi.org/10.1007/s001090050186

Goldstein SA, Bockenhauer D, O'Kelly I, Zilberberg N (2001) Potassium leak channels and the KCNK family of two-P-domain subunits. Nat Rev Neurosci 2:175–184. https://doi.org/10.1038/35058574

Gotter AL, Santarelli VP, Doran SM, Tannenbaum PL, Kraus RL, Rosahl TW, Meziane H, Montial M, Reiss DR, Wessner K, McCampbell A, Stevens J, Brunner JI, Fox SV, Uebele VN, Bayliss DA, Winrow CJ, Renger JJ (2011) TASK-3 as a potential antidepressant target. Brain Res 1416:69–79. https://doi.org/10.1016/j.brainres.2011.08.021

Guyenet PG, Stornetta RL, Souza G, Abbott SBG, Shi Y, Bayliss DA (2019) The retrotrapezoid nucleus: central chemoreceptor and regulator of breathing automaticity. Trends Neurosci 42:807–824. https://doi.org/10.1016/j.tins.2019.09.002

Hay RT (2005) SUMO: a history of modification. Mol Cell 18:1–12. https://doi.org/10.1016/j.molcel.2005.03.012

Hodgkin AL, Huxley AF (1952) A quantitative description of membrane current and its application to conduction and excitation in nerve. J Physiol 117:500

Honore E (2007) The neuronal background K2P channels: focus on TREK1. Nat Rev Neurosci 8:251–261. https://doi.org/10.1038/nrn2117

Honore E, Maingret F, Lazdunski M, Patel AJ (2002) An intracellular proton sensor commands lipid- and mechano-gating of the K(+) channel TREK-1. EMBO J 21:2968–2976. https://doi.org/10.1093/emboj/cdf288

Hoshi T, Zagotta WN, Aldrich RW (1991) Two types of inactivation in shaker K+ channels: effects of alterations in the carboxy-terminal region. Neuron 7:547–556

Hwang EM, Kim E, Yarishkin O, Woo DH, Han K-S, Park N, Bae Y, Woo J, Kim D, Park M (2014) A disulphide-linked heterodimer of TWIK-1 and TREK-1 mediates passive conductance in astrocytes. Nat Commun 5:1–15

Hwang HY, Zhang E, Park S, Chung W, Lee S, Kim DW, Ko Y, Lee W (2015) TWIK-related spinal cord K+ channel expression is increased in the spinal dorsal horn after spinal nerve ligation. Yonsei Med J 56:1307–1315

Kang D, Han J, Talley EM, Bayliss DA, Kim D (2004) Functional expression of TASK-1/TASK-3 heteromers in cerebellar granule cells. J Physiol 554:64–77. https://doi.org/10.1113/jphysiol.2003.054387

Kang D, Choe C, Kim D (2005) Thermosensitivity of the two-pore domain K+ channels TREK-2 and TRAAK. J Physiol 564:103–116. https://doi.org/10.1113/jphysiol.2004.081059

Ketchum KA, Joiner WJ, Sellers AJ, Kaczmarek LK, Goldstein SA (1995) A new family of outwardly rectifying potassium channel proteins with two pore domains in tandem. Nature 376:690–695

Kim D, Gnatenco C (2001) TASK-5, a new member of the tandem-pore K(+) channel family. Biochem Biophys Res Commun 284:923–930. https://doi.org/10.1006/bbrc.2001.5064

Kim Y, Bang H, Kim D (2000) TASK-3, a new member of the tandem pore K(+) channel family. J Biol Chem 275:9340–9347. https://doi.org/10.1074/jbc.275.13.9340

Koh SD, Monaghan K, Sergeant GP, Ro S, Walker RL, Sanders KM, Horowitz B (2001) TREK-1 regulation by nitric oxide and cGMP-dependent protein kinase. An essential role in smooth muscle inhibitory neurotransmission. J Biol Chem 276:44338–44346. https://doi.org/10.1074/jbc.M108125200

Kollewe A, Lau AY, Sullivan A, Roux B, Goldstein SA (2009) A structural model for K2P potassium channels based on 23 pairs of interacting sites and continuum electrostatics. J Gen Physiol 134:53–68. https://doi.org/10.1085/jgp.200910235

Kuang Q, Purhonen P, Hebert H (2015) Structure of potassium channels. Cell Mol Life Sci 72:3677–3693

Lafrenière RG, Cader MZ, Poulin J-F, Andres-Enguix I, Simoneau M, Gupta N, Boisvert K, Lafrenière F, McLaughlan S, Dubé M-P (2010) A dominant-negative mutation in the TRESK potassium channel is linked to familial migraine with aura. Nat Med 16:1157–1160

Lengyel M, Czirjak G, Enyedi P (2018) TRESK background potassium channel is not gated at the helix bundle crossing near the cytoplasmic end of the pore. PLoS One 13:e0197622. https://doi.org/10.1371/journal.pone.0197622

Lesage F (2003) Pharmacology of neuronal background potassium channels. Neuropharmacology 44:1–7. https://doi.org/10.1016/s0028-3908(02)00339-8

Lesage F, Barhanin J (2011) Molecular physiology of pH-sensitive background K2P channels. Physiology 26:424–437

Lesage F, Lazdunski M (2000) Molecular and functional properties of two-pore-domain potassium channels. Am J Physiol Renal Physiol 279:F793–F801. https://doi.org/10.1152/ajprenal.2000.279.5.F793

Lesage F, Guillemare E, Fink M, Duprat F, Lazdunski M, Romey G, Barhanin J (1996a) A pH-sensitive yeast outward rectifier K channel with two pore domains and novel gating properties. J Biol Chem 271:4183–4187

Lesage F, Guillemare E, Fink M, Duprat F, Lazdunski M, Romey G, Barhanin J (1996b) TWIK-1, a ubiquitous human weakly inward rectifying K+ channel with a novel structure. EMBO J 15:1004–1011

Lesage F, Terrenoire C, Romey G, Lazdunski M (2000) Human TREK2, a 2P domain mechano-sensitive K+ channel with multiple regulations by polyunsaturated fatty acids, lysophospholipids, and Gs, Gi, and Gq protein-coupled receptors. J Biol Chem 275:28398–28405. https://doi.org/10.1074/jbc.M002822200

Lewis A, McCrossan ZA, Manville RW, Popa MO, Cuello LG, Goldstein SA (2020) TOK channels use the two gates in classical K+ channels to achieve outward rectification. FASEB J. https://doi.org/10.1096/fj.202000545R

Liu C, Au JD, Zou HL, Cotten JF, Yost CS (2004) Potent activation of the human tandem pore domain K channel TRESK with clinical concentrations of volatile anesthetics. Anesth Analg 99:1715–1722. https://doi.org/10.1213/01.ANE.0000136849.07384.44

Lolicato M, Riegelhaupt PM, Arrigoni C, Clark KA, Minor DL Jr (2014) Transmembrane helix straightening and buckling underlies activation of mechanosensitive and thermosensitive K (2P) channels. Neuron 84:1198–1212. https://doi.org/10.1016/j.neuron.2014.11.017
Lolicato M, Arrigoni C, Mori T, Sekioka Y, Bryant C, Clark KA, Minor DL Jr (2017) K2P2.1 (TREK-1)-activator complexes reveal a cryptic selectivity filter binding site. Nature 547:364–368. https://doi.org/10.1038/nature22988
Lolicato M, Natale A, Aberemane-Ali F, Crottès D, Capponi S, Duman R, Wagner A, Rosenberg JM, Grabe M, Minor DL (2020) K2P channel C-type gating involves asymmetric selectivity filter order-disorder transitions. bioRxiv 6(44):eabc9174
Lopes CM, Gallagher PG, Buck ME, Butler MH, Goldstein SA (2000) Proton block and voltage gating are potassium-dependent in the cardiac leak channel Kcnk3. J Biol Chem 275:16969–16978
Lopes CM, Zilberberg N, Goldstein SA (2001) Block of Kcnk3 by protons evidence that 2-P-domain potassium channel subunits function as homodimers. J Biol Chem 276:24449–24452
Lotshaw DP (2007) Biophysical, pharmacological, and functional characteristics of cloned and native mammalian two-pore domain K+ channels. Cell Biochem Biophys 47:209–256. https://doi.org/10.1007/s12013-007-0007-8
Maingret F, Patel AJ, Lesage F, Lazdunski M, Honore E (1999) Mechano- or acid stimulation, two interactive modes of activation of the TREK-1 potassium channel. J Biol Chem 274:26691–26696. https://doi.org/10.1074/jbc.274.38.26691
Maingret F, Lauritzen I, Patel AJ, Heurteaux C, Reyes R, Lesage F, Lazdunski M, Honore E (2000) TREK-1 is a heat-activated background K(+) channel. EMBO J 19:2483–2491. https://doi.org/10.1093/emboj/19.11.2483
Mathie A, Veale EL (2015) Two-pore domain potassium channels: potential therapeutic targets for the treatment of pain. Pflugers Arch 467:931–943. https://doi.org/10.1007/s00424-014-1655-3
Meadows HJ, Benham CD, Cairns W, Gloger I, Jennings C, Medhurst AD, Murdock P, Chapman CG (2000) Cloning, localisation and functional expression of the human orthologue of the TREK-1 potassium channel. Pflugers Arch 439:714–722. https://doi.org/10.1007/s004249900235
Medhurst AD, Rennie G, Chapman CG, Meadows H, Duckworth MD, Kelsell RE, Gloger II, Pangalos MN (2001) Distribution analysis of human two pore domain potassium channels in tissues of the central nervous system and periphery. Brain Res Mol Brain Res 86:101–114. https://doi.org/10.1016/s0169-328x(00)00263-1
Miller AN, Long SB (2012) Crystal structure of the human two-pore domain potassium channel K2P1. Science 335:432–436. https://doi.org/10.1126/science.1213274
Nie X, Arrighi I, Kaissling B, Pfaff I, Mann J, Barhanin J, Vallon V (2005) Expression and insights on function of potassium channel TWIK-1 in mouse kidney. Pflugers Arch 451:479–488
Niemeyer MI, Gonzalez-Nilo FD, Zuniga L, Gonzalez W, Cid LP, Sepulveda FV (2007) Neutralization of a single arginine residue gates open a two-pore domain, alkali-activated K+ channel. Proc Natl Acad Sci U S A 104:666–671. https://doi.org/10.1073/pnas.0606173104
Niemeyer MI, Cid LP, Pena-Munzenmayer G, Sepulveda FV (2010) Separate gating mechanisms mediate the regulation of K2P potassium channel TASK-2 by intra- and extracellular pH. J Biol Chem 285:16467–16475. https://doi.org/10.1074/jbc.M110.107060
Ozaita A, Vega-Saenz de Miera E (2002) Cloning of two transcripts, HKT4.1a and HKT4.1b, from the human two-pore K+ channel gene KCNK4. Chromosomal localization, tissue distribution and functional expression. Brain Res Mol Brain Res 102:18–27. https://doi.org/10.1016/s0169-328x(02)00157-2
Patel AJ, Honoré E, Lesage F, Fink M, Romey G, Lazdunski M (1999) Inhalational anesthetics activate two-pore-domain background K+ channels. Nat Neurosci 2:422–426
Pergel E, Lengyel M, Enyedi P, Czirják G (2019) TRESK (K2P18. 1) background potassium channel is activated by novel-type protein kinase C via dephosphorylation. Mol Pharmacol 95:661–672

Piechotta PL, Rapedius M, Stansfeld PJ, Bollepalli MK, Ehrlich G, Andres-Enguix I, Fritzenschaft H, Decher N, Sansom MS, Tucker SJ, Baukrowitz T (2011) The pore structure and gating mechanism of K2P channels. EMBO J 30:3607–3619. https://doi.org/10.1038/emboj.2011.268

Plant LD (2012) A role for K2P channels in the operation of somatosensory nociceptors. Front Mol Neurosci 5:21. https://doi.org/10.3389/fnmol.2012.00021

Plant LD, Dementieva IS, Kollewe A, Olikara S, Marks JD, Goldstein SA (2010) One SUMO is sufficient to silence the dimeric potassium channel K2P1. Proc Natl Acad Sci 107:10743–10748

Plant LD, Zuniga L, Araki D, Marks JD, Goldstein SA (2012) SUMOylation silences heterodimeric TASK potassium channels containing K2P1 subunits in cerebellar granule neurons. Sci Signal 5:ra84

Pollema-Mays SL, Centeno MV, Ashford CJ, Apkarian AV, Martina M (2013) Expression of background potassium channels in rat DRG is cell-specific and down-regulated in a neuropathic pain model. Mol Cell Neurosci 57:1–9

Pope L, Lolicato M, Minor DL Jr (2020) Polynuclear ruthenium amines inhibit K2P channels via a "Finger in the Dam" mechanism. Cell Chem Biol 27(5):511–524.e4

Pountney DJ, Gulkarov I, Vega-Saenz de Miera E, Holmes D, Saganich M, Rudy B, Artman M, Coetzee WA (1999) Identification and cloning of TWIK-originated similarity sequence (TOSS): a novel human 2-pore K+ channel principal subunit. FEBS Lett 450:191–196

Rajan S, Wischmeyer E, Xin Liu G, Preisig-Muller R, Daut J, Karschin A, Derst C (2000) TASK-3, a novel tandem pore domain acid-sensitive K+ channel. An extracellular histiding as pH sensor. J Biol Chem 275:16650–16657. https://doi.org/10.1074/jbc.M000030200

Rajan S, Wischmeyer E, Karschin C, Preisig-Muller R, Grzeschik KH, Daut J, Karschin A, Derst C (2001) THIK-1 and THIK-2, a novel subfamily of tandem pore domain K+ channels. J Biol Chem 276:7302–7311. https://doi.org/10.1074/jbc.M008985200

Rajan S, Plant LD, Rabin ML, Butler MH, Goldstein SA (2005) Sumoylation silences the plasma membrane leak K+ channel K2P1. Cell 121:37–47

Renigunta V, Zou X, Kling S, Schlichthorl G, Daut J (2014) Breaking the silence: functional expression of the two-pore-domain potassium channel THIK-2. Pflugers Arch 466:1735–1745. https://doi.org/10.1007/s00424-013-1404-z

Reyes R, Duprat F, Lesage F, Fink M, Salinas M, Farman N, Lazdunski M (1998) Cloning and expression of a novel pH-sensitive two pore domain K+ channel from human kidney. J Biol Chem 273:30863–30869. https://doi.org/10.1074/jbc.273.47.30863

Rödström KE, Kiper AK, Zhang W, Rinné S, Pike AC, Goldstein M, Conrad LJ, Delbeck M, Hahn MG, Meier H (2020) A lower X-gate in TASK channels traps inhibitors within the vestibule. Nature:1–5

Sano Y, Inamura K, Miyake A, Mochizuki S, Kitada C, Yokoi H, Nozawa K, Okada H, Matsushime H, Furuichi K (2003) A novel two-pore domain K+ channel, TRESK, is localized in the spinal cord. J Biol Chem 278:27406–27412

Schewe M, Nematian-Ardestani E, Sun H, Musinszki M, Cordeiro S, Bucci G, de Groot BL, Tucker SJ, Rapedius M, Baukrowitz T (2016) A non-canonical voltage-sensing mechanism controls gating in K2P K(+) channels. Cell 164:937–949. https://doi.org/10.1016/j.cell.2016.02.002

Sesti F, Shih TM, Nikolaeva N, Goldstein SA (2001) Immunity to K1 killer toxin: internal TOK1 blockade. Cell 105:637–644. https://doi.org/10.1016/s0092-8674(01)00376-2

Şterbuleac D (2019) Molecular determinants of chemical modulation of two-pore domain potassium channels. Chem Biol Drug Des 94:1596–1614

Talley EM, Solorzano G, Lei Q, Kim D, Bayliss DA (2001) Cns distribution of members of the two-pore-domain (KCNK) potassium channel family. J Neurosci 21:7491–7505

Thomas D, Plant LD, Wilkens CM, McCrossan ZA, Goldstein SA (2008) Alternative translation initiation in rat brain yields K2P2. 1 potassium channels permeable to sodium. Neuron 58:859–870

Tian C, Zhu R, Zhu L, Qiu T, Cao Z, Kang T (2014) Potassium channels: structures, diseases, and modulators. Chem Biol Drug Des 83:1–26

Tulleuda A, Cokic B, Callejo G, Saiani B, Serra J, Gasull X (2011) TRESK channel contribution to nociceptive sensory neurons excitability: modulation by nerve injury. Mol Pain 7:30. https://doi.org/10.1186/1744-8069-7-30

Weir GA, Pettingill P, Wu Y, Duggal G, Ilie AS, Akerman CJ, Cader MZ (2019) The role of TRESK in discrete sensory neuron populations and somatosensory processing. Front Mol Neurosci 12:170. https://doi.org/10.3389/fnmol.2019.00170

Woo DH, Han K-S, Shim JW, Yoon B-E, Kim E, Bae JY, Oh S-J, Hwang EM, Marmorstein AD, Bae YC (2012) TREK-1 and Best1 channels mediate fast and slow glutamate release in astrocytes upon GPCR activation. Cell 151:25–40

Yang S-B, Jan LY (2013) Potassium channels: their physiological and molecular diversity. In: Roberts GCK (ed) Encyclopedia of biophysics. Springer, Berlin, pp 1933–1941

Yellen G (1998) The moving parts of voltage-gated ion channels. Q Rev Biophys 31:239–295

Zhou X-L, Vaillant B, Loukin SH, Kung C, Saimi Y (1995) YKC1 encodes the depolarization-activated K+ channel in the plasma membrane of yeast. FEBS Lett 373:170–176

Zhou J, Yang CX, Zhong JY, Wang HB (2013) Intrathecal TRESK gene recombinant adenovirus attenuates spared nerve injury-induced neuropathic pain in rats. Neuroreport 24:131–136. https://doi.org/10.1097/WNR.0b013e32835d8431

Zilberberg N, Ilan N, Goldstein SA (2001) KCNKO: opening and closing the 2-P-domain potassium leak channel entails "C-type" gating of the outer pore. Neuron 32:635–648. https://doi.org/10.1016/s0896-6273(01)00503-7

Zuniga L, Zuniga R (2016) Understanding the cap structure in K2P channels. Front Physiol 7:228. https://doi.org/10.3389/fphys.2016.00228

Control of Biophysical and Pharmacological Properties of Potassium Channels by Ancillary Subunits

Geoffrey W. Abbott

Contents

G. W. Abbott (✉)
Bioelectricity Laboratory, Department of Physiology and Biophysics, School of Medicine, University of California, Irvine, CA, USA
e-mail: abbottg@hs.uci.edu

N. Gamper, K. Wang (eds.), *Pharmacology of Potassium Channels*,
Handbook of Experimental Pharmacology 267, https://doi.org/10.1007/164_2021_512

Abstract

Potassium channels facilitate and regulate physiological processes as diverse as electrical signaling, ion, solute and hormone secretion, fluid homeostasis, hearing, pain sensation, muscular contraction, and the heartbeat. Potassium channels are each formed by either a tetramer or dimer of pore-forming α subunits that co-assemble to create a multimer with a K^+-selective pore that in most cases is capable of functioning as a discrete unit to pass K^+ ions across the cell membrane. The reality in vivo, however, is that the potassium channel α subunit multimers co-assemble with ancillary subunits to serve specific physiological functions. The ancillary subunits impart specific physiological properties that are often required for a particular activity in vivo; in addition, ancillary subunit interaction often alters the pharmacology of the resultant complex. In this chapter the modes of action of ancillary subunits on K^+ channel physiology and pharmacology are described and categorized into various mechanistic classes.

Keywords

K2P · KCNE · KCNQ · Kir · Kv7 · Long QT syndrome · Voltage-gated potassium channel

1 Introduction

The potassium channels are one of the most numerous and diverse groups of ion channels, reflecting their many roles within specialized cell types in a range of tissues. The diversity is generated by five main mechanisms. First there are a high number of K^+ channel pore-forming α subunit isoforms – for example, the voltage-gated potassium (Kv) channel family numbers 40 in the human genome. Second, K^+ channels are composed of either tetrameric (the majority) or dimeric (the K2P channels) assemblies of α subunits. In turn, in many K^+ channel subfamilies, different α subunit isoforms can heteromultimerize to generate channels with properties distinct from homomeric relatives. Third, most K^+ channels are composed not just of α subunits, but also of ancillary subunits that co-assemble with the α subunits to impart often unique physiological and pharmacological properties. Fourth, many α and ancillary subunits exhibit splice variation with consequent altered properties, and fifth, further added functional complexity arises from post-translational modifications that can dynamically alter activity, which can also include the modulatory effects of signaling cascades. This great diversity of molecular identity and composition of K^+ channels gives rise as one might expect to a wide functional diversity when it comes to the currents generated.

The potassium ion-selective channels can be divided by sequence and structural delineations into three to five families: the voltage-gated potassium (Kv) channels, the inward rectifier (K_{ir}) channels, the two-pore domain potassium (KCNK or K2P) channels, and two families of calcium-activated potassium (KCa) channels (SK/IK and BK) (Atkinson et al. 1991; Fakler et al. 1995; Goldstein et al. 1996; Hille et al.

1999; Ketchum et al. 1995; Kubo et al. 1993; MacKinnon 2003; Tempel et al. 1987, 1988). Ca^{2+}-activated K^+ channels sometimes are combined with Kv channels into a larger family of 6-transmembrane domain K^+ channels.

The Kir channels comprise tetramers of α subunits each containing two transmembrane (TM) domains and a single pore (P) loop; in contrast, mammalian K2P channels form as dimers of 4-TM subunits each with two P-loops. Kv channels form as tetramers of 6-TM α subunits each with a P-loop. SK/IK architecture resembles that of Kv channels, while BK channels contain an extra (7th) TM domain at the N-terminal end. Aside from the extra TM domain of BK channels, the N-terminal end of which is exposed to the extracellular side of the plasma membrane, all K^+ channel α subunit N- and C-termini are instead exposed at the intracellular face of the cell membrane.

For the purposes of this chapter, ancillary subunits are defined as proteins for which the principal activity is to regulate specific ion channels. Therefore, other proteins involved in generalized protein trafficking or localization, or post-translational modifications such as kinases and phosphorylases, are generally not included (although specific examples are described where informative).

Rather than a comprehensive list based on the different channel or ancillary subunit classes (which has been described in, e.g., (Abbott 2014, 2015, 2016b, c; Pongs and Schwarz 2010; Gonzalez-Perez and Lingle 2019)), here the focus is on the different modes of regulation of K^+ channel function and pharmacology and, accordingly, this chapter is organized based on the modes of action of various ancillary subunits to give an overview, using examples, of how they regulate channel function and pharmacology.

Effects on channel function are separated into:

1. obligate, essential for channel activity,
2. not required for channel function but essential for a known physiological role,
3. inhibitory to channel activity.

Effects on channel pharmacology are separated into:

1. drug sensitization by providing a novel binding site absent on the channel α subunit,
2. drug sensitization by increasing the affinity of a small molecule that also binds to and modulates α subunit-only channels,
3. drug desensitization by decreasing the affinity of a small molecule that binds to and modulates α subunit-only channels.

2 Ancillary Subunit Modulation of K^+ Channel Function

2.1 Obligate Ancillary Subunit Interactions, Essential for Channel Activity

Perhaps the best known of the obligate ancillary subunits among the K^+ channel regulators are the sulfonylurea (SUR) receptors, essential components of a specific class of inward rectifier potassium (Kir) channels. Inwardly rectifying potassium (Kir) channel α subunits exhibit a relatively simple architecture of two TM domains and a P-loop, organized into tetramers to form functional channels (in some cases with the necessity of additional regulatory subunits – see below). Inward rectification of Kir channels, which tends to limit outward versus inward K^+ current under experimental conditions across the entire measurable voltage range, is caused by blockade by specific intracellular molecules that limit outward current. The most important of these molecules are Mg^{2+} ions and polyamines (of which spermine is particularly potent) (Kubo et al. 1993; Nishida and MacKinnon 2002; Ruppersberg 2000).

SURs are TM proteins that co-assemble via their cytoplasmic domains in a 4:4 stoichiometry with Kir6.x α subunits to form K_{ATP} channels, which sense cellular metabolic state, coupling this to cellular excitability (Chan et al. 2003). K_{ATP} channels are essential for regulation of insulin secretion in pancreatic β-cells, and for modulating cardiac myocyte and neuronal excitability, and vascular smooth muscle tone, among other functions. SURs are a form of ATP binding cassette (ABC) transporter that no longer transports nutrients and other molecules into cells, and instead now serves as a regulatory subunit for K_{ATP} channels. While Kir6.x α subunits directly bind ATP (in the absence of Mg^{2+}) to inhibit K_{ATP} channels, binding of Mg^{2+}-adenosine nucleotides (ATP or ADP) to SURs activates K_{ATP} channels. Thus, K_{ATP} channels sense changes in the ATP:ADP ratio, with a decrease in this ratio activating the channel and thereby suppressing excitability (because positive to the K^+ equilibrium potential, which is generally around −65 to −80 mV under physiological conditions, K^+ diffuses out of the cell when K^+ channel pores open – hyperpolarizing the cell). SURs are essential to Kir6 channel function because in their absence, Kir6.1 or 6.2 α subunits fail to reach the cell surface and are therefore nonfunctional. This is because of the RKR motif, an endoplasmic reticulum (ER) retention signal that lies within each of the Kir6 and SUR subunits (Zerangue et al. 1999), which ensures ER retention of the homomeric subunits. Co-expression of the two subunit types permits each to reciprocally mask the RKR motifs of the other, permitting maturation and surface trafficking of the entire K_{ATP} channel complex. The Kir6.2 ER retention signal lies within the last 26 residues of the distal C-terminus; accordingly, deletion of this portion of the α subunit permits surface expression and channel activity (Tucker et al. 1997; Zerangue et al. 1999). Thus, SURs are essential modulators in K_{ATP} channels by virtue of an ER retention motif that has evolved to prevent the α subunits from reaching the cell surface alone. For in-depth coverage of physiology and pharmacology of K_{ATP} channels, please see chapters "Kir Channel Molecular Physiology, Pharmacology and Therapeutic Implications" and "The Pharmacology of ATP-Sensitive K^+ Channels (K_{ATP})".

2.2 Ancillary Subunit Interactions Not Necessary for Channel Function But Essential for a Known Physiological Role

2.2.1 KCNE1 Effects on KCNQ1 Function

Ancillary subunits bearing a single TM segment are a common theme among various K^+ channel types and are typically not required for channel function per se but can be essential for a known physiological role. These often-diminutive subunits can exert profound effects on K^+ channel function, from early maturation and trafficking all the way to function at the cell surface and endocytosis for recycling or degradation. The KCNE subunits are a very highly studied class and their roles as examples in the various categories of α subunit-ancillary subunit interactions are discussed in detail throughout this chapter.

The KCNE subunits are single-pass TM proteins best known for their regulation of Kv channel α subunits, but also able to regulate other ion channel classes (Abbott and Goldstein 1998; Abbott et al. 1999, 2001b). The first member of this gene family, which has 5 members in the human genome, was cloned in 1988 by expression cloning of rat kidney RNA fractions in the *Xenopus laevis* oocyte expression system (Takumi et al. 1988). The founding member was termed IsK or MinK (minimal potassium channel), and is now more commonly referred to as KCNE1, similar to its gene name, *KCNE1*. KCNE1 was first thought to generate a slow-activating Kv current on its own but it was subsequently realized that KCNE1 co-assembles with endogenous *Xenopus* oocyte KCNQ1 to augment its current, also slowing KCNQ1 activation. Co-assembly of KCNE1 with human KCNQ1 recapitulates properties of human cardiac I_{Ks} (slow-activating K^+ current), a Kv current that is important in human ventricular myocyte repolarization (Barhanin et al. 1996; Sanguinetti et al. 1996). Mutations in either KCNQ1 or KCNE1 cause inherited variants of long QT syndrome, a surface electrocardiogram abnormality that reflects a delay in ventricular cardiac myocyte repolarization and can lead to the potentially life-threatening *torsades de pointes* arrhythmia, which can in turn degenerate intro ventricular fibrillation and sudden cardiac death (Splawski et al. 1997a, b; Wang et al. 1996) (see also chapter "Cardiac K^+ Channels and Channelopathies").

KCNQ1 is capable of forming functional Kv channels in the absence of KCNE subunits (Barhanin et al. 1996; Sanguinetti et al. 1996), although it requires calmodulin (CaM) for folding and assembly, so in this manner CaM could be considered as an essential regulatory subunit for KCNQ1 (and in fact all the KCNQ isoforms, 1–5) (Ghosh et al. 2006; Shamgar et al. 2006; Sun and MacKinnon 2017; Wiener et al. 2008). However, thus far no native currents have been discovered that are generated by KCNQ1 in the absence of KCNE subunits. Functional ventricular I_{Ks} channels contain four KCNQ1 α subunits and also require KCNE1 in the complex; thus, KCNE1 is not absolutely required for KCNQ1 channel function but is essential for a known physiological role of KCNQ1. The stoichiometry of KCNE1 in complexes with 4 KCNQ1 α subunits is still under discussion, with some contending variable stoichiometry (between 1 and 4 KCNE1 subunits) and others contending there is a fixed stoichiometry of 4:2 (KCNQ1:KCNE1) (Morin and Kobertz 2008; Nakajo et al. 2010; Plant et al. 2014; Wang and Kass 2012).

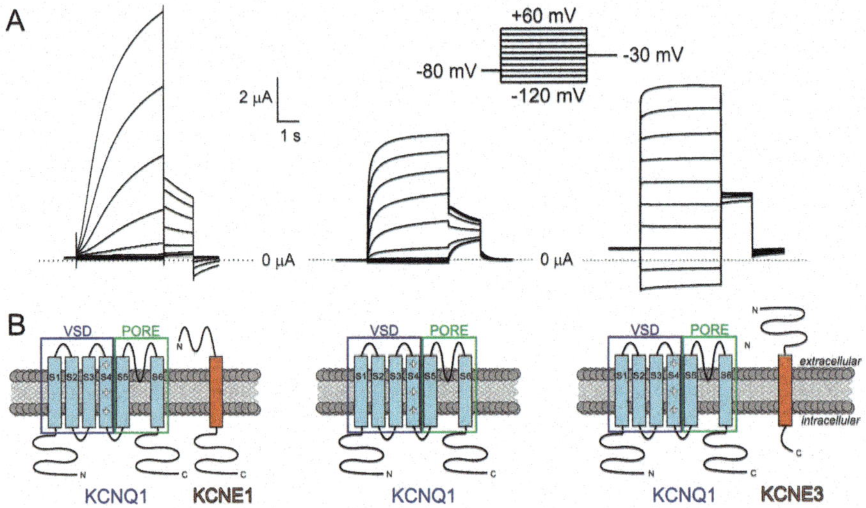

Fig. 1 Regulation of KCNQ1 activity by KCNE1 and KCNE3. (**a**) Representative electrophysiological recordings from *Xenopus* oocytes expressing KCNQ1 alone (center), or with KCNE1 (left) or KCNE3 (right). Voltage protocol inset. (**b**) Subunit topologies shown beneath corresponding traces from (**a**). *VSD* voltage-sensing domain. Adapted from Abbott (2017b)

In addition to slowing KCNQ1 activation five to tenfold, KCNE1 shifts the KCNQ1 voltage dependence of activation towards more positive voltages and increases its unitary conductance several fold (Barhanin et al. 1996; Sanguinetti et al. 1996; Sesti and Goldstein 1998) (Fig. 1). KCNE1 also eliminates KCNQ1 inactivation – a process that occurs in most Kv channels (but not in other members of the KCNQ/Kv7 family), in which the channel remains "open," but the inactivation gate is closed, so no current passes, despite the membrane being depolarized. Inactivation in homomeric KCNQ1 channels is typically difficult to observe during a depolarizing pulse because the activation process is slow and is followed in turn by moderate inactivation, causing an equilibrium to be reached at the macroscopic (whole-cell) scale usually with little noticeable current decay. The inactivation process becomes evident, however, during a subsequent repolarizing tail pulse, in which a current "hook" caused by recovery from inactivation is observed for KCNQ1 but not for KCNQ1-KCNE1 channels (Tristani-Firouzi and Sanguinetti 1998).

The mechanism for the dramatic slowing of KCNQ1 activation imposed by KCNE1 is debated. Some studies suggest that KCNE1 directly slows KCNQ1 pore opening, whereas others favor KCNE1 slowing activation of the voltage-sensing domain (VSD) and thus indirectly slowing pore opening. Various lines of evidence suggest that KCNQ1 channels can, unusually for Kv channels, open when only one of the four VSDs is activated; in contrast, KCNQ1-KCNE1 pore opening may require multiple, and even all four VSDS to activate – similar to the conventional model for Kv channels outside the KCNQ family (Gofman et al. 2012; Nakajo

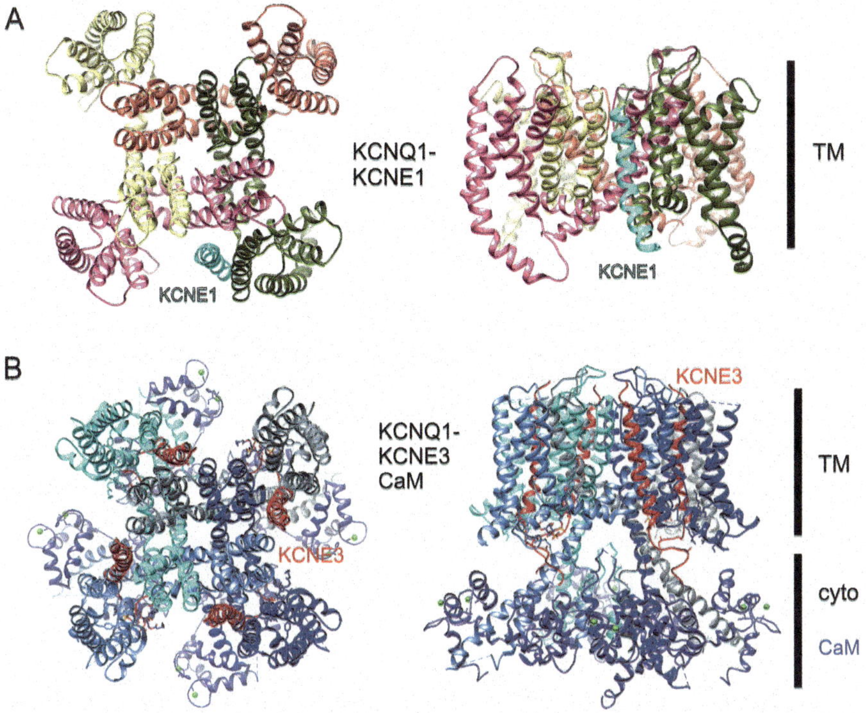

Fig. 2 KCNQ1-KCNE1 model and KCNQ1-KCNE3-CaM structure. (**a**) Structural model of KCNQ1-KCNE1 – image plotted from coordinates from Kang et al. (2008). (**b**) High-resolution cryo-EM structure of KCNQ1-KCNE3-CaM – image plotted from coordinates from Sun and MacKinnon (2020)

and Kubo 2007; Osteen et al. 2010a, b, 2012; Ruscic et al. 2013). Interaction of KCNE1 with the VSD of KCNQ1 could also impose the positive shift in voltage dependence of activation. This might occur by modifying the manner in which S4 (the positively charged segment within the VSD that is primarily responsible for responding to changes in membrane potential) "senses" stabilizing acidic residues in other, nearby KCNQ1 TM domains (Strutz-Seebohm et al. 2011; Wu et al. 2010a, b).

The multiple gating effects of KCNE1 on KCNQ1 suggest a tight physical coupling to multiple KCNQ1 domains and this is borne out by structural models (built from NMR structural analysis of KCNE1, high-resolution structures of other channels and incorporating mutagenesis results from functional studies of KCNQ1-KCNE1 complexes) showing KCNE1 lying in a groove between the pore module and VSD, where it can impact multiple functionally important gating processes (Fig. 2a) (Sahu et al. 2014; Smith et al. 2007; Vanoye et al. 2009). Although no direct structure determination exists for KCNQ1-KCNE1 complexes, the structures of KCNQ1-CaM and KCNQ1-KCNE3-CaM have been solved and they also support the positioning of KCNEs in the aforementioned groove (Fig. 2b) (Sun and MacKinnon 2017, 2020).

2.2.2 KCNE1 Effects on KCNQ1 Trafficking

KCNQ1 α subunits require CaM for folding and assembly, prerequisites for trafficking to the cell surface to become functional channels (Ghosh et al. 2006; Shamgar et al. 2006; Sun and MacKinnon 2017; Wiener et al. 2008). Once at the plasma membrane, as for all membrane proteins, KCNQ1 channels have a finite lifetime at the cell surface and must be recycled at some point either to be returned to the cell surface when needed again or degraded. Both KCNQ1 and KCNQ1-KCNE1 channels are internalized by Nedd4/Nedd4-like and Rab-5 processes (Jespersen et al. 2007; Krzystanek et al. 2012). There is also an alternative, KCNE1-dependent internalization process, in which KCNE1 orchestrates clathrin-mediated endocytosis of KCNQ1-KCNE1 complexes, either when heterologously expressed in immortal cell lines, or when studying native I_{Ks} channels in guinea-pig ventricular myocytes. This internalization requires the dynamin GTPase, which pinches off the clathrin-lined vesicle from the plasma membrane so that it can be internalized along with membrane-associated proteins in the vesicle. Clathrin-mediated endocytosis of KCNQ1-KCNE1 can be disrupted by mutagenesis of either dynamin (the K44A dominant-negative dynamin mutation increases I_{Ks} current density by blocking KCNQ1-KCNE1 internalization) or KCNE1. Three motifs are necessary and sufficient for KCNE1-facilitated KCNQ1-KCNE1 internalization. First KCNE1 contains a DPFNVY motif in the C-terminal domain (residues 76–81) similar to a motif that is highly conserved in G-protein-coupled receptors (GPCRs), (D/N)PX_{2-3}Y, which is implicated in GPCR internalization; the FNVY motif within this stretch in KCNE1 is also similar to the YXXϕ motif required in other clathrin-internalized proteins for binding of the AP-2 adaptor complex, which helps to internalize cargo for endocytosis (and also present in KCNE2 and KCNE3). Accordingly, mutagenic disruption of this motif in KCNE1 reduces KCNQ1-KCNE1 internalization by about half. KCNE1 also contains a consensus SH3-binding domain (PSP) in the distal C-terminus, mutation of which blocks over half the KCNQ1-KCNE1 internalization (Xu et al. 2009). Finally, S102 in KCNE1 is a consensus PKC phosphorylation site, and had been shown in a prior study to mediate PKC regulation of I_{Ks} current magnitude by an unknown mechanism (Varnum et al. 1993; Zhang et al. 1994). S102 is conserved in all 5 KCNE proteins; mutation of S102 to alanine also blocked around half of KCNQ1-KCNE1 channel internalization. Mutation of all three sites simultaneously eliminated clathrin-mediated endocytosis of the channel. Further studies showed that PKC dynamically regulated I_{Ks} by inducing KCNQ1-KCNE1 endocytosis via S102 phosphorylation (Kanda et al. 2011c). Because PKC phosphorylation is a dynamic post-translational modification, this process could potentially act to dynamically alter I_{Ks} current properties in vivo as needed; in addition to decreasing current magnitude, selective removal by this process of solely KCNQ1-KCNE1 complexes could alter I_{Ks} gating kinetics by leaving at the membrane homomeric KCNQ1 channels lacking KCNE1 (Kanda et al. 2011c).

2.2.3 Effects of KCNE2 on KCNQ1 Function and Trafficking

Over a decade after the discovery of KCNE1, we and others discovered the remaining four members of the family, KCNE2–5, also named MinK-related peptides (MiRPs) 1–4; KCNE5 was originally named KCNE1L (KCNE1-like) (Abbott et al. 1999; Piccini et al. 1999). KCNE2–5 also all regulate KCNQ1, but with widely different outcomes compared to KCNE1, and for different physiological purposes (Abbott 2014).

KCNE2 confers upon KCNQ1 constitutive activation in the physiological voltage range, together with much-reduced current density compared to homomeric KCNQ1 – giving a shallow but linear IV curve, in contrast to the outward rectifying properties displayed by KCNQ1 or KCNQ1-KCNE1 (Tinel et al. 2000). The mechanism for reduced current has not been investigated at the single-channel level but there may be a contribution from altered trafficking due to KCNE2 interaction (Hu et al. 2019). KCNQ1-KCNE2 complexes serve essential functions in a range of non-excitable epithelial cell types, including in gastric parietal cells, thyroid epithelial and choroid plexus epithelial cells (Roepke et al. 2006, 2009, 2010, 2011a). In parietal cells, KCNQ1-KCNE2 channels facilitate gastric acid secretion by the gastric H^+/K^+-ATPase by providing a conduit through which K^+ ions can return to the stomach lumen to feed the ATPase to ensure K^+/proton exchange through it to acidify the oxyntic pits. Highlighting the importance of KCNE2 in gastric complexes with KCNQ1, germline deletion of *Kcne2* in mice results in complete abolition of gastric acid secretion and ultimately development of gastritis cystica profunda and gastric neoplasia, likely due to both bacterial overgrowth and inflammation because of the lack of gastric acid secretion. There are also separate effects of *Kcne2* depletion on gastric cellular proliferation. Reduced gastric KCNE2 expression is similarly associated with human gastritis cystica profunda, gastric neoplasia, and resistance to 5-fluorouracil-based chemotherapy or with gastric tumor proliferation outpacing this therapy (Abbott and Roepke 2016; Li et al. 2016; Roepke et al. 2006, 2009, 2010, 2011a).

In addition to permitting KCNQ1 to remain open (and non-inactivated) at the moderately polarized membrane potentials in the parietal cells (−20 to −40 mV), KCNE2 bestows at least two other crucial properties on KCNQ1 for its gastric role. First, homomeric KCNQ1 (and KCNQ1-KCNE1) channels are inhibited by extracellular low pH, and therefore would be nonfunctional when at the apical side of the parietal cell and facing the stomach lumen, where they are required to ensure K^+ return to the lumen. KCNE2 reverses this property such that KCNQ1-KCNE2 activity is augmented by extracellular low pH. This switch in pH sensitivity requires KCNE2 residues in its extracellular N-terminus and N-terminal portion of the TM segment, and is not observed in KCNQ1-KCNE3 channels, which are constitutively active (see below) yet pH-insensitive (Heitzmann et al. 2007).

Second, in the absence of KCNE2, KCNQ1 traffics to the basolateral membrane instead of the apical membrane of parietal cells, where it cannot serve its function of returning K^+ to the stomach lumen. Using single- and double knockouts of the *Kcne2* and *Kcne3* genes in mice, we showed that in the absence of Kcne2, Kcne3 – normally more highly expressed in the colon and intestine than in the stomach – is upregulated in the stomach and hijacks Kcnq1, taking it to the basolateral side of parietal cells. Knockout of both *Kcne2* and *Kcne3* restored Kcnq1 to the apical side, but the mice still lacked gastric acid secretion – likely because lacking in Kcne2 their gastric Kcnq1 channels were inhibited by low pH (Roepke et al. 2011b).

KCNE2 is not required for KCNQ1 channel function per se but is essential for several of its known physiological roles in the stomach, thyroid, and choroid plexus. Germline deletion of *Kcne2* in mice also causes hypothyroidism, because Kcnq1-Kcne2 complexes are required at the basolateral side of thyroid epithelial cells for efficient function of the sodium-iodide symporter (NIS). Without Kcne2, NIS is unable to sequester sufficient iodide into thyroid cells, an essential ion for production of thyroid hormone, at key times in the mouse life cycle, including in early development, during gestation and lactation, and in old age (Roepke et al. 2009). We found that this was an uptake defect and not a storage defect, and that the effects of *Kcne2* deletion on thyroid iodide uptake deficiency could be recapitulated using chromanol 293B, a blocker of KCNQ1-containing channels, in vitro and in vivo (Purtell et al. 2012). Studies using positron emission tomography to track I^{124} showed that lactating dams could not produce sufficient milk to feed their pups, and that the iodide that did reach their pups via milk was inefficiently sequestered by their thyroid glands, leading to developmental delay, cardiac hypertrophy, and alopecia. These defects could be prevented by surrogacy of *Kcne2* knockout pups with wild-type dams (and vice-versa), indicating a prominent role for maternal *Kcne2* status in the pups' phenotype (Roepke et al. 2009). While the molecular mechanism underlying the need for Kcnq1-Kcne2 channels to facilitate efficient iodide uptake by NIS is not fully understood, it is not simply a requirement for maintenance of a whole-cell transmembrane electrochemical gradient to permit sodium-dependent solute transporter function, as the sodium-coupled monocarboxylate transporter (SMCT) was able to function normally in the absence of Kcne2 (Purtell et al. 2012).

Kcne2 deletion also causes increased handling-related seizures in mice, increased susceptibility to seizures initiated by the chemoconvulsant pentylenetetrazole (PTZ), and reduced depressive/despair-type behavior – each suggesting increased excitability in the nervous system. These changes can be traced to another role of Kcne2 complexes, in the apical membrane of the choroid plexus epithelium. As the primary site of cerebrospinal fluid (CSF) production and secretion, the choroid plexus plays an important role in regulating the environment of the nervous system. *Kcne2* deletion in mice uncovered its regulation of two different Kv channel isoforms in the choroid plexus, Kcnq1 and Kcna3 (Kv1.3). Compared to wild-type choroid plexus epithelia cells, *Kcne2* knockout cells exhibited increased XE991-sensitive outward current but reduced XE991-sensitive inward current at hyperpolarized membrane potentials. This profile matches what one would expect for loss of Kcne2 from Kcnq1 complexes, XE991 being a relatively KCNQ-specific inhibitor; KCNE2 being known to decrease KCNQ1 outward current but permit its constitutive activation at hyperpolarized voltages. In addition, *Kcne2* deletion increased Margatoxin (MgTx)-sensitive currents, while leaving Dendrotoxin-sensitive currents unchanged, indicative of altered Kcna3 (Kv1.3) but not Kcnq1 currents, respectively. Accordingly, in heterologous co-expression experiments, human KCNE2 was found to increase KCNA3 current but not alter KCNA1 current. Co-assembly of mouse Kcne2 with Kcnq1 and Kcne3 in choroid plexus epithelium was also supported by co-localization and co-immunoprecipitation experiments (Abbott et al. 2014; Roepke et al. 2011a).

Kcne2 deletion hyperpolarized the choroid plexus epithelial cell membrane potential, as would be predicted from the above effects, and also increased CSF [Cl^-], but a mechanism for increased seizure susceptibility was not immediately apparent (Roepke et al. 2011a). Metabolomics analysis subsequently revealed that *Kcne2* deletion caused a reduction in CSF *myo*-inositol, and restoration of normal *myo*-inositol levels by mega-dosing in the water supply also restored normal seizure susceptibility and depressive behavior. The reason for reduced CSF *myo*-inositol was uncovered by further analysis – KCNQ1-KCNE2 channels negatively regulate activity of the SMIT1 sodium-dependent myo-inositol transporter, which is expressed at both the apical and basolateral membranes of the choroid plexus epithelium and which regulates movement of myo-inositol between the blood and CSF. Loss of Kcne2 in mice resulted in loss of the brake upon uptake of *myo*-inositol from the CSF back into the choroid plexus, suggesting a plausible mechanism for depleted CSF *myo*-inositol. This work represented the discovery of direct physical interaction of a K^+ channel with a sodium-coupled solute transporter and has since led to many other examples of this type of interaction (Abbott 2016a; Abbott et al. 2014; Neverisky and Abbott 2015, 2017).

2.2.4 Effects of KCNE3 on KCNQ1 Function

KCNE3, like KCNE2, induces constitutive activation in channels formed with KCNQ1, by greatly negative-shifting the voltage dependence of activation and thus dramatically reducing the time-and voltage dependence of the current across the physiological voltage range (Schroeder et al. 2000). This is achieved by KCNE3 locking the KCNQ1 voltage sensor in the activated state, thus indirectly locking open the pore (Nakajo and Kubo 2007; Panaghie and Abbott 2007). Unlike KCNE2, KCNE3 does not concurrently reduce overall current magnitude of KCNQ1, so the resultant channels are much easier to study and the majority of mechanistic analyses on loss of voltage dependence have been conducted in KCNQ1-KCNE3 complexes (reviewed in Abbott 2016b). The constitutive activation of KCNQ1-KCNE3 channels, which essentially provides a continuous K^+ selective current that does not inactivate, permits the utility of KCNQ1-KCNE3 channels in non-excitable cells such as those in the colonic epithelium, where they regulate membrane potential and in turn, cAMP-stimulated chloride secretion (Schroeder et al. 2000), and in non-excitable cells of the airway (Grahammer et al. 2001) and mammary epithelia (vanTol et al. 2007).

KCNE3 also endows KCNQ1 with sensitivity to estrogen. Increased estrogen causes downregulation of KCNE3, which leaves homomeric KCNQ1 alone to function in the non-excitable colonic cell membrane without KCNE3 to lock open its voltage sensor and keep it activated. Consequently, during proestrus, when estrogen levels are high, female mammals exhibit water retention partly because

KCNQ1 is less able than KCNQ1-KCNE3 to promote cAMP-stimulated chloride secretion. In addition, male rats have higher colonic crypt KCNE3 expression and enriched KCNQ1-KCNE3 channel versus homomeric KCNQ1, compared to female rats. Estrogen regulation of KCNE3 requires KCNE3-S82 because it requires a signaling cascade that includes phosphorylation of S82 by PKC (Alzamora et al. 2011; Kroncke et al. 2016; O'Mahony et al. 2009; Rapetti-Mauss et al. 2013). Phosphorylation of KCNE3-S82 is also required for KCNE3 to induce subthreshold activation in Kv3.4-KCNE3 complexes, which are important in skeletal muscle physiology (Abbott et al. 2001a, 2006; King et al. 2017).

2.2.5 SK Regulatory Proteins and BK γ Subunits

Separate from the classic Kv channels, the calcium-activated potassium channels have superficially similar architecture but different functional properties (Berkefeld and Fakler 2013; Berkefeld et al. 2010); see also chapter "Calcium-Activated K^+ Channels (K_{Ca}) and Therapeutic Implications"). The majority of calcium-activated potassium channels fall into one of two main groups. The BK (big conductance) calcium-activated potassium channels form one group, while the SK (small conductance) and IK (intermediate conductance) calcium-activated potassium channels fall into the other group. SK channels have a relatively small unitary conductance (~10 pS) and share somewhat similar tetrameric architecture and topology to that of Kv channels, with each α subunit containing 6 TM domains and a P-loop. The major difference to Kv channels is that the S4 segment of SK channels is only weakly charged (only two basic residues compared to seven for typical Kv channels) and SK channels are accordingly relatively insensitive to voltage, instead being activated by Ca^{2+}. There are three SK channel α subunit isoforms (SK1–3), encoded by *KCNN1–3*, and the isoforms are expressed throughout the nervous system in higher animals, where they regulate neuronal excitability (especially the medium afterhyperpolarizing potential, mAHP) and participate in calcium signaling and synaptic plasticity (Bond et al. 1999).

SK channels are not known to possess any channel-specific ancillary subunits, but they participate in functionally essential interactions with generalized protein modulators calmodulin (CaM, which endows calcium sensitivity), casein kinase 2 (CK2, which phosphorylates T80 of channel-bound CaM) and protein phosphatase 2 (PP2A, which dephosphorylates CaM-T80). IK channels exhibit unitary conductances of ~35 pS; these are to date the least studied of the KCa channels. There is one IK isoform in the human genome, KCa3.1 (a.k.a. IKCa1, SK4), encoded by *KCNN4*. KCa3.1 exhibits similar TM topology and activation mechanism to that of the SK channels, including the voltage insensitivity and the dependence on CaM for Ca^{2+} binding and sensitivity. IK channels are expressed in epithelia, blood cells, and in peripheral neurons (Lujan et al. 2009; Weatherall et al. 2010).

BK (also MaxiK or Slo) channels – encoded by *KCNMA* genes – belong in a different class to the SK and IK channels for several reasons. First, BK channels, as the name suggests, have a much larger unitary conductance, of 200–300 pS. Second, BK channels contain an additional TM helix, termed S0, at the N-terminal end, that distinguishes them from both SK/IK channels and from Kv channels and in fact all the classic voltage-dependent S4 family (repeating units of Nav and Cav channels each contain 6 TM segments). Third, BK channels possess intrinsic voltage sensitivity, via positive charges in their voltage-sensing domain (VSD), primarily TM segments S2-S4. Fourth, BK channels also exhibit intrinsic Ca^{2+} sensitivity, with two types of Ca^{2+} binding site present on the channel α subunits themselves – high-affinity "regulator for conductance of potassium" (RCK) domains within the α subunit cytosolic C-terminal (RCK1 and RCK2) and a lower affinity binding site that binds Ca^{2+} (and other divalent metal ions such as Mg^{2+}) located between the TM segments and the gating ring (Kshatri et al. 2018) (Vergara et al. 1998). Fifth, BK channels are regulated by channel-specific ancillary subunits. Specifically, BK channels are modulated by physical interaction with ancillary proteins termed β subunits, here referred to as BKβ subunits to distinguish them from Kvβ subunits and γ subunits (Li and Yan 2016).

The BK γ (gamma) subunits are represented as four isoforms (BKγ1–4, encoded by *LRRC26, 52, 55* and *38*, respectively) in the human genome. LRRC is an acronym for "leucine-rich repeat containing" and refers to the extracellular LRR domain. BKγ1 is widely expressed including in the brain, various glandular epithelia, aorta and mucosa; BKγ2 in skeletal muscle, testis placenta and sperm, BKγ3 also in sperm and in the brain, but also in the liver, spleen and olfactory bulb, and BKγ4 in muscle, testis and sperm, cerebellum, adrenal gland, and thymus. They are each potent regulators of BK channel activation, shifting the voltage dependence of activation by −100 to −140 mV across the Ca^{2+} concentration range, such that the γ-containing BK channels are constitutively active at resting membrane potentials, even in the absence of elevated intracellular Ca^{2+}, with approximately half activation at −80 mV in 100 μM intracellular free Ca^{2+} (Yan and Aldrich 2010). Each γ subunit also speeds activation and slows deactivation of BK channels, without inducing inactivation. It is thought that γ subunits exhibit variable stoichiometry, from 1–4 in a complex of 4 α subunits; increasing the number of γ subunits per complex increases the negative-shift in voltage dependence of BK channel activation.

2.2.6 BK β Subunits

Not as common as those possessing a single TM span, some ion channel regulatory subunits possess a 2-TM topology (Sun et al. 2012). In the K^+ channels, the BKβ subunits are the exemplar, and they are another example of ancillary subunits that modify channel function to enable specific physiological roles but are not essential for channel activity per se (Castillo et al. 2015). The BKβ subunits each possess two TM domains, and there are four isoforms (BKβ1–4). BKβ1 and BKβ2 each shift the voltage dependence of human BK channel activation (with elevated intracellular Ca^{2+}) by −70 and −50 mV, respectively; BKβ2 also induces inactivation that is absent from homomeric BK α subunit channels. BKβ3 hyperpolarizes murine BK channel activation across the Ca^{2+} concentration range (by −30 mV) but does not alter voltage dependence of human BK channel activation; however, BKβ3 induces inactivation in both human and murine BK. BKβ4 is able to differentially alter BK

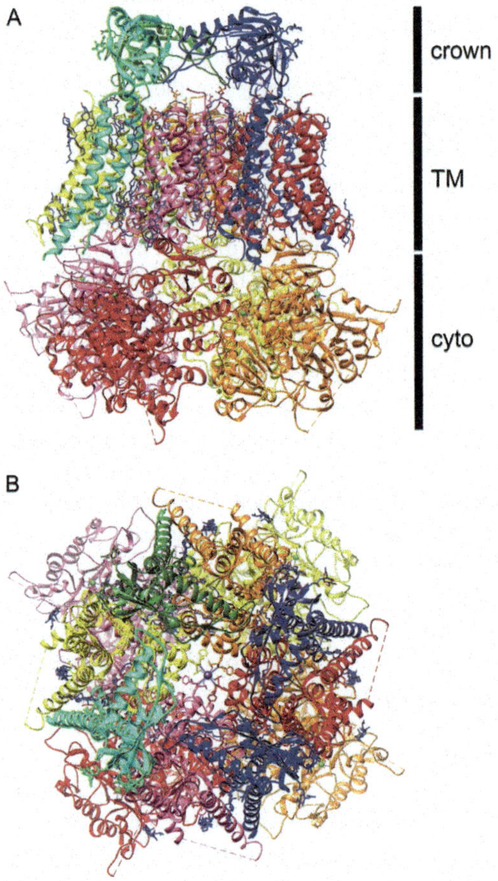

Fig. 3 Structure of human Slo1-β4 complex. (**a**) Side view of Ca^{2+}-bound human Slo1-β4 complex, solved by cryo-EM, plotted from coordinates in Tao and MacKinnon (2019). Slo1 (BK) α subunit colored pink, red, orange, yellow. B4 subunits colored purple, blue, cyan, green. Crown is an extracellular region formed from the β4 subunits. (**b**) View of Slo1-β4 complex as above but from extracellular face

activation voltage dependence depending on Ca^{2+} levels: a 20-mV positive shift in low Ca^{2+}, and a somewhat greater negative shift in high Ca^{2+}; these effects are coupled with reduced activation and deactivation rate. BKβ1 and BKβ4 do not induce BK channel inactivation. As for γ subunits, β subunits are reported to exhibit variable stoichiometry, from 1–4 in a complex of 4 α subunits; increasing the number of BKβ in a channel increases their relative effect on the voltage dependence of BK activation (Castillo et al. 2015) (Fig. 3).

2.2.7 Cytosolic Subunits that Add Novel Functionality to Kv Channels

An array of cytosolic ancillary subunits regulate K^+ channels, typically by interaction with the cytosolic portions of the K+ channel and sometimes adding completely unique functionality not exhibited by the α subunits alone. One example is the Kvβ subunit family. Some but not all Kvβ subunits contain a fast inactivation domain that can plug the pore of slow-inactivating delayed rectifier α subunits to induce fast inactivation, that is, rapid current decay following voltage-dependent activation. The structural basis of Kvβ-mediated fast inactivation has been suggested by solving of the structure of the prokaryotic KcsA K^+ channel containing tetrabutylantimony, an electron-dense quaternary ammonium ion that resembles the distal end of the inactivation peptide and adopts a deep pore binding site. In addition, solving of the octameric complex of Kvβ2 with the cytoplasmic tetramerization (T1) domain of KCNA1 suggested that lateral negatively charged fenestrations above this octameric structure could permit inactivation peptide access to the pore in the intact α subunit/β subunit complex (Gulbis et al. 2000; Zhou et al. 2001).

An earlier crystallographic analysis revealed that the Kvβ2 subunit displays structural homology to aldo-keto reductases, proteins that contain a bound NADP (+) co-factor (as does Kvβ2) and can sense the redox state of the cell (Gulbis et al. 1999). Further functional analyses have indeed shown that Kvβ2 can couple channel activity and other characteristics to the redox state of the cell, and that this occurs in vivo. For example, using *Kcnab2* knockout mice for comparison, it was found that Kvβ2 is important for adequate expression of cardiac transient outward current (I_{to}, generated by Kv1.4, Kv4.2, and Kv4.3) and $I_{K,slow}$ (generated by Kv1.5) and that this ancillary subunit also coupled myocyte repolarization to metabolic state via its interaction with these Kv α subunits (Campomanes et al. 2002; Kilfoil et al. 2019).

Another class of cytoplasmic subunits, the K^+ channel interacting proteins (KChIPs) endows Kv channels with calcium sensitivity. KChIPs are best known for regulating Kv4 α subunits, inducing large increases in current magnitude upon co-expression; knockout of the cardiac KChIP isoform, KChIP2, in mice leads to loss of cardiac myocyte I_{to} because of impaired Kv4 channel activity. KChIPs increase the surface density of Kv4 α subunits, negative-shift the voltage dependence of activation, slow inactivation and speed recovery from inactivation. KChIPs are also important for regulating what are termed A-type current or I_{SA} (fast-inactivating Kv4-generated currents) in the brain. KChIPs contain four EF-hand-like domains and are calcium-binding proteins related to neurocalcins and neuronal calcium sensor-1 (NCS-1). While there have been conflicting observations on the effects Ca^{2+} in Kv4-KChIP complex formation, stoichiometry, and gating, it does appear that changes in intracellular $[Ca^{2+}]$ can modulate Kv4-KChIP gating properties. Increasing intracellular Ca^{2+}, but not Co^{2+}, via patch pipette was found to increase I_{SA} current magnitude, thought to be expressed by Kv4-KChIP complexes, in cultured rat cerebellar granule cells. In another study conducted using heterologous expression in HEK-293 cells it was concluded that direct binding of Ca^{2+} to KChIP2 EF hand motifs EF3 and EF4 likely accelerates inactivation recovery of Kv4.3/KChIP2 channels (Abbott 2017a; Abbott et al. 2007; Amadi et al. 2007; An et al. 2000; Bahring 2018; Bahring et al. 2001; Foeger et al. 2013; Gong et al. 2006; Takimoto and Ren 2002).

2.2.8 SUMO and 14-3-3 Interactions with K2P Channels

K2P channels are the molecular correlates of many potassium-selective leak currents in the cells of mammals and other organisms. K2P channels tend to exhibit linear current-voltage (IV) relationships with symmetrical $[K^+]$ on either side of the plasma membrane, while behaving as outward (open) rectifiers under physiological

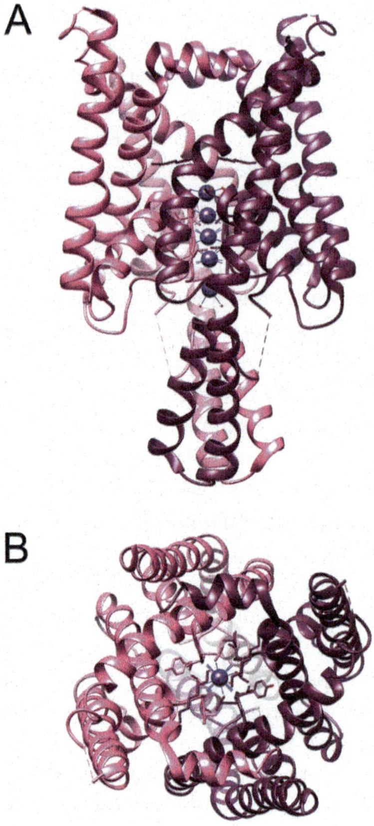

Fig. 4 Structure of human K2P1 channel. (**a**) Side view of human K2P1 channel, solved by X-ray crystallography, plotted from coordinates in Miller and Long (2012). The two α subunits are colored pink and purple. (**b**) View of human K2P1 channels as above but from extracellular face

conditions (low K^+ outside, high K^+ inside the cell). K2P channels form as dimers each with 4 P-loops (two contributed from each subunit) and can form as heterodimers with unique properties (Fig. 4). There are at least 18 K2P isoforms in the human genome, and most of them are categorized into four subfamilies: TALK, TASK, TREK, and TWIK (Duprat et al. 2007; Goldstein et al. 1996, 2001; Ketchum et al. 1995; Kollewe et al. 2009; Mathie et al. 2010); see also chapter "The Pharmacology of Two-Pore Domain Potassium Channels".

At the time of writing, K2P-specific ancillary subunits have yet to be uncovered; given that the first K2P α subunits were cloned a quarter century ago (in 1995), this suggests that K2P-specific regulatory subunits may be uncommon or absent. However, there are some notable examples of K2P channels being regulated by more general modulatory proteins. First, 14-3-3β (a cytoplasmic protein that forms as dimers and is composed of 9 α-helices per monomer in an anti-parallel organization) is required for maturation and expression of K2P3 and K2P9 at the cell surface. This is because β-COP, which belongs to the coatamer class of proteins, binds to K2P3 and K2P9 to prevent them from exiting the endoplasmic reticulum after translation. 14-3-3β binding to the K2Ps requires phosphorylation at a single residue on the channel C-terminus; once this occurs, β-COP binding is inhibited, permitting channel maturation and surface expression. 14-3-3 also serves to facilitate folding and maturation of other, non-channel proteins, using a similar mechanism (O'Kelly et al. 2002; O'Kelly and Goldstein 2008).

While 14-3-3 rescues K2P channel surface expression and activity, another protein – the small ubiquitin-related modifier protein (SUMO) – modifies K2P channels by silencing them when it is covalently attached to, e.g., K2P1 via interaction of the channel with the SUMO conjugating enzyme, Ubc-9 ligase. Removal of SUMO by the SUMO-specific protease, SENP1, restores K2P1 activity. Covalent modification by addition of SUMO occurs via a single residue, K2P1-K274 (Plant et al. 2010; Rajan et al. 2005). Before this discovery, SUMO was previously thought to not regulate proteins at the plasma membrane and instead perform solely nuclear functions; it is now known to regulate other plasma membrane proteins in addition to K2P1, including the cardiac KCNQ1-KCNE1 potassium channel and the neuronal and cardiac Kv2.1 potassium channel (Plant et al. 2011, 2014).

2.3 Ancillary Subunit Interactions Inhibitory to Channel Activity

While it is not always completely understood why this occurs, some ancillary subunits strongly inhibit channel activity. KCNQ1 activity is strongly inhibited by the KCNE4 and KCNE5 subunits. KCNE4 is the largest of the KCNE subunits, with a recent N-terminal extension not previously detected bringing the total length of the long version of human KCNE4 (hKCNE4L) to 221 residues, still with a single TM segment. The shorter (170 amino acid) hKCNE4S isoform inhibits KCNQ1 by as much as 90% in heterologous co-expression studies in *Xenopus laevis* oocytes (and hKCNE4S also strongly inhibits KCNQ1 in COS cells), while hKCNE4L inhibits by only 40% in *Xenopus* oocytes (Abbott 2016d). The mechanistic studies of KCNE4 inhibition were all performed using KCNE4S. KCNE4S does not alter the surface expression of KCNQ1 protein, and the mechanism by which it inhibits KCNQ1 involves binding to both KCNQ1 and also directly to CaM, which is essential for KCNQ1 activity. A tetraleucine motif on the membrane-proximal region of the KCNE4 intracellular C-terminus is required for interaction with CaM. Further, KCNQ1 inhibition by KCNE4S is calcium-sensitive, with lower intracellular calcium levels decreasing inhibition. Thus, KCNE4, in concert with CaM, provides a mechanism to tune KCNQ1 current to intracellular calcium, and this may provide a reason for the existence of KCNE4 as a KCNQ1 inhibitory subunit (Abbott et al. 1999; Bendahhou et al. 2005; Ciampa et al. 2011; Grunnet et al. 2002, 2005; Lundquist et al. 2005; Manderfield et al. 2009; Manderfield and George 2008). KCNE4 also inhibits Kv1.1 and Kv1.3 channel activity (Grunnet et al. 2003).

KCNE5 also inhibits KCNQ1, shifting its voltage dependence by +140 mV such that it does not activate at physiologically relevant membrane potentials. The mechanism does not appear to be CaM-dependent and KCNE5 lacks the tetraleucine motif important for KCNE4 interaction with CaM. The physiological role of this inhibition is not known, although it is of note that both KCNE4 and KCNE5 can inhibit I_{Ks} (KCNQ1-KCNE1) complexes and all are expressed in human heart. KCNE4 is reportedly the highest expressed of all the KCNEs in human heart, which raises the question, still unanswered, of its role there – could it be to negatively regulate I_{Ks}, or render it calcium-sensitive? Perhaps KCNQ1 channels can gain or lose KCNE subunits while at the cell surface, or in endocytic pits, providing a dynamic mechanism for regulation of I_{Ks} kinetics, voltage dependence, and other properties (Angelo et al. 2002; Bendahhou et al. 2005; Piccini et al. 1999).

While KCNE1 and KCNE2 have opposite effects to one another on KCNQ1 (one slowing activation and increasing outward current, the other inducing constitutive activation and reducing outward current), they each exert a similar inhibitory effect on N-type inactivating (hereafter referred to as N-type) α subunits of the Kv1 and Kv3 (KCNA and KCNC) subfamilies. N-type subunits are fast-inactivating by virtue of a cytoplasmic "ball" domain at the end of the α subunit N-terminal domain, which plugs the pore from the intracellular side after channel activation, to induce inactivation, or block of the activated pore. This produces currents with rapid decay after rapid activation. In the Kv1 subfamily, Kv1.4 is the N-type subunit, while Kv1.1 and Kv1.2, for example, are delayed rectifier subunits (fast activating, slow-inactivating by a C-type mechanism involving pore collapse). In the Kv3 subfamily, Kv3.3 and Kv3.4 are the N-type subunits, while Kv3.1 and Kv3.2 are delayed rectifiers.

Within the same subfamily, co-expression of fast-inactivating N-type subunits with slow-inactivating delayed rectifier subunits produces heteromeric channels (if co-assembly is permissible) with intermediate inactivation kinetics (Coleman et al. 1999; Isacoff et al. 1990; Ruppersberg et al. 1990). This is because the rate of N-type inactivation is proportional to the number of N-type inactivation domains in the tetrameric complex, and any one of the up to 4 inactivation domains is necessary and sufficient to induce fast inactivation; a higher number increases the chance of this occurring in a given time period (Malysiak and Grzywna 2008). The rate of N-type inactivation is an important factor in determining the excitability and action potential frequency of cells such as neurons in which Kv1 and Kv3 α subunits are expressed. Therefore, it follows that co-assembly of N-type with delayed rectifier α subunits should be tightly regulated, to control inactivation rate. KCNE1 and KCNE2 were unexpectedly discovered to provide such a mechanism. When either of these subunits is co-expressed with any of the N-type subunits, Kv1.4, Kv3.3, or Kv3.4, it results in strong inhibition of channel activity. This is due to the KCNE subunits retaining the N-type subunits early in the secretory pathway, probably in the endoplasmic reticulum. Co-expression of KCNE1 or KCNE2 with the delayed rectifier subunits Kv1.1 or Kv3.1 did not result in inhibition. Crucially, triple expression of KCNE1 with both Kv1.1 and Kv1.4 prevented inhibition by KCNE1, permitting surface expression and channel activity of heteromeric Kv1.1-Kv1.4 complexes. A similar effect was observed with Kv3,1, which prevented intracellular retention of Kv3.4 by KCNE1 (and indeed prevented interaction of KCNE1 with Kv3.4). Furthermore, Kv3.2 was also able to rescue Kv3.4 from intracellular retention by KCNE1, while Kv2.1 – which can interact with KCNE1 but not Kv3.4 – was unable to rescue Kv3.4 from retention by KCNE1 (Kanda et al. 2011a, b). Thus, KCNE1 and KCNE2 can act as checkpoints that prevent surface expression of homomeric N-type channels and instead permit surface expression only of same-subfamily heteromeric N-type/delayed rectifier channels that have intermediate inactivation rates.

3 Ancillary Subunit Modulation of K^+ Channel Pharmacology

3.1 Sensitization by Providing a Novel Binding Site Absent from the Channel α Subunit to Directly Impact Channel Function

Perhaps the most profound mode of pharmacomodulation by an ancillary subunit is when it introduces a completely novel site to endow sensitivity to a drug or other small molecule to which the α subunit-alone channel is insensitive. Here we divide these into two classes depending on whether the novel small molecule effect is on function or protein folding/assembly/trafficking. BKβ subunits, for example, act in the first category, by providing a novel binding site and thereby sensitizing BK channels to activation by a range of steroids and steroid-like compounds. BKβ1 sensitizes BK channels to dehydrosoyasaponin-I (DHS-I), lithocholate, tamoxifen, 17β-estradiol and xenoestrogen; likewise, BKβ2 and BKβ4 sensitize BK channels to dehydroepiandrosterone and corticosterone, respectively. Several residues (T169, L172, and L173) in the BKβ1 TM2 segment form a steroid-sensing site essential for binding of, e.g., lithocholate. Neither homomeric BK α subunit channels, nor BK channels formed with any of the other β subunits, are activated by this steroid (Bukiya et al. 2007, 2009).

Sulfonylurea receptor (SUR) subunits participate in both forms of modulation. SURs are so-called because they bind sulphonylureas, which are used to treat type 2 diabetes mellitus and also some types of neonatal diabetes. Sulfonylurea binding to the SUR subunits inhibits K_{ATP} channels, thereby depolarizing pancreatic β-cells, which leads to voltage-gated calcium channel activation, and insulin secretion. Another important pharmacological role for SURs is in mediating the action of pharmacochaperones. This term describes drugs that can rescue K_{ATP} channel folding or assembly mutants, i.e., channels carrying mutations that prevent K_{ATP} channel surface expression, causing β-cell hyperexcitability and resultant congenital hyperinsulinism. Recent structural analyses utilizing cryo-electron microscopy (cryo-EM) showed that glibenclamide, repaglinide, and carbamazepine bind within a common binding pocket located in the SUR1 subunit of Kir6.2-SUR1 channels, which are the pancreatic β-cell K_{ATP} channel isoform. Because this site is located close to the Kir6.2 N-terminus, binding of the drugs to the SUR receptor stabilizes the Kir6.2 N-terminus, providing a solid base for channel folding and assembly, rescuing the mutant K_{ATP} channel activity by facilitating their surface expression, even though the pharmacochaperones are also channel inhibitors (Martin et al. 2017, 2019, 2020) (Fig. 5). Thus, SURs provide a novel binding site absent on the channel α subunit, to directly impact channel folding/assembly.

The Seebohm laboratory discovered a structural basis for drug sensitization of KCNQ1 by KCNE1, involving introduction of windows, or fenestrations, by

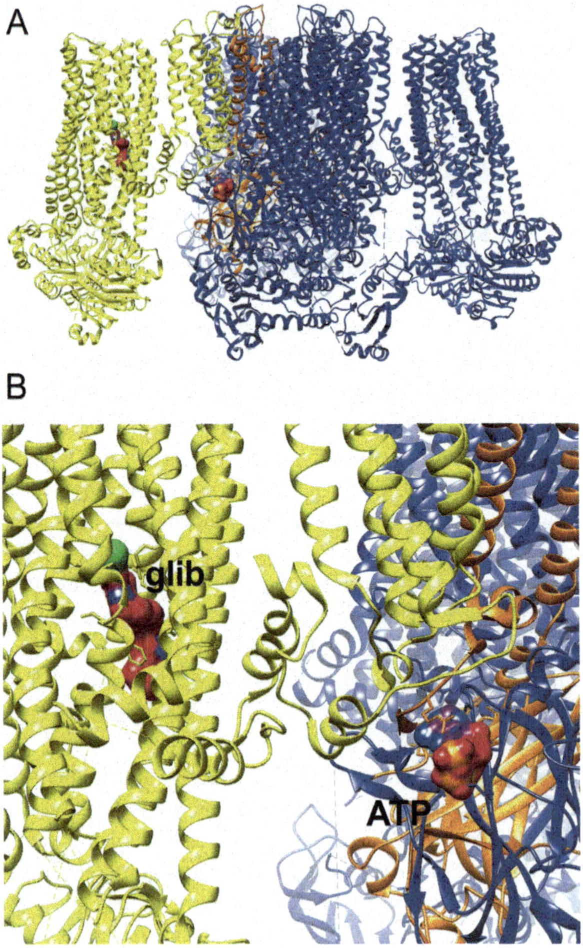

Fig. 5 Structure of Syrian hamster SUR1-rat Kir6.2 with glibenclamide and ATP bound. (**a**) Side view of Syrian hamster SUR1-rat Kir6.2 with glibenclamide and ATP bound solver by cryo-EM, plotted from coordinates in Martin et al. (2017). One SUR1 subunit is colored yellow, one Kir6.2 subunit is colored orange. (**b**) Close-up of structure in panel A, with glibenclamide (glib) and ATP labeled

KCNE1 that endow novel sensitivity of KCNQ1 to adamantane compounds. While KCNQ1-KCNE1 channels display an IC_{50} of 78.4 nM for inhibition by 2-(4-chlorophenoxy)-2-methyl-*N*-[5-[(methylsulfonyl)amino]tricyclo[3.3.1.13,7] dec-2-yl]-propanamide (also named JNJ303), homomeric KCNQ1 channels, or those formed by KCNQ1 with isoforms KCNE2–5, are completely insensitive to 1 μM JNJ303, which inhibits KCNQ1-KCNE1 by 80%. It is thought that rather than blocking the KCNQ1-KCNE1 pore, adamantane compounds instead stabilize a non-conducting closed state. By introducing a fenestration in the KCNQ1 pore, KCNE1 reportedly introduces a novel binding site that permits JNJ303 and other adamantane compounds to stabilize the closed state; hence, homomeric KCNQ1 is unaffected by these compounds (Wrobel et al. 2016). Pharmacological exploitation of this phenomenon has the potential to produce highly subtype-specific I_{Ks} modulators.

KCNQ channels can also form reciprocally regulating complexes with sodium-coupled solute transporters. In these cases, each protein can be considered as the others' ancillary subunit (Abbott 2016a; Abbott et al. 2014; Neverisky and Abbott 2015, 2017). We found that KCNQ channels function as chemosensors to permit sodium-coupled *myo*-inositol transporters to respond to stimuli to which they are typically insensitive, including the neurotransmitter GABA, related metabolites β-hydroxybutyrate (an important ketone body) and the KCNQ2/3-targeting anticonvulsant, retigabine – all of which activate KCNQ2/3 channels, do not act on homomeric SMITs, but inhibit myo-inositol transport activity of SMITs in physical complexes with KCNQ2/3 channels (Manville and Abbott 2020). Reciprocally, SMITs alter the response of KCNQs to some of these molecules. KCNQ2/3 channels are the primary molecular correlate of the neuronal M-current, which regulates neuronal excitability. They are sensitive to GABA and β-hydroxybutyrate by virtue of a binding pocket on their KCNQ3 subunits formed by residues in the S4–5 linker region and S5 (Manville et al. 2018, 2020). Homomeric KCNQ channels are activated by both GABA and β-hydroxybutyrate, while homomeric KCNQ2 is insensitive to both. In contrast, when in complexes with SMIT1, KCNQ2 channels are still insensitive to GABA, but can sense and are able to facilitate inhibition of SMIT1 by β-hydroxybutyrate. Therefore, SMIT1 modulates KCNQ2 pharmacology and introduces a novel sensitivity of the channel α subunit to β-hydroxybutyrate (Manville and Abbott 2020). A related example is that for Kv4.3 (or Kv4.2) channel complexes to become sensitive to the drug NS5806, the KChiP3 ancillary subunit is required, because it provides the NS5806 binding site. In addition, the mode of action of NS5806 (potentiation or inhibition) depends on the presence of yet another ancillary subunit – DPP6 (Zhang et al. 2020).

3.2 Sensitization by Increasing the Affinity of a Small Molecule That Also Binds to and Modulates α Subunit-Only Channels

3.2.1 Modulation of KCNQ1 Pharmacology by KCNEs

As described above, KCNQ1 is differentially regulated by a molecular toolkit of KCNE subunits to permit it to function in a wide range of cellular environments and overcome challenges such as extracellular acidification (in gastric parietal cells), lack of excitability (various epithelia cells), and diverse localization requirements (e.g., apical for KCNQ1-KCNE2 in parietal cells, basolateral for KCNQ1-KCNE3 in the lower digestive tract). KCNE subunits have long been known to alter the pharmacology of small molecules that modulate KCNQ1 (Panaghie and Abbott 2006). Therefore, there is the possibility that the differences in pharmacology of the different KCNQ1-KCNE complexes could 1 day be exploited to increase selectivity and target the desired tissue type while sparing others.

Unlike JNJ303, described in the previous section, Chromanol 293B is a classic inhibitor of both KCNQ1 and I_{Ks}, but it blocks with considerable differential sensitivity dependent on the presence of KCNE subunits. In particular, KCNE3 highly sensitizes KCNQ1 to inhibition by chromanol 293B, decreasing the IC_{50}

100-fold from 65 to 0.5 μM; KCNQ1-KCNE1 complexes lie in the middle, with an IC_{50} of 15 μM. One might suspect that this effect could arise from the widely different gating attributes of the three channel types, in particular that KCNQ1-KCNE3 has a much higher open probability at more negative voltages than the others. However, it was found that KCNQ1-KCNE3-V72T channels, which exhibit time- and voltage-dependent activation more akin to KCNQ1-KCNE1 channels, showed similar chromanol 293B sensitivity to that of wild-type KCNQ1-KCNE3, thus suggesting a change in the drug binding site itself rather than solely an indirect effect linked to altered gating (Bett et al. 2006).

3.2.2 Wild-Type and Inherited Mutant KCNE2 Effects on hERG Pharmacology

The hERG (human ether-à-go-go related gene product) potassium channel (a.k.a. Kv11.1; KCNH2) is essential for human ventricular myocyte repolarization. It generates I_{Kr} (the rapidly activating potassium current) in ventricular myocardium of humans, guinea pigs, rabbits, rats, but is absent in mice. Inherited loss-of-function mutations in the *KCNH2* gene that encodes hERG account for as many as 40–45% of sequenced inherited Long QT syndrome cases (termed LQT2 if they involve hERG), roughly equal to the proportion for *KCNQ1* (LQT1), with sodium channel Nav1.5 gene *SCN5A* gain-of-function mutations (LQT3) reaching around 10%. The remaining small fraction of sequenced LQTS cases mostly involve ancillary subunits or other channels (Abbott 2013) (see also chapter "Cardiac K^+ Channels and Channelopathies").

hERG is regulated by both KCNE1 (originally termed MinK, IsK) and KCNE2 (MiRP1), although the relative importance of homomeric hERG, versus hERG-KCNE1 or hERG-KCNE2, channels in the myocardia of humans and the various other species is not fully understood (Abbott 2015; Abbott et al. 1999; McDonald et al. 1997). There is biochemical evidence for each type of heteromeric complex in cardiac tissue and also evidence from human genetics studies. While KCNE1 doubles hERG current by an unclear mechanism, KCNE2 reduces hERG current by 40%, partly by reducing unitary conductance. Other effects include speeding of deactivation, right-shifting of the voltage dependence of activation, reduction of sensitivity to external K^+, and endowing hERG with a biphasic response to the inhibitor E-4031, more closely matching that of native I_{Kr} (Abbott 2015; Abbott et al. 1999; McDonald et al. 1997).

Inherited *KCNE1* mutations are known to cause LQTS (the LQT5 form) but this is primarily considered to occur because of dysfunction of KCNQ1-KCNE1 (I_{Ks}) complexes in cardiac myocytes. A recessive inherited disorder, Jervell Lange-Nielsen syndrome (JLNS), is well known to cause both LQTS and sensorineural deafness. This reflects the importance of KCNQ1-KCNE1 channels in the inner ear, where they regulate K^+ secretion into the endolymph. Without this channel functioning correctly, the inner ear develops abnormally resulting in deafness. Mutations in either *KCNE1* or *KCNQ1* cause JLNS; typically, both alleles carry a loss-of-function mutation to cause JLNS although this is not always the case (Schulze-Bahr et al. 1997; Tranebjaerg et al. 1993; Tyson et al. 1997). KCNQ1 has been found to form complexes with hERG in vitro and in vivo, a surprising result given that it is generally considered that Kv channels can only heteromultimerize with α subunits from their own subfamily (e.g., Kv1.1 with Kv1.4; Kv3.1 with Kv3.4). It is therefore possible that KCNE1 (or KCNE2) mutations could reduce myocyte repolarization reserve by disrupting I_{Ks}, I_{Kr}, or coupled I_{Ks}-I_{Kr} complexes (Organ-Darling et al. 2013; Ren et al. 2010).

For inherited *KCNE2* gene variants, several associations between specific mutations and polymorphisms have been found to associate with LQTS, a caveat being that in most cases, a second insult is likely required to cause arrhythmia. hERG is known to be highly susceptible to inhibition by a wide range of small molecules, many of which are or were FDA-approved and in clinical use for disorders unrelated to cardiac function, e.g., antihistamines, antibiotics. Paradoxically, some Class I antiarrhythmics (main antiarrhythmic action is sodium channel block) are also proarrhythmic in some settings, because they also block hERG channels. This has resulted in restricted use of some anti-arrhythmics based on the diagnosis and susceptibility to acquired (drug-induced) arrhythmia, and also drugs being completely withdrawn from clinical use. Thus, avoiding hERG block is seen as an important step during drug development, to help avoid cardiotoxicity (Guo et al. 2009; Recanatini et al. 2005; Roy et al. 1996; Testai et al. 2004) (see also chapter "Cardiac hERG K^+ Channel as Safety and Pharmacological Target").

The relevance with respect to KCNE2 is that some mutations or even common polymorphisms in the *KCNE2* gene can increase drug susceptibility of hERG-KCNE2 channels and therefore predispose to acquired arrhythmia (a combined inherited/acquired form of LQTS). There are several different categories of KCNE2 gene variant in this regard, three of which are found in gene variants that occur in a cluster on the extracellular N-terminal of KCNE2. The first is represented by the T8A polymorphism, which is present in around 1.6% of Caucasians studied in the USA and was not found in African-Americans. T8A does not alter channel function at baseline. However, it increases susceptibility of hERG-KCNE2 channels to block by the antibiotic sulfamethoxazole, shifting the K_i almost twofold, from 380 to 210 μg/ml. In addition, sulfamethoxazole only speeded deactivation of T8A, and not wild-type hERG-KCNE2 channels. T8A was discovered in a patient with sulfamethoxazole-induced arrhythmia and appears to predispose to this condition based on the clinical and cellular electrophysiology findings. Thus, T8A is a silent polymorphism at baseline, its effects on function only uncovered in the context of drug block (Abbott et al. 1999; Sesti et al. 2000). Interestingly, the mechanism for enhanced drug block was found to be loss of a consensus glycosylation site, preventing addition of a carbohydrate moiety that normally helps to shield hERG-KCNE2 channels from sulfamethoxazole block (Park et al. 2003).

These effects contrast with that of KCNE2-Q9E, which causes both altered hERG-KCNE2 channel function at baseline (+9 mV shift in the voltage dependence of activation and an 80% increase in the fast time constant of deactivation, with no change in the slow component) and threefold-increased sensitivity to the macrolide antibiotic, clarithromycin. The Q9E variant was originally discovered in an African-American woman who presented with clarithromycin-induced LQTS after receiving intravenous clarithromycin in hospital; Q9E was subsequently discovered to be present in 3% of African-Americans and absent from Caucasians in the USA (Abbott et al. 1999; Ackerman et al. 2003).

A third gene variant in this N-terminal cluster, KCNE2-T10M, was found in a woman who had a history of episodes of auditory-induced syncope (a hallmark of *KCNH2*-linked LQTS) and who presented with prolonged QT interval and ventricular fibrillation with hypomagnesemia and hypocalcemia after running the New York Marathon. She also exhibited later ventricular fibrillation concurrent with hypokalemia. Other family members carried the variant and did not exhibit LQTS; the subject did not harbor mutations in KCNH2 or other known LQTS genes. In cellular electrophysiology experiments, we found that T10M reduced hERG-KCNE2 currents by up to 80%, left-shifts the voltage dependence of inactivation, slows inactivation and recovery from inactivation, but did not alter sensitivity to extracellular cations. It was concluded that the T10M variant is a loss-of-function mutation that is tolerable until overlapping with another insult, in this case perturbation of extracellular cations such as K^+. Increased extracellular $[K^+]$ paradoxically augments hERG and hERG-KCNE2 outward currents, against driving force (up to 10 mM extracellular K^+), because of beneficial interactions with the outer pore. Thus, the hypokalemia experienced by the patient, exacerbated by lower repolarization reserve due to the T10M variant, likely contributed to their arrhythmia (Gordon et al. 2008).

Three other variants, M54T, I57T, and A116V, were found in individuals with procainamide-, oxatomide-, and quinidine-induced LQTS, respectively (Abbott et al. 1999; Sesti et al. 2000). None of these mutations altered the affinity of hERG-KCNE2 channels to block by the respective drugs, but all impaired channel function at baseline (by ~30% in terms of peak tail current inhibition). These then would fall into the category, like T10M of KCNE2 variants that predispose to acquired arrhythmia by reducing repolarization reserve enough that superimposition of another I_{Kr}-impairing environmental agent (e.g., drug, electrolyte imbalance) causes pathogenic I_{Kr} reduction. We also identified M54T and I57T in subjects with LQTS with no known environmental I_{Kr} reducing factors, so it is possible that in rare cases a drug or other agent is not needed for KCNE2-associated LQTS. Overall, given its effects on hERG sensitivity to electrolytes, E-4031, sulfamethoxazole and clarithromycin, it is clear that KCNE2 (either in wild-type or mutant form) falls into the category of an ancillary subunit that modulates hERG sensitivity to drugs that also act on the homomeric channel (Abbott et al. 1999; Sesti et al. 2000).

3.2.3 G Protein βγ (Gβγ) Subunits

Aside from the SURs (described in Sect. 2.1), the other most important K_{ir} channel ancillary subunits are the membrane-associated G protein βγ (Gβγ) subunits. G protein-coupled inwardly rectifying potassium channels (GIRKs) are activated by a combination of phosphatidylinositol 4,5-bisphosphate (PIP_2) and by GPCR stimulation, e.g., by neurotransmitters. GPCR stimulation causes Gβγ subunits to be released, dissociating from the Gα subunit. Gβγ subunits are then free to directly bind to and activate Kir3.1–3.4 (GIRK1–4) α subunits; the resultant heteromeric channel complexes are referred to as Kir3.x or GIRK channels. Kir3.1–3.3 are expressed predominantly in the nervous system, while Kir3.4 is primarily a cardiac subunit. The minor membrane phospholipid component but prominent signaling molecule, PIP_2, is able to directly bind to and activate homomeric Kir3 channels in the absence of Gβγ subunits, but PIP_2 is a much more potent activator of heteromeric Kir3-Gβγ channels. In contrast, Gβγ subunits cannot activate Kir3 channels in the absence of PIP_2. Thus, Gβγ subunits regulate Kir3 channels by sensitizing them to PIP_2, probably by increasing the affinity of Kir3 channels for PIP_2 (Glaaser and Slesinger 2015; Hibino et al. 2010; Sadja et al. 2003; Yamada et al. 1998).

3.3 Desensitization by Decreasing the Affinity of a Small Molecule That Binds to and Modulates α Subunit-Only K^+ Channels

Receptor desensitization by ancillary subunits is an important consideration in channel pharmacology because it can potentially be leveraged to provide increased specificity if the α subunit-ancillary subunit expression and interaction profiles are well understood across different cell and tissue types. Several of the TM ancillary subunits desensitize their α subunit partners to pharmacological agents and other types of channel regulators, and we consider various key examples below.

The BKβ subunits can act to decrease the sensitivity of BK channels to some molecules that regulate homomeric BK α subunit channels. Thus, BKβ1 and more so BKβ2- β4 reduce BK channel sensitivity to block by charybodotoxin (ChTx) and/or Iberiotoxin (IbTx), with β4 inducing almost complete resistance. The extracellular loop of the BKβ subunits mediates the altered sensitivity to ChTx, specifically L90, Y91, T93, and E94 in β1, and a lysine-rich ring comprising K137, 141, 147, and 150 in β2 (Li and Yan 2016; Torres et al. 2014). In BKβ4, which is a more powerful protector against toxin inhibition, K120, R121 and K125 together with some BK α subunit residues, form a shield against toxin blockade (Gan et al. 2008) (Fig. 3).

The KCNE3 regulatory subunit, which regulates KCNQ1 in colonic epithelium, also regulates the Kv3.4 potassium channel α subunit, in skeletal muscle. An R83H gene variant in KCNE3 predisposes to human periodic paralysis and reduces the ability of Kv3.4-KCNE3 channels to activate at subthreshold membrane potentials, thereby shifting the membrane potential of C2C12 skeletal muscle cells more positive (Abbott et al. 2001a, 2006). Mice with germline deletion of *Kcne3* exhibit skeletal muscle downregulation of hERG (also a partner of KCNE3), Kv3.4 and the K2P channel, KCNK4 (TRAAK). *Kcne3* knockout gastrocnemius muscle exhibits transcript remodeling and – consistent with increased oxidative metabolic activity – increased type IIa fast twitch oxidative muscle tissue. *Kcne3* knockout also causes abnormal hind limb clasping upon tail suspension, loss of the normal biphasic decline in contractile force upon repetitive hind limb muscle stimulation, and impaired skeletal myoblast K^+ currents, indicating a primary role for Kv3.4-KCNE3 channels in skeletal muscle (King et al. 2017). Strikingly, and as observed for the toxin response in BK channels containing BKβ subunits (see above), KCNE3 greatly desensitizes Kv3.4 channels to block by the sea anemone toxin, blood-depressing substance II (BDS-II), assessed by quantifying open probability at the single-channel level. KCNE3 reduced the BDS-II Ki of Kv3.4 30-fold, from 260 nM to 7.2 μM, recapitulating the BDS-II sensitivity of native C2C12 muscle cell currents (6.9 μM). It is likely that KCNE3 forms a protective shield that disrupts toxin binding to a site on the α subunit, as observed for BK channels (Abbott et al. 2001a).

Finally, as mentioned above, homomeric KCNQ3 channels are sensitive to and are activated by GABA and β-hydroxybutyrate (Manville et al. 2018, 2020). Heteromeric KCNQ2/3 channels are also sensitive and can transmit this sensitivity to physically coupled SMIT1 *myo*-inositol transporter, resulting in inhibition of transport activity. However, when KCNQ3 (in the absence of KCNQ2) forms physical complexes with SMIT1, the complex loses sensitivity to β-hydroxybutyrate, while retaining GABA sensitivity. An arginine in the S4–5 linker (R242 in KCNQ3) is required for communicating GABA binding to SMIT1 and is also known to be important for GABA binding; mutation of this residue also results in dramatic current inhibition upon SMIT1 binding to the channel. GABA induces a novel pore conformation in KCNQ3 and KCNQ5 that represents an activated state (GABA also binds to KCNQ2 and KCNQ4 but neither alters the pore conformation nor activates these two isoforms) and SMIT1 itself binds to the KCNQ pore module and induces a similar conformational shift (Manville and Abbott 2020). In sum, it appears that SMIT1 co-assembly alters the conformation of the S5/S4–5 binding site that coordinates binding of GABA, β-hydroxybutyrate and other small molecules such as retigabine.

SMIT1 also reduces sensitivity of KCNQ2/3 to the classic K^+ channel pore-blocking quaternary ammonium ion, tetraethylammonium (TEA) (by fourfold, increasing the EC_{50} from 7 to 29 mM at 0 mV). SMIT1 also decreased KCNQ1 TEA sensitivity by fourfold at 0 mV (EC_{50} shifted from 29 to 121 mV) and reversed the voltage dependence of KCNQ1 inhibition (Manville et al. 2017).

4 Conclusions

The ion channel structural biology revolution that started with X-ray crystallography and that has now benefitted from rapid advances in cryo-EM is enabling structural biologists to solve high-resolution structures of ion channel complexes with both their α and ancillary subunits, and in some cases drugs and other regulatory small molecules bound. These data can be understood in the context of advances in our knowledge of the essential roles of ancillary subunits in specific tissues and physiological roles, and with understanding of the effects of ancillary subunits on native ion channel function, derived from cellular electrophysiology, transgenic mouse and human genetics studies. Emerging from this are realistic prospects of future drugs that achieve specificity by targeting specific α-ancillary subunit complexes while leaving others untouched.

Acknowledgements GWA is grateful for financial support from the National Institutes of Health, National Institute of General Medical Sciences (GM130377 to GWA).

References

Abbott GW (2013) KCNE genetics and pharmacogenomics in cardiac arrhythmias: much ado about nothing? Expert Rev Clin Pharmacol 6:49–60. https://doi.org/10.1586/ecp.12.76

Abbott GW (2014) Biology of the KCNQ1 potassium channel. New J Sci 2014:26. https://doi.org/10.1155/2014/237431

Abbott GW (2015) The KCNE2 K(+) channel regulatory subunit: ubiquitous influence, complex pathobiology. Gene 569:162–172. https://doi.org/10.1016/j.gene.2015.06.061

Abbott GW (2016a) Channel-transporter complexes: an emerging theme in cell signaling. Biochem J 473:3759–3763. https://doi.org/10.1042/BCJ20160685C

Abbott GW (2016b) KCNE1 and KCNE3: the yin and yang of voltage-gated K(+) channel regulation. Gene 576:1–13. https://doi.org/10.1016/j.gene.2015.09.059

Abbott GW (2016c) KCNE4 and KCNE5: K(+) channel regulation and cardiac arrhythmogenesis. Gene 593:249–260. https://doi.org/10.1016/j.gene.2016.07.069

Abbott GW (2016d) Novel exon 1 protein-coding regions N-terminally extend human KCNE3 and KCNE4. FASEB J 30:2959–2969. https://doi.org/10.1096/fj.201600467R

Abbott GW (2017a) Beta subunits functionally differentiate human Kv4.3 potassium channel splice variants. Front Physiol 8:66. https://doi.org/10.3389/fphys.2017.00066

Abbott GW (2017b) Chansporter complexes in cell signaling. FEBS Lett 591:2556–2576. https://doi.org/10.1002/1873-3468.12755

Abbott GW, Goldstein SA (1998) A superfamily of small potassium channel subunits: form and function of the MinK-related peptides (MiRPs). Q Rev Biophys 31:357–398

Abbott GW, Roepke TK (2016) KCNE2 and gastric cancer: bench to bedside. Oncotarget 7:17286–17287. https://doi.org/10.18632/oncotarget.7921

Abbott GW, Sesti F, Splawski I, Buck ME, Lehmann MH, Timothy KW, Keating MT, Goldstein SA (1999) MiRP1 forms IKr potassium channels with HERG and is associated with cardiac arrhythmia. Cell 97:175–187

Abbott GW, Butler MH, Bendahhou S, Dalakas MC, Ptacek LJ, Goldstein SA (2001a) MiRP2 forms potassium channels in skeletal muscle with Kv3.4 and is associated with periodic paralysis. Cell 104:217–231

Abbott GW, Goldstein SA, Sesti F (2001b) Do all voltage-gated potassium channels use MiRPs? Circ Res 88:981–983

Abbott GW, Butler MH, Goldstein SA (2006) Phosphorylation and protonation of neighboring MiRP2 sites: function and pathophysiology of MiRP2-Kv3.4 potassium channels in periodic paralysis. FASEB J 20:293–301. https://doi.org/10.1096/fj.05-5070com

Abbott GW, Xu X, Roepke TK (2007) Impact of ancillary subunits on ventricular repolarization. J Electrocardiol 40:S42–S46. https://doi.org/10.1016/j.jelectrocard.2007.05.021

Abbott GW, Tai KK, Neverisky DL, Hansler A, Hu Z, Roepke TK, Lerner DJ, Chen Q, Liu L, Zupan B, Toth M, Haynes R, Huang X, Demirbas D, Buccafusca R, Gross SS, Kanda VA, Berry GT (2014) KCNQ1, KCNE2, and Na+−coupled solute transporters form reciprocally regulating complexes that affect neuronal excitability. Sci Signal 7:ra22. https://doi.org/10.1126/scisignal.2005025

Ackerman MJ, Tester DJ, Jones GS, Will ML, Burrow CR, Curran ME (2003) Ethnic differences in cardiac potassium channel variants: implications for genetic susceptibility to sudden cardiac death and genetic testing for congenital long QT syndrome. Mayo Clin Proc 78:1479–1487. https://doi.org/10.4065/78.12.1479

Alzamora R, O'Mahony F, Bustos V, Rapetti-Mauss R, Urbach V, Cid LP, Sepulveda FV, Harvey BJ (2011) Sexual dimorphism and oestrogen regulation of KCNE3 expression modulates the functional properties of KCNQ1 K(+) channels. J Physiol 589:5091–5107. https://doi.org/10.1113/jphysiol.2011.215772

Amadi CC, Brust RD, Skerritt MR, Campbell DL (2007) Regulation of Kv4.3 closed state inactivation and recovery by extracellular potassium and intracellular KChIP2b. Channels 1:305–314

An WF, Bowlby MR, Betty M, Cao J, Ling HP, Mendoza G, Hinson JW, Mattsson KI, Strassle BW, Trimmer JS, Rhodes KJ (2000) Modulation of A-type potassium channels by a family of calcium sensors. Nature 403:553–556. https://doi.org/10.1038/35000592

Angelo K, Jespersen T, Grunnet M, Nielsen MS, Klaerke DA, Olesen SP (2002) KCNE5 induces time- and voltage-dependent modulation of the KCNQ1 current. Biophys J 83:1997–2006. https://doi.org/10.1016/S0006-3495(02)73961-1

Atkinson NS, Robertson GA, Ganetzky B (1991) A component of calcium-activated potassium channels encoded by the Drosophila slo locus. Science 253:551–555. https://doi.org/10.1126/science.1857984

Bahring R (2018) Kv channel-interacting proteins as neuronal and non-neuronal calcium sensors. Channels (Austin) 12:187–200. https://doi.org/10.1080/19336950.2018.1491243

Bahring R, Dannenberg J, Peters HC, Leicher T, Pongs O, Isbrandt D (2001) Conserved Kv4 N-terminal domain critical for effects of Kv channel-interacting protein 2.2 on channel expression and gating. J Biol Chem 276:23888–23894. https://doi.org/10.1074/jbc.M101320200

Barhanin J, Lesage F, Guillemare E, Fink M, Lazdunski M, Romey G (1996) K(V)LQT1 and lsK (minK) proteins associate to form the I(Ks) cardiac potassium current. Nature 384:78–80. https://doi.org/10.1038/384078a0

Bendahhou S, Marionneau C, Haurogne K, Larroque MM, Derand R, Szuts V, Escande D, Demolombe S, Barhanin J (2005) In vitro molecular interactions and distribution of KCNE family with KCNQ1 in the human heart. Cardiovasc Res 67:529–538. https://doi.org/10.1016/j.cardiores.2005.02.014

Berkefeld H, Fakler B (2013) Ligand-gating by Ca2+ is rate limiting for physiological operation of BK(ca) channels. J Neurosci 33:7358–7367. https://doi.org/10.1523/JNEUROSCI.5443-12.2013

Berkefeld H, Fakler B, Schulte U (2010) Ca2+-activated K+ channels: from protein complexes to function. Physiol Rev 90:1437–1459. https://doi.org/10.1152/physrev.00049.2009

Bett GC, Morales MJ, Beahm DL, Duffey ME, Rasmusson RL (2006) Ancillary subunits and stimulation frequency determine the potency of chromanol 293B block of the KCNQ1 potassium channel. J Physiol 576:755–767. https://doi.org/10.1113/jphysiol.2006.116012

Bond CT, Maylie J, Adelman JP (1999) Small-conductance calcium-activated potassium channels. Ann N Y Acad Sci 868:370–378. https://doi.org/10.1111/j.1749-6632.1999.tb11298.x

Bukiya AN, Liu J, Toro L, Dopico AM (2007) Beta1 (KCNMB1) subunits mediate lithocholate activation of large-conductance Ca2+-activated K+ channels and dilation in small, resistance-size arteries. Mol Pharmacol 72:359–369. https://doi.org/10.1124/mol.107.034330

Bukiya AN, Vaithianathan T, Toro L, Dopico AM (2009) Channel beta2-4 subunits fail to substitute for beta1 in sensitizing BK channels to lithocholate. Biochem Biophys Res Commun 390:995–1000. https://doi.org/10.1016/j.bbrc.2009.10.091

Campomanes CR, Carroll KI, Manganas LN, Hershberger ME, Gong B, Antonucci DE, Rhodes KJ, Trimmer JS (2002) Kv beta subunit oxidoreductase activity and Kv1 potassium channel trafficking. J Biol Chem 277:8298–8305. https://doi.org/10.1074/jbc.M110276200

Castillo K, Contreras GF, Pupo A, Torres YP, Neely A, Gonzalez C, Latorre R (2015) Molecular mechanism underlying beta1 regulation in voltage- and calcium-activated potassium (BK) channels. Proc Natl Acad Sci U S A 112:4809–4814. https://doi.org/10.1073/pnas.1504378112

Chan KW, Zhang H, Logothetis DE (2003) N-terminal transmembrane domain of the SUR controls trafficking and gating of Kir6 channel subunits. EMBO J 22:3833–3843. https://doi.org/10.1093/emboj/cdg376

Ciampa EJ, Welch RC, Vanoye CG, George AL Jr (2011) KCNE4 juxtamembrane region is required for interaction with calmodulin and for functional suppression of KCNQ1. J Biol Chem 286:4141–4149. https://doi.org/10.1074/jbc.M110.158865

Coleman SK, Newcombe J, Pryke J, Dolly JO (1999) Subunit composition of Kv1 channels in human CNS. J Neurochem 73:849–858. https://doi.org/10.1046/j.1471-4159.1999.0730849.x

Duprat F, Lauritzen I, Patel A, Honore E (2007) The TASK background K2P channels: chemo- and nutrient sensors. Trends Neurosci 30:573–580. https://doi.org/10.1016/j.tins.2007.08.003

Fakler B, Brandle U, Glowatzki E, Weidemann S, Zenner HP, Ruppersberg JP (1995) Strong voltage-dependent inward rectification of inward rectifier K+ channels is caused by intracellular spermine. Cell 80:149–154

Foeger NC, Wang W, Mellor RL, Nerbonne JM (2013) Stabilization of Kv4 protein by the accessory K(+) channel interacting protein 2 (KChIP2) subunit is required for the generation of native myocardial fast transient outward K(+) currents. J Physiol 591:4149–4166. https://doi.org/10.1113/jphysiol.2013.255836

Gan G, Yi H, Chen M, Sun L, Li W, Wu Y, Ding J (2008) Structural basis for toxin resistance of beta4-associated calcium-activated potassium (BK) channels. J Biol Chem 283:24177–24184. https://doi.org/10.1074/jbc.M800179200

Ghosh S, Nunziato DA, Pitt GS (2006) KCNQ1 assembly and function is blocked by long-QT syndrome mutations that disrupt interaction with calmodulin. Circ Res 98:1048–1054. https://doi.org/10.1161/01.RES.0000218863.44140.f2

Glaaser IW, Slesinger PA (2015) Structural insights into GIRK channel function. Int Rev Neurobiol 123:117–160. https://doi.org/10.1016/bs.irn.2015.05.014

Gofman Y, Shats S, Attali B, Haliloglu T, Ben-Tal N (2012) How does KCNE1 regulate the Kv7.1 potassium channel? Model-structure, mutations, and dynamics of the Kv7.1-KCNE1 complex. Structure 20:1343–1352. https://doi.org/10.1016/j.str.2012.05.016

Goldstein SA, Price LA, Rosenthal DN, Pausch MH (1996) ORK1, a potassium-selective leak channel with two pore domains cloned from *Drosophila melanogaster* by expression in *Saccharomyces cerevisiae*. Proc Natl Acad Sci U S A 93:13256–13261

Goldstein SA, Bockenhauer D, O'Kelly I, Zilberberg N (2001) Potassium leak channels and the KCNK family of two-P-domain subunits. Nat Rev Neurosci 2:175–184. https://doi.org/10.1038/35058574

Gong N, Bodi I, Zobel C, Schwartz A, Molkentin JD, Backx PH (2006) Calcineurin increases cardiac transient outward K+ currents via transcriptional up-regulation of Kv4.2 channel subunits. J Biol Chem 281:38498–38506. https://doi.org/10.1074/jbc.M607774200

Gonzalez-Perez V, Lingle CJ (2019) Regulation of BK channels by beta and gamma subunits. Annu Rev Physiol 81:113–137. https://doi.org/10.1146/annurev-physiol-022516-034038

Gordon E, Panaghie G, Deng L, Bee KJ, Roepke TK, Krogh-Madsen T, Christini DJ, Ostrer H, Basson CT, Chung W, Abbott GW (2008) A KCNE2 mutation in a patient with cardiac arrhythmia induced by auditory stimuli and serum electrolyte imbalance. Cardiovasc Res 77:98–106. https://doi.org/10.1093/cvr/cvm030

Grahammer F, Warth R, Barhanin J, Bleich M, Hug MJ (2001) The small conductance K+ channel, KCNQ1: expression, function, and subunit composition in murine trachea. J Biol Chem 276:42268–42275. https://doi.org/10.1074/jbc.M105014200

Grunnet M, Jespersen T, Rasmussen HB, Ljungstrom T, Jorgensen NK, Olesen SP, Klaerke DA (2002) KCNE4 is an inhibitory subunit to the KCNQ1 channel. J Physiol 542:119–130

Grunnet M, Rasmussen HB, Hay-Schmidt A, Rosenstierne M, Klaerke DA, Olesen SP, Jespersen T (2003) KCNE4 is an inhibitory subunit to Kv1.1 and Kv1.3 potassium channels. Biophys J 85:1525–1537. https://doi.org/10.1016/S0006-3495(03)74585-8

Grunnet M, Olesen SP, Klaerke DA, Jespersen T (2005) hKCNE4 inhibits the hKCNQ1 potassium current without affecting the activation kinetics. Biochem Biophys Res Commun 328:1146–1153. https://doi.org/10.1016/j.bbrc.2005.01.071

Gulbis JM, Mann S, MacKinnon R (1999) Structure of a voltage-dependent K+ channel beta subunit. Cell 97:943–952

Gulbis JM, Zhou M, Mann S, MacKinnon R (2000) Structure of the cytoplasmic beta subunit-T1 assembly of voltage-dependent K+ channels. Science 289:123–127

Guo D, Klaasse E, de Vries H, Brussee J, Nalos L, Rook MB, Vos MA, van der Heyden MA, Ijzerman AP (2009) Exploring chemical substructures essential for HERG k(+) channel

blockade by synthesis and biological evaluation of dofetilide analogues. ChemMedChem 4:1722–1732. https://doi.org/10.1002/cmdc.200900203
Heitzmann D, Koren V, Wagner M, Sterner C, Reichold M, Tegtmeier I, Volk T, Warth R (2007) KCNE beta subunits determine pH sensitivity of KCNQ1 potassium channels. Cell Physiol Biochem 19:21–32. https://doi.org/10.1159/000099189
Hibino H, Inanobe A, Furutani K, Murakami S, Findlay I, Kurachi Y (2010) Inwardly rectifying potassium channels: their structure, function, and physiological roles. Physiol Rev 90:291–366. https://doi.org/10.1152/physrev.00021.2009
Hille B, Armstrong CM, MacKinnon R (1999) Ion channels: from idea to reality. Nat Med 5:1105–1109. https://doi.org/10.1038/13415
Hu B, Zeng WP, Li X, Al-Sheikh U, Chen SY, Ding J (2019) A conserved arginine/lysine-based motif promotes ER export of KCNE1 and KCNE2 to regulate KCNQ1 channel activity. Channels 13:483–497. https://doi.org/10.1080/19336950.2019.1685626
Isacoff EY, Jan YN, Jan LY (1990) Evidence for the formation of heteromultimeric potassium channels in Xenopus oocytes. Nature 345:530–534. https://doi.org/10.1038/345530a0
Jespersen T, Membrez M, Nicolas CS, Pitard B, Staub O, Olesen SP, Baro I, Abriel H (2007) The KCNQ1 potassium channel is down-regulated by ubiquitylating enzymes of the Nedd4/Nedd4-like family. Cardiovasc Res 74:64–74. https://doi.org/10.1016/j.cardiores.2007.01.008
Kanda VA, Lewis A, Xu X, Abbott GW (2011a) KCNE1 and KCNE2 inhibit forward trafficking of homomeric N-type voltage-gated potassium channels. Biophys J 101:1354–1363. https://doi.org/10.1016/j.bpj.2011.08.015
Kanda VA, Lewis A, Xu X, Abbott GW (2011b) KCNE1 and KCNE2 provide a checkpoint governing voltage-gated potassium channel alpha-subunit composition. Biophys J 101:1364–1375. https://doi.org/10.1016/j.bpj.2011.08.014
Kanda VA, Purtell K, Abbott GW (2011c) Protein kinase C downregulates I(Ks) by stimulating KCNQ1-KCNE1 potassium channel endocytosis. Heart Rhythm 8:1641–1647. https://doi.org/10.1016/j.hrthm.2011.04.034
Kang C, Tian C, Sonnichsen FD, Smith JA, Meiler J, George AL Jr, Vanoye CG, Kim HJ, Sanders CR (2008) Structure of KCNE1 and implications for how it modulates the KCNQ1 potassium channel. Biochemistry 47:7999–8006. https://doi.org/10.1021/bi800875q
Ketchum KA, Joiner WJ, Sellers AJ, Kaczmarek LK, Goldstein SA (1995) A new family of outwardly rectifying potassium channel proteins with two pore domains in tandem. Nature 376:690–695. https://doi.org/10.1038/376690a0
Kilfoil PJ, Chapalamadugu KC, Hu X, Zhang D, Raucci FJ Jr, Tur J, Brittian KR, Jones SP, Bhatnagar A, Tipparaju SM, Nystoriak MA (2019) Metabolic regulation of Kv channels and cardiac repolarization by Kvbeta2 subunits. J Mol Cell Cardiol 137:93–106. https://doi.org/10.1016/j.yjmcc.2019.09.013
King EC, Patel V, Anand M, Zhao X, Crump SM, Hu Z, Weisleder N, Abbott GW (2017) Targeted deletion of Kcne3 impairs skeletal muscle function in mice. FASEB J 31:2937–2947. https://doi.org/10.1096/fj.201600965RR
Kollewe A, Lau AY, Sullivan A, Roux B, Goldstein SA (2009) A structural model for K2P potassium channels based on 23 pairs of interacting sites and continuum electrostatics. J Gen Physiol 134:53–68. https://doi.org/10.1085/jgp.200910235
Kroncke BM, Van Horn WD, Smith J, Kang C, Welch RC, Song Y, Nannemann DP, Taylor KC, Sisco NJ, George AL Jr, Meiler J, Vanoye CG, Sanders CR (2016) Structural basis for KCNE3 modulation of potassium recycling in epithelia. Sci Adv 2:e1501228. https://doi.org/10.1126/sciadv.1501228
Krzystanek K, Rasmussen HB, Grunnet M, Staub O, Olesen SP, Abriel H, Jespersen T (2012) Deubiquitylating enzyme USP2 counteracts Nedd4-2-mediated downregulation of KCNQ1 potassium channels. Heart Rhythm 9:440–448. https://doi.org/10.1016/j.hrthm.2011.10.026
Kshatri AS, Gonzalez-Hernandez A, Giraldez T (2018) Physiological roles and therapeutic potential of Ca(2+) activated potassium channels in the nervous system. Front Mol Neurosci 11:258. https://doi.org/10.3389/fnmol.2018.00258

Kubo Y, Baldwin TJ, Jan YN, Jan LY (1993) Primary structure and functional expression of a mouse inward rectifier potassium channel. Nature 362:127–133. https://doi.org/10.1038/362127a0

Li Q, Yan J (2016) Modulation of BK channel function by auxiliary beta and gamma subunits. Int Rev Neurobiol 128:51–90. https://doi.org/10.1016/bs.irn.2016.03.015

Li X, Cai H, Zheng W, Tong M, Li H, Ao L, Li J, Hong G, Li M, Guan Q, Yang S, Yang D, Lin X, Guo Z (2016) An individualized prognostic signature for gastric cancer patients treated with 5-fluorouracil-based chemotherapy and distinct multi-omics characteristics of prognostic groups. Oncotarget 7:8743. https://doi.org/10.18632/oncotarget.7087

Lujan R, Maylie J, Adelman JP (2009) New sites of action for GIRK and SK channels. Nat Rev Neurosci 10:475–480. https://doi.org/10.1038/nrn2668

Lundquist AL, Manderfield LJ, Vanoye CG, Rogers CS, Donahue BS, Chang PA, Drinkwater DC, Murray KT, George AL Jr (2005) Expression of multiple KCNE genes in human heart may enable variable modulation of I(Ks). J Mol Cell Cardiol 38:277–287. https://doi.org/10.1016/j.yjmcc.2004.11.012

MacKinnon R (2003) Potassium channels. FEBS Lett 555:62–65

Malysiak K, Grzywna ZJ (2008) On the possible methods for the mathematical description of the ball and chain model of ion channel inactivation. Cell Mol Biol Lett 13:535–552. https://doi.org/10.2478/s11658-008-0015-8

Manderfield LJ, George AL Jr (2008) KCNE4 can co-associate with the I(Ks) (KCNQ1-KCNE1) channel complex. FEBS J 275:1336–1349. https://doi.org/10.1111/j.1742-4658.2008.06294.x

Manderfield LJ, Daniels MA, Vanoye CG, George AL Jr (2009) KCNE4 domains required for inhibition of KCNQ1. J Physiol 587:303–314. https://doi.org/10.1113/jphysiol.2008.161281

Manville RW, Abbott GW (2020) Potassium channels act as chemosensors for solute transporters. Commun Biol 3:90. https://doi.org/10.1038/s42003-020-0820-9

Manville RW, Neverisky DL, Abbott GW (2017) SMIT1 modifies KCNQ channel function and pharmacology by physical interaction with the pore. Biophys J 113:613–626. https://doi.org/10.1016/j.bpj.2017.06.055

Manville RW, Papanikolaou M, Abbott GW (2018) Direct neurotransmitter activation of voltage-gated potassium channels. Nat Commun 9:1847. https://doi.org/10.1038/s41467-018-04266-w

Manville RW, Papanikolaou M, Abbott GW (2020) M-channel activation contributes to the anticonvulsant action of the ketone body beta-hydroxybutyrate. J Pharmacol Exp Ther 372:148–156. https://doi.org/10.1124/jpet.119.263350

Martin GM, Kandasamy B, DiMaio F, Yoshioka C, Shyng SL (2017) Anti-diabetic drug binding site in a mammalian KATP channel revealed by Cryo-EM. Elife 6:e31054. https://doi.org/10.7554/eLife.31054

Martin GM, Sung MW, Yang Z, Innes LM, Kandasamy B, David LL, Yoshioka C, Shyng SL (2019) Mechanism of pharmacochaperoning in a mammalian KATP channel revealed by cryo-EM. Elife 8:e46417. https://doi.org/10.7554/eLife.46417

Martin GM, Sung MW, Shyng SL (2020) Pharmacological chaperones of ATP-sensitive potassium channels: mechanistic insight from cryoEM structures. Mol Cell Endocrinol 502:110667. https://doi.org/10.1016/j.mce.2019.110667

Mathie A, Al-Moubarak E, Veale EL (2010) Gating of two pore domain potassium channels. J Physiol 588:3149–3156. https://doi.org/10.1113/jphysiol.2010.192344

McDonald TV, Yu Z, Ming Z, Palma E, Meyers MB, Wang KW, Goldstein SA, Fishman GI (1997) A minK-HERG complex regulates the cardiac potassium current I(Kr). Nature 388:289–292. https://doi.org/10.1038/40882

Miller AN, Long SB (2012) Crystal structure of the human two-pore domain potassium channel K2P1. Science 335:432–436. https://doi.org/10.1126/science.1213274

Morin TJ, Kobertz WR (2008) Counting membrane-embedded KCNE beta-subunits in functioning K+ channel complexes. Proc Natl Acad Sci U S A 105:1478–1482. https://doi.org/10.1073/pnas.0710366105

Nakajo K, Kubo Y (2007) KCNE1 and KCNE3 stabilize and/or slow voltage sensing S4 segment of KCNQ1 channel. J Gen Physiol 130:269–281. https://doi.org/10.1085/jgp.200709805
Nakajo K, Ulbrich MH, Kubo Y, Isacoff EY (2010) Stoichiometry of the KCNQ1 – KCNE1 ion channel complex. Proc Natl Acad Sci U S A 107:18862–18867. https://doi.org/10.1073/pnas.1010354107
Neverisky DL, Abbott GW (2015) Ion channel-transporter interactions. Crit Rev Biochem Mol Biol 51:257–267. https://doi.org/10.3109/10409238.2016.1172553
Neverisky DL, Abbott GW (2017) KCNQ-SMIT complex formation facilitates ion channel-solute transporter cross talk. FASEB J 31:2828–2838. https://doi.org/10.1096/fj.201601334R
Nishida M, MacKinnon R (2002) Structural basis of inward rectification: cytoplasmic pore of the G protein-gated inward rectifier GIRK1 at 1.8 A resolution. Cell 111:957–965
O'Kelly I, Goldstein SA (2008) Forward transport of K2p3.1: mediation by 14-3-3 and COPI, modulation by p11. Traffic 9:72–78. https://doi.org/10.1111/j.1600-0854.2007.00663.x
O'Kelly I, Butler MH, Zilberberg N, Goldstein SA (2002) Forward transport. 14-3-3 binding overcomes retention in endoplasmic reticulum by dibasic signals. Cell 111:577–588
O'Mahony F, Thomas W, Harvey BJ (2009) Novel female sex-dependent actions of oestrogen in the intestine. J Physiol 587:5039–5044. https://doi.org/10.1113/jphysiol.2009.177972
Organ-Darling LE, Vernon AN, Giovanniello JR, Lu Y, Moshal K, Roder K, Li W, Koren G (2013) Interactions between hERG and KCNQ1 alpha-subunits are mediated by their COOH termini and modulated by cAMP. Am J Physiol Heart Circ Physiol 304:H589–H599. https://doi.org/10.1152/ajpheart.00385.2012
Osteen JD, Gonzalez C, Sampson KJ, Iyer V, Rebolledo S, Larsson HP, Kass RS (2010a) KCNE1 alters the voltage sensor movements necessary to open the KCNQ1 channel gate. Proc Natl Acad Sci U S A 107:22710–22715. https://doi.org/10.1073/pnas.1016300108
Osteen JD, Sampson KJ, Kass RS (2010b) The cardiac IKs channel, complex indeed. Proc Natl Acad Sci U S A 107:18751–18752. https://doi.org/10.1073/pnas.1014150107
Osteen JD, Barro-Soria R, Robey S, Sampson KJ, Kass RS, Larsson HP (2012) Allosteric gating mechanism underlies the flexible gating of KCNQ1 potassium channels. Proc Natl Acad Sci U S A 109:7103–7108. https://doi.org/10.1073/pnas.1201582109
Panaghie G, Abbott GW (2006) The impact of ancillary subunits on small-molecule interactions with voltage-gated potassium channels. Curr Pharm Des 12:2285–2302
Panaghie G, Abbott GW (2007) The role of S4 charges in voltage-dependent and voltage-independent KCNQ1 potassium channel complexes. J Gen Physiol 129:121–133. https://doi.org/10.1085/jgp.200609612
Park KH, Kwok SM, Sharon C, Baerga R, Sesti F (2003) N-glycosylation-dependent block is a novel mechanism for drug-induced cardiac arrhythmia. FASEB J 17:2308–2309. https://doi.org/10.1096/fj.03-0577fje
Piccini M, Vitelli F, Seri M, Galietta LJ, Moran O, Bulfone A, Banfi S, Pober B, Renieri A (1999) KCNE1-like gene is deleted in AMME contiguous gene syndrome: identification and characterization of the human and mouse homologs. Genomics 60:251–257. https://doi.org/10.1006/geno.1999.5904
Plant LD, Dementieva IS, Kollewe A, Olikara S, Marks JD, Goldstein SA (2010) One SUMO is sufficient to silence the dimeric potassium channel K2P1. Proc Natl Acad Sci U S A 107:10743–10748. https://doi.org/10.1073/pnas.1004712107
Plant LD, Dowdell EJ, Dementieva IS, Marks JD, Goldstein SA (2011) SUMO modification of cell surface Kv2.1 potassium channels regulates the activity of rat hippocampal neurons. J Gen Physiol 137:441–454. https://doi.org/10.1085/jgp.201110604
Plant LD, Xiong D, Dai H, Goldstein SA (2014) Individual IKs channels at the surface of mammalian cells contain two KCNE1 accessory subunits. Proc Natl Acad Sci U S A 111: E1438–E1446. https://doi.org/10.1073/pnas.1323548111
Pongs O, Schwarz JR (2010) Ancillary subunits associated with voltage-dependent K+ channels. Physiol Rev 90:755–796. https://doi.org/10.1152/physrev.00020.2009

Purtell K, Paroder-Belenitsky M, Reyna-Neyra A, Nicola JP, Koba W, Fine E, Carrasco N, Abbott GW (2012) The KCNQ1-KCNE2 K(+) channel is required for adequate thyroid I(−) uptake. FASEB J 26:3252–3259. https://doi.org/10.1096/fj.12-206110

Rajan S, Plant LD, Rabin ML, Butler MH, Goldstein SA (2005) Sumoylation silences the plasma membrane leak K+ channel K2P1. Cell 121:37–47. https://doi.org/10.1016/j.cell.2005.01.019

Rapetti-Mauss R, O'Mahony F, Sepulveda FV, Urbach V, Harvey BJ (2013) Oestrogen promotes KCNQ1 potassium channel endocytosis and postendocytic trafficking in colonic epithelium. J Physiol 591:2813–2831. https://doi.org/10.1113/jphysiol.2013.251678

Recanatini M, Poluzzi E, Masetti M, Cavalli A, De Ponti F (2005) QT prolongation through hERG K(+) channel blockade: current knowledge and strategies for the early prediction during drug development. Med Res Rev 25:133–166. https://doi.org/10.1002/med.20019

Ren XQ, Liu GX, Organ-Darling LE, Zheng R, Roder K, Jindal HK, Centracchio J, McDonald TV, Koren G (2010) Pore mutants of HERG and KvLQT1 downregulate the reciprocal currents in stable cell lines. Am J Physiol Heart Circ Physiol 299:H1525–H1534. https://doi.org/10.1152/ajpheart.00479.2009

Roepke TK, Anantharam A, Kirchhoff P, Busque SM, Young JB, Geibel JP, Lerner DJ, Abbott GW (2006) The KCNE2 potassium channel ancillary subunit is essential for gastric acid secretion. J Biol Chem 281:23740–23747. https://doi.org/10.1074/jbc.M604155200

Roepke TK, King EC, Reyna-Neyra A, Paroder M, Purtell K, Koba W, Fine E, Lerner DJ, Carrasco N, Abbott GW (2009) Kcne2 deletion uncovers its crucial role in thyroid hormone biosynthesis. Nat Med 15:1186–1194. https://doi.org/10.1038/nm.2029

Roepke TK, Purtell K, King EC, La Perle KM, Lerner DJ, Abbott GW (2010) Targeted deletion of Kcne2 causes gastritis cystica profunda and gastric neoplasia. PLoS One 5:e11451. https://doi.org/10.1371/journal.pone.0011451

Roepke TK, Kanda VA, Purtell K, King EC, Lerner DJ, Abbott GW (2011a) KCNE2 forms potassium channels with KCNA3 and KCNQ1 in the choroid plexus epithelium. FASEB J 25:4264–4273

Roepke TK, King EC, Purtell K, Kanda VA, Lerner DJ, Abbott GW (2011b) Genetic dissection reveals unexpected influence of beta subunits on KCNQ1 K+ channel polarized trafficking in vivo. FASEB J 25:727–736. https://doi.org/10.1096/fj.10-173682

Roy M, Dumaine R, Brown AM (1996) HERG, a primary human ventricular target of the nonsedating antihistamine terfenadine. Circulation 94:817–823

Ruppersberg JP (2000) Intracellular regulation of inward rectifier K+ channels. Pflugers Arch 441:1–11

Ruppersberg JP, Schroter KH, Sakmann B, Stocker M, Sewing S, Pongs O (1990) Heteromultimeric channels formed by rat brain potassium-channel proteins. Nature 345:535–537. https://doi.org/10.1038/345535a0

Ruscic KJ, Miceli F, Villalba-Galea CA, Dai H, Mishina Y, Bezanilla F, Goldstein SA (2013) IKs channels open slowly because KCNE1 accessory subunits slow the movement of S4 voltage sensors in KCNQ1 pore-forming subunits. Proc Natl Acad Sci U S A 110:E559–E566. https://doi.org/10.1073/pnas.1222616110

Sadja R, Alagem N, Reuveny E (2003) Gating of GIRK channels: details of an intricate, membrane-delimited signaling complex. Neuron 39:9–12. https://doi.org/10.1016/s0896-6273(03)00402-1

Sahu ID, Kroncke BM, Zhang R, Dunagan MM, Smith HJ, Craig A, McCarrick RM, Sanders CR, Lorigan GA (2014) Structural investigation of the transmembrane domain of KCNE1 in proteoliposomes. Biochemistry 53:6392–6401. https://doi.org/10.1021/bi500943p

Sanguinetti MC, Curran ME, Zou A, Shen J, Spector PS, Atkinson DL, Keating MT (1996) Coassembly of K(V)LQT1 and minK (IsK) proteins to form cardiac I(Ks) potassium channel. Nature 384:80–83. https://doi.org/10.1038/384080a0

Schroeder BC, Waldegger S, Fehr S, Bleich M, Warth R, Greger R, Jentsch TJ (2000) A constitutively open potassium channel formed by KCNQ1 and KCNE3. Nature 403:196–199. https://doi.org/10.1038/35003200

Schulze-Bahr E, Wang Q, Wedekind H, Haverkamp W, Chen Q, Sun Y, Rubie C, Hordt M, Towbin JA, Borggrefe M, Assmann G, Qu X, Somberg JC, Breithardt G, Oberti C, Funke H (1997) KCNE1 mutations cause jervell and Lange-Nielsen syndrome. Nat Genet 17:267–268. https://doi.org/10.1038/ng1197-267

Sesti F, Goldstein SA (1998) Single-channel characteristics of wild-type IKs channels and channels formed with two minK mutants that cause long QT syndrome. J Gen Physiol 112:651–663

Sesti F, Abbott GW, Wei J, Murray KT, Saksena S, Schwartz PJ, Priori SG, Roden DM, George AL Jr, Goldstein SA (2000) A common polymorphism associated with antibiotic-induced cardiac arrhythmia. Proc Natl Acad Sci U S A 97:10613–10618. https://doi.org/10.1073/pnas.180223197

Shamgar L, Ma L, Schmitt N, Haitin Y, Peretz A, Wiener R, Hirsch J, Pongs O, Attali B (2006) Calmodulin is essential for cardiac IKS channel gating and assembly: impaired function in long-QT mutations. Circ Res 98:1055–1063. https://doi.org/10.1161/01.RES.0000218979.40770.69

Smith JA, Vanoye CG, George AL Jr, Meiler J, Sanders CR (2007) Structural models for the KCNQ1 voltage-gated potassium channel. Biochemistry 46:14141–14152. https://doi.org/10.1021/bi701597s

Splawski I, Timothy KW, Vincent GM, Atkinson DL, Keating MT (1997a) Molecular basis of the long-QT syndrome associated with deafness. N Engl J Med 336:1562–1567. https://doi.org/10.1056/NEJM199705293362204

Splawski I, Tristani-Firouzi M, Lehmann MH, Sanguinetti MC, Keating MT (1997b) Mutations in the hminK gene cause long QT syndrome and suppress IKs function. Nat Genet 17:338–340. https://doi.org/10.1038/ng1197-338

Strutz-Seebohm N, Pusch M, Wolf S, Stoll R, Tapken D, Gerwert K, Attali B, Seebohm G (2011) Structural basis of slow activation gating in the cardiac I Ks channel complex. Cell Physiol Biochem 27:443–452. https://doi.org/10.1159/000329965

Sun J, MacKinnon R (2017) Cryo-EM structure of a KCNQ1/CaM complex reveals insights into congenital long QT syndrome. Cell 169:1042–1050 e9. https://doi.org/10.1016/j.cell.2017.05.019

Sun J, MacKinnon R (2020) Structural basis of human KCNQ1 modulation and gating. Cell 180:340–347 e9. https://doi.org/10.1016/j.cell.2019.12.003

Sun X, Zaydman MA, Cui J (2012) Regulation of voltage-activated K(+) channel gating by transmembrane beta subunits. Front Pharmacol 3:63. https://doi.org/10.3389/fphar.2012.00063

Takimoto K, Ren X (2002) KChIPs (Kv channel-interacting proteins)--a few surprises and another. J Physiol 545:3. https://doi.org/10.1113/jphysiol.2002.033993

Takumi T, Ohkubo H, Nakanishi S (1988) Cloning of a membrane protein that induces a slow voltage-gated potassium current. Science 242:1042–1045

Tao X, MacKinnon R (2019) Molecular structures of the human Slo1 K(+) channel in complex with beta4. Elife 8:e51409. https://doi.org/10.7554/eLife.51409

Tempel BL, Papazian DM, Schwarz TL, Jan YN, Jan LY (1987) Sequence of a probable potassium channel component encoded at Shaker locus of Drosophila. Science 237:770–775

Tempel BL, Jan YN, Jan LY (1988) Cloning of a probable potassium channel gene from mouse brain. Nature 332:837–839. https://doi.org/10.1038/332837a0

Testai L, Bianucci AM, Massarelli I, Breschi MC, Martinotti E, Calderone V (2004) Torsadogenic cardiotoxicity of antipsychotic drugs: a structural feature, potentially involved in the interaction with cardiac HERG potassium channels. Curr Med Chem 11:2691–2706

Tinel N, Diochot S, Borsotto M, Lazdunski M, Barhanin J (2000) KCNE2 confers background current characteristics to the cardiac KCNQ1 potassium channel. EMBO J 19:6326–6330. https://doi.org/10.1093/emboj/19.23.6326

Torres YP, Granados ST, Latorre R (2014) Pharmacological consequences of the coexpression of BK channel alpha and auxiliary beta subunits. Front Physiol 5:383. https://doi.org/10.3389/fphys.2014.00383

Tranebjaerg L, Samson RA, Green GE (1993) Jervell and Lange-Nielsen syndrome. In: Pagon RA, Adam MP, Ardinger HH, Bird TD, Dolan CR, Fong CT, Smith RJH, Stephens K (eds) Genetic hearing loss. GeneReviews(R), Seattle

Tristani-Firouzi M, Sanguinetti MC (1998) Voltage-dependent inactivation of the human K+ channel KvLQT1 is eliminated by association with minimal K+ channel (minK) subunits. J Physiol 510(Pt 1):37–45

Tucker SJ, Gribble FM, Zhao C, Trapp S, Ashcroft FM (1997) Truncation of Kir6.2 produces ATP-sensitive K+ channels in the absence of the sulphonylurea receptor. Nature 387:179–183. https://doi.org/10.1038/387179a0

Tyson J, Tranebjaerg L, Bellman S, Wren C, Taylor JF, Bathen J, Aslaksen B, Sorland SJ, Lund O, Malcolm S, Pembrey M, Bhattacharya S, Bitner-Glindzicz M (1997) IsK and KvLQT1: mutation in either of the two subunits of the slow component of the delayed rectifier potassium channel can cause Jervell and Lange-Nielsen syndrome. Hum Mol Genet 6:2179–2185

Vanoye CG, Welch RC, Daniels MA, Manderfield LJ, Tapper AR, Sanders CR, George AL Jr (2009) Distinct subdomains of the KCNQ1 S6 segment determine channel modulation by different KCNE subunits. J Gen Physiol 134:207–217. https://doi.org/10.1085/jgp.200910234

vanTol BL, Missan S, Crack J, Moser S, Baldridge WH, Linsdell P, Cowley EA (2007) Contribution of KCNQ1 to the regulatory volume decrease in the human mammary epithelial cell line MCF-7. Am J Physiol Cell Physiol 293:C1010–C1019. https://doi.org/10.1152/ajpcell.00071.2007

Varnum MD, Busch AE, Bond CT, Maylie J, Adelman JP (1993) The min K channel underlies the cardiac potassium current IKs and mediates species-specific responses to protein kinase C. Proc Natl Acad Sci U S A 90:11528–11532

Vergara C, Latorre R, Marrion NV, Adelman JP (1998) Calcium-activated potassium channels. Curr Opin Neurobiol 8:321–329

Wang M, Kass RS (2012) Stoichiometry of the slow I(ks) potassium channel in human embryonic stem cell-derived myocytes. Pediatr Cardiol 33:938–942. https://doi.org/10.1007/s00246-012-0255-2

Wang Q, Curran ME, Splawski I, Burn TC, Millholland JM, VanRaay TJ, Shen J, Timothy KW, Vincent GM, de Jager T, Schwartz PJ, Toubin JA, Moss AJ, Atkinson DL, Landes GM, Connors TD, Keating MT (1996) Positional cloning of a novel potassium channel gene: KVLQT1 mutations cause cardiac arrhythmias. Nat Genet 12:17–23. https://doi.org/10.1038/ng0196-17

Weatherall KL, Goodchild SJ, Jane DE, Marrion NV (2010) Small conductance calcium-activated potassium channels: from structure to function. Prog Neurobiol 91:242–255. https://doi.org/10.1016/j.pneurobio.2010.03.002

Wiener R, Haitin Y, Shamgar L, Fernandez-Alonso MC, Martos A, Chomsky-Hecht O, Rivas G, Attali B, Hirsch JA (2008) The KCNQ1 (Kv7.1) COOH terminus, a multitiered scaffold for subunit assembly and protein interaction. J Biol Chem 283:5815–5830. https://doi.org/10.1074/jbc.M707541200

Wrobel E, Rothenberg I, Krisp C, Hundt F, Fraenzel B, Eckey K, Linders JT, Gallacher DJ, Towart R, Pott L, Pusch M, Yang T, Roden DM, Kurata HT, Schulze-Bahr E, Strutz-Seebohm-N, Wolters D, Seebohm G (2016) KCNE1 induces fenestration in the Kv7.1/KCNE1 channel complex that allows for highly specific pharmacological targeting. Nat Commun 7:12795. https://doi.org/10.1038/ncomms12795

Wu D, Delaloye K, Zaydman MA, Nekouzadeh A, Rudy Y, Cui J (2010a) State-dependent electrostatic interactions of S4 arginines with E1 in S2 during Kv7.1 activation. J Gen Physiol 135:595–606. https://doi.org/10.1085/jgp.201010408

Wu D, Pan H, Delaloye K, Cui J (2010b) KCNE1 remodels the voltage sensor of Kv7.1 to modulate channel function. Biophys J 99:3599–3608. https://doi.org/10.1016/j.bpj.2010.10.018

Xu X, Kanda VA, Choi E, Panaghie G, Roepke TK, Gaeta SA, Christini DJ, Lerner DJ, Abbott GW (2009) MinK-dependent internalization of the IKs potassium channel. Cardiovasc Res 82:430–438. https://doi.org/10.1093/cvr/cvp047

Yamada M, Inanobe A, Kurachi Y (1998) G protein regulation of potassium ion channels. Pharmacol Rev 50:723–760

Yan J, Aldrich RW (2010) LRRC26 auxiliary protein allows BK channel activation at resting voltage without calcium. Nature 466:513–516. https://doi.org/10.1038/nature09162

Zerangue N, Schwappach B, Jan YN, Jan LY (1999) A new ER trafficking signal regulates the subunit stoichiometry of plasma membrane K(ATP) channels. Neuron 22:537–548

Zhang ZJ, Jurkiewicz NK, Folander K, Lazarides E, Salata JJ, Swanson R (1994) K+ currents expressed from the Guinea pig cardiac IsK protein are enhanced by activators of protein kinase C. Proc Natl Acad Sci U S A 91:1766–1770

Zhang H, Zhang H, Wang C, Wang Y, Zou R, Shi C, Guan B, Gamper N, Xu Y (2020) Auxiliary subunits control biophysical properties and response to compound NS5806 of the Kv4 potassium channel complex. FASEB J 34:807–821. https://doi.org/10.1096/fj.201902010RR

Zhou M, Morais-Cabral JH, Mann S, MacKinnon R (2001) Potassium channel receptor site for the inactivation gate and quaternary amine inhibitors. Nature 411:657–661. https://doi.org/10.1038/35079500

Peptide Toxins Targeting K_V Channels

Kazuki Matsumura, Mariko Yokogawa, and Masanori Osawa

Contents

Abstract

A number of peptide toxins isolated from animals target potassium ion (K^+) channels. Many of them are particularly known to inhibit voltage-gated K^+ (K_V) channels and are mainly classified into pore-blocking toxins or gating-modifier toxins. Pore-blocking toxins directly bind to the ion permeation pores of K_V channels, thereby physically occluding them. In contrast, gating-modifier toxins bind to the voltage-sensor domains of K_V channels, modulating their voltage-dependent conformational changes. These peptide toxins are useful molecular tools in revealing the structure-function relationship of K_V channels

K. Matsumura · M. Yokogawa · M. Osawa (✉)
Graduate School of Pharmaceutical Sciences, Keio University, Tokyo, Japan
e-mail: osawa-ms@pha.keio.ac.jp

N. Gamper, K. Wang (eds.), *Pharmacology of Potassium Channels*,
Handbook of Experimental Pharmacology 267, https://doi.org/10.1007/164_2021_500

and have potential for novel treatments for diseases related to K_V channels. This review focuses on the inhibition mechanism of pore-blocking and gating-modifier toxins that target K_V channels.

Keywords

Gating-modifier toxin · Peptide toxin · Pore-blocking toxin · Voltage-gated potassium channel

1 Introduction

Peptide toxins are found in the venoms of poisonous organisms, such as scorpions, sea anemones, spiders, cone snails, snakes, bees, and centipedes. The biological roles of the toxins in these animals are self-defence against natural enemies and predation (Finol-Urdaneta et al. 2020; Harvey and Robertson 2004; Jiménez-Vargas et al. 2017). Some peptide toxins target potassium (K^+) channels in humans and other higher organisms. In the various types of cells, tissues, and organs, K^+ channels play essential roles in life functions and are also related to some diseases. Therefore, the mechanisms of K^+ channel modulation by peptide toxins have attracted much attention because these toxins can be used as effective molecular tools for elucidating the structure-function relationship of K^+ channels and for developing novel drugs.

2 Classification of Peptide Toxins Targeting K^+ Channels

K^+ channels are classified into four types: inward-rectifier K^+ (K_{ir}) channels, tandem-pore domain K^+ (K_{2P}) channels, voltage-gated K^+ (K_V) channels, and calcium-activated K^+ (K_{Ca}) channels (Fig. 1a). Many peptide toxins that act on K^+ channels target K_V and K_{Ca} channels (Jiménez-Vargas et al. 2017), whereas some peptide toxins target K_{ir} channels (Doupnik 2017; Ramu and Lu 2019; Ramu et al. 2018). To date, no peptide toxins targeting K_{2P} channels have been reported.

Based on their mode of inhibition, peptide toxins targeting K^+ channels include pore-blocking toxins and gating-modifier toxins (Tables 1 and 2). Pore-blocking toxins bind to the pore domain (PD) of K_V, K_{Ca}, and K_{ir} channels and physically occlude their ion permeation pores, which reduces ionic currents (Fig. 1b, c; Banerjee et al. 2013; MacKinnon and Miller 1988). Gating-modifier toxins bind to the voltage-sensing domain (VSD) of K_V channels and modulate their voltage-dependent conformational changes that allosterically control ion permeation in the PD (Fig. 1b; Swartz 2007). Many gating-modifier toxins inhibit K_V channels in a voltage-dependent manner, shifting the conductance-voltage relationship towards the positive potential (Fig. 1d; Lee et al. 2003; Swartz and MacKinnon 1997a). In this chapter, we mainly describe the inhibition mechanism of K_V channels by pore-blocking and gating-modifier toxins.

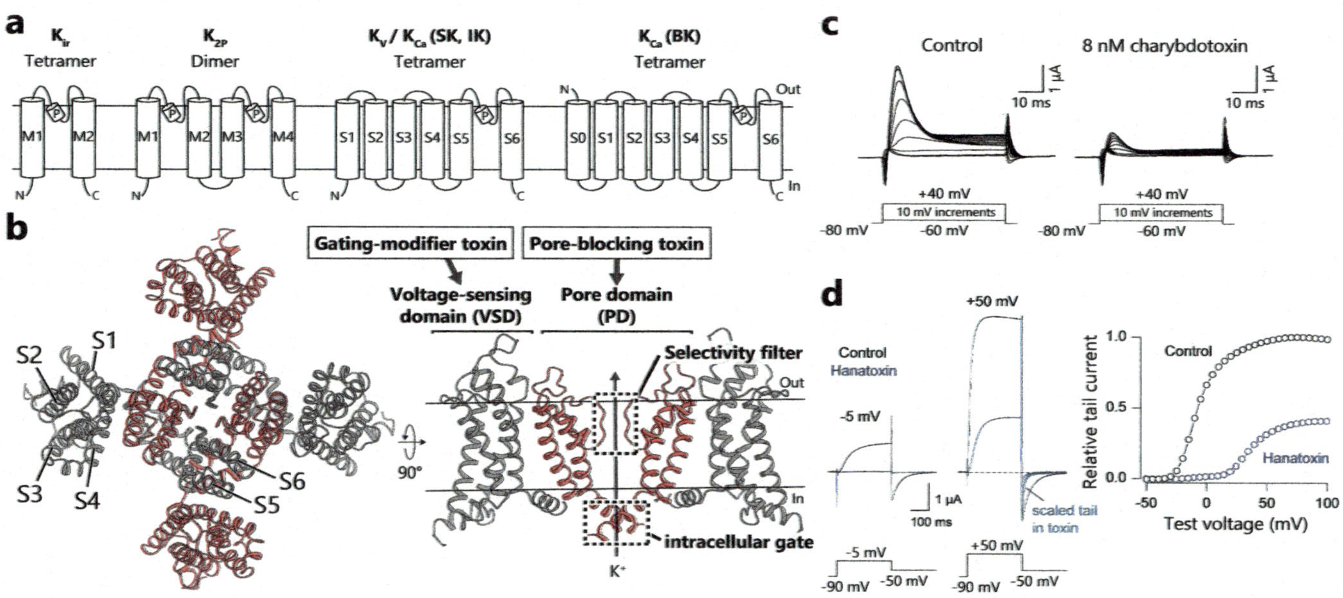

Fig. 1 Structural and functional features of peptide toxins binding to K^+ channels. (**a**) Membrane topologies of K_{ir}, K_{2P}, K_V, and K_{Ca} channels. The transmembrane helices (M1-M4 or S0-S6) and the pore helix (P) are indicated. (**b**) Top view (left) and side view (right) of the structure of the $K_V1.2$-$K_V2.1$ chimera are represented with ribbons (PDB code: 2R9R; Long et al. 2007). The two opposing subunits are shown in grey or red. In the side view, the VSDs and PDs are only shown in the diagonal position. (**c**) Shaker channel (the *Drosophila* ortholog of mammalian K_V1 channels) currents before and after addition of 8 nM charybdotoxin (figure modified from Fig. 1 in MacKinnon et al. 1988 with permission). (**d**) $K_V2.1$ currents elicited by a depolarising pulse of −5 and +50 mV before and after addition of 5 μM hanatoxin 1 (left). Conductance-voltage relationship of $K_V2.1$ in the presence or absence of 5 μM hanatoxin 1 (right) (figure modified from Fig. 2 in Lee et al. 2003 with permission)

Table 1 Pore-blocking toxins targeting K^+ channels[a]

Name	Species	Length	UniProt code	Target and IC_{50}, K_d, or K_i
α-KTx 1.1 (charybdotoxin: CTX)	*Leiurus quinquestriatus*	37	P13487	$K_V1.2$: $K_d = 14$ nM (Grissmer et al. 1994) $K_V1.3$: $K_d = 2.6$ nM (Grissmer et al. 1994) $K_V1.6$: $K_i = 22$ nM (Garcia et al. 1994) $K_{Ca}1.1$: $K_i = 3.5$ nM (Takacs et al. 2009) $K_{Ca}3.1$: $K_d = 5$ nM (Rauer et al. 2000)
α-KTx 1.2 (charybdotoxin-Lq2: Lq2)	*Leiurus quinquestriatus*	37	P45628	$K_{ir}1.1$: $K_i = 0.41$ μM (Lu and MacKinnon 1997)
α-KTx 1.3 (iberiotoxin: IbTx)	*Hottentotta tamulus*	37	P24663	$K_{Ca}1.1$: $K_d = 1.07$ nM (Candia et al. 1992)
α-KTx 2.1 (noxiustoxin: NTX)	*Centruroides noxius*	39	P08815	$K_V1.2$: $K_d = 2$ nM (Grissmer et al. 1994) $K_V1.3$: $K_d = 1$ nM (Grissmer et al. 1994)
α-KTx 2.2 (margatoxin: MgTX)	*Centruroides margaritatus*	39	P40755	$K_V1.2$: $K_d = 6.4$ pM (Bartok et al. 2014) $K_V1.3$: $K_d = 11.7$ pM (Bartok et al. 2014)
α-KTx 3.1 (kaliotoxin-1: KTX1)	*Androctonus mauritanicus*	38	P24662	$K_V1.1$: $K_d = 41$ nM (Grissmer et al. 1994) $K_V1.2$: $K_i = 20$ nM (Takacs et al. 2009) $K_V1.3$: $K_d = 0.65$ nM (Grissmer et al. 1994) $K_{Ca}1.1$: $K_d = 20$ nM (Crest et al. 1992)
α-KTx 3.2 (agitoxin-2: AgTx2)	*Leiurus quinquestriatus*	38	P46111	Shaker: $K_i = 0.64$ nM (Garcia et al. 1994) $K_V1.1$: $K_i = 0.044$ nM (Garcia et al. 1994) $K_V1.2$: $K_i = 3.4$ nM (Takacs et al. 2009) $K_V1.3$: $K_i = 0.004$ nM (Garcia et al. 1994) $K_V1.6$: $K_i = 37$ pM (Garcia et al. 1994)
BgK	*Bunodosoma granuliferum*	37	P29186	$K_V1.1$: $K_d = 6$ nM (Cotton et al. 1997) $K_V1.2$: $K_d = 15$ nM (Cotton et al. 1997) $K_V1.3$: $K_d = 39$ nM (Rauer et al. 1999)

(continued)

Table 1 (continued)

Name	Species	Length	UniProt code	Target and IC_{50}, K_d, or K_i
				$K_V1.6$ (Cotton et al. 1997; Racapé et al. 2002) $K_{Ca}3.1$: $K_d = 172$ nM (Rauer et al. 1999)
ShK	*Stichodactyla helianthus*	35	P29187	Shaker: $K_i = 2$ nM (Zhao et al. 2015) $K_V1.1$: $IC_{50} = 16$ pM (Kalman et al. 1998) $K_V1.2$: $IC_{50} = 9{,}000$ pM (Kalman et al. 1998) $K_V1.3$: $IC_{50} = 11$ pM (Kalman et al. 1998) $K_V1.6$: $IC_{50} = 165$ pM (Kalman et al. 1998) $K_V3.2$: $IC_{50} = 6$ nM (Yan et al. 2005) $K_{Ca}3.1$: $K_d = 30$ nM (Rauer et al. 1999)
HmK	*Heteractis magnifica*	35	O16846	Shaker: $K_i = 1.0$ nM (Zhao et al. 2015) $K_V1.2$: $K_i = 2.5$ nM (Zhao et al. 2015) $K_V1.3$: $K_i = 3.1$ nM (Zhao et al. 2015)
κO-conotoxin PVIIA (PVIIA)	*Conus purpurascens*	27	P56633	Shaker: $IC_{50} = 60$ nM/ 70 nM (Terlau et al. 1996)
κM-conotoxin RIIIJ (RIIIJ)	*Conus radiatus*	25	P0CG45	$K_V1.1$: $IC_{50} = 2.8$ μM (Cordeiro et al. 2019) $K_V1.2$: $IC_{50} = 0.32$ μM (Cordeiro et al. 2019) $K_V1.3$: $IC_{50} = 1.96$ μM (Cordeiro et al. 2019) $K_V1.6$: $IC_{50} = 4.4$ μM (Cordeiro et al. 2019) $K_V1.7$: $IC_{50} = 7.41$ μM (Cordeiro et al. 2019)
Tertiapin-Q	*Apis mellifera*	21	P56587 A0A087ZPM1	$K_{Ca}1.1$: $IC_{50} = 5.8$ nM (Kanjhan et al. 2005) $K_{ir}1.1$: $K_i = 2$ nM (Jin and Lu 1998) $K_{ir}3.1$/Kir3.4: $K_i = 8$ nM (Jin and Lu 1998)
SsTx	*Scolopendra subspinipes mutilans*	53	A0A2L0ART2	$K_V1.3$: $IC_{50} = 5.26$ μM (Du et al. 2019) $K_V7.1$: $IC_{50} = 2.8$ μM (Luo et al. 2018)

(continued)

Table 1 (continued)

Name	Species	Length	UniProt code	Target and IC_{50}, K_d, or K_i
				$K_V7.2$: $IC_{50} = 2.7$ μM (Luo et al. 2018) $K_V7.4$: $IC_{50} = 2.5$ μM (Luo et al. 2018) $K_V7.5$: $IC_{50} = 2.7$ μM (Luo et al. 2018) $K_{ir}6.2$: $K_d = 180$ nM (Ramu and Lu 2019) Shaker, Shal (Yang et al. 2020)

[a]More detailed and comprehensive information is compiled in the following databases: Venom zone, https://venomzone.expasy.org/, Kalium, https://kaliumdb.org/ (Tabakmakher et al. 2019); ConoServer, http://www.conoserver.org/ (Kaas et al. 2008)

3 Pore-Blocking Toxin

3.1 Overview

Pore-blocking toxins bind to the extracellular side of the PD of K^+ channels, physically occluding a K^+-permeation pore (Fig. 2a; Banerjee et al. 2013). These toxins include the α-KTx toxins from scorpions, ShK-motif toxins from sea anemones, conotoxins from cone snails, etc. (Fig. 2b, c). Although the sequences of these toxins are highly diverse, these toxins possess a "functional dyad", a pair of a pore-plugging basic residue (usually lysine) and a hydrophobic residue (usually tyrosine or phenylalanine) within close proximity, which is crucial for the inhibition of K^+ channels (Fig. 2b, c; Dauplais et al. 1997; Mouhat et al. 2005). The pore-plugging mechanism using the functional dyad is common among many peptide toxins derived from various organisms (Mouhat et al. 2008), as observed in the following structures of toxin-channel complexes: charybdotoxin (CTX)-$K_V1.2$ (Banerjee et al. 2013); agitoxin-2 (AgTx2)-KcsA (Takeuchi et al. 2003); kaliotoxin (KTX1)-$K_V1.3$ (Lange et al. 2006); BgK-$K_V1.1$ (Gilquin et al. 2005); ShK- and HmK-$K_V1.3$ (Zhao et al. 2020); PVIIA-Shaker (Huang et al. 2005); RIIIJ-K_V1 (Cordeiro et al. 2019); and SsTx-Shal (Yang et al. 2020). In contrast, there are some peptide toxins that bind to the PD of K^+ channels, inhibiting it allosterically rather than in pore-plugging mechanism (Frénal et al. 2004; Imredy and MacKinnon 2000; Karbat et al. 2019; Lamy et al. 2010; Xu et al. 2003; Zhang et al. 2003). Here, we mainly describe the interaction mode of pore-blocking toxins and K^+ channels, focusing on the scorpion α-KTx toxin, CTX (Fig. 2a; Banerjee et al. 2013) as an example.

Table 2 Gating-modifier toxins targeting K^+ channels[a]

Name	Species	Length	UniProt code	Target and IC_{50} or K_d
Hanatoxin 1 (HaTx1)	*Grammostola spatulata*	35	P56852	$K_V2.1$: $K_d = 19$ nM/102 nM (Swartz and MacKinnon 1997a) $K_V4.2$ (Swartz and MacKinnon 1995) Shaker: activator (Milescu et al. 2013)
SGTx1	*Scodra griseipes*	34	P56855	$K_V2.1$: $K_d = 2.7$ μM (Lee et al. 2004; Wang et al. 2004)
Heteropodatoxin 2 (HpTx2)	*Heteropoda venatoria*	30	P58426	$K_V4.1$: $K_d = 7.1$ μM (DeSimone et al. 2011) $K_V4.2$: $IC_{50} = 100$ nM (Escoubas et al. 2002) $K_V4.3$: $K_d = 2.3$ μM (DeSimone et al. 2011)
Phrixotoxin 1 (PaTx1)	*Phrixotrichus auratus*	29	P61230	$K_V2.1$: (Diochot et al. 1999) $K_V2.2$ (Diochot et al. 1999) $K_V4.1$ (Escoubas et al. 2002) $K_V4.2$: $IC_{50} = 5$ nM (Escoubas et al. 2002) $K_V4.3$: $IC_{50} = 28$ nM (Escoubas et al. 2002)
Stromatoxin 1 (ScTx1)	*Stromatopelma calceata*	34	P60991	$K_V2.1$: $IC_{50} = 12.7$ nM (Escoubas et al. 2002) $K_V2.2$: $IC_{50} = 21.4$ nM (Escoubas et al. 2002) $K_V4.2$: $IC_{50} = 1.2$ nM (Escoubas et al. 2002) $K_V2.1$/Kv9.3: $IC_{50} = 7.2$ nM (Escoubas et al. 2002)
Heteroscodratoxin 1 (HmTx1)	*Heteroscodra maculata*	35	P60992	$K_V2.1$ (Escoubas et al. 2002) $K_V2.2$ (Escoubas et al. 2002) $K_V4.1$: $IC_{50} = 280$ nM (Escoubas et al. 2002) $K_V4.2$ (Escoubas et al. 2002) $K_V4.3$ (Escoubas et al. 2002)
Voltage-sensor toxin 1 (VSTx1)	*Grammostola spatulata*	34	P60980	K_VAP (Ruta et al. 2003; Ruta and MacKinnon 2004) hERG: $IC_{50} = 7.4$ μM (Redaelli et al. 2010)
Theraphosa leblondi toxin 1 (TLTx1)	*Theraphosa leblondi*	35	P83745	$K_V4.2$: $IC_{50} = 193$ nM (Ebbinghaus et al. 2004)
Guangixitoxin-1E (GxTx-1E)	*Plesiophrictus guangxiensis*	36	P84835	$K_V2.1$: $K_d = 2.6$ nM/15.1 nM (Herrington et al. 2006) $K_V2.2$: $K_d = 2.6$ nM (Herrington et al. 2006)

(continued)

Table 2 (continued)

Name	Species	Length	UniProt code	Target and IC_{50} or K_d
				K_V4.3: IC_{50} = 24 nM (Herrington et al. 2006)
SNX-482	*Hysterocrates gigas*	41	P56854	K_V4.2 (Kimm and Bean 2014) K_V4.3 (Kimm and Bean 2014)
Jingzhaotoxin-I (JZTX-I)	*Chilobrachys jingzhao*	33	P83974	K_V2.1: K_d = 12.2 μM (Yuan et al. 2007b) Kv4.1 (Yuan et al. 2007b)
Jingzhaotoxin-III (JZTX-III)	*Chilobrachys jingzhao*	36	P62520	K_V2.1: K_d = 0.43 μM (Yuan et al. 2007b)
Jingzhaotoxin-V (JZTX-V)	*Chilobrachys jingzhao*	29	Q2PAY4	K_V2.1: K_d = 14.3 μM (Yuan et al. 2007b) K_V4.2: IC_{50} = 13 nM (Zhang et al. 2019)
Jingzhaotoxin-XI (JZTX-XI)	*Chilobrachys jingzhao*	34	P0C247	K_V2.1: IC_{50} = 0.39 μM (Tao et al. 2016)
Jingzhaotoxin-XII (JZTX-XII)	*Chilobrachys jingzhao*	29	P0C5X7	K_V4.1: IC_{50} = 0.363 μM (Yuan et al. 2007a)
Jingzhaotoxin-35 (JZTX-35)	*Chilobrachys jingzhao*	36	B1P1F9	K_V2.1: IC_{50} = 3.62 μM (Wei et al. 2014)
Blood depressing substance-I (BDS-I)	*Anemonia sulcata*	43	P11494	K_V3.1: IC_{50} = 220 nM (Yeung et al. 2005) K_V3.2 (Yeung et al. 2005) K_V3.4: IC_{50} = 47 nM (Diochot et al. 1998)
Blood depressing substance-II (BDS-II)	*Anemonia sulcata*	43	P59084	K_V3.1: IC_{50} = 750 nM (Yeung et al. 2005) K_V3.2 (Yeung et al. 2005) K_V3.4: IC_{50} = 56 nM (Diochot et al. 1998)
APETx1	*Anthopleura elegantissima*	42	P61541	hERG: K_d = 87 nM/141.1 nM/16.3 nM (Diochot et al. 2003; Zhang et al. 2007)
APETx2	*Anthopleura elegantissima*	42	P61542	hERG: IC_{50} = 1.21 μM (Jensen et al. 2014)
APETx4	*Anthopleura elegantissima*	42	C0HL40	K_V10.1: IC_{50} = 1.1 μM (Moreels et al. 2017)

[a]More detailed and comprehensive information is compiled in the following databases: Venom zone, https://venomzone.expasy.org/, Kalium, https://kaliumdb.org/ (Tabakmakher et al. 2019)

3.2 K^+ Channel Inhibition by CTX in the Pore-Plugging Mechanism

Charybdotoxin (CTX: α-KTx 1.1) is a 37-residue peptide toxin isolated from *Leiurus quinquestriatus*, and one of the most well-characterised scorpion α-KTx-family toxins (Fig. 2b; Miller 1995; Miller et al. 1985). CTX comprises an

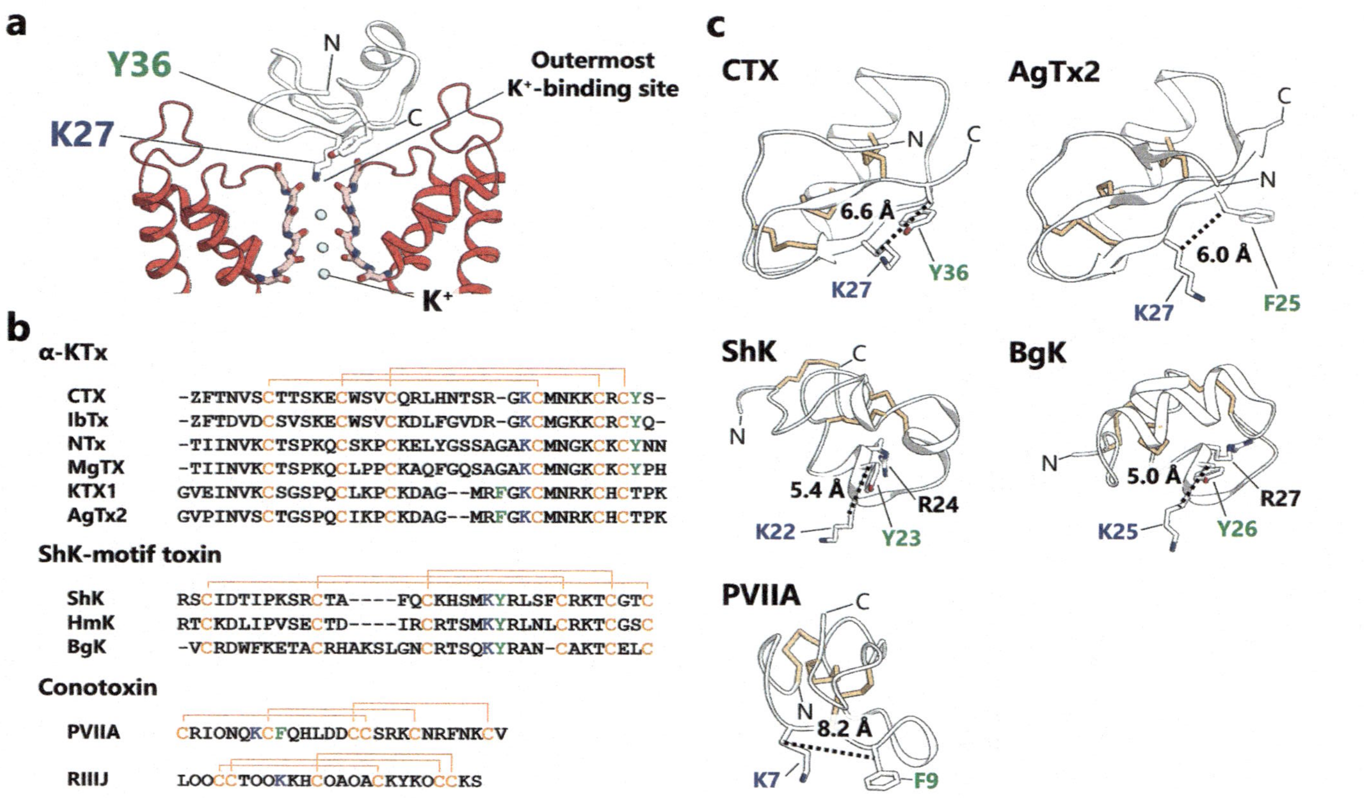

Fig. 2 Pore-blocking toxins. (**a**) The structure of the extracellular region of the PD of the $K_V1.2$-$K_V2.1$ chimera with charybdotoxin (PDB code: 4JTA; Banerjee et al. 2013). CTX is represented as a white ribbon, and its functional dyad residues are represented as sticks. The PDs of the two opposing subunits are

α-helix and a two-stranded β-sheet, which are bridged by three disulphide bonds (cysteine-stabilised α/β motif: CS-α/β motif) similar to other α-KTx-family toxins (Fig. 2c; Bontems et al. 1992; Rodríguez de la Vega and Possani 2004). CTX inhibits Shaker-type K_V (K_V1), $K_{Ca}1.1$ (BK, Slo1), and $K_{Ca}3.1$ (IK, SK4) channels at low nanomolar concentrations (Table 1; Garcia et al. 1994; Grissmer et al. 1994; Rauer et al. 2000; Takacs et al. 2009).

An X-ray crystallographic study provided an obscure electron density map of CTX bound to the $K_V1.2$-$K_V2.1$ chimera (K_V chimera) because the asymmetric CTX molecule binds to the fourfold symmetric K_V chimeric tetramer (Banerjee et al. 2013). This technical challenge was overcome by using three CTX derivatives with distinct sites labelled with heavy atoms, and a structural model obtained in this way demonstrates good complementarity between the shapes of the toxin and the channel (Fig. 3a; Banerjee et al. 2013). This model effectively explains the observations in the functional analysis, as described below.

The crystal structure shows that CTX binds to the PD of the K_V chimera as in a "lock and key" model (Fig. 2a; Banerjee et al. 2013). The positively charged ε-amino group of CTX K27, which is the basic residue of the functional dyad, was positioned near the outermost ion-binding site in the selectivity filter of the K_V chimera (Fig. 2a; Banerjee et al. 2013). The ε-amino group of CTX K27 causes electrostatic repulsion with permeant ions. Thus, CTX-binding made the outermost ion-binding site in the selectivity filter empty (Banerjee et al. 2013), whereas K^+ density was observed in the crystal structure of the K_V chimera alone (Long et al. 2007). In the electrophysiological analysis, the CTX dissociation from the Shaker channel is voltage-dependent because the permeant ions from the intracellular side destabilise the toxin-channel complex (Goldstein and Miller 1993). This effect is called the "trans-enhanced dissociation effect" (MacKinnon and Miller 1988) and was eliminated by the charge-neutralisation mutations of CTX K27 (Goldstein and Miller 1993), suggesting that K27 is the crucial pore-plugging residue in K^+ channel inhibition.

In addition to K27, the CTX active residues contributing to the inhibition of the Shaker channel were identified by mutational analysis (Fig. 3b; Goldstein and Miller 1992; Goldstein et al. 1994; Naranjo and Miller 1996; Stocker and Miller 1994) and were found to be located at the toxin-channel interface of the crystal structure

---

Fig. 2 (continued) shown. The backbones in the signature sequence (TTVGYG) of the selectivity filter are represented as sticks, and K^+ ions are represented as cyan balls. (**b**) The sequence alignments of pore-blocking toxins: α-KTx-family toxins from scorpions, ShK-motif toxins from sea anemones, and conotoxins from cone snails. The one letter code of unnatural amino acids are as follows: Z, pyroglutamic acid; O, hydroxyproline. The pairs of cysteine residues forming disulphide bonds are indicated with orange lines. The basic and hydrophobic residues of the functional dyad are coloured with blue and green, respectively. (**c**) The structures of pore-blocking toxins: charybdotoxin, CTX (PDB code: 2CRD; Bontems et al. 1992); agitoxin 2, AgTx2 (PDB code: 1AGT; Krezel et al. 1995); ShK (PDB code: 1ROO; Tudor et al. 1996); BgK (PDB code: 1BGK; Dauplais et al. 1997); and κM-conotoxin PVIIA (PDB code: 1AV3; Scanlon et al. 1997). The distances between Cβ atoms of the functional dyad residues are shown with dotted lines

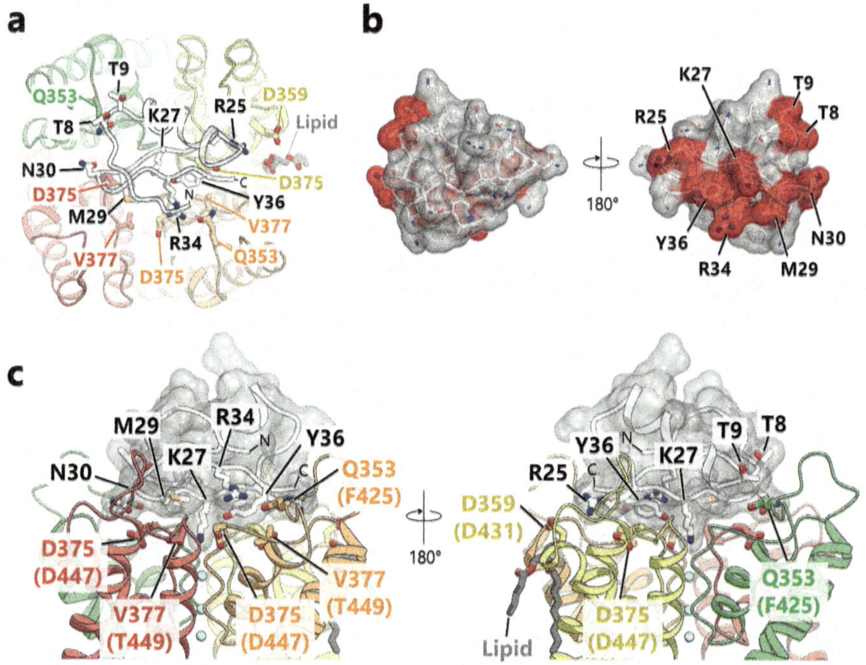

Fig. 3 Residues forming the interaction between CTX and K_V chimera (PDB code: 4JTA; Banerjee et al. 2013). (**a**) Extracellular view of the PD of the K_V chimera in complex with CTX. CTX is represented by a white semi-transparent ribbon. The K_V chimera is coloured according to each subunit. The interacting residues of both molecules, deduced by functional analysis, are shown as sticks representing the side chains. (**b**) The molecular surface of CTX is represented as the semi-transparent surface with the sticks. The orientation of CTX on the left panel is the same as that in panel (**a**). The CTX residues involved in inhibiting the Shaker channel are shown as a red surface. (**c**) Side view of the complex. The residue numbers of the K_V chimera are indicated and the corresponding residues of the Shaker channel are given in parenthesis

(Fig. 3a; Banerjee et al. 2013). CTX Y36, the hydrophobic residue of the functional dyad, contacts both D375 and V377 of the K_V chimera (Fig. 3a, c). A conservative Y36F mutation was relatively tolerated, whereas the other mutations in this position remarkably attenuated the inhibition, which indicates that an aromatic ring in this position of CTX is crucial (Goldstein et al. 1994). CTX M29 is also located in close proximity to D375 and V377 in another subunit of the K_V chimera (Fig. 3c, left). The M29Q, M29E, and M29G mutations drastically reduced inhibition (Goldstein et al. 1994). Moreover, mutant cycle analysis showed that CTX M29 is energetically coupled with T449 in the Shaker channel corresponding to V377 in the K_V chimera (Naranjo and Miller 1996).

CTX R25 is positioned possibly to form electrostatic interactions with D359 of the K_V chimera and the head group of the lipid observed in the crystal structure (Fig. 3c, right), whereas CTX R34 is located close enough to form electrostatic interactions with D375 and a hydrogen bond with Q353 (Fig. 3; left). Charge-reversal mutations, R25D and R34D/E of CTX, decreased the inhibition of the Shaker channel, and a conservative R34K mutation did not affect the inhibition (Goldstein et al. 1994), which indicates that these basic residues are important for inhibition.

CTX N30 is located close to D375 of the K_V chimera (Fig. 3c, left). The N30D/E mutations remarkably weakened the inhibition of the Shaker channel (Goldstein et al. 1994), which suggests that CTX N30 interacts with D375 of the K_V chimera by forming a hydrogen bond. In close proximity, the backbones of T8 and T9 of CTX contact D353 of the Kv chimera (Fig. 3c, right). A residue corresponding to D353 of $K_V1.2$ is F425 in the Shaker channel, whose substitution with glycine increased the affinity for CTX by 1,900-fold (Goldstein and Miller 1992). Moreover, complementary mutational analysis suggested that F425 is physically close to T8 and T9 in CTX (Goldstein et al. 1994).

As mentioned above, the crystal structure of the CTX-K_V complex suggests a "lock and key" model, in which the toxin binds to the channel without spending binding free energy on the conformational change (Banerjee et al. 2013). In contrast, it has also been proposed that α-KTx family toxins, including CTX, inhibit K^+ channels with dynamic interactions. Molecular dynamics simulation combined with electrophysiological analysis shows the wobbling of CTX on the Shaker channels, which suggests transitional states in dissociation pathways (Moldenhauer et al. 2019). Another α-KTx family toxin, kaliotoxin-1 (α-KTx 3.1, KTX-1) has been proposed to bind to the $K_V1.3$ chimera channel with both molecules undergoing conformational changes, based on the chemical shift changes in solid-state nuclear magnetic resonance spectroscopy (Lange et al. 2006). Moreover, high-speed atomic force microscopy revealed that the binding affinity of agitoxin-2 (α-KTx 3.2, AgTx2) to the prokaryotic K^+ channel, KcsA, increases during persistent binding, which supports the induced-fit model (Sumino et al. 2019). These studies propose that the inhibition of the K^+ channel by pore-blocking toxins is attained by dynamic interactions, not by a simple "lock and key" interaction.

3.3 Pore-Blocking Toxin with a Distinct Structural Fold

Pore-blocking toxins targeting K_V channels have been discovered in organisms other than scorpions (Table 1). They form distinct structural folds but possess a common functional dyad (Fig. 2b, c; Dauplais et al. 1997; Mouhat et al. 2005). Sea anemone ShK-motif toxins, BgK, ShK, and HmK, are characterised by homologous sequences and structures and are considered to inhibit K_V1 channels by a pore-plugging mechanism (Fig. 2b, c; Dauplais et al. 1997). Three consecutive residues (Lys-Tyr-Arg) are conserved, and the lysine and tyrosine residues (K22 and Y23 in ShK) mainly contribute to the inhibition of K_V1 channels (Dauplais et al. 1997; Finol-Urdaneta et al. 2020; Pennington et al. 1996). Recently, it was proposed that ShK inhibits $K_V1.3$ by an arginine-dependent blocking mechanism using R24, whereas HmK inhibits $K_V1.3$ by a conventional lysine-dependent blocking mechanism using K22, suggesting that the binding of the ShK-motif toxins is multimodal

(Zhao et al. 2020). A cone snail toxin, κO-conotoxin PVIIA, contains the inhibitor-cystine-knot (ICK) fold (Fig. 2b, c) and inhibits the Shaker channel by plugging K7 into the pore (Huang et al. 2005). Another cone snail toxin, κM-conotoxin RIIIJ, which shows no sequence homology with PVIIA (Fig. 2b), is proposed to bind to the asymmetric K_V1 heterotetramer, inserting K9 into the pore (Cordeiro et al. 2019). Recently, several centipede toxins targeting K^+ channels have been identified (Chu et al. 2020), and one of them, SsTx, is likely to utilise different basic residues to inhibit $K_V1.3$, $K_V7.4$, and Shal K^+ channel (Du et al. 2019; Luo et al. 2018; Yang et al. 2020).

3.4 Peptide Toxins That Bind to the PD with Indirect Inhibition Mechanisms

Despite binding to the PD, several peptide toxins inhibit K^+ channels through indirect inhibition mechanisms rather than by pore-plugging mechanisms. Scorpion γ-KTx family toxins, including ErgTx1 (γ-KTx 1.1) and BeKm-1 (γ-KTx 2.1), inhibit *ether-ά-go-go*-related gene (ERG) K^+ channels (Gurrola et al. 1999; Korolkova et al. 2001). These toxins share a CS-αβ motif with α-KTx family toxins but bind to the linker connecting S5 and the pore helix (turret; Fig. 1b) of ERG channels rather than the selectivity filter, indirectly modulating gating kinetics (Hill et al. 2007; Tseng et al. 2007; Xu et al. 2003). δ-Dendrotoxin (δ-DTX), a snake Kunitz-type-fold toxin, is also proposed to bind to the turret region rather than the centre of the pore in $K_V1.1$, based on the findings of mutant cycle analysis (Imredy and MacKinnon 2000). Cone snail conkunitzin-S1 (Conk-S1) possesses a Kunitz-type fold and inhibits the Shaker channel by binding to the turret regions and causing asymmetric conformational changes in the selectivity filter (Karbat et al. 2019). In the inactivation-deficient mutants of the Shaker channel, Conk-S1 retains its binding capacity to the channels but shows no inhibition, which suggests that Conk-S1 is a "pore-modulating toxin" that collapses the pore by exploiting a slow inactivation mechanism (Karbat et al. 2019). Apamin, an 18-residue small peptide, is a component of bee venom inhibiting $K_{Ca}2.1$-2.3 (SK1-3) channels with high affinity (Brown et al. 2020). It is suggested that apamin is an allosteric inhibitor, which acts by binding to an extracellular region distinct from that of tetraethylammonium, which is a small molecule that binds to the centre of the pore (Lamy et al. 2010). Apamin interacts with not only the PD but also the S3-S4 loop of the VSD (Weatherall et al. 2011).

4 Gating-Modifier Toxins

4.1 Overview

Gating-modifier toxins bind to the VSD of voltage-dependent ion channels to modify the voltage dependence of the channels by trapping a specific functional state; this mechanism is called voltage-sensor-trapping mechanism (Catterall et al. 2007; Cestèle et al. 1998; Swartz 2007). Most gating-modifier toxins targeting K_V channels were isolated from spiders (mostly tarantula) or sea anemones (Table 2; Swartz 2007). Spider gating-modifier toxins include hanatoxin 1 (HaTx1) of *Grammostola spatulata* (Swartz and MacKinnon 1995), SGTx1 of *Scodra griseipes* (Marvin et al. 1999), and guangxitoxin-1E (GxTx-1E) of *Plesiophrictus guangxiensis* (Herrington et al. 2006). These toxins form ICK folds containing an anti-parallel β-sheet with three disulphide bonds, of which a bond connecting positions III and VI is passed through the others (Fig. 4a; Lee et al. 2004, 2010; Takahashi et al. 2000). Sea anemone gating-modifier toxins, such as blood depressing substance-I (BDS-I) of *Anthopleura sulcata* (Diochot et al. 1998; Yeung et al. 2005) and APETx1 of *Anthopleura elegantissima* (Diochot et al. 2003; Zhang et al. 2007), form a β-defensin fold, in which the pattern of disulphide bonds is distinct from that in the spider ICK-fold toxins, as shown in Fig. 4b (Chagot et al. 2005; Driscoll et al. 1989a, b). Gating-modifier toxins possess a cluster of hydrophobic residues called hydrophobic patches on their molecular surfaces (Fig. 4a, b). The hydrophobic patch plays an important role in the inhibition of K_V channels by interacting with K_V channels directly and partitioning into the lipid membrane (Lee and MacKinnon 2004; Milescu et al. 2007, 2009).

The inhibition mechanism of K_V channels by gating-modifier toxins has been well characterised in spider toxins (Swartz 2007). HaTx1, the first isolated spider toxin targeting K^+ channels, stabilises the VSDs of K_V channels in the resting state, which is an impermeable state of the channel under resting membrane potential (Lee et al. 2003; Phillips et al. 2005; Swartz and MacKinnon 1997a). Another spider toxin, VSTx1, is also a gating-modifier toxin that binds to the depolarised conformation of the VSD of K_V channels, detaining the inactivated state (Schmidt et al. 2009). Here, we describe the inhibition modes of K_V channels by gating-modifier toxins, which were mainly revealed by functional analysis.

4.2 The S3-S4 Region of VSD Is the Binding Site of Gating-Modifier Toxins

HaTx1 inhibits $K_V2.1$ at nanomolar concentrations (Table 2; Swartz and MacKinnon 1995, 1997a). Electrophysiological analysis demonstrated that HaTx1 binding is not competitive with the pore-blocking toxin AgTx2, suggesting that its binding site is on the outside of the pore (Swartz and MacKinnon 1997b). Moreover, multiple HaTx1 molecules can bind to a single K_V channel (Swartz and MacKinnon 1997a). Alanine-scanning mutational analysis of $K_V2.1$ revealed that three S3b residues,

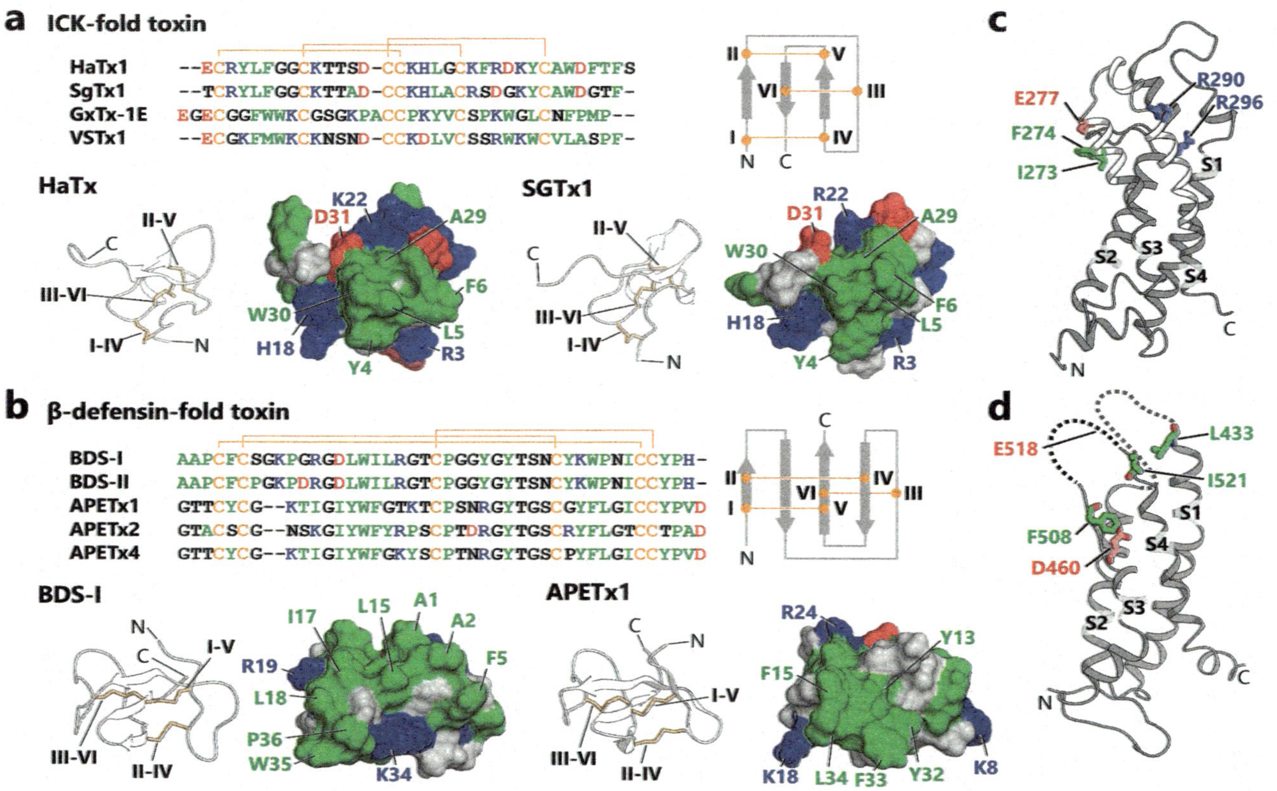

Fig. 4 Gating-modifier toxins. Sequences and structures are coloured according to the residue: acidic residues (Asp and Glu), red; basic residues (Lys, Arg, and His), blue; hydrophobic residues (Ala, Val, Leu, Ile, Pro, Met, Phe, Tyr, and Trp), green; others (Gly, Cys, Ser, Thr, Asn, and Gln), grey. (**a**) Sequence

I273, F274, and E277, contribute to the inhibition by HaTx1 and, to a lesser extent, two S4 residues, R290 and R296, suggesting that HaTx1 binds to the VSD (Fig. 4c; Li-Smerin and Swartz 2000; Swartz and MacKinnon 1997b). It has also been reported that the S3b-S4 region of K_V channels plays a critical role in the binding of other spider gating-modifier toxins (Alabi et al. 2007; DeSimone et al. 2009; Li-Smerin and Swartz 1998; Milescu et al. 2009; Tao et al. 2013a, b; 2016; Zhang et al. 2019).

Sea anemone gating-modifier toxins are also considered to interact with the S3b-S4 region of VSDs in K_V channels. Two related sea anemone toxins, BDS-I and BDS-II, inhibit K_V3 channels (Diochot et al. 1998), and several mutations in the S3b-S4 region of $K_V3.2$ reduce the sensitivity of these toxins (Yeung et al. 2005). Other sea anemone toxins, APETx1 and APETx2, inhibit hERG (Diochot et al. 2003; Jensen et al. 2014). Three hERG S3-S4 mutations, F508A, E518C, and I521A, drastically decreased the effect of APETx1, whereas the two S1-S2 mutations, L433A and D460A, also modestly decreased it (Fig. 4d; Matsumura et al. 2021; Zhang et al. 2007). These results suggest that sea anemone gating-modifier toxins mainly bind to the S3-S4 regions of K_V channels similar to that of the spider gating-modifier toxins.

4.3 Hydrophobic Surface Contributing to the Inhibition of Gating-Modifier Toxins

SGTx1 shares 80% sequence identity with HaTx1, and its structure resembles that of HaTx1 (Fig. 4a; Lee et al. 2004; Takahashi et al. 2000). SGTx1 inhibits $K_V2.1$, but its affinity is reported to be 20-fold weaker than that of HaTx1 (Table 2; Lee et al. 2004; Wang et al. 2004). SGTx1 was used to identify an active surface of spider gating-modifier toxins instead of HaTx1, which is difficult to prepare using peptide synthesis or recombinant expression systems because of the multiple conformers generated during the folding process (Lee et al. 2004; Milescu et al. 2007; Wang et al. 2004). Alanine-scanning mutational analysis of SGTx1 revealed that five hydrophobic residues (Y4, L5, F6, A29, and F30) and four charged residues (R3,

Fig. 4 (continued) alignments and molecular structure of ICK-fold toxins from spiders: HaTx1 (PDB code: 1D1H; Takahashi et al. 2000), SGTx1 (PDB code: 1LA4; Lee et al. 2004). (**b**) Sequence alignments and molecular structure of β-defensin-fold toxins from sea anemones: BDS-I (PDB code: 1BDS; Driscoll et al. 1989a, b), APETx1 (PDB code: 1WQK; Chagot et al. 2005). (**c**) Mapping the residues contributing to the inhibition of $K_V2.1$ by HaTx1 on the VSD of the K_V chimera (PDB code: 2R9R; Li-Smerin and Swartz 2000; Long et al. 2007; Swartz and MacKinnon 1997b). The VSD structure of the K_V chimera is represented by a ribbon coloured according to the region of $K_V1.2$ (grey) and $K_V2.1$ (white). (**d**) Mapping the residues contributing to hERG inhibition by APETx1 on the VSD of hERG (PDB code: 5VA2; Matsumura et al. 2021; Wang and MacKinnon 2017; Zhang et al. 2007). The S1-S2 and S3-S4 loops, missing in the cryo-EM structure, are shown with dotted lines

H18, R22, and D31) contribute to $K_V2.1$ inhibition (Fig. 4a; Wang et al. 2004). These active residues are conserved between SGTx1 and HaTx1, except for a conservative substitution from R22 in SGTx1 to K22 in HaTx1. The active residues are localised on the molecular surface of the toxins, and a cluster of hydrophobic residues form a hydrophobic patch (Fig. 4a; Wang et al. 2004). Other spider gating-modifier toxins are also known to possess hydrophobic patches, contributing to their inhibition of K_V channels (Tao et al. 2013a; Zhang et al. 2019).

The amphipathic surface of the spider gating-modifier toxins is important for partitioning into the lipid membrane to inhibit K_V channels (Lee and MacKinnon 2004; Milescu et al. 2007, 2009; Phillips et al. 2005). The molecular surfaces of SGTx1 that interact with $K_V2.1$ and the lipid membrane are overlapped (Milescu et al. 2007; Wang et al. 2004). The HaTx1-binding site on $K_V2.1$ S3b is located at the interface with the lipid membrane (Fig. 4c; Long et al. 2007; Swartz and MacKinnon 1997b). Moreover, the modification in lipid composition affects the inhibition of $K_V2.1$ by gating-modifier toxins, suggesting that the local lipid membrane environment around the S3b-S4 regions is involved in the stability of the toxin-channel complex (Milescu et al. 2007, 2009).

A hydrophobic patch is also observed on the molecular surface of sea anemone gating-modifier toxins (Fig. 4b). Mutational analysis of APETx1 showed that the four hydrophobic residues, F15, Y32, F33, and L34, which are located on the molecular surface, are crucial for hERG inhibition (Matsumura et al. 2021). These residues form a hydrophobic patch of APETx1 (Fig. 4b; Chagot et al. 2005; Matsumura et al. 2021). APETx2 has also been proposed to use hydrophobic residues to inhibit hERG (Jensen et al. 2014). These reports suggest that the hydrophobic patch plays a critical role in inhibition by the gating-modifier toxins of spiders and sea anemones.

4.4 Voltage-Dependent Inhibition of K_V Channel by Gating-Modifier Toxins

Most spider and sea anemone gating-modifier toxins targeting K_V channels strongly inhibit ionic currents at lower depolarised voltages (e.g. 0 mV) rather than higher ones (e.g. +100 mV), shifting the activation voltage towards the positive direction (Fig. 1d; Swartz 2007). HaTx1 and GxTx-1E also inhibit the gating currents of $K_V2.1$ (Lee et al. 2003; Tilley et al. 2019) and BDS-II inhibits that of $K_V3.2$ (Yeung et al. 2005). These results show that the gating-modifier toxins inhibit K_V channels by trapping the VSDs in the resting state, in which the intracellular gate of K_V channels is closed. This resting state trapping mechanism of gating-modifier toxins is corroborated by kinetic analyses that measured the onset and recovery rates of inhibition (Lee et al. 2003; Phillips et al. 2005; Swartz and MacKinnon 1997a; Tilley et al. 2014).

Kinetic analyses of inhibition demonstrate that the binding of HaTx1 and GxTx-1E to $K_V2.1$ becomes weaker when the channel is activated at a depolarised voltage (Phillips et al. 2005; Tilley et al. 2014). However, the kinetics in recovery from the inhibition by HaTx1 are much slower than the channel gating kinetics, suggesting that the channel can be activated with HaTx1 bound to the VSDs (Phillips et al. 2005). This continuous toxin binding during channel activation explains the observation that many gating-modifier toxins accelerate the channel closing (deactivation) kinetics reflected in the attenuation of the tail current (Ebbinghaus et al. 2004; Herrington et al. 2006; Lee et al. 2003, 2004; Li-Smerin and Swartz 1998; Liao et al. 2006; Matsumura et al. 2021; Phillips et al. 2005; Swartz 2007; Swartz and MacKinnon 1997a; Tao et al. 2016; Yuan et al. 2007b).

As mentioned above, GxTx-1E stabilises $K_V2.1$ in the resting state (Herrington et al. 2006; Tilley et al. 2014). Recently, the inhibitory effect of GxTx-1E on $K_V2.1$ was evaluated by a single channel recording and gating current analysis (Tilley et al. 2019). The single channel recording revealed that GxTx-1E reduces macroscopic currents of $K_V2.1$ by lowering its open probability without affecting its unitary conductance. Furthermore, GxTx-1E inhibited the movement of all gating charges in $K_V2.1$, indicating that GxTx-1E detains $K_V2.1$ in the early resting state by stabilising the deepest resting conformation of the VSD. In contrast, HaTx1 shows the fractional inhibition of gating currents of $K_V2.1$, suggesting that HaTx1 inhibits later voltage-dependent movements of the VSD (Lee et al. 2003).

4.5 Gating-Modifier Toxins Binding to the Depolarised Conformation of VSD

As described above, many gating-modifier toxins targeting K_V channels stabilise the resting conformation of VSD (Swartz 2007). However, one of the well-known spider gating-modifier toxins, VSTx1, inhibits K_VAP by binding to the depolarised conformation of VSD to stabilise the inactivated state (Schmidt et al. 2009). To date, two distinct structural models of the VSTx1-VSD complex in the depolarised conformation are proposed based on interaction analyses using solution nuclear magnetic resonance spectroscopy (Lau et al. 2016; Ozawa et al. 2015).

In contrast to the inhibitory effect on $K_V2.1$, HaTx1 acts on the Shaker channel as an activator that shifts the activation voltage towards the negative direction and increases the macroscopic current (Milescu et al. 2013). Although the activating effect on the Shaker channel by HaTx1 was weak, this effect was drastically enhanced by introducing three mutations into the S3b of the Shaker channel (Milescu et al. 2013). HaTx1 not only stabilised the depolarised conformation of the VSD, but also facilitated the final pore-opening transition. Unlike HaTx1, another inhibitor of $K_V2.1$, GxTx-1E, robustly inhibited a Shaker channel mutant with five mutations on S3b, positively shifting its activation voltage. These results suggest that the complementarity of the toxin-channel interface determines the activating or inhibitory effect of gating-modifier toxins (Milescu et al. 2013).

5 Conclusion

To date, though the crystal structure of the K_V channel in complex with CTX has been reported (Banerjee et al. 2013), other structures of K^+ channels in complex with peptide toxins remain elusive. Recently, membrane proteins embedded in nanodiscs and liposomes mimicking physiological lipid membrane environments have been used for structural analysis (Gao et al. 2020; Matthies et al. 2018; Yao et al. 2020), and these methods may be applicable to the analysis of toxin-channel complexes. In the future, further analysis would reveal the structural mechanism of the inhibition of K^+ channels by peptide toxins, which would provide a structural basis for the development of ligands and drugs with novel mechanisms of action for K^+ channels.

Acknowledgements All molecular graphics images were prepared using CueMol (http://www.cuemol.org). The authors are grateful to the Japan Society for the Promotion of Science KAKENHI Grant Numbers JP17H03978 and JP19H04973 (to M.O.), a grant from The Vehicle Racing Commemorative Foundation (to M.O.), and a grant from Takeda Science Foundation (to M.Y. and M.O.).

References

Alabi AA, Bahamonde MI, Jung HJ, Kim JI, Swartz KJ (2007) Portability of paddle motif function and pharmacology in voltage sensors. Nature 450:370–375

Banerjee A, Lee A, Campbell E, Mackinnon R (2013) Structure of a pore-blocking toxin in complex with a eukaryotic voltage-dependent K^+ channel. Elife 2:e00594

Bartok A, Toth A, Somodi S, Szanto TG, Hajdu P, Panyi G, Varga Z (2014) Margatoxin is a non-selective inhibitor of human Kv1.3 K^+ channels. Toxicon 87:6–16

Bontems F, Gilquin B, Roumestand C, Ménez A, Toma F (1992) Analysis of side-chain organization on a refined model of charybdotoxin: structural and functional implications. Biochemistry 31:7756–7764

Brown BM, Shim H, Christophersen P, Wulff H (2020) Pharmacology of small- and intermediate-conductance calcium-activated potassium channels. Annu Rev Pharmacol Toxicol 60:219–240

Candia S, Garcia ML, Latorre R (1992) Mode of action of iberiotoxin, a potent blocker of the large conductance Ca^{2+}-activated K^+ channel. Biophys J 63:583–590

Catterall WA, Cestèle S, Yarov-Yarovoy V, Yu FH, Konoki K, Scheuer T (2007) Voltage-gated ion channels and gating modifier toxins. Toxicon 49:124–141

Cestèle S, Qu Y, Rogers JC, Rochat H, Scheuer T, Catterall WA (1998) Voltage sensor-trapping: enhanced activation of sodium channels by β-scorpion toxin bound to the S3-S4 loop in domain II. Neuron 21:919–931

Chagot B, Diochot S, Pimentel C, Lazdunski M, Darbon H (2005) Solution structure of APETx1 from the sea anemone *Anthopleura elegantissima*: a new fold for an HERG toxin. Proteins 59:380–386

Chu Y, Qiu P, Yu R (2020) Centipede venom peptides acting on ion channels. Toxins (Basel) 12

Cordeiro S, Finol-Urdaneta RK, Köpfer D, Markushina A, Song J, French RJ, Kopec W, de Groot BL, Giacobassi MJ, Leavitt LS, Raghuraman S, Teichert RW, Olivera BM, Terlau H (2019) Conotoxin κM-RIIIJ, a tool targeting asymmetric heteromeric K_V1 channels. Proc Natl Acad Sci U S A 116:1059–1064

Cotton J, Crest M, Bouet F, Alessandri N, Gola M, Forest E, Karlsson E, Castañeda O, Harvey AL, Vita C, Ménez A (1997) A potassium-channel toxin from the sea anemone *Bunodosoma granulifera*, an inhibitor for Kv1 channels. Revision of the amino acid sequence, disulfide-bridge assignment, chemical synthesis, and biological activity. Eur J Biochem 244:192–202

Crest M, Jacquet G, Gola M, Zerrouk H, Benslimane A, Rochat H, Mansuelle P, Martin-Eauclaire MF (1992) Kaliotoxin, a novel peptidyl inhibitor of neuronal BK-type Ca^{2+}-activated K^+ channels characterized from *Androctonus mauretanicus mauretanicus* venom. J Biol Chem 267:1640–1647

Dauplais M, Lecoq A, Song J, Cotton J, Jamin N, Gilquin B, Roumestand C, Vita C, de Medeiros CL, Rowan EG, Harvey AL, Ménez A (1997) On the convergent evolution of animal toxins. Conservation of a diad of functional residues in potassium channel-blocking toxins with unrelated structures. J Biol Chem 272:4302–4309

DeSimone CV, Lu Y, Bondarenko VE, Morales MJ (2009) S3b amino acid substitutions and ancillary subunits alter the affinity of *Heteropoda venatoria* toxin 2 for Kv4.3. Mol Pharmacol 76:125–133

DeSimone CV, Zarayskiy VV, Bondarenko VE, Morales MJ (2011) Heteropoda toxin 2 interaction with Kv4.3 and Kv4.1 reveals differences in gating modification. Mol Pharmacol 80:345–355

Diochot S, Schweitz H, Béress L, Lazdunski M (1998) Sea anemone peptides with a specific blocking activity against the fast inactivating potassium channel Kv3.4. J Biol Chem 273:6744–6749

Diochot S, Drici MD, Moinier D, Fink M, Lazdunski M (1999) Effects of phrixotoxins on the Kv4 family of potassium channels and implications for the role of I_{to1} in cardiac electrogenesis. Br J Pharmacol 126:251–263

Diochot S, Loret E, Bruhn T, Beress L, Lazdunski M (2003) APETx1, a new toxin from the sea Anemone *Anthopleura elegantissima*, blocks voltage-gated human *Ether-a-go-go*-related gene potassium channels. Mol Pharmacol 64:59–69

Doupnik CA (2017) Venom-derived peptides inhibiting Kir channels: past, present, and future. Neuropharmacology 127:161–172

Driscoll PC, Clore GM, Beress L, Gronenborn AM (1989a) A proton nuclear magnetic resonance study of the antihypertensive and antiviral protein BDS-I from the sea anemone *Anemonia sulcata*: sequential and stereospecific resonance assignment and secondary structure. Biochemistry 28:2178–2187

Driscoll PC, Gronenborn AM, Clore GM (1989b) The influence of stereospecific assignments on the determination of three-dimensional structures of proteins by nuclear magnetic resonance spectroscopy. Application to the sea anemone protein BDS-I. FEBS Lett 243:223–233

Du C, Li J, Shao Z, Mwangi J, Xu R, Tian H, Mo G, Lai R, Yang S (2019) Centipede KCNQ Inhibitor SsTx also targets $K_V1.3$. Toxins (Basel) 11

Ebbinghaus J, Legros C, Nolting A, Guette C, Celerier ML, Pongs O, Bähring R (2004) Modulation of Kv4.2 channels by a peptide isolated from the venom of the giant bird-eating tarantula *Theraphosa leblondi*. Toxicon 43:923–932

Escoubas P, Diochot S, Célérier ML, Nakajima T, Lazdunski M (2002) Novel tarantula toxins for subtypes of voltage-dependent potassium channels in the Kv2 and Kv4 subfamilies. Mol Pharmacol 62:48–57

Finol-Urdaneta RK, Belovanovic A, Micic-Vicovac M, Kinsella GK, McArthur JR, Al-Sabi A (2020) Marine toxins targeting Kv1 channels: pharmacological tools and therapeutic scaffolds. Mar Drugs 18

Frénal K, Xu CQ, Wolff N, Wecker K, Gurrola GB, Zhu SY, Chi CW, Possani LD, Tytgat J, Delepierre M (2004) Exploring structural features of the interaction between the scorpion toxin CnErg1 and ERG K^+ channels. Proteins 56:367–375

Gao S, Valinsky WC, On NC, Houlihan PR, Qu Q, Liu L, Pan X, Clapham DE, Yan N (2020) Employing NaChBac for cryo-EM analysis of toxin action on voltage-gated Na^+ channels in nanodisc. Proc Natl Acad Sci U S A 117:14187–14193

Garcia ML, Garcia-Calvo M, Hidalgo P, Lee A, MacKinnon R (1994) Purification and characterization of three inhibitors of voltage-dependent K^+ channels from *Leiurus quinquestriatus* var. *hebraeus* venom. Biochemistry 33:6834–6839

Gilquin B, Braud S, Eriksson MA, Roux B, Bailey TD, Priest BT, Garcia ML, Ménez A, Gasparini S (2005) A variable residue in the pore of Kv1 channels is critical for the high affinity of blockers from sea anemones and scorpions. J Biol Chem 280:27093–27102

Goldstein SA, Miller C (1992) A point mutation in a Shaker K^+ channel changes its charybdotoxin binding site from low to high affinity. Biophys J 62:5–7

Goldstein SA, Miller C (1993) Mechanism of charybdotoxin block of a voltage-gated K^+ channel. Biophys J 65:1613–1619

Goldstein SA, Pheasant DJ, Miller C (1994) The charybdotoxin receptor of a Shaker K^+ channel: peptide and channel residues mediating molecular recognition. Neuron 12:1377–1388

Grissmer S, Nguyen AN, Aiyar J, Hanson DC, Mather RJ, Gutman GA, Karmilowicz MJ, Auperin DD, Chandy KG (1994) Pharmacological characterization of five cloned voltage-gated K+ channels, types Kv1.1, 1.2, 1.3, 1.5, and 3.1, stably expressed in mammalian cell lines. Mol Pharmacol 45:1227–1234

Gurrola GB, Rosati B, Rocchetti M, Pimienta G, Zaza A, Arcangeli A, Olivotto M, Possani LD, Wanke E (1999) A toxin to nervous, cardiac, and endocrine ERG K^+ channels isolated from Centruroides noxius scorpion venom. FASEB J 13:953–962

Harvey AL, Robertson B (2004) Dendrotoxins: structure-activity relationships and effects on potassium ion channels. Curr Med Chem 11:3065–3072

Herrington J, Zhou YP, Bugianesi RM, Dulski PM, Feng Y, Warren VA, Smith MM, Kohler MG, Garsky VM, Sanchez M, Wagner M, Raphaelli K, Banerjee P, Ahaghotu C, Wunderler D, Priest BT, Mehl JT, Garcia ML, McManus OB, Kaczorowski GJ, Slaughter RS (2006) Blockers of the delayed-rectifier potassium current in pancreatic β-cells enhance glucose-dependent insulin secretion. Diabetes 55:1034–1042

Hill AP, Sunde M, Campbell TJ, Vandenberg JI (2007) Mechanism of block of the hERG K^+ channel by the scorpion toxin CnErg1. Biophys J 92:3915–3929

Huang X, Dong F, Zhou HX (2005) Electrostatic recognition and induced fit in the κ-PVIIA toxin binding to Shaker potassium channel. J Am Chem Soc 127:6836–6849

Imredy JP, MacKinnon R (2000) Energetic and structural interactions between delta-dendrotoxin and a voltage-gated potassium channel. J Mol Biol 296:1283–1294

Jensen JE, Cristofori-Armstrong B, Anangi R, Rosengren KJ, Lau CHY, Mobli M, Brust A, Alewood PF, King GF, Rash LD (2014) Understanding the molecular basis of toxin promiscuity: the analgesic sea anemone peptide APETx2 interacts with acid-sensing ion channel 3 and hERG channels via overlapping pharmacophores. J Med Chem 57:9195–9203

Jiménez-Vargas JM, Possani LD, Luna-Ramírez K (2017) Arthropod toxins acting on neuronal potassium channels. Neuropharmacology 127:139–160

Jin W, Lu Z (1998) A novel high-affinity inhibitor for inward-rectifier K^+ channels. Biochemistry 37:13291–13299

Kaas Q, Westermann JC, Halai R, Wang CK, Craik DJ (2008) ConoServer, a database for conopeptide sequences and structures. Bioinformatics 24:445–446

Kalman K, Pennington MW, Lanigan MD, Nguyen A, Rauer H, Mahnir V, Paschetto K, Kem WR, Grissmer S, Gutman GA, Christian EP, Cahalan MD, Norton RS, Chandy KG (1998) ShK-Dap22, a potent Kv1.3-specific immunosuppressive polypeptide. J Biol Chem 273:32697–32707

Kanjhan R, Coulson EJ, Adams DJ, Bellingham MC (2005) Tertiapin-Q blocks recombinant and native large conductance K^+ channels in a use-dependent manner. J Pharmacol Exp Ther 314:1353–1361

Karbat I, Altman-Gueta H, Fine S, Szanto T, Hamer-Rogotner S, Dym O, Frolow F, Gordon D, Panyi G, Gurevitz M, Reuveny E (2019) Pore-modulating toxins exploit inherent slow inactivation to block K^+ channels. Proc Natl Acad Sci U S A 116:18700–18709

Kimm T, Bean BP (2014) Inhibition of A-type potassium current by the peptide toxin SNX-482. J Neurosci 34:9182–9189

Korolkova YV, Kozlov SA, Lipkin AV, Pluzhnikov KA, Hadley JK, Filippov AK, Brown DA, Angelo K, Strøbaek D, Jespersen T, Olesen SP, Jensen BS, Grishin EV (2001) An ERG channel inhibitor from the scorpion *Buthus eupeus*. J Biol Chem 276:9868–9876
Krezel AM, Kasibhatla C, Hidalgo P, MacKinnon R, Wagner G (1995) Solution structure of the potassium channel inhibitor agitoxin 2: caliper for probing channel geometry. Protein Sci 4:1478–1489
Lamy C, Goodchild SJ, Weatherall KL, Jane DE, Liégeois JF, Seutin V, Marrion NV (2010) Allosteric block of $K_{Ca}2$ channels by apamin. J Biol Chem 285:27067–27077
Lange A, Giller K, Hornig S, Martin-Eauclaire MF, Pongs O, Becker S, Baldus M (2006) Toxin-induced conformational changes in a potassium channel revealed by solid-state NMR. Nature 440:959–962
Lau CHY, King GF, Mobli M (2016) Molecular basis of the interaction between gating modifier spider toxins and the voltage sensor of voltage-gated ion channels. Sci Rep 6:34333
Lee SY, MacKinnon R (2004) A membrane-access mechanism of ion channel inhibition by voltage sensor toxins from spider venom. Nature 430:232–235
Lee HC, Wang JM, Swartz KJ (2003) Interaction between extracellular Hanatoxin and the resting conformation of the voltage-sensor paddle in Kv channels. Neuron 40:527–536
Lee CW, Kim S, Roh SH, Endoh H, Kodera Y, Maeda T, Kohno T, Wang JM, Swartz KJ, Kim JI (2004) Solution structure and functional characterization of SGTx1, a modifier of Kv2.1 channel gating. Biochemistry 43:890–897
Lee S, Milescu M, Jung HH, Lee JY, Bae CH, Lee CW, Kim HH, Swartz KJ, Kim JI (2010) Solution structure of GxTX-1E, a high-affinity tarantula toxin interacting with voltage sensors in Kv2.1 potassium channels. Biochemistry 49:5134–5142
Liao Z, Yuan C, Deng M, Li J, Chen J, Yang Y, Hu W, Liang S (2006) Solution structure and functional characterization of jingzhaotoxin-XI: a novel gating modifier of both potassium and sodium channels. Biochemistry 45:15591–15600
Li-Smerin Y, Swartz KJ (1998) Gating modifier toxins reveal a conserved structural motif in voltage-gated Ca^{2+} and K^+ channels. Proc Natl Acad Sci U S A 95:8585–8589
Li-Smerin Y, Swartz KJ (2000) Localization and molecular determinants of the hanatoxin receptors on the voltage-sensing domains of a K^+ channel. J Gen Physiol 115:673–684
Long SB, Tao X, Campbell EB, MacKinnon R (2007) Atomic structure of a voltage-dependent K^+ channel in a lipid membrane-like environment. Nature 450:376–382
Lu Z, MacKinnon R (1997) Purification, characterization, and synthesis of an inward-rectifier K^+ channel inhibitor from scorpion venom. Biochemistry 36:6936–6940
Luo L, Li B, Wang S, Wu F, Wang X, Liang P, Ombati R, Chen J, Lu X, Cui J, Lu Q, Zhang L, Zhou M, Tian C, Yang S, Lai R (2018) Centipedes subdue giant prey by blocking KCNQ channels. Proc Natl Acad Sci U S A 115:1646–1651
MacKinnon R, Miller C (1988) Mechanism of charybdotoxin block of the high-conductance, Ca^{2+}-activated K^+ channel. J Gen Physiol 91:335–349
MacKinnon R, Reinhart PH, White MM (1988) Charybdotoxin block of Shaker K^+ channels suggests that different types of K^+ channels share common structural features. Neuron 1:997–1001
Marvin L, De E, Cosette P, Gagnon J, Molle G, Lange C (1999) Isolation, amino acid sequence and functional assays of SGTx1. The first toxin purified from the venom of the spider *scodra griseipes*. Eur J Biochem 265:572–579
Matsumura K, Shimomura T, Kubo Y, Oka T, Kobayashi N, Imai S, Yanase N, Akimoto M, Fukuda M, Yokogawa M, Ikeda K, Kurita J-I, Nishimura Y, Shimada I, Osawa M (2021) Mechanism of hERG inhibition by gating-modifier toxin, APETx1, deduced by functional characterization. BMC Mol Cell Biol 22(1):3
Matthies D, Bae C, Toombes GE, Fox T, Bartesaghi A, Subramaniam S, Swartz KJ (2018) Single-particle cryo-EM structure of a voltage-activated potassium channel in lipid nanodiscs. Elife 7
Milescu M, Vobecky J, Roh SH, Kim SH, Jung HJ, Kim JI, Swartz KJ (2007) Tarantula toxins interact with voltage sensors within lipid membranes. J Gen Physiol 130:497–511

Milescu M, Bosmans F, Lee S, Alabi AA, Kim JI, Swartz KJ (2009) Interactions between lipids and voltage sensor paddles detected with tarantula toxins. Nat Struct Mol Biol 16:1080–1085
Milescu M, Lee HC, Bae CH, Kim JI, Swartz KJ (2013) Opening the Shaker K^+ channel with hanatoxin. J Gen Physiol 141:203–216
Miller C (1995) The charybdotoxin family of K^+ channel-blocking peptides. Neuron 15:5–10
Miller C, Moczydlowski E, Latorre R, Phillips M (1985) Charybdotoxin, a protein inhibitor of single Ca^{2+}-activated K^+ channels from mammalian skeletal muscle. Nature 313:316–318
Moldenhauer H, Díaz-Franulic I, Poblete H, Naranjo D (2019) Trans-toxin ion-sensitivity of charybdotoxin-blocked potassium-channels reveals unbinding transitional states. Elife 8
Moreels L, Peigneur S, Galan DT, De Pauw E, Béress L, Waelkens E, Pardo LA, Quinton L, Tytgat J (2017) APETx4, a Novel Sea anemone toxin and a modulator of the cancer-relevant potassium channel $K_V10.1$. Mar Drugs 15
Mouhat S, De Waard M, Sabatier JM (2005) Contribution of the functional dyad of animal toxins acting on voltage-gated Kv1-type channels. J Pept Sci 11:65–68
Mouhat S, Andreotti N, Jouirou B, Sabatier JM (2008) Animal toxins acting on voltage-gated potassium channels. Curr Pharm Des 14:2503–2518
Naranjo D, Miller C (1996) A strongly interacting pair of residues on the contact surface of charybdotoxin and a Shaker K^+ channel. Neuron 16:123–130
Ozawa S, Kimura T, Nozaki T, Harada H, Shimada I, Osawa M (2015) Structural basis for the inhibition of voltage-dependent K^+ channel by gating modifier toxin. Sci Rep 5:14226
Pennington MW, Mahnir VM, Khaytin I, Zaydenberg I, Byrnes ME, Kem WR (1996) An essential binding surface for ShK toxin interaction with rat brain potassium channels. Biochemistry 35:16407–16411
Phillips LR, Milescu M, Li-Smerin Y, Mindell JA, Kim JI, Swartz KJ (2005) Voltage-sensor activation with a tarantula toxin as cargo. Nature 436:857–860
Racapé J, Lecoq A, Romi-Lebrun R, Liu J, Kohler M, Garcia ML, Ménez A, Gasparini S (2002) Characterization of a novel radiolabeled peptide selective for a subpopulation of voltage-gated potassium channels in mammalian brain. J Biol Chem 277:3886–3893
Ramu Y, Lu Z (2019) A family of orthologous proteins from centipede venoms inhibit the hKir6.2 channel. Sci Rep 9:14088
Ramu Y, Xu Y, Lu Z (2018) A novel high-affinity inhibitor against the human ATP-sensitive Kir6.2 channel. J Gen Physiol 150:969–976
Rauer H, Pennington M, Cahalan M, Chandy KG (1999) Structural conservation of the pores of calcium-activated and voltage-gated potassium channels determined by a sea anemone toxin. J Biol Chem 274:21885–21892
Rauer H, Lanigan MD, Pennington MW, Aiyar J, Ghanshani S, Cahalan MD, Norton RS, Chandy KG (2000) Structure-guided transformation of charybdotoxin yields an analog that selectively targets Ca^{2+}-activated over voltage-gated K^+ channels. J Biol Chem 275:1201–1208
Redaelli E, Cassulini RR, Silva DF, Clement H, Schiavon E, Zamudio FZ, Odell G, Arcangeli A, Clare JJ, Alagón A, de la Vega RC, Possani LD, Wanke E (2010) Target promiscuity and heterogeneous effects of tarantula venom peptides affecting Na^+ and K^+ ion channels. J Biol Chem 285:4130–4142
Rodríguez de la Vega RC, Possani LD (2004) Current views on scorpion toxins specific for K^+-channels. Toxicon 43:865–875
Ruta V, MacKinnon R (2004) Localization of the voltage-sensor toxin receptor on KvAP. Biochemistry 43:10071–10079
Ruta V, Jiang Y, Lee A, Chen J, MacKinnon R (2003) Functional analysis of an archaebacterial voltage-dependent K^+ channel. Nature 422:180–185
Scanlon MJ, Naranjo D, Thomas L, Alewood PF, Lewis RJ, Craik DJ (1997) Solution structure and proposed binding mechanism of a novel potassium channel toxin κ-conotoxin PVIIA. Structure 5:1585–1597
Schmidt D, Cross SR, MacKinnon R (2009) A gating model for the archeal voltage-dependent K^+ channel KvAP in DPhPC and POPE:POPG decane lipid bilayers. J Mol Biol 390:902–912

Stocker M, Miller C (1994) Electrostatic distance geometry in a K^+ channel vestibule. Proc Natl Acad Sci U S A 91:9509–9513

Sumino A, Sumikama T, Uchihashi T, Oiki S (2019) High-speed AFM reveals accelerated binding of agitoxin-2 to a K^+ channel by induced fit. Sci Adv 5:eaax0495

Swartz KJ (2007) Tarantula toxins interacting with voltage sensors in potassium channels. Toxicon 49:213–230

Swartz KJ, MacKinnon R (1995) An inhibitor of the Kv2.1 potassium channel isolated from the venom of a Chilean tarantula. Neuron 15:941–949

Swartz KJ, MacKinnon R (1997a) Hanatoxin modifies the gating of a voltage-dependent K^+ channel through multiple binding sites. Neuron 18:665–673

Swartz KJ, MacKinnon R (1997b) Mapping the receptor site for hanatoxin, a gating modifier of voltage-dependent K^+ channels. Neuron 18:675–682

Tabakmakher VM, Krylov NA, Kuzmenkov AI, Efremov RG, Vassilevski AA (2019) Kalium 2.0, a comprehensive database of polypeptide ligands of potassium channels. Sci Data 6:73

Takacs Z, Toups M, Kollewe A, Johnson E, Cuello LG, Driessens G, Biancalana M, Koide A, Ponte CG, Perozo E, Gajewski TF, Suarez-Kurtz G, Koide S, Goldstein SA (2009) A designer ligand specific for Kv1.3 channels from a scorpion neurotoxin-based library. Proc Natl Acad Sci U S A 106:22211–22216

Takahashi H, Kim JI, Min HJ, Sato K, Swartz KJ, Shimada I (2000) Solution structure of hanatoxin1, a gating modifier of voltage-dependent K^+ channels: common surface features of gating modifier toxins. J Mol Biol 297:771–780

Takeuchi K, Yokogawa M, Matsuda T, Sugai M, Kawano S, Kohno T, Nakamura H, Takahashi H, Shimada I (2003) Structural basis of the KcsA K^+ channel and agitoxin2 pore-blocking toxin interaction by using the transferred cross-saturation method. Structure 11:1381–1392

Tao H, Chen JJ, Xiao YC, Wu YY, Su HB, Li D, Wang HY, Deng MC, Wang MC, Liu ZH, Liang SP (2013a) Analysis of the interaction of tarantula toxin Jingzhaotoxin-III (β-TRTX-Cj1α) with the voltage sensor of Kv2.1 uncovers the molecular basis for cross-activities on Kv2.1 and Nav1.5 channels. Biochemistry 52:7439–7448

Tao H, Wu Y, Deng M, He J, Wang M, Xiao Y, Liang S (2013b) Molecular determinants for the tarantula toxin jingzhaotoxin-I interacting with potassium channel Kv2.1. Toxicon 63:129–136

Tao H, Chen X, Deng M, Xiao Y, Wu Y, Liu Z, Zhou S, He Y, Liang S (2016) Interaction site for the inhibition of tarantula Jingzhaotoxin-XI on voltage-gated potassium channel Kv2.1. Toxicon 124:8–14

Terlau H, Shon KJ, Grilley M, Stocker M, Stühmer W, Olivera BM (1996) Strategy for rapid immobilization of prey by a fish-hunting marine snail. Nature 381:148–151

Tilley DC, Eum KS, Fletcher-Taylor S, Austin DC, Dupré C, Patrón LA, Garcia RL, Lam K, Yarov-Yarovoy V, Cohen BE, Sack JT (2014) Chemoselective tarantula toxins report voltage activation of wild-type ion channels in live cells. Proc Natl Acad Sci U S A 111:E4789–E4796

Tilley DC, Angueyra JM, Eum KS, Kim H, Chao LH, Peng AW, Sack JT (2019) The tarantula toxin GxTx detains K^+ channel gating charges in their resting conformation. J Gen Physiol 151:292–315

Tseng GN, Sonawane KD, Korolkova YV, Zhang M, Liu J, Grishin EV, Guy HR (2007) Probing the outer mouth structure of the HERG channel with peptide toxin footprinting and molecular modeling. Biophys J 92:3524–3540

Tudor JE, Pallaghy PK, Pennington MW, Norton RS (1996) Solution structure of ShK toxin, a novel potassium channel inhibitor from a sea anemone. Nat Struct Biol 3:317–320

Wang W, MacKinnon R (2017) Cryo-EM structure of the open human *Ether-à-go-go*-related K^+ channel hERG. Cell 169:422–430.e10

Wang JM, Roh SH, Kim S, Lee CW, Kim JI, Swartz KJ (2004) Molecular surface of tarantula toxins interacting with voltage sensors in K_V channels. J Gen Physiol 123:455–467

Weatherall KL, Seutin V, Liégeois JF, Marrion NV (2011) Crucial role of a shared extracellular loop in apamin sensitivity and maintenance of pore shape of small-conductance calcium-activated potassium (SK) channels. Proc Natl Acad Sci U S A 108:18494–18499

Wei P, Xu C, Wu Q, Huang L, Liang S, Yuan C (2014) Jingzhaotoxin-35, a novel gating-modifier toxin targeting both Nav1.5 and Kv2.1 channels. Toxicon 92:90–96

Xu CQ, Zhu SY, Chi CW, Tytgat J (2003) Turret and pore block of K^+ channels: what is the difference? Trends Pharmacol Sci 24:446–448. author reply 448–9

Yan L, Herrington J, Goldberg E, Dulski PM, Bugianesi RM, Slaughter RS, Banerjee P, Brochu RM, Priest BT, Kaczorowski GJ, Rudy B, Garcia ML (2005) *Stichodactyla helianthus* peptide, a pharmacological tool for studying Kv3.2 channels. Mol Pharmacol 67:1513–1521

Yang S, Wang Y, Wang L, Kamau P, Zhang H, Luo A, Lu X, Lai R (2020) Target switch of centipede toxins for antagonistic switch. Sci Adv 6:eabb5734

Yao X, Fan X, Yan N (2020) Cryo-EM analysis of a membrane protein embedded in the liposome. Proc Natl Acad Sci U S A 117:18497–18503

Yeung SY, Thompson D, Wang Z, Fedida D, Robertson B (2005) Modulation of Kv3 subfamily potassium currents by the sea anemone toxin BDS: significance for CNS and biophysical studies. J Neurosci 25:8735–8745

Yuan C, Liao Z, Zeng X, Dai L, Kuang F, Liang S (2007a) Jingzhaotoxin-XII, a gating modifier specific for Kv4.1 channels. Toxicon 50:646–652

Yuan C, Yang S, Liao Z, Liang S (2007b) Effects and mechanism of Chinese tarantula toxins on the Kv2.1 potassium channels. Biochem Biophys Res Commun 352:799–804

Zhang M, Korolkova YV, Liu J, Jiang M, Grishin EV, Tseng GN (2003) BeKm-1 is a HERG-specific toxin that shares the structure with ChTx but the mechanism of action with ErgTx1. Biophys J 84:3022–3036

Zhang M, Liu XS, Diochot S, Lazdunski M, Tseng GN (2007) APETx1 from sea Anemone *Anthopleura elegantissima* is a gating modifier peptide toxin of the human *Ether-a-go-go-*related Potassium Channel. Mol Pharmacol 72:259–268

Zhang Y, Luo J, He J, Rong M, Zeng X (2019) JZTX-V targets the voltage sensor in Kv4.2 to inhibit I_{to} potassium channels in cardiomyocytes. Front Pharmacol 10:357

Zhao R, Dai H, Mendelman N, Cuello LG, Chill JH, Goldstein SA (2015) Designer and natural peptide toxin blockers of the KcsA potassium channel identified by phage display. Proc Natl Acad Sci U S A 112:E7013–E7021

Zhao R, Dai H, Mendelman N, Chill JH, Goldstein SAN (2020) Tethered peptide neurotoxins display two blocking mechanisms in the K^+ channel pore as do their untethered analogs. Sci Adv 6:eaaz3439

Therapeutic Antibodies Targeting Potassium Ion Channels

Janna Bednenko, Paul Colussi, Sunyia Hussain, Yihui Zhang, and Theodore Clark

Contents

J. Bednenko · P. Colussi · S. Hussain · Y. Zhang
TetraGenetics Inc, Arlington, MA, USA
e-mail: jbednenko@tetragenetics.com; pcolussi@tetragenetics.com; shussain@tetragenetics.com; yzhang@tetragenetics.com

T. Clark (✉)
TetraGenetics Inc, Arlington, MA, USA

Department of Microbiology and Microbiology, Cornell University, Ithaca, NY, USA
e-mail: tgc3@cornell.edu

N. Gamper, K. Wang (eds.), *Pharmacology of Potassium Channels*,
Handbook of Experimental Pharmacology 267, https://doi.org/10.1007/164_2021_464

Abstract

Monoclonal antibodies combine specificity and high affinity binding with excellent pharmacokinetic properties and are rapidly being developed for a wide range of drug targets including clinically important potassium ion channels. Nonetheless, while therapeutic antibodies come with great promise, K^+ channels represent particularly difficult targets for biologics development for a variety of reasons that include their dynamic structures and relatively small extracellular loops, their high degree of sequence conservation (leading to immune tolerance), and their generally low-level expression in vivo. The process is made all the more difficult when large numbers of antibody candidates must be screened for a given target, or when lead candidates fail to cross-react with orthologous channels in animal disease models due to their highly selective binding properties. While the number of antibodies targeting potassium channels in preclinical or clinical development is still modest, significant advances in the areas of protein expression and antibody screening are converging to open the field to an avalanche of new drugs. Here, the opportunities and constraints associated with the discovery of antibodies against K^+ channels are discussed, with an emphasis on novel technologies that are opening the field to exciting new possibilities for biologics development.

Keywords

Biologic · Ion channel · Kv1.3 · Potassium channel · Therapeutic antibody

1 K^+ Channels as Targets for Therapeutic Antibody Development

As detailed throughout this volume, potassium-selective ion channels are involved in a wide range of physiological processes important in human health and disease. While not all K^+ channels are validated drug targets, many are, and the past decade has seen notable successes in the discovery and development of both small and large molecule drugs that can modulate the activity of a number of important disease-related K^+ channel targets. Antibodies that can recognize and modulate the activities of human Kv1.3, Kv10.1, Kv11.1, and TASK3 have now been described and their potential use in the context of disease is discussed in detail at the end of the chapter (Sects. 4.1–4.4).

First, we consider the advantages of antibodies as therapeutic compounds, the mechanisms by which they achieve their effect, the strategies underlying their discovery and development (Fig. 1; Box 1) along with the practical aspects of antigen design, protein production, immunization, and screening to identify potential lead compounds that can move forward through clinical trials.

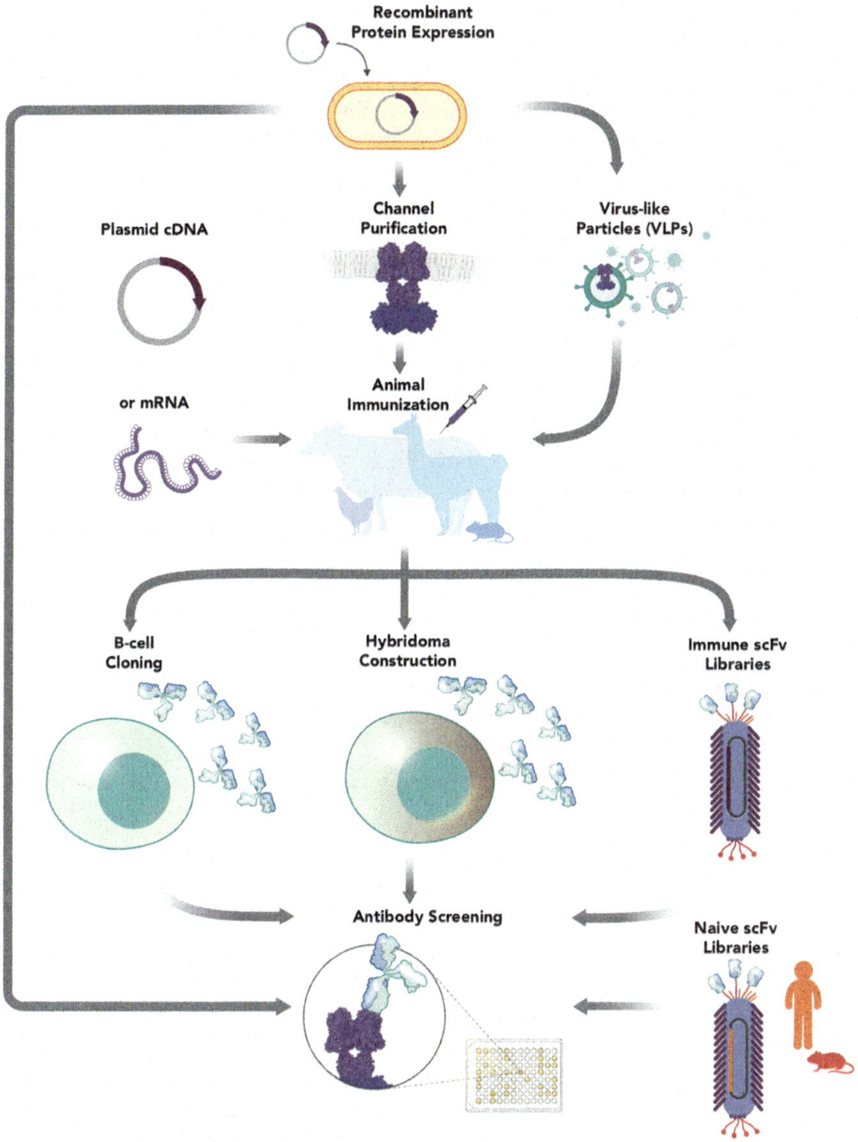

Fig. 1 Alternative routes toward antibody discovery

Box 1 Antibody Discovery

Obtaining therapeutic antibodies that can modulate the activity of K^+ channels can proceed via different routes depending, to some extent, on the source

(continued)

Box 1 (continued)
material for antibody generation and screening. As shown in Fig. 1, traditional approaches involve immunizing animals with purified channel proteins (or, in some cases, virus-like particles containing channel proteins) that are expressed recombinantly in heterologous systems. Alternatively, genetic immunization with nucleic acids encoding relevant proteins (either in the form of cDNA or mRNA) offers yet another way to induce channel-specific antibody production in animal hosts. Following immunization, culture supernatants from cloned B-cells or immortalized hybridoma cell lines that secrete antibodies are then screened for binding to channels of interest. As an alternative to screening culture supernatants, B-cells from immunized animals can be harvested and used to construct phage or yeast libraries that display single-chain antibody fragments (scFvs) on their surface as a way to indirectly screen for antibodies that bind and modulate channel activity. In fact, animal immunization can be precluded altogether using single-domain antibody display libraries constructed from B-cells of naïve animals or human patients (bottom right). These libraries are often of sufficient depth that they can be mined for rare antibodies that might never be produced in response to immunization but are valuable therapeutic candidates, nonetheless. Finally, regardless of the source of antibodies used for discovery, recombinant channel proteins that are either purified or expressed in mammalian cell lines for functional (patch-clamp) studies, provide necessary material for antibody screening.

2 Antibody Discovery and Development: Challenges and Opportunities

Based on their long half-lives and exquisite target selectivity, antibodies have distinct advantages relative to both small molecule drugs, which often lack selectivity, and peptide toxins that can be highly selective but are rapidly cleared from the bloodstream (Hutchings et al. 2019; Posner et al. 2019; Wulff et al. 2019). With regard to the latter, approaches toward increasing the half-lives of peptide toxins (including the use of antibody scaffolds (see Sect. 2.1)) are being actively pursued, however, maintaining the native binding activity of re-engineered toxins remains a significant issue (Lau and Dunn 2018; Murray et al. 2019).

In the case of small molecules, the drug-binding pockets where these entities typically exert their effect are often conserved among members of a given ion channel family, have relatively small surface area (~300 $Å^2$ (Coleman and Sharp 2010)), and suffer from general promiscuity at the level of both the pocket and ligand. Conversely, the large solvent accessible surface area of the paratope/epitope interaction site formed between the complementarity-determining regions of immunoglobulins and their cognate antigens (~1,500–2,000 $Å^2$ (Ramaraj et al.

2012)) drives high specificity binding for a given target and is one of the principal advantages of antibodies compared to small molecule drugs.

High-affinity binding is clearly a hallmark of therapeutic antibodies, however, in order to achieve clinical relevance, antibodies targeting K^+ channels must not only bind the target, they must also alter a specific physiological process in vivo either by blocking or potentiating channel activity, and/or drive the elimination of disease-causing cells by engaging immune effector mechanisms. This has been a difficult problem to solve for reasons that are both theoretical and practical.

2.1 Theoretical Considerations

Although much is known about the gating properties of K^+ channels, there are relatively few examples of antibodies that can modulate channel activity and there is little-to-no structural information that can inform us as to how such antibodies either inhibit or potentiate channel activity. Thus, in the absence of a solid framework for understanding how antibodies alter the functional activity of a channel, a rational approach toward generating such antibodies remains elusive.

That being said, among the handful of antibodies that can modulate K^+ channel activity, a majority are directed against the small extracellular loop structures of their cognate channel proteins. While this may seem obvious given that these loop structures are the sole regions on K^+ channels available for antibody binding, an antibody directed against TASK3 appears to inhibit the channel indirectly by driving receptor-mediated endocytosis of the channel itself. Inhibitory antibodies to Kv1.3 and Kv10.1, on the other hand, appear to block their respective channels more directly.

In the absence of detailed structural studies, one can only speculate that antibodies in this latter group act through either steric hindrance (that is, blockage of the pore) or allosteric interactions that interfere with the normal cycling between conformationally distinct (resting, open, and inactivated) states of the channel in each case (Kuang et al. 2015). In the second instance, antibodies that recognize unique epitopes associated with a particular state could potentially stabilize or lock a channel into a given conformation leaving it irreversibly opened or closed. Alternatively, it is possible that binding to an epitope (s) shared by all conformational states could prevent cycling or force a channel into a particular state. Either way, such antibodies could be useful from a therapeutic standpoint and, in the case of true "state-dependency," could offer an additional layer of selectivity on target binding. Indeed, recent evidence has suggested that a lead antibody targeting Kv1.3 is capable of blocking channel activity in a use-dependent manner by recognizing an epitope associated with the open-state of the channel accessible in activated human T_{EM} cells but not in "resting" Kv1.3 expressing CHO cells (Colussi, personal communication).

The idea that K^+ channels may expose state-dependent epitopes is clearly consistent with voluminous studies showing that these ion channels are structurally dynamic and that conductance changes are generally accompanied by changes in the 3-dimensional conformation of the channel itself (Tombola et al. 2006; Gupta et al. 2010; Vargas et al. 2012; Kuang et al. 2015; Islas 2016). Although it is unclear whether such conformational shifts generate novel B-cell epitopes, studies using stabilized GPCR structures (Hutchings et al. 2017; Soave et al. 2018) suggest possible strategies for screening antibodies that recognize unique determinants associated with particular conformational states of a given K^+ channel. Additionally, active consideration is being given to the use of biparatopic and bispecific antibodies that can bind different epitopes on the same or different proteins (in the case of homomeric or heteromeric channels, respectively) as a way to prevent dynamic structural changes thereby blocking channel activity. Finally, it should be noted that antibodies that can modulate K^+ channel activity have been identified using purified full-length proteins in an undetermined conformational state as the starting immunogen (Colussi, unpublished).

As noted above, inhibition of channel activity by antibodies may result from steric blockage of the pore. In this case, single-domain antibodies (nanobodies) from camelids or sharks may be particularly useful due to their small size and apparent ability to bind recessed epitopes (Henry and Mackenzie 2018). While a precise mechanism (s) is lacking, several camelid single-domain constructs have been shown to inhibit K1.3 (see Sect. 4.1). Along the same lines, antibodies produced in some species have ultra-long hypervariable regions in their heavy chains. This is perhaps best exemplified in cows where the complementarity-determining region 3 (CDRH3) of the heavy chain folds into an extended "stalk" and disulfide-bonded "knob" structure. Wang and colleagues took advantage of this structure to build chimeric humanized antibodies containing the extended bovine CDRH3 stalk domain but substituting the knob region with the peptide toxins Moka-1 or Vm24 that bind to the outer vestibule of the Kv1.3 channel pore (Wang et al. 2016). In each case, the chimeric antibodies blocked Kv1.3 channel activity, presumably by interfering with conductance through the pore. While such chimera are highly engineered constructs, the fact that the ultra-long bovine CDRH3 can gain access to the outer regions of the pore suggests that cows and other species that generate antibodies with novel properties could be ideal platforms for producing pore-blocking antibodies.

2.2 Practical Considerations

From a practical standpoint, immunization and screening campaigns to identify modulating antibodies typically require large amounts of high-quality starting material to achieve success. Nevertheless, K^+ channels tend to be expressed at low levels in their native contexts and, as multi-span transmembrane proteins that can be toxic when overexpressed, are often difficult to produce as recombinant proteins. This problem is compounded when considering heteromeric K^+ channels where more than a single polypeptide must be expressed and assembled at the proper stoichiometric ratios before antibody generation can begin. Despite this, a number of successful approaches are being used to generate sufficient levels of antigen for animal immunization including the use of novel expression systems (Sect. 3.2).

A more basic problem relative to antibody generation is the high level of sequence conservation between K^+ channels of different species. Tolerance mechanisms during B- and T-cell development dampen the immune response to autoantigens and can be difficult to overcome when challenging a given species with antigens that are evolutionarily conserved and therefore similar or identical to self. The most relevant regions of the K^+ channels in this case are the small extracellular loops where antibodies are expected to bind. As illustrated for the Kv1.3 channel (Table 1), these regions show varying degrees of sequence conservation in orthologs from humans and mice and, as expected, greater divergence in more distantly related species (specifically, chickens). Interestingly, when comparing homologous K^+ channel family members of a given species, extracellular loop structures tend to be less conserved than the transmembrane domains separating them (Table 1) and, theoretically, provide sufficient sequence diversity to elicit production of highly selective antibodies in a given animal host.

It is also worth noting that while the number of potential B-cell epitopes associated with the extracellular loops is somewhat limited due to their small size (roughly 10–20% of total protein for the annotated human K^+ channels (Table 1)), the proteins themselves exist as multimers of the same or different subunits. As a result, surface epitopes are displayed in a repetitive fashion on native proteins. Such repetitive features may be conducive to antibody production via crosslinking of the Ig receptor on B-cells and may provide some advantage for antibody screening based on binding avidity. Certainly, the ability to generate native loop structures on recombinant proteins together with the use of strong adjuvants and evolutionarily diverse species for immunization offer logical approaches to antibody generation (Sects. 3.2 and 3.4). And, while peptide constructs corresponding to loop structures tend to be weakly immunogenic on their own, methods are being developed to enhance their potency including multimerization of antigen constructs. A notable example in this case is mAb56, which blocks the Kv10.1 channel (Sect. 4.2) and was generated by linking a peptide from the third extracellular loop of the channel to a tetramerization domain from the C-terminal region of the same protein (Hartung et al. 2011).

Beyond the need to generate antibodies, screening to identify not only binders, but also antibodies that can modulate channel activity in vivo has its own unique challenges. This is a complex task that involves surveying vast repertoires of antibody molecules for antigen binding, recovering genetic sequences for antibodies that bind, and screening potentially hundreds of candidates for their ability to modulate channel activity. While these steps are often fraught, a variety of novel technologies are being married with more traditional approaches to screen and identify candidates of interest. A more detailed description of these technologies is provided in Sect. 3.4.

The preceding discussion has focused on the discovery of antibodies that can modify channel function by blocking or potentiating current flow. While this is arguably the most difficult goal to achieve in the development of therapeutic antibodies against K^+ channels, it is worth noting that modulating channel activity is not the sole mechanism for achieving both specificity and biological effect. As

Table 1 Average percent identities calculated from percent identity matrices generated by Clustal Omega alignments

Channel family	Members (n)	Whole protein	Average % identity														ECL (% total protein)
			TMD						Re-entrant loop (RL)		Extracellular						
			S1	S2	S3	S4	S5	S6	RL	RL2	ECL1	ECL2	ECL-RL	RL-ECL	ECL-RL2	RL2-ECL	
KCa2.x	3	68.1	96.7	85.0	74.6	87.3	96.8	100.0	100	NA	56.0	80.7	89.5	91.7	NA	NA	8.2
KNav1.x	2	76.7	85.7	60.0	61.1	100	76.2	81.3	81.0	NA	35.3	100.0	55.6	50.0	NA	NA	3.9
Kv1.x	8	67.0	79.1	73.5	78.0	96.8	98.9	97.4	89.5	NA	35.9	49.3	64.5	78.6	NA	NA	15.1
Kv2.x	2	67.0	90.9	100.0	94.7	100.0	100.0	100.0	100.0	NA	75.0	100.0	84.6	100.0	NA	NA	6.2
Kv3.x	4	73.5	90.4	76.7	93.9	100.0	100.0	100.0	100.0	NA	49.1	95.5	69.1	90.5	NA	NA	13.5
Kv4.x	3	70.0	96.8	79.4	82.5	100.0	100.0	100.0	96.7	NA	65.7	81.5	74.4	90.5	NA	NA	8.5
Kv6.x	4	49.5	53.8	50.8	53.2	76.2	65.9	62.9	82.5	NA	43.3	34.9	36.7	66.7	NA	NA	11.9
Kv7.x	5	47.8	63.8	46.2	52.9	84.6	80.5	73.2	85.2	NA	42.0	77.5	35.8	63.3	NA	NA	4.9
Kv8.x	2	39.4	40.9	38.1	52.4	100.0	52.4	42.9	61.9	NA	16.7	33.3	38.5	33.3	NA	NA	8.6
Kv9.x	3	51.6	48.5	46.4	59.8	84.1	71.2	78.2	83.3	NA	41.5	10.0	38.5	76.2	NA	NA	10.5
Kv10.x	2	76.8	90.5	90.5	94.7	100.0	91.3	81.0	81.8	NA	71.4	100.0	68.8	71.4	NA	NA	7.8
Kv11.x	3	63.4	96.8	85.7	93.7	100.0	93.7	96.8	100.0	NA	33.3	77.8	67.6	100.0	NA	NA	6.4
Kv12.x	3	56.5	81.0	69.8	76.2	90.5	81.0	96.8	87.3	NA	22.2	41.7	58.5	86.7	NA	NA	7.4
Kir2.x	4	64.4	64.0	67.4	NA	NA	NA	NA	79.0	NA	45.3	77.8	NA	NA	NA	NA	8.3
Kir3.x	4	63.2	62.0	76.5	NA	NA	NA	NA	85.1	NA	54.9	83.3	NA	NA	NA	NA	7.7
Kir4.x	2	61.0	68.0	77.3	NA	NA	NA	NA	94.7	NA	48.0	66.7	NA	NA	NA	NA	9.0
Kir6.x	2	69.7	80.0	77.3	NA	NA	NA	NA	94.7	NA	43.5	88.9	NA	NA	NA	NA	8.9
K2P	15	28.9	41.5	46.0	24.5	37.8	NA	NA	57.2	56.9	NA	NA	22.9	45.5	40.7	16.9	22.7

Approximate boundaries of transmembrane domains (TMD: S1-S6), Re-entrant loops (RL), Extracellular loops (ECL) and ECLs either N-terminal (ECL-RL) or C-terminal (RL-ECL) of re-entrant loops were based on annotation in UniProtKB, Clustal alignments and where necessary manual alignment. Kv5.1 is the only member of the Kv5 family and therefore it was not analyzed. *NA* not applicable noted above, a monoclonal antibody targeting the TASK3 channel inhibits human lung cancer xenografts and murine breast cancer metastasis in mice through internalization of the channel/antibody complex (Sun et al. 2016). Furthermore, Fc-mediated effector functions such as antibody-dependent cellular cytotoxicity (ADCC), antibody-dependent cellular phagocytosis (ADCP), and complement-dependent cytotoxicity (CDC) offer additional mechanisms by which depletion of diseased cells could achieve therapeutic benefit (de Taeye et al. 2019; Graziano and Engelhardt 2019). In the same vein, engineering approaches involving optimization of Fc domain structure (Lobner et al. 2016; Saunders 2019), the use of monovalent, bi- and multi-valent formats (Brinkmann and Kontermann 2017; Husain and Ellerman 2018), antibody drug conjugates (Joubert et al. 2020), as well as cellular approaches that engage chimeric antigen receptors on engineered T cells (Golubovskaya and Wu 2016) represent additional opportunities for K^+ channel antibody development.

3 Workflows in Antibody Production and Screening

As outlined in Fig. 1, having chosen a particular target, a wide range of options is available in the path toward therapeutic antibody development. This begins with decisions around whether to immunize animals or rely on surface display technologies to select for antibodies (typically in single-chain scFv or Fab formats in the case of phage, yeast, or ribosome display) that can bind the target of interest (Fig. 1; Box 1). Since therapeutic antibodies must contain humanized or fully human heavy and light chain scaffolds, one advantage of surface display is the potential to screen commercially available libraries prepared from human DNA. Indeed, such libraries have been constructed from both naïve and autoimmune patient samples and are of sufficient depth that one can mine the entire B-cell receptor repertoire of a given individual, albeit, without native heavy and light chain pairings (Lee et al. 2018; Rouet et al. 2018). Alternatively, one can choose to immunize animals and create de novo surface display libraries, which, theoretically, could increase the likelihood of selecting for antibodies of interest, particularly in cases where a given channel protein is highly conserved across species boundaries and pre-existing libraries might lack strong binders.

Regardless of the direction one chooses, success in antibody discovery ultimately depends on having an antigen to screen with (Fig. 1). If the initial choice is to immunize animals with protein antigens, the immunogen itself can be used for screening purposes in combination with any of a variety of binding assays including conventional ELISA and bead-based assays that allow colorimetric or fluorescence read-outs. Similarly, if one opts for nucleic acid immunization, or relies solely on surface display to select for antibodies of interest, the screening entity can be a peptide, a protein fragment, or a full-length protein purified directly from native tissue or produced recombinantly and bound to a solid support (see Sect. 3.1). Cell-based flow cytometry offers an additional method of screening but suffers a number of pitfalls including low resolution (if the channel is not abundantly expressed) and the potential to miss antibodies that bind a particular conformational state. Despite these problems, flow cytometry has the major advantage that it can potentially select for only those antibodies that bind surface exposed loops on channel proteins and is especially useful at later stages of screening when binders have already been identified.

Among the most powerful technologies now being applied to antibody discovery are manual and automated high-throughput B-cell cloning methods that allow one to screen the secreted products of thousands to hundreds-of-thousands of individual plasma cells from naive or immunized animals. These include the optofluidic BEACON platform created by Berkeley Lights, Inc. (Emeryville, CA, USA), the Genovac Nano platform (Aldevron LLC, Fargo, ND, USA), and Abcellera's microfluidic platform (Vancouver, Canada) among others. A somewhat different approach involves the capture of individual B-cells in porous microspheres that can be directly screened under a microscope using diffusible fluorescent probes (Izquierdo et al. 2016). Of course, the power of these single-cell technologies is that cDNA sequences encoding the heavy and light chains of potentially every B-cell clone in their native pairings can be readily recovered by standard RT-PCR and sequencing to regenerate the antibody of interest either as a native immunoglobulin or a chimeric antibody containing the variable regions of the original clone within a human immunoglobulin scaffold. Although not routinely available to academic laboratories, the platforms themselves can be accessed through commercial sources (Aldevron, LLC, Fargo, NC; Abcellera, Vancouver, Canada; Ligand Pharmaceuticals Inc., San Diego, CA).

Lastly, antibodies identified in binding assays must be further screened for their ability to modulate K^+ channel activity and/or induce a biological effect in vitro and in vivo. Depending on the channel and target cell, these assays will differ but assuming the desired effect is achieved, the antibody will then be evaluated for binding affinity, selectivity, cross-reactivity with non-human orthologs, potency, manufacturability (expression yield, solubility, thermal and long-term stability), and immunogenicity and, if necessary, subjected to additional engineering prior to selection of a final lead and initiation of preclinical development. While these latter criteria are beyond the scope of this chapter, the sections that follow provide additional detail on key aspects of antibody generation and screening beginning with the selection of immunogens.

3.1 Choice of Immunogen

For programs involving animal immunization, antibody development begins with the choice of immunogen, which are grouped here into four categories, namely, (1) peptides and soluble protein fragments; (2) native full-length proteins; (3) recombinant full-length proteins; and, (4) nucleic acids encoding antigens of interest. The use of peptides and protein fragments enables production of antibodies against targeted regions of the channel that are accessible at or near the cell surface and likely to play a role in ion flux. When larger fragments or full-length proteins are used as immunogens, screening methods designed to preferentially select those antibodies that can bind extracellular regions of the channel (either on a solid support or on target cells) are preferred (Bednenko et al. 2018).

Although peptides and small protein fragments have been used to generate antibodies against a number of K^+ channels (Hemmerlein et al. 2006; Gómez-Varela et al. 2007; Sun et al. 2016; Hartung et al. 2020; Fan et al. 2020), they are weakly immunogenic and lack the full spectrum of multi-domain, conformational epitopes present in the full-length, correctly folded and functional channel proteins (Dodd et al. 2018). Efforts to improve their immunogenicity and solubility include linking such fragments to sequences that promote oligomerization (Hemmerlein et al. 2006; Gómez-Varela et al. 2007; Hartung et al. 2020).

By comparison, full-length proteins provide all possible B- and T-cell epitopes associated with a given target and, when available, are the antigens of choice for generating inhibitory antibodies against channel proteins; the principal caveat being that K^+ channels are not abundantly expressed in their native cell types and their overexpression in heterologous cells can be toxic resulting in limited yields. Modulating antibodies against a small number of ion channels (the calcium channel, Orai1 (Lin et al. 2013); the ligand-gated ion channel P2X7 (Buell et al. 1998); and the cation channel TRPA1 (Lee et al. 2014)) have been generated using proteins derived from mammalian expression systems. While this has not been demonstrated in the case of the K^+ channels, alternative approaches using DNA immunization (Stortelers et al. 2018), peptide/carrier protein conjugates (Fan et al. 2020), virus-like particles (Adam et al. 2014; Doms et al. 2014) and proteins produced in protozoan expression systems (Bednenko et al. 2018) have had success targeting the important potassium channel, Kv1.3.

Lastly, vector-encoded antigens (in the case of DNA immunization) and, more recently, mRNA immunization (Pardi et al. 2018; Zhang et al. 2019) obviate the need to generate and purify proteins altogether and are discussed in more detail below (Sect. 3.2).

3.2 Expression Platforms for Antigen Generation

While expression in *Escherichia coli* and other bacteria are among the most robust platforms in terms of protein yield, they are generally not recommended for mammalian membrane proteins due to their reducing cytosolic environments along with significant differences between prokaryotic and eukaryotic protein processing and membrane insertion mechanisms (Pandey et al. 2016). However, several eukaryotic systems have demonstrated consistently strong performance in expression of membrane protein drug targets. We took advantage of the Protein Data Bank (PDB; rcsb.org) to obtain information on production of eukaryotic full-length K^+ channels for structure analysis (Table 2). Of the 21 potassium channels in this archive, 9 were expressed in human embryonic kidney HEK293 cells using BacMam technology, 6 in methylotrophic yeast *Pichia pastoris*, 5 in insect cells (*Spodoptera frugiperda* Sf9 and *Trichoplusia ni* High Five cells), and 1 in rat insulinoma INS-1832/13 cells.

Table 2 Potassium ion channels that were expressed as full-length proteins for structural analysis (source: Protein Data Bank, rcsb.org)

Potassium channel	Species	Expression system	PDB accession number	References
Kv1.2	*Rattus norvegicus*	*Pichia pastoris*	2A79	Long et al. (2005)
Kv7.1	*Homo sapiens*	Human HEK293S GnTI⁻ cells (BacMam technology)	6UZZ	Sun and MacKinnon (2020)
Kv7.1	*Xenopus laevis*	Human HEK293S GnTI⁻ cells (BacMam technology)	5VMS	Sun and MacKinnon (2017)
Kv10.1	*Rattus norvegicus*	Human HEK293S GnTI⁻ cells (BacMam technology)	6PBY 6PBX 5K7L	Whicher and MacKinnon (2016), (2019)
Kv11.1	*Homo sapiens*	Human HEK293S GnTI⁻ cells (BacMam technology)	5VA1 5VA2 5VA3	Wang and MacKinnon (2017)
KCa1.1	*Homo sapiens*	Human HEK293S GnTI⁻ cells (BacMam technology)	6V3G 6V38 6V35 6V22	Tao and MacKinnon (2019)
KCa1.1	*Aplysia californica*	*Trichoplusia ni* (High Five cells)	5TJI 5TJ6	Hite et al. (2017), Tao et al. (2017)
KCa3.1	*Homo sapiens*	Human HEK293S GnTI⁻ cells (BacMam technology)	6CMN 6CNN 6CNO	Lee and MacKinnon (2018)
KCa4.1	*Gallus gallus*	*Spodoptera frugiperda* Sf9 cells	5A6E 5U70 5U76	Hite et al. (2015), Hite and MacKinnon (2017)
Kir2.2	*Gallus gallus*	*Pichia pastoris*	3JYC 3SPC 3SPG 3SPH 3SPI 3SPJ 5KUK 5KUM 6M84 6M85 6M86	Tao et al. (2009), Hansen et al. (2011), Lee et al. (2016b), Zangerl-Plessl et al. (2020)
Kir3.2	*Mus musculus*	*Pichia pastoris*	3SYA 3SYC 3SYO 3SYP	Whorton and MacKinnon (2011), (2013)

(continued)

Table 2 (continued)

Potassium channel	Species	Expression system	PDB accession number	References
			3SYQ 4KFM	
Kir6.2	*Homo sapiens*	Human HEK293S GnTI⁻ cells (BacMam technology)	6C3O 6C3P	Lee et al. (2017)
Kir6.2	*Mus musculus*	Human Free-style HEK293-F cells (BacMam technology)	5YKE 5YKF 5YKG 5YW8 5YW9 5YWA 5YWB 5YWC 6JB1	Wu et al. (2018), Ding et al. (2019)
Kir6.2	*Rat norvegicus*	Rat insulinoma INS-1 832/13 cells transduced with adenovirus	5TWV 6BAA	Martin et al. (2017a, b, 2019)
Kir6.2	*Mus musculus*	HEK293S GnTI⁻ cells (BacMam technology)	5WUA	Li et al. (2017)
TWIK1	*Homo sapiens*	*Pichia pastoris*	3UKM	Miller and Long (2012)
TREK1	*Homo sapiens*	*Spodoptera frugiperda*	4TWK	To be published
TREK2	*Homo sapiens*	*Spodoptera frugiperda* Sf9 cells	4BW5 4XDJ 4XDK 4XDL	Dong et al. (2015)
TRAAK	*Homo sapiens*	*Pichia pastoris*	3UM7 4I9W 4RUE 4RUF 4WFE 4WFF 4WFG 4WFH	Brohawn et al. (2012), (2013), (2014), Lolicato et al. (2014)
TRAAK	*Mus musculus*	*Pichia pastoris*	6PIS	Brohawn et al. (2019)
TASK1	*Homo sapiens*	*Spodoptera frugiperda* Sf9 cells	6RV2 6RV3 6RV4	Rödström et al. (2020)

Both BacMam and insect cell platforms utilize modified double-stranded DNA baculoviruses as vehicles for gene delivery. Baculoviruses do not replicate in mammalian cells, and high-level transient expression in BacMam system is achieved by the presence of a mammalian expression cassette containing a strong CMV promoter (Goehring et al. 2014).

While insect cells are capable of generating high protein yields, mammalian cells offer additional advantages, most importantly, native N-glycosylation and membrane lipid composition (e.g., higher cholesterol content), which increases the likelihood of correct protein folding (Goehring et al. 2014). However, overexpression of recombinant ion channels can be toxic for mammalian cells, presumably due to unregulated activity of the channel in a heterologous environment (Claire 2006). To avoid cytotoxicity, expression can be carried out in eukaryotic or prokaryotic cell-free systems that may include detergents and/or lipids to enhance membrane protein solubility. Both Kv1.1 and Kv1.3 have been successfully expressed using cell-free technologies (Renauld et al. 2017; Cortes et al. 2018), although it is important to note that cell-free systems typically produce only moderate amounts of protein, are relatively expensive and often difficult to scale up.

As an alternative to mammalian and insect cells, fast-growing and efficient microbial eukaryotes such as yeast (especially, *Pichia pastoris* and *Saccharomyces cerevisiae*) or the ciliated protist, *Tetrahymena thermophila*, have also been explored for the expression of recombinant potassium channels (see Table 3 for comparison of *P. pastoris* and *T. thermophila*). *P. pastoris* owes its success in large part to a very powerful and tightly regulated methanol-inducible AOX1 promoter as well as to its ability to reach very high cell densities in culture (Vogl and Glieder 2013; Guyot et al. 2020). Growth of *P. pastoris* either in shake flasks or bioreactors at the liter scale can be sufficient to produce milligram quantities of purified membrane protein to support animal immunization and screening trials (Guyot et al. 2020).

Tetrahymena, on the other hand, is a free-living protist with biological properties that make it ideal for the production of eukaryotic membrane proteins. These properties include the absence of a cell wall, the ability to introduce foreign genes at very high copy number (~18,000 copies per cell), near uniform N-glycosylation (primarily 3Man2GlcNAc) and a metabolism that is geared toward the production of membrane and secreted proteins (Nusblat et al. 2012; Guerrier et al. 2017). Additionally, the *T. thermophila* genome is characterized by expanded gene families encoding proteins involved in membrane dynamics and transport, including more than 300 potassium voltage-gated ion channels (Eisen et al. 2006). Typically, it takes 1–2 weeks to obtain *Tetrahymena* transformants with confirmed recombinant protein expression, and 2–3 more weeks to accumulate sufficient biomass for production of 2–10 mg of purified membrane protein. TetraGenetics Inc. (Arlington, MA, USA) has employed this system for production of a number of human K^+ channels being targeted for therapeutic antibody development in the treatment of cancer, autoimmune and fibroproliferative disorders (Bednenko et al. 2018, Shim et al. 2020; Bednenko and Colussi, unpublished).

Some antibody development technologies do not require immunogen purification. For example, Integral Molecular (Philadelphia, PA, USA) produced

Table 3 *Pichia pastoris* and *Tetrahymena thermophila*

	Pichia pastoris	*Tetrahymena thermophila*
Cell size	4–6 μm	30–60 × 50–100 μm
Cell wall	Yes	No
Typical cell density at the time of induction of protein expression	10^8–10^9 cells/ml	10^6 cells/ml
Examples of high membrane protein expression yield	90 mg purified human aquaporin 1 from 1 l of culture (Nyblom et al. 2007) 13 mg purified mouse P-glycoprotein from 100 g of cells (Bai et al. 2011)	3 mg purified human Kv1.3 from 1 l of culture (Bednenko et al. 2018)
Inducible promoter	AOX1, methanol-inducible	MTT, cadmium-inducible
Typical growth temperature	27–30°C	30–37°C
Typical induction temperature	20–27°C	24–37°C
Doubling time	2–3.5 h (1–3 h)	2–3 h
Induction time	16–60 h	2–24 h
N-glycosylation	8–14 mannose residues per side chain GlycoSwitch technology: engineer your own strains to obtain controlled, human-like N-glycosylation (Jacobs et al. 2009)	Primarily $Man_{2-5}GlcNAc_2$ (Calow et al. 2016)
O-glycosylation	Very little	None
Membrane lipid composition	Ergosterol is the major sterol. Strains producing cholesterol have been engineered (Hirz et al. 2013)	No sterol synthesis, tetrahymanol as a sterol surrogate. Membrane lipid composition can be modified by addition of cholesterol and other lipids into a growth media (Nusblat et al. 2012)

Kv1.3-containing murine leukemia virus-like particles (VLPs) in HEK293T cells, with the yield of 400 pmol Kv1.3/mg total VLP protein, or ~ 2.6% w/w (Adam et al. 2014; Doms et al. 2014). Due to their repetitive surface structures, VLPs are known to elicit robust humoral and cellular immune responses (Mohsen et al. 2017).

As noted above, nucleic acid immunization technologies short-circuit the need for protein antigens altogether (Liu et al. 2016). Typically, DNA expression vectors encoding the antigen of choice are delivered into animal tissues by intradermal microparticle bombardment or by intramuscular injection, while liposome formulations and/or electroporation are used to increase transfection efficiency. Ablynx (Ghent, Belgium) immunized llamas with Kv1.3-encoding plasmid DNA to develop functional anti-Kv1.3 single-domain nanobodies (Stortelers et al. 2018). The caveat with DNA immunization strategies is that they often yield low antibody titers and may need to be supplemented with a purified recombinant protein for animal boosting and/or antibody screening (Bednenko et al. 2018). Along with DNA, in vitro transcribed messenger RNA (mRNA) can serve as a template for antigen synthesis in vivo following cell entry. Chemically formulated mRNA vaccines have demonstrated encouraging results in the field of vaccine development and may become an indispensable tool in the fight against emerging infectious diseases and cancer (Pardi et al. 2018; Corey et al. 2020; Espeseth et al. 2020).

3.3 Purification and Formulation of Target Immunogens

An important consideration when designing purification strategies for K^+ channels is maintenance of physiological activity, which in the end can be validated using one or more analytical methods depending on the formulation (e.g., ligand binding, ion-flux, electrophysiology). This is particularly important when modifying native sequences with various tags to aid in purification (6-10xHis, GFP, antibody epitopes such as FLAG, Rho1D4, and others), or when truncations are introduced to the N- and C-termini to limit aggregation and impart stability (Wang and MacKinnon 2017; Whicher and MacKinnon 2019; Sun and MacKinnon 2020). While intracellular domains are often irrelevant for therapeutic antibody discovery, care should be taken to retain regions of the protein that are responsible for oligomerization and membrane trafficking (Jenke et al. 2003). An additional consideration for immunogen design is whether auxiliary proteins such as beta or gamma subunits (Tao and MacKinnon 2019), or calmodulin for some Kv and KCa channels (Wang and MacKinnon 2017; Lee and MacKinnon 2018; Whicher and MacKinnon 2019; Sun and MacKinnon 2020) need to be co-expressed and purified along with the primary alpha channel subunit.

Extraction of K^+ channels from membranes typically requires solubilization of the lipid bilayer with detergents, preferably those deemed more gentle and likely to stabilize the channel such as the maltoside series of non-ionic detergents (e.g., DM, DDM, Cymal-series). Additionally, incorporation of cholesterol analogs (e.g., cholesteryl hemisuccinate), phospholipids (alone or in mixtures, e.g., POPC, POPG, POPE, POPA, POPS) can also help preserve activity of detergent-stabilized K^+ channels (Hansen et al. 2011; Whorton and MacKinnon 2011; Sun and MacKinnon 2020).

While some antibody discovery programs have utilized ion channels in detergent solution (Brohawn et al. 2013; Shcherbatko et al. 2016), reconstituting the target in a membrane-like environment is the preferred option to stabilize the channel in preparation for immunization, screening, and binding assays. Liposomes, which are composed of synthetic or natural lipids that are sonicated or extruded to obtain spherical unilamellar membranes, are the most commonly used formulations for this purpose. Following reconstitution into liposomes (Seddon et al. 2004), the resulting proteoliposomes can be used directly for immunization, as well for screening antibody libraries and testing channel activity in flux assays (Wang and Sigworth 2009; Bednenko et al. 2018; Lee and MacKinnon 2018). Nevertheless, a drawback of liposome formulations is the difficulty in controlling the orientation of a reconstituted membrane protein which is dependent on various empirical factors, including the reconstitution method (Yanagisawa et al. 2011), lipid composition (Hickey and Buhr 2011), and protein structure. This issue can be addressed via engineered tags located on intracellular loops or termini that can tether a protein to a solid support prior to reconstitution into lipids, thereby forcing the channels to orient with the extracellular loops facing out (Sumino et al. 2017). Such approaches can be effective for screening phage or yeast display libraries to preferentially recover antibody fragments that recognize surface epitopes.

In recent years, numerous surfactants have been developed for keeping membrane proteins stable in detergent-free solutions. Such surfactants offer additional options for presenting K^+ channels during immunization and screening and fall into two general categories: (1) polymer-based surfactants such as amphipols (Tribet et al. 1996; Popot et al. 2003) along with styrene-maleic acid (SMA) copolymers (Lee et al. 2016a); and (2) peptide-based surfactants that are variations on membrane-interacting proteins such as saposins or apolipoproteins (Frauenfeld et al. 2016; Denisov et al. 2004; Denisov and Sligar 2016). Each of these platforms results in disc-like membrane protein-surfactant complexes that theoretically expose both the intracellular and extracellular domains of the protein in the same complex. Currently, the most established surfactant technology is the protein-based nanodisc (Denisov et al. 2004; Denisov and Sligar 2016) in which a dimer of an engineered apoA1 construct (Membrane Scaffold Protein, or MSP) encircles a lipid membrane containing an embedded membrane protein. This approach has been used successfully in the reconstitution of a number of K^+ channels (Xu et al. 2015; Matthies et al. 2018; Shenkarev et al. 2018; Sun and MacKinnon 2020). Supporting their utility in antibody discovery, membrane protein nanodiscs have been used to identify Fab fragments from phage display libraries (Dominik et al. 2016). Amphipols, such as the polyacrylate derivative A8–35 and PMAL-C8, have been used successfully to stabilize K^+ channels (Spear et al. 2015; Lee et al. 2017) and amphipol formulations have been employed to generate monoclonal antibodies against other membrane protein antigens in rodents (Agosto et al. 2014; Vij et al. 2018; Storek et al. 2019). SMA lipoparticles (SMALPs) have also been demonstrated to solubilize K^+ channels (Dörr et al. 2014), including human channels expressed in mammalian cells (Karlova et al. 2019) but, to date, there have been no reports of their use in antibody discovery.

3.4 Antibody Platforms and Initial Phases of Screening

As noted above, the decision tree leading to antibody discovery (Fig. 1) requires a fundamental choice between the use of animal immunization, surface display libraries, or some combination of the two. Each of these approaches has advantages and disadvantages and, with sufficient resources, an argument can be made to pursue parallel paths.

In the case of animal immunization, one has the opportunity to generate antibodies with high affinity that arise naturally through a process of reiterative B-cell selection following somatic hypermutation of heavy and light chain genes in the germinal centers of the spleen and secondary lymph nodes of vaccinated animals. The principal downsides of animal immunization are the potential for tolerance mechanisms to interfere with antibody production and the possibility that the immune system will focus the B-cell response on so-called immunodominant epitopes (to the exclusion of others) that are irrelevant with respect to therapeutic potential of resulting antibodies. Additionally, while antibodies produced against a given target may, in fact, be excellent drug candidates, downstream manufacturing of those antibodies requires linking phenotype (that is, antibody binding and/or modulating activity) to genotype (or more precisely, the sequences of the hypervariable regions of the heavy and light chain genes that produce the antibody). Accomplishing this requires identification and isolation of individual B-cell clones responsible for antibody production, which, until recently, relied on standard mouse hybridoma technology that has since been expanded to other species (Parray et al. 2020). The ability to immortalize human B-cells (Kwakkenbos et al. 2016), along with single-cell methods that permit high-throughput screening of secreted antibodies from large numbers of B-cell clones is proving to be extremely powerful as well and may eventually supplant hybridoma technology altogether (see below).

In choosing to immunize, a variety of factors including animal species/strain, antigen dose, choice of adjuvant, route of injection, vaccination schedule must all be considered before initiating the process. With regard to the animal host, specialized mouse strains, as well as evolutionary distant species provide opportunities to overcome tolerance mechanisms limiting the response to highly conserved channel proteins (Sect. 2.2; Table 1). At the same time, antibodies produced in llamas, cows, and sharks have unique structures that may be well suited to blocking channel activity (see below).

Although standard mouse lines (especially BALB/c) have long been used for monoclonal antibody production, the development of hyperimmune strains that produce autoantibodies such as DiversimAb (Abveris, abveris.com, Canton, MA, USA) and others (Perry et al. 2011; Lee et al. 2012), as well as mice that overexpress the neonatal Fc receptor (Cervenak et al. 2015; Schneider et al. 2015) has proven to be useful in generating immune responses to antigens with high homology to endogenous mouse proteins. Additionally, tolerance issues should be precluded in knock-out mice that do not express channels of interest (Hrabovska et al. 2010) and strains lacking *KCNA3*, *KCNA5,* and *KCNN4* genes all appear to be viable (London et al. 2001; Koni et al. 2003; Begenisich et al. 2004). Finally, in addition to engineered mouse strains, evolutionary distant species provide another option to overcome tolerance mechanisms. Chickens, in particular, are capable of generating high-affinity antibodies against broad sets of epitopes on mammalian proteins (Abdiche et al. 2016; Bednenko et al. 2018). Indeed, Bednenko and co-workers identified 9 functional anti-Kv1.3 mAbs following immunization of a single animal with full-length Kv1.3 expressed recombinantly in *T. thermophila* (Bednenko et al. 2018) (Sect. 4.1).

Along with chickens and mice, other species have garnered attention from ion channel researchers based on their ability to generate antibodies with unusual structures. In cows, for example, roughly 10% of antibodies contain ultra-long CDRH3 heavy chain loops with protruding "stalk-and-knob" structures that extend away from the main antibody scaffold and play a predominant role in antigen binding (Wang et al. 2013; Dong et al. 2019; Stanfield et al. 2018, 2020). Researchers have explored the use of these unique structures in the development of broadly neutralizing therapeutic antibodies against HIV (Sok et al. 2017). Furthermore, the "knob" domains of cow antibodies have been modified to incorporate various polypeptides including protease inhibitors and cytokines, as well as peptide toxins that have been shown to block the activity of Kv1.3 (Zhang et al. 2013a, b; Liu et al. 2015; Wang et al. 2016).

In camelids (i.e., llamas, alpacas, and camels) and cartilaginous fish (sharks), a subset of immunoglobulins (or immunoglobulin-like molecules in the case of sharks) is entirely devoid of light chains (Könning et al. 2017). The variable domains of these heavy chain-only isotypes (referred to as VHHs, or nanobodies, in camelids and V-NARs, for New Antigen Receptors, in sharks) represent autonomous antigen-binding units. While having different evolutionary origins (Flajnik et al. 2011; English et al. 2020), camelid nanobodies and V-NARs share an immunoglobulin-like fold along with a number of other features including small size (12–15 kDa), protruding CDR3 loops (up to 40 amino acids in length), enormous structural diversity, modular design, and high stability and solubility (Könning et al. 2017; Mitchell and Colwell 2018; Feng et al. 2019). Camelid VHH genes are highly homologous to human VH genes while shark V-NARs arose from a non-immunoglobulin lineage. Generally, camelid nanobodies exhibit low immunogenicity in humans but nevertheless require humanization (Vincke et al. 2009), a process which is far more challenging for V-NAR sequences (Steven et al. 2017).

Over the past decade, camelid nanobodies have dominated the single-domain antibody discovery field and are viewed as next generation tools for use in medical and biotechnological applications (Hoey et al. 2019). Nanobody-derived products are at various stages of clinical development and at least one (caplacizumab; Cablivi) received FDA approval in 2019 for treatment of acquired Thrombotic Thrombocytopenia Purpura (aTTP) (Hanlon and Metjian 2020; Jovčevska and Muyldermans 2020; Kaplon et al. 2020). A number of promising shark single-domain antibodies, including a V-NAR against the ion channel, P2X3, are currently in preclinical development (English et al. 2020; www.ossianix.co.uk).

As with bovine antibodies, single-domain entities can access buried epitopes, such as enzyme active sites and, potentially, the transmembrane pores of ion channels. Small size, high tissue/tumor penetration, and fast renal clearance make single-domain antibodies particularly useful for in vivo diagnostics. To enable half-life extension (HLE) for therapeutic purposes, nanobodies have either been PEGylated, engineered in a bivalent or multivalent format with an albumin-binding domain, or fused to human antibody scaffolds containing the Fc domain. For example, trivalent anti-P2X7 and anti-Kv1.3 nanobodies with extended half-lives have been developed that contain two identical channel blocking domains linked to an anti-albumin binding domain (Danquah et al. 2016; Stortelers et al. 2018). Other examples of engineered multivalent nanobodies include a biparatopic dimer targeting distinct epitopes of CXCR2 (Bradley et al. 2015) and an engineered fusion of the anti-mGluR1 IgG with the blood brain barrier-penetrating nanobody FC5 (Webster et al. 2016).

As indicated above, regardless of the species, antibodies developed in animals must be humanized in order to limit immunogenicity in therapeutic applications. This is a complex process and often needs to be coupled to affinity maturation. An alternative strategy relies on the immunization of transgenic animals carrying human immunoglobulin loci (Green 2014). The first rodent strains that utilized this approach, the HuMAbMouse® and XenoMouse® (Lonberg et al. 1994; Mendez et al. 1997), produced fully human antibodies, while subsequent versions (the OmniRat and VelocImmune mouse) focused on substituting human variable domains for the rodent versions without modifying rodent constant regions allowing for a more efficient immune response in immunized animals (Osborn et al. 2013; Murphy et al. 2014). Following hit identification, human constant regions are appended to fully human variable regions using standard molecular biology tools. To date, more than 20 monoclonal antibodies discovered in transgenic mice have been approved by the FDA (Lu et al. 2020) and many different species – rats, chickens, cows – have been transformed to create human antibody-producing factories. Some humanized commercial platforms not listed above include IntelliSelect (kymab.com), OmniMouse, OmniChicken, OmniFlic, OmniClic (omniab.com), H2L2 Mouse and HCAB (harbourantibodies.com), Trianni mouse (Trianni) and Tc-bovines (sabbiotherapeutics.com).

As a way around animal immunization, investigators have the option to screen combinatorial libraries that display variable regions of antibodies on the surfaces of either phage particles or microbial cells (bacteria or yeast) to screen for potential binders to K^+ channels of interest. Surface display technologies have been in widespread use for more than two decades and have seen enormous advancement in both the underlying display platforms and types of molecules that are displayed, and in the case of antibodies, allow the user to mine vast numbers (on the order of 10^{12} or more) of heavy and light chain pairings from naïve or immune animals, while at the same time, have access to their sequence information.

Variations on the types of surface display platforms and their relative advantages and disadvantages are too great to adequately cover in this chapter, although several excellent reviews are available in the literature (Almagro et al. 2019; Kunamneni et al. 2020; Ministro et al. 2020). Briefly, the libraries themselves are classified as immune, naïve, synthetic and semi-synthetic depending on the source material and the way in which the libraries are constructed. Naïve libraries are antigen independent and are constructed from non-rearranged V gene segments from humans or animals, as well as synthetic and shuffled V genes. In some respects, they are considered "universal" libraries since they are used to screen for binders to essentially any target. Immune libraries, on the other hand, are constructed from humans or animals that have been immunized or are expected to contain antibodies against specific proteins as a result of a medically relevant condition. While these are typically more circumscribed in their overall sequence coverage, V_H and V_L regions have already undergone gene rearrangement and somatic hypermutation in vivo, thereby increasing the likelihood of identifying high affinity binders to the antigen of interest. By contrast, identification of antibody fragments with high binding activity displayed on naïve, synthetic, or semi-synthetic libraries often depends on the size of the library being screened. As suggested by their names, much of the diversity of V_H and V_L domains in synthetic and semi-synthetic libraries is generated artificially through oligonucleotide synthesis and/or random in vitro assembly of V and DJ gene segments. In either case, such libraries are commercially available and can be mined in-house or on a fee-for-service basis using commercial suppliers.

One of the inherent issues with combinatorial antibody display libraries has been the non-native pairing of heavy and light chains generated during library construction that can give rise to antibodies having lower affinity and selectivity than bona fide immunoglobulins produced in response to immunization. Additionally, non-natively paired immunoglobulin can be prone to less than desirable biophysical properties (e.g., aggregation, etc.) although more recent engineering approaches to library design and construction have mitigated these drawbacks. Using a novel single-cell emulsion and oligo-dT mRNA capture approach, Wang and co-workers have recently succeeded in retaining the native V_H and V_L pairings in scFv expressing yeast display libraries prepared from human patient plasmablasts (Wang et al. 2018). This approach yielded broadly neutralizing antibodies against HIV, as well as high-affinity antibodies to the Ebola virus glycoprotein and influenza HA (Wang et al. 2018) and will likely have numerous applications in antibody discovery in the future.

3.5 Functional Screening Assays

Following generation of anti-K^+ channel antibodies in immunized animals, or, alternatively, interrogation of surface display antibody libraries, screening to identify mAbs with desired characteristics typically requires a multi-tiered approach. Generally, initial antibody screening establishes specificity and identifies binders that recognize the target K^+ channel. Protein-based binding assays such as ELISA and HTRF-FRET are highly sensitive, amenable to high throughput and can be used to screen unpurified samples such as hybridoma supernatants and bacterial periplasmic extracts. However, cell-based assays (e.g., flow cytometry) using either endogenous or recombinant cell lines expressing the K^+ channel of interest in a more physiologically relevant conformation can offer a significantly more robust method to identify surface binders that are more likely to modulate channel function. Nevertheless, while the latter is typically viewed as superior to the former, this is not always the case where, for example, antibodies bind an exposed epitope on a purified protein preparation in an ELISA format that is not necessarily exposed on cells owing to the particular state of the channel.

Identification of antibody binders is typically followed by functional screening to isolate clones capable of modulating K^+ channel function. Analytical methods used for functional screens can be broadly divided into ion flux-based assays and electrophysiology that directly measures electrical currents produced by potassium ion movement through the channel. Ion flux assays include methods that use radioactive tracers such as ^{86}Rb that can pass through K^+ channels and can be detected by atomic absorption spectrometry (Karczewski et al. 2009). However, hazards associated with radioactivity can make these assays inconvenient to perform (Gill et al. 2003; Yu et al. 2016) and alternative methods using ion sensitive dyes have been developed and are preferable for some laboratories. For example, K^+ channels are also permeable to thallium (Tl^+) ions, and Tl^+-sensitive dyes have been developed and applied to K^+ channel screens (Weaver et al. 2004). In Tl^+-based ion flux assays, cells expressing the K^+ channel of interest are initially loaded with a non-fluorescent, Tl^+-specific, and membrane permeant reporter dye such as the BTC-AM or FLUXOR. Tl^+ is added to the medium and, following stimulation, flows through the open channel, activating the dye, thereby generating a fluorescent signal that can be measured by imaging plate-readers (Beacham et al. 2010). Tl^+ flux assays have been used to identify modulators of multiple types of K+ channels including voltage-gated (Cheung et al. 2012; Yu et al. 2016) inward rectifier (Wang et al. 2011; Kaufmann et al. 2013; Wydeven et al. 2014; Swale et al. 2016), two-pore domain (Sun et al. 2016; Wright et al. 2017), and K_{Ca} channels (Jørgensen et al. 2013).

An alternative method of detecting ion flux is based on monitoring the change of membrane potential induced by movement of ions across the membrane using a voltage-sensitive dye. Multiple voltage-sensitive dyes utilizing different mechanisms have been developed (Miller 2016) and include: (1) electrochromic probes such as amino-naphthyl ethenyl-pyridinium (ANEP) dyes whose fluorescence spectra change in response to changes in the surrounding electric field. ANEP dyes can detect sub-millisecond membrane potential changes, but typically suffer from low sensitivity; (2) lipophilic redistribution dyes (e.g., oxonol dye DiBAC4) that have superior sensitivity, but slower response times; and, (3) fluorescence resonance energy transfer (FRET) based probes which typically combine high sensitivity with fast response times (Yuan et al. 2013). Additionally, a new type of voltage-sensing dye has been developed in recent years that utilizes a photo-induced electron transfer (PET) mechanism (Kulkarni et al. 2016).

The direct measurement of electrical currents by patch-clamp electrophysiology is still generally considered the gold standard for ion channel research, with the whole-cell patch-clamp configuration being widely used for investigating ion channels and membrane potential. The whole-cell patch-clamp technique uses a glass micropipette electrode to form a high resistance seal with the cell membrane allowing it to record and manipulate ionic current across the whole-cell membrane (Liu et al. 2019). The inherent flexibility of manual patch recording makes it ideal for characterizing ion channel functional properties as well as investigating mechanisms of action of modulating compounds. Indeed, several modulating antibodies targeting K^+ channels have been successfully characterized using manual patch clamp techniques including Kv1.3 (Bednenko et al. 2018) and Eag1 (Gómez-Varela et al. 2007). For screening purposes, manual patch clamp is hampered by several factors including the fact that it is labor-intensive, can only be performed by highly trained experimentalists, and has relatively low throughput.

To meet the demand for large-scale screening of drug candidates, several automated patch clamp systems have been developed that can be divided into two types, based on their patching system (Dunlop et al. 2008; Annecchino and Schultz 2018). The first type inherited the glass micropipette of conventional patch clamp and includes Flyscreen, AutoPatch, and roboPatch for mammalian cells, and the robocyte17 and OpusXpress 6000A for Xenopus oocytes. More popular automated platforms use micro-fabricated planar electrode-based patch clamp that replaces the glass micropipette in conventional patch clamp and include IonWorks Barracuda and PatchXpress from Molecular Devices (Xu et al. 2003; Kuryshev et al. 2014; Cerne et al. 2016), Patchliner and Syncropatch 384/786i from Nanion (Farre et al. 2009; Obergrussberger et al. 2018), and QPatch and Qube 384 from Sophion (Asmild et al. 2003; Chambers et al. 2016).

Compared to manual patch clamp, the appeal of automated platforms is their semi-high throughput capabilities that enable compound library screening or acquisition of multiple data points in a single experiment. Nevertheless, automated systems suffer from some limitations including compatible cell types (typically recombinant as opposed to primary), inability to establish long recording times and generally lower quality recordings compared to manual patch (Annecchino and Schultz 2018), though the latter continues to improve in association with platform advancements (Bell and Dallas 2018). Based on their respective advantages and disadvantages, automated platforms are typically involved in early drug discovery while confirmation and in-depth characterization of compounds of interest rely on manual patch clamp techniques.

It is important to note that whichever approach is used, screening antibodies has at least two unique challenges. First, any functional effect bestowed by an antibody can reasonably be expected to be directly related to the kinetics of association and dissociation from the target on a timescale that will likely be represented in minutes not seconds as is typically observed with small molecule modulators. Second, a number of antibodies targeting K+ channels demonstrate some level of state-dependency (e.g., Kv1.3 [Colussi, personal communication], Eag1 [Gómez-Varela et al. 2007]) that should be considered when designing screening experiments in order to mitigate the potential of false-negative results. Such state-dependency may depend on expression of the channel in its native context and be missed completely in recombinant cell lines.

4 Current Status of the Field

Despite the challenges associated with antibody generation and screening, successful attempts to produce antibodies that can modulate the activities of K^+ channels have been made. While these antibodies are at various stages of clinical development, we fully anticipate that therapeutic antibodies targeting K^+ channels will find their way into clinical use for a wide range of human diseases in the not too distant future.

4.1 Kv1.3

Based on its essential role in the activation and proliferation of chronically stimulated effector memory T-cells (T_{EM} cells), the voltage-gated Kv1.3 channel is a well-established drug development target for autoimmune and inflammatory diseases such as type 1 diabetes, psoriasis, and rheumatoid arthritis (reviewed in Feske et al. 2015; Chandy and Norton 2017). Additionally, Kv1.3 is expressed in a variety of other blood cells including platelets (McCloskey et al. 2010; Feske et al. 2015; Fan et al. 2020) where it is thought to play a role in platelet-dependent thrombosis (see below). It is also considered a key therapeutic target for neuroinflammatory diseases (Wang et al. 2020) and there is accumulating evidence that inhibition of the Kv1.3 channel may be beneficial in cancer therapy (Teisseyre et al. 2019). While a variety of small molecules, peptides, and peptide/antibody fusions have been reported to downregulate Kv1.3 activity (Schmitz et al. 2005; Wang et al. 2016; Chandy and Norton 2017), there remains a substantial need for selective anti-Kv1.3 monoclonal antibodies that can be used therapeutically in humans.

To develop antibodies against the Kv1.3 channel, Bednenko et al. utilized a powerful *Tetrahymena thermophila* protein expression system to express and purify multi-milligram quantities of Kv1.3 formulated as particulate antigens on polysterene beads, magnetic beads, or proteoliposomes (Bednenko et al. 2018). Two phylogenetically diverse species, chickens and llamas, were used to generate antibodies. Chickens were immunized with Kv1.3-containing proteoliposomes, and individual B-cell clones were screened for the production of antibodies against Kv1.3 displayed on magnetic or polystyrene beads using gel-encapsulated microenvironment (GEM) assays (Izquierdo et al. 2016). Llamas were immunized with a DNA vaccine encoding Kv1.3 and boosted with Kv1.3 proteoliposomes, followed by construction of phage libraries displaying single-chain variable-region antibody fragments (scFv) with non-native heavy and light chain pairings. Libraries were then screened using recombinant Kv1.3 linked to beads to identify scFv clones with binding activity. Finally, chicken antibodies or llama scFv fragments were used to construct chimeric scFv-Fc clones that were analyzed by whole-cell patch clamp electrophysiology with L929 mouse fibroblasts transiently expressing human Kv1.3. Nine chicken and one llama antibody were found to inhibit Kv1.3 activity by 44–82% within 10 min of antibody addition. Select antagonist mAbs demonstrated high potency (IC50 $<$ 10 nM), cross-reactivity to rat and cynomolgus monkey Kv1.3, and significant selectivity over related Kv1.x family members, hERG and Nav1.5.

In independent studies with llamas, genetic immunization enabled the production of nanobodies that successfully blocked Kv1.3 function as demonstrated by electrophysiology (Steeland et al. 2016; Stortelers et al. 2018). These nanobodies recognized the first extracellular loop (ECL1) of Kv1.3 and cross-reacted with rat and cynomolgus monkey Kv1.3 channels. In addition, they displayed functionality in primary T cell assays. Formatting monovalent nanobodies into bivalent and trivalent constructs resulted in increased potency. Additionally, a construct comprised of two monovalent nanobody components and one albumin-binding nanobody component (Nb12–12-alb11) was created, and it demonstrated efficacy in vivo in a rat delayed-type hypersensitivity model.

Recently, Fan and co-workers utilized mouse hybridoma technology to develop a monoclonal antibody (6E12#15) against a 14 amino acid peptide EADDPTSGFSSIPD corresponding to a portion of the extracellular loop 3 (ECL3) of human Kv1.3 (Fan et al. 2020). This antibody recognized both the human and mouse Kv1.3 channels and inhibited their activity in transfected HEK293 cells. Consistent with the important role Kv1.3 plays in the regulation of membrane potential and calcium signaling in platelets (McCloskey et al. 2010), mAb 6E12#15 inhibited aggregation, adhesion, and activation of platelets isolated from humans and wild-type mice but not mice lacking the *KCNA3* gene. Moreover, 6E12#15 impaired platelet-driven thrombus formation when injected into wild-type, but not *KCNA3* knock-out mice, opening up the possibility for developing therapeutics for the treatment of platelet-dependent thrombosis.

4.2 Kv10.1

Kv10.1, or Eag1, belongs to the ether-á-go-go (EAG) subfamily of voltage-gated K^+ channels and it is recognized for its pathological overexpression in solid tumors (Cázares-Ordoñez and Pardo 2017; Barros et al. 2020). In healthy human tissues, Kv10.1 is primarily expressed in the brain. Kv10.1 is implicated in a variety of cellular processes, including proliferation, migration, and adhesion, as well as cell cycle progression. In addition, Kv10.1 has been genetically linked to two developmental disorders, Temple-Baraitser syndrome and Zimmermann-Laband syndrome (Kortüm et al. 2015; Simons et al. 2015).

In the search for specific inhibitors of Kv10.1 that did not interfere with the function of other EAG family members such as Kv10.2 and Kv11.1 (hERG), Gómez-Varela et al. developed a monoclonal antibody, mAb56, using standard mouse hybridoma technology (Gómez-Varela et al. 2007). A 79 residue ECL3 sequence fused to the C-terminal Kv10.1 tetramerization domain was used to immunize mice (Hemmerlein et al. 2006). Following hybridoma construction, mAb56 was selected based on its ability to bind Kv10.1 and to inhibit Kv10.1 activity, but not hERG currents in whole-cell patch clamp experiments with Kv10.1-expressing HEK293 cells (~40% inhibition of current amplitude with an IC50 of ~73 nM). mAb56 displayed cross-reactivity with rat and mouse Kv10.1, and its cognate epitope has been mapped to a linear peptide, GSGSGKWEG, near the middle of ECL3. Consistent with a role for Kv10.1 in cancer, mAb56 has been shown to decrease proliferation of cancer cells in vitro and to reduce tumor growth in mouse xenograft models (e.g., MDA-MB-435S human breast cancer cell xenografts; PAXF1657 pancreatic cancer patient-derived xenografts). The mAb56 antibody is commercially available through Sigma-Aldrich (catalog # MABN378) and has been used as a highly selective Kv10.1 inhibitor for research purposes (Chen et al. 2011; Hernández-Reséndiz et al. 2020).

Ideally, therapeutics against cancer should not only reduce tumor growth, but also mediate cancer cell death. Toward that end, Hartung et al. designed an anti-Kv10.1 antibody fusion with TRAIL (tumor necrosis factor-related apoptosis-inducing ligand) with the goal of inducing apoptosis in Kv10.1-positive cancer cells (Hartung et al. 2011, 2020; Hartung and Pardo 2016). In the first set of studies, mAb62 antibody, which was selected in the same immunization campaign as mAb56 (Hemmerlein et al. 2006; Gómez-Varela et al. 2007), was used as a template to generate the single-chain variable fragment, scFv62, and then a trimeric scFv62-TRAIL fusion. mAb62 has been shown to interact with Kv10.1, but not with its closest relative Kv10.2. Unlike mAb56, however, mAb62 did not inhibit Kv10.1 K^+ currents, and its linear epitope, DYEIFDEDT, was located within the N-terminal portion of ECL3, proximal to the mAb56 epitope. The scFv62-TRAIL fusion induced apoptosis in cancer cells when used in combination with cytotoxic drugs, such as doxorubicin (Hartung et al. 2011; Hartung and Pardo 2016).

A third antibody with high affinity for ECL3 of Kv10.1, VHH-D9, was developed by immunization of llamas with a Kv10.1 ECL3-containing construct similar to the one used in the mAb56 and mAb62 development campaigns. This single-domain antibody (nanobody) was then fused to three TRAIL-encoding sequences in tandem to produce the potent apoptosis-inducing construct VHH-D9-scTRAIL (Hartung et al. 2020). Together, these studies suggest a promising approach to cancer therapy that utilizes antibody fusion constructs to target pro-apoptotic factors (in this case TRAIL) to the surface of tumor cells.

4.3 Kv11.1 (hERG)

Another member of the EAG subfamily, Kv11.1 (in humans, referred to as hERG), plays a fundamental role in cardiac repolarization. While hERG is also ectopically expressed in many types of cancer cells and may play a role in cancer progression (He et al. 2020), suppression of Kv11.1 function in vivo, either by genetic mutation or, in some cases, by administration of small molecule drugs, can lead to the potentially fatal long QT syndrome (Barros et al. 2020; Garrido et al. 2020). Since the vast majority of drugs associated with QT prolongation have been shown to interact with hERG, lead compounds are counter-screened against human Kv11.1 early in drug discovery to rule out potential interactors.

Despite this, antibodies against hERG that specifically recognize intracellular epitopes within the N-terminal portion of the protein have been developed as biochemical, immunofluorescence, and electrophysiological tools (Hausammann et al. 2013; Harley et al. 2016). In the first instance (Hausammann et al. 2013), mice were immunized with the purified full-length hERG expressed in *Spodoptera frugiperda* Sf9 cells. Twelve hERG-specific monoclonal antibodies were identified based on their ability to bind hERG in detergent-permeabilized hERG-expressing CHO cells. Most antibodies in this series recognized linear epitopes within the intracellular region spanning amino acid residues 130 through 320 corresponding to the disordered linker region between the N-terminal Per-Arnt-Sim (PAS) domain and the voltage-sensor domain. Antibodies developed in this study were utilized to develop sensitive ELISA-based protocols for quantification of hERG in cell lysates and membrane preparations.

Harley and colleagues developed and characterized single-chain variable fragments recognizing the PAS domain of Kv11.1 (Harley et al. 2016). In this case, a phage display library containing scFv fragments derived from chickens immunized with the N-terminal region of the protein (amino acid resides 1–135) was screened with the purified PAS domain. Following biochemical and functional characterization of various scFv fragments, two clones, designated scFv 2.10 and scFv 2.12, were selected. Both were able to modulate hERG activity in whole-cell patch electrophysiology assays when delivered into the cytoplasm of either hERG-expressing HEK293 cells or stem cell-derived human cardiomyocytes via the patch pipette. Interestingly, both antibodies appeared to augment hERG channel activity although by different mechanisms, consistent with them interacting with two different regions in the PAS domain.

4.4 TASK3

The TWIK-related Acid-Sensitive K^+ channel 3 (TASK3) is a member of the K2P family of potassium channels expressed primarily in the central nervous system (Bittner et al. 2010; Feliciangeli et al. 2015). TASK3 is also overexpressed in cancer cells. Expression of the TASK3-encoding *KCNK9* gene was found to be amplified threefold to tenfold in ~10% of breast tumor samples (24 out of 247) and *KCNK9* mRNA expression was elevated between 2 and >100-fold in 36% of lung tumors and in 44% of breast tumors (Mu et al. 2003). Furthermore, wild-type TASK3, but not the G95E mutant with minimal-to-no channel activity, promotes tumorigenesis in mice, presumably by enhancing cancer cell survival under hypoxic or serum-deprived conditions (Pei et al. 2003). Based on these findings, TASK3 has generated interest as a potassium channel target for cancer therapy development.

To develop a monoclonal antibody against TASK3, Sun and colleagues (Sun et al. 2016) took advantage of a specific structural feature of the K2P channels, a large extracellular cap that extends above the selectivity filter (Brohawn et al. 2012; Miller and Long 2012; Dong et al. 2015; Rödström et al. 2020). A 59 amino acid region of the human TASK3 extracellular loop 1 corresponding to the cap-forming domain was expressed in mammalian cells as a fusion protein with the Fc domain of mouse IgG2a and then used to immunize mice (Sun et al. 2016). The most potent monoclonal antibody, Y4 (an IgG1), demonstrated binding to human TASK3 with sub-nanomolar affinity, cross-reactivity with mouse TASK3, and selectivity when tested against other human K2P family members. Although no inhibitory effect on channel activity was observed in electrophysiology experiments after relatively short periods of incubation and recording (20 min), Y4 functionality was demonstrated in ion flux assays using human TASK3-expressing HEK293 cells after preincubation for >6 h. The delayed effect of Y4 is presumably due to antibody-induced internalization of the channel which has been shown to occur in both recombinant TASK3-

expressing HEK293 cells as well as in selected human and murine cancer cell lines (Sun et al. 2016). Importantly, following prolonged incubation (24–72 h), the Y4 antibody resulted in decreased viability of TASK3-expressing cancer cells, an effect exacerbated when the cells were grown in low serum. Furthermore, Y4 suppressed tumor growth and metastasis in vivo when administered to mice (Sun et al. 2016).

5 Concluding Remarks

Targeting potassium channels with antibodies remains a challenging endeavor. However, the ability to achieve high selectivity, potency, and favorable pharmacokinetics using biologics that can modulate the activity of a wide range of K^+ channels implicated human disease make this approach well worth the effort. Indeed, the growing list of functionally active K^+ channel antibodies identified to date, along with concomitant advances in target expression, formulation and methods for antibody discovery, suggests that this burgeoning field will continue to expand leading ultimately to novel and effective antibody therapeutics.

References

Abdiche YN, Harriman R, Deng X et al (2016) Assessing kinetic and epitopic diversity across orthogonal monoclonal antibody generation platforms. MAbs 8:264–277. https://doi.org/10.1080/19420862.2015.1118596

Adam SV, Banik SSR, Doranz BJ (2014) Membrane protein solutions for antibody discovery. Genet Technol Bioeng News 34(8). https://www.integralmolecular.com/wp-content/uploads/2020/02/2014_MPS-Discovery_GEN.pdf

Agosto MA, Zhang Z, He F et al (2014) Oligomeric state of purified transient receptor potential melastatin-1 (TRPM1), a protein essential for dim light vision. J Biol Chem 289:27019–27033. https://doi.org/10.1074/jbc.M114.593780

Almagro JC, Pedraza-Escalona M, Arrieta HI et al (2019) Phage display libraries for antibody therapeutic discovery and development. Antibodies (Basel) 8:44. https://doi.org/10.3390/antib8030044

Annecchino LA, Schultz SR (2018) Progress in automating patch clamp cellular physiology. Brain Neurosci Adv 2:2398212818776561. https://doi.org/10.1177/2398212818776561

Asmild M, Oswald N, Krzywkowski KM et al (2003) Upscaling and automation of electrophysiology: toward high throughput screening in ion channel drug discovery. Recept Channels 9:49–58

Bai J, Swartz DJ, Protasevich II et al (2011) A gene optimization strategy that enhances production of fully functional P-glycoprotein in Pichia pastoris. PLoS One 6:e22577. https://doi.org/10.1371/journal.pone.0022577

Barros F, de la Peña P, Domínguez P et al (2020) The EAG voltage-dependent K^+ channel subfamily: similarities and differences in structural organization and gating. Front Pharmacol 11:411. https://doi.org/10.3389/fphar.2020.0041

Beacham DW, Blackmer T, O'Grady M et al (2010) Cell-based potassium ion channel screening using the FluxOR assay. J Biomol Screen 15:441–446. https://doi.org/10.1177/1087057109359807

Bednenko J, Harriman R, Mariën L et al (2018) A multiplatform strategy for the discovery of conventional monoclonal antibodies that inhibit the voltage-gated potassium channel Kv1.3. MAbs 10:636–650. https://doi.org/10.1080/19420862.2018.1445451

Begenisich T, Nakamoto T, Ovitt CE et al (2004) Physiological roles of the intermediate conductance, Ca2+−activated potassium channel Kcnn4. J Biol Chem 279:47681–47687. https://doi.org/10.1074/jbc.M409627200

Bell DC, Dallas ML (2018) Using automated patch clamp electrophysiology platforms in pain-related ion channel research: insights from industry and academia. Br J Pharmacol 175:2312–2321. https://doi.org/10.1111/bph.13916

Bittner S, Budde T, Wiendl H et al (2010) From the background to the spotlight: TASK channels in pathological conditions. Brain Pathol 20:999–1009. https://doi.org/10.1111/j.1750-3639.2010.00407.x

Bradley ME, Dombrecht B, Manini J et al (2015) Potent and efficacious inhibition of CXCR2 signaling by biparatopic nanobodies combining two distinct modes of action. Mol Pharmacol 87:251–262. https://doi.org/10.1124/mol.114.094821

Brinkmann U, Kontermann RE (2017) The making of bispecific antibodies. MAbs 9:182–212. https://doi.org/10.1080/19420862.2016.1268307

Brohawn SG, del Mármol J, MacKinnon R (2012) Crystal structure of the human K2P TRAAK, a lipid- and mechano-sensitive K+ ion channel. Science 335:436–441. https://doi.org/10.1126/science.1213808

Brohawn SG, Campbell EB, MacKinnon R (2013) Domain-swapped chain connectivity and gated membrane access in a Fab-mediated crystal of the human TRAAK K+ channel. Proc Natl Acad Sci U S A 110:2129–2134. https://doi.org/10.1073/pnas.1218950110

Brohawn SG, Campbell EB, MacKinnon R (2014) Physical mechanism for gating and mechanosensitivity of the human TRAAK K+ channel. Nature 516:126–130. https://doi.org/10.1038/nature14013

Brohawn SG, Wang W, Handler A et al (2019) The mechanosensitive ion channel TRAAK is localized to the mammalian node of Ranvier. eLife 8:e50403. https://doi.org/10.7554/eLife.50403

Buell G, Chessell IP, Michel AD et al (1998) Blockade of human P2X7 receptor function with a monoclonal antibody. Blood 92:3521–3528

Calow J, Behrens AJ, Mader S et al (2016) Antibody production using a ciliate generates unusual antibody glycoforms displaying enhanced cell-killing activity. MAbs 8:1498–1511. https://doi.org/10.1080/19420862.2016.1228504

Cázares-Ordoñez V, Pardo LA (2017) Kv10.1 potassium channel: from the brain to the tumors. Biochem Cell Biol 95:531–536. https://doi.org/10.1139/bcb-2017-0062

Cerne R, Wakulchik M, Li B et al (2016) Optimization of a high-throughput assay for calcium channel modulators on IonWorks Barracuda. Assay Drug Dev Technol 14:75–83. https://doi.org/10.1089/adt.2015.678

Cervenak J, Kurrle R, Kacskovics I (2015) Accelerating antibody discovery using transgenic animals overexpressing the neonatal Fc receptor as a result of augmented humoral immunity. Immunol Rev 268:269–287. https://doi.org/10.1111/imr.12364

Chambers C, Witton I, Adams C et al (2016) High-throughput screening of Na(V)1.7 modulators using a giga-seal automated patch clamp instrument. Assay Drug Dev Technol 14:93–108. https://doi.org/10.1089/adt.2016.700

Chandy KG, Norton RS (2017) Peptide blockers of K_v1.3 channels in T cells as therapeutics for autoimmune disease. Curr Opin Chem Biol 38:97–107. https://doi.org/10.1016/j.cbpa.2017.02.015

Chen Y, Sánchez A, Rubio ME et al (2011) Functional K(v)10.1 channels localize to the inner nuclear membrane. PLoS One 6:e19257. https://doi.org/10.1371/journal.pone.0019257

Cheung Y-Y, Yu H, Xu K et al (2012) Discovery of a series of 2-phenyl-N-(2-(pyrrolidin-1-yl)phenyl)acetamides as novel molecular switches that modulate modes of K(v)7.2 (KCNQ2) channel pharmacology: identification of (S)-2-phenyl-N-(2-(pyrrolidin-1-yl)phenyl)butanamide (ML252) as a potent, brain penetrant K(v)7.2 channel inhibitor. J Med Chem 55:6975–6979. https://doi.org/10.1021/jm300700v

Claire JJ (2006) Functional expression of ion channels in mammalian systems. In: Clare JJ, Trezise DJ (eds) Expression and analysis of recombinant ions channels. Wiley-VCH, Weinheim, pp 79–109. https://doi.org/10.1002/9783527610754.tr06

Coleman RG, Sharp KA (2010) Shape and evolution of thermostable protein structure. Proteins 78:420–433. https://doi.org/10.1002/prot.22558

Corey L, Mascola JR, Fauci AS et al (2020) A strategic approach to COVID-19 vaccine R&D. Science 368:948–950. https://doi.org/10.1126/science.abc5312

Cortes S, Barette C, Beroud R et al (2018) Functional characterization of cell-free expressed Kv1.3 channel using a voltage-sensitive fluorescent dye. Protein Expr Purif 145:94–99. https://doi.org/10.1016/j.pep.2018.01.006

Danquah W, Meyer-Schwesinger C, Rissiek B et al (2016) Nanobodies that block gating of the P2X7 ion channel ameliorate inflammation. Sci Transl Med 8:366ra162. https://doi.org/10.1126/scitranslmed.aaf8463

de Taeye SW, Rispens T, Vidarsson G (2019) The ligands for human IgG and their effector functions. Antibodies (Basel) 8:30. https://doi.org/10.3390/antib8020030

Denisov IG, Sligar SG (2016) Nanodiscs for structural and functional studies of membrane proteins. Nat Struct Mol Biol 23:481–486. https://doi.org/10.1038/nsmb.3195

Denisov IG, Grinkova YV, Lazarides AA et al (2004) Directed self-assembly of monodisperse phospholipid bilayer Nanodiscs with controlled size. J Am Chem Soc 126:3477–3487. https://doi.org/10.1021/ja0393574

Ding D, Wang M, Wu JX et al (2019) The structural basis for the binding of repaglinide to the pancreatic K_{ATP} channel. Cell Rep 27:1848–1857.e4. https://doi.org/10.1016/j.celrep.2019.04.050

Dodd RB, Wilkinson T, Schofield DJ (2018) Therapeutic monoclonal ntibodies to complex membrane protein targets: antigen generation and antibody discovery strategies. BioDrugs 32:339–355. https://doi.org/10.1007/s40259-018-0289-y

Dominik PK, Borowska MT, Dalmas O et al (2016) Conformational chaperones for structural studies of membrane proteins using antibody phage display with nanodiscs. Structure 24:300–309. https://doi.org/10.1016/j.str.2015.11.014

Doms R, Rucker J, Hoffman TL et al (2014) Method for production of antibodies that bind to multiple membrane spanning proteins. US Patent 8,680,244

Dong YY, Pike AC, Mackenzie A et al (2015) K2P channel gating mechanisms revealed by structures of TREK-2 and a complex with Prozac. Science 347:1256–1259. https://doi.org/10.1126/science.1261512

Dong J, Finn JA, Larsen PA et al (2019) Structural diversity of ultralong CDRH3s in seven bovine antibody heavy chains. Front Immunol 10:558. https://doi.org/10.3389/fimmu.2019.00558

Dörr JM, Koorengevel MC, Schäfer M et al (2014) Detergent-free isolation, characterization, and functional reconstitution of a tetrameric K+ channel: the power of native nanodiscs. Proc Natl Acad Sci U S A 111:18607–18612. https://doi.org/10.1073/pnas.1416205112

Dunlop J, Bowlby M, Peri R et al (2008) High-throughput electrophysiology: an emerging paradigm for ion-channel screening and physiology. Nat Rev Drug Discov 7:358–368. https://doi.org/10.1038/nrd2552

Eisen JA, Coyne RS, Wu M et al (2006) Macronuclear genome sequence of the ciliate Tetrahymena thermophila, a model eukaryote. PLoS Biol 4:e286. https://doi.org/10.1371/journal.pbio.0040286

English H, Hong J, Ho M (2020) Ancient species offers contemporary therapeutics: an update on shark V_{NAR} single domain antibody sequences, phage libraries and potential clinical applications. Antib Ther 3:1–9. https://doi.org/10.1093/abt/tbaa001

Espeseth AS, Cejas PJ, Citron MP et al (2020) Modified mRNA/lipid nanoparticle-based vaccines expressing respiratory syncytial virus F protein variants are immunogenic and protective in rodent models of RSV infection. NPJ Vaccines 5:16. https://doi.org/10.1038/s41541-020-0163-z

Fan C, Yang X, Wang WW et al (2020) Role of Kv1.3 channels in platelet functions and thrombus formation. Arterioscler Thromb Vasc Biol 40:2360–2375. https://doi.org/10.1161/ATVBAHA.120.314278

Farre C, Haythornthwaite A, Haarmann C et al (2009) Port-a-patch and patchliner: high fidelity electrophysiology for secondary screening and safety pharmacology. Comb Chem High Throughput Screen 12:24–37. https://doi.org/10.2174/138620709787047966

Feliciangeli S, Chatelain FC, Bichet D et al (2015) The family of K2P channels: salient structural and functional properties. J Physiol 593:2587–2603. https://doi.org/10.1113/jphysiol.2014.287268

Feng M, Bian H, Wu X et al (2019) Construction and next-generation sequencing analysis of a large phage-displayed V_{NAR} single-domain antibody library from six naïve nurse sharks. Antib Ther 2:1–11. https://doi.org/10.1093/abt/tby011

Feske S, Wulff H, Skolnik EY (2015) Ion channels in innate and adaptive immunity. Annu Rev Immunol 33:291–353. https://doi.org/10.1146/annurev-immunol-032414-112212

Flajnik MF, Deschacht N, Muyldermans S (2011) A case of convergence: why did a simple alternative to canonical antibodies arise in sharks and camels? PLoS Biol 9:e1001120. https://doi.org/10.1371/journal.pbio.1001120

Frauenfeld J, Löving R, Armache JP et al (2016) A saposin-lipoprotein nanoparticle system for membrane proteins. Nat Methods 13:345–351. https://doi.org/10.1038/nmeth.3801

Garrido A, Lepailleur A, Mignani SM et al (2020) hERG toxicity assessment: useful guidelines for drug design. Eur J Med Chem 195:112290. https://doi.org/10.1016/j.ejmech.2020.112290

Gill S, Gill R, Lee SS et al (2003) Flux assays in high throughput screening of ion channels in drug discovery. Assay Drug Dev Technol 1:709–717. https://doi.org/10.1089/154065803770381066

Goehring A, Lee CH, Wang KH et al (2014) Screening and large-scale expression of membrane proteins in mammalian cells for structural studies. Nat Protoc 9:2574–2585. https://doi.org/10.1038/nprot.2014.173

Golubovskaya V, Wu L (2016) Different subsets of T cells, memory, effector functions, and CAR-T immunotherapy. Cancers (Basel) 8:36. https://doi.org/10.3390/cancers8030036

Gómez-Varela D, Zwick-Wallasch E, Knötgen H et al (2007) Monoclonal antibody blockade of the human Eag1 potassium channel function exerts antitumor activity. Cancer Res 67:7343–7349. https://doi.org/10.1158/0008-5472.CAN-07-0107

Graziano RF, Engelhardt JJ (2019) Role of FcγRs in antibody-based cancer therapy. Curr Top Microbiol Immunol 423:13–34. https://doi.org/10.1007/82_2019_150

Green LL (2014) Transgenic mouse strains as platforms for the successful discovery and development of human therapeutic monoclonal antibodies. Curr Drug Discov Technol 11:74–84. https://doi.org/10.2174/1570163811310999003 8

Guerrier S, Plattner H, Richardson E et al (2017) An evolutionary balance: conservation vs innovation in ciliate membrane trafficking. Traffic 18:18–28. https://doi.org/10.1111/tra.12450

Gupta S, Bavro VN, D'Mello R et al (2010) Conformational changes during the gating of a potassium channel revealed by structural mass spectrometry. Structure 18:839–846. https://doi.org/10.1016/j.str.2010.04.012

Guyot L, Hartmann L, Mohammed-Bouteben S et al (2020) Preparation of recombinant membrane proteins from pichia pastoris for molecular investigations. Curr Protoc Protein Sci 100:e104. https://doi.org/10.1002/cpps.104

Hanlon A, Metjian A (2020) Caplacizumab in adult patients with acquired thrombotic thrombocytopenic purpura. Ther Adv Hematol 11:2040620720902904. https://doi.org/10.1177/2040620720902904

Hansen SB, Tao X, MacKinnon R (2011) Structural basis of PIP2 activation of the classical inward rectifier K+ channel Kir2.2. Nature 477:495–498. https://doi.org/10.1038/nature10370

Harley CA, Starek G, Jones DK et al (2016) Enhancement of hERG channel activity by scFv antibody fragments targeted to the PAS domain. Proc Natl Acad Sci U S A 113:9916–9921. https://doi.org/10.1073/pnas.1601116113

Hartung F, Pardo LA (2016) Guiding TRAIL to cancer cells through Kv10.1 potassium channel overcomes resistance to doxorubicin. Eur Biophys J 45:709–719. https://doi.org/10.1007/s00249-016-1149-7

Hartung F, Stühmer W, Pardo LA (2011) Tumor cell-selective apoptosis induction through targeting of K(V)10.1 via bifunctional TRAIL antibody. Mol Cancer 10:109. https://doi.org/10.1186/1476-4598-10-109

Hartung F, Krüwel T, Shi X et al (2020) A novel anti-Kv10.1 nanobody fused to single-chain TRAIL enhances apoptosis induction in cancer cells. Front Pharmacol 11:686. https://doi.org/10.3389/fphar.2020.00686

Hausammann GJ, Heitkamp T, Matile H et al (2013) Generation of an antibody toolbox to characterize hERG. Biochem Biophys Res Commun 431(1):70–75. https://doi.org/10.1016/j.bbrc.2012.12.089

He S, Moutaoufik MT, Islam S et al (2020) HERG channel and cancer: a mechanistic review of carcinogenic processes and therapeutic potential. Biochim Biophys Acta Rev Cancer 1873:188355. https://doi.org/10.1016/j.bbcan.2020.188355

Hemmerlein B, Weseloh RM, Mello de Queiroz F et al (2006) Overexpression of Eag1 potassium channels in clinical tumours. Mol Cancer 5:41. https://doi.org/10.1186/1476-4598-5-41

Henry KA, MacKenzie CR (2018) Antigen recognition by single-domain antibodies: structural latitudes and constraints. MAbs 10:815–826. https://doi.org/10.1080/19420862.2018.1489633

Hernández-Reséndiz I, Pacheu-Grau D, Sánchez A et al (2020) Inhibition of Kv10.1 channels sensitizes mitochondria of cancer cells to antimetabolic agents. Cancers (Basel) 12:920. https://doi.org/10.3390/cancers12040920

Hickey KD, Buhr MM (2011) Lipid bilayer composition affects transmembrane protein orientation and function. J Lipids 2011:208457. https://doi.org/10.1155/2011/208457

Hirz M, Richter G, Leitner E et al (2013) A novel cholesterol-producing Pichia pastoris strain is an ideal host for functional expression of human Na,K-ATPase a3b1 isoform. Appl Microbiol Biotechnol 97:9465–9478. https://doi.org/10.1007/s00253-013-5156-7

Hite RK, MacKinnon R (2017) Structural titration of Slo2.2, a Na^+-dependent K^+ channel. Cell 168:390–399.e11. https://doi.org/10.1016/j.cell.2016.12.030

Hite RK, Yuan P, Li Z et al (2015) Cryo-electron microscopy structure of the Slo2.2 Na(+)-activated K(+) channel. Nature 527:198–203. https://doi.org/10.1038/nature14958

Hite RK, Tao X, MacKinnon R (2017) Structural basis for gating the high-conductance Ca^{2+}-activated K^+ channel. Nature 541:52–57. https://doi.org/10.1038/nature20775

Hoey RJ, Eom H, Horn JR (2019) Structure and development of single domain antibodies as modules for therapeutics and diagnostics. Exp Biol Med (Maywood) 244:1568–1576. https://doi.org/10.1177/1535370219881129

Hrabovska A, Bernard V, Krejci E (2010) A novel system for the efficient generation of antibodies following immunization of unique knockout mouse strains. PLoS One 5:e12892. https://doi.org/10.1371/journal.pone.0012892

Husain B, Ellerman D (2018) Expanding the boundaries of biotherapeutics with bispecific antibodies. BioDrugs 32:441–464. https://doi.org/10.1007/s40259-018-0299-9

Hutchings CJ, Koglin M, Olson WC et al (2017) Opportunities for therapeutic antibodies directed at G-protein-coupled receptors. Nat Rev Drug Discov 16:787–810. https://doi.org/10.1038/nrd.2017.91

Hutchings CJ, Colussi P, Clark TG (2019) Ion channels as therapeutic antibody targets. MAbs 11:265–296. https://doi.org/10.1080/19420862.2018.1548232

Islas LD (2016) Functional diversity of potassium channel voltage-sensing domains. Channels (Austin) 10:202–213. https://doi.org/10.1080/19336950.2016.1141842

Izquierdo SM, Varela S, Park M et al (2016) High-efficiency antibody discovery achieved with multiplexed microscopy. Microscopy 65:341–352. https://doi.org/10.1093/jmicro/dfw014

Jacobs PP, Geysens S, Vervecken W (2009) Engineering complex-type N-glycosylation in Pichia pastoris using GlycoSwitch technology. Nat Protoc 4:58–70. https://doi.org/10.1038/nprot.2008.213

Jenke M, Sánchez A, Monje F et al (2003) C-terminal domains implicated in the functional surface expression of potassium channels. EMBO J 22:395–403. https://doi.org/10.1093/emboj/cdg035

Jørgensen S, Dyhring T, Brown DT et al (2013) A high-throughput screening campaign for detection of ca(2+)-activated k(+) channel activators and inhibitors using a fluorometric imaging plate reader-based tl(+)-influx assay. Assay Drug Dev Technol 11:163–172. https://doi.org/10.1089/adt.2012.479

Joubert N, Beck A, Dumontet C et al (2020) Antibody-drug conjugates: the last decade. Pharmaceuticals (Basel) 13:245. https://doi.org/10.3390/ph13090245

Jovčevska I, Muyldermans S (2020) The therapeutic potential of nanobodies. BioDrugs 34:11–26. https://doi.org/10.1007/s40259-019-00392-z

Kaplon H, Muralidharan M, Schneider Z (2020) Antibodies to watch in 2020. MAbs 12:1703531. https://doi.org/10.1080/19420862.2019.1703531

Karczewski J, Kiss L, Kane SA et al (2009) High-throughput analysis of drug binding interactions for the human cardiac channel, Kv1.5. Biochem Pharmacol 77:177–185. https://doi.org/10.1016/j.bcp.2008.09.035

Karlova MG, Voskoboynikova N, Gluhov GS et al (2019) Detergent-free solubilization of human Kv channels expressed in mammalian cells. Chem Phys Lipids 219:50–57. https://doi.org/10.1016/j.chemphyslip.2019.01.013

Kaufmann K, Romaine I, Days E et al (2013) ML297 (VU0456810), the first potent and selective activator of the GIRK potassium channel, displays antiepileptic properties in mice. ACS Chem Neurosci 4:1278–1286. https://doi.org/10.1021/cn400062a

Koni PA, Khanna R, Chang MC et al (2003) Compensatory anion currents in Kv1.3 channel-deficient thymocytes. J Biol Chem 278:39443–39451. https://doi.org/10.1074/jbc.M304879200

Könning D, Zielonka S, Grzeschik J et al (2017) Camelid and shark single domain antibodies: structural features and therapeutic potential. Curr Opin Struct Biol 45:10–16. https://doi.org/10.1016/j.sbi.2016.10.019

Kortüm F, Caputo V, Bauer CK et al (2015) Mutations in KCNH1 and ATP6V1B2 cause Zimmermann-Laband syndrome. Nat Genet 247:661–667. https://doi.org/10.1038/ng.3282

Kuang Q, Purhonen P, Hebert H (2015) Structure of potassium channels. Cell Mol Life Sci 72:3677–3693. https://doi.org/10.1007/s00018-015-1948-5

Kulkarni RU, Yin H, Pourmandi N et al (2016) A rationally designed, general strategy for membrane orientation of photoinduced electron transfer-based voltage-sensitive dyes. ACS Chem Biol 12:407–413. https://doi.org/10.1021/acschembio.6b00981

Kunamneni A, Ogaugwu C, Bradfute S et al (2020) Ribosome display technology: applications in disease diagnosis and control. Antibodies (Basel) 9:28. https://doi.org/10.3390/antib9030028

Kuryshev YA, Brown AM, Duzic E et al (2014) Evaluating state dependence and subtype selectivity of calcium channel modulators in automated electrophysiology assays. Assay Drug Dev Technol 12:110–119. https://doi.org/10.1089/adt.2013.552

Kwakkenbos MJ, van Helden PM, Beaumont T et al (2016) Stable long-term cultures of self-renewing B cells and their applications. Immunol Rev 270:65–77. https://doi.org/10.1111/imr.12395

Lau JL, Dunn MK (2018) Therapeutic peptides: historical perspectives, current development trends, and future directions. Bioorg Med Chem 26:2700–2707. https://doi.org/10.1016/j.bmc.2017.06.052

Lee CH, MacKinnon R (2018) Activation mechanism of a human SK-calmodulin channel complex elucidated by cryo-EM structures. Science 360:508–513. https://doi.org/10.1126/science.aas9466

Lee BH, Gauna AE, Pauley KM et al (2012) Animal models in autoimmune diseases: lessons learned from mouse models for Sjögren's syndrome. Clin Rev Allergy Immunol 42:35–44. https://doi.org/10.1007/s12016-011-8288-5

Lee KJ, Wang W, Padaki R et al (2014) Mouse monoclonal antibodies to transient receptor potential ankyrin 1 act as antagonists of multiple modes of channel activation. J Pharmacol Exp Ther 350:223–231. https://doi.org/10.1124/jpet.114.215574

Lee SC, Knowles TJ, Postis VL et al (2016a) A method for detergent-free isolation of membrane proteins in their local lipid environment. Nat Protoc 11:1149–1162. https://doi.org/10.1038/nprot.2016.070

Lee SJ, Ren F, Zangerl-Plessl EM et al (2016b) Structural basis of control of inward rectifier Kir2 channel gating by bulk anionic phospholipids. J Gen Physiol 148:227–237. https://doi.org/10.1085/jgp.201611616

Lee KPK, Chen J, MacKinnon R (2017) Molecular structure of human KATP in complex with ATP and ADP. eLife 6:e32481. https://doi.org/10.7554/eLife.32481

Lee NK, Bidlingmaier S, Su Y et al (2018) Modular construction of large non-immune human antibody phage-display libraries from variable heavy and light chain gene cassettes. Methods Mol Biol 1701:61–82. https://doi.org/10.1007/978-1-4939-7447-4_4

Li N, Wu JX, Ding D et al (2017) Structure of a pancreatic ATP-sensitive potassium channel. Cell 168:101–110.e10. https://doi.org/10.1016/j.cell.2016.12.028

Lin FF, Elliott R, Colombero A et al (2013) Generation and characterization of fully human monoclonal antibodies against human Orai1 for autoimmune disease. J Pharmacol Exp Ther 345:225–238. https://doi.org/10.1124/jpet.112.202788

Liu T, Fu G, Luo X et al (2015) Rational design of antibody protease inhibitors. J Am Chem Soc 137:4042–4045. https://doi.org/10.1021/ja5130786

Liu S, Wang S, Lu S (2016) DNA immunization as a technology platform for monoclonal antibody induction. Emerg Microbes Infect 5:e33. https://doi.org/10.1038/emi.2016.27

Liu C, Li T, Chen J (2019) Role of high-throughput electrophysiology in drug discovery. Curr Protoc Pharmacol 87:e69. https://doi.org/10.1002/cpph.69

Lobner E, Traxlmayr MW, Obinger C et al (2016) Engineered IgG1-Fc--one fragment to bind them all. Immunol Rev 270:113–131. https://doi.org/10.1111/imr.12385

Lolicato M, Riegelhaupt PM, Arrigoni C et al (2014) Transmembrane helix straightening and buckling underlies activation of mechanosensitive and thermosensitive K(2P) channels. Neuron 84:1198–1212. https://doi.org/10.1016/j.neuron.2014.11.017

Lonberg N, Taylor LD, Harding FA et al (1994) Antigen-specific human antibodies from mice comprising four distinct genetic modifications. Nature 368:856–859. https://doi.org/10.1038/368856a0

London B, Guo W, Pan X et al (2001) Targeted replacement of KV1.5 in the mouse leads to loss of the 4-aminopyridine-sensitive component of I(K,slow) and resistance to drug-induced qt prolongation. Circ Res 88:940–946. https://doi.org/10.1161/hh0901.090929

Long SB, Campbell EB, Mackinnon R (2005) Crystal structure of a mammalian voltage-dependent Shaker family K+ channel. Science 309:897–903. https://doi.org/10.1126/science.1116269

Lu RM, Hwang YC, Liu IJ et al (2020) Development of therapeutic antibodies for the treatment of diseases. J Biomed Sci 27:1. https://doi.org/10.1186/s12929-019-0592-z

Martin GM, Kandasamy B, DiMaio F et al (2017a) Anti-diabetic drug binding site in a mammalian K_{ATP} channel revealed by Cryo-EM. eLife 6:e31054. https://doi.org/10.7554/eLife.31054

Martin GM, Yoshioka C, Rex EA et al (2017b) Cryo-EM structure of the ATP-sensitive potassium channel illuminates mechanisms of assembly and gating. eLife 6:e24149. https://doi.org/10.7554/eLife.24149

Martin GM, Sung MW, Yang Z et al (2019) Mechanism of pharmacochaperoning in a mammalian K_{ATP} channel revealed by cryo-EM. eLife 8:e46417. https://doi.org/10.7554/eLife.46417

Matthies D, Bae C, Toombes GE et al (2018) Single-particle cryo-EM structure of a voltage-activated potassium channel in lipid nanodiscs. eLife 7:e37558. https://doi.org/10.7554/eLife.37558

McCloskey C, Jones S, Amisten S et al (2010) Kv1.3 is the exclusive voltage-gated K+ channel of platelets and megakaryocytes: roles in membrane potential, Ca2+ signalling and platelet count. J Physiol 588:1399–1406. https://doi.org/10.1113/jphysiol.2010.188136

Mendez MJ, Green LL, Corvalan JR et al (1997) Functional transplant of megabase human immunoglobulin loci recapitulates human antibody response in mice. Nat Genet 15:146–156. https://doi.org/10.1038/ng0297-146

Miller EW (2016) Small molecule fluorescent voltage indicators for studying membrane potential. Curr Opin Chem Biol 33:74–80. https://doi.org/10.1016/j.cbpa.2016.06.003

Miller AN, Long SB (2012) Crystal structure of the human two-pore domain potassium channel K2P1. Science 335:432–436. https://doi.org/10.1126/science.1213274

Ministro J, Manuel AM, Goncalves J (2020) Therapeutic antibody engineering and selection strategies. Adv Biochem Eng Biotechnol 171:55–86. https://doi.org/10.1007/10_2019_116

Mitchell LS, Colwell LJ (2018) Comparative analysis of nanobody sequence and structure data. Proteins 86:697–706. https://doi.org/10.1002/prot.25497

Mohsen MO, Zha L, Cabral-Miranda G (2017) Major findings and recent advances in virus-like particle (VLP)-based vaccines. Semin Immunol 34:123–132. https://doi.org/10.1016/j.smim.2017.08.014

Mu D, Chen L, Zhang X et al (2003) Genomic amplification and oncogenic properties of the KCNK9 potassium channel gene. Cancer Cell 3:297–302. https://doi.org/10.1016/s1535-6108(03)00054-0

Murphy AJ, Macdonald LE, Stevens S et al (2014) Mice with megabase humanization of their immunoglobulin genes generate antibodies as efficiently as normal mice. Proc Natl Acad Sci U S A 111:5153–5158. https://doi.org/10.1073/pnas.1324022111

Murray JK, Wu B, Tegley CM et al (2019) Engineering $Na_V1.7$ inhibitory JzTx-V peptides with a potency and basicity profile suitable for antibody conjugation to enhance pharmacokinetics. ACS Chem Biol 14:806–818. https://doi.org/10.1021/acschembio.9b00183

Nusblat AD, Bright LJ, Turkewitz AP (2012) Conservation and innovation in Tetrahymena membrane traffic: proteins, lipids, and compartments. Methods Cell Biol 109:141–175. https://doi.org/10.1016/B978-0-12-385967-9.00006-2

Nyblom M, Oberg F, Lindkvist-Petersson K et al (2007) Exceptional overproduction of a functional human membrane protein. Protein Expr Purif 56:110–120. https://doi.org/10.1016/j.pep.2007.07.007

Obergrussberger A, Goetze TA, Brinkwirth N et al (2018) An update on the advancing high-throughput screening techniques for patch clamp-based ion channel screens: implications for drug discovery. ExpertOpin Drug Discov 13:269–277. https://doi.org/10.1080/17460441.2018.1428555

Osborn MJ, Ma B, Avis S et al (2013) High-affinity IgG antibodies develop naturally in Ig-knockout rats carrying germline human IgH/Igκ/Igλ loci bearing the rat CH region. J Immunol 190:1481–1490. https://doi.org/10.4049/jimmunol.1203041

Pandey A, Shin K, Patterson RE et al (2016) Current strategies for protein production and purification enabling membrane protein structural biology. Biochem Cell Biol 94:507–527. https://doi.org/10.1139/bcb-2015-0143

Pardi N, Hogan MJ, Porter FW et al (2018) mRNA vaccines – a new era in vaccinology. Nat Rev Drug Discov 17:261–279. https://doi.org/10.1038/nrd.2017.243

Parray HA, Shukla S, Samal S et al (2020) Hybridoma technology a versatile method for isolation of monoclonal antibodies, its applicability across species, limitations, advancement and future perspectives. Int Immunopharmacol 85:106639. https://doi.org/10.1016/j.intimp.2020.106639

Pei L, Wiser O, Slavin A et al (2003) Oncogenic potential of TASK3 (Kcnk9) depends on K+ channel function. Proc Natl Acad Sci U S A 100:7803–7807. https://doi.org/10.1073/pnas.1232448100

Perry D, Sang A, Yin Y et al (2011) Murine models of systemic lupus erythematosus. J Biomed Biotechnol 2011:271694. https://doi.org/10.1155/2011/271694

Popot JL, Berry EA, Charvolin D et al (2003) Amphipols: polymeric surfactants for membrane biology research. Cell Mol Life Sci 60:1559–1574. https://doi.org/10.1007/s00018-003-3169-6

Posner J, Barrington P, Brier T et al (2019) Monoclonal antibodies: past, present and future. Handb Exp Pharmacol 260:81–141. https://doi.org/10.1007/164_2019_323

Ramaraj T, Angel T, Dratz EA et al (2012) Antigen-antibody interface properties: composition, residue interactions, and features of 53 non-redundant structures. Biochim Biophys Acta 1824:520–532. https://doi.org/10.1016/j.bbapap.2011.12.007

Renauld S, Cortes S, Bersch B et al (2017) Functional reconstitution of cell-free synthesized purified K_v channels. Biochim Biophys Acta Biomembr 1859:2373–2380. https://doi.org/10.1016/j.bbamem.2017.09.002

Rödström KEJ, Kiper AK, Zhang W et al (2020) A lower X-gate in TASK channels traps inhibitors within the vestibule. Nature 582:443–447. https://doi.org/10.1038/s41586-020-2250-8

Rouet R, Jackson KJL, Langley DB et al (2018) Next-generation sequencing of antibody display repertoires. Front Immunol 9:118. https://doi.org/10.3389/fimmu.2018.00118

Saunders KO (2019) Conceptual approaches to modulating antibody effector functions and circulation half-life. Front Immunol 10:1296. https://doi.org/10.3389/fimmu.2019.01296

Schmitz A, Sankaranarayanan A, Azam P et al (2005) Design of PAP-1, a selective small molecule Kv1.3 blocker, for the suppression of effector memory T cells in autoimmune diseases. Mol Pharmacol 68:1254–1270. https://doi.org/10.1124/mol.105.015669

Schneider Z, Jani PK, Szikora B et al (2015) Overexpression of bovine FcRn in mice enhances T-dependent immune responses by amplifying T helper cell frequency and germinal center enlargement in the spleen. Front Immunol 6:357. https://doi.org/10.3389/fimmu.2015.00357

Seddon AM, Curnow P, Booth PJ (2004) Membrane proteins, lipids and detergents: not just a soap opera. Biochim Biophys Acta 1666:105–117. https://doi.org/10.1016/j.bbamem.2004.04.011

Shcherbatko A, Foletti D, Poulsen K et al (2016) Modulation of P2X3 and P2X2/3 receptors by monoclonal antibodies. J Biol Chem 291:12254–12270. https://doi.org/10.1074/jbc.M116.722330

Shenkarev ZO, Karlova MG, Kulbatskii DS et al (2018) Recombinant production, reconstruction in lipid-protein nanodiscs, and electron microscopy of full-length α-subunit of human potassium channel Kv7.1. Biochemistry (Mosc) 83:562–573. https://doi.org/10.1134/S0006297918050097

Shim H, Nguyen H, Cui Y et al (2020) Search for new KCa3.1-targeting small molecules and monoclonal antibodies. In: Experimental Biology 2020 meeting abstracts. FASEB J 34:1. https://doi.org/10.1096/fasebj.2020.34.s1.03756

Simons C, Rash LD, Crawford J et al (2015) Mutations in the voltage-gated potassium channel gene KCNH1 cause Temple-Baraitser syndrome and epilepsy. Nat Genet 47:73–77. https://doi.org/10.1038/ng.3153

Soave M, Cseke G, Hutchings CJ et al (2018) A monoclonal antibody raised against a thermo-stabilised β_1-adrenoceptor interacts with extracellular loop 2 and acts as a negative allosteric modulator of a sub-set of β_1-adrenoceptors expressed in stable cell lines. Biochem Pharmacol 147:38–54. https://doi.org/10.1016/j.bcp.2017.10.015

Sok D, Le KM, Vadnais M et al (2017) Rapid elicitation of broadly neutralizing antibodies to HIV by immunization in cows. Nature 548:108–111. https://doi.org/10.1038/nature23301

Spear JM, Koborssy DA, Schwartz AB et al (2015) Kv1.3 contains an alternative C-terminal ER exit motif and is recruited into COPII vesicles by Sec24a. BMC Biochem 16:16. https://doi.org/10.1186/s12858-015-0045-6

Stanfield RL, Haakenson J, Deiss TC et al (2018) The unusual genetics and biochemistry of bovine immunoglobulins. Adv Immunol 137:135–164. https://doi.org/10.1016/bs.ai.2017.12.004

Stanfield RL, Berndsen ZT, Huang R et al (2020) Structural basis of broad HIV neutralization by a vaccine-induced cow antibody. Sci Adv 6:eaba0468. https://doi.org/10.1126/sciadv.aba0468

Steeland S, Vandenbroucke RE, Libert C (2016) Nanobodies as therapeutics: big opportunities for small antibodies. Drug Discov Today 21:1076–1113. https://doi.org/10.1016/j.drudis.2016.04.003

Steven J, Müller MR, Carvalho MF et al (2017) In vitro maturation of a humanized shark VNAR domain to improve its biophysical properties to facilitate clinical development. Front Immunol 8:1361. https://doi.org/10.3389/fimmu.2017.01361

Storek KM, Chan J, Vij R et al (2019) Massive antibody discovery used to probe structure-function relationships of the essential outer membrane protein LptD. eLife 8:e46258. https://doi.org/10.7554/eLife.46258

Stortelers C, Pinto-Espinoza C, Van Hoorick D et al (2018) Modulating ion channel function with antibodies and nanobodies. Curr Opin Immunol 52:18–26. https://doi.org/10.1016/j.coi.2018.02.003

Sumino A, Uchihashi T, Oiki S (2017) Oriented reconstitution of the full-length KcsA potassium channel in a lipid bilayer for AFM imaging. J Phys Chem Lett 8:785–793. https://doi.org/10.1021/acs.jpclett.6b03058

Sun J, MacKinnon R (2017) Cryo-EM structure of a KCNQ1/CaM complex reveals insights into congenital long QT syndrome. Cell 169:1042–1050.e9. https://doi.org/10.1016/j.cell.2017.05.019

Sun J, MacKinnon R (2020) Structural basis of human KCNQ1 modulation and gating. Cell 180:340–347.e9. https://doi.org/10.1016/j.cell.2019.12.003

Sun H, Luo L, Lal B et al (2016) A monoclonal antibody against KCNK9 K(+) channel extracellular domain inhibits tumour growth and metastasis. Nat Commun 7:10339. https://doi.org/10.1038/ncomms10339

Swale DR, Kurata H, Kharade SV et al (2016) ML418: the first selective, sub-micromolar pore blocker of Kir7.1 potassium channels. ACS Chem Neurosci 7:1013–1023. https://doi.org/10.1021/acschemneuro.6b00111

Tao X, MacKinnon R (2019) Molecular structures of the human Slo1 K^+ channel in complex with β4. eLife 8:e51409. https://doi.org/10.7554/eLife.51409

Tao X, Avalos JL, Chen J et al (2009) Crystal structure of the eukaryotic strong inward-rectifier K+ channel Kir2.2 at 3.1 A resolution. Science 326:1668–1674. https://doi.org/10.1126/science.1180310

Tao X, Hite RK, MacKinnon R (2017) Cryo-EM structure of the open high-conductance Ca^{2+}-activated K^+ channel. Nature 541:46–51. https://doi.org/10.1038/nature20608

Teisseyre A, Palko-Labuz A, Sroda-Pomianek K et al (2019) Voltage-gated potassium channel Kv1.3 as a target in therapy of cancer. Front Oncol 9:933. https://doi.org/10.3389/fonc.2019.00933

Tombola F, Pathak MM, Isacoff EY (2006) How does voltage open an ion channel? Annu Rev Cell Dev Biol 22:23–52. https://doi.org/10.1146/annurev.cellbio.21.020404.145837

Tribet C, Audebert R, Popot JL (1996) Amphipols: polymers that keep membrane proteins soluble in aqueous solutions. Proc Natl Acad Sci U S A 93:15047–15050. https://doi.org/10.1073/pnas.93.26.15047

Vargas E, Yarov-Yarovoy V, Khalili-Araghi F (2012) An emerging consensus on voltage-dependent gating from computational modeling and molecular dynamics simulations. J Gen Physiol 140:587–594. https://doi.org/10.1085/jgp.201210873

Vij R, Lin Z, Chiang N et al (2018) A targeted boost-and-sort immunization strategy using Escherichia coli BamA identifies rare growth inhibitory antibodies. Sci Rep 8:7136. https://doi.org/10.1038/s41598-018-25609-z

Vincke C, Loris R, Saerens D (2009) General strategy to humanize a camelid single-domain antibody and identification of a universal humanized nanobody scaffold. J Biol Chem 284:3273–3284. https://doi.org/10.1074/jbc.M806889200

Vogl T, Glieder A (2013) Regulation of Pichia pastoris promoters and its consequences for protein production. New Biotechnol 30:385–404. https://doi.org/10.1016/j.nbt.2012.11.010

Wang W, MacKinnon R (2017) Cryo-EM structure of the open human ether-à-go-go-related K^+ channel hERG. Cell 169:422–430.e10. https://doi.org/10.1016/j.cell.2017.03.048

Wang L, Sigworth FJ (2009) Structure of the BK potassium channel in a lipid membrane from electron cryomicroscopy. Nature 461:292–295. https://doi.org/10.1038/nature08291

Wang HR, Wu M, Yu H et al (2011) Selective inhibition of the K(ir)2 family of inward rectifier potassium channels by a small molecule probe: the discovery, SAR, and pharmacological characterization of ML133. ACS Chem Biol 6:845–856. https://doi.org/10.1021/cb200146a

Wang F, Ekiert DC, Ahmad I et al (2013) Reshaping antibody diversity. Cell 153:1379–1393. https://doi.org/10.1016/j.cell.2013.04.049

Wang RE, Wang Y, Zhang Y et al (2016) Rational design of a Kv1.3 channel-blocking antibody as a selective immunosuppressant. Proc Natl Acad Sci U S A 113:11501–11506. https://doi.org/10.1073/pnas.1612803113

Wang B, DeKosky BJ, Timm MR et al (2018) Functional interrogation and mining of natively paired human V_H:V_L antibody repertoires. Nat Biotechnol 36:152–155. https://doi.org/10.1038/nbt.4052

Wang X, Li G, Guo J et al (2020) Kv1.3 channel as a key therapeutic target for neuroinflammatory diseases: state of the art and beyond. Front Neurosci 13:1393. https://doi.org/10.3389/fnins.2019.01393

Weaver CD, Harden D, Dworetzky SI et al (2004) A thallium-sensitive, fluorescence-based assay for detecting and characterizing potassium channel modulators in mammalian cells. J Biomol Screen 9:671–677. https://doi.org/10.1177/1087057104268749

Webster CI, Caram-Salas N, Haqqani AS et al (2016) Brain penetration, target engagement, and disposition of the blood-brain barrier-crossing bispecific antibody antagonist of metabotropic glutamate receptor type 1. FASEB J 30:1927–1940. https://doi.org/10.1096/fj.201500078

Whicher JR, MacKinnon R (2016) Structure of the voltage-gated K^+ channel Eag1 reveals an alternative voltage sensing mechanism. Science 353:664–669. https://doi.org/10.1126/science.aaf8070

Whicher JR, MacKinnon R (2019) Regulation of Eag1 gating by its intracellular domains. eLife 8: e49188. https://doi.org/10.7554/eLife.49188

Whorton MR, MacKinnon R (2011) Crystal structure of the mammalian GIRK2 K+ channel and gating regulation by G proteins, PIP2, and sodium. Cell 147:199–208. https://doi.org/10.1016/j.cell.2011.07.046

Whorton MR, MacKinnon R (2013) X-ray structure of the mammalian GIRK2-βγ G-protein complex. Nature 498:190–197. https://doi.org/10.1038/nature12241

Wright PD, Veale EL, McCoull D et al (2017) Terbinafine is a novel and selective activator of the two-pore domain potassium channel TASK3. Biochem Biophys Res Commun 493:444–450. https://doi.org/10.1016/j.bbrc.2017.09.002

Wu JX, Ding D, Wang M et al (2018) Ligand binding and conformational changes of SUR1 subunit in pancreatic ATP-sensitive potassium channels. Protein Cell 9:553–567. https://doi.org/10.1007/s13238-018-0530-y

Wulff H, Christophersen P, Colussi P et al (2019) Antibodies and venom peptides: new modalities for ion channels. Nat Rev Drug Discov 18:339–357. https://doi.org/10.1038/s41573-019-0013-8

Wydeven N, Marron Fernandez de Velasco E, Du Y et al (2014) Mechanisms underlying the activation of G-protein-gated inwardly rectifying K+ (GIRK) channels by the novel anxiolytic drug, ML297. Proc Natl Acad Sci U S A 111:10755–10760. https://doi.org/10.1073/pnas.1405190111

Xu J, Guia A, Rothwarf D et al (2003) A benchmark study with sealchip planar patch-clamp technology. Assay Drug Dev Technol 1:675–684. https://doi.org/10.1089/154065803770381039

Xu H, Hill JJ, Michelsen K et al (2015) Characterization of the direct interaction between KcsA-Kv1.3 and its inhibitors. Biochim Biophys Acta 1848:1974–1980. https://doi.org/10.1016/j.bbamem.2015.06.011

Yanagisawa M, Iwamoto M, Kato A et al (2011) Oriented reconstitution of a membrane protein in a giant unilamellar vesicle: experimental verification with the potassium channel KcsA. J Am Chem Soc 133:11774–11779. https://doi.org/10.1021/ja2040859

Yu H, Li M, Wang W et al (2016) High throughput screening technologies for ion channels. Acta Pharmacol Sin 37:34–43. https://doi.org/10.1038/aps.2015.108

Yuan LIN, Lin W, Zheng K (2013) FRET-based small-molecule fluorescent probes: rational design and bioimaging applications. Acc Chem Res 46:1462–1473. https://doi.org/10.1021/ar300273v

Zangerl-Plessl EM, Lee SJ, Maksaev G et al (2020) Atomistic basis of opening and conduction in mammalian inward rectifier potassium (Kir2.2) channels. J Gen Physiol 152:e201912422. https://doi.org/10.1085/jgp.201912422
Zhang Y, Wang D, de Lichtervelde L et al (2013a) Functional antibody CDR3 fusion proteins with enhanced pharmacological properties. Angew Chem Int Ed Engl 52:8295–8298. https://doi.org/10.1002/anie.201303656
Zhang Y, Wang D, Welzel G et al (2013b) An antibody CDR3-erythropoietin fusion protein. ACS Chem Biol 8:2117–2121. https://doi.org/10.1021/cb4004749
Zhang C, Maruggi G, Shan H et al (2019) Advances in mRNA vaccines for infectious diseases. Front Immunol 10:594. https://doi.org/10.3389/fimmu.2019.00594

Printed by Printforce, United Kingdom